MECHATRONICS

SECOND EDITION

MECHATRONICS

with Experiments

SABRI CETINKUNT

University of Illinois at Chicago, USA

This edition first published 2015
© 2015 John Wiley & Sons Ltd

Registered office
John Wiley & Sons Ltd, The Atrium, Southern Gate, Chichester, West Sussex, PO19 8SQ, United Kingdom
For details of our global editorial offices, for customer services and for information about how to apply for
permission to reuse the copyright material in this book please see our website at www.wiley.com.

Library of Congress Cataloging-in-Publication Data

Cetinkunt, Sabri.
  [Mechatronics]
  Mechatronics with experiments / Sabri Cetinkunt. – Second edition.
    pages cm
  Revised edition of Mechatronics / Sabri Cetinkunt. 2007
  Includes bibliographical references and index.
  ISBN 978-1-118-80246-5 (cloth)
 1. Mechatronics.   I. Title.
  TJ163.12.C43 2015
  621.381–dc23

                                                        2014032267

A catalogue record for this book is available from the British Library.

ISBN: 9781118802465

Set in 10/12pt Times by Aptara Inc., New Delhi, India

3   2015

CONTENTS

PREFACE

This second edition of the textbook has the following modifications compared to the first edition:

- Twelve experiments have been added. The experiments require building of electronic interface circuits between the microcontroller and the electromechanical system, writing of real-time control code in C language, and testing and debugging the complete system to make it work.

- All of the chapters have been edited and more examples have been added where appropriate.

- A brief tutorial on MATLAB®/Simulink®/Stateflow is included.

I would like to thank Paul Petralia, Tom Carter and Anne Hunt [Acquisitions Editor, Project Editor and Associate Commissioning Editor, respectively] at John Wiley and Sons for their patience and kind guidance throughout the process of writing this edition of the book.

Sabri Cetinkunt
Chicago, Illinois, USA
March 19, 2014

ABOUT THE COMPANION WEBSITE

This book has a companion website:

www.wiley.com/go/cetinkunt/mechatronics

The website includes:

- A solutions manual

CHAPTER 1

INTRODUCTION

THE **MECHATRONICS** field consists of the synergistic integration of three distinct traditional engineering fields for system level design processes. These three fields are

1. mechanical engineering where the word "mecha" is taken from,
2. electrical or electronics engineering, where "tronics" is taken from,
3. computer science.

The file of mechatronics is not simply the sum of these three major areas, but can be defined as the intersection of these areas when taken in the context of systems design (Figure 1.1). It is the current state of evolutionary change of the engineering fields that deal with the design of controlled electromechanical systems. A mechatronic system is a computer controlled mechanical system. Quite often, it is an *embedded computer*, not a general purpose computer, that is used for control decisions. The word mechatronics was first coined by engineers at Yaskawa Electric Company [1,2]. Virtually every modern electromechanical system has an embedded computer controller. Therefore, computer hardware and software issues (in terms of their application to the control of electromechanical systems) are part of the field of mechatronics. Had it not been for the widespread availability of low cost microcontrollers for the mass market, the field of mechatronics as we know it today would not exist. The availability of embedded microprocessors for the mass market at ever reducing cost and increasing performance makes the use of computer control in thousands of consumer products possible.

The old model for an electromechanical product design team included

1. engineer(s) who design the mechanical components of a product,
2. engineer(s) who design the electrical components, such as actuators, sensors, amplifiers and so on, as well as the control logic and algorithms,
3. engineer(s) who design the computer hardware and software implementation to control the product in real-time.

A mechatronics engineer is trained to do all of these three functions. In addition, the design process is not sequential with mechanical design followed by electrical and computer control system design, but rather all aspects (mechanical, electrical, and computer control) of design are carried out simultaneously for optimal product design. Clearly, mechatronics is not a new engineering discipline, but the current state of the evolutionary process of the engineering disciplines needed for design of electromechanical systems. The end product of a mechatronics engineer's work is a working prototype of an embedded computer controlled electromechanical device or system. This book covers the fundamental

Mechatronics with Experiments, Second Edition. Sabri Cetinkunt.
© 2015 John Wiley & Sons, Ltd. Published 2015 by John Wiley & Sons, Ltd.
Companion Website: www.wiley.com/go/cetinkunt/mechatronics

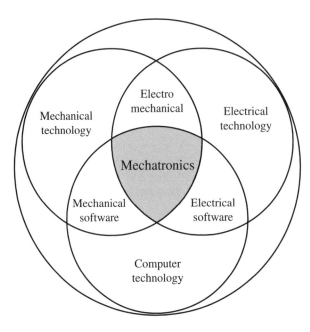

FIGURE 1.1: The field of mechatronics: intersection of mechanical engineering, electrical engineering, and computer science.

technical topics required to enable an engineer to accomplish such designs. We define the word *device* as a stand-alone product that serves a function, such as a microwave oven, whereas a *system* may be a collection of multiple devices, such as an automated robotic assembly line.

As a result, this book has sections on mechanical design of various mechanisms used in automated machines and robotic applications. Such mechanisms are designs over a century old and these basic designs are still used in modern applications. Mechanical design forms the "skeleton" of the electromechanical product, upon which the rest of the functionalities are built (such as "eyes," "muscles," "brains"). These mechanisms are discussed in terms of their functionality and common design parameters. Detailed stress or force analysis of them is omitted as these are covered in traditional stress analysis and machine design courses.

The analogy between a human controlled system and computer control system is shown in Figure 1.2. If a process is controlled and powered by a human operator, the operator observes the behavior of the system (i.e., using visual observation), then makes a decision regarding what action to take, then using his muscular power takes a particular control action. One could view the outcome of the decision making process as a low power control or decision signal, and the action of the muscles as the actuator signal which is the amplified version of the control (or decision) signal. The same functionalities of a control system can be automated by use of a digital computer as shown in the same figure.

The sensors replace the eyes, the actuators replace the muscles, and the computer replaces the human brain. Every computer controlled system has these four basic functional blocks:

1. process to be controlled,
2. actuators,
3. sensors,
4. controller (i.e., digital computer).

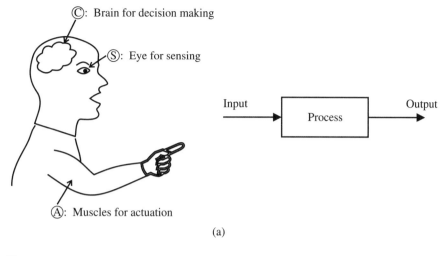

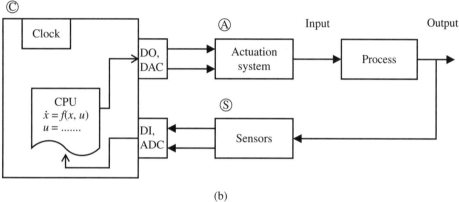

FIGURE 1.2: Manual and automatic control system analogy: (a) human controlled, (b) computer controlled.

The microprocessor (μP) and digital signal processing (DSP) technology had two impacts on control world,

1. it replaced the *existing* analog controllers,

2. prompted *new* products and designs such as fuel injection systems, active suspension, home temperature control, microwave ovens, and auto-focus cameras, just to name a few.

Every mechatronic system has some sensors to measure the status of the process variables. The sensors are the "eyes" of a computer controlled system. We study most common types of sensors used in electromechanical systems for the measurement of temperature, pressure, force, stress, position, speed, acceleration, flow, and so on (Figure 1.3). This list does not attempt to cover every conceivable sensor available in the current state of the art, but rather makes an attempt to cover all major sensor categories, their working principles and typical applications in design.

Actuators are the "muscles" of a computer controlled system. We focus in depth on the actuation devices that provide high performance control as opposed to simple ON/OFF actuation devices. In particular, we discuss hydraulic and electric power actuators in detail. Pneumatic power (compressed air power) actuation systems are not discussed.

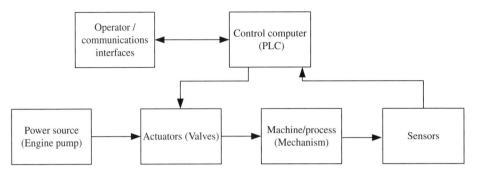

FIGURE 1.3: Main components of any mechatronic system: mechanical structure, sensors, actuators, decision making component (microcontroller), power source, human/supervisory interfaces.

They are typically used in low performance, ON/OFF type control applications (although, with advanced computer control algorithms, even they are starting to be used in high performance systems). The component functionalities of pneumatic systems are similar to those of hydraulic systems. However, the construction detail of each is quite different. For instance, both hydraulic and pneumatic systems need a component to pressurize the fluid (pump or compressor), a valve to control the direction, amount, and pressure of the fluid flow in the pipes, and translation cylinders to convert the pressurized fluid flow to motion. The pumps, valves, and cylinders used in hydraulic systems are quite different to those used in pneumatic systems.

Hardware and software fundamentals for embedded computers, microprocessors, and digital signal processors (DSP), are covered with applications to the control of electrome-chanical devices in mind. Hardware I/O interfaces, microprocessor hardware architectures, and software concepts are discussed. The basic electronic circuit components are discussed since they form the foundation of the interface between the digital world of computers and the analog real world. It is important to note that the hardware interfaces and embed-ded controller hardware aspects are largely standard and do not vary greatly from one application to another. On the other hand, the software aspects of mechatronics designs are different for every product. The development tools used may be same, but the final software created for the product (also called the application software) is different for each product. It is not uncommon that over 80% of engineering effort in the development of a mechatronic product is spent on the software aspects alone. Therefore, the importance of software, especially as it applies to embedded systems, cannot be over emphasized.

Mechatronic devices and systems are the natural evolution of automated systems. We can view this evolution as having three major phases:

1. completely mechanical automatic systems (before and early 1900s),
2. automatic devices with electronic components such as relays, transistors, op-amps (early 1900s to 1970s),
3. computer controlled automatic systems (1970s–present)

Early automatic control systems performed their automated function solely through mechanical means. For instance, a water level regulator for a water tank uses a float connected to a valve via a linkage (Figure 1.4). The desired water level in the tank is set by the adjustment of the float height or the linkage arm length connecting it to the valve. The float opens and closes the valve in order to maintain the desired water level. All the functionalities of a closed loop control system ("sensing-comparison-corrective actuation"

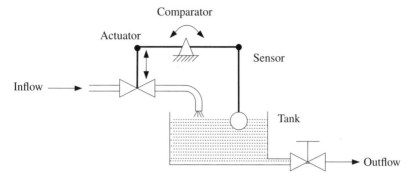

FIGURE 1.4: A completely mechanical closed loop control system for liquid level regulation.

or "sensor-logic-actuation") may be embedded in one component by design, as is the case in this example.

Another classic automatic control system that is made of completely mechanical components (no electronics) is Watt's flyball governor, which is used to regulate the speed of an engine (Figure 1.5). The same concept is still used in some engines today. The engine speed is regulated by controlling the fuel control valve on the fuel supply line. The valve is controlled by a mechanism that has a desired speed setting using the bias in the spring in the flywheel mechanism. The actual speed is measured by the flyball mechanism. The higher the speed of the engine is, the more the flyballs move out due to centrifugal force. The difference between the desired speed and actual speed is turned into control action by the movement of the valve, which controls a small cylinder which is then used to control the fuel control valve. In today's engines, the fuel rate is controlled directly by an electrically actuated injector. The actual speed of the engine is sensed by an electrical sensor (i.e., tachometer, pulse counter, encoder) and an embedded computer controller decides on how

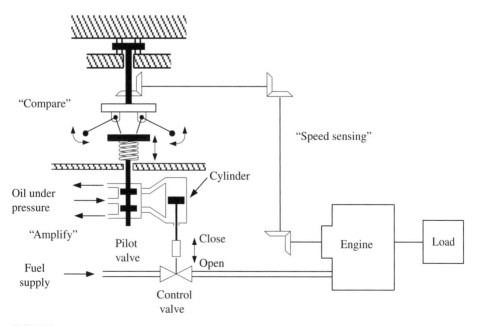

FIGURE 1.5: Mechanical "governor" concept for automatic engine speed control using all mechanical components.

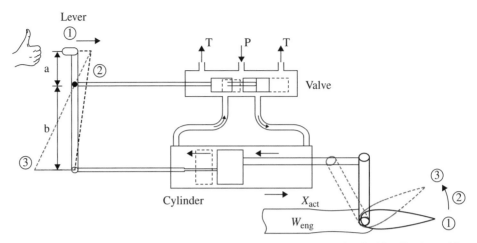

FIGURE 1.6: Closed loop cylinder position control system with mechanical feedback used in the actuation of the main valve.

much fuel to inject based on the difference between the desired and actual engine speed (Figure 1.9).

Figure 1.6 shows a closed loop cylinder position control system where the position feedback is mechanical. The command signal is the desired cylinder position and is generated by the motion of the lever moved by the pilot, and converted to the actuation power to the valve spool displacement through the mechanical linkage. The position feedback is provided by the mechanical linkage connection between the cylinder rod and the lever arm. When the operator moves the lever to a new position, it is the desired cylinder position (position 1 to position 2 in the figure). Initially, that opens the valve, and the fluid flow to the cylinder makes the piston move. As the piston moves, it also moves the linkage connected to the lever. This in turn moves the valve spool (position 2 to position 3 in the figure) to neutral position where the flow through the valve stops when the cylinder position is proportional to the lever displacement. In steady-state, when the cylinder reaches the desired position, it will push the lever such that the valve will be closed again (i.e., when the error is zero, the actuation signal is zero). The proportional control decision based on error is implemented hydro-mechanically without any electronic components.

$$x_{\text{valve}}(t) = \frac{1}{a} \cdot x_{\text{cmd}}(t) - \frac{1}{b} \cdot x_{\text{actual}}(t) \tag{1.1}$$

Analog servo controllers using operational amplifiers led to the second major change in mechatronic systems. As a result, automated systems no longer had to be all mechanical. An operational amplifier is used to compare a desired response (presented as an analog voltage) and a measured response by an electrical sensor (also presented as an analog voltage) and send a command signal to actuate an electrical device (solenoid or electric motor) based on the difference. This brought about many electromechanical servo control systems (Figures 1.7, 1.8). Figure 1.7 shows a web handling machine with tension control. The wind-off roll runs at a speed that may vary. The wind-up roll is to run such that no matter what the speed of the web motion is, a certain tension is maintained on the web. Therefore, a displacement sensor on the web is used to indirectly sense the web tension since the sensor measures the displacement of a spring. The measured tension is then compared to the desired tension (command signal in the figure) by an operational amplifier. The operational amplifier sends a speed or current command to the amplifier of the motor based on the tension error. Modern tension control systems use a digital computer controller in place of the analog operational amplifier controller. In addition, the digital controller may

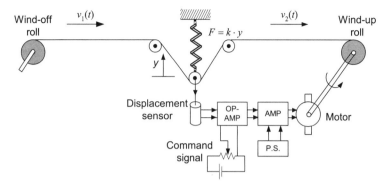

FIGURE 1.7: A web handling motion control system. The web is moved at high speed while maintaining the desired tension. The tension control system can be considered a mechatronic system, where the control decision is made by an analog op-amp, not a digital computer.

use a speed sensor from the wind-off roll or from the web on the incoming side in order to react to tension changes faster and improve the dynamic performance of the system.

Figure 1.8 shows a temperature control system that can be used to heat a room or oven. The heat is generated by the electric heater. Heat is lost to the outside through the walls. A thermometer is used to measure the temperature. An analog controller has the desired temperature setting. Based on the difference between the set and measured temperature, the op-amp turns ON or OFF the relay which turns the heater ON/OFF. In order to make sure

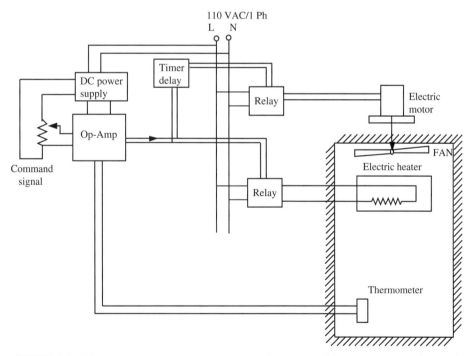

FIGURE 1.8: A furnace or room temperature control system and its components using analog op-amp as the controller. Notice that a fan driven by an electric motor is used to force the air circulation from the heater to the room. A timer is used to delay the turn ON and turn OFF time of the fan motor by a specified amount of time after the heater is turned ON or OFF. A microcontroller-based digital controller can replace the op-amp and timer components.

the relay does not turn ON and OFF due to small variations around the set temperature, the op-amp would normally have a *hysteresis* functionality implemented on its circuit. More details on the relay control with hysteresis will be discussed in later chapters.

Finally, with the introduction of microprocessors into the control world in the late 1970s, programmable control and intelligent decision making were introduced to automatic devices and systems. Digital computers not only duplicated the automatic control functionality of previous mechanical and electromechanical devices, but also brought about new possibilities for device designs that were not possible before. The control functions incorporated into the designs included not only the servo control capabilities but also many operational logic, fault diagnostics, component health monitoring, network communication, nonlinear, optimal, and adaptive control strategies (Figure 1.3). Many such functions were practically impossible to implement using analog op-amp circuits. With digital controllers, such functions are rather easy to implement. It is only a matter of coding these functionalities in software. The difficulty is in knowing what to code that works.

The automotive industry, the largest industry in the world, has transformed itself both in terms of its products (the content of the cars) and the production methods of its products since the introduction of microprocessors. Use of microprocessor-based embedded controllers significantly increased the robotics-based programmable manufacturing processes, such as assembly lines, CNC machine tools, and material handling. This changed the way the cars are made, reducing the necessary labor and increasing the productivity. The product itself, cars, has also changed significantly. Before the widespread introduction of 8-bit and 16-bit microcontrollers into the embedded control mass market, the only electrical components in a car were the radio, starter, alternator, and battery charging system. Engine, transmission, and brake subsystems were all controlled by mechanical or hydro-mechanical means. Today, the engine in a modern car has a dedicated embedded microcontroller that controls the timing and amount of fuel injection in an optimized manner based on the load, speed, temperaturem and pressure sensors in real time. Thus, it improves the fuel efficiency, reduces emissions, and increases performance (Figure 1.9). Similarly, automatic transmission is controlled by an embedded controller. The braking system includes ABS (anti-lock braking system), TCS (traction-control system), DVSC (dynamic vehicle

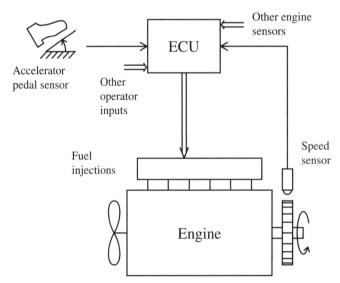

FIGURE 1.9: Electronic "governor" concept for engine control using embedded microcontrollers. The electronic control unit decides on fuel injection timing and amount in real time based on sensor information.

stability control) systems which use dedicated microcontrollers to modulate the control of brake, transmission and engine in order to maintain better control of the vehicle. It is estimated that an average car today has over 30 embedded microprocessor-based controllers on board. This number continues to increase as more intelligent functions are added to cars, such as the autonomous self driving cars by Google Inc and others. It is clear that the traditionally all-mechanical devices in cars have now become computer controlled electromechanical devices, which we call mechatronic devices. Therefore, the new generation of engineers must be well versed in the technologies that are needed in the design of modern electromechanical devices and systems. The field of mechatronics is defined as the integration of these areas to serve this type of modern design process.

Robotic manipulator is a good example of a mechatronic system. The low-cost, high computational power, and wide availability of digital signal processors (DSP) and microprocessors energized the robotics industry in late 1970s and early 1980s. The robotic manipulators, the reconfigurable, programmable, multi degrees of freedom motion mechanisms, have been applied in many manufacturing processes and many more applications are being developed, including robotic assisted surgery. The main sub-systems of a robotic manipulator serve as a good example of mechatronic system. A robotic manipulator has four major sub-systems (Figure 1.3), and every modern mechatronic system has the same sub-system functionalities:

1. a mechanism to transmit motion from actuator to tool,
2. an actuator (i.e., a motor and power amplifier, a hydraulic cylinder and valve) and power source (i.e., DC power supply, internal combustion engine and pump),
3. sensors to measure the motion variables,
4. a controller (DSP or microprocessor) along with operator user interface devices and communication capabilities to other intelligent devices.

Let us consider an electric servo motor-driven robotic manipulator with three axes. The robot would have a predefined mechanical structure, for example Cartesian, cylindrical, spherical, SCARA type robot (Figures 1.10, 1.11, 1.12). Each of the three electric servo motors (i.e., brush-type DC motor with integrally mounted position sensor such as an encoder or stepper motor with position sensor) drives one of the axes. There is a separate power amplifier for each motor which controls the current (hence torque) of the motor. A DC power supply provides a DC bus at a constant voltage and derives it from a standard AC line. The DC power supply is sized to support all three motor-amplifiers.

The power supply, amplifier, and motor combination forms the actuator sub-system of a motion system. The sensors in this case are used to measure the position and velocity

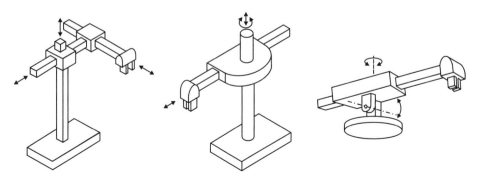

FIGURE 1.10: Three major robotic manipulator mechanisms: Cartesian, cylindrical, spherical coordinate axes.

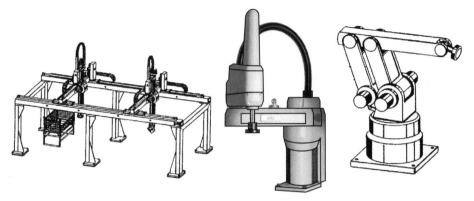

FIGURE 1.11: Gantry, SCARA, and parallel linkage drive robotic manipulators.

of each motor so that this information is used by the axis controller to control the motor through the power amplifier in a closed loop configuration. Other external sensors not directly linked to the actuator motions, such as a vision sensors or a force sensors or various proximity sensors, are used by the supervisory controller to coordinate the robot motion with other events. While each axis has a dedicated closed loop control algorithm, there has to be a supervisory controller that coordinates the motion of the three motors in order to generate a coordinated motion by the robot, that is straight line motion, and so on circular motion etc. The hardware platform to implement the coordinated and axis level controls can be based on a single DSP/microprocessor or it may be distributed over multiple processors as shown. Figure 1.12 shows the components of a robotic manipulator in block diagram form. The control functions can be implemented on a single DSP hardware or a distributed DSP hardware. Finally, just as no man is an island, no robotic manipulator is an

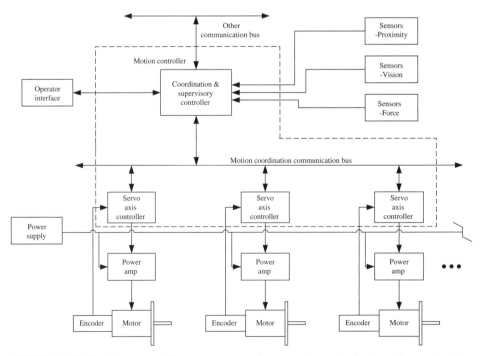

FIGURE 1.12: Block diagram of the components of a computer controlled robotic manipulator.

island. A robotic manipulator must communicate with a user and other intelligent devices to coordinate its motion with the rest of the manufacturing cell. Therefore, it has one or more other communication interfaces, typically over a common fieldbus (i.e., DeviceNET, CAN, ProfiBus, Ethernet). The capabilities of a robotic manipulator are quantified by the following;

1. workspace: volume and envelope that the manipulator end effector can reach,
2. number of degrees of freedom that determines the positioning and orientation capabilities of the manipulator,
3. maximum load capacity, determined by the actuator, transmission components, and structural component sizing,
4. maximum speed (top speed) and small motion bandwidth,
5. repeatability and accuracy of end effector positioning,
6. manipulator's physical size (weight and volume it takes).

Figure 1.13 shows a computer numeric controlled (CNC) machine tool. A multi axis vertical milling machine is shown in this figure. There are three axes of motion controlled precisely (i.e., within 1/1000 in or 25 micron = 25/1000 mm accuracy) in x, y and z directions by closed loop controlled servo motors. The rotary motion of each of the servo motors is converted to linear motion of the table by the ball-screw or lead-screw

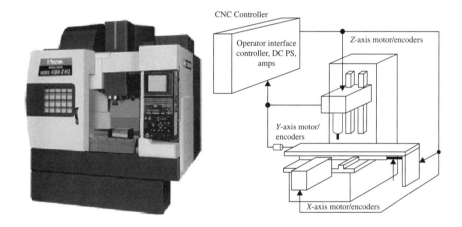

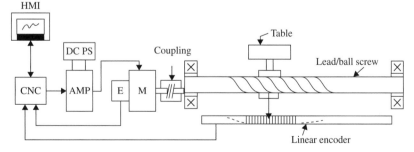

FIGURE 1.13: Computer numeric controlled (CNC) machine tool: (a) picture of a vertical CNC machine tools, reproduced with permission from Yamazaki Mazak Corporation, (b) x-y-z axes of motion, actuated by servo motors, (c) closed loop control system block diagram for one of the axis motion control system, where two position sensors per axis (motor-connected and load-connected) are shown (also known as dual position feedback).

mechanism in each axis. The fourth motion axis is the spindle rotation which typically runs at a constant speed. Each axis has its own servo motor (i.e., brushless DC motor with position feedback), amplifier and DC power supply. In high precision machine tools, in addition to the position sensors integral to the servo motor, there are also linear position sensors (i.e., linear encoders) attached to the moving part of the table on each axis in order to measure the translational position of the table directly. Using this measurement, the controller can compensate for position errors due to backlash and mechanical transmission errors in the lead-screw/ball-screw. The CNC controller implements the desired motion commands for each axis in order to generate the desired cut-shape, as well as the closed loop position control algorithm such as a PID controller. When two position sensors are used for one degree of motion (one located at the actuator point (on the motor shaft) and one located at the actuated-tool point (table)), it is referred to as *dual position feedback control system*. A typical control logic in dual-position feedback system is to use the motor-based encoder feedback in velocity loop, and load-based encoder feedback in position loop control. Current state of the art technology in CAD/CAM and CNC control is such that a desired part is designed in CAD software, then the motion control software to run on the CNC controller (i.e., G-code or similar code which defines the sequence of desired motion profiles for each axis) is automatically generated from the CAD file of the part, downloaded to the CNC controller, which then controls each motion axis of the machine in closed loop to cut the desired shape.

Figure 1.14 shows the power flow in a modern construction equipment. The power source in most mobile equipment is an internal combustion engine, which is a diesel engine in large power applications. The power is hydro-mechanically transmitted from engine to transmission, brake, steering, implement, and cooling fan. All sub-systems get their power in hydraulic form from a group of pumps mechanically connected to the engine. These pumps convert mechanical power to hydraulic power. In automotive type designs, the power from engine to transmission gear mechanism is linked via a torque converter. In other designs, the transmission may be a hydrostatic design where the mechanical power is converted to hydraulic power by a pump and then back to mechanical power by hydraulic motors. This is the case in most excavator designs. Notice that each major sub-system has its own electronic control module (ECM). Each ECM deals with the control of the sub-system and possibly communicates with a machine level master controller. For instance, ECM for engines deals with maintaining an engine speed commanded by the operator pedal. As the load increases and the engine needs more power, the ECM automatically commands more fuel to the engine to regulate the desired speed. The transmission ECM deals with the control of a set of solenoid actuated pressure valves which then controls a set of clutch and brakes in order to select the desired gear ratio. Steering ECM controls a valve which controls the flow rate to a steering cylinder. Similarly, other sub-system ECMs controls electrically controlled valves and other actuation devices to modulate the power used in that sub-system.

The agricultural industry uses harvesting equipment where the equipment technology has the same basic components used in the automotive and construction equipment industry. Therefore, automotive technology feeds and benefits agricultural technology. Using global positioning systems (GPS) and land mapping for optimal utilization, large scale farming has started to be done by autonomous harvesters where the machine is automatically guided and steered by GPS systems. Farm lands are fertilized in an optimal manner based on previously collected satellite maps. For instance, the planning and execution of an earth moving job, such as road building or a construction site preparation or farming, can be done completely under the control of GPSs and autonomously driven machines without any human operators on the machine. However, safety concerns have so far delayed the introduction of such autonomous machine operations. The underlying technologies are

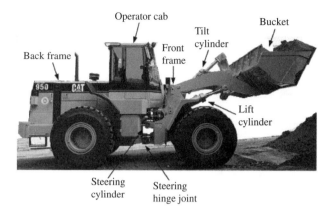

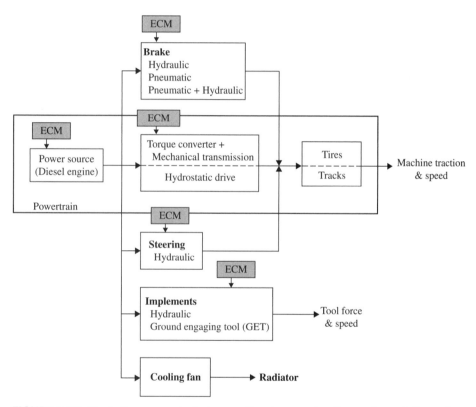

FIGURE 1.14: Block diagram controlled power flow in a construction equipment. Power flow in automotive applications is similar. Notice that modern construction equipment has electronic control modules (ECMs) for most major sub-systems such as engine, transmission, brake, steering, implement sub-systems.

relatively mature for autonomous construction equipment and farm equipment operation (Figure 1.15).

The chemical process industry involves many large scale computer controlled plants. The early application of computer controlled plants was based on a large central computer controlling most of the activities. This is called the *centralized control* model. In recent years, as microcontrollers became more powerful and low cost, the control systems for large plants have been designed using many layers of hierarchy of controllers. In other words,

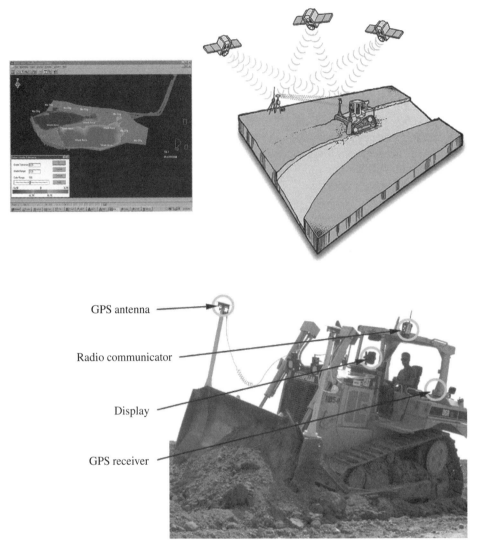

FIGURE 1.15: Semi-autonomous construction equipment operation using global positioning system (GPS), local sensors and on-vehicle sensors for closed loop sub-system control.

the control logic is distributed physically to many microcomputers. Each microcomputer is physically closer to the sensors and actuators it is responsible for. Distributed controllers communicate with each other and higher level controllers over a standard communication network. There may be a separate communication network at each layer of the hierarchical control system. The typical variables of control in process industry are fluid flow rate, temperature, pressure, mixture ratio, fluid level in tank, and humidity.

Energy management and control of large buildings is a growing field of application of optimized computer control. Home appliances are more and more microprocessor controlled, instead of being just an electromechanical appliances. For instance, old ovens used relays and analog temperature controllers to control the electric heater in the oven. The new ovens use a microcontroller to control the temperature and timing of the oven operation. Similar changes have occurred in many other appliances used in homes, such as washers and driers.

Micro electromechanical systems (MEMS) and MEMS devices incorporate all of the computer control, electrical and mechanical aspects of the design directly on the silicon substrate in such a way that it is impossible to discretely identify each functional component. Finally, the application of mechatronic design in medical devices, such as surgery assistive devices, robotic surgery, and intelligent drills, is perhaps one of the most promising field in this century.

Computer controlled medical devices (implant and external assistive, rehabilitation equipment) have been experiencing exponential growth as the physical size of sensing and computing devices becomes very tiny such that they can be integrated with small actuators as implant devices for human body. The basic principle of the sensing-decision-actuation is being put to many uses in embedded computer controlled medical devices (also called bio-mechatronic devices, Figure 1.16). In time these devices will be able to integrate a growing set of tiny sensors, and make more sophisticated real-time decisions about what (if any) intervention action to take to assist the functioning of the human body. For instance, implant defibrillators and pace-makers for heart patients are examples of such devices. A pace-maker is a heart implant device that provides electrical pulses to the heart muscles to regulate its rate when it senses that the heart rate has fallen below a critical level. The

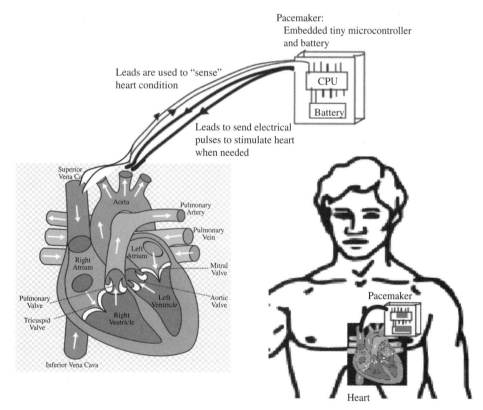

FIGURE 1.16: Example of an embedded computer controlled medical device: a bio-mechatronic device. The pulse generator houses the battery (electrical power source) and a tiny embedded computer. The electrical wires between the heart and the pulse generator (pace-maker) are for both sensing the heart condition (sensor cables) and actuating the heart beat by electrical pulse signal shocks to the heart muscle. The sensing-decision-actuation functions are integrated via the pulse generator and electrical signal leads. Wapcaplet, Yaddah [GFDL (www.gnu.org/copyleft/fdl.html) or CC-BY-SA-3.0 (http://creativecommons.org/licenses/by-sa/3.0/)], via Wikimedia.

pace-maker senses the heart rate, and if the heart rate is below a critical rate, it sends electrical pulses to the heart in order to increase the heart rate. The sensing and actuation components are interfaced to the heart through electrical wires. The embedded computer, battery (electrical power source for the pulse power) and pulse generator circuit is one integrated unit which is implanted under the skin somewhere close to the heart.

In addition, many computer controlled orthopeadic devices are in the process of development as implanted aid devices as well as rehabilitative devices. For example, in artificial hand devices, the embedded computer senses the desired motion signals in the remaining muscles which are sent from the brain, then interprets them to actuate the mechanical hand like it would function in a natural hand. The compact electromechanical design of the hand mechanism, its integrated actuation and sensing (position and force sensors) devices are *electromechanical design problems*. Measurement and interpretation of the desired motion signals from human brain to the residual muscles and, based on that information, determining the desired motion of the hand is an *intelligent signal processing and control problem* (see National Geographic Magazine, ... issue).

1.1 CASE STUDY: MODELING AND CONTROL OF COMBUSTION ENGINES

The internal combustion engine is the power source for most of the mobile equipment applications including automotive, construction, and agricultural machinery. As a result, it is an essential component in most mobile equipment applications. Here, we discuss the modeling and basic control concepts of internal combustion engines from a mechatronics engineering point of view. This case study may serve as an example of how a dynamic model and a control system should be developed for a computer controlled electromechanical system. Basic modeling and control of any dynamic system invariably involves use of Laplace transforms. As a result, detailed analysis using Laplace transforms is minimized here in this introductory chapter.

We will discuss the basic characteristics of a diesel engine from a mechatronics engineer's point of view. Any modeling and control study should start with a good physical understanding of how a system works. We identify the main components and sub-systems. Then each component is considered in terms of its input and output relationship in modeling. For control system design purposes, we identify the necessary sensors and controlled actuators. With this guidance, we study

1. engine components – basic mechanical components of the engine,
2. operating principles and performance – how energy is produced (converted from chemical energy to mechanical energy) through the combustion process,
3. electronic control system components: actuators, sensors, and electronic control module (ECM),
4. dynamic models of the engine from a mechatronics engineer's point of view,
5. control algorithms – basic control algorithms and various extensions in order to meet fuel efficiency and emission requirements.

An engine converts the chemical energy of fuel to mechanical energy through the combustion process. In a mobile equipment, sub-systems derive their power from the engine. There are two major categories of internal combustion engines: (i) Clerk (two-stroke) cycle engine; (ii) Otto (four-stroke) cycle engine. In a two-stroke cycle engine, there is a combustion in each cylinder once per revolution of the crankshaft. In a four-stroke cycle engine, there is a combustion in each cylinder once every two revolutions of the crankshaft. Only four-stroke cycle engines are discussed below.

Four-stroke cycle internal combustion engines are also divided into two major categories: (i) gasoline engines; (ii) diesel engines. The fundamental difference between them is in the way the combustion is ignited every cycle in each cylinder. Gasoline engines use a spark plug to start the combustion, whereas the combustion is self-ignited in diesel engines as a result of the high temperature rise (typical temperature levels in the cylinder towards the end of the compression cylce is around 700–900 °C range) due to the large compression ratio. If the ambient air temperature is very low (i.e., extremely cold conditions), the temperature rise in the cylinder of a diesel engine due to the compression of air–fuel mixture may not be high enough for self-ignition. Therefore, diesel engines have electric heaters to pre-heat the engine block before starting the engine in a very cold environment.

The basic mechanical design and size of the engine defines an envelope of maximum performance (speed, torque, power, and fuel consumption). The specific performance of an engine within the envelope of maximum performance is customized by the engine controller. The decision block between the sensory data and fuel injection defines a particular performance within the bounds defined by the mechanical size of the engine. This decision block includes considerations of speed regulation, fuel efficiency, and emission control.

1.1.1 Diesel Engine Components

The main mechanical components of a diesel engine are located on the engine block (Figure 1.17). The engine block provides the frame for the combustion chambers where each combustion chamber is made of a cylinder, a piston, one or more intake valves and

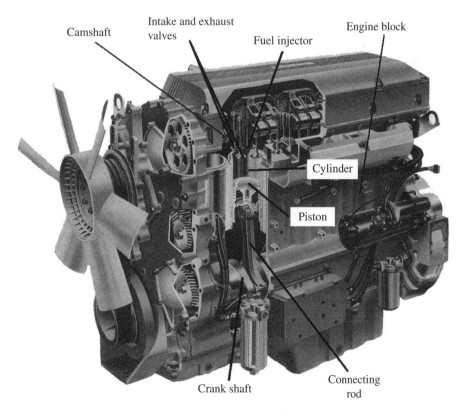

FIGURE 1.17: Mechanical components of an engine block: 1. engine block, 2. cylinder, 3. piston, 4. connecting rod, 5. crankshaft, 6. cam-shaft, 7. intake valve, 8. exhaust valve, 9. fuel injector.

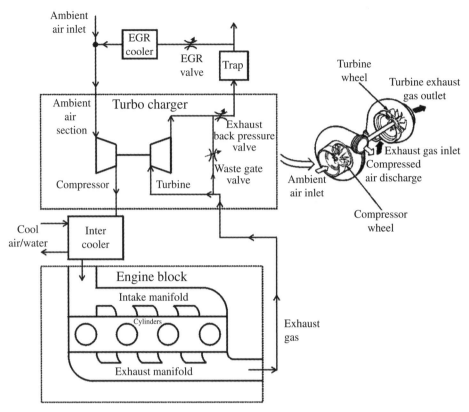

FIGURE 1.18: Engine and its surrounding sub-systems: intake manifold, exhaust manifold, turbo charger with waste-gate valve, charge (inter) cooler, exhaust gas recirculation (EGR), trap or catalytic converter.

exhaust valves, and a fuel injector. The power obtained from the combustion process is converted to the reciprocating linear motion of the piston. The linear motion of the piston is converted to a unidirectional continuous rotation of the crankshaft through the connecting rod. In the case of a spark ignited engine (gasoline engine), there would also be a spark plug to generate ignition. The compression ratio of diesel engines is in the range of 1:14 to 1:24, while gasoline engine compression ratio range is about half of that.

Normally, there are multiple cylinders (i.e., 4, 6, 8, 12) where each cylinder operates with a different crankshaft phase angle from each other in order to provide non-pulsating power. An engine's power capacity is determined primarily by the number of cylinders, volume of each cylinder (piston diameter and stroke length), and compression ratio. Figure 1.18 shows the engine block and its surrounding sub-systems: throttle, intake manifold, exhaust manifold, turbo charger, charge cooler. In most diesel engines there is not a physical throttle valve. A typical diesel engine does not control the inlet air, it takes the available air and controls the injected fuel rate, while some diesel engines control both the inlet air (via the throttle valve) and the injected fuel rate.

The surrounding sub-systems support the preparation of the air and fuel mixture before the combustion and exhausting of it. Timing of the intake valve, exhaust valve, and injector is controlled either by mechanical means or by electrical means. In completely mechanically controlled engines, a mechanical *camshaft* coupled to the *crankshaft* by a timing belt with a 2:1 gear ratio is used to control the timing of these components which is periodic with two revolutions of the crankshaft. *Variable valve control* systems incorporate a mechanically controlled lever which adjusts the phase of the camshaft sections in order to vary the timing of the valves. Similar phase adjustment mechanisms are designed into individual fuel injectors as well. In electronically controlled engines some or all of these components (fuel injectors, intake and exhaust valves) are each controlled by electrical actuators (i.e. solenoid actuated valves). In today's diesel engines, the injectors are electronically controlled, while the intake valves and exhaust valves are controlled by the mechanical camshaft. In fully electronically controlled engines, also called *camless engines*, the intake and exhaust valves are also electronically controlled.

Turbo chargers (also called *super chargers*) and *charge coolers* (also called *inter coolers*, or *after coolers*) are passive mechanical devices that assist in the efficiency and maximum power output of the engine. The turbo charger increases the amount of air pumped ("charged") into the cylinders. It gets the necessary energy to perform the pumping function from the exhaust gas. The turbo charger has two main components: a turbine and compressor, which are connected to the same shaft. Exhaust gas rotates the turbine, and it in turn rotates the compressor which performs the pumping action. By making partial use of the otherwise wasted energy in the exhaust gas, the turbo charger pumps more air, which in turn means more fuel can be injected for a given cylinder size. Therefore, an engine can generate more power from a given cylinder size using a turbo charger. An engine without turbo charger is called a *naturally aspirated engine*. The turbo charger gain is a function of the turbine speed, which is related to the engine speed. Therefore, some turbo chargers have variable blade orientation or a moving nozzle (called *variable geometry turbochargers (VGT)*) to increase the turbine gain at low speed and reduce it at high speed (Figure 1.18).

While the main purpose of the turbo charger is to increase the amount of inlet air pumped into the cylinders, it is not desirable to increase the inlet boost pressure beyond a maximum value. Some turbo charger designs incorporate a *waste-gate* valve for that purpose. When the boost pressure sensor indicates that the pressure is above a certain level, the electronic control unit opens a solenoid actuated butterfly type valve at the waste-gate. This routes the exhaust gas to bypass the turbine to the exhaust line. Hence the name "waste-gate" since it wastes the exhaust gas energy. This reduces the speed of the turbine and the compressor. When the boost pressure drops below a certain value, the waste-gate valve is closed again and the turbo charger operates in its normal mode.

Another feature of some turbo chargers is the *exhaust back pressure device*. Using a butterfly type valve, the exhaust gas flow is restricted and hence the exhaust back pressure is increased. As a result, the engine experiences larger exhaust pressure resistance. This leads to faster heating of the engine block. This is used for rapid warming of the engine under cold starting conditions.

In some designs of turbo chargers, in order to reduce the cylinder temperature, a *charge cooler* (also called *inter cooler*) is used between the turbo charger's compressor output and the intake manifold. The turbo charger's compressor outputs air with temperatures as high as 150 °C. The ideal temperature for inlet air for a diesel engine is around 35–40 °C. The charge cooler performs the cooling function of the intake air so that air density can be increased. A high air temperature reduces the density of the air (hence the air–fuel ratio) as well as increases the wear in the combustion chamber components.

The exhaust gas recirculation (EGR) mixes the intake air with a controlled amount of exhaust gas for combustion. The main advantage of the EGR is the reduction of NO_x content in the emission. However, EGR results in more engine wear, and increases smoke and particulate content in the emission.

Fuel is injected in to the cylinder by cam-actuated (mechanically controlled) or solenoid actuated (electrically controlled) injectors. The solenoid actuation force is amplified by hydraulic means in order to provide the necessary force for the injectors. Figure 1.19 shows an electrically controlled fuel injector system where a hydraulic oil pressure line is

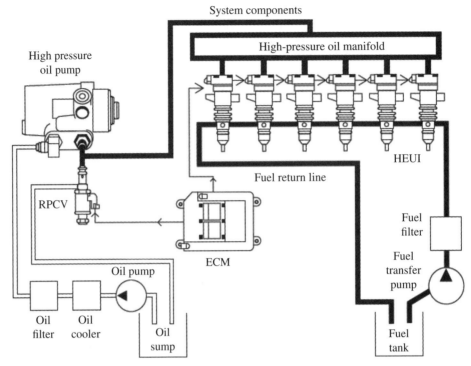

FIGURE 1.19: An example of an electronically controlled fuel injection system used in diesel engines: HEUI fuel injection system by Caterpillar Inc and Navistar Inc.

used as the amplifier stage between the solenoid signal and injection force. Notice that the main components of the fuel delivery system are

1. fuel tank,
2. fuel filter,
3. fuel pump,
4. pressure regulator valve,
5. high pressure oil pump,
6. fuel injectors.

The fuel pump maintains a constant fuel pressure line for the injectors. In common-rail (CR) and electronic unit injector (EUI) type fuel injection systems (such as the hydraulic electronic unit injector (HEUI) by Caterpillar and Navistar Inc), a high pressure oil pump in conjunction with a pressure regulating valve is used to provide the high pressure oil line to act as an amplifier line for the injectors. Hydraulic oil is the same oil used for engine lubrication. The injectors are controlled by the low power solenoid signals coming from the electronic control module (ECM). The motion of the solenoid plunger is amplified by the high pressure oil line to provide the higher power levels needed for the fuel injector.

There are four different fluids involved in any internal combustion engine:

1. fuel for combustion,
2. air for combustion and cooling,
3. oil for lubrication,
4. water–coolant mixture for cooling.

Each fluid (liquid or gas) circuit has a component to store, condition (filter, heat or cool), move (pump), and direct (valve) it within the engine.

The cooling and lubrication systems are closed circuit systems which derive their power from the crankshaft via a gearing arrangement to the coolant pump and the oil pump. Reservoir, filter, pump, valve, and circulation lines are very similar to other fluid circuits. The main component of the cooling system is the radiator. It is a heat exchanger where the heat from the coolant is removed to the air through the a series of convective tubes or cores. The coolant is used not only to remove heat from the engine block, but also to remove heat from the intake air at after-cooler (inter-cooler) as well as to remove heat from the lubrication oil. Finally the heat is dissipated out to the environment at the radiator. The radiator fan provides forced air for higher heat exhange capacity. Typically, the cooling system includes a temperature regulator valve which directs the coolant flow path when the engine is cold in order to help it warm up quickly.

The purpose of lubrication is to reduce the mechanical friction between two surfaces. As the friction is reduced, the friction related heat is reduced. The lubrication oil forms a thin film between any two moving surfaces (i.e., bearings). The oil is sucked from the oil pan by the oil pump, passed through an oil filter and cooler, then guided to the cylinder block, piston, connecting rod, and crankshaft bearings. The lubrication oil temperature must be kept around 105–115 °C. Too high a temperature reduces the load handling capacity, whereas too low a temperature increases viscosity and reduces lubrication capability. A pressure regulator keeps the lube oil pressure around a nominal value (40 to 50 psi range).

The fuel pump, lubrication oil pump, cooling fan, and coolant pump all derive their power from the crank with gear and belt couplings. The current trend in engine design is to use electric generators to transfer power from the engine to the electric motor-driven pumps for the sub-systems. That is, instead of using mechanical gears and belts to transmit and distribute power, the new designs use electrical generators and motors.

Diesel Engine Operating Principles   Let us consider one of the cylinders in a four-stroke cycle diesel engine (Figure 1.20). Other cylinders go through the same sequence of cycles except offset by a crankshaft phase angle. In a four-cylinder diesel engine, each cylinder goes through the same sequence of four-stroke cycles offset by 180° of crankshaft angle. Similarly, this phase angle is 120° for a six-cylinder engine, and 90° for an eight-cylinder engine. The phase angle between cylinders is (720°)/(number of cylinders). During the intake stroke, the intake valve opens and the exhaust valve closes. As the piston moves down, the air is sucked into the cylinder until the piston reaches the bottom dead center (BDC). The next stroke is the compression stroke during which the intake valve closes and, as the piston moves up, the air is compressed. The fuel injection (and spark ignition in the SI engine) is started at some position before the piston reaches the top dead center (TDC).

The combustion, and the resulting energy conversion to mechanical energy, are accomplished during the expansion stroke. During that stroke, the intake valve and exhaust valve are closed. Finally, when the piston reaches the BDC position and starts to move up, the exhaust valve opens to evacuate the burned gas. This is called the exhaust stroke. The cycle ends when the piston reaches the TDC position.

This four-stroke cycle repeats for each cylinder. Note that each cylinder is in one of these strokes at any given time. For the purpose of illustrating the basic operating principle, we stated above that the intake and exhaust valve open and close at the end or beginning of each cycle. In an actual engine, the exact opening and closing position of these valves, as well as the fuel injection timing and duration, are a little different than the BDC or TDC positions of the piston.

It is indeed these intake and exhaust valve timings as well as the fuel injection timing (start time, duration, and injection pulse shape) decisions that are made by the

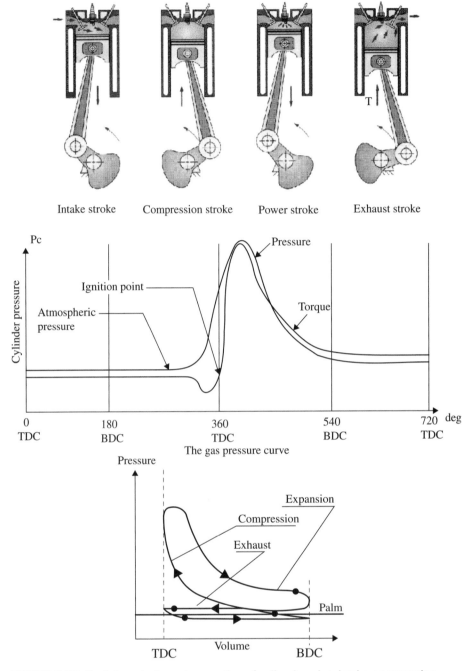

FIGURE 1.20: Basic four-stroke cycle operation of a diesel engine: intake, compression, expansion, exhaust stroke. The pressure in a cylinder is a function of crankshaft position. Other cylinders have identical pressures as a function of crankshaft angle except with a phase angle. Compression ratio is the ratio of the cylinder volume at BDC (V_{BDC}) to the cylinder volume at TDC (V_{TDC}). Notice that during the compression stroke, the cylinder pressure opposes the crank motion, hence the effective torque is negative. During the expansion stoke, the cylinder pressure supports the crank motion, hence the effective torque is positive.

electronic control module (ECM) in real time. The control decisions are made relative to the crankshaft angular position based on a number of sensory data. The timing relative to the crankshaft position may be varied as a function of engine speed in order to optimize the engine performance. The delay from the time current pulse sent to the injector and the time that combustion is fully developed is in the order of 15 degrees of crankshaft angle. The typical shape of the pressure in the cylinder during a four-stroke cycle are also shown in the Figure 1.20. The maximum combustion pressure is in the range of 30 bar (3 Mpa) to 160 bar (16 MPa). Notice that even though the pressure in the cylinder is positive, the torque contribution of each cylinder as a result of this pressure is positive during the cycle when the piston is moving down (the pressure is helping the motion and the net torque contribution to the crankshaft is positive) and negative during the cycle when the piston if moving up (the pressure is opposing the motion and the net torque contribution to the crankshaft is negative). As result, the net torque generated by each cylinder oscillates as function of crankshaft angle with a period of two revolutions. The mean value of that generated torque by all cylinders is the value used for characterizing the performance of the engine.

In electronically controlled engines, the fuel injection timing relative to the crankshaft position is varied as a function of engine speed in order to give enough time for the combustion to develop. This is called the *variable timing fuel injection* control. As the engine speed increases, the injection time is advanced, that is, fuel is injected earlier relative to the TDC of the cylinder during the compression cycle. Injection timing has a significant effect on the combustion efficiency, hence the torque produced, as well as the emission content.

It is standard in the literature to look at the cylinder pressure versus the combustion chamber volume during the four-stroke cycle. The so-called p-v diagram shape in general looks like as is shown in Figure 1.20. The net energy developed by the combustion process is proportional to the area enclosed by the p-v diagram. In order to understand the shape of the torque generated by the engine, let us look at the pressure curve as a function of the crankshaft (Figure 1.20) and superimpose the same pressure curve for other cylinders with the appropriate crankshaft phase angle. The sum of the pressure contribution from each cylinder is the total pressure curve generated by the engine (Figure 1.20). The pressure multiplied by the piston top surface area is the net force generated. The effective moment arm of the connecting rod multiplied by the force gives the torque generated at the crankshaft.

The net change in the acceleration is the net torque divided by the inertia. If the inertia is large, the transient variations in the net torque will result in smaller acceleration changes, and hence smaller speed changes. At the same time, it takes a longer time to accelerate or decelerate the engine to a different speed. These are the advantages and disadvantages of the *flywheel* used on the crankshaft.

1.1.2 Engine Control System Components

There are three groups of components of an engine control system: (1) sensors, (2) actuators, (3) electronic control module (ECM) (Figure 1.21). The number and type of sensors used in an electronic engine controller varies from manufacturer to manufacturer. The following is a typical list of sensors used:

1. accelerator pedal position sensor,
2. throttle position sensor (if the engine has throttle),
3. engine speed sensor,
4. air mass flow rate sensor,
5. intake manifold (boost) absolute pressure sensor,

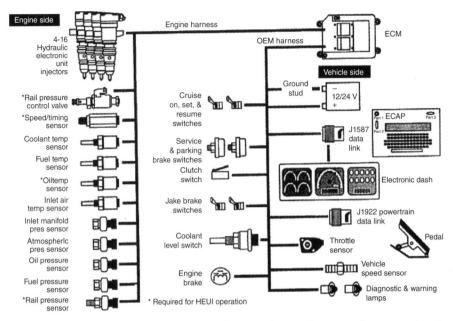

FIGURE 1.21: Control system components for a modern engine controller: sensor inputs, ECM (electronic control module), outputs to actuators. Reprinted Courtesy of Caterpillar Inc.

  6. atmospheric pressure sensor,

  7. manifold temperature sensor,

  8. ambient air temperature sensor,

  9. exhaust gas oxygen (EGO) sensor,

  10. knock detector sensor (piezo-accelerometer sensor)

The controlled actuators in an engine (outputs) are:

  1. fuel injector actuation: injection timing, duration (injected fuel amount, also called fuel ratio) and pulse shape control,

  2. ignition sparks: timing (only in spark-ignited gasoline engines, not used in diesel engines),

  3. exhaust gas recirculation (EGR) valve,

  4. idle air control (IAC) valve, which may not be present in all engine designs,

  5. waste-gate valve,

  6. exhaust back pressure valve, and

  7. turbo charger.

ECM is the digital computer hardware which has the interface circuitry for the sensors and actuators and runs the engine control algorithm (engine control software) in real-time. The engine control algorithm implements the control logic that defines the relationship between the sensor signals and actuator control signals. The main objectives of engine control are

  1. engine speed control,

  2. fuel efficiency, and

  3. emission concerns.

In its simplest form, the engine control algorithm controls the fuel injector in order to maintain a desired speed set by the accelerator pedal position sensor.

The actively controlled variables by ECM are the injectors (when and how much fuel to inject – an analog signal per injector), and the RPCV valve which is used to regulate the pressure of the amplification oil line. As a result, for a six-cylinder, four-stroke cycle diesel engine, the engine controller has seven control outputs: six outputs (one for each injector solenoid) and one output for the RPCV valve (Figure 1.19). Notice that at 3000 rpm engine speed, 36° of crankshaft rotation takes only about 2.0 ms, which is about the window of opportunity to complete the fuel injection. Controlling the injection start time with an accuracy of 1° of crankshaft position requires about 55.5 microsecond repeatability in the fuel-injection control system timing. Therefore, accuracy in controlling the injection start time and duration at different engine speeds is clearly very important. Since we know that the combustion and injection processes have their own inherent delay due to natural physics, we can anticipate these delays in a real-time control algorithm, and advance or retard the injection timing as a function of the engine speed. This is called *variable injection timing* in engine control.

Solenoid actuated fuel injectors are digitally controlled, thereby making the injection start-time and duration changeable in real-time based on various sensory and command data. The injection start-time and duration are controlled by the signal sent to the solenoid. The solenoid motion is amplified to high pressure injection levels via high pressure hydraulic lines (i.e., in the case of HEUI injectors by Caterpillar Inc.) or by cam-driven push rod arms (i.e., in the case of EUI injectors by Caterpillar Inc.).

The intake manifold absolute pressure is closely related to the load on the engine – as the load increases, this pressure increases. The engine control algorithm uses this sensor to estimate the load. Some engines also include a high bandwidth acceleration sensor (i.e., piezoelectric accelerometer) on the engine cylinder head to detect the "knock" condition in the engine. Knock condition is the result of excessive combustion pressures in the cylinders (usually under loaded conditions of engine) as a result of premature and unusually fast propagation of ignition of the air–fuel mixture. The higher the compression ratio is, the more likely the knock condition is. The accelerometer signal is digitally filtered and evaluated for knock condition by the control algorithm. Once the control algorithm has determined which cylinders have knock condition, the fuel injection timing is retarded until the knock is eliminated in the cyclinders in which it has been detected.

In diesel engines with electronic governors, the operator sets the desired speed with the pedal which defines the desired speed as a percentage of maximum speed. Then the electronic controller modulates the fuel rate up to the maximum rate in order to maintain that speed. The engine operates along the vertical line between the desired speed and the lug curve (Figure 1.22). If the load at that speed happens to be larger than the maximum torque the engine can provide at that speed, the engine speed drops and torque increases until the balance between load torque and engine torque is achieved. In most gasoline engines, the operator pedal command is a desired engine torque. The driver closes the loop on the engine speed by observing and reacting to the vehicle speed. When "cruise control" is activated, than the electronic controller regulates the engine fuel rate in order to maintain the desired vehicle speed.

1.1.3 Engine Modeling with Lug Curve

If we neglect the transient response delays in the engine performance and the oscillations of engine torque within one cycle (two revolutions of crank angle), the steady-state performance of an engine can be described in terms of its mean (average) torque per cycle, power, and fuel efficiency as a function of engine speed (Figure 1.22). The most important

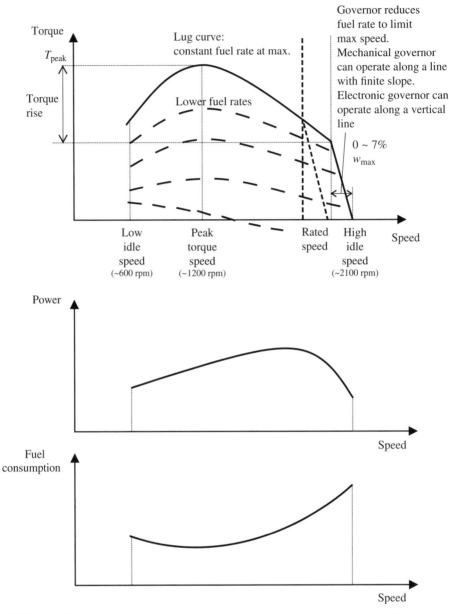

FIGURE 1.22: Steady-state engine performance: torque (lug), power and fuel efficiency as function of engine speed.

of these three curves that defines the capabilities of an engine is the torque-speed curve. This curve is also called the "lug curve" due to its shape. As the speed is reduced down from the rated speed, the mean torque generated by the engine increases under constant fuel rate conditions. Hence, if the load increases to slow down the engine, the engine inherently increases torque to overcome the load. In order to define the lug curve model for an engine, we need a table of torque versus the engine speed for maximum fuel rate. A linear interpolation between intermediate points is satisfactory for initial analysis. The table should have, at minimum, the low idle, high idle, peak torque, and rated speed points (four data points). The points under that curve are achieved by lower fuel rates. As the fuel injection rate decreases from its maximum value, the lug curve does not necessarily scale linearly with it. At very low fuel rates, the combustion process works in such a way that the shape of the lug curve becomes quite different (Figure 1.22). The best fuel rate is accomplished when the engine speed is around the middle of low and high idle speeds.

The torque defined by the lug curve is the *mean effective* steady-state torque capacity of the engine as a function of speed at maximum fuel rate. This curve does not consider the cyclic oscillations of net torque as a result of the combustion process. The actual torque output of the engine is oscillatory as a function of the crankshaft angle within each four-stoke (intake-compression-combustion-exhaust) per two-revolution cycle. The role of the flywheel is to reduce the oscillations of engine speed. Each cylinder has one cycle of net torque contribution (which has both positive and negative portions) per two revolutions, and the torque functions of each cyclinder are phase shifted from each other.

A given mechanical engine size defines the boundaries of this curve:

1. maximum engine speed (high idle speed) determined by the friction in the bearings and combustion time needed,

2. lug portion of the curve corresponding to torque-speed relation when the maximum fuel rate is injected to the engine. This is determined by the heat capacity of the engine as well as the injector size.

$$T_{eng}^{lug} = f_0^{lug}(w_{eng}) \quad \text{for} \quad u_{fuel} = u_{fuel}^{max} \tag{1.2}$$

Once the limits of the engine capacity are defined by mechanical design and sizing of the components, we can operate the engine at any point under that curve with the governor – the engine controller. Any point under the engine lug curve corresponds to a fuel rate and engine speed. Notice that the left side of the lug curve (speeds below the peak torque) is unstable since the engine torque capacity decreases. At speeds below this point, if the load increases as the speed decreases, the engine will stall. The slope of the high idle speed setting can be 3–7% of speed change in mechanically controlled engines, or it can be made perpendicular (almost zero speed change) in electronically controlled engines. The governor (engine controller) cuts down the fuel rate in order to limit the maximum speed of the engine. The sharp, almost vertical, line on the lug curve (steady-state torque-speed curve) is the result of the governor action to limit the engine speed by changing the fuel rate from its maximum value at the rated speed down to smaller values. Hence, electronically controlled engines can maintain the same speed as the load conditions vary from zero to maximum torque capacity of the engine.

The parameterized version of the lug curve (torque versus engine speed), where there is a curve defined for each value of fuel rate, is called the *torque map*. The engine is modeled as a torque source (output: torque) as function of two input variables: fuel rate and engine speed. Such a model assumes that the necessary air flow is provided in order to satisfy the fuel rate to torque transfer function (Figures 1.23, 1.22).

$$T_{eng} = f_0(u_{fuel}, w_{eng}) \tag{1.3}$$

Notice that, when $u_{fuel} = u_{fuel}^{max}$, the function $f_0(\cdot, \cdot)$ represents the lug curve. In order to define the points below the lug curve, data points need to be specified for different values of fuel rate (Figure 1.22). An analytical representation of such a model can be expressed as

$$T_{eng} = \left(\frac{u_{fuel}}{u_{fuel}^{max}}\right) \cdot f_0\left(u_{fuel}^{max}, w_{eng}\right) \tag{1.4}$$

Clearly, if the engine capacity limits are to be imposed on the model, the lug curve limits can be added as a nonlinear block in this model. The standard "lug curve" defines the

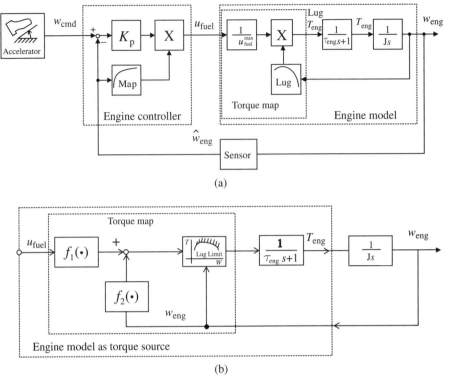

(a)

(b)

FIGURE 1.23: Engine models and closed loop control: (a) Engine is modeled as a relationship between the fuel rate and generated torque. The time delay and filtering effects due to combustion and injection are taken into account with a first-order filter dynamics with a time constant of τ_{eng}. Torque pulsation due to individual cylinder combustion is also neglected, however it can be included by using such a model for individual cylinders. This model assumes that the lug curve scales linearly with the fuel injection rate. (b) This model assumes that the torque can be expressed as the difference of two functions where one function output depends on engine speed and the other on the fuel rate.

steady-state torque-speed relationship for an engine when the maximum fuel rate is injected. This is also called the *rack stop* limit of the torque-speed curve. Under the governor control (mechanical or electronic engine controller), any point (speed, torque) under that curve can be accomplished by a fuel rate that is less than the maximum fuel rate. In order to impose the engine lug curve limits, the following must be defined

1. torque-speed (the lug curve) for maximum fuel rate ($T_{eng}(w_{eng})$ when $u_{fuel} = u_{fuel}^{max}$),
2. the parameterized torque-speed curves for different fuel rates less than the maximum fuel rate.

This model and the lug curve model do not include the transient behavior due to the combustion process. The simplest transient (dynamic) model to account for the delay between the fuel rate and generated torque can be included in the form of a first-order filter. In other words, there is a filtering type delay between the torque obtained from the lug curve and the actual torque developed by the engine for a given fuel rate and engine speed,

$$\tau_e \frac{dT_{eng}(t)}{dt} + T_{eng}(t) = T_{eng}^{lug}(t) \tag{1.5}$$

and its Laplace transform (readers who are not familiar with Laplace transforms can skip related material in the rest of this section without loss of continuity),

$$T_{eng}(s) = \frac{1}{(\tau_{eng} \cdot s + 1)} \cdot T_{eng}^{lug}(s) \qquad (1.6)$$

where τ_{eng} represents the time constant of the combustion to torque generation process, T_{eng}^{lug} is the torque prediction based on lug-curve, and T_{eng} is the torque produced including the filtering delay.

Special Case: Simple Engine Model   Simpler versions of this model can be used to represent engine steady-state dynamics as follows,

$$T_{eng} = f_1(u_{fuel}) - f_2(w_{eng}) \qquad (1.7)$$

where T_{eng} is the torque generated by the engine, u_{fuel} is the injected fuel rate, w_{eng} is engine speed, $f_0(\cdot,\cdot)$ represents the nonlinear mapping function between the two independent variables (fuel rate and engine speed) and the generated torque. The function $f_1(\cdot)$ represent the fuel rate to torque generation through the combustion process, $f_2(\cdot)$ represents the load torque due to friction in the engine as a function of engine speed (Figure 1.23).

1.1.4   Engine Control Algorithms: Engine Speed Regulation using Fuel Map and a Proportional Control Algorithm

A very simple engine control algorithm may decide on the fuel rate based on the accelerator pedal position and engine speed sensors as follows (Figure 1.23),

$$u_{fuel} = g_1(w_{eng}) \cdot K_p \cdot (w_{cmd} - w_{eng}) \qquad (1.8)$$

where $g_1(\cdot)$ is fuel rate look-up table as a function of engine speed and K_p is a gain multiplying the speed error.

An actual engine control algorithm is more complicated, uses more sensory data and embedded engine data in the form of look-up tables, estimators, and various logic functions such as cruise control mode and cold start mode. In addition, the control algorithm decides not only on the fuel rate (u_{fuel}) but also on the injection timing relative to the crank shaft position. Other controlled variables include the exhaust gas recirculation (EGR) and idle air control valves. However, simple engine control algorithms like this are useful in various stages of control system development in vehicle applications.

Example: Electronic Governor for Engine Control   The word "*electronic governor*" is an industry standard name used to define the digital closed-loop controller (or the embedded control module or electronic control module (ECM)) which is used to control ("regulate", "govern") the engine speed (Figure 1.24). It is simply the digital implementation of the mechanical governor, except that in addition to controlling speed, the digital controller can take many other factors into consideration when deciding on "how much" and "when" and "how" (injection pulse pattern) to inject fuel to control the engine speed.

Let us consider the steady-state torque-speed relationship for an engine, that is the lug curve (Fig. 1.19). Under a properly designed electronic governor, we can have the engine operate at any point under the lug curve. At *low idle speed*, which is the minimum speed allowed, the torque-speed curve is generally maintained to be a straight vertical line. If the throttle pedal sensor output is zero, which means the throttle pedal is not pressed at all, the desired engine speed is interpreted as the low idle speed, and the engine

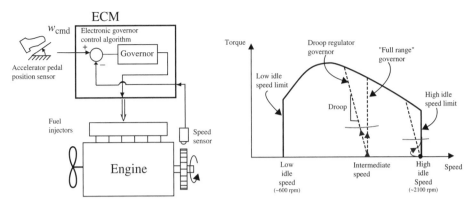

FIGURE 1.24: Electronic governors: steady-state operation and performance description on lug curve.

controller is suppose to control the fuel rate from zero to maximum fuel rate range in order to maintain the engine speed equal to the low idle speed. As the load torque on the engine changes (i.e., due to friction, temperature condition, or other parasitic loads such as heating and air-conditioner load), the controller will increase or decrease the fuel rate along a vertical line on the low idle side, from zero (or minimum) fuel rate to maximum fuel rate range.

On the *high idle speed*, that is when throttle pedal is pressed to maximum position, the electronic governor again tries to maintain the maximum desired high idle speed by selecting the proper fuel rate from zero to maximum fuel rate. Let us consider that the engine is at the high idle speed (throttle is pressed to maximum position) and the load torque is zero. After some time, let us assume the engine load torque increases to a value that is less than the maximum torque capacity of the engine at the high idle speed. This will result in a drop in engine speed in transient response. But, in steady-state the engine speed should recover to the high idle speed. In other words, engine speed is regulated along a vertical line at high idle speed. Mechanical governors are able to operate along a line that has a finite slope instead of being vertical line. If we desired to emulate mechanical governor behavior in our electronic governor, then the control algorithm for the electronic governor may be modified as follows. The commanded (desired) speed (w_{cmd}) is determined first from the throttle position sensor, then it is modified (w^*_{cmd}) based on the estimated load torque. Let us assume that we have estimated load torque (T_l) information or measurement in real-time.

$$w_{cmd} = f(\theta_{throttle}); \quad \textit{throttle or accelerator pedal sensor position} \quad (1.9)$$

$$= w_{max}; \quad \textit{for high idle condition} \quad (1.10)$$

$$w^*_{cmd} = w_{cmd} - \frac{1}{K_{tw}} \cdot T_l \quad (1.11)$$

With this modified command signal to the closed loop control algorithm, the net result of the high idle speed regulation turns into a line with slope of $-K_{tw}$ instead of being a vertical line.

Likewise, at any intermediate speed between low idle and high idle, the engine can be controlled to maintain a desired speed exactly, regardless of the load torque amount as long as it is less than the maximum torque the engine can provide at that speed using maximum fuel rate. This torque is defined by the lug curve for each speed. If we want to achieve a vertical line of speed regulation (ability to maintain speed despite load torque as long as

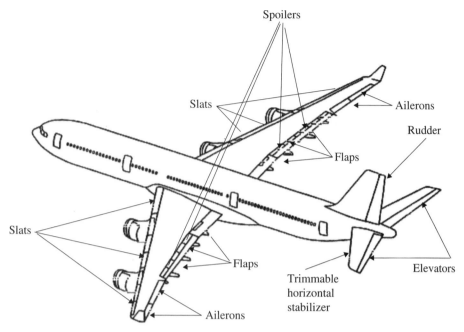

FIGURE 1.25: A modern commercial airplane (Boeing 777) and its flight control surfaces.

load torque is less than maximum available), the commanded speed will be only a function of the pedal position.

$$w_{\text{cmd}} = f(\theta_{\text{throttle}}); \quad \textit{throttle or accelerator pedal sensor position} \quad (1.12)$$

Emission Issues One of the fundamental challenges in engine control comes from emission requirements. There are five major emission concerns in the exhaust smoke:

1. CO_2 content due to global warming effects,
2. CO – health concerns (colorless, odorless, tasteless, but when inhaled in 0.3% of air can result in death within 30 minutes),
3. NO_x content (NO and NO_2),
4. HC when combined with NO_2 becomes harmful to health,
5. PM (particulate matter) solid and liquids present in the exhaust gas.

The fuel injection timing has a major influence on the quality of combustion and hence the exhaust gas composition. The challenge is to design engine and controllers which will create efficient combustion while reducing all of the undesirable emission components in the exhaust gas.

1.2 EXAMPLE: ELECTRO-HYDRAULIC FLIGHT CONTROL SYSTEMS FOR COMMERCIAL AIRPLANES

Figure 1.25 shows an example of a modern commercial airplane widely used in passenger and cargo transportation, the Boeing 777, and its the flight control surfaces.[1] The forward

---

[1] This section was written with significant assistance from Dr. Olaf Cochoy.

propulsion force for the airplane is provided by jet engines (or propeller engines in smaller planes). There are four major groups of forces acting on an airplane (Figure 1.26):

1. Gravity force (F_g).
2. Thrust force generated by the engines (F_t).
3. Lift force generated by the movable wings (flight control surfaces) (F_l).
4. Drag force due to friction between the air and aircraft body (F_d).

When the thrust force is in the horizontal direction, the vertical acceleration of the airplane is determined by the difference between lift force and gravitational force, whereas the horizontal acceleration is determined by the difference between the thrust force and drag force. Gravitational force is always in the z-direction of the global coordinate system. Thrust force direction is a function of the attitude (pitch angle) of the airplane. The lift and drag forces are aerodynamic forces and are strongly dependent on the relative speed between the airplane body and air flow, as well as the aerodynamic shape of the airplane, that is the configuration of the movable wing sections.

There are six coordinate variables to be controlled in an airplane motion: x_G, y_G, z_G position coordinates of the center of mass of the airplane relative to a reference on earth, the orientation orientation θ, β, ϕ (*yaw, roll, pitch*) of the airplane relative to a reference frame.

By convention, airplane motion can be described in two groups:

1. *longitudinal motion* that includes the motion along axial direction x and vertical direction z as well as pitch rotation motion about the y axis,
2. *lateral motion* that includes the rotational motions of roll and yaw.

The longitudinal and lateral motion variables are lightly coupled to each other, that is, motions in the lateral plane result in small motions in the longitudinal plane, and vice versa. The position coordinates are obtained by the integral of the speed vector of the airplane along its flight path. The linear speed vector of the airplane is determined by the integral of the acceleration vector which is determined by the net force vector acting on the center of mass of the airplane.

For simplicity, if we assume the thrust force and the drag force are in the x-direction, and lift and gravitational forces are in the z-direction, the dynamic relationship for motion of the airplane in the x and z directions can be expressed as (Figure 1.26),

$$m \cdot \ddot{x}_G(t) = F_t(t) - F_d(t) \tag{1.13}$$

$$m \cdot \ddot{z}_G(t) = F_l(t) - F_g(t) \tag{1.14}$$

where $x_G(t), y_G(t), z_G(t)$ are the coordinates of the center of mass of the airplane with respect to a fixed reference frame, F_t, F_d, F_l, F_g are thrust, drag, lift and gravity forces. For simplicity we show them as if they act as center of mass of the airplane as single vectors. In reality, they are distributed over the whole plane and depend on the current position of the control surface. The motion in y-direction is negligable since the accelerations in the y-direction are due to the small aerodynamic drag forces resulting from the relative speed of the airplane and air in the y-direction.

$$m \cdot \ddot{y}_G(t) = -F_{dy}(t) \tag{1.15}$$

where $F_{dy}(t)$ is the drag force in the y-direction. However, there are finite drag forces in the y-direction, and hence the resulting motion in y-direction. The airplane path and orientation is modified to correct for the unintended motion in the y-direction. Reference [4] gives a more accurate dynamic relationship between aerodynamic forces generated by the control surfaces, thrust force, and aircraft motion.

Orientation speeds are similarly obtained by integral of the angular accelerations which are determined by the net moments about the respective axes by all of the above referenced forces.

$$\frac{d}{dt}(\vec{H}_G(t)) = \vec{T}_G(t) \tag{1.16}$$

$$I_{3x3} \cdot \vec{\alpha}(t) + \vec{\omega}(t) \times \vec{H}_G(t) = \vec{T}_G(t) \tag{1.17}$$

where

$$\vec{H}_G(t) = H_{Gx}(t) \cdot \vec{i} + H_{Gy}(t) \cdot \vec{j} + H_{Gz}(t) \cdot \vec{k} \tag{1.18}$$

$$= I_{3x3} \cdot \vec{w}(t) \tag{1.19}$$

$$\vec{T}_G(t) = T_{Gx}(t) \cdot \vec{i} + T_{Gy}(t) \cdot \vec{j} + T_{Gz}(t) \cdot \vec{k} \tag{1.20}$$

$$\vec{w}(t) = w_x(t) \cdot \vec{i} + w_y(t) \cdot \vec{j} + w_z(t) \cdot \vec{k} \tag{1.21}$$

$$\vec{\alpha}(t) = \frac{dw_x(t)}{dt} \cdot \vec{i} + \frac{dw_y(t)}{dt} \cdot \vec{j} + \frac{dw_z(t)}{dt} \cdot \vec{k} \tag{1.22}$$

$$= \alpha_x(t) \cdot \vec{i} + \alpha_y(t) \cdot \vec{j} + \alpha_k(t) \cdot \vec{k} \tag{1.23}$$

$$I_{3x3} = \begin{bmatrix} I_{xx} & -I_{xy} & -I_{xz} \\ -I_{yx} & I_{yy} & -I_{yz} \\ -I_{zx} & -I_{zy} & I_{zz} \end{bmatrix} \tag{1.24}$$

$$H_{Gx}(t) = I_{xx} \cdot w_x(t) - I_{xy} \cdot w_y(t) - I_{xz} \cdot w_z(t) \tag{1.25}$$

$$H_{Gy}(t) = -I_{yx} \cdot w_x(t) + I_{yy} \cdot w_y(t) - I_{yz} \cdot w_z(t) \tag{1.26}$$

$$H_{Gz}(t) = -I_{zx} \cdot w_x(t) - I_{zy} \cdot w_y(t) + I_{zz} \cdot w_z(t) \tag{1.27}$$

where $\vec{H}_G(t)$ is the angular momentum vector of the airplane with respect to a coordinate frame fixed to the center of mass of the airplane.

$\vec{T}_G(t)$ is the net moment vector about the axes of the coordinate frame due to all the forces acting on the airplane (gravity, thrust, lift (forces generated by flight control surfaces), drag forces).

I_{3x3} is the moment of inertia matrix which is symmetric ($I_{xy} = I_{yx}, I_{xz} = I_{zx}, I_{yz} = I_{zy}$), and $\vec{w}(t)$ is the angular velocity vector, $\vec{\alpha}(t)$ is the angular acceleration vector, of the airplane, with respect to the same coordinate frame. The coordinate frame xyz is fixed to the airplane center of mass and moves with it. Hence if the weight distribution of the airplane does not change, the I_{3x3} matrix is constant. However, due to consumed fuel and movement of passengers during the flight, the inertia matrix (I_{3x3}) changes slowly. In addition, the center of mass coordinates of the airplane also change during flight due to the same reasons. Fuel stored in the wings and other parts of the airplane body is pumped to the engines in such a way that the change in the center of mass location is smooth and slow as a function of time.

The movable surfaces on the two wings and on the tail of the airplane are used to effect the aerodynamic forces, hence the lift and the orientation of the airplane (Figures 1.25 and 1.26). Aerodynamic forces (lift and drag forces) acting on the airplane are functions of

1. relative speed between the airplane and the air (hence the speed of the airplane, as well as the air speed (direction and magnitude) and turbulance conditions),

2. shape of the airplane, where some of the shapes are adjustable during flight, such as the control surfaces on the wings and the tail.

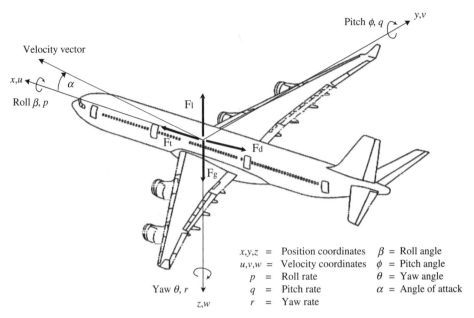

FIGURE 1.26: Motion variables of an airplane: three position coordinate variables x, y, z of a coordinate frame origin attached to the airplane, and the orientation coordinate variables roll, yaw, and pitch (θ, β, α) of a coordinate frame attached to the airplane relative to a reference frame. Three main sources of force on the airplane: 1. trust force generated by the engines, 2. gravitational force, 3. aerodynamic force (drag and lift) which are functions of the relative speed of the airplane with respect to surrounding air, shape of the plane geometry (i.e., changing shape via control surfaces), and air density.

The propulsion force vector is generated by the jet engines. The direction of this force relative to the aircraft body is fixed, whereas the magnitude of it is controlled by the throttle level of the engine. When the airplane has non-zero velocity, the orientation of the airplane about three coordinate axes can be affected by moving a set of control surfaces.

It is important to recognize that the airplane must be moving at a relatively high speed in order to generate sufficient aerodynamic lift forces with the aid of control surfaces. For a given airplane frame and speed, the net direction and magnitude of aerodynamic forces are controlled by the movable flight control surfaces. These aerodynamic forces affect the orientation and lift of the airplane. As the orientation of the airplane changes, the direction of the propulsion forces also change since they are generated by the engines connected to the airplane frame. For instance, when the pitch of the airplane is positive, the thrust force generated by the engines helps the airplane gain altitude in the z-direction. In modern fighter jets, the engine trust vector is controllable. That is, not only the magnitude of the trust force, but also its vector direction relative to the aircraft frame is controllable. In such airplanes, the trust force is controlled as a vector quantity (both magnitude and direction). Hence, the trust force can be actively used to aid the lift force for the airplane.

The flight control surfaces deal with controlling the orientation angles of the airplane (yaw, roll, and pitch) during take-off and landing, as well as during flight. The flight control surfaces and their control systems are grouped into two categories:

1. primary flight control systems,
2. secondary flight control systems (also called high-lift flight control systems).

The secondary (high-lift) flight control surfaces are used only during the take-off and landing phases of the flight. As the name implies, these surfaces are designed to provide increased lift for the airplane while it is at a relatively low speed. Secondary flight control surfaces include the *left and right flaps*, *left and right slats*, and *spoilers*. The *flaps* and *slats* are positioned to predefined positions during take-off and landing. Spoilers are primarily used to reduce the lift and are generally used during landing, hence the name "spoilers" which "spoil" the lift. Slats and flaps increase the lift coefficient of the airplane by providing a larger lift surface area and by providing a guided airflow path for larger lift coefficient. They are typically moved to a desired position and held at that position during the take-off and landing phase of the flight until that phase is completed. Then these surfaces are moved back to their original fold positions during flight. The secondary surfaces require a relatively low bandwidth control system since they move to a predefined position and stay there. Slats and flaps typically contribute to the lift by increasing the effective angle of attack and lift surface area for a given airplane (Figure 1.25a). During take-off, the highest possible angle of attack is desired, hence slats are fully deployed and flaps might be partially deployed at a certain position. During landing, the lowest possible speed is desired, as opposed to maximum attack angle during take-off, hence the flaps are fully deployed while slats might only be partially deployed.

Primary flight control surfaces are used during the flight to maintain the orientation of the airplane. The primary flight control surfaces include the *left and right ailerons, right and left elevators, and rudder*. At any given time, the airplane has a commanded orientation in terms of pitch, yaw, and roll angles. An on-board orientation measurement sensor (an *electromechanical gyroscope* with a rotor spinning at a constant high speed by an electric motor or laser based *ring-laser gyroscope*) is used to measure the actual orientation of the airplane relative a fixed coordinate frame. Then the primary flight control surfaces are actuated based on closed-loop controls in order to maintain the desired orientation. The bandwidth of the closed loop control system for primary flight control surfaces must be fast and well damped in order to maintain a very smooth flight condition. Whereas the bandwidth requirements of the secondary flight control systems is much smaller. An auto-pilot generates the desired orientation and engine thrust signals during the flight. A pilot can over-ride these command signals from the auto-pilot and command them manually with a joystick.

Although the motion of the primary control surfaces affects multiple orientation variables, they are mainly related in a one-to-one relationship as follows:

1. *Left and right ailerons* are always actuated asymmetrically (in the opposite direction and equal amount) to generate opposite aerodynamic forces, hence torque, for *roll* motion. In some conditions, the roll motion is augmented by the motion of *spoilers* if decreasing the lift, hence reducing the altitude of the plane, is desired during the roll motion. This is typically needed during the approach for landing.

2. *Left and right elevators* are always operated symmetrically (in the same direction and equal amount) to generate aerodynamic forces, hence torque about center of mass, for the *pitch* motion. During flight, the center of mass of the airplane shifts as fuel is used and passengers move around. In order to balance the forces and maintain the pitch angle of the plane, *trimmable horizontal stabilizer* (THS) is used. The trimmable horizontal stabilizer pivots up and down, hence moving the whole tail section up and down. For instance, if during flight a change in cruise altitude is desired, the elevators are used to induce a pitch moment and change the angle of attack of the airplane. On the other hand, during a constant altitude cruising flight condition, the pitch moment about the center of gravity of the airplane has to be balanced. As the fuel is used during the flight and passengers move around, the center of gravity changes. This balancing

is done by the trimmable horizontal stabilizer which changes the angle of attack slightly, while allowing elevators to remain in their neutral position. Technically, the role of trimmable horizontal stabilizer can be performed by the elevators. However, that would result in increased drag.

3. The rudder is used to generate aerodynamic forces, hence torque about center of mass, for the *yaw* motion.

Quite often, the orientation motions happen simultaneously, that is the airplane makes a roll motion while at the same time making a change in its yaw and/or pitch orientation.

In modern large civilian aircrafts, the flight control surfaces are actuated by hydraulic power, where the delivered hydraulic power is controlled by electrical signals. The total hydraulic power generated by the hydraulic pumps, driven by the jet engines, on a Boeing 747, is about 300 kW, on a Boeing 777 it is about 400 kW, and on an Airbus 380 it is about 800 kW. The electrical control signals are delivered to hydraulic components by-wire, that is an electrical current signal delivered to a servo valve. As a result, the name "fly-by-wire" is used to describe flight control systems based on electro-hydraulic systems. Hydraulic power is the most widely used power type for motion on large aircraft flight control surfaces. In order to take advantage of digital computer control, the control of hydraulic power is carried out electrically by digital computers. A closed loop fly-by-wire system means that the control signal to the actuator is generated by the flight control computer based on the error between a command signal (generated either by auto-pilot software or by the pilot via the joystick) and the on-board sensors, just like any other closed loop control system. If the control signal to the actuators is generated only based on the command signal, and no feedback sensor is used, then it is an open loop fly-by-wire system.

The design of a fly-by-wire electrohydraulic flight control system is dictated by requirements in

1. safety and reliability,

2. power,

3. weight,

4. operational and functional performance.

Safety and reliability are the utmost considerations in flight control systems. Failure modes, such as loss of hydraulic pressure or control surface runaway, are classified in different categories (minor, major, hazardous, catastrophic) by their effect on the function of the plane. The highest requirement is for catastrophic failures, leading to possible loss of the aircraft, for which it has to be proven that the catastrophic failure condition has a probability to occur in less than one in 10^9 per flight hour. Redundant systems are the key for improved safety. That means designing double or triple redundancy for a given function so that there are two or three more alternatives in actuating and controlling a surface if one or two of them fails. The redundancy must be provided at the level of

1. the power source (i.e., engine driven pump, electric generator driven pump where generator is driven by engine, ram air driven generator in case of all engine failure),

2. the power distribution and metering elements (redundant hydraulic pipe lines and valves),

3. the power delivery (redundant hydraulic cylinders and motors),

4. the sensing (redundant position, orientation, pressure sensors), and

5. the control computer (redundant control signal lines to actuators as well as redundant flight control computers).

For redundancy purposes, all hydraulic actuators (electro-hydraulic servo actuators (EHSA) and motors) are powered by one of three independent hydraulic power supply systems. A typical hydraulic power supply system is of constant pressure type in aerospace applications. The constant pressure may be supplied by a fixed displacement pump with a pressure limiting unload-valve or a output pressure regulated variable displacement pump. The second option is more energy efficient in converting mechanical engine power to desired hydraulic power.

These hydraulic power supply systems are named Blue, Green, and Yellow. In order to meet reliability requirements for the supply systems, several independent power sources are used. The main power sources for aircrafts are the engines. They not only create the necessary thrust for flight, but also supply power to drive generators (integrated drive generator, IDG) and hydraulic pumps (engine driven pump (EDP)). An additional electrically driven pump is connected to the yellow system. A backup for the blue system is the ram air turbine (RAT), which is activated (roughly located below the cockpit) in case both engines fail. The green and the yellow hydraulic systems are connected by a so-called power transfer unit (PTU), comprising of two hydraulic machines (one fixed, one variable displacement), both of which can operate as hydraulic pumps or motors. Thus, energy can be transferred between both supply systems. The PTU might be used during normal operation in case the power demand from one of the two hydraulic systems is too high (e.g., during deployment of slats and flaps). The "hydraulic power consumers" (actuators) are connected to the hydraulic supply systems with different priorities controlled by a priority valve, with the primary flight control actuators at highest priority (Figure 1.27).

Every actuator in flight control surfaces uses one of the constant supply pressure lines as a hydraulic power source, and meters (controls) the flow to the rotary hydraulic

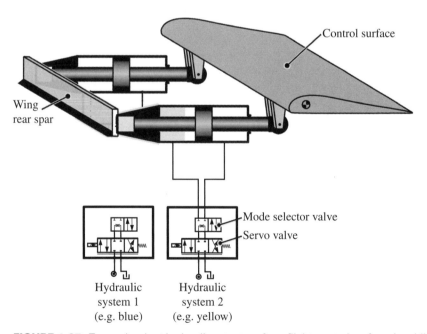

FIGURE 1.27: Two redundant hydraulic actuators for a flight control surface. In addition to redundant cylinder actuators, there are redundant flow control valves as well as redundant hydraulic power source (blue and yellow supply). As shown in this example, redundancy for all three main components of a hydraulic control system is designed into the hardware in the design: redundant pumps, redundant valves, and redundant actuators (hydraulic cylinders or motors).

motor or a hydraulic cylinder using a servo valve. The actuator output shaft drives a linkage connected to the moveable control surface. The nominal hydraulic supply line pressure in commercial airplanes today is in the range of 3000–5000 psi. In order to increase the hydraulic power density (hydraulic power delivered/mass of the hydraulic components), the future operating pressures are expected to get higher.

The current trend in the aerospace industry is towards the so-called "more-electric aircraft" (or ultimately, the "all-electric aircraft"). As this name suggests, it is desired to replace hydraulic powered systems by electrically powered ones. Historically, flight control systems have evolved through four different stages (Figure 1.28) in terms of the way the aerodynamic surfaces are *controlled* and *powered*,

1. mechanically control signaled (via mechanical linkages and cables), hydraulically powered flight control systems of the past (Figure 1.28a),

2. pilot-hydraulically control signaled, hydraulically powered flight control systems of the past (Figure 1.28b),

3. electrically control signaled, hydraulically powered flight control systems of the present (Figure 1.28c),

4. electrically (or optically) control signaled, electrically powered flight control systems of the future (Figure 1.28d).

In the so called hydro-mechanical control systems (Figure 1.28b), there are no electronic components. All of the control decisions and sensing are accomplished via mechanical and hydraulic means (Figure 1.6).

The main reasons for the push for more electric aircraft lies in:

1. easier installation and maintainability of electric wiring compared to hydraulic piping,

2. improved efficiency of power usage (at generation, transmission, and actuation stages) due to on-going improvement in electric motor and power electronics technology.

1.3 EMBEDDED CONTROL SOFTWARE DEVELOPMENT FOR MECHATRONIC SYSTEMS

The trend in industrial practice is that the embedded control software development part of modern mechatronics engineering is done involving three phases (Figures 1.29, 1.30, 1.31):

- Phase 1: Control software development and simulation in non-real-time environment.
- Phase 2: Hardware in-the-loop (HIL) simulation and testing in real-time environment.
- Phase 3: Testing and validation on actual machine.

In phase 1, the control software is developed by using graphical software tools, such as Simulink® and Stateflow, simulated and analyzed on a non-real-time computer environment (Figure 1.29). The "plant model," which is the computer model of the machine to be controlled, is a non-real-time detailed dynamic model. Simulations and analysis are done in this non-real-time environment.

In phase 2, the "same control software" is tested on a target embedded control module (ECM). That "same control software" is a C-code which is auto-generated from the graphical diagrams of Simulink® and Stateflow using auto-code generation tools such as Simulink® Coder, Embedded Coder, and MATLAB® Coder. That real-time controller software is run on the target embedded controller module (ECM) hardware in real-time, which can be connected to another computer which simulates the controlled process dynamics in real-time. This case is called the *hardware in-the-loop (HIL)* simulation (Figure 1.30).

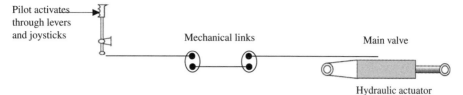

(a) Mechanically control signaled, hydraulically powered

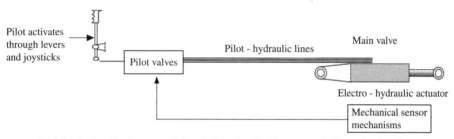

(b) Pilot hydraulically control signaled, hydraulically powered (fly-by-hydraulics)

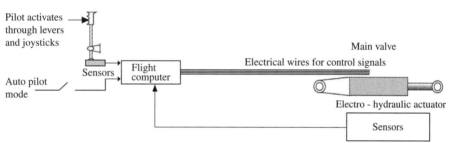

(c) Electrically control signaled, hydraulically powered (fly-by-wire)

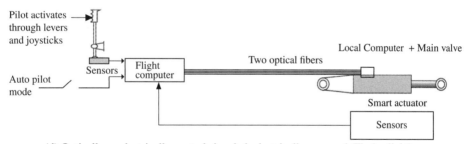

(d) Optically or electrically control signaled, electrically powered (Fly-by-light)

FIGURE 1.28: Four different flight control concepts in historical order of development: (a) mechanically signaled, hydraulically powered, (b) pilot-hydraulically signaled, hydraulically powered, (c) electrically signaled, hydraulically powered, (d) electrically signaled, electrically actuated.

This process allows the engineer to test the control software on the actual ECM hardware in real-time. The computer which simulates the plant model in real-time provides the simulated I/O connections to the ECM. The fundamental challenge in HIL simulations is the need to find a balance between the model accuracy (hence more complex and detailed models) and the need for real-time simulation. As real-time modeling capabilities improve, virtual dynamic testing and validation of complete machines using HIL tools will become a

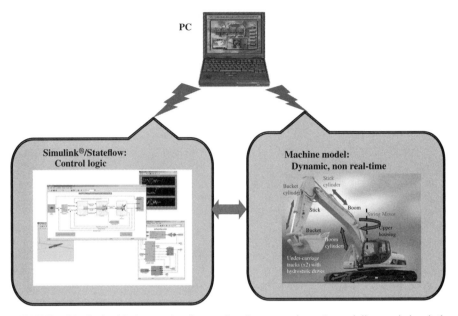

FIGURE 1.29: Embedded control software development phase 1: modeling and simulation in non-real-time.

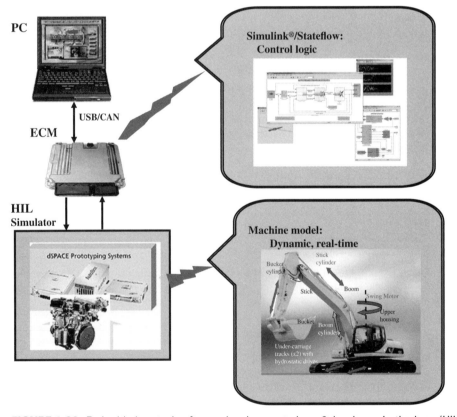

FIGURE 1.30: Embedded control software development phase 2: hardware-in-the-loop (HIL) real-time simulation and testing.

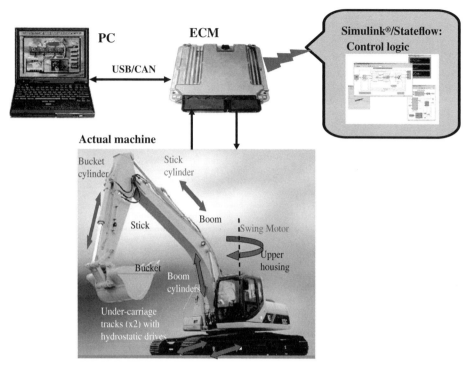

FIGURE 1.31: Embedded control software development phase 3: on a target embedded control module (ECM, also called electronic control module) and testing on actual machine.

reality (Figures 1.29, 1.30) in engineering design and development processes for embedded control systems.

In phase 3, the ECM with the control code is tested on the actual prototype machine (Figure 1.31). First, all of the I/O hardware is verified for proper operation. The sensors and actuators (i.e., solenoid drives, amplifiers) are *calibrated*. The software logic is tested to make sure all contingencies for fault conditions are taken into account. Then, the control algorithm parameters are *tuned* to obtain the best possible dynamic performance based on expert operator and end-user comments. The performance and reliability of the machine is tested, compared to benchmark results, and documented in preparation for production release.

HIL is the testing and validation engineering process between pure software simulation (100% software, Figure 1.29) and pure hardware testing (100% actual machine with all its hardware and embedded software, Figure 1.31), where some of the components are actual hardware and some are simulated in real-time software. The pure software based simulation cannot capture sufficiently the real-time conditions to provide sufficient confidence in the overall system functionality and reliability (Figure 1.29). Whereas pure hardware testing is quite often too expensive due to the cost of actual hardware, its custom instrumentation for testing purposes, and the team of engineers and operators involved in the testing (Figure 1.31). Furthermore, some tests (especially failure modes) can not be tested (or are very difficult to test, i.e., flight control systems) on the actual hardware. HIL simulation testing is an engineering process that is in the middle between pure software and pure hardware testing (Figure 1.30).

HIL tools have been developed rapidly in recent years in that some of the hardware components of the control system are included as actual hardware (such as electronic control

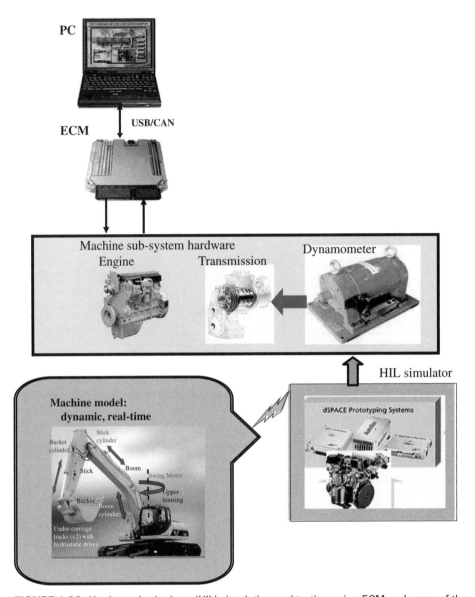

FIGURE 1.32: Hardware-in-the-loop (HIL) simulation and testing using ECM and some of the actual hardware in a powertrain control application and some of the hardware simulated in real-time: engine and transmission physically exist, the lower powertrain and load conditions are simulated via the dynomameter based on the real-time dynamic model.

unit (ECU), engine, transmission, dynamometer), and some of the components are present in software form running in real-time and its results are reflected on the control system by a generic simulator (i.e., a dynamometer which represents the load on the powertrain based on the machine dynamics and operating conditions, (Figure 1.32). Early versions of HIL simulations were used to test only the static input output behavior of the ECM running the intended real-time control code, where I/O behavior is tested with a static I/O simulator. Modern HIL simulation and testing are performed for dynamic testing, as well as static testing, where the I/O to ECM is driven by dynamic and detailed models of the actual machine.

1.4 PROBLEMS

1. Consider the mechanical closed loop control system for the liquid level shown in Figure 1.4. a) draw the block diagram of the whole system showing how it works to maintain a desired liquid level, b) modify this system with an electromechanical control system involving a digital controller. Show the components in the modified system and explain how they would function. c) Draw the block diagram of the digital control system version.

2. Consider the mechanical governor used in regulating the engine speed in Figure 1.5. a) Draw the block diagram of the system and explain how it works. b) Assume you have a speed sensor on the crank shaft, an electrically actuated valve in place of the valve which is actuated by the fly-ball mechanism, and a microcontroller. Modify the system components for digital control version and draw the new block diagram.

3. Consider the web tension control system shown in Figure 1.7. a) Draw the block diagram of the system and explain how it works. b) Assume the analog electronic circuit of the op-amp and command signal source is replaced by a microcontroller. Draw the new components and block diagram of the system. Explain how the new digital control system would work. c) Discuss what would be different in the real-time control algorithm if the microcontroller was to control the speed of the wind-off roll's motor instead of the wind-up roll's motor. Assume wind-up motor speed is constant.

4. Consider the room temperature control system show in Figure 1.8. The electric heater and fan motor is turned ON or OFF depending on the actual room temperature and desired room temperature. a) Draw the block diagram of the control system and explain how it works. b) Replace the op-amp, command signal source, and timer components with a microcontroller. Explain using pseudo-code the main logic of the real-time software that must run on the microcontroller.

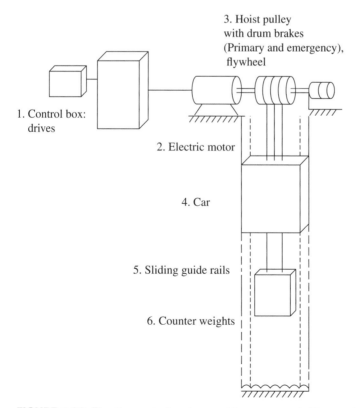

FIGURE 1.33: Elevator control system and its components (Source: http://www.smartelevatorsco.com/ropedelevators.html).

5. The ASIMO humanoid robot, developed by Honda Motor Co., is capable of autonomous walking, navigation, and performing various service tasks such as carrying a tray of objects (Google.com, Asimo Images). It can run at 6 km/hr speed. ASIMO's weight is about 54 kg and its height is 130 cm. The robot is equipped with digital vision cameras, voice recognition sensors, ultra sonic range finding sensors, as well as joint sensors at its legs and arms which are colocated with the motion actuators. Assume that each leg has four motorized joints ($2 \times 4 = 8$ degrees of freedom (DOF) for two legs), each arm has four motorized joints ($2 \times 4 = 8\,DOF$ for arms), and each finger has three motorized joints ($3x10 = 30\,DOF$ for fingers). In addition, the robot has three rotational degrees of freedom at the waist ($3\,DOF$) as well as at its head ($3\,DOF$). This results in a total of $8 + 8 + 30 + 3 + 3 = 52\,DOF$ motion. Design a block diagram representation of a conceptual control system for this robot. The control system should include a block for actuators (motors), sensors, microcontrollers, and power sources for the robot motion. Dedicate a separate microcontroller for the vision system, for the voice recognition system, and the individual closed loop motion control of each leg and arm. Note that the problem has many possible solutions. It is intended to give the student an idea about the number of I/O and distributed embedded controllers involved in such a robotic device.

6. Typical high rise buildings have in the range of 60 to 80 floors. Moving people and loads in high rise buildings safely and efficiently is increasingly important. Since the elevators are used by general public, safety is very important. In addition, for efficient operation of society, the operation must be as fast as possible (i.e., 30 km/h) while maintaining the comfort of the passengers. Electric motor driven elevators are the norm in the modern high rise buildings. Figure 1.33 shows the basic configuration of an electric motor driven elevator.

Here, we ask the student to do a little research to determine the following aspects of the modern elevator control systems,

1. *The mechanism* or the basic mechanical or electromechanical system, analogous to skeleton of human body.
2. *The actuators* (similar to muscles in human body analogy), including its power supply and amplification component which is used to control how much power is allowed to flow into the actuator from the power supply.
3. *The sensors* (similar to eyes in human body analogy) used to measure various variables such as position, velocity, weight, and so on.
4. *The controller* (similar to the brain in the human body) used to make control decisions.
5. *The control logic and simulation:* closed loop control model and simulation: model the dynamics of the elevator, and a PID position loop controller. Simulate a motion from one level to another under a gravity load disturbance, focusing on the effect of "I" (integral) action of the PID controller to eliminate the positioning error.

CHAPTER 2

CLOSED LOOP CONTROL

THIS CHAPTER contains the fundamental material on closed loop control systems. Before one uses feedback, that is closed loop control, one should explore the option of open loop control. We will address the following questions:

- What are the advantages and disadvantages of closed loop control versus open loop control?
- Why should we use feedback control instead of open loop control?
- In what cases may open loop control be better than closed loop control?

A control system is designed to *make a system do what we want it to do*. Therefore, a control system designer needs to know the desired behavior or performance expected from the system. The performance specifications of a control system must cover certain fundamental characteristics, such as stability, quality of response, and robustness. Despite the great variety and richness of control theory, more than 90% of the feedback controllers in practice are of the proportional-integral-derivative (PID) type. Due to its wide usage in practice, PID control is considered a fundamental controller type. PID control is discussed in the last section of this chapter.

The control decisions can be made either by an analog control circuit, in which case the controller is called an *analog controller*, or by a digital computer, in which case the controller is called a *digital controller*. In analog control, the control decision rules are designed into the analog circuit hardware. In digital control, the control decision rules are coded in software. This software code implementing the control decisions is called the *digital control algorithm*.

The main advantages of digital control over analog control are as follows:

1. Increased flexibility: changing the control algorithm is a matter of changing the software. Making software changes in digital control is much easier than changing analog circuit design in analog control.

2. Increased level of decision making capability: implementing nonlinear control functions, logical decision functions, conditional actions to be taken, and learning from experience can all be programmed in software. Building analog controllers with these capabilities would be a prohibitive task, if not impossible.

It is important to identify the place of the control of dynamic systems in the big picture of control systems. Real-world control systems involve many discrete event controls using sequencing and logic decisions. Discrete event control refers to the control logic based on sensors and uses actuators which have only two level states, ON/OFF, (i.e., pneumatic cylinders controlled by an ON/OFF solenoid, relays). The sequence controllers use sensors and actuators which have only an ON/OFF state, and the control algorithm is a logic

---

Mechatronics with Experiments, Second Edition. Sabri Cetinkunt.
© 2015 John Wiley & Sons, Ltd. Published 2015 by John Wiley & Sons, Ltd.
Companion Website: www.wiley.com/go/cetinkunt/mechatronics

between the sensors and ON/OFF actuators. Such controls are generally implemented using programmable logic controllers (PLC) in the automation industry. Servo control loops may be part of such a control system. Closed loop servo control is often a sub-system of the logic control systems where servo and logic control are hierarchically organized.

A control system is called *closed loop* if the control decisions are made based on some sensor signals. If the control decisions do not take any sensor signal of the controlled variables into account and decisions are made based on some pre-defined sequence or operator commands, such a control system is called *open loop*. It has been long recognized that using feedback information (sensor signals) about the controlled variable in determining the control action provides robustness against changing conditions and disturbances.

2.1 COMPONENTS OF A DIGITAL CONTROL SYSTEM

Let us consider the control of a process using (i) analog control (Figure 2.1), and (ii) digital control (Figure 2.2). The only difference is the controller box. In analog control, all of the signals are continuous, whereas in digital control the sensor signals must be converted to digital form, and the digital control decisions must be converted to analog signals to send to the actuation system.

The basic components of a digital controller are shown in Figure 2.2:

1. a central processing unit (CPU) for implementing the logic and mathematical control algorithms (decision making process)

2. discrete state input and output devices (i.e., for interfacing switches and lamps),

3. an analog to digital converter (A/D or PWM input) to convert the sensor signals from analog to digital signals,

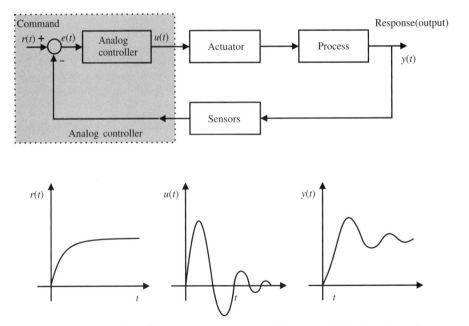

FIGURE 2.1: Analog closed loop control system and the nature of the signals involved.

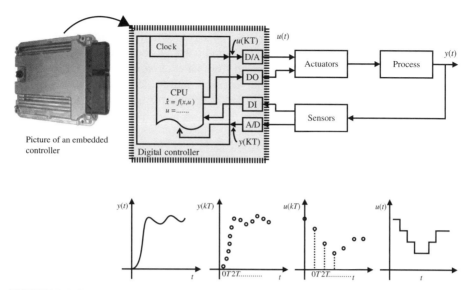

FIGURE 2.2: Digital closed loop control system and the nature of the signals involved.

4. a digital to analog converter (D/A or PWM output) to convert the control decisions made by the control algorithm in the central processing unit (CPU) to the analog signal form so that it can be commanded to the actuation system for amplification,

5. a clock for controlling the operation of the digital computer. The digital computer is a discrete device and its speed of operations are controlled by the clock cycle. The clock is to the computer what the heart is to a body.

In Figure 2.2, it is shown that the signals travel from sensors to the control computer in analog form. Similarly, the control signals from the controller to the amplifier/actuator travel in analog form. The conversion of the signal from analog to digital form (A/D converter) occurs at the control computer end. Likewise, the conversion of the digital signal to analog signal occurs at the control computer end (D/A converter) and travels to the actuators in analog form. Recent trends in computer controlled systems are such that the analog to digital, and digital to analog, conversion occurs at the sensor and actuator point. Such sensors and actuators are marketed as "smart sensors" and "smart actuators." In this approach, the signal travels from sensor point to control computer, and from control computer to the actuator point, in digital form. Especially, the use of a fiber optic transmission medium provides very high signal transmission speed with high noise immunity. It also simplifies the interface problems between the computers, the sensors, and the amplifiers. In either case, digital input and output (DI/DO), A/D, and D/A operations are needed in a computer control system as an interface between the digital world of computers and the analog world of real systems. The exact location of the digital and analog interface functions can vary from application to application.

Let us consider the operations performed by the components of a digital control computer and their implications compared to analog control:

1. time delay associated with signal conversion (at A/D and D/A) and processing (at CPU),

2. sampling,

3. quantization,

4. reconstruction.

A digital computer is a discrete-event device. It can work with finite samples of signals. The sampling rate can be programmed based on the clock frequency. Every sampling period, the sensor signals are converted to digital form by the A/D converter (the sampling operation). If the command signals are generated from an external analog device, this also must be sampled. During the same sampling period, control calculations must be performed, and the results must be sent out through the D/A converter. The A/D and D/A conversions are finite precision operations. Therefore there is always a quantization error.

2.2 THE SAMPLING OPERATION AND SIGNAL RECONSTRUCTION

Due to the fact that the controller is a digital computer, the following problems are introduced in a closed loop control system: time delay associated with signal conversion and processing, sampling, quantization error due to finite precision, and reconstruction of signals.

2.2.1 Sampling: A/D Operation

In this section, we will focus on the sampling only and its implications. We will consider the sampling operation in the following order:

1. physical circuit of the sampler,
2. mathematical model of sampling,
3. implications of sampling.

2.2.2 Sampling Circuit

Consider the sample and hold circuit shown in Figure 2.3. When the switch is turned ON, the output will track the input signal. This is the sampling operation. When the switch is turned OFF, the output will stay constant at the last value. This is the hold operation.

While the switch is ON, the output voltage is

$$\bar{y}(t) = \frac{1}{C} \int_0^t i(\tau)d\tau \qquad (2.1)$$

where

$$i(t) = \frac{y(t) - \bar{y}(t)}{R} \qquad (2.2)$$

Taking the Laplace transforms of the differential equations, and substituting the value of i from the second equation gives the input–output transfer function of the sample and hold circuit.

$$\bar{y}(s) = \frac{1}{(RCs + 1)} y(s) \qquad (2.3)$$

While the switch is OFF $i(t) = 0$; $\bar{y}$ remains constant (hold operation).

Let T be sampling period, T_0 is the portion of T for which the switch stays ON, T_1 is the remaining portion of the sampling period during which the switch stays OFF (Figure 2.4). If the input signal $y(t)$ changes as a step function, the output signal will track it according to the solution of the transfer function in response to the step input. Figure 2.4 shows the typical response of a realistic sample and hold circuit of an A/D converter.

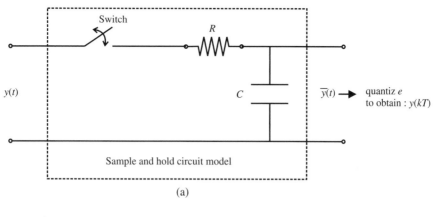

(a)

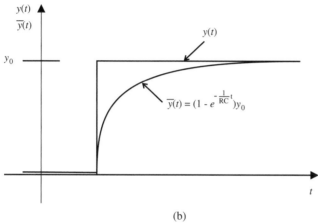

(b)

FIGURE 2.3: (a) Sample and hold circuit model and (b) response of the sampled voltage output.

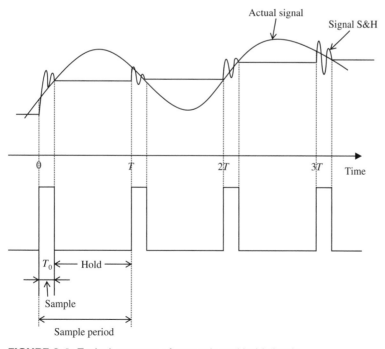

FIGURE 2.4: Typical response of a sample and hold circuit.

2.2.3 Mathematical Idealization of the Sampling Circuit

Let us consider the limiting case of the sampling circuit as a mathematical idealization for further analysis. Let us consider that the RC value goes to zero.

$$RC \to 0; \ \frac{1}{RC} \to \infty$$

This means that as soon as the switch is turned ON, $\bar{y}(t)$ will reach the value of the $y(t)$. Therefore, there is no need to keep the switch ON any more than an infinitesimally small period of time. The ON time of the switch can go to zero, $T_0 \to 0^+$ (Figure 2.4).

$$y(t) \simeq y(kT) \tag{2.4}$$

With this idealization in mind, the sampling operation can be viewed as a sequence of periodic impulse functions.

$$\sum_{k=-\infty}^{+\infty} \delta(t - kT)$$

This also says that the sampling operation acts as a "comb" function (Figure 2.5). If we represent the sequence of samples of the signal with $\{y(kT)\}$, the following relationship holds,

$$\{y(kT)\} = \sum_{k=-\infty}^{\infty} \delta(t - kT)]y(t) \tag{2.5}$$

Now, we will consider the following three questions concerning a continuous time signal, $y(t)$, and its samples, $y(kT)$ that is sampled at a sampling frequency $w_s = 2\pi/T$ by an A/D converter (Figure 2.6).

- **Question 1:** what is the relationship between the Laplace transform of the samples and the Laplace transform of the original continuous signal?

$$L\{y(kT)\} \ ? \ L\{y(t)\}$$

- **Question 2:** what is the relationship between the Fourier transform of the samples and the Fourier transform of the original continuous signal? Shannon's sampling theorem provides the answer to this question.

$$F\{y(kT)\} \ ? \ F\{y(t)\}$$

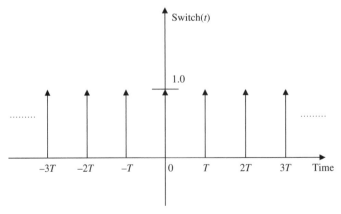

FIGURE 2.5: Idealized mathematical model of the sampling operation via a "comb" function.

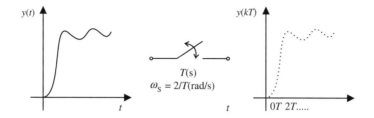

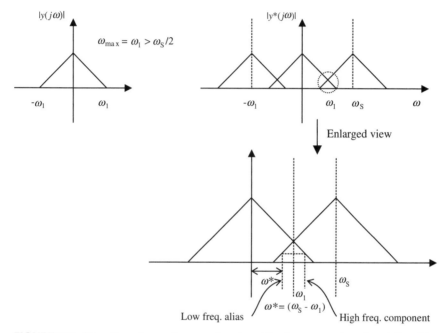

FIGURE 2.6: Sampling of a continuous signal, and the frequency domain relationship between the original signal and the sampled signal.

- **Question 3:** Point out a least three implications of the sampling theorem that come out of the relationship derived in question 2.

Let us address each one of these questions in order.

Question 1 Notice that since the sampling "comb" function is periodic, it can be expressed as a sum of Fourier series,

$$\sum_{k=-\infty}^{\infty} \delta(t - kT) = \sum_{n=-\infty}^{\infty} C_n e^{j\frac{2\pi}{T} nt} \tag{2.6}$$

where,

$$C_n = \frac{1}{T} \int_{-\frac{T}{2}}^{\frac{T}{2}} \sum \delta(t - kT) e^{-j\frac{2\pi}{T} \cdot n \cdot t} dt = \frac{1}{T} \tag{2.7}$$

Therefore,

$$\sum_{k=-\infty}^{\infty} \delta(t - kT) = \frac{1}{T} \sum_{n=-\infty}^{\infty} e^{j(\frac{2\pi}{T})n\cdot t} \tag{2.8}$$

The Laplace transform of the sampled signal is a two-sided Laplace transform, see [5]

$$
\begin{aligned}
L\{y(kT)\} &= \int_{-\infty}^{\infty} y(t) \sum_{n=-\infty}^{\infty} \delta(t - kt)e^{-st}dt \\
&= \int_{-\infty}^{\infty} y(t)\frac{1}{T} \sum_{n=-\infty}^{\infty} e^{j\frac{2\pi}{T}nt} e^{-st}dt \\
&= \frac{1}{T} \sum_{n=-\infty}^{\infty} \int_{-\infty}^{\infty} y(t)e^{-(s-j\frac{2\pi}{T}n)t}dt
\end{aligned} \tag{2.9}
$$

The relationship between the Laplace transform of the sampled signal and the Laplace transform of the original continuous signal is

$$Y^*(s) = \frac{1}{T} \sum_{n=-\infty}^{\infty} Y(s - jw_s n) \tag{2.10}$$

where $w_s = \frac{2\pi}{T}$ is the sampling frequency and T is the sampling period.

Question 2   The Fourier transform of a signal can be obtained from the Laplace transform by substituting jw in place of s in the Laplace transform of the function. Therefore, we obtain the following relationship between the Fourier transforms of the sampled and continuous signal (Figure 2.6),

$$Y^*(jw) = \frac{1}{T} \sum_{n=-\infty}^{\infty} Y(jw - jw_s n) \tag{2.11}$$

where $Y^*(jw)$ is the Fourier transform of sampled signal, and $Y(jw)$ is the Fourier transform of the original signal.

The frequency content of the sampled signal is the frequency content of the original signal plus the same content shifted in the frequency axis by integer multiples of the sampling frequency. In addition, the magnitude of the frequency content is scaled by the sampling period. The physical interpretation of the above relation is the famous sampling theorem, also called the Shannon's sampling theorem.

Sampling Theorem   In order to recover the original signal from its samples, the sampling frequency, w_s, must be at least two times the highest frequency content, w_{max}, of the signal,

$$w_s \geq 2 \cdot w_{max} \tag{2.12}$$

Question 3   We now consider various implications of the sampling operation.

   (i) Aliasing: Aliasing is the result of violating the sampling theorem, that is

$$w_s < 2 \cdot w_{max} \tag{2.13}$$

The high frequency components of the original signal show up in the sampled signal as if they are low frequency components (Figure 2.12). This is called the *aliasing*. The aliasing frequency that shows up on the samples of the signal is

$$w^* = \left| \left(w_1 + \frac{w_s}{2} \right) \mathrm{mod} \, (w_s) - \left(\frac{w_s}{2} \right) \right| \tag{2.14}$$

where w_1 is the frequency content of the original signal. For simplicity, we consider a specific frequency content for the original signal. If the sampling theorem is violated (the sampling frequency is less than twice the highest frequency content of the original signal),

1. reconstruction of the original signal from its samples is impossible,
2. high frequency components look like low frequency components.

Let us consider two sinusoidal signals with frequencies 0.1 Hz and 0.9 Hz. If we sample both of the signals at $w_s = 1$ Hz, the sampling theorem is not violated in sampling the first signal, but it is violated in sampling the second signal. Due to the aliasing, the samples of the 0.9 Hz sinusoidal signal will look like the samples of the 0.1 Hz signal (Figure 2.6). Figure 2.12 shows the two cases,

1. continuous signal $sin(2\pi(0.9)t)$ and sampling frequency $w_s = 1.0$ Hz,
2. continuous signal $sin(2\pi(0.1)t)$ and sampling frequency $w_s = 1.0$ Hz.

The high frequency signal 0.9 Hz looks like a 0.1 Hz signal when sampled at the 1.0 Hz rate as a result of the aliasing,

$$w^* = \left| \left(w_1 + \frac{w_s}{2} \right) \mathrm{mod}(w_s) - \left(\frac{w_s}{2} \right) \right|$$
$$w^* = |(0.9 + 0.5)\mathrm{mod}(1.) - 0.5| = |0.4 - 0.5|$$
$$w^* = 0.1 \text{ Hz} \tag{2.15}$$

Another example is a sinusoidal signal with $w_1 = 3$ Hz, and sampled values of it at $w_s = 4$ Hz. The sampling theorem is violated. The samples will show a 1 Hz oscillation which does not exist in the original signal.

In summary, if the sampling theorem is violated, high frequency content of a signal shows up as low frequency (aliasing frequency) content in the sampled signal as a result of aliasing. The aliasing frequency is given by Eqn. 2.14.

(ii) Hidden oscillations: If the original signal has a frequency content which is an exact integer multiple of the sampling frequency (sampling theorem is violated), then there could be hidden oscillations. In other words, the original signal would have high frequency oscillations, whereas the sampled signal would not show them at all (Figure 2.13). If $w_{signal} = n \cdot w_s; n = 1, 2, \dots$.

When $n = \frac{1}{2}$; with correct phase of the sampling time to the oscillation frequency, hidden oscillations are also possible.

(iii) Beat Phenomenon: The beat phenomenon is observed when two signals with very close frequency content with very close magnitudes are added. The result looks like two signals (one slowly varying, the other fast varying) are multiplied.

This phenomenon occurs as a result of the sampling operation when the sampling frequency is just a little larger than twice the highest frequency content of the signal (Figure 2.14). Notice that the sampling theorem is not violated. Consider the following signal,

$$u(t) = A \cdot (\cos(w_1 t) + \cos(w_2 t)) \tag{2.16}$$

$$= 2A \cdot \cos \left(\frac{w_2 - w_1}{2} t \right) \cdot \cos \left(\frac{w_1 + w_2}{2} t \right) \tag{2.17}$$

$$u(t) = 2A \cdot \cos(w_{beat} t) \cdot \cos(w_{ave} t) \tag{2.18}$$

where $w_{ave} = (w_1 + w_2)/2$, $w_{beat} = (w_2 - w_1)/2$. Let us show the above equality directly using trigonometric relations. The following trigonometric relations are used in the derivation,

$$\cos(\alpha + \beta) = \cos(\alpha)\cos(\beta) - \sin(\alpha)\sin(\beta) \qquad (2.19)$$

$$\sin^2\theta + \cos^2\theta = 1 \qquad (2.20)$$

$$\cos^2\theta = \frac{1 + \cos 2\theta}{2} \qquad (2.21)$$

Then,

$$u(t) = 2A \cdot \cos\left(\frac{w_2 - w_1}{2}t\right) \cdot \cos\left(\frac{w_1 + w_2}{2}t\right) \qquad (2.22)$$

$$= 2A \cdot \left(\cos\frac{w_2 t}{2}\cos\frac{w_1 t}{2} + \sin\frac{w_2 t}{2}\sin\frac{w_1 t}{2}+ \right. \qquad (2.23)$$

$$\left. \cdot \left(\cos\frac{w_2 t}{2}\cos\frac{w_1 t}{2} - \sin\frac{w_2 t}{2}\sin\frac{w_1 t}{2}\right)\right) \qquad (2.24)$$

$$= 2A \cdot \left(\cos^2\frac{w_2 t}{2}\cos^2\frac{w_1 t}{2} - \sin^2\frac{w_2 t}{2}\sin^2\frac{w_1 t}{2}\right) \qquad (2.25)$$

$$= 2A \cdot \left(\cos^2\frac{w_2 t}{2}\cos^2\frac{w_1 t}{2} - \left(1 - \cos^2\frac{w_2 t}{2}\right)\left(1 - \cos^2\frac{w_1 t}{2}\right)\right) \qquad (2.26)$$

$$= 2A \cdot \left(\cos^2\frac{w_2 t}{2}\cos^2\frac{w_1 t}{2} - \right. \qquad (2.27)$$

$$\left(1 - \cos^2\frac{w_2 t}{2} - \cos^2\frac{w_1 t}{2} + \cos^2\frac{w_2 t}{2}\cos^2\frac{w_1 t}{2}\right)\right) \qquad (2.28)$$

$$= 2A \cdot \left(-\left(1 - \cos^2\frac{w_2 t}{2} - \cos^2\frac{w_1 t}{2}\right)\right) \qquad (2.29)$$

$$= 2A \cdot \left(\frac{1 + \cos w_2 t}{2} + \frac{1 + \cos w_1 t}{2} - 1\right) \qquad (2.30)$$

$$= 2A \cdot \left(\frac{\cos w_2 t}{2} + \frac{\cos w_1 t}{2}\right) \qquad (2.31)$$

$$= A \cdot (\cos w_2 t + \cos w_1 t) \qquad (2.32)$$

which shows that the *addition* of two cosine functions of two discrete frequencies can also be expressed as *multiplication* of two cosine functions with two frequencies that are the average and difference of the original two frequencies. When the two frequencies are close to each other, then this results in the *beat phenomenon*.

If w_1 is very close to w_2, that is $w_1 = 5$, $w_2 = 5.2$, then the so called beat phenomenon occurs,

$$u(t) = A\cos(w_{beat}t) \cdot \cos(w_{ave}t) \qquad (2.33)$$

where $w_{beat} = (w_2 - w_1)/2$, $w_{ave} = (w_2 + w_1)/2$. The same effect occurs when we sample a signal with a frequency which is very closed to the minimum sampling frequency required, yet the sampling theorem is not violated.

Let us consider a sinusoidal signal,

$$y(t) = \sin(2\pi 0.9t) \qquad (2.34)$$

and sample it with a sampling frequency of $w_s = 1.9$ Hz, which satisfies the sampling theorem requirements. However, since sampling a signal effectively shifts the original signal frequency content in the frequency domain in integer multiples of sampling frequency, it results in adding two very close frequency contents. The result is that the sampled signal shows the beat phenomenon (Figure 2.14). Another example of the same phenomenon is

$$y(t) = \sin(2\pi 4.9t) \tag{2.35}$$

and sample it with a sampling frequency of $w_s = 10.0$ Hz.

In Figure 2.14, the top-left figure shows the frequency content of the original signal, and the top-right figure shows the frequency content of the sampled signal. If we approximate the sampled signal with the lowest two frequency contents,

$$y^*(t) \approx \frac{1}{T}(\sin(2\pi\, 0.9t) + \sin(2\pi\, 1.0t)) \tag{2.36}$$

which is the addition of two sinusoidal signals with very close frequencies, as discussed above. Therefore, the sampling is expected to have (and the above equation explains) the beat phenomenon observed in the time domain plots of the sampled signal.

2.2.4 Signal Reconstruction: D/A Operation

A D/A converter is used to convert a digital number to an analog voltage signal. It is also referred to as the signal reconstruction device. Let us consider that we sample a continuous signal through an A/D converter, then send that signal out without any modification through a D/A converter. The difference between the original analog signal (input to the A/D converter) and the analog signal output from the D/A converter is an undesired distortion due to sampling, quantization, time delay, and reconstruction errors (Figure 2.1). For instance, in communication systems, the analog voice signal is sampled, transmitted digitally over the phone lines, and converted back to the analog voice signal at the other end of the phone line. The goal there is to be able to reconstruct the original signal as accurately as possible.

We know that the sampled signal frequency content is the original frequency content plus the same content shifted in the frequency axis by integer multiples of the sampling frequency. In order to recover the frequency content of the original signal, we need an ideal filter which has a square gain and zero phase transfer function (Figure 2.7). Clearly, if there

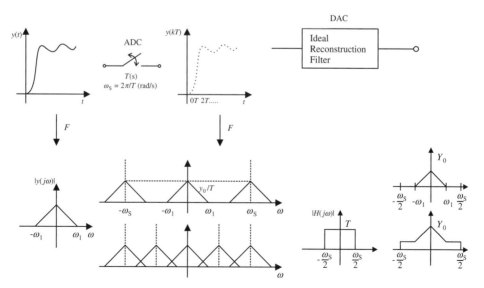

FIGURE 2.7: Signal reconstruction with an ideal D/A converter.

is aliasing, it is impossible to recover the original signal even with an ideal re-construction filter. Here we will consider the following two resconstruction D/A converters:

1. Ideal reconstruction filter (Shannon's Reconstruction).
2. ZOH (Zero-Order-Hold).

In order to focus on D/A functionality and its ability to reconstruct the original signal from its samples, let us assume that the time delay and quantization errors are negligable and that the sampled signal is sent out to the D/A converter without any modification.

(i) Ideal Reconstruction Filter D/A Converter

In order to recover the original signal from the sampled signal, we need to recover the original frequency content of the signal from the frequency content of the sampled signal (Figure 2.7). Therefore, we need an ideal filter, an *ideal reconstruction filter*, which has the following frequency response,

$$
\begin{cases}
|H(j\omega)| = \begin{cases} T; & w \in \left[-\frac{\omega_s}{2}, \frac{\omega_s}{2}\right] \\ 0; & otherwise \end{cases} \\
\angle H(j\omega) = 0; \ \forall \ \omega
\end{cases}
\tag{2.37}
$$

Let us take the inverse Fourier transform of this filter transfer function to determine what kind of impulse response such a filter would have.

$$
F^{-1}(H(jw)) = \frac{1}{2\pi} \int_{-\frac{\pi}{T}}^{\frac{\pi}{T}} T \cdot e^{jwt} \cdot dw
$$

$$
= \frac{T}{2\pi} \int_{-\frac{\pi}{T}}^{\frac{\pi}{T}} e^{j\omega t} \cdot dw
\tag{2.38}
$$

$$
h(t) = \frac{2}{w_s t} \cdot \sin\left(\frac{w_s t}{2}\right)
\tag{2.39}
$$

This is the impulse response of an ideal reconstruction filter. Notice that the impulse response of the ideal reconstruction filter is non-causal (Figure 2.8).

In order to practically implement it, one must introduce a time delay into the system that is large enough compared to sampling period. It cannot be implemented in closed loop control systems due to the stability problem the time delay would cause. The original signal could, in theory if not in practical applications, be reconstructed from its samples as follows,

$$
y(t) = \sum_{k=-\infty}^{\infty} y(kT) \cdot \sin c \frac{\pi(t - kT)}{T}
\tag{2.40}
$$

(ii) D/A: Zero-Order-Hold (ZOH)

The great majority of D/A converters operate as zero-order-hold functions. The signal is kept at the last value until a new value comes. The change between the two values is a step change. Let us try to obtain a transfer function for a zero-order-hold (ZOH) D/A converter. To this end, consider that a single unit pulse is sent to the ZOH D/A (Figure 2.9). The output of the ZOH D/A would be a single pulse with unit magnitude

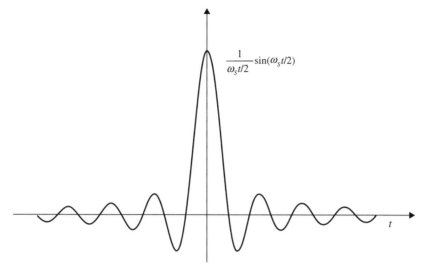

FIGURE 2.8: Impulse response of an ideal reconstruction D/A filter. Notice that it is a non-causal filter.

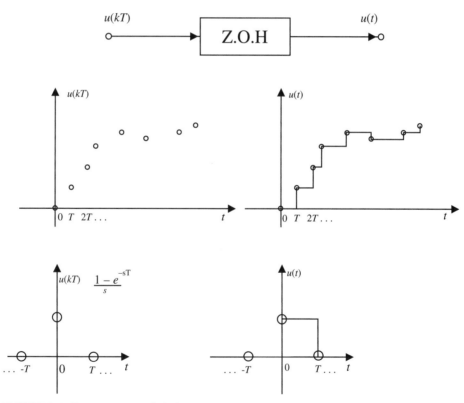

FIGURE 2.9: Zero order hold (ZOH) type practical D/A converter, its operation and transfer function.

and duration of a single sampling period. The transfer function of it is an integrator response to an impulse minus the same response delayed by a sampling period,

$$\text{ZOH}(s) = \left(\frac{1 - e^{-sT}}{s} \right) \tag{2.41}$$

The frequency response (filter transfer function) of the ZOH D/A can be obtained from the above equation by substituting jw for the s variable, and after some algebraic manipulations it can be shown that the frequency response of a ZOH D/A is

$$
\begin{aligned}
\text{ZOH}(j\omega) &= \left(\frac{1 - e^{-j\omega T}}{j\omega} \right) \\
&= e^{-\frac{j\omega T}{2}} \left(\frac{e^{\frac{j\omega T}{2}} - e^{-\frac{j\omega T}{2}}}{jw} \right) \\
&= T \cdot e^{-\frac{j\omega T}{2}} \frac{\sin \frac{\omega T}{2}}{\frac{\omega T}{2}} \\
&= e^{-\frac{j\omega T}{2}} \cdot T \cdot \left(\sin c \frac{\omega T}{2} \right)
\end{aligned} \tag{2.42}
$$

Clearly, compared to Equation 2.37 which represents the ideal reconstruction filter transfer function, the transfer function of the ZOH type D/A converter is different than the ideal case, but it is a practical one.

2.2.5 Real-time Control Update Methods and Time Delay

Time delay is an important issue in control systems. Time delay in the feedback loop can cause instability since it introduces phase lag. There is inherent time delay in digital control. The A/D and D/A conversion takes a finite amount of time to complete. The execution of the control calculations takes a finite amount of time. The sampling period is determined by the sum of the time periods that these operations take. The sampling period is a good indication of the time delay introduced into the loop due to the digital implementation. If the sampling frequency is much higher than the bandwidth of the closed loop system (i.e., 50 times faster), the influence of time delay due to the digital sampling period will not be significant. As the sampling frequency gets closer to the control system bandwidth (i.e., 2 times), the time delay associated with sampling rate can create very serious stability and performance problems. Figure 2.10 shows two different implementations of a closed loop system in terms of sampling and control update timing.

A control system will have periodic sampling intervals. The sampling period can be programmed using a clock. After every sampling period is passed, an interrupt is generated. The real-time control software can be divided into two groups,

1. foreground program,
2. background program.

Normally, the CPU runs the background program handling operator input/output operations, checks error and alarm conditions, and checks other process inputs and outputs (I/O) not used in closing the control loop but used for other logic and sequencing functions. The foreground program is the one that is executed every time the sampling clock generates an interrupt. When a new interrupt is generated every sampling period, the CPU saves the status of what it is doing in the background program, and jumps to the foreground program

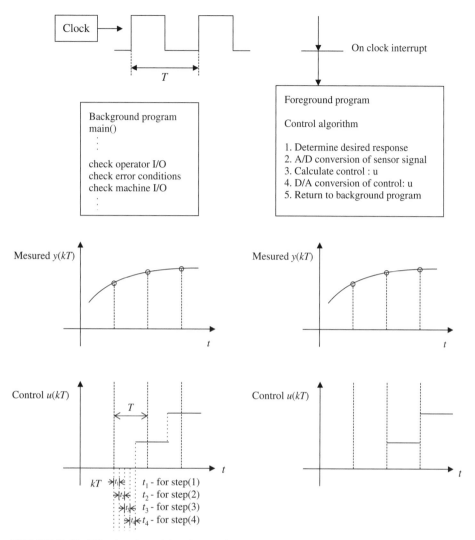

FIGURE 2.10: Effective time delay due to signal conversion, processing, and update methods in a digital control system.

as soon as possible. In the foreground program, it handles the closed loop control update for the current sampling period. This involves the following tasks,

1. determine the desired response (sample from A/D if necessary),

2. A/D sample of the sensor signal,

3. calculate the control action,

4. send the control action to D/A,

5. return to the background program.

This implementation and timing of the sequence of operations are shown in Figure 2.10 on the left-hand side. Note that A/D conversion is an iterative process and the conversion times may vary from one cycle to the next. As a result, the effective update period for control signals can be a little different from one cycle to the next due to the variations in the A/D conversion. The period for which the control signal is held constant is

$$T_u = T - T_{c_k} + T_{c_{k+1}} \tag{2.43}$$

where $T_{c_k} = (t_1 + t_2 + t_3 + t_4)_k$ and $T_{c_{k+1}} = (t_1 + t_2 + t_3 + t_4)_{k+1}$ are the total time spent in the foreground program execution during sampling interval k, and $k + 1$, respectively. Due to the variations in the A/D conversion time, t_2, the effective update period T_u for control action may vary from one sample to the next. Hence, the control system may not have a truly constant sampling period. This is a potential problem which may or may not be significant depending on the application. However, this implementation minimizes the time delay between the measured sensor signal and the corresponding control action since the control action is updated as soon as it is available. If minimizing the time delay associated with the digital controller is more important than maintaining a truly constant sampling frequency, this implementation is appropriate to use. Another implementation is shown on the right side of Figure 2.10. The sequence of operations in the foreground program is a little different:

1. send the control action calculated from the previous period to D/A,
2. determine the desired response (sample from A/D if necessary),
3. A/D sample of the sensor signal,
4. calculate the control action and keep it for the next sampling period,
5. return to the background program.

The difference is that sending out the control signal to the D/A converter is the first thing done every sampling period. The control signal sent out is the signal calculated during the previous sampling period. The signal calculated during this period is sent to the D/A converter at the beginning of the next sampling period. The advantage of this implementation is that the control signal is updated in a truly fixed sampling period. As long as the sampling period is long enough to complete the foreground program every sampling period, the update of the control signal is done at a fixed frequency. The disadvantage is that the effective time delay associated with digital processing of the control signal is longer than the previous implementation. The time delay is at most one sampling period long. If the sampling frequency is much larger than the bandwidth of the closed loop system, that should not be a serious problem. As the sampling frequency gets closer to the bandwidth of the closed loop system, the larger time delay in the second implementation compared to the first implementation may have serious performance degrading consequences.

2.2.6 Filtering and Bandwidth Issues

The sampling theorem requires that the sampling frequency must be at least twice the highest frequency content of the signal being sampled. However, real-world signals always have some level of noise in them. The frequency of the noise component of the signal is generally very high. Therefore, it would not be practical to use very fast sampling rates in order to handle the noise and avoid the aliasing problem. Furthermore, the noise content of the signal is not something we would like to capture. It is an unwanted component. Therefore, signals are generally passed through anti-aliasing filters before sampling at the A/D circuit. This is called *pre-filtering*. The purpose of anti-aliasing filters is to attenuate the high frequency noise components, and pass the low frequency components of the signal (Figure 2.11). The bandwidth of the anti-aliasing filter (also called the noise filter) should be such that it should not pass much of the signal beyond $1/2$ of the sampling frequency. An

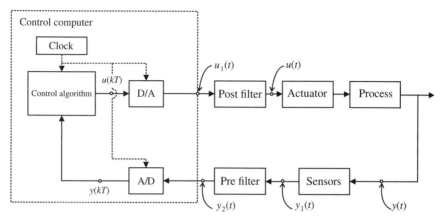

FIGURE 2.11: Filtering and bandwidth issues in a digital closed loop control system.

ideal filter would have frequency response characteristics similar to the ideal reconstruction filter. It would pass identically all the frequency components up to 1/2 of the sampling frequency, and cancel the rest. However, it is not practical to use such filters in control and data acqusition systems.

Generally, a second order or higher order filter is used as a noise filter. Typically the filter transfer function for a second-order filter is

$$G_{\mathrm{F}}(s) = \frac{w_{\mathrm{n}}^2}{s^2 + 2\xi_{\mathrm{n}} w_{\mathrm{n}} s + w_{\mathrm{n}}^2}$$

If a higher order filter is desired, multiple second-order filters can be cascaded in series. The exact parameters of the noise filter ($\xi_{\mathrm{n}}, w_{\mathrm{n}}$) are selected based on the class of the filter. The Butterworth, ITAE, and Bessel filter are popular filter parameters used for that purpose.

Similarly, since the output of the D/A converter is a sequence of step changes, it may be useful to smooth the control signal before it is applied to the amplification stage. The same type of noise filters can be used as a *post-filtering* device to reduce the high frequency content of the control signal.

The pre-filters and post-filters add time delay into the closed loop system due to their finite bandwidth. In order to use the pre- and post-filters for noise cancellation and smoothing purposes without significantly affecting the closed loop system bandwidth, the following general guidelines should be followed. We have four frequencies of interest:

1. closed loop system bandwidth, w_{cls},
2. sampling frequency, w_{s},
3. pre- and post-filter bandwidth, w_{filter},
4. the maximum frequency content of the signal presented to the sampling and hold circuit of A/D converter, w_{signal}.

In order to make sure that the pre- and post-filtering does not affect the closed loop system bandwidth, the filters must have about 10 times or more higher bandwidth than the closed loop system bandwidth. The filter bandwidth is also a good estimate of the highest frequency content allowed to enter the sampling circuit.

$$w_{\mathrm{filter}} \approx w_{\mathrm{signal}} \approx 5 \; to \; 10 * w_{\mathrm{cls}}$$

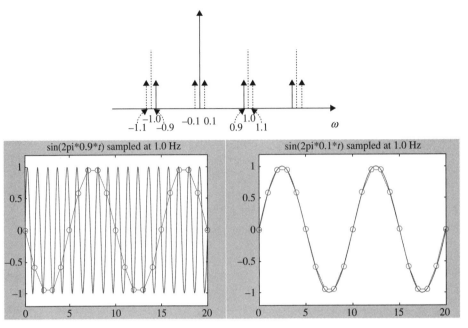

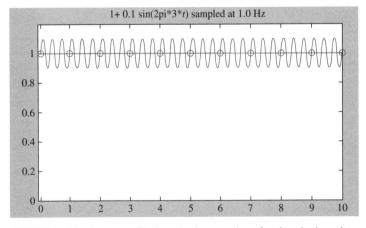

FIGURE 2.12: Example of aliasing as a result of sampling operation and the misleading picture of the sampled signal.

From the sampling theorem,

$$w_s \geq 2 * w_{\text{signal}}$$

In practice, the sampling frequency should be much larger than the minimum requirement imposed by the sampling theorem, that is

$$w_s \approx 5 \, to \, 20 * w_{\text{signal}}$$

Therefore, the magnitude relation between the four frequencies of interest is

$$w_s \approx 5 \, to \, 20 * w_{\text{signal}} \approx 5 \, to \, 20 * w_{\text{filter}} \approx 25 \, to \, 200 * w_{\text{cls}} \quad (2.44)$$

FIGURE 2.13: Hidden oscillations in the samples of a signal when the sampling rate is an integer multiple of a frequency content in the continuous signal.

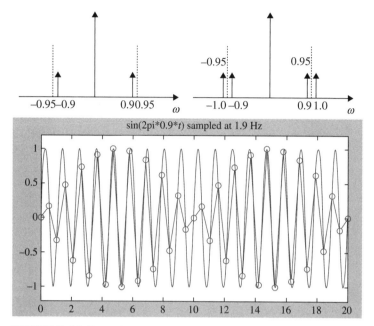

FIGURE 2.14: Beat phenomenon and its explanation in the frequency and time domain. Notice that the sampling theorem is not violated. The sampling frequency is very close to the minimum requirement.

2.3 OPEN LOOP CONTROL VERSUS CLOSED LOOP CONTROL

Open loop control means control decisions are made without making use of any measurement of the actual response of the system. An open loop control decision does not need a sensor. Closed loop control means control decisions are made based on the measurements of the actual system response. The actual response is fed back to the controller, and the control decision is made based on this feedback signal and the desired response. In comparison, we will take the real-world issues into consideration, namely, *disturbances, changes in process dynamics, and sensor noise.* These are common real-world problems faced to different degrees of significance in every control system. Consider a general dynamic process nominally modeled by $G(s)$, controlled by an open loop and a closed loop controller (Figure 2.15).

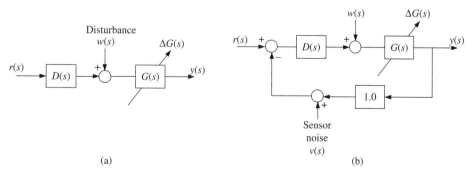

FIGURE 2.15: Open loop versus closed loop control system comparison.

The response for the open loop control case is

$$y(s) = D(s)G(s)\,r(s) + G(s)\,w(s)$$

and for the closed loop control case is

$$y(s) = \frac{D(s)G(s)}{1 + D(s)G(s)}\,r(s) + \frac{G(s)}{1 + D(s)G(s)}\,w(s) - \frac{D(s)G(s)}{1 + D(s)G(s)}\,v(s)$$

In real-world systems, the following issues exist and must be addressed:

1. Disturbances ($w(s)$): there are always disturbances which are not under our control. They exist and cause error in the system response. For instance, the wind acts as a disturbance on an airplane changing its flight direction. Low outside temperature and the heat loss due to it from the walls of a heated house acts as a disturbance on the control system since the outside temperature is not under our control, yet it affects the temperature of the house.

2. Variations in process dynamics ($\Delta G(s)$): the dynamics of the process may change structurally or parametrically. Structural changes in the dynamics imply drastic significant changes, such as the change in the dynamics of an aircraft due to loss of an engine or a wing. Whereas parametric changes imply less significant, more smooth, non-drastic changes, such as the change in the weight of an aircraft as the fuel is being consumed, or due to opening of the wing control surfaces.

3. Sensor noise ($v(s)$): closed loop control requires the measurement of the actual response (the controlled variable). The sensor signals always have some noise in the measurement. The noise is included in the control decisions and hence affects the overall performance of the system.

Let us consider the effect of these three groups of real-world problems in system performance under open loop and closed loop control (Figure 2.14a). In open loop control, the effect of disturbance is,

$$y_w(s) = G(s)\,w(s)$$

the effect of process dynamic variations is,

$$\begin{aligned} y(s) &= (G_0(s) + \Delta G(s))\,r(s) \\ &= G_0(s)r(s) + \Delta G(s)\,r(s) \end{aligned}$$

Since no feedback sensors are needed in open loop control, there is no sensor noise problem.

The responses due to disturbance $G(s) \cdot w(s)$ and due to process dynamic variations $\Delta G(s) \cdot r(s)$ are not wanted and are considered errors. Open loop control has no mechanism to correct for these errors. The error is proportional to the disturbance magnitude and the changes in the process dynamics. On the one hand, if the process dynamics is well known and no variations occur, and there is no disturbance or the nature of disturbance is well known, the open loop control can provide excellent performance. On the other hand, if process dynamics vary or there are disturbances whose nature is not known or not repeatable, open loop control has no mechanism to reduce their effect.

In closed loop control, there is an added component, the sensor, to provide feedback measurement about the actual output of the process. By definition, feedback control action is generated based on the error between desired and actual output. Hence, error is inherent part of such a design.

Let us consider the output of the system shown in Figure 2.14b.

$$y(s) = \frac{D(s)G(s)}{1 + D(s)G(s)} r(s) + \frac{G(s)}{1 + D(s)G(s)} w(s) - \frac{D(s)G(s)}{1 + D(s)G(s)} v(s)$$

Ideally we would like $y(s) = r(s)$ and no output be caused as a result of $w(s)$, variations in $G(s)$, and $v(s)$. Let us consider the effects of disturbance, dynamic process variations, and sensor noise.

1. Disturbance effect: $w(s)$

$$y_w(s) = \frac{G(s)}{1 + D(s)G(s)} w(s)$$

The response due to the disturbance, $y_w(s)$, is desired to be small or zero if possible. If we can make $D(s)G(s) \gg G(s)$ and $D(s)G(s) \gg 1$ for the frequency range where $w(s)$ is significant, then the response due to disturbances in that frequency range would be small. If the response due to disturbances is small or zero, the control system is said to *have good disturbance rejection* or to *be insensitive to the disturbances*. Another way of stating this result is that it is desirable to have a controller with large gain.

2. Variation of process dynamics: $G(s) = G_0(s) + \Delta G(s)$. Consider only the command signal and process dynamic variations, and let us analyze the effect of the variations in the process dynamics on the response.

$$y_r(s) = \frac{D(s)(G_0(s) + \Delta G(s))}{1 + D(s)(G_0(s) + \Delta G(s))} r(s)$$

$$= \frac{D(s)G_0(s)}{1 + D(s)(G_0(s) + \Delta G(s))} r(s) + \frac{D(s)\Delta G(s)}{1 + D(s)(G_0(s) + \Delta G(s))} r(s)$$

The second term is the main contribution of process dynamic variations to the output. In order to make the effect of this on the system response small, the following condition must hold,

$$D(s)G(s) \gg D(s), \quad \text{and} \quad D(s)G(s) \gg 1$$

If the response due to changes in the process dynamics is small, the control system is called *insensitive to the variations in process dynamics* which is a desired property.

So far disturbance rejection capability requires,

$$D(s)G(s) \gg G(s), \quad \text{and} \quad D(s)G(s) \gg 1$$

and insensitivity to process dynamic variations requires,

$$D(s)G(s) \gg D(s), \quad \text{and} \quad D(s)G(s) \gg 1$$

Therefore, the conditions $\{D(s)G(s) \gg G(s), D(s)G(s) \gg D(s), \text{ and } D(s)G(s) \gg 1\}$ mean that the loop gain must be well balanced between the controller and the process in order to have good disturbance rejection and be insensitive to process dynamics variations.

3. Sensor noise effect: sensor noise is a problem for closed loop control systems only. The open loop control does not need feedback sensors and therefore it does not have a sensor noise problem. Let us consider the response, $y_v(s)$, of a closed loop system due to sensor noise, $v(s)$,

$$y_v(s) = -\frac{D(s)G(s}{1 + D(s)G(s)} v(s)$$

In order to make $y_v(s)$ small, $D(s)G(s) \ll 1$ must be, which is contradictory to the loop gain requirements for the previous two properties: disturbance rejection and insensitivity to variations in process dynamics. *This is the fundamental design conflict of feedback control systems.* Robustness against disturbances and variations in process dynamics require large loop gain balanced between controller and process, $D(s)G(s) \gg 1$, while robustness against sensor noise requires $D(s)G(s) \ll 1$. These are conflicting requirements and both cannot be satisfied at the same time for all frequencies.

In practice, the control engineering problems are generally such that disturbances and variations in process dynamics are slowly varying and have low frequency content. Whereas sensor noise has high frequency content. If a given control problem has this frequency separation property between various uncertainties, then a controller can be designed such that $D(s)G(s) \gg 1$, $D(s)$, $G(s)$ for a low frequency range so that the system has good robustness against disturbances and variations in process dynamics, and $D(s)G(s) \ll 1$ for a high frequency range so that sensor noise is also rejected. *This is the basic feedback control system design compromise.* If there is no such frequency separation between disturbance, variations in process dynamics, and sensor noise, no feedback controller can be designed to provide robustness against all of these real-world problems.

In summary, the loop transfer function of a typical well designed control system has the following desired shape as a function of frequency: it should be as large as possible at low frequencies to provide robustness against disturbance and variations in process dynamics, and it should be as small as possible at high frequencies to reject sensor noise (Figure 2.16). Furthermore, it should cross the 0 db magnitude by about -20 db/decade slope in $20\log_{10}|D(jw)G(jw)|$ versus $\log_{10} w$ plot in order to have a good stability margin.

So far, we have compared the advantages and disadvantages of closed loop control versus open loop control. The main advantage of feedback control over open loop control is that it increases the robustness of the system against the disturbances and variations in the process dynamics. The general characteristics of control systems are discussed in terms of the shape of the loop transfer function in order to provide good robustness against these undesirable real-world problems of control systems. However, sensor noise or sensor failure can make a closed loop system unstable. If the process dynamics does not vary much and the disturbances are well known, open loop control may be a better choice than closed loop control. Open loop control does not suffer from the potential stability problems associated with sensor failures.

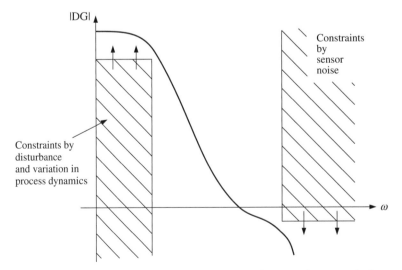

FIGURE 2.16: Desired performance specification for control systems in frequency domain.

2.4 PERFORMANCE SPECIFICATIONS FOR CONTROL SYSTEMS

The performance desired from a control system can be described under three groups:

1. response quality:
   (a) transient response,
   (b) steady-state response,
2. stability,
3. robustness of the system stability and response quality against various uncertainties such as disturbances, process dynamic variations, sensor noise.

The main advantage of feedback control over open loop control is its ability to reduce the effect of disturbances and process dynamic variations on the quality of system response. In other words, the main advantage of feedback control is the robustness it provides against various uncertainties.

A basic feedback control system and typical uncertainties associated with it are shown in Figure 2.14b. As we discussed in the previous section, the total response of the system due to command, disturbance, and sensor noise (for $H = 1$ case) is

$$y = \frac{DG}{1 + DG} r + \frac{G}{1 + DG} w - \frac{DG}{1 + DG} v$$

The goal of the control is to make $y(t)$ equal to $r(t)$. Therefore $DG \gg 1$ should be in general. If $DG \gg G$, the effect of disturbance, w, is reduced. However, the sensor noise directly contributes to the output, y. In order to track r and reject disturbance, w, we want $DG \gg 1$ (large), but in order to reject sensor noise we want $DG \ll 1$ (small). This is the basic dillema of feedback control design. A compromise is reached by the following engineering judgment: disturbance, $w(t)$ is generally of low frequency content, whereas sensor noise $v(t)$ is high frequency content. Therefore, if we design a controller such that $DG \gg 1$ around the low frequency region to reject disturbances, and $DG \ll 1$ around the high frequency region to reject sensor noise, the closed loop system has good robustness against uncertainties.

The robustness of the closed loop system (CLS) is closely related to the gain of loop transfer function as a function of frequency (Figure 2.16). Therefore the robustness properties are best conveyed in the frequency domain. In general, a loop transfer function should have a large gain at low frequency in order to reject low frequency disturbances and slow variations in process dynamics, and low loop gain at high frequency in order to reject sensor noise. The s-plane pole-zero representation of a transfer function does not convey gain information. Hence, robustness properties are not well conveyed by the s-plane pole-zero structure of the transfer function.

Stability requirements are equally well described in the s-plane as well as frequency domain. In the s-plane, all the CLS poles must be on the left-hand plane. In the frequency domain, the gain margin and phase margin must be large enough to provide a sufficient stability margin. The desired relative stability margin from a CLS can be expressed either in terms of gain and phase margin in frequency domain, or in terms of the distance of CLS poles from the imaginary axis in the s-plane.

Finally, the response quality must be specified. The response of a dynamic system can be divided into two parts: (i) transient response part, (ii) steady-state response part.

Transient response is the immediate response of the system when it is commanded for new desired output. The steady-state response is the response of the system after a

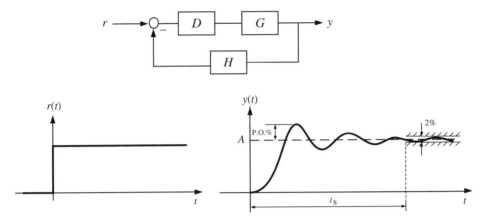

FIGURE 2.17: A typical closed loop control system step response shape. Quantification of transient response characteristics to step input can done using percent overshoot and settling time.

sufficient amount of time has passed. The steady-state response quality is quantified by the error between the desired and actual output after a long enough time has passed for the system to respond. Clearly, the response of a dynamic system depends on its input. The standard input signal used in defining the transient response characteristics of a dynamic system is the step input. By using a standard test signal, various competing controller and process designs can be compared in terms of their performances (Figure 2.17).

In general a CLS step response looks like the response shown in (Figure 2.17). The transient response to the step command can be characterized by a few quantitative measures of the response, such as the maximum percent overshoot or the time it takes for the response to settle down within certain percentage of the final value.

The transient response to a step input is typically described by maximum percent overshoot, PO%, and settling time, t_s, the time it takes for the output to settle down to within $\pm 2\%$ or $\pm 1\%$ of the desired output.

For a second-order system, there is a one to one relationship between (PO%, t_s) and the damping ratio, and natural frequency (ξ, ω_n) of the second order system (Figures 2.17 and 2.18). It can be shown that

$$PO\% = e^{\frac{-\pi\xi}{\sqrt{1-\xi^2}}}$$

$$t_s = \begin{cases} \dfrac{4}{\xi\omega_n}, & \pm 2\% \\ \dfrac{4.6}{\xi\omega_n}, & \pm 1\% \end{cases}$$

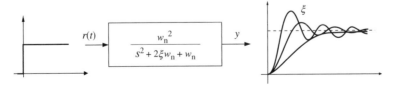

FIGURE 2.18: A standard second-order system model and its step response. The step response is determined by the damping ratio (ξ) and natural frequency (w_n).

Therefore, if it is desired that the closed loop system (CLS) step response should not have more than a certain amount of PO%, and should settle down within ±2% of the final value within t_s (seconds), the designer must seek a controller which will make the closed loop system behave like (or similar to) a second-order system whose two poles are given by the above relationships.

2.5 TIME DOMAIN AND *S*-DOMAIN CORRELATION OF SIGNALS

The response of a linear time invariant dynamic system, $y(t)$, as a result of an input signal, $r(t)$ can be calculated using the Laplace transforms

$$y(s) = G(s)\, r(s)$$

The impulse response $y(s)$ can be expanded to its partial fraction expansion (PFE) which has the general form as follows,

$$y(s) = \frac{1}{s} + \sum_{i=1}^{m} \frac{A_i}{s + \sigma_i} + \sum_{j=1}^{m} \frac{B_j}{s^2 + 2\alpha_j s + \left(\alpha_j^2 + \omega_i^2\right)}$$

$$A_i,\ B_j - \text{residue of PFE of } G(s)\, r(s)$$

The time domain response, $y(t)$ can be obtained by taking the inverse Laplace transform of each term of the PFE,

$$y(t) = 1(t) + \sum_{i=1}^{m} A_i e^{-\sigma_i t} + \sum_{j=1}^{m} B_j e^{-\alpha_j t} \left(\frac{1}{\omega_j} \sin \omega_j t\right)$$

The correlation between the time domain response (impulse response) and the poles of the transfer function is shown in Figure 2.19.

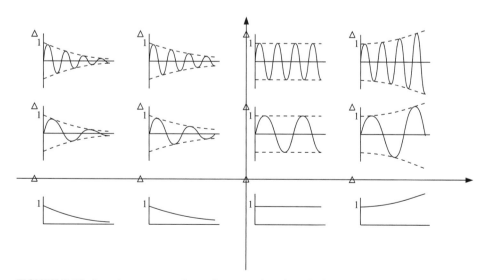

FIGURE 2.19: Impulse response for various root locations in the *s*-plane.

2.6 TRANSIENT RESPONSE SPECIFICATIONS: SELECTION OF POLE LOCATIONS

The feedback control changes the dynamics of the open loop system to the desired form by closed loop control action. For linear systems, closed loop control changes the locations of the poles. The control effort required is proportional to the amount of movement of the pole locations (the difference between the open loop and closed loop pole locations). Large pole movements will require unnecessarily large actuators. The desired pole locations can be selected to approximate the step response behavior of a dominant second-order model or the pole locations of some standard filters such as Bessel and Butterworth filters. The second-order system parameters (ω_n, ξ) can be selected fairly accurately to satisfy t_{settling}, and PO% specifications by designing the CLS such that it has a dominant second-order system poles and the rest of poles are further to the left in the s-plane (Figure 2.20).

2.6.1 Step Response of a Second-Order System

Step response is the standard signal used in evaluating the transient response characteristics of a control system. Specifically, the step response behavior is summarized by the maximum percent overshoot (PO%), and the amount of time it takes for the output to settle to within 1 or 2% of the commanded step output (settling time, t_s), the time it takes for the output to reach 90% of command rise time t_r, and to reach maximum value (peak time, t_p). The (PO%, t_s, t_r, t_p) all are related to the pole locations of a second-order system. We will consider the step response of a second-order system of the form

$$\frac{\omega_n^2}{s^2 + 2\xi\omega_n s + \omega_n^2}$$

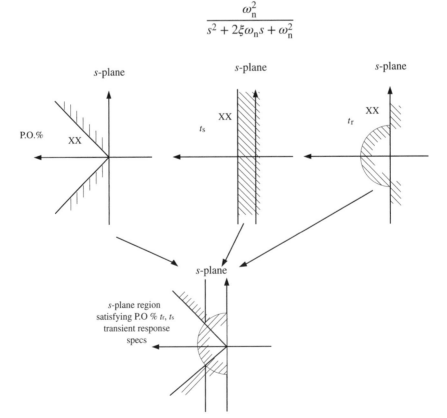

FIGURE 2.20: Desired response performance specifications at s-domain pole locations.

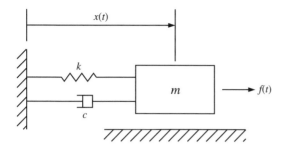

FIGURE 2.21: Mass-spring system dynamics and its position control.

Further, we will consider the step response when there is an additional pole and zero

$$\frac{\left(\frac{s}{a}+1\right)}{\left(\frac{s}{b}+1\right)} \frac{\omega_n^2}{s^2 + \xi\omega_n s + \omega_n^2}$$

The utility of this is that given transient response specifications for a control system ($PO\%, t_s$) we can determine where the dominant poles should be in order to meet the specifications. Even though our system may not be second order, second-order system pole-zero locations can provide a good starting point in design, especially if the higher order system can be made to have a dominant second-order dynamics.

Consider the step response of a second-order system (Figure 2.21);

$$m\ddot{x} + c\dot{x} + kx = f$$

let $f(t) = kr(t)$

$$\ddot{x} + \frac{c}{m}\dot{x} + \frac{k}{m}x = \frac{k}{m}r$$

Let $\frac{c}{m} = 2\xi\omega_n$, $\frac{k}{m} = \omega_n^2$ and take the Laplace transform of the differential equation with zero initial conditions,

$$\frac{x(s)}{r(s)} = \frac{\omega_n^2}{s^2 + 2\xi\omega_n s + \omega_n^2}$$

If $r(t)$ is a step input, $r(s) = \frac{1}{s}$, the response $x(s)$ is given by

$$x(s) = \frac{\omega_n^2}{s^2 + 2\xi\omega_n s + \omega_n^2} \cdot \frac{1}{s}$$

Using the partial fraction expansions and taking the inverse Laplace transform, the response can be found as

$$x(t) = 1 - e^{-\xi\omega_n t}\left(\cos\sqrt{1-\xi^2}\omega_n t + \frac{\xi}{\sqrt{1-\xi^2}}\sin\sqrt{1-\xi^2}\omega_n t\right) \quad \text{for } 0 \le \xi < 1 \text{ range}$$

The maximum overshoot occurs at the first time instant, t_p, where the derivative of x is zero. Once t_p is found, then the maximum value of response at that time can be evaluated and the percent overshoot can be determined.

$$\frac{dx(t)}{dt} = 0 \quad \Rightarrow \quad \text{find} \quad t_p: \quad t_p = \frac{2.5}{\omega_n}$$

Then

$$PO\% = \frac{x(t_p) - 1}{1} \times 100 = e^{\frac{-\pi\xi}{\sqrt{1-\xi^2}}}$$

The settling time is the time it takes for the response to settle within ± 1 or $\pm 2\%$ of the final value, and it can be shown that

$$t_s = \begin{cases} \frac{4.6}{\xi\omega_n}; & \pm 1\% \\ \frac{4.0}{\xi\omega_n}; & \pm 2\% \end{cases}$$

Therefore, given a $(PO\%, t_s)$ specification, the corresponding second-order system pole locations $(-\xi w_n \pm \sqrt{(1-\xi^2)}w_n$ can be directly obtained.

Effect of an additional zero   Let us consider a second-order system with a real zero (Figure 2.22). The system is the same as a standard second-order system with two complex conjugate poles and d.c. gain of 1, with an additional zero on the real axis.

$$G(s) = \left(\frac{s}{a} + 1\right) \frac{\omega_n^2}{s^2 + 2\xi\omega_n s + \omega_n^2}$$

Let $\omega_n = 1$; $a = \alpha \xi\omega_n$, the transfer function can be expressed as

$$G(s) = \frac{\left(\frac{s}{\alpha\xi} + 1\right)}{s^2 + 2\xi s + 1} = \frac{1}{s^2 + 2\xi s + 1} + \frac{1}{\alpha\xi} \frac{s}{s^2 + 2\xi s + 1}$$

Notice that the effect of zero is to add the derivative of the step response to the second-order system response by an amount proportional to $\frac{1}{\alpha\xi}$. Clearly if α is large, a is to the left of $\xi\omega_n$, and the influence of the addition of zero is not much. As α gets smaller, a gets closer to $\xi\omega_n$ area, and $(1/\alpha\xi)$ grows. Hence, the influence of zero on the response increases. The main effect of zero as it gets close to the $\xi\omega_n$ value is to increase the percent overshoot. If the zero is on the right half s-plane (non-mininum phase transfer function) the initial value of step response goes in the opposite direction. This is illustrated in Figures 2.23 and 2.24.

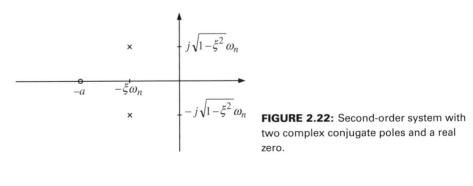

FIGURE 2.22: Second-order system with two complex conjugate poles and a real zero.

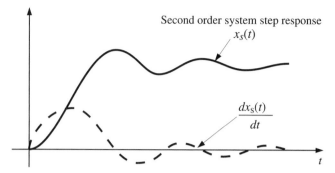

FIGURE 2.23: Step response of a second-order system and the derivative of its response.

The step response of a second-order system with a zero on the left half plane has the form

$$x_{\text{step}}(t) + \frac{1}{\alpha\xi}\dot{x}_{\text{step}}(t)$$

and response with a zero on the right half s-plane (make the $(s/a + 1)$ term $(s/a - 1)$ in the above equation and the result that follows (Figure 2.24) has the form

$$x_{\text{step}}(t) - \frac{1}{\alpha\xi}\dot{x}_{\text{step}}(t)$$

The Effect of an Additional Pole The system becomes third order when an additional pole exists in addition to the two complex conjugate poles.

$$\left(\frac{1}{\frac{s}{b} + 1}\right)\left(\frac{1}{\frac{s^2}{\omega_n^2} + \frac{2\xi}{\omega_n}s + 1}\right)$$

If the $b = \alpha\xi\omega_n$ is 3 to 5 times to the left of $\xi\omega_n$, the effect of an additional pole on the step response is negligable. As b gets close to $\xi\omega_n$, the effect is to increase the rise time and hence slow the response of the system (Figure 2.24).

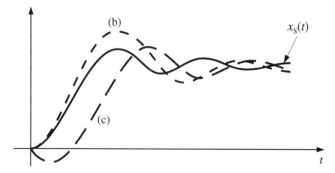

FIGURE 2.24: Step response of a second-order system and the effect of a zero: (a) without zero, (b) zero on the left hand plane, (c) zero on the right hand plane.

FIGURE 2.25: Second-order filters.

2.6.2 Standard Filters

Most dynamic systems are higher order than second order. However, a higher order system behaves very similar to a standard second-order system if the poles are distributed on the s-plane such that there are two dominant poles and the rest of the poles and zeros are far (3 to 5 times) to the left of these dominant poles. Instead of choosing a dominant second-order system model as a design goal, we can choose other higher-order system models. Among the popular standard filters, which can be used as the goal for a control system performance, are Bessel, Butterworth, and ITAE filters Figure 2.25.

The step response and frequency response characteristics of these filters are well tabulated in standard textbooks on control systems and digital signal processing fields. They may be used as a reference to describe the desired performance (hence desired pole-zero locations) for a closed loop control design problem. Note that the more the open loop poles must be moved to desired closed loop pole locations, the larger the control action requirements. This may quickly saturate existing actuators and result in poor transient performance or require unnecessarily large actuators on the system. The dominant closed loop system poles (also called the bandwidth) should be as large as possible with sufficient damping on the s-plane, but it should be a balanced choice between desired speed of response and required actuator size and control effort.

2.7 STEADY-STATE RESPONSE SPECIFICATIONS

Steady-state response is usually characterized by the steady-state error between a desired output and the actual output.

$$y(\cdot) = \frac{D(\cdot)G(\cdot)}{1 + D(\cdot)G(\cdot)} r(\cdot)$$

The error between the desired and actual response is

$$e(\cdot) = r(\cdot) - y(\cdot)$$
$$= \left(1 - \frac{D(\cdot)G(\cdot)}{1 + D(\cdot)G(\cdot)}\right) r(\cdot)$$
$$e(\cdot) = \frac{1}{1 + D(\cdot)G(\cdot)} r(\cdot)$$

In the s-domain the steady-state error (for continuous time systems)

$$e(s) = \frac{1}{1 + D(s)G(s)} r(s)$$

The steady-state value of the error as time goes to infinity can be determined using the final value theorem of Laplace transforms, provided that the $e(s)$ has stable poles and at most have one pole at the origin, $s = 0$,

$$\lim_{t \to \infty} e(t) = \lim_{s \to 0} s\, e(s) = \lim_{s \to 0} s \cdot \frac{1}{1 + D(s)G(s)} r(s)$$

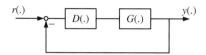

FIGURE 2.26: Block diagram of a standard feedback control system.

Let us consider the following cases:

1. the loop transfer function, $D(s)G(s)$, has N number of poles at the origin $s = 0$,

$$D(s)\,G(s) = \frac{\prod_{i=1}^{m}(s+z_i)}{s^N \prod_{i=1}^{n}(s+p_i)}; \ \ N = 0, 1, 2, ...$$

2. the commanded signal is a step, ramp, or parabolic signal (Figure 2.26),

$$r(s) = \frac{1}{s}, \frac{A}{s^2}, \frac{2B}{s^3}$$

Now we will consider the steady-state error of a closed loop system in response to a step, ramp, and parabolic command signal where the loop transfer function $D(s)G(s)$ has $N(N = 0, 1, 2)$ poles at the origin (Figure 2.27).

1. $N = 0$;

(a) $\lim_{t\to\infty} e_{step}(t) = \lim_{s\to 0} s \dfrac{1}{1 + \dfrac{\prod(s+z_i)}{\prod(s+p_i)}} \dfrac{1}{s} = \dfrac{1}{1 + D(0)G(0)} = \dfrac{1}{1 + K_p}$

(b) $\lim_{t\to\infty} e_{ramp}(t) = \lim_{s\to 0} s \dfrac{1}{1 + \dfrac{\prod(s+z_i)}{\prod(s+p_i)}} \dfrac{A}{s^2} = \dfrac{A}{0} \Rightarrow \infty$

(c) $\lim_{t\to\infty} e_{parab}(t) = \lim_{s\to 0} s \dfrac{1}{1 + \dfrac{\prod(s+z_i)}{\prod(s+p_i)}} \dfrac{2B}{s^3} = \dfrac{1}{0} \Rightarrow \infty$

$r(t)$ \ N	0	1	2
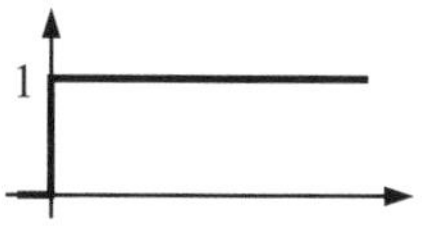 (step)	$\dfrac{1}{1+K_p}$	0	0
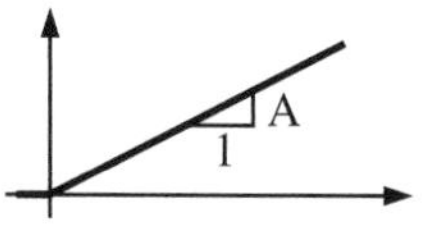 (ramp)	∞	$\dfrac{A}{K_v}$	0
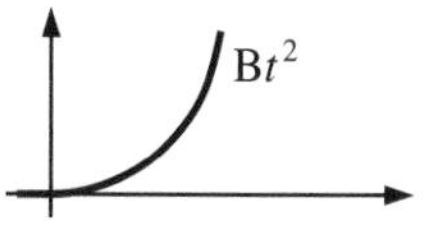 (parabola Bt^2)	∞	∞	$\dfrac{2B}{K_a}$

FIGURE 2.27: The steady-state error of a feedback control system in response to various command signals depends on the number of poles at the origin of the loop transfer function.

2. $N = 1$

(a) $\lim_{t \to \infty} e_{\text{step}}(t) = \lim_{s \to 0} s \dfrac{1}{1 + \dfrac{1}{s} \dfrac{\prod(s + z_i)}{\prod(s + p_i)}} \dfrac{1}{s} = 0$

(b) $\lim_{t \to \infty} e_{\text{ramp}}(t) = \lim_{s \to 0} s \dfrac{1}{1 + \dfrac{1}{s} \dfrac{\prod(s + z_i)}{\prod(s + p_i)}} \dfrac{A}{s^2} = \lim_{s \to 0} \dfrac{A}{sD(s)G(s)} = \dfrac{A}{K_v}$

(c) $\lim_{t \to \infty} e_{\text{parab}}(t) = \lim_{s \to 0} s \dfrac{1}{1 + \dfrac{1}{s} \dfrac{\prod(s + z_i)}{\prod(s + p_i)}} \dfrac{2B}{s^3} = \dfrac{1}{0} \Rightarrow \infty$

3. $N = 2$

(a) $\lim_{t \to \infty} e_{\text{step}}(t) = \lim_{s \to 0} s \dfrac{1}{1 + \dfrac{1}{s^2} \dfrac{\prod(s + z_i)}{\prod(s + p_i)}} \dfrac{1}{s} = 0$

(b) $\lim_{t \to \infty} e_{\text{ramp}}(t) = \lim_{s \to 0} s \dfrac{1}{1 + \dfrac{1}{s^2} \dfrac{\prod(s + z_i)}{\prod(s + p_i)}} \dfrac{A}{s^2} = 0$

(c) $\lim_{t \to \infty} e_{\text{parab}}(t) = \lim_{s \to 0} s \dfrac{1}{1 + \dfrac{1}{s^2} \dfrac{\prod(s + z_i)}{\prod(s + p_i)}} \dfrac{2B}{s^3} \lim_{s \to 0} \dfrac{2B}{s^2 D(s)G(s)} = \dfrac{2B}{K_a}$

Notice that the DC gain $(D(0)G(0))$ and the number of poles that the loop transfer function has at the origin are important factors in determining the steady-state error. It is convenient to define three constants to describe the steady-state error behavior of a closed loop system: K_p the position error constant, K_v velocity error constant, and K_a acceleration error constant.

$$K_p = \lim_{s \to 0} D(s) G(s)$$

$$K_v = \lim_{s \to 0} s D(s) G(s)$$

$$K_a = \lim_{s \to 0} s^2 D(s) G(s)$$

2.8 STABILITY OF DYNAMIC SYSTEMS

Stability of a control system is always a fundamental requirement. In fact, not only is it required that the system be stable, but it must also be stable against uncertainties and reasonable variations in the system dynamics. In other words, it must have a good stability robustness. The stability of a dynamic system can be defined by two general terms (Figure 2.28):

1. in terms of input–output magnitudes,
2. in terms of the stability around an equilibrium point.

A dynamic system is said to be bounded input–bounded output (BIBO) stable if the response of the system stays bounded for every bounded input. This definition is referred to as *input–output stability* or *BIBO stability*. The stability of a dynamic system can also be defined just in terms of its equilibrium points and initial conditions without any reference to input. This definition is called the *stability in the sense of Lyapunov* or *Lyapunov stability*.

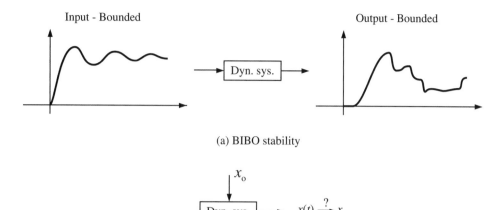

(a) BIBO stability

(b) Stability about an equilibrium point

FIGURE 2.28: Definition of two different stability notions.

2.8.1 Bounded Input–Bounded Output Stability

Definition: a dynamic system (linear or nonlinear) is said to be bounded input–bounded output (BIBO) stable if for every bounded input, the output is bounded.

For a linear time invariant (LTI) system, the output to any input can be calculated as,

$$y(t) = \int_{-\infty}^{t} h(\tau)u(t - \tau)d\tau$$

where $h(t)$ is the impulse response of the LTI system. If the input is bounded, there must exist a constant M such that

$$|u(t)| \leq M < \infty$$

Hence

$$|y(t)| = \left| \int_{-\infty}^{t} h(\tau)u(t - \tau)d\tau \right|$$

$$\leq \int_{-\infty}^{t} |h(\tau)||u(t - \tau)|d\tau$$

$$\leq M \int_{-\infty}^{t} |h(\tau)|d\tau$$

For the LTI system to be BIBO stable, $y(t)$ must be bounded for all t as $t \to \infty$. Therefore, for $y(t)$

$$\lim_{t \to \infty} |y(t)| \leq M \int_{-\infty}^{\infty} |h(z)|dz$$

to be bounded, the following expression must be bounded,

$$\int_{-\infty}^{\infty} |h(z)|dz$$

This requires that

$$h(t) \to 0 \ as \ t \to \infty$$

In conclusion, if a LTI system is BIBO stable, this means that its impulse response goes to zero as time goes to infinity. The opposite is also true. If the impulse response of a LTI system goes to zero as time goes to infinity, the LTI system is BIBO stable (Figure 2.29).

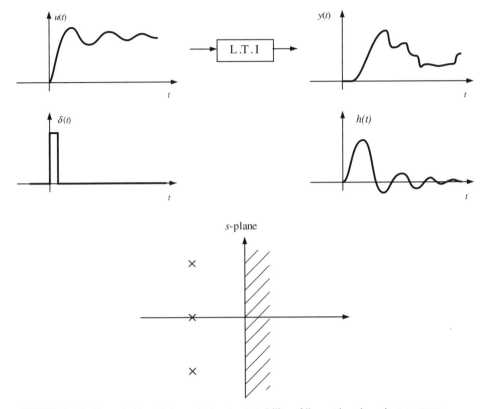

FIGURE 2.29: Bounded input–bounded output stability of linear time invariant systems.

Remark The impulse response going to zero as time goes to infinity means that all the poles of the dynamic system are on the left half of the s-plane. Therefore, for LTI systems, BIBO stability means that all the poles have a negative real part on the s-plane. The following expression summarizes the BIBO stability for LTI systems.

$$\{BIBO \quad stable\} \quad \Leftrightarrow \quad \{h(t) \quad \rightarrow \quad 0 \quad as \quad t \rightarrow \infty\} \quad \Leftrightarrow \quad \{\forall R_e(p_i) < 0\}$$

2.9 EXPERIMENTAL DETERMINATION OF FREQUENCY RESPONSE

Consider the dynamics system shown in Figure 2.16. It is excited by an input signal in the range such that the dynamic system behaves as an LTI system. Let us assume that we can set the magnitude and phase of the input signal, and that we can measure the response magnitude and phase.

The experimental procedure to determine the frequency response is:

1. Select A, and $w = w_0$ (i.e., $w_0 = 0.001$).
2. Apply input signal: $u(t) = A \sin(wt)$.
3. Wait long enough so that the output reaches the steady-state response and the transients die out.
4. Measure B and ψ of the response in $y(t) = B \cdot \sin(wt + \psi)$.
5. Record $w, B/A, \psi$.

6. Repeat until $w = w_{\max}$ where $w_{\max}$ is the maximum frequency of interest, by incrementing $w = w + \Delta w$. Δw is the increment of frequency as the experiment sweeps the frequency range from w_0 to $w_{\max}$.

7. Plot $B/A, \psi$ versus w.

8. Curve fit to B/A and ψ as function of w and obtain a mathematical expression for the frequency response as a rational function.

2.9.1 Graphical Representation of Frequency Response

The frequency response of a dynamic system is conveniently represented by a complex function of frequency. The complex function can be graphically represented in many different ways. In control systems studies, the three most commonly known representations are:

1. Bode Plots: plot $20 \log_{10} |G(jw)|$ v.s. $\log_{10}(w)$ and $Phase(G(jw))$ v.s. $\log_{10}(w)$.

2. Nyquist Plots (polar plots): plot $Re(G(jw))$ v.s. $Im(G(jw))$ on the complex $G(jw)$ plane where w-frequency is parameterized along the curve.

3. Log Magnitude versus Phase Plot: plot the $20 \log_{10}(G(jw))$ (y-axis) versus $Phase(G(jw)$ (x-axis) and w-frequency is parameterized along the curve.

One can choose to graphically plot the complex frequency response function in many other ways. The above three representations are the most common ones. With the aid of CAD-tools, it is a very simple task to plot a given frequency response in any one of the above forms. However, the ability to plot basic building blocks of transfer functions by hand sketches still remains a powerful tool in design.

Therefore, we will discuss the manual plotting of various transfer functions next. Let us consider a general transfer function which has

1. DC gain,

2. zeros and poles at the origin,

3. first-order zeros and poles,

4. second-order zeros and poles.

$$G(s) = K_0 \frac{\Pi(s/z_i + 1)\Pi((s/w_{zi})^2 + 2\xi_{zi}(s/w_{zi}) + 1)}{s^{\pm N}\Pi(s/p_i + 1)\Pi((s/w_{pi})^2 + 2\xi_{pi}(s/w_{pi}) + 1)} \qquad (2.45)$$

The rest of the graphical plotting discussions will consider this general form of the transfer function.

Bode Plots   Given a frequency response data either in explicit mathematical form as $G(jw)$ or as raw experimental data as magnitude and phase information $B/A = |G(jw)|$ and $\psi(w)$, one possible graphical represenation is as two plots:

- Plot 1, y-axis: $20 \log_{10} |G(jw)|$, x-axis: $\log_{10} w$,
- Plot 2, y-axis: $Phase(G(jw))$, x-axis: $\log_{10} w$.

Such graphical representation is called a *Bode plot*. The Bode plot is the most commonly used graphical representation of frequency response information. Let us consider

$$G(s)|_{s=jw} = G(jw) = |G(jw)|e^{j\psi(w)} \qquad (2.46)$$

The most general form of a transfer function and its frequency domain representation,

$$G(s) = K_0 \frac{\Pi(s/z_i + 1)\Pi[(s/w_{zi})^2 + 2\xi_{zi}(s/w_{zi}) + 1]}{s^{\pm N}\Pi(s/p_i + 1)\Pi[(s/w_{pi})^2 + 2\xi_{pi}(s/w_{pi}) + 1]} \tag{2.47}$$

$$G(jw) = K_0 \frac{\Pi(jw/z_i + 1)\Pi[(jw/w_{zi})^2 + 2\xi_{zi}(jw/w_{zi}) + 1]}{jw^{\pm N}\Pi(jw/p_i + 1)\Pi[(jw/w_{pi})^2 + 2\xi_{pi}(jw/w_{pi}) + 1]} \tag{2.48}$$

$$G(jw) = K_0 \frac{\Pi[1 + (w/z_i)^2]^{1/2}e^{j\tan^{-1}(w/z_i)}\Pi[(1 - w^2/w_{zi}^2)^2 + (2\xi_{zi}w/w_{zi})^2]^{1/2}e^{j\tan^{-1}(\frac{(2\xi_{zi}w/w_{zi})}{(1-w^2/w_{zi}^2)})}}{w^{\pm N}e^{\pm jN90}\Pi[1 + (w/p_i)^2]^{1/2}e^{j\tan^{-1}(w/p_i)}\Pi[(1 - w^2/w_{pi}^2)^2 + (2\xi_{pi}w/w_{pi})^2]^{1/2}e^{j\tan^{-1}(\frac{(2\xi_{pi}w/w_{pi})}{(1-w^2/w_{pi}^2)})}} \tag{2.49}$$

$$= |G(jw)|e^{j\psi(w)} \tag{2.50}$$

Now, let us express the magnitude and phase information separately, and take the logarithm of the magnitude information. Further, let us multiply the logarithm of the magnitude by 20 in order to express the magnitude information in dB (decibel) units.

$$20\log_{10}|G(jw)| = 20\log_{10}K_0 + \sum 20\log_{10}[1 + (w/z_i)^2]^{1/2} \tag{2.51}$$

$$+ \sum 20\log_{10}[(1 - w^2/w_{zi}^2)^2 + (2\xi_{zi}w/w_{zi})^2]^{1/2} \tag{2.52}$$

$$- 20(\pm N)\log_{10}w - \sum \log_{10}[1 + (w/p_i)^2]^{1/2} \tag{2.53}$$

$$- \sum \log_{10}[(1 - w^2/w_{pi}^2)^2 + (2\xi_{pi}w/w_{pi})^2]^{1/2} \tag{2.54}$$

$$\psi(w) = \angle G(jw) = \sum \tan^{-1}(w/z_i) + \sum \tan^{-1}\frac{(2\xi_{zi}w/w_{zi})}{(1 - w^2/w_{zi}^2)} \tag{2.55}$$

$$- \sum \pm N90 - \sum \tan^{-1}(w/p_i) - \sum \tan^{-1}\frac{(2\xi_{pi}w/w_{pi})}{(1 - w^2/w_{pi}^2)} \tag{2.56}$$

A Bode plot of the frequency response $G(jw)$ is the two plots of the above two equations versus the $\log_{10}w$.

The implication of taking the logarithm of the magnitude information is that the contribution of gain, zeros, and poles to the overall gain plot becomes additive. The phase information is already additive. When designing compensators, the additive nature of frequency response in Bode plots is very helpful. As we try different controllers, we do not have to replot the open loop system frequency response. Logarithmic scale in frequency allows us to capture the behavior of the system at very low frequencies as well as very high frequencies while using a reasonable size for the x-axis.

A Bode plot of any frequency response which can be expressed as a rational polynomial can be drawn as a linear summation of magnitude and phase contribution of (i) gain, (ii) zero/pole at origin, (iii) first-order zero/pole, (iv) second-order zero/pole. Quite often, the asymptotic sketches of the contribution of each of these dynamic components are more useful than their exact plots due to the fact that the asymptotic approximate sketches can be plotted rather quickly by hand.

Bode Plots of Standard Elements of a Transfer Function

1. *Constant Gain, K_0 :* A constant gain will have a constant logarithmic magnitude as a function of frequency, and a zero phase. If the sign of the gain is negative, the phase will be $-180°$,

$$20 \log_{10} | \cdot | = 20 \log_{10} K_0 \tag{2.57}$$

$$\angle(.) = \tan^{-1} \frac{Im(K_0)}{Re(K_0)} = \tan^{-1} \frac{0}{Re(K_0)} = 0 \tag{2.58}$$

$$= \tan^{-1} \frac{Im(K_0)}{Re(K_0)} = \tan^{-1} \frac{0}{Re(K_0)} = -180°; \text{ for } K_0 < 0 \tag{2.59}$$

2. *Pole/zero at the origin:* Pole at the origin

$$\frac{1}{s^N}|_{s=jw} = \frac{1}{(jw)^N} = \frac{1}{w^N e^{jN90}} = \frac{1}{w^N} \cdot e^{-jN90} \tag{2.60}$$

The magnitude and phase in Bode plots is given by

$$20 \log_{10} \left| \frac{1}{(jw)^N} \right| = 20 \log_{10} \frac{1}{w^N} = -20 N \log_{10} w \tag{2.61}$$

$$\angle \left(\frac{1}{(jw)^N} \right) = -N \cdot 90° \tag{2.62}$$

Similarly, for zero(s) at the origin, the Bode plot is

$$s^N|_{s=jw} = (jw)^N = w^N e^{jN90} = w^N e^{jN90} \tag{2.63}$$

and

$$20 \log_{10} |(jw)^N| = 20 \log_{10} w^N = 20 N \log_{10} w \tag{2.64}$$

$$\angle(jw)^N = N \cdot 90° \tag{2.65}$$

The Bode plots of gain, pole(s), and zero(s) at the origin are shown in Figure 2.30.

3. *First-order pole and zero:* Let us consider a pole on a real axis,

$$\frac{1}{(s/p_i + 1)}|_{s=jw} = \frac{1}{(jw/p_i + 1)} \tag{2.66}$$

$$= \frac{1}{[1 + (w/p_i)^2]^{1/2} e^{j \tan^{-1}(w/p_i)}} \tag{2.67}$$

$$= \frac{1}{[1 + (w/p_i)^2]^{1/2}} e^{-j \tan^{-1}(w/p_i)} \tag{2.68}$$

The magnitude and phase as a function of frequency are given by

$$20 \log_{10} \left| \frac{1}{jw/p_i + 1} \right| = 20 \log \frac{1}{[1 + (w/p_i)^2]^{1/2}} = -20 \log [1 + (w/p_i)^2]^{1/2} \tag{2.69}$$

$$\approx -20 \log 1 = 0; \quad \text{for } w/p_i \ll 1 \tag{2.70}$$

$$\approx -20 \log(w/p_i); \quad \text{for } w/p_i \gg 1 \tag{2.71}$$

$$\angle \frac{1}{(jw/p_i + 1)} = -\tan^{-1}(w/p_i) \tag{2.72}$$

Similiar algebraic calculations can be carried out for a zero on the real axis, and the Bode plots are as follows,

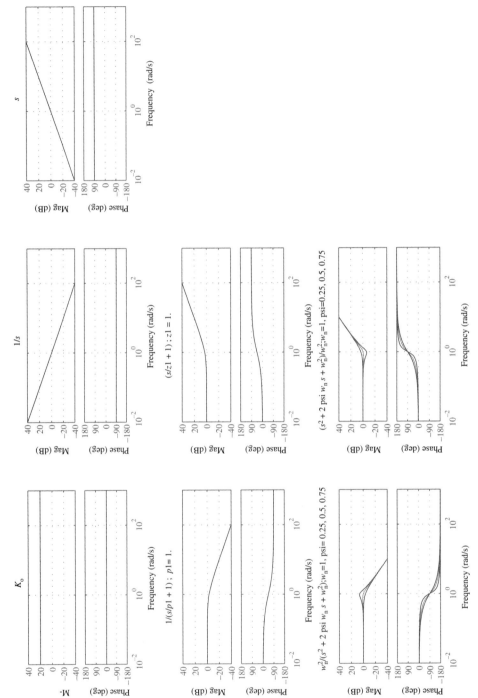

FIGURE 2.30: Bode plots of basic transfer functions: gain, pole/zero at the origin, first-order pole/zero, second-order pole/zero (for different damping coefficients).

$$(s/z_i + 1)|_{s=jw} = (jw/z_i + 1) = [1 + (w/z_i)^2]^{1/2} e^{j \tan^{-1}(w/z_i)} \tag{2.73}$$

$$20 \log |jw/z_i + 1| = 20 \log [1 + (w/z_i)^2]^{1/2} = 20 \log [1 + (w/z_i)^2]^{1/2} \tag{2.74}$$

$$= \approx 20 \log 1 = 0; \quad \text{for } w/z_i \ll 1 \tag{2.75}$$

$$= \approx 20 \log(w/z_i); \quad \text{for } w/z_i \gg 1 \tag{2.76}$$

$$\angle (jw/z_i + 1) = \tan^{-1}(w/z_i) \tag{2.77}$$

The Bode plots of first-order (real) poles and zeros are also shown in Figure 2.30.

4. *Second-order (complex conjugate) poles and zeros:* Consider a complex conjugate pole pair, and its frequency response

$$\frac{1}{(s/w_i)^2 + 2\xi(s/w_i) + 1}\Big|_{s=jw} = \frac{1}{[(1 - w^2/w_i^2)^2 + (2\xi w/w_i)^2]^{1/2} e^{j \tan^{-1}\left(\frac{2\xi w/w_i}{1-(w^2/w_i^2)}\right)}} \tag{2.78}$$

The magnitude and phase as function of frequency are given by

$$20 \log | \cdot | = -20 \log[(1 - w^2/w_i^2)^2 + (2\xi w/w_i)^2]^{1/2} \tag{2.79}$$

$$\approx 0; \quad w/w_i \ll 1 \tag{2.80}$$

$$\approx -40 \log(w/w_i); \quad w/w_i \gg 1 \tag{2.81}$$

$$\angle (\cdot) = -\tan^{-1} \left(\frac{2\xi(w/w_i)}{(1 - w^2/w_i^2)} \right) \tag{2.82}$$

Similarly, the Bode plot of a complex conjugate zero and its asymptotic plot can be found as (Figure 2.30).

$$(s/w_i)^2 + 2\xi(s/w_i) + 1|_{s=jw} = [(1 - (w^2/w_i^2))^2 + (2\xi w/w_i)^2]^{1/2} e^{j \tan^{-1}\left(\frac{2\xi w/w_i}{1-(w^2/w_i^2)}\right)} \tag{2.83}$$

$$20 \log | \cdot | = 20 \log[(1 - (w^2/w_i^2))^2 + (2\xi w/w_i)^2]^{1/2} \tag{2.84}$$

$$\approx 0; \quad w/w_i \ll 1 \tag{2.85}$$

$$\approx 40 \log(w/w_i); \quad w/w_i \gg 1 \tag{2.86}$$

$$\angle (.) = \tan^{-1} \left(\frac{2\xi(w/w_i)}{1 - (w^2/w_i^2)} \right) \tag{2.87}$$

Nyquist (Polar) Plots of Standard Elements of a Transfer Function

Nyquist plots (also called polar plots) are the graphical representation of the frequency response data on a complex plane where the y-axis is the imaginary part, and the x-axis is the real part of the frequency response. The frequency is parameterized along the curve.

$$G(jw) = G(s)|_{s=jw} = Re(G(jw)) + jIm(G(jw)) = X(w) + jY(w) \tag{2.88}$$

For every point along the imaginary axis of the s-plane, $s = jw$, there is a point on the curve plotted on the $(G(jw))$-plane. Nyquist plots of various standard elements are shown in Figure 2.31.

Log Magnitude versus Phase Plots of Standard Elements of a Transfer Function

Frequency response is conveniently represented by a complex function. Graphical representation of a complex function must convey the real and imaginary part or magnitude and phase information. Another possible way of plotting the frequency response

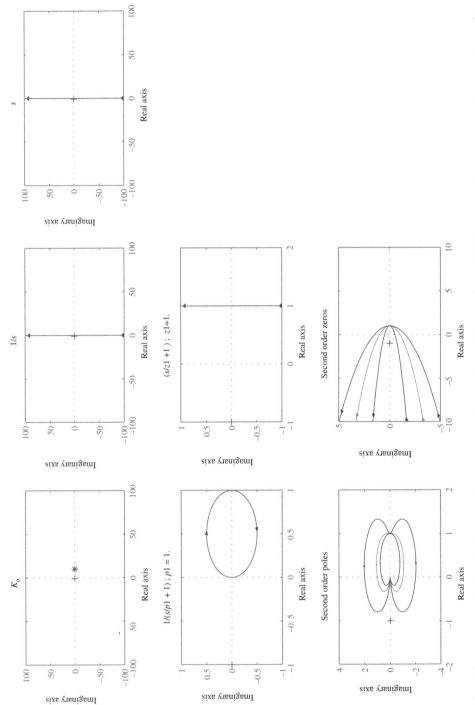

FIGURE 2.31: Nyquist (polar) plots of basic transfer functions: gain, pole/zero at the origin, first-order pole/zero, second-order pole/zero, second-order pole/zero (for different damping coefficients).

data is to plot the $20 \log_{10} |G(jw)|$ as y-axis versus $\angle(G(jw))$ as the x-axis, where w is parameterized along the plot. A frequency response can be expressed as

$$G(jw) = |G(jw)|\angle(G(jw)) \tag{2.89}$$

Log magnitude ($20 \log_{10} |G(jw)|$)versus phase angle ($\angle(G(jw))$) plots of various elements are shown in Figure 2.32. A MATLAB® program to generate the plot for Figures 2.30–2.32 is shown below.

```
K0 = 10.0 ;
z1 = 1.0  ;
p1 = 1.0  ;

num1 = [K0]        ; den1 = [1]      ; sys1=tf(num1,den1) ;  % Ko
num2 = [1]         ; den2 = [1  0]   ; sys2=tf(num2,den2) ;  % 1/s
num3 = [1 0]       ; den3 = [1]      ; sys3=tf(num3,den3) ;  % s
num4= [1]          ; den4 = [1/p1  1]; sys4=tf(num4,den4) ;  % 1/(s/p1+1)
num5= [1/z1 1]     ; den5 = [1]      ; sys5=tf(num5,den5) ;  % (s/z1+1)

psi = 0.25;  w_n = 1.0 ;
num6=[w_n^2] ; den6=[1   2*psi*w_n   w_n^2]; sys6=tf(num6,den6) ;
psi = 0.5;  w_n = 1.0 ;
num7=[w_n^2] ; den7=[1   2*psi*w_n   w_n^2]; sys7=tf(num7,den7) ;
psi = 0.75;  w_n = 1.0 ;
num8=[w_n^2] ; den8=[1   2*psi*w_n   w_n^2]; sys8=tf(num8,den8) ;
psi = 0.25;  w_n = 1.0 ;
num9=[1  2*psi*w_n  w_n^2];   den9=[w_n^2] ; sys9=tf(num9,den9) ;
psi = 0.5;  w_n = 1.0 ;
num10=[1  2*psi*w_n  w_n^2];   den10=[w_n^2] ;  sys10=tf(num10,den10) ;
psi = 0.75;  w_n = 1.0 ;
num11=[1  2*psi*w_n  w_n^2];   den11=[w_n^2] ; sys11=tf(num11,den11) ;

figure(1) ; grid on;                    % Bode Plots
subplot(3,3,1) ;  bode(sys1); grid on;
subplot(3,3,2) ;  bode(sys2); grid on;
subplot(3,3,3) ;  bode(sys3); grid on;
subplot(3,3,4) ;  bode(sys4); grid on;
subplot(3,3,5) ;  bode(sys5); grid on;
subplot(3,3,7) ;  bode(sys6);  hold on; bode(sys7);  hold on;
                  bode(sys8);  hold on;grid on;
subplot(3,3,8) ;  bode(sys9);  hold on; bode(sys10);  hold on;
                  bode(sys11); hold on;grid on;

figure(2) ; grid off;                   % Nyquist (Polar) plots
subplot(3,3,1) ;  nyquist(sys1); grid off;
subplot(3,3,2) ;  nyquist(sys2); grid off;
subplot(3,3,3) ;  nyquist(sys3); grid off;
subplot(3,3,4) ;  nyquist(sys4); grid off;
subplot(3,3,5) ;  nyquist(sys5); grid off;
subplot(3,3,7) ;  nyquist(sys6);  hold on; nyquist(sys7);  hold on;
                  nyquist(sys8);  hold on; grid off;
subplot(3,3,8) ;  nyquist(sys9);  hold on; nyquist(sys10);  hold on;
                  nyquist(sys11); hold on; grid off;

figure(3) ; grid off;                   % Log Magnitude versus Phase plots
subplot(3,3,1) ;  nichols(sys1); grid off;
subplot(3,3,2) ;  nichols(sys2); grid off;
subplot(3,3,3) ;  nichols(sys3); grid off;
subplot(3,3,4) ;  nichols(sys4); grid off;
subplot(3,3,5) ;  nichols(sys5); grid off;
```

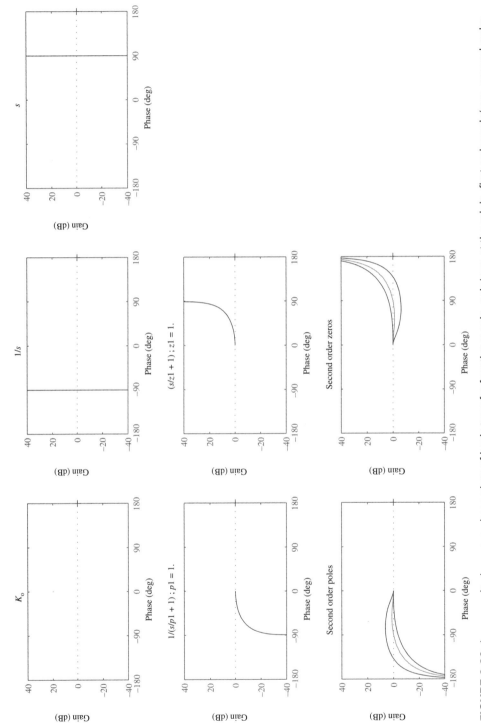

FIGURE 2.32: Log magnitude versus phase plots of basic transfer functions: gain, pole/zero at the origin, first-order pole/zero, second-order pole/zero (for different damping coefficients). Notice that the plot for constant gain K_0 is only a point in this case since it has one magnitude value and zero phase (a point on this plot) for all frequencies.

```
subplot(3,3,7) ;  nichols(sys6);  hold on; nichols(sys7);  hold on;
                  nichols(sys8);  hold on; grid off;
subplot(3,3,8) ;  nichols(sys9);  hold on; nichols(sys10);  hold on;
                  nichols(sys11); hold on; grid off;
```

2.9.2 Stability Analysis in the Frequency Domain: Nyquist Stability Criteria

Frequency domain methods are the most commonly used control system design methods in practice. Therefore, it is important to be able to evaluate the stability of a dynamic system in the frequency domain.

The stability of a linear time invariant dynamic system can be determined in the frequency domain using the *Nyquist stability criteria*. Furthermore, *relative stability* can be quantified (if stable, how far is the system from being unstable, or if unstable how far is the system from being stable) using the *gain margin* and *phase margin* measures.

Consider the feedback control system shown in the Figure 2.17. The question is "how can we determine if the closed loop system is stable" using frequency domain methods.

$$CLS\,poles: \quad 1 + G(s)H(s) = 0 \tag{2.90}$$

Are there any roots of this equation on the right-half of the s-plane?

The Nyquist stability criteria answers that question by using the frequency response data of the loop transfer function, $G(s)H(s)$ or $G(s)$ if the sensor dynamics is included in the loop transfer function. In the s-domain, we can find the roots of this equation. If any roots are on the right-half s-plane, then the CLS is unstable. However, we would like to determine if the CLS has poles on the RHP using frequency domain methods without explicitly solving for the roots of the closed loop characteristic equation.

The Nyquist stability criteria is derived as a special case of the mapping theorem of complex variables. Consider the mapping shown in Figure 2.33a. A contour C_1 from the s-plane is mapped to $F(s)$-plane by the function $F(s)$. The closed contour C_1 will be mapped to another closed contour C_1' on the $F(s)$-plane. The important point to note in this mapping is the number of poles and zeros of the mapping function $F(s)$ inside the C_1 contour and its relationship to the number of encirclements of the origin in the $F(s)$-plane by the C_1' contour. Notice that as a phasor from a zero inside C_1 traverses clockwise (CW) over the C_1, the C_1' will encircle the origin in the CW direction (Figure 2.33b). Similarly, as a phasor from a pole inside C_1 traverses clockwise (CW) over the C_1, the C_1' will encircle the origin in the counter clockwise (CCW) direction (Figure 2.33c). If there are no poles or zeros of $F(s)$ inside the contour C_1, then the mapped contour C_1' does not encircle the origin in the $F(s)$-plane. If there are two poles inside the contour C_1, then the C_1' contour encircles the origin CCW two times (Figure 2.33d). If there are two zeros inside the contour C_1, then the C_1' contour encircles the origin CW two times.

In summary, as the mapping $F(s)$ traverses for s variable values over the C_1 contour in CW direction, the corresponding contour in $F(s)$-plane, C_1', will encircle the origin in the CW direction based on the following relationship:

$$N = Z - P \tag{2.91}$$

where,

N is the number of CW encirclements of the origin by C_1',

Z is the number of zeros of $F(s)$ inside C_1,

P is the number of poles of $F(s)$ inside C_1.

Notice that encirclements of origin in CW direction are counted as positive, and the encirclements of the origin in the CCW direction are counted as negative (Figure 2.33a–d).

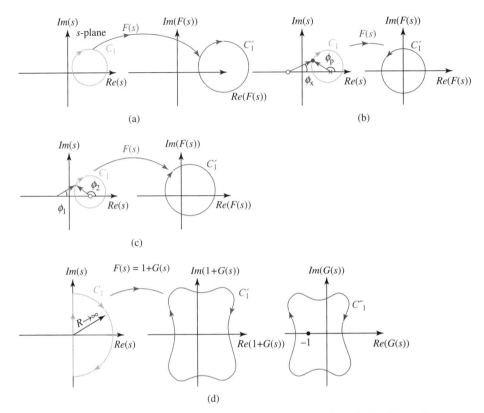

FIGURE 2.33: Stability analysis in the frequency domain using Nyquist Stability criteria: (a) mapping of a closed contour C_1 from s-plane to $F(s)$-plane, where $F(s)$ has no pole nor zero inside the contour C_1 in the s-plane, (b) same as (a) except $F(s)$ has a pole inside C_1, (c) same as (a) except $F(s)$ a zero inside the C_1 contour, (d) mapping of right-hand s-plane as a special choice of C_1 contour in s-plane and $F(s) = 1 + GH(s)$, typical selection of C_1 contour to exclude poles and zeros on jw axis. If C_1 is chosen to include them, Nyquist stability criteria would still give us the correct answer about the number of unstable poles, Nyquist stability criteria and Nyquist (polar) plot where only $s = jw$ for $w = 0 \longrightarrow \infty$ is mapped. Use of Nyquist stability criteria to determine relative stability: does the closed loop system have poles inside this contour C_1, if so, how many?

The Nyquist stability criteria is an application of the mapping theorem to determine the stability of a closed loop LTI dynamic system. Let us consider that the C_1 contour is a contour containing the RHP in s-plane, and that $F(s) = 1 + G(s)$ (Figure 2.33d). Then, the number of CW encirclements of the origin in $(1 + G(s))$ plane by the C_1' contour is equal to the number of zeros of $(1 + G(s))$ in the RHP (which is the number of unstable closed loop poles) minus the number of poles of $(1 + G(s))$ in the RHP (which is the number of unstable open loop poles). Finally, instead of $(1 + G(s))$ mapping, we can consider the $G(s)$ mapping alone, and revise the above conclusion for $(-1, 0)$ point encirclement instead of the origin $(0, 0)$.

Nyquist Stability Criteria: The number of unstable closed loop poles (Z) of the system shown in Figure 2.33, is equal to the number of CW encirclements of the $(-1, 0)$ point plus the number of unstable poles of the open loop system,

$$Z = N + P \qquad (2.92)$$

Notice that if the open loop system is stable, $P = 0$, then $Z = N$.

In most engineering systems, the transfer function has more poles than zeros. Therefore, as we map the C_1 contour with $G(s)$,

- The half circle arc (as s goes to infinity) will map to zero magnitude for $deg(d(s)) > deg(n(s))$ or a finite value is $deg(d(s)) = deg(n(s))$ (Figure 2.33d).

- The mapping of the lower half of the imaginary axis will be symmetric to the mapping of the upper half of the imaginary axis (Figure 2.33d).

Hence, the Nyquist plot can be determined only from the mapping of the positive jw axis.

If there are poles or zeros on the imaginary axis, the contour C_1 should either include them or exclude them from being inside the C_1 contour. Either approach would give the same final conclusion regarding closed loop stability. It is customary to exclude the poles and zeros on the imaginary axis from the C_1 contour's inside.

Relative Stability   The relative stability of a CLS is quantified by the use of the distance of the Nyquist plot from the $(-1, 0)$ point. That is "how far is the closed loop system from the stability boundary."

There are two quantities defined for relative stability (Figure 2.34): the *gain margin* (GM) and the *phase margin* (PM). The gain margin is the inverse of the magnitude of the loop transfer function when the phase is 180°. It indicates how much the loop gain can be increased before the system reaches the stability boundary. The phase margin is the phase angle difference between the loop transfer function and $-180°$ when the magnitude of the loop transfer function is 1. The PM indicates the amount of phase lag that can be introduced into the loop transfer function before it reaches the stability boundary. The measurement of GM and PM on Bode and Nyquist plots is shown in Figure 2.34.

2.10   THE ROOT LOCUS METHOD

The root locus method is a graphical method for plotting the roots of an algebraic equation as one or more parameters vary. It is used primarily in studying the effect of variations in one parameter of a control system on the locations of closed loop system poles. In the next section, we will use the root locus method in order to understand the characteristics of PID type closed loop controllers. The basic mathematical functionality is to *find* and *plot* the roots of an algebraic equation for various values of a parameter. The solution of algebraic equations can be easily done by numerical means using a digital computer. Solving it for various values of one or more parameters is nothing more than implementing the numerical procedure in an iteration loop, that is FOR or DO loop in a high level programming language. This is certainly a tool which became more and more effective with the availablity of CAD tools for control system design. A control engineer must, however, always keep in mind that the basic principle about computers is *garbage in – garbage out*. Therefore, it is very important that a designer should be able to quickly verify in general terms a computer calculation with hand calculations or analysis. The graphical hand sketching rules of the root locus method provides such a tool. Understanding the graphical root locus method not only provides a way of quickly checking the results of a computer calculations, but also develops very valuable insights into the control system design.

Let us consider an algebraic equation, that is a polynomial of degree n,

$$a_0 s^n + a_1 s^{n-1} + \ldots + a_{n-1} s + a_n = 0 \qquad (2.93)$$

which has n roots $\{s_1, s_2, \ldots, s_n\}$. If the value of any of a_i changes, there will be a new set of n roots $\{s_1, s_2, \ldots, s_n\}$. If a particular parameter a_i varies from one minimum value to

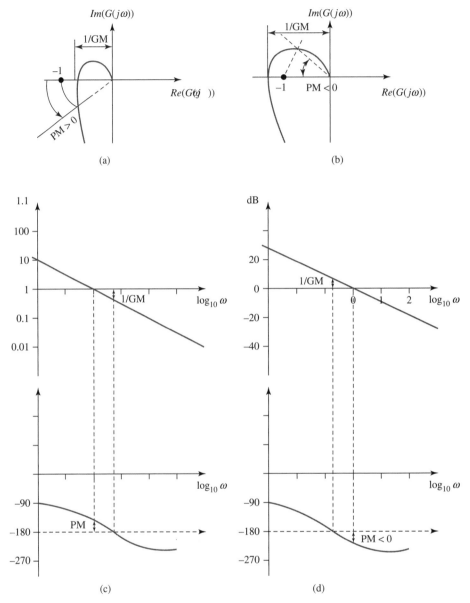

FIGURE 2.34: Gain margin and phase margin measures in Nyquist plots (a and b) and Bode plots (c and d).

another value, an n set of roots can be solved for every value of the parameter a_i. If we plot the roots on the complex s-plane, we end up with the graphical representation of the *locus of the roots* of the algebraic equation as one of the parameters in the equation varies. This is the basic functionality of the root locus method. Clearly, this can be done numerically by a computer algorithm.

Let us consider the feedback control system shown in Figure 2.35. The closed loop system transfer function is

$$\frac{y(s)}{r(s)} = \frac{KG(s)}{1 + KG(s)} \qquad (2.94)$$

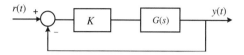

FIGURE 2.35: Basic root locus problem.

where the poles of the closed loop system are given by the roots of the denominator,

$$\Delta_{cls}(s) = 1 + KG(s) \tag{2.95}$$

The standard root locus analysis problem involves the sketch of the locus of the roots of this equation as K varies from zero to infinity. From the above general discussion, it is clear that the root locus method is not limited to that type of problem. It can address any problem which involves finding roots of an algebraic equation as any one or more parameters vary. Consider another example as shown in Figure 2.36, where a is a parameter. We would like to study the locations of closed loop system poles as the parameter a varies from zero to infinity.

$$\frac{y(s)}{r(s)} = \frac{1}{1 + s(s+a)} = \frac{1}{s^2 + as + 1} \tag{2.96}$$

The characteristic equation is

$$\Delta_{cls}(s) = s^2 + as + 1 = 0 \tag{2.97}$$

which can be expressed in standard root locus formulation form suitable for graphical sketching as

$$1 + a \cdot \frac{s}{s^2 + 1} = 0 \tag{2.98}$$

The graphical root locus method rules are developed for sketching the roots of a polynomial equation as one parameter varies (Figure 2.37). The polynomial equation is always expressed in the form of

$$1 + (parameter) \cdot \frac{numerator}{denominator} = 0 \tag{2.99}$$

Therefore, the locus of roots can be studied as a function of any parameter in the closed loop system, not just the gain of the loop transfer function. Let us assume that we are interested in studying the locus of the roots of the following equation as parameter b varies from zero to infinity,

$$s^3 + 6s^2 + bs + 8 = 0 \tag{2.100}$$

This problem can be expressed in a form suitable for the application of the root locus sketching rules as follows,

$$1 + b \cdot \frac{s}{s^3 + 6s^2 + 8} = 0 \tag{2.101}$$

MATLAB® provides the *rlocus(...)* function for root locus. The *rlocus()* function is overloaded and can accept different parameters. In principle, it takes the loop transfer

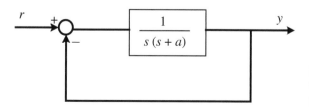

FIGURE 2.36: An example: closed loop transfer function poles as parameter a varies.

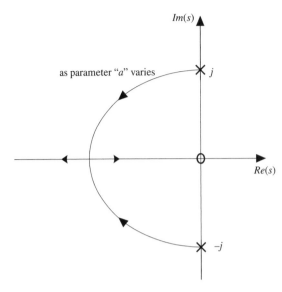

as parameter "*a*" varies

FIGURE 2.37: Value of gain is parameterized along the root locus curves. The particular value of the gain can be calculated in order to be at a selected point on the root locus.

function and assumes that a parameter K is in the feedback loop. Therefore, in order to use the *rlocus()* function to analyze the roots of a transfer function or algebraic equation as one parameter varies, the equivalent root locus problem must be formed before calling the *rlocus()* function. Below are some samples of calls to *rlocus()* function.

```
sys = tf(num,den) ;   /* sys can be formed by tf, ss, zpk function calls */
sys = zpk(z,p,k)  ;   /* sys= (K (s-z1)(s-z2).../)(s-p1).(s-p2)....) */
sys = ss(A,B,C,D) ;   /* G(s) = C (sI-A)^-1 B + D */

rlocus(sys) ;       /* given LTI system, plot closed loop poles as K varies
                        from 0 to infinity
rlocus(sys, K);     /*..............for the values of the parameter given
                        in the vector K*/
[R]=rlocus(sys,K); /* Stores the closed loop roots in the R for numerical
                        reference. */
rltool(sys) ;       /* Interactive graphical tools for plotting root locus */
```

Root Locus Sketching Rules   In this section we list the rules used in approximate hand sketching of the root locus. The derivation of the rules is given in many classic textbooks on control systems.

1. Put the problem in the $1 + KG(s) = 0$ form, where K is the varying parameter. Mark the poles of $G(s) = n(s)/d(s)$ by x and the zeros by o on the s-plane. Notice that root locus begins at x's for $K = 0$ and ends at o's as $K \longrightarrow \infty$. The x's and o's represent the asymptotic location of closed loop pole locations. If K is indeed the loop transfer function gain, the o's are also the zeros of the open and closed loop system. Otherwise, they only represent the asymptotic location of closed loop system poles.

2. Mark the part of the real axis to the left of odd number of poles and zeros as part of the root locus. All the points on the real axis to the left of odd number of poles and zeros satisfy the angle criteria, hence are part of the root locus. Notice that the

root locus is the collection of points on the s-plane that makes the phase angle of the transfer function equal to $180°$, since

$$K = -\frac{1}{G(s)};$$ (2.102)

Since, K is a positive real number, the $G(s)$ complex function will have negative real values at the collection of s-points that makes up the root locus. In other words, the phase angle of $G(s)$ at all points that are part of the root locus is $180°$.

$$G(s) = -\frac{1}{K} = \left|\frac{1}{K}\right| \cdot e^{j180°}$$ (2.103)

3. If there are n poles, and m zeros, m of the poles end up at the m zero locations as parameter K goes to infinity. The remaining $n - m$ poles go to infinity along asymptotes. If $n - m = 1$, then the one extra pole goes to infinity along the negative real axis. If $n - m \geq 2$, then they go to infinity along the asymptotes defined by the asymptote center and angles as follows:

$$\sigma = \frac{\sum p_i - \sum z_i}{n - m}$$ (2.104)

$$\psi_l = \frac{180 + l \cdot 360}{n - m}; \quad l = 0, 1, 2, \ldots, n - m - 1.$$ (2.105)

where p_i's are the pole locations z_i's are the zero locations of the $G(s)$,

$$G(s) = \frac{\Pi_{i=1}^m (s - z_i)}{\Pi_{i=1}^n (s - p_i)}$$ (2.106)

The quick hand sketches of the root locus allow the designer to quickly check the computer analysis results for correctness and provide valuable insight in controller design. Note that when the excessive number of poles than zeros is $n - m = 0, 1, 2, 3, 4$, the angles of asymptotes are *none*, 180, $\{90, 270\}$, $\{60, 180, 300\}$, $\{45, 135, 225, 315\}$, respectively.

2.11 CORRELATION BETWEEN TIME DOMAIN AND FREQUENCY DOMAIN INFORMATION

Three major groups of events which are not under our control and affect the system performance are:

1. variations in the process parameters and dynamics,
2. disturbances,
3. sensor noise.

A desired performance specification for any CLS includes specifications regarding

1. stability,
2. response quality (transient and steady state),
3. robustness of stability and response quality despite real-world imperfections, that is variations in the process dynamics, disturbances, and sensor noise.

The *stability* of CLS requires that all of the CLS poles be in the left half of the s-plane (LHP). If a certain degree of relative stability is required, then we can further impose conditions that CLS poles must have a real part smaller than a negative real value

$$Re(p_i) < 0 \quad or \quad -a \tag{2.107}$$

or in the frequency domain it can be specified in terms of gain and phase margins,

$$GM > GM_{min}, \quad PM > PM_{min} \tag{2.108}$$

The *response quality* is generally divided into two groups: transient response and steady-state response. The transient response is generally specified as the step response. The step response is quantified using percent overshoot, settling time, and rise time specifications: $(PO\%, t_s, t_r)$. For a *second-order closed loop system*, the PO% and t_s uniquely determine the closed loop system poles with damping ratio (ξ) and natural frequency (w_n):

$$p_{1,2} = -\xi w_n \pm j\sqrt{1-\xi^2}w_n \tag{2.109}$$

$$t_s = \frac{4.0}{\xi w_n}; \tag{2.110}$$

$$PO = e^{-\xi\pi/\sqrt{1-\xi^2}}; \quad for\ 0 \le \xi < 1.0 \tag{2.111}$$

In the frequency domain, the cross-over frequency (w_{cr}) of the loop transfer function and bandwidth (w_{bw}) of the closed loop transfer function, along with the phase margin of the loop transfer function, correlate well with the transient response. Cross-over frequency is defined for the loop transfer function where the magnitude of the loop transfer function is 1 (or 0 dB). Bandwidth frequency is defined for the closed loop transfer function where the magnitude is 0.707 or (−3 dB). Bandwidth closely relates to the speed of response (t_s), and phase margin (PM) is closely related to the damping ratio. For **a second-order closed loop system** as shown in Figure 2.38, it can be shown [6] that

$$PM = \tan^{-1}\frac{2\xi}{\sqrt{\sqrt{1+4\xi^2}-2\xi^2}} \tag{2.112}$$

$$\approx 100\,\xi \ for\ \xi \le 0.6 \tag{2.113}$$

$$w_{cr} = \sqrt{-2\xi^2 + \sqrt{4\xi^2+1}} \cdot w_n; \ w = w_{cr}\ at\ |G(jw)| = 1.0 \tag{2.114}$$

$$; 20\log_{10}|G(jw)| = 0.0\ \text{dB} \tag{2.115}$$

$$w_{bw} = w_n \cdot \sqrt{(1-2\xi^2) + \sqrt{(2\xi^2-1)^2 + 1}} \tag{2.116}$$

$$; w = w_{bw}\ at\ |G(jw)/(1+G(jw))| = 0.707 \tag{2.117}$$

$$; 20\log_{10}|G(jw)/(1+G(jw))| = -3\ \text{dB} \tag{2.118}$$

$$w_p = w_n\sqrt{1 - 2 \cdot \xi^2}; \ where\ |G(jw)/(1+G(jw))|\ is\ maximum \tag{2.119}$$

$$; w = w_p\ at\ \frac{d}{dw}(|G(jw)/(1+G(jw))|) = 0.0 \tag{2.120}$$

$$; for\ 0 \le \xi \le 0.707 \tag{2.121}$$

$$M_p = \max|G(jw)/(1+G(jw))| \tag{2.122}$$

$$= \frac{1}{2\xi\sqrt{1-\xi^2}}; for\ 0 \le \xi \le 0.707 \tag{2.123}$$

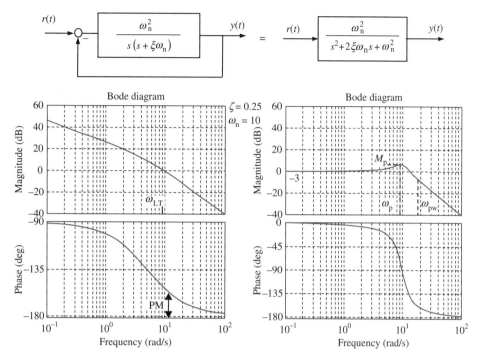

FIGURE 2.38: Second-order linear system: Bode plot of the loop transfer function and the Bode plot of the closed loop transfer function, along with some transient response related performance measures.

where settling time t_s is defined as the time it takes for the step response to settle within 2% of the commanded signal, PO is the maximum percent overshoot of the step response. In general, it can be shown that $w_{cr} \leq w_p \leq 2 \cdot w_{cr}$.

A Bode plot of the loop transfer function and closed loop transfer function of a second-order system can be generated for specific values of the parameters ξ and w_n as shown in Figure 2.38.

```
psi = 0.25 ;
wn  = 10.0 ;
s=tf('s') ;
G1 = wn^2/(s^2+2*psi*wn*s) ; % Loop tranfer function
G2 = G1/(1+G1) ;            % Closed loop transfer function
figure(1) ; grid on ;
subplot(1,2,1) ;  bode(G1,'r') ; grid on ;
subplot(1,2,2) ;  bode(G2,'r') ; grid on ;
```

The typical desired gain margin and phase margins are; $GM \geq 6$ dB and PM around 30° to 60°. For any stable minimum phase system, $G(s)$, (a dynamic system whose zeros and poles are on the LHP), the phase of $G(jw)$ is uniquely related to the magnitude of $G(jw)$. The approximate phase is the slope of the magnitude curve (n, slope of magnitude curve) times 90° at any frequency,

$$\angle G(jw) \approx \pm n \times 90° \tag{2.124}$$

Therefore, around cross-over frequency, $w \approx w_{cr}$, $|G(jw)| \approx 1$, if $n = -1$, that is the slope of the magnitude curve is -20 dB/decade, then $\angle G(jw_{cr}) \approx -90°$, and hence the $PM \approx 90°$. If $n = -2$, $\angle G(jw_{cr}) \approx 180°$ and the $PM \approx 0$. Therefore, the slope of the magnitude curve

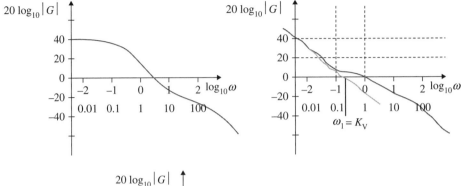

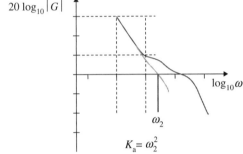

FIGURE 2.39: Correlation between the time domain steady-state response and frequency response (Bode plot) characteristics.

on the Bode plot should have about -20 dB/decade slope around the cross-over frequency so that the the system has about $90°$ phase margin.

The steady-state response is specified in terms of the steady-state error of the CLS in response to three standard test command signals: step, ramp, and parabolic signals. It can be shown that the steady-state error in response to step, ramp, and parabolic command signals are defined as

$$e_{\text{step}}(\infty) = \frac{1}{1 + K_p}; K_p = \lim_{s \to 0} D(s)G(s) \qquad (2.125)$$

$$e_{\text{ramp}}(\infty) = \frac{1}{K_v}; K_v = \lim_{s \to 0} sD(s)G(s) \qquad (2.126)$$

$$e_{\text{parabolic}}(\infty) = \frac{1}{K_a}; K_a = \lim_{s \to 0} s^2 D(s)G(s) \qquad (2.127)$$

The error constants K_p, K_v, K_a can be directly determined from the asymptotic behavior of the frequency response plots using Bode or Nyquist plots (Figure 2.39). Let us consider a frequency response equation in the following form

$$G(jw) = K \frac{(1 + jw/z_1)(1 + jw/z_2)...}{(jw)^N(1 + jw/p_1)(1 + jw/p_2)...} \qquad (2.128)$$

Clearly, if we would like to estimate the low frequency gain in order to get the error constants, let $w \longrightarrow 0$ and the frequency response can be approximated as

$$G(jw) \approx \frac{K}{(jw)^N} \qquad (2.129)$$

- If $N = 0$, then $K_p = K$.
- If $N = 1$, then $|G(jw)| = K/|jw| = 1$ or $K_v = w_1$ where $20 \log |G(jw)|_{w=w_1} = 0$ dB.
- Similarly, If $N = 2$, then $|G(jw)| = 1$, then, $K_a = w_2^2$ where w_2 is such that $20 \log |G(jw)|_{w=w_2} = 0$.

The type of the loop transfer function can also be immediately determined from the slope of the magnitude curve as frequency goes to zero or from the phase plot at low frequencies.

The robustness specification deals with the sensitivity of the system. The most important advantage of feedback control over open loop control is that the feedback improves the robustness of the system performance against the variations in process dynamics and disturbances. The closed loop system should not only be stable and have good response quality for the nominal parameters of the operating conditions, but also should stay stable and have good response quality despite the real-world imperfections.

In summary, the correlation between the time-domain specifications and frequency domain behavior is as follows

- Good stability means a large gain margin and phase margin. In order to have a reasonably good P.M., the slope of the magnitude curve should be about -20 dB/decade around the cross-over frequency.
- Larger loop gain at low frequencies results in lower steady-state errors, and good disturbance rejection against low frequency disturbances.
- Low loop gain and a fast decaying rate at the high frequency region increase the ability to reject the effect of high frequency noise.

Overall, the stability, steady-state error, and robustness characteristics of a CLS is well represented in the frequency response of the loop transfer function, whereas the transient response is not represented with the same accuracy. The s-domain pole-zero representation of a CLS correlates to the transient response behavior well, but does not give information about disturbance and sensor noise rejection ability. Therefore, frequency response (i.e., Bode plots) and s-domain methods (i.e., root locus method) complement each other in the graphical information they display regarding the control system characteristics (i.e., transient and steady-state response).

2.12 BASIC FEEDBACK CONTROL TYPES

Figure 2.40 shows the three basic feedback control actions: proportional, integral, and derivative control actions. Figure 2.41 shows the input–output behavior of these control types. In practical terms, proportional control action is generated based on the current error, the integral control action is generated based on the past error, and the derivative control action is generated based on the anticipated future error. The integral of the error can be interpreted as the past information about it. The derivative of the error can be interpreted as as a measure of future error to come. Assume that the error signal entering the control blocks has a trapeziodal form. The control actions generated by the proportional, integral, and derivative actions are shown in Figure 2.41. Proportional - integral - derivative (PID) control has control decision blocks which take into account the past, current, and future error. In a way, it covers all the history of error. Therefore, most practical feedback controllers are either a form of the PID controller or have the properties of a PID controller.

The block diagram of a textbook standard PID controller is shown in Figure 2.42. The control algorithm can be expressed in both the continuous (analog) time domain (which can be implemented using op-amps) and in the discrete (digital) time domain (which can be implemented using a digital computer in software). At any given time t, the control signal $u(t)$ is determined as function,

$$u(t) = K_{\mathrm{p}}e(t) + K_{\mathrm{I}} \int_0^t e(\tau)d\tau + K_{\mathrm{D}}\dot{e}(t) \tag{2.130}$$

Proportional control (P)

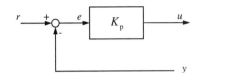

$$u(t) = K_p e(t)$$

$$D(s) = \frac{u(s)}{e(s)} = K_p$$

Integral control (I)

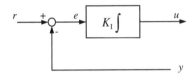

$$u(t) = K_I \int_0^t e(\tau)d\tau$$

$$D(s) = \frac{u(s)}{e(s)} = \frac{K_I}{s}$$

Derivative control (D)

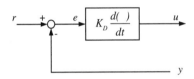

$$u(t) = K_D \frac{d}{dt}(e(t))$$

$$D(s) = \frac{u(s)}{e(s)} = K_D s$$

FIGURE 2.40: Basic feedback control actions: proportional control, integral control, derivative control.

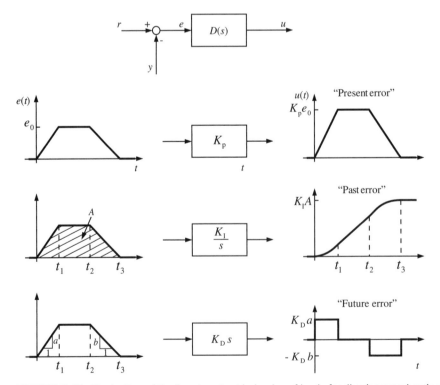

FIGURE 2.41: Illustration of the input–output behavior of basic feedback control actions: proportional, integral, derivative control.

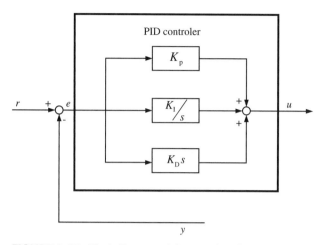

FIGURE 2.42: Block diagram of the standard PID controller.

which shows that the control signal is function of the error between the commanded and measured output signals, $e(t)$ at time t, as well as the derivative of the error signal $\dot{e}(t)$ and the integral of the error signal since the control loop is enabled ($t = 0$), $\int_0^t e(\tau)d\tau$.

The discrete time approximation of the PID control algorithm can be implemented by finite difference approximation to the derivative and integral functions. In digital implementation, the control signal can be updated at periodic intervals, T, also called the sampling period. The control is updated at integer multiples of the sampling period. The value of the signal is kept constant between each update period. At any update instant k, time is $t = kT$, and the previous update instant is $t - T = kT - T$, the next update instant is $t = kT + T$ and so on, the control signal can be expressed as $u(t) = u(kT)$,

$$u(kT) = K_{\mathrm{p}} \cdot e(kT) + K_{\mathrm{I}} \cdot u_{\mathrm{I}}(kT) + K_{\mathrm{D}}\frac{(e(kT) - e(kT - T))}{T}) \qquad (2.131)$$

where

$$u_{\mathrm{I}}(kT) = u_{\mathrm{I}}(kT - T) + e(kT) \cdot T \qquad (2.132)$$

$$u_{\mathrm{I}}(0) = 0.0; \quad \textit{at the initialization} \qquad (2.133)$$

Let us take the Laplace transform of the continuous time domain (analog) version of the PID control and analyze its effect on a controlled system. Basically, the same results apply for the discrete time version (digital implementation) provided the sampling period is short enough (high sampling frequency) relative to the bandwidth of the closed loop system.

$$u(s) = \left(k_{\mathrm{p}} + K_{\mathrm{I}}\frac{1}{s} + K_{\mathrm{D}}s\right)e(s) \qquad (2.134)$$

$$D(s) = K_{\mathrm{p}} + K_{\mathrm{I}}\frac{1}{s} + K_{\mathrm{D}}s \qquad (2.135)$$

$$= K_{\mathrm{p}}\left(1 + \frac{1}{T_{\mathrm{I}}s} + T_{\mathrm{D}}s\right) \qquad (2.136)$$

$$K_{\mathrm{I}} = \frac{K_{\mathrm{p}}}{T_{\mathrm{I}}}, \quad K_{\mathrm{D}} = K_{\mathrm{p}}T_{\mathrm{D}}$$

Consider a second order mass-force system to study its behavior under various forms of PID control (Figure 2.43).

$$m\ddot{x}(t) = f(t) - f_{\mathrm{d}}(t)$$

$$\ddot{x}(t) = \frac{1}{m}f(t) - \frac{1}{m}f_{\mathrm{d}}(t)$$

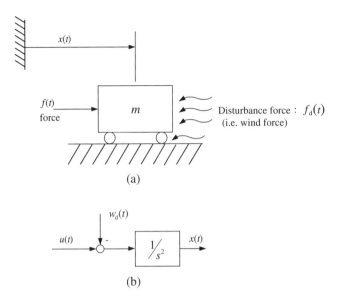

(a)

(b)

FIGURE 2.43: Mass-force system model (a) model components, (b) block diagram.

$$\ddot{x}(t) = u(t) - w_d(t)$$
$$s^2 x(s) = u(s) - w_d(s)$$
$$x(s) = \frac{1}{s^2}u(s) - \frac{1}{s^2}w_d(s)$$

2.12.1 Proportional Control

Let us consider the input and output relationship, and do not consider disturbance for the purpose of studying proportional control properties only (Figure 2.44),

$$\frac{x(s)}{u(s)} = \frac{1}{s^2}$$

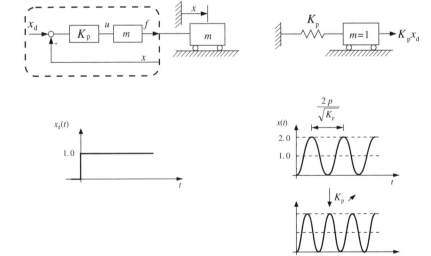

FIGURE 2.44: Mass–force system with position feedback control and its step response.

If the control action $u(t)$ is decided upon by a proportional control based on the error between the desired position, $x_d(t)$ and the actual measured position, $x(t)$,

$$u(t) = K_p(x_d(t) - x(t))$$
$$u(s) = K_p(x_d(s) - x(s))$$

CLS transfer function from the commanded position to the actual position under the proportional control is

$$x(s) = \frac{K_p}{s^2 + K_p} x_d(s)$$

The proportional control alone on a mass-force system is equivalent to adding a spring to the system where the spring constant is equal to the proportional feedback gain, K_p (Figure 2.44). The response of this system to a commanded step change in position is shown in Figure 2.44. Figure 2.46a shows the CLS root locus as K_p varies from zero to infinity. The steady-state error due to constant disturbance is

$$X(s) = -\frac{1}{m}\frac{1}{s^2 + K_p} \cdot F_d(s) \qquad (2.144)$$

$$\lim_{t \to \infty} x(t) = \lim_{s \to 0} sX(s) = -\frac{1}{m}\frac{1}{K_p} \qquad (2.145)$$

2.12.2 Derivative Control

Let us consider only the derivative control on the same mass-force system. Assume that the control is proportional to the derivative of position which means proportional to the velocity,

$$u(t) = -K_D\dot{x}$$
$$u(s) = -K_D s x(s)$$
$$s^2 x(s) = -K_D s x(s)$$
$$s(s + K_D)x(s) = 0$$

If we consider disturbance in the model, the transfer function from the disturbance (i.e., wind force) to the position of the mass can be determined as

$$m\ddot{x} = f(t) - f_d(t)$$
$$\ddot{x} = u(t) - w_d(t)$$
$$s(s + K_D)x(s) = -w_d(s)$$
$$x(s) = \frac{1}{s(s + K_D)}(-w_d(s))$$

Let us consider the case that the disturbance is a constant step function, $w_d = \frac{1}{s}$ and the resultant response is

$$x(s) = -\frac{1}{s}\frac{1}{s(s + K_D)}$$
$$= \frac{a_1}{s^2} + \frac{a_2}{s} + \frac{a_3}{(s + K_D)}$$

where

$$a_1 = \lim_{s \to 0} s^2 x(s) = -\frac{1}{K_D} \tag{2.146}$$

$$a_2 = \lim_{s \to 0} \frac{d}{ds} [s^2 x(s)] = \frac{1}{K_D^2} \tag{2.147}$$

$$a_3 = \lim_{s \to -K_D} (s + K_D) x(s) = -\frac{1}{K_D^2} \tag{2.148}$$

$$x(t) = -\frac{1}{K_D} \cdot t + \frac{1}{K_D^2} \cdot 1(t) - \frac{1}{K_D^2} \cdot e^{-K_D t} \tag{2.149}$$

The position of mass under the derivative control due to a constant disturbance force is

$$x(t) = \frac{1}{K_D} t - \frac{1}{K_D^2} 1(t) + \frac{1}{K_D^2} e^{-K_D t}$$

In steady state,

$$\dot{x}(t) = -\frac{1}{K_D} \tag{2.150}$$

Since $\dot{x}_d(t) = 0.0$ (desired velocity is zero), the error in steady-state

$$\dot{e}(t) = \dot{x}_d(t) - \dot{x}(t) = \frac{1}{K_D} \tag{2.151}$$

which means that in steady state, the velocity error due to a step input constant force disturbance will result in constant velocity of the mass, even though the desired velocity is zero. The velocity error is inversely proportional to the velocity feedback gain.

This example shows that the derivative feedback control alone would not be able to reject a constant disturbance acting on a second-order mass–force system. Derivative feedback introduces damping into the closed loop system poles (Figure 2.45) and increases the stability margin.

2.12.3 Integral Control

Now let's consider the case where the control action is based on the integral of position error,

$$u(t) = K_I \int_0^t [x_d(\tau) - x(\tau)] d\tau$$

$$u(s) = \frac{K_I}{s} [x_d(s) - x(s)]$$

Substituting this into mass-force model

$$s^2 x(s) = u(s) = \frac{K_I}{s} [x_d(s) - x(s)]$$

$$s^3 x(s) + K_I x(s) = K_I x_d(s)$$

$$(s^3 + K_I) x(s) = K_I x_d(s)$$

The closed loop system poles are given by (Figure 2.46);

$$\Delta_{cl}(s) = 1 + K_I \frac{1}{s^3}$$

Figure 2.46b shows the locus of CLS poles for various values of K_I as it takes on values from 0 to ∞. The integral control alone would result in an unstable mass-force system. It tends to destabilize the system. However, the main purpose of integral control

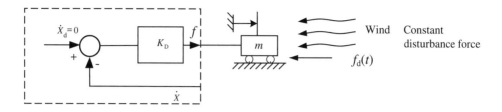

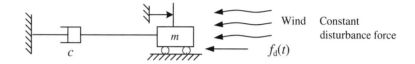

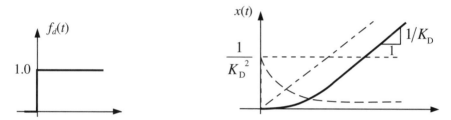

FIGURE 2.45: Mass force system under velocity feedback control, and response to a constant disturbance.

is to reject the disturbances and reduce the steady-state error, as will be shown in the next section.

2.12.4 PI Control

Let us consider the mass-force system under a proportional plus integral (PI) control. The control algorithm is

$$u(t) = K_p(x_d(t) - x(t)) + K_I \int_0^t (x_d(\tau) - x(\tau))d\tau$$

and its Laplace transform is

$$u(s) = K_p(x_d(s) - x(s)) + K_I \frac{1}{s}(x_d(s) - x(s))$$

$$= \left(K_p + \frac{K_I}{s} \right) (x_d(s) - x(s))$$

Let's take the Laplace transform of the mass-force system and substitute the PI controller for $u(s)$

$$\ddot{x} = u(t) - w_d(t)$$

$$s^2 x(s) = u(s) - w_d(s) = \left(K_p + \frac{K_I}{s} \right) (x_d(s) - x(s)) - w_d(s)$$

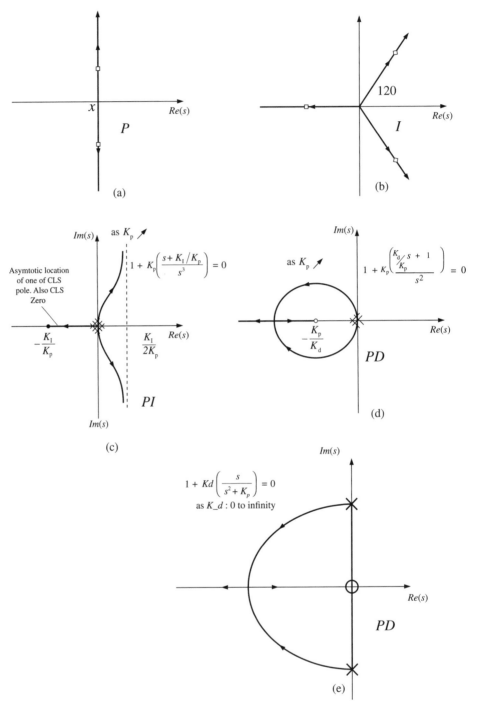

FIGURE 2.46: Locus of the poles of the system under P, I, PI, and PD control (closed loop system poles) as the gain increases from zero to infinity.

After a few algebraic manipulations, the transfer function between position, desired position, and disturbance force is found as

$$x(s) = \frac{K_\mathrm{p}s + K_\mathrm{I}}{s^3 + K_\mathrm{p}s + K_\mathrm{I}} x_\mathrm{d}(s) - \frac{s}{s^3 + K_\mathrm{p}s + K_\mathrm{I}} w_\mathrm{d}(s)$$

Consider the case that the commanded position is zero, $x_\mathrm{d}(t) = 0$ and there is a constant step disturbance, $w_\mathrm{d}(s) = \frac{1}{s}$. Any non-zero response due to the disturbance would be an error.

$$x(s) = -\frac{s}{s^3 + K_\mathrm{p}s + K_\mathrm{I}} \frac{1}{s}$$

$$x(s) = -\frac{1}{s^3 + K_\mathrm{p}s + K_\mathrm{I}}$$

If $\Delta_\mathrm{cls}(s) = s^3 + K_\mathrm{p}s + K_\mathrm{I}$ has stable roots, the response of the system will be zero despite a constant disturbance.

Using the final value theorem,

$$\lim_{t \to \infty} e(t) = e_\mathrm{ss}(\infty) = \lim_{s \to 0} se(s)$$

$$= \lim_{s \to 0} s \frac{1}{s^3 + K_\mathrm{p}s + K_\mathrm{I}}$$

$$e_\mathrm{ss}(\infty) = 0$$

The steady-state error due to a constant disturbance is zero under the PI type control. If there is no integral control action, $K_\mathrm{I} = 0$, the steady-state error would have been

$$e(s) = \frac{s}{s(s^2 + K_\mathrm{p})} \frac{1}{s}$$

$$\lim_{s \to 0} se(s) = \frac{1}{s^2 + K_\mathrm{p}} = \frac{1}{K_\mathrm{p}} \to \neq 0$$

Therefore it is clear that it is the integral of position error used in feedback control which enables the control system to reject the constant disturbance and keep $x(t) = x_\mathrm{d}(t)$ in steady state. The transient response to a step command change in desired position under no disturbance condition,

$$w_\mathrm{d}(t) = 0$$

$$x_\mathrm{d}(t) = 1(t)$$

$$x(s) = \frac{K_\mathrm{p}s + K_\mathrm{I}}{s^3 + K_\mathrm{p}s + K_\mathrm{I}} x_\mathrm{d}(s)$$

The closed loop system has a zero at

$$-\frac{K_\mathrm{I}}{K_\mathrm{p}}$$

and three poles at the roots of the following equation,

$$s^3 + K_p s + K_I = 0$$

$$1 + \frac{K_p}{s^3}\left(s + \frac{K_I}{K_p}\right) = 0$$

$$1 + K_p \frac{s + \frac{K_I}{K_p}}{s^3} = 0$$

Let us study the locus of the roots of this equation for various values of K_p, K_I (Figure 2.46c). Closed loop system dominant poles are such that one of them is on the negative real axis, but the other two have positive real parts, limited to $\frac{1}{2}\frac{K_I}{K_p}$, and are unstable. If we want to stabilize the system, we must introduce more damping using derivative (D) control, which is considered next. The pure mass-force system is unstable under a PI controller for any positive values of the PI gains (K_p, K_I). Many real systems have some inherent damping in open loop. In other words, the loop force-position transfer function is $\frac{1}{s(s+c)}$ instead of $\frac{1}{s^2}$. If the open loop damping is large enough, the closed loop system under PI control would be stable for a finite range of K_p, K_I gains. That is the reason why many physical second-order systems are stable and well controlled by a PI controller alone, without the derivative control action.

2.12.5 PD Control

Now, we will consider the characteristics of the mass-force of a system under proportional plus derivative (PD) control. The PD control algorithm is given by,

$$u(t) = K_p(x_d(t) - x(t)) - K_D\dot{x}(t)$$
$$u(s) = K_p(x_d(s) - x(s)) - K_D\dot{x}(s)$$

Substituting this into the mass-force model,

$$s^2 x(s) = u(s) - w_d(s)$$
$$(s^2 + K_D s + K_p)x(s) = K_p x_d(s) - w_d(s)$$

We will consider the dominant response to a step command in the desired position, and the steady-state response to a constant disturbance.

(i) Transient response in the s-domain

$$x(s) = \frac{K_p}{s^2 + K_D s + K_p}(1/s)$$

and the time domain response is found by taking the inverse Laplace transform,

$$x(t) = 1 - e^{-\xi\omega_n t}\frac{1}{\sqrt{1-\xi^2}}\sin(\sqrt{1-\xi^2}\omega_n t + \phi)$$

where

$$K_p = \omega_n^2$$
$$K_D = 2\xi\omega_n$$
$$\phi = \tan^{-1}\left(\frac{\sqrt{1-\xi^2}}{\xi}\right)$$

Since PD control gains determine the natural frequency and the damping ratio of the closed loop system, PD control can be efficient in shaping transient response.

Figures 2.46d and 2.46e show the closed loop system pole locations as K_p varies from zero to infinity. The poles of the closed loop transfer function can be expressed as

$$\Delta_{cls}(s) = s^2 + K_D s + K_p = 0 \tag{2.152}$$

$$= 1 + K_p \frac{(K_D/K_p)s + 1}{s^2} = 0 \tag{2.153}$$

$$= 1 + K_p \frac{1}{s(s + K_D)} = 0 \tag{2.154}$$

If we sketch the root locus, it is clear that the closed loop poles will always be stable for any positive values of K_p, K_D (Figure 2.46d). If we plot the root locus for a constant value of K_D/K_p as K_p varies from zero to infinity, we have the root locus as shown in Figure 2.46d. If we plot the root locus of the last form of the equation for a given constant value of K_D as K_p varies from zero to infinity, we have the root locus shown in Figure 2.46e. Although the shape of the root locuses are different for the same closed loop system in question, the Figure 2.46d is a root locus as both K_p and K_D vary such that K_D/K_p is constant, whereas Figure 2.46e is a root locus as K_D varies for a constant K_p. For a given value of K_p, K_D, both root locuses predict the same closed loop root locations, as they should.

(ii) Zero command, constant step disturbance case - $x_d = 0$; $w_d \neq 0$; $w_d(s) = \frac{1}{s}$. The response due to step disturbance is given by

$$x(s) = -\frac{1}{s^2 + K_D s + K_p} \frac{1}{s}$$

$$x(s) = e(s) = -\frac{1}{s^2 + K_D s + K_p} \frac{1}{s}$$

Any non-zero response due to the disturbance is indeed an error. The magnitude of the error under PD control is

$$\lim_{t \to \infty} e(t) = \lim_{s \to 0} s(s)$$

$$= \lim_{s \to 0} (-s) \frac{1}{s^2 + K_D s + K_p} \frac{1}{s}$$

$$= \frac{1}{K_p}$$

Therefore, the PD control alone cannot provide zero steady-state error in the presence of constant disturbance.

2.12.6 PID Control

PID control is basically a PD control plus PI control. It combines the capabilities of PD and PI control. PD control is primarily used to shape transient response and stabilize the system. The D (derivative) action introduces damping into the closed loop system. If the steady-state error is constant, hence its derivative is zero, the derivative action has no influence on the steady-state response. PI control is used to reduce the steady state error and improve disturbance rejection capability. Almost all practical controllers exhibit the features of PID control. They have control action components which deal with the present error (proportional – P control), past error using the integral of error (integral – I control), and the future error using the anticipatory nature of derivative (D-control). There are many different implementations of PID control. One possible implementation of PID control is shown below. In this implementation, the derivative action is only applied to the feedback signal, not on the error. Sometimes, this may be preferable if the command signal has jump discontiniuties such as step changes.

$$u(t) = K_p(x_d(t) - x(t)) + K_I(\int_0^t (x_d(\tau) - x(\tau))d\tau - K_D \dot{x}(t)$$

$$u(s) = K_p(x_d(s) - x(s)) + \frac{K_I}{s}(x_d(s) - x(s)) - K_D s x(s)$$

$$= -\left(K_p + \frac{K_I}{s} + K_D s\right) x(s) + \left(K_p + \frac{K_I}{s}\right) x_d(s)$$

If we use the PID controller for position control of the mass-force system,

$$s^2 x(s) = u(s) - w_d(s)$$

$$\left\{ s^2 + \left(K_p + \frac{K_I}{s} + K_D s\right) \right\} x(s) = \left(K_p + \frac{K_I}{s}\right) x_d(s) - w_d(s)$$

$$(s^3 + K_D s^2 + K_p s + K_I) x(s) = (K_p s + K_I) x_d(s) - s w_d(s)$$

The closed loop system transfer function for the mass-force system under the PID control is

$$x(s) = \frac{(K_p s + K_I)}{(s^3 + K_D s^2 + K_p s + K_I)} x_d(s) - \frac{s}{(s^3 + K_D s^2 + K_p s + K_I)} w_d(s)$$

Let us consider the behavior of this system for two different conditions: (i) commanded input is zero, but there is a constant unit magnitude disturbance, $x_d(t) = 0$; $w_d(t) = 1(t)$, (ii) there is a unit magnitude step command, but no disturbance, $x_d(t) = 1(t)$; $w_d(t) = 0$.

(i) $x_d(t) = 0.0$; $w_d(t) = 1(t)$: The non-zero response due to disturbance is an unwanted response and can be considered an error,

$$x(t) = e(t)$$
$$x(s) = e(s)$$

$$= -\frac{s}{(s^3 + K_D s^2 + K_p s + K_I)} \frac{1}{s}$$

In steady state, the position of the mass-force system under the PID control

$$\lim_{t \to \infty} x(t) = \lim_{s \to 0} x(s) = -\frac{s}{(s^3 + K_D s^2 + K_p s + K_I)} = 0$$

is zero as commanded despite the constant magnitude disturbance force, If there is no integral action, $K_I = 0$, the response is

$$e(\infty) = \frac{1}{K_p}$$

finite and inversely proportional to the proportional feedback gain. Notice the importance of the integral (I) action in rejecting the disturbance. The integral action makes the system type I with respect to the disturbance entering the system after the controller block.

(ii), $x_d(t) = 1(t)$; $w_d(t) = 0$: Let us study the step response of the system when there is no disturbance,

$$x_d(t) = 1(t); \quad \text{step function}$$

$$x(s) = \frac{K_p s + K_I}{(s^3 + K_D s^2 + K_p s + K_I)} \frac{1}{s}$$

The PID controller can be designed as a cascade of PD and PI controllers. Design PD control first to set the shape of the transient response, then design the PI control to shape the steady-state response. The PD control introduces a zero to the open and closed loop transfer function. Therefore, it has a tendency to pull the root locus to the left side of s-plane, and hence has a stabilizing effect on the closed loop system. PI control introduces a zero close to the origin and a pole at the origin. Generally, the PI controller zero is placed closer to the origin relative to the other poles and zeros of the system. The result of placing the zero of PI control closer to the pole at the origin is that it will not influence the transient response much, which was shaped by the PD control, but will still increase the *type* of the loop transfer function by one. Therefore, the PI controller primarily influences the steady-state error. The resultant PID control parameters as a function of PI and PD controller parameters can be found as follows,

$$D(s) = K_p \left(1 + \frac{1}{T_I s} + T_D s \right)$$

$$= K_p^* \left(1 + \frac{1}{T_I^* s} \right) (1 + T_D^* s)$$

where

$$K_p = K_p^* (1 + T_D^*/T_I^*)$$
$$T_I = T_I^* + T_D^*$$
$$T_D = (T_I^* T_D^*)/(T_I^* + T_D^*)$$

Understanding the PID control components in terms of their frequency domain representation is useful. Figure 2.47 shows the Bode plots of the P, D, I, PD, PI, and PID controllers. The Bode plot is obtained by replacing $s = jw$ in the transfer function, and plotting the magnitude and the phase of the transfer function as a function of frequency in logarithmic scale.

$$D(jw) = D(s)|_{s=jw} \tag{2.155}$$

$$= |D(jw)| \cdot e^{j\psi} \tag{2.156}$$

$$= [(Re(D(jw)))^2 + (Im(D(jw)))^2]^{1/2} \tag{2.157}$$

$$\cdot e^{j \tan^{-1}(Im(D(jw))/Re(D(jw)))} \tag{2.158}$$

$$|D(jw)| = [(Re(D(jw)))^2 + (Im(D(jw)))^2]^{1/2} \tag{2.159}$$

$$20 \log_{10} |D(jw)| = 20 \log_{10}([(Re(D(jw)))^2 + (Im(D(jw)))^2]^{1/2}) \tag{2.160}$$

$$\psi = \tan^{-1} \frac{Im(D(jw))}{Re(D(jw))} \tag{2.161}$$

The following MATLAB® code generates the Bode plots of various PID controllers versions for the gains $K_p = 10.0, K_d = 1.0, K_i = 1.0$,

```
Kp = 10.0 ;
Kd = 1.0  ;
Ki = 1.0  ;

num1=[Kp] ; den1=[1] ;   % P- control
num2=[Kd 0]; den2 = [1]; % D-control
num3=[Ki]; den3=[1 0];   % I - control
num4=[Kd Kp]; den4 = [1]; % PD - control
num5=[Kp Ki]; den5=[1 0]; % PI - control
num6=[(Kp*Kd) (Kp*Kp+Ki*Kd)  Ki*Kp]; den6=[1 0]; % PID =
     PD * PI - control
```

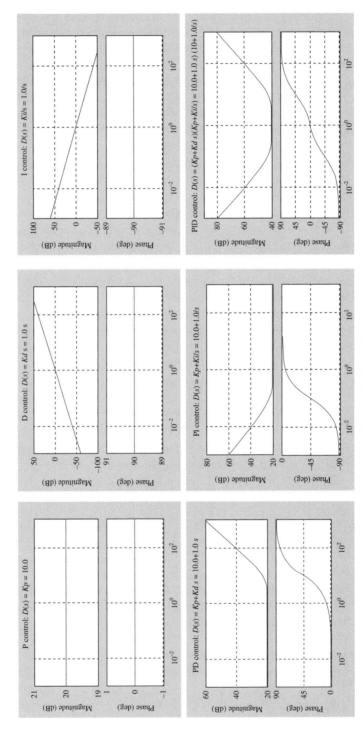

FIGURE 2.47: Bode plots of PID controller components: P, D, I, PI, PD, and PID.

```
figure(1); bode(num1,den1); title('P control:
          D(s) = Kp = 10.0'); grid on;
figure(2); bode(num2,den2); title('D control:
          D(s) = Kd s = 1.0 s'); grid on;
figure(3); bode(num3,den3); title('I control:
          D(s) = Ki/s = 1.0/s');grid on;
figure(4); bode(num4,den4); title('PD control:
          D(s) = Kp+Kd s  = 10.0+1.0 s');grid on;
figure(5); bode(num5,den5); title('PI control:
          D(s) = Kp+Ki/s  = 10.0+1.0/s'); grid on;
figure(6); bode(num6,den6); title('PID :D(s)=(Kp+Kds)
          (Kp+Ki/s)=(10.0+1.0s)(10+1.0/s)');grid on;
```

Using frequency domain methods, the PID controller gains can be selected to shape the Bode plots (frequency response) of the controller in order to give the desired shape to the combined Bode plot of the controlled process and the controller. The desired shape of the combined Bode plot, that is the frequency response of the loop transfer function including the controller and the process, is primarily characterized by

1. stability: the gain margin (the magnitude of the plot at the frequency when the phase is $-180°$), the phase margin (the difference of the phase of the frequency response from the $-180°$ at the frequency that the magnitude crosses 0 dB), i.e., if the phase of the frequency response at that frequency is $-150°$, the phase margin is $+30°$),

2. speed of response (bandwidth): cross-over frequency (the frequency at which the magnitude plot crosses 0 dB) which determines the speed of response (bandwidth) of the closed loop system,

3. steady-state response and disturbance rejection: gain of the loop transfer function at the low frequency range (the higher the low frequency gain is, the smaller the steady-state error and the better the disturbance rejection against low frequency disturbances),

4. noise rejection: gain of the loop transfer function at the high frequency range should be low in order to reject high frequency noise.

2.12.7 Practical Implementation Issues of PID Control

Anti-Windup Integral Control   When integral control is used in a control system which has actuator saturation, the integral control can very adversely affect the transient response and the stability of the closed loop system. Almost all practical control systems have actuators with saturation. All physical actuation components have limits on the output they can provide, that is a valve can be fully open, an electric motor can provide a known maximum torque or force, an amplifier output voltage would be limited to a maximum supply voltage. When an actuator saturates, the control output no longer changes, effectively rendering an open loop control system. During that time, if the error sign does not change, the output of the integrator continues to increase even though the actuator is saturated. When the error sign changes, and the control signal sign should change, the integral control component may prevent that due to its large contribution to the control signal that is a result of the "accumulation" (integral) of the past errors. This is called *integrator windup*. It results in poor transient response and possibly in stability problems. The solution is then to add a component that will have an *anti-windup* function on the integrator. In digital control, the easiest and most common method of implementing the *integrator anti-windup* is to stop

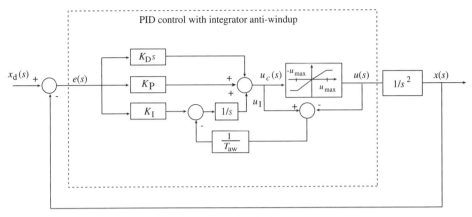

FIGURE 2.48: PID controller with integrator anti-windup.

executing the integral term whenever the actuator saturates. The following code segment shows an example of a digital PID control with integrator anti-windup.

```
%  PID Control with Intergator Anti-windup:  Version 1
%     T is sampling period.
%     K_p, K_d, K_i are PID gains,
%     e and edot are error signal and its time derivative
...
...
if  abs(u_c) <  u_max
    u_i = u_i + (K_i * e) * T   ;
elseif abs(u_c) >= u_max
    u_i = u_i ;
endif
u_c = K_p * e + K_d * edot + u_i ;
...
```

This means that during the actuator saturation, the integration function of the controller is turned off.

Let the parameters of the control system shown in Figure 2.48 be as follows:

$$G(s) = 1/s^2 \tag{2.162}$$

$$u_{\text{max}} = \pm 2 \tag{2.163}$$

$$K_{\text{p}} = 10.0, K_{\text{D}} = 3.0, K_{\text{I}} = 2.0 \tag{2.164}$$

and the commanded position is a unit step change function starting at time $t = 0.5$ s. The simulation results for commanded position, actual position, total control signal, and integral action control signal are shown for PID control alone and PID control with integrator anti-windup. It is clear that the anti-windup implementation of stopping the integral operation as soon as the actuator saturates improves the transient response (Figure 2.49).

Derivative Control The derivative control takes the derivative of the input signal (i.e., error signal) and multiplies it with a gain. The derivative control has the advantage of adding damping to the system, hence increasing its stability. However, it also amplifies the high frequency noise content in the signal. Therefore, if the signal has high frequency noise content, instead of pure derivative control an approximation to it with a low pass filter in

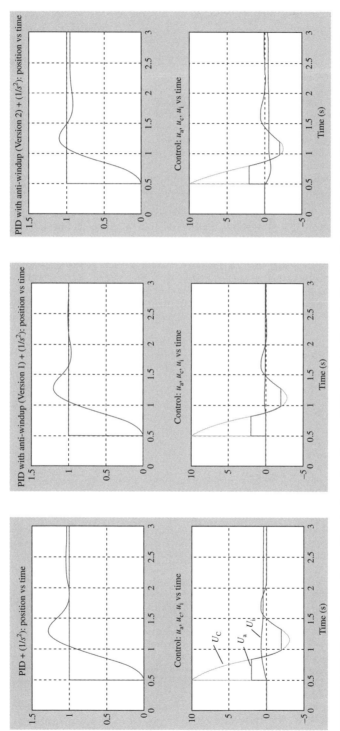

FIGURE 2.49: Mass-force system with actuator saturation and PID control with and without anti-windup logic: (a) PID control only, (b) PID control with anti-windup implemented as stopping the integral control whenever the actuator saturates, (c) PID control with anti-windup implemented as the digital equivalent of the analog control block diagram shown in Figure 2.48.

series should be implemented. In other words, the derivative control may be modified as

$$u_d(s) = K_D \, s \, e(s) \qquad (2.165)$$

$$u_d(t) = K_D \cdot \frac{de(t)}{dt} \qquad (2.166)$$

which should be modified to reduce its high frequency gain, and hence reduce the noise amplification of it,

$$u_d(s) = K_D \, s \, \frac{1}{(K_D/N)s + 1} \, e(s) \qquad (2.167)$$

$$(K_D/N) \cdot \frac{du_d(t)}{dt} = -u_d(t) + K_D \cdot \frac{de(t)}{dt} \qquad (2.168)$$

The modified derivative gain implementation basically adds a low pass filter to the pure derivative control. The cross-over frequency of the low pass filter should be selected as high as possible so as not to significantly change the derivative function, but also low enough not to amplify high frequency noise. The two versions of the derivative control can be compared in terms of their effect on the loop transfer function by considering their added pole-zero on the s-plane and their Bode plots in the frequency domain (Figure 2.50). The

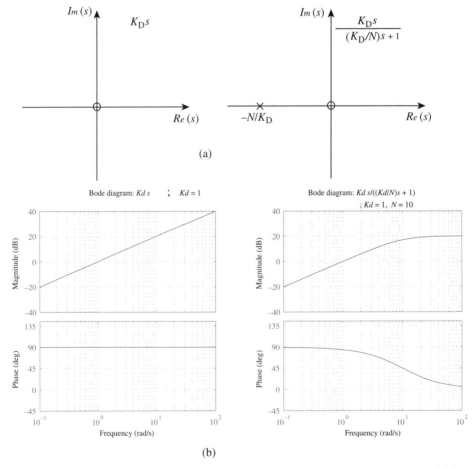

FIGURE 2.50: A practical implementation of "Derivative" control in order to reduce its high frequency noise amplification: (a) pole-zero structure in s-plane, (b) Bode plots.

pure derivative control adds a zero at the origin, whereas the modified derivative control adds a zero at the origin and a pole on the negative real axis to the loop transfer function. Similarly, the Bode plot of pure derivative control has a gain that increases to infinity as the frequency increases to infinity. Whereas, the gain of the modified derivative control has a finite gain as frequency goes to infinity (Figure 2.49). The MATLAB® code segment to obtain the Bode plots of the pure derivative control and derivative with a first-order pole control is given below.

```
Kd= 1.0 ;
N = 10.0 ;
num1 = [Kd   0]    ; den1 = [1]         ; sys1=tf(num1,den1) ;  % Kds
num2 = [Kd   0]    ; den2 = [Kd/N   1]  ; sys2=tf(num2,den2) ;  % 1/s
figure(1) ; grid on;              % Bode Plots
subplot(1,2,1) ;  bode(sys1); grid on;
subplot(1,2,2) ;  bode(sys2); grid on;
```

The high frequency gain of the modified derivative control can be made almost zero by adding a second first-order low pass filter in series with the first one.

$$u_d(s) = K_D s \cdot \frac{1}{(K_D/N_1)s + 1} \cdot \frac{1}{(K_D/N_1)s + 1} \cdot e(s) \tag{2.169}$$

Another way to implement a practical PD (proportional and derivative) control is to implement it as a phase-lead compensator, that is

$$u(s) = (K_D s + K_p) \cdot e(s) \tag{2.170}$$

$$u(s) = \frac{(K_D s + K_p)}{(T_f s + 1)} \cdot e(s) \tag{2.171}$$

where the last equation is an approximation to a PD control. This type of filter is called a phase-lead filter since for $T_f < (K_D/K_p)$, that is $T_f = 0.1 \cdot (K_D/K_p)$, the phase of the filter is positive, which is why it is called a *phase-lead* filter.

Another common modification to derivative control is to use it only on the feedback signal (output sensor signal) instead of the error signal if the command signal has discontinuities, that is step changes in the command signal where the derivative at the discontinuity is infinite. With this modification, the transient response of the system is improved due to avoided discontinuous and large variations in the control signal, such as

$$e(s) = r(s) - y(s) \tag{2.172}$$

$$u_d(s) = K_D s \cdot \frac{1}{(K_D/N)s + 1} \cdot (-y(s)) \tag{2.173}$$

Other Practical Variations of PID Control Figure 2.51 shows variations of the PID control algorithm in real-world applications. However, it should be noted that all of these variations are not necessarily implemented in one given application. So far, we have discussed the P-I-D (proportional, integral, and derivative) gains of the PID controller. In addition we also discussed the integral anti-windup to deal with actuator saturation, and a more practical application of derivative control to reduce effects of noise in the error signal. In this figure we illustrate the following practical modifications to the PID controller.

1. Velocity and acceleration feedforward terms are used to improve the transient response. The velocity feedforward term (K_{vf}) helps improve the overshoot and raise time by effectively commanding a larger control signal before the velocity error

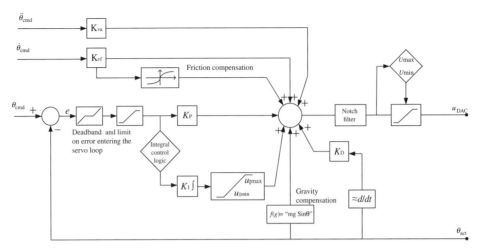

FIGURE 2.51: Practical implementation and various modifications of PID control algorithm: (i) velocity and acceleration feedforward, (ii) deadband and limit on error passed on to the servo loop, (iii) enable/disable integral action, (iv) friction compensation, (v) gravity compensation, (vi) notch filters, (vii) control signal limit.

gets large. The acceleration feedforward (K_{af}) term approximately tries to cancel or reduce the inertial effects on the dynamic response.

2. Deadband, applied on the error that is allowed to enter into the servo control loop, is sometimes useful in some applications, when a certain amount of error is acceptable and it is more important to hold the output steady instead of trying to achieve perfect zero error. Similarly, limiting the maximum value of the error passed on to the servo loop that is multiplied with the PID gains further might improve transient response and reduce overshoot.

3. In some cases, integral control maybe completely enabled or disabled. For instance, we may disable the integral control while in transient response phase, that is derivative of the command signal is not zero, and then enable it in steady-state condition when derivative of command signal is zero. In motion control applications, if the closed loop control variable is the position, then the derivative of the command signal is the commanded velocity.

```
. . .

if ( (dot_theta_cmd) != 0)
    Ki = 0.0
else
    Ki = Ki_original
end

u_i = u_i + K_i * e * T_sampling

. . .
```

4. Friction is a common problem in motion control systems. If the friction force/torque is known approximately as a function of speed and/or position, it can be compensated for in the feedforward or feedback path. Compensating in feedforward is more reliable in terms of stability. Compensating in the feedback block may lead to instability due to the stick-slip type discontinuous nature and variations in the actual friction conditions.

5. Gravitational load can be compensated for as a function of the measured position and inertia of the of the controlled axis. In nonlinear mechanisms and varying load conditions, a conservative value of effective inertia may be used or a more detailed inertia estimation algorithm may be used to support this control function.

6. Notch filters are used to reduce the likelihood of exciting the structural resonance frequencies of the controlled sytem by minimizing the frequency content of the control signal in the vicinity of the resonant frequencies. If there is only one resonant frequency, then only a single second-order digital Notch filter would be sufficient. If there are multiple resonant frequencies of concern, then multiple Notch filters in series should be used.

7. Finally, the output magnitude of the control signal to the amplifier may be limited at the controller level by a saturation function. In some cases, the maximum and minimum values of the control signal may need to be different, that is the absolute value of maximum and minimum values do not have to be same (i.e., $u_{max} = 10.0$ and $u_{min} = -5.0$).

8. Other practical aspects of PID control implementation are at the higher (supervisory) level where an algorithm would set the maximum and minimum allowed limits on the commanded position, commanded velocity, commanded acceleration, following error to send a warning message to the user, following error to stop the closed loop control (fatal error). The command signals would not be allowed to exceed these limits. In addition, the maximum output control signal (u_{max}) and root-mean-square value (u_{RMS}) value of the output signal can be monitored to protect the amplifier and actuator from overheating. If these values are exceeded, the control logic may reduce the control signal and/or send a warning signal to the user or shut-down the closed loop control algorithm as a fault condition (i.e., when the maximum control signal level is exceeded or the RMS control signal level is exceeded).

2.12.8 Time Delay in Control Systems

Time delay is a common problem in control systems. It typically occurs due to transport delay or actuator response delay. A *pure time delay* example is in a fluid or thermal control system. Consider a system where fluid moves with speed V, an actuator (heater) adds heat at some location, and the measurement of the temperature is taken at some downstream location at a distance l from where the actuator is located. Clearly, any effect of the actuator action will be measured only after a time delay of $t_d = \frac{l}{V}$.

Pure time delay can be represented mathematically as follows (Figure 2.52),

$$u_2(t) = u_1(t - t_d) \tag{2.174}$$

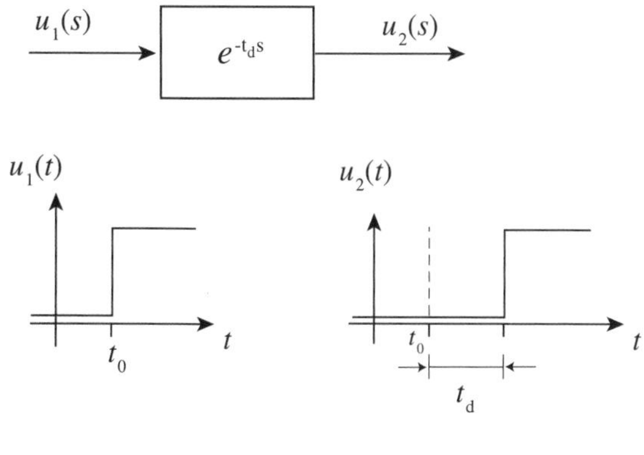

FIGURE 2.52: Time delay in control systems: pure time delay.

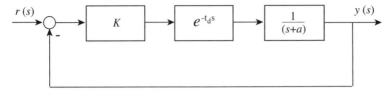

FIGURE 2.53: A closed loop control system with a process which has pure time delay.

which states that the $u_2(t)$ is simply t_d time period delayed version of $u_1(t)$. Let the Laplace transform of $u_1(t)$ be $U_1(s)$, and the Laplace transform of $u_2(t)$ be $U_2(s)$; it can be easily shown from the application of Laplace transform equation that

$$L\{u_2(t)\} = L\{u_1(t - t_d)\} \tag{2.175}$$

$$= e^{-t_d s} \cdot U_1(s) \tag{2.176}$$

$$\frac{U_2(s)}{U_1(s)} = e^{-t_d s} \tag{2.177}$$

which shows that the transfer function of *pure time delay* is $e^{-t_d s}$ where t_d is the magnitude of the time delay.

Consider the closed loop control system shown in Figure 2.53. When $t_d = 0.0$, there is no time delay in the loop. The closed loop transfer function is

$$\frac{Y(s)}{R(s)} = \frac{Ke^{-t_d s}}{(s + a) + Ke^{-t_d s}} \tag{2.178}$$

and the closed loop system characteristic equation which determines the closed loop pole locations is

$$\Delta_{cls}(s) = (s + a) + Ke^{-t_d s} = 0 \tag{2.179}$$

$$= 1 + K\frac{e^{-t_d s}}{s + a} = 0 \tag{2.180}$$

Figure 2.54 shows the root locus of the closed loop system poles for $t_d = 0.0$, and two different approximations to the pure time delay: one with a first-order filter and one with a second-order filter,

$$e^{-t_d s} \approx \frac{1}{t_d s + 1}; \quad \textit{Approximation 1} \tag{2.181}$$

$$e^{-t_d s} \approx \frac{1}{(t_d s + 1)^2}; \quad \textit{Approximation 2} \tag{2.182}$$

Clearly, the time delay approximations show that it has destabilizing effect on the closed loop pole locations. When the time delay does not exist ($t_d = 0.0$), the closed loop system is stable for all values of $K : 0 \longrightarrow \infty$. When a two-pole filter approximation is made to a non-zero time delay, the closed loop system is stable only for the range $K : 0 \longrightarrow K^*$, where

$$K^* : 1 + K^* \frac{1}{(t_d s + 1)^2} \frac{1}{s + a}|_{s=jw} = 0.0 \tag{2.183}$$

and unstable for $K > K^*$.

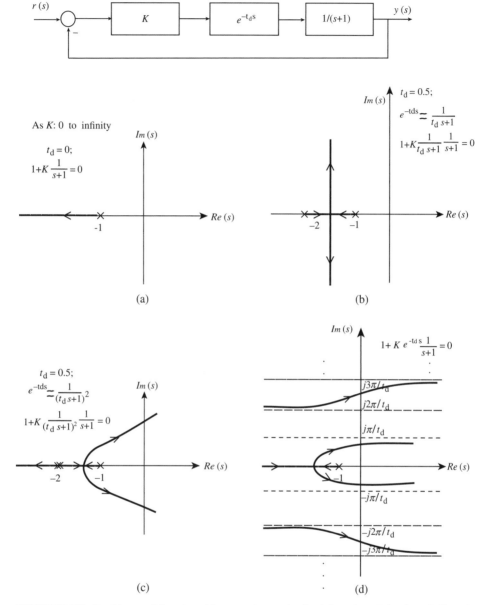

FIGURE 2.54: Root locus of the closed loop system poles for (a) no time delay ($t_d = 0.0$), and (b) and (c) two different approximations to the pure time delay ($t_d \neq 0.0$): one with a first-order filter and one with a second-order filter, and (d) pure time delay accurately taken into account without approximation.

A more detailed analysis of the root locus without approximation to the time delay shows that the closed loop root locus would have infinite number of branches and all of them would eventually go to the right hand plane for large values of K. The nature of the infinite number of branches of the root locus can be observed from the fact that

$$1 + K \cdot e^{-t_d s} \cdot \frac{1}{(s + a)} = 0 \qquad (2.184)$$

where

$$e^{-t_d s} = e^{-t_d Re(s)} \cdot e^{-jt_d Im(s)} \tag{2.185}$$

$$= (Real\ Number) \cdot e^{-jt_d w} \tag{2.186}$$

$$e^{-jt_d w} = \cos(t_d w) - j\sin(t_d w) \tag{2.187}$$

where the $\cos(t_d w)$ and $\sin(t_d w)$ terms have an infinite number of roots for w. Each solution for w corresponds to a point on a separate branch of the root locus. The most significant branch of the infinite number of root locus branches is the one closest to the origin. It is that branch that goes unstable before the other branches, and hence dominates the transient response and stability of the closed loop system.

While accurate analysis of a pure time delay in the root locus is rather difficult due to the infinite number of branches, the analysis in the frequency domain is rather easy. The Bode plot of a pure time delay is simply a unit magnitude and linear phase angle as a function of frequency. However, when we plot the phase angle in logarithmic scale of the frequency, it looks nonlinear (Figure 2.55a).

$$G_1(jw) = e^{-t_d s}|_{s=jw} \tag{2.188}$$

$$= e^{-t_d jw} \tag{2.189}$$

$$= 1.0 \cdot e^{-j(t_d w))} \tag{2.190}$$

$$|G_1(jw)| = 1.0 \tag{2.191}$$

$$\psi[\text{rad}] = -t_d\,\text{s} \cdot w\,[\text{rad/s}] \tag{2.192}$$

$$\psi[\text{deg}] = -57.3\,\text{deg/s}\,t_d\,\text{s} \cdot w\,[\text{rad/s}] \tag{2.193}$$

The Bode plot of the pure time delay is shown in Figure 2.55a.

```
s=tf('s');
G1 = exp(-1.0*s) ; % Time delay
G2 = 1/(s+1) ;
G3 = G1 * G2;

figure(1) ; grid on;
subplot(2,1,1) ; bode(G1,'b'); grid on;  % Bode plot of time delay only: e^(-td s)
subplot(2,1,2) ; bode(G1,'b',G2,'g',G3,'r'); grid on;
                % Bode plot of, e^(-td s), 1/(s+1) and e^(-td s) / (s+1)
```

Notice that the effect of the time delay on the stability (through phase margin and gain margin measures) of the closed loop system is obvious: it adds phase lag to the loop transfer function and reduces its stability margins. Eventually, it will make the closed loop system unstable as the closed loop gain gets larger. Figure 2.55b shows the Bode plot of a closed loop system (Figure 2.53), with and without time delay in the loop transfer function, for the following parameters $a = 1.0$, $t_d = 1.0\,\text{s}$. Clearly, the closed loop system is stable for all values of the gain K when there is no time delay, as was confirmed by the root locus as well. However, when the time delay is taken into account, the closed loop system eventually goes unstable due to added phase lag by the time delay (the same result was also confirmed by the root locus analysis above). The closed loop system becomes unstable at a value of gain K where the phase margin becomes zero. This gain value can be determined from the Bode plot as follows:

1. Read the value of frequency when the total phase of the loop transfer function (including the time delay) is $-180°$: $w = w^*$.

2. At that frequency, read the value of the loop transfer function gain, $|G(jw^*)|$.

3. The closed loop gain which defines the stability margin is: $K^* = \frac{1}{|G(jw^*)|}$.

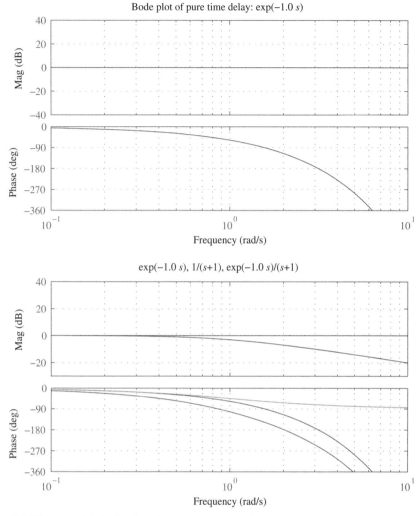

FIGURE 2.55: (a) Bode plot of pure time delay. Magnitude is constant
(0 dB = 1.0 Output Unit/Input Unit) as a function of frequency. The phase angle is linear as
function of frequency, but looks nonlinear in logarithmic scaled x-axis. (b) Effect of pure time
delay on the stability and performance of a closed loop control system, as displayed by the
Bode plot comparisons.

Notice that these calculations can be directly read from the Bode diagram due to the additive
nature of logarithmic scale, since

$$20 \log_{10} K^* = 20 \log_{10} \frac{1}{|G(jw^*)|} \tag{2.194}$$

$$20 \log_{10} K^* = 20 \log_{10} 1 - 20 \log_{10} |G(jw^*)| \tag{2.195}$$

$$20 \log_{10} K^* = 0 - 20 \log_{10} |G(jw^*)| \tag{2.196}$$

$$20 \log_{10} K^* = -20 \log_{10} |G(jw^*)| \tag{2.197}$$

$$K^* = \frac{1}{|G(jw^*)|} \tag{2.198}$$

which means that, in dB scale, the gain value that defines the stability limit is the negative
value of the loop transfer function's magnitude in logarithmic scale (the inverse of the loop

transfer function gain in linear scale) at the frequency where the loop transfer function has a phase angle of $-180°$.

Time delay is a serious problem in closed loop control systems in that it can make an otherwise stable system unstable. In order to keep the system stable, loop gain must be limited. Therefore, the overall price paid (to keep the system stable despite the time delay) is lower bandwidth due to lower loop gain. If the time delay is known in advance accurately, one way to deal with its destabilizing effects is to use the *Otto-Smith regulator* type controller (Figure 2.56a). The Otto-Smith regulator accomplishes closed loop stability behavior as if there is no time delay, while maintaining the same time delay in the closed loop system as the open loop system time delay. In other words, the Otto-Smith regulator does not try to get rid of or change the time delay. It keeps the time delay, plus adds the desired closed loop poles as if there was no time delay. It can be shown that Figure 2.56a is equivalent to Figure 2.56b. However, the Otto-Smith regulator can be successfully implemented only if

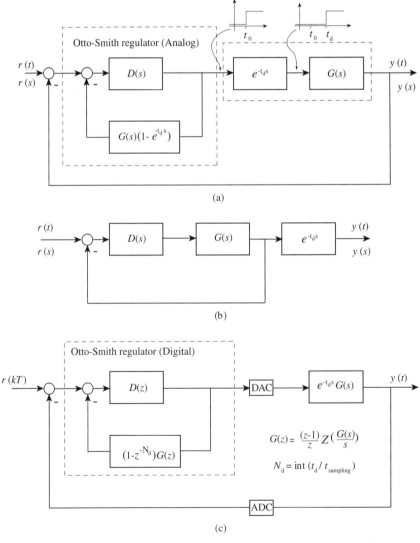

FIGURE 2.56: (a) Otto-Smith regulator implemented as analog controller, (b) equivalent implementation as analog controller, (c) digital implementation.

the time delay is known in advance accurately. Digital implementation of the time delay logic in the control loop (Figure 2.56c) requires $N_d = t_d / T_{sample}$ number of sampling time delays in the processed signal, where T_{sample} is the sampling period of the digital controller, that is

$$e^{-t_d s} \approx z^{-N_d} \tag{2.199}$$

that is, $t_d = 2.0$, $T_{sample} = 0.1$, then $N_d = 20$. In order to implement this time delay in a digital Otto-Smith regular controller, we would have to keep track of the past 20 samples of the control signals.

Example: PID Control of a Motion System with Friction

Many mechanical motion systems have friction opposing the motion. When two surfaces move against each other, the resistance force (or torque) before the relative motion starts is larger than the resistance force once the motion has started. This friction resists motion. For the relative motion to start, the applied external force must be larger than the friction force. Furthermore, that initial friction force, so called "stiction" (Coulomb) friction, is not constant and varies for the same system due to the condition of the surfaces and lubrication levels. The dynamic model of such a basic motion system is a mass-force system with friction as follows,

$$m \ddot{x}(t) = f_{control}(t) - f_{friction}(t) \tag{2.200}$$

where $f_{friction}(t)$ represents the friction in the system. If the mass is at rest, the only way for the motion to start is for the control force to be larger than the friction force. Once the motion starts, the net force (difference between control force and friction force) determines the acceleration and deceleration of the mass. Again, once it stops, the motion can start again only if the control force is larger than the friction force.

Without the integral control action, that is the PD controller, a closed loop position control system which has stiction friction will result in a finite steady-state positioning error. The reason is that when the control force, determined by the PD controller, is less than the friction force, the motion of the mass will start to decelerate and come to zero velocity. At that point, if the actual position is a finite value (the likelihood of the actual position being exactly the desired position is a random possibility) the control signal from the PD controller will be only due to the proportional controller since the mass speed is zero, and that control signal value would be smaller than friction force. As a result, the mass cannot move and will be stuck at that position with the finite steady-state position error. The following equations describe this condition,

$$f_{control}(t) = K_p e(t) + K_d \dot{e}(t) \tag{2.201}$$

$$f_{control}(t) < f_{friction}(t) \longrightarrow \ddot{x}(t) < 0.0 \tag{2.202}$$

$$\longrightarrow \dot{x}(t) = 0.0 \; eventually \tag{2.203}$$

$$f_{control}(t) = K_p e(t) \quad when \quad \dot{x}(t) = 0.0 \tag{2.204}$$

$$f_{control}(t) < f_{friction}(t) \longrightarrow \ddot{x}(t) = 0.0 \; and \; \dot{x}(t) = 0.0 \; no \; motion. \tag{2.205}$$

With integral control action, that is a PID controller, a closed loop position system will tend to oscillate about the target position which is referred to as *limit cycle oscillations*. This is a fundamental condition that is common in many closed loop motion control systems.

This behavior can be explained by simply adding the effect of the integral control action to the PD control behavior above. When the position of the inertia stops at a finite error, integral control will eventually build the control signal to be larger than the friction and will be able to start the motion again.

$$f_{control}(t) = K_p e(t) + K_d \dot{e}(t) + \int e(\tau)d\tau \qquad (2.206)$$

But, in the process of doing so, it will result in back and forth overshoot behavior, and hence result in oscillations about the desired position.

The remedy is either to design and maintain the system to minimize the stiction friction such that the resulting position error is acceptably small, or to predict it and compensate for it in the control algorithm explicitly based on the friction prediction. For a given stiction friction level, the positioning error cannot be guaranteed to be zero by any PID type controller unless there is a way to eliminate the source of the stiction friction. Friction prediction and compensation in real-time to achieve zero positioning error is not practically realistic due to the random and highly varying nature of friction physics.

The simulation results and physical explanations are shown in Figure 2.57.

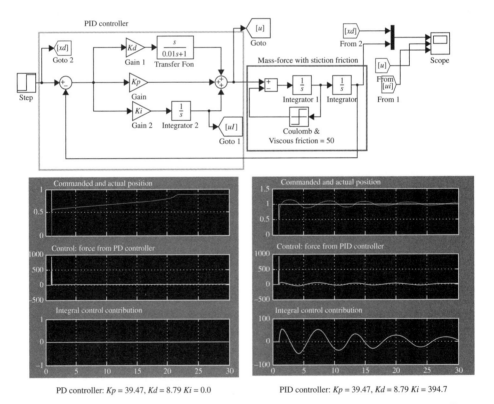

PD controller: $Kp = 39.47$, $Kd = 8.79$ $Ki = 0.0$ PID controller: $Kp = 39.47$, $Kd = 8.79$ $Ki = 394.7$

FIGURE 2.57: Open loop system which has stiction friction and a PID closed loop controller. Due to the stiction friction, the integral action of the PID controller causes a cycle of oscillation, as seen on the simulation results for the PID controller on the right. Without the integral control, there will be a finite steady-state error, as seen on the simulation results for the PD controller on the left.

2.13 TRANSLATION OF ANALOG CONTROL TO DIGITAL CONTROL

A controller can be completely analyzed and designed using continuous time methods. The resultant controller is an analog controller which can also be implemented in hardware using op-amps. The controller can be approximated with a digital controller which would be implemented using a digital computer. The fundamental tool is the approximation of differentiation by finite differences.

The basic problem is the following: given $G_c(s)$ (analog controller transfer function) find $H_c(z)$ (digital controller transfer function) such that the closed loop system (CLS) under a digital controller performs as close as possible to the CLS under an analog controller (Figure 2.58).

The finite-difference approximations considered are

- forward difference approximation,
- backward difference approximation,
- trapezoidal approximation.

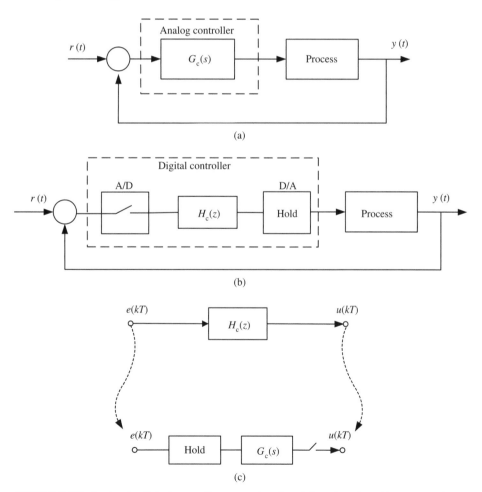

FIGURE 2.58: Analog to digital controller approximation: (a) analog closed loop control system, (b) digital closed loop control system, (c) digital controller approximation to the analog controller.

There are other methods such as *trapezoidal approximation with frequency pre-warping, zero order hold (ZOH) equivalent approximation, pole-zero mapping,* and *first-order equivalence* which are not discussed here.

It should be quickly noted that as the sampling rate gets very large relative to the bandwidth of the controller (i.e., 20 to 50 times larger), the differences between different approximation methods becomes insignificant. Likewise, if the sampling frequency is not very large relative to the bandwidth of the controller (i.e., 2 to 4 times larger than the controller bandwidth), the differences between different approximations become significant.

2.13.1 Finite Difference Approximations

The basic concept in approximation of analog filters by digital filters is the finite difference approximation of differentiation and integration. Let us consider an error signal, $e(t)$ and its differentiation and integration, and the samples of the error signal $\{\ldots, e(kT - T), e(kT), e(kT + T), \ldots\}$,

$$\frac{d}{dt}[e(t)], \left(\int_{t_0}^{t} e(\tau)d\tau\right) \iff (e(kT), e(kT - T), \ldots) \tag{2.207}$$

Consider a first-order transfer function example,

$$\frac{u(s)}{e(s)} = G(s) = \frac{a}{s + a} \tag{2.208}$$

$$\dot{u}(t) + au(t) = ae(t) \tag{2.209}$$

$$u(t)|_{t=kT} = \int_{0}^{kT} [-au(\tau) + ae(\tau)]d\tau \tag{2.210}$$

Discretize the integration

$$u(kT) = \int_{0}^{kT-T} [-au(\tau) + ae(\tau)]d\tau + \int_{kT-T}^{kT} [-au(\tau) + ae(\tau)]d\tau \tag{2.211}$$

$$u(kT) = u(kT - T) + \int_{kT-T}^{kT} [-au(\tau) + ae(\tau)]d\tau \tag{2.212}$$

Now, we consider three different finite difference approximations where each one makes a different approximation to the integration term in the above equation.

(i) *Forward difference approximation:*

$$u(kT) = u(kT - T) + T[-au(kT - T) + ae(kT - T)] \tag{2.213}$$

$$u(kT) = (1 - aT)u(kT - T) + aTe(kT - T) \tag{2.214}$$

Notice that this equation can be easily implemented in software on a digital computer. At every control sampling period, all that is needed is the value of the output at the previous cycle and the error. The algorithm involves two multiplication and one addition operation.

In order to develop a more generic relationship between the analog and digital controller approximate conversion, we take the Z-transform of the above difference equation,

$$[1 - (1 - aT)z^{-1}]u(z) = aTz^{-1}e(z) \tag{2.215}$$

Notice that a single sampling period of delay in a signal adds a z^{-1} to the transform of the signal. Likewise, an advance of a single sampling period in a signal adds a z to the transform of a signal. Using that principle, it is easy to take the Z-transform of difference equations and inverse Z-transform of z-domain transfer functions to obtain difference equations. For real-time algorithmic implementation, we need the difference equation form of the controller.

$$\frac{u(z)}{e(z)} = \frac{aTz^{-1}}{1 - (1 - aT)z^{-1}}$$

$$= \frac{a.T}{z - (1 - aT)}$$

$$= \frac{a}{((z - 1)/T) + a} \tag{2.216}$$

Notice the substitution relationship between s and z using this approximation

$$\frac{a}{s + a} \longrightarrow \frac{a}{\frac{z-1}{T} + a} \tag{2.217}$$

$$s \longrightarrow \frac{z - 1}{T}; \quad z = sT + 1 \tag{2.218}$$

(ii) *Backward difference approximation:*
Another possible approximation is to use the backward difference rule

$$u(kT) = u(kT - T) + T[-a\,u(kT) + a\,e(kT)] \tag{2.219}$$

Again, the above equation is in a form suitable for real-time implementation in software. In order to obtain a more generic relationship for this type of approximation, let us take the Z-transform of the above equation,

$$u(z) = z^{-1}u(z) - T\,a\,u(z) + T\,a\,e(z) \tag{2.220}$$

$$(1 + Ta - z^{-1})u(z) = T\,a\,e(z) \tag{2.221}$$

$$\frac{u(z)}{e(z)} = \frac{Ta}{1 + Ta - z^{-1}} = \frac{zTa}{z - 1 + Taz}$$

$$= \frac{a}{\frac{z-1}{zT} + a} \tag{2.222}$$

The backward approximation is equivalent to the following substitution between s and z,

$$\frac{a}{s + a} \longrightarrow \frac{a}{\frac{z-1}{Tz} + a} \tag{2.223}$$

$$s \longrightarrow \frac{z - 1}{Tz} \tag{2.224}$$

(iii) *Trapezoidal approximation (Tustin's method, bilinear transformation)*
Finally, we will consider the trapezoidal rule approximation among the finite difference approximations to the integration,

$$u(kT) = u(kT - T) + \frac{T}{2}[-a[u(kT - T) + u(kT)] + [a(e(kT - T) + e(kT)]] \tag{2.225}$$

Similarly, we take the Z-transform of the above equation,

$$zu(z) = u(z) + \frac{T}{2}[-a(1 + z)u(z) + a(1 + z)e(z)] \tag{2.226}$$

$$\left[z + \frac{Ta}{2}(1 + z) - 1\right]u(z) = \frac{T \cdot a}{2}(1 + z)e(z) \tag{2.227}$$

$$\frac{u(z)}{e(z)} = \frac{\frac{T}{2} \cdot a(1+z)}{(z-1) + \frac{T \cdot a}{2}(1+z)}$$

$$= \frac{a}{\frac{2}{T}\frac{z-1}{z+1} + a} \tag{2.228}$$

The equivalent substitution relationship between s and z is

$$\frac{a}{s+a} \longleftrightarrow \frac{a}{\frac{2}{T}\frac{z-1}{z+1} + a} \tag{2.229}$$

$$s \longleftrightarrow \frac{2}{T}\frac{z-1}{z+1} \tag{2.230}$$

A summary of finite difference based digital appoximation of analog filters is given below.

method	approximation
$FWD - rule$	$s \longleftarrow \dfrac{z-1}{T}$
$BWD - rule$	$s \longleftarrow \dfrac{z-1}{Tz}$
$Trapezoidal - rule$	$s \longleftarrow \dfrac{2}{T}\dfrac{z-1}{z+1}$

$$\tag{2.231}$$

2.14 PROBLEMS

1. Consider a time domain signal $y(t) = 1.0 \cdot \sin(2\pi \cdot 10t)$ which is periodic. Select proper sampling frequencies to illustrate the following sampling effects.

   1. A sampling period that is fast enough for accurate sampling and does not violate the sampling theorem.
   2. A sampling period that illustrates aliasing problem as a result of sampling.
   3. A sampling period that illustrates the beat phenomenon as a result of sampling.
   4. A sampling period that illustrates the hidden oscillations problem as a result of sampling.

Explain your time domain results with the frequency content of the sampled signals. Plot the original signal and sample signals in the time domain. Plot the magnitude component of the Fourier series of the original signal and the Fourier transform of the sampled signals as a function of frequency.

2. Consider a mass-force system. Let $m = 5$ kg. A controller decides on the force using a PD controller on position error. Select the PD controller gains such that the step response of the closed loop system has no more than 5% overshoot, and the settling time is less than 2.0 s. Confirm your results with a Simulink® or MATLAB® simulation.

3. What is the effect of switching the sign on the controller, that is $u(s) = (K_p + K_d s)(x(s) - x_{cmd}(s))$ (this can easily happen in practice by swapping the signal input lines to the analog PD controller between the command signal and the sensor signal) or by switching the polarity of the controller output (sign change once in the controller)? What happens if we switch the polarity of both the command signal and sensor signal as well as the output polarity of the controller (sign change twice in the controller)? Use a simple mass-force system model under a PD control algorithm to simulate your analysis and present results.

4. Consider a mass-force system. Let $m = 5$ kg. A controller decides on the force using a PID type controller on *velocity* error. Let velocity commands be $r(t) = 10 \cdot t$. It is assumed that the feedback sensor provides the velocity measurement of the mass. The controller acts on the velocity error.

   1. What is the steady-state error when a P-type controller is used?

2. What is the steady-state error when a PD-type controller is used?

3. What is the steady-state error when a PI-type controller is used?

Confirm your results with a Simulink® or MATLAB® simulation.

5. Consider the dynamics of a DC motor speed control system and its current mode amplifier. Consider the the motor torque–speed transfer function is a first-order filter and the current mode amplifier input–output dynamics is also a first-order filter.

$$\frac{w(s)}{T(s)} = \frac{100}{0.02s + 1} \tag{2.232}$$

$$T(s) = 10 \cdot i(s) \tag{2.233}$$

$$\frac{i(s)}{i_{cmd}(s)} = \frac{10}{0.005s + 1} \tag{2.234}$$

Consider that the current command is generated by an analog controller (PID type) using motor speed as feedback signal and commanded velocity signals.

(a) Ignore the filtering effect of the current amplifier, and determine the locus of closed loop poles (root locus) of the closed loop control system under three different controllers: P, PD, PI. In each case vary proportional gain as the root locus parameter and select different values of derivative and integral gains.

$$i_{cmd}(s) = K_p \cdot (w_{cmd}(s) - w(s)); \quad P \ control \tag{2.235}$$

$$= (K_p + K_d s) \cdot (w_{cmd}(s) - w(s)); \quad PD \ control \tag{2.236}$$

$$= \left(K_p + \frac{K_i}{s}\right) \cdot (w_{cmd}(s) - w(s)); \quad PI \ control \tag{2.237}$$

(b) Repeat the same root locus analysis by including the amplifier dynamics in the analysis. Discuss your results in terms of the effect of selecting different controller types (P, PD, PI), their gains, and the effect of additional filter type dynamics in the control loop.

(c) Consider the same DC motor control problem, except this time consider the closed loop position control problem. All other components are same, except that the commanded signal is the desired position signal and the feedback signal is the position signal. As a result, the only difference in the model is the torque–position transfer function, which has one additional integrator term compared to the speed control problem,

$$\frac{\theta(s)}{T(s)} = \frac{100}{s(0.02s + 1)} \tag{2.238}$$

Repeat the analysis for part (a) and (b) for the closed loop position control of a DC motor problem.

6. Consider the dynamics of a DC motor speed control system and its current mode amplifier. Consider the the motor torque–speed transfer function is a first-order filter and the current mode amplifier input–output dynamics is also a first-order filter.

$$\frac{w(s)}{T(s)} = \frac{K_m}{\tau_m s + 1} \tag{2.239}$$

$$T(s) = K_T \cdot i(s) \tag{2.240}$$

$$\frac{i(s)}{i_{cmd}(s)} = \frac{K_a}{\tau_a s + 1} \tag{2.241}$$

Assume that there is a load torque acting as a disturbance on the motor and that it is in the form of (i) step function, (ii) a ramp function. If we want to make sure that the steady-state speed error due to the load torque is zero, what kind of controller is required in order to deal with each type of disturbance (load) torque? Why can't a PD type controller do the job?

7. Given an analog PD controller,

$$G_c(s) = K_p + K_d \cdot s \tag{2.242}$$

(a) obtain its digital PD controller equivalent for sampling period T using all of the digital approximation methods discussed above.

(b) Let $K_p = 1.0$, $K_d = 0.7$. Plot the Bode diagrams of the analog and the digital approximations for three different values of $T = 1/5\,\text{s}$, $1/50\,\text{s}$, $1/500\,\text{s}$.

8. Given an analog PI controller,

$$G_c(s) = K_p + K_i \cdot (1/s) \tag{2.243}$$

(a) obtain its digital PI controller equivalent for sampling period T using all of the digital approximation methods discussed above.

(b) Let $K_p = 1.0$, $K_i = 0.1$. Plot the Bode diagrams of the analog and the digital approximations for three different values of $T = 1/5\,\text{s}$, $1/50\,\text{s}$, $1/500\,\text{s}$.

9. Given an analog PID controller,

$$G_c(s) = K_p + K_i \cdot (1/s) + K_d \cdot s \tag{2.244}$$

(a) obtain its digital PID controller equivalent for sampling period T using all of the digital approximation methods discussed above.

(b) Let $K_p = 1.0$, $K_i = 0.1$, $K_d = 0.7$. Plot the Bode diagrams of the analog and the digital approximations for three different values of $T = 1/5\,\text{s}$, $1/50\,\text{s}$, $1/500\,\text{s}$.

10. Consider a DC motor, its amplifier, a closed loop position controller, and load which is connected to the motor via a gear reducer (Figure 2.59). Consider an analog controller of type PI,

$$D(s) = \frac{u(s)}{e(s)} = K_p + K_i \frac{1}{s} \tag{2.245}$$

$$e(s) = \theta_d(s) - \theta(s) \tag{2.246}$$

the amplifier model,

$$i(s) = K_a \cdot u(s) \tag{2.247}$$

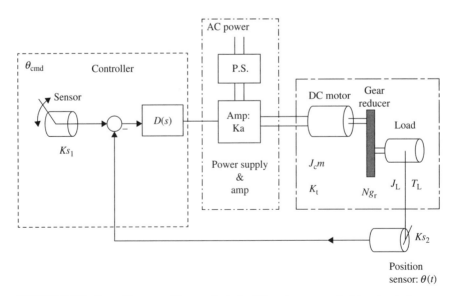

FIGURE 2.59: Closed loop position control of a DC motor connected to a load via a gear reducer.

the DC motor model,

$$T_m(s) = K_t \cdot i(s) \tag{2.248}$$

$$\theta(s) = \frac{1}{(J_t s^2 + cs)} \cdot (T_m(s) - T_l(s)) \tag{2.249}$$

Let the parameters of the components be as follows: $K_a = 1.0$, $K_t = 1.0$, $c = 10.0$, $J_t = J_1 + \frac{1}{N^2} J_2$. $J_1 = 0.1$, $N = 5$, $J_2 = 1.0$.

1. Find the closed loop system transfer function between the commanded position signal, load torque, and actual position signal. Select the gains of the PI controller K_p and K_i so that the step response setting time is about 100 ms with overshoot of about no more than 5%, and that the steady state error in response to a step disturbance load, $T_l(t) = 1.0 \, Step(t)$, should be zero (assuming no actuator saturation).

2. Implement the PI controller digitally. Use the trapezoidal rule for the digital approximation. Neglect the effect of quantization errors due to finite resolution of ADC and DAC. Let sampling period to be $T_{sample} = 1.0 \, ms = 0.001 \, s$. Obtain the digital version of the PI controller in the z-domain as a transfer function, and in the time domain as a difference equation expression.

3. Simulate in the time domain the response of the closed loop system to a step input command and disturbance torque condition,

$$\theta_d(t) = 1.0 \; for \; t > 1.0 \, s \tag{2.250}$$

$$T_l(t) = 1.0 \; for \; t > 2.0 \, s \tag{2.251}$$

for both analog controller implementation and digital controller implementation.

CHAPTER *3*

MECHANISMS FOR MOTION TRANSMISSION

3.1 INTRODUCTION

Every computer controlled mechanical system that involves motion is built around a basic frame of a mechanism which is used to transmit the motion generated by the actuators to the tool. The actuators provide purely rotary or purely linear motion. For instance, a rotary electric motor can be viewed as a rotary motion source while a hydraulic or pneumatic cylinder can be viewed as a linear motion source. In general, it is not practical to place the actuator exactly at the location where the motion of a tool is needed. Therefore, a motion transmission mechanism is needed between the actuator and the tool. Motion transmission mechanisms perform two different roles,

1. they transmit motion from actuator to tool when the actuator cannot be designed into the same location as the tool with the desired motion type,

2. they increase or reduce torque and speed between input and output shafts while maintaining the power conservation between input and output (output power is input power minus the power losses).

The most common motion transmission mechanisms fit into one of three major categories:

1. rotary to rotary motion transmission mechanisms (gears, belts, and pulleys),

2. rotary to translational motion transmission mechanisms (lead-screw, rack-pinion, belt-pulley),

3. cyclic motion transmission mechanisms (linkages and cams).

Common to all of these mechanisms is that an input shaft displacement is related to the output shaft displacement with a fixed mechanical relationship. During the conversion, there is inevitable loss of power due to friction. However, for analysis purposes here, we will assume ideal motion transmission mechanisms with 100% efficiency.

The efficiency of a motion transmission mechanism is defined as the ratio between the output power and input power,

$$\eta = \frac{P_{\text{out}}}{P_{\text{in}}} \tag{3.1}$$

Mechatronics with Experiments, Second Edition. Sabri Cetinkunt.
© 2015 John Wiley & Sons, Ltd. Published 2015 by John Wiley & Sons, Ltd.
Companion Website: www.wiley.com/go/cetinkunt/mechatronics

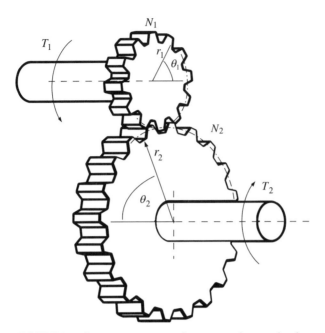

FIGURE 3.1: Rotary to rotary motion conversion mechanisms: gear mechanism.

The efficiency can vary from 75 to 95% for different types of motion transmission mechanisms. If we assume perfect efficiency, then

$$P_{in} = P_{out} \tag{3.2}$$

which is a convenient relationship in determining the input–output relationships.

The mechanical construction of the mechanism determines the ratio of input displacement to output displacement, which is called the *effective gear ratio*. The effective gear ratio is not influenced by efficiency. If a mechanism is not 100% efficient, the loss is a percentage of the torque or force transmitted. In other words, let us consider a simple gear arrangement (Figure 3.1) with a gear ratio of $N = \Delta\theta_{in}/\Delta\theta_{out}$,

$$P_{out} = \eta \cdot P_{in} \tag{3.3}$$
$$T_{out} \cdot \dot{\theta}_{out} = \eta \cdot T_{in} \cdot \dot{\theta}_{in} \tag{3.4}$$

and regardless of the efficiency,

$$N = \frac{\dot{\theta}_{in}}{\dot{\theta}_{out}} \tag{3.5}$$

Hence,

$$T_{out} = \eta \cdot N \cdot T_{in} \tag{3.6}$$

where the torque output is reduced by the efficiency of the mechanism.

The effective gear ratio of a mechanism is determined using the energy equations. The kinetic energy of the tool on the output is expressed in terms of output speed. Then, the output speed is expressed as a function of the input speed. Since both expressions represent the same kinetic energy, the effective gear ratio is obtained. For the following discussion, we refer to the load inertia (J_1) and load torque (T_1) as being applied on the output shaft. In other words, $J_1 = J_{out}$, $T_1 = T_{out}$. The reflected values of these on the input shaft side are referred to as $J_{in,eff} = J_{in}$, $T_{in,eff} = T_{in}$. For instance, for a rotary gear reducer, let KE_1

be the kinetic energy of the load with inertia J_1 and output speed $\dot{\theta}_{out}$. In order to provide such an energy, the actuator must provide a kinetic energy plus the losses. Hence,

$$KE_1 = \eta \cdot KE_{in} \tag{3.7}$$

$$KE_1 = \frac{1}{2} \cdot J_1 \cdot \dot{\theta}_{out}^2 \tag{3.8}$$

$$= \eta \cdot \frac{1}{2} \cdot J_{in,eff} \cdot \dot{\theta}_{in}^2 \tag{3.9}$$

$$= \frac{1}{2} \cdot J_1 \cdot (\dot{\theta}_{in}/N)^2 \tag{3.10}$$

Hence, the effective reflected inertia is

$$J_{in,eff} = \frac{1}{\eta \cdot N^2} \cdot J_1 \tag{3.11}$$

In summary, an ideal motion transmission mechanism (efficiency is 100%, $\eta = 1.0$) has the following reflection properties between its input and output shafts

$$\dot{\theta}_{in} = N \cdot \dot{\theta}_{out} \tag{3.12}$$

$$J_{in,eff} = \frac{1}{N^2} \cdot J_1 \tag{3.13}$$

$$T_{in,eff} = \frac{1}{N} \cdot T_1 \tag{3.14}$$

where N is the effective gear ratio. The efficiency factor of the motion transmission mechanism is often taken into account by a relatively large safety factor. If the efficiency factor is to be explicitly included in the actuator sizing calculations, then the following relations hold,

$$\dot{\theta}_{in} = N \cdot \dot{\theta}_{out} \tag{3.15}$$

$$J_{in,eff} = \frac{1}{N^2 \cdot \eta} \cdot J_1 \tag{3.16}$$

$$T_{in,eff} = \frac{1}{N \cdot \eta} \cdot T_1 \tag{3.17}$$

It is important to note that in either direction of power transmission, the efficiency factor is in the denominator of the equations which indicates loss of power due to transmission efficiency in either direction.

A motion transmission mechanism is characterized by the following parameters:

1. The main characteristic of a motion transmission mechanism is its gear ratio. This is sometimes called the *effective gear ratio* since the motion conversion may not necessarily be performed by gears.

2. *Efficiency:* efficiency of a real gear ratio is always less than 100%. For most gear mechanisms, forward and back drive efficiencies are same except for the lead-screw and ball-screw type mechanisms. In such mechanisms, it is appropriate to talk about rotary to linear motion conversion efficiency, η_f, and linear to rotary motion conversion efficiency, η_b. For ball-screw, the typical values of the efficiency coefficients are $\eta_f = 0.9$, $\eta_b = 0.8$, and for lead-screws they vary as a function of the lead angle. The lead angle is defined as the angle the lead helix makes with a line perpendicular to the axis of rotation. As the lead (linear distance traveled per rotation) increases, so does the lead angle, hence the efficiency.

3. *Backlash:* there is always an effective *backlash* in motion transmission mechanisms. The backlash is given in units of *arc minutes* $= 1/60\ degrees$ in rotary mechanisms, and in inches or mm units in translational mechanisms. Notice that backlash directly affects the positioning accuracy. If the position sensor is connected to the motor, not to the load, it will not be able to measure the positioning error accurately due to backlash. Therefore, if backlash is large enough to be a concern for positioning accuracy, there has to be a position sensor connected to the load in order to measure the true position, including the effect of backlash. In such systems, it is generally necessary to use two position sensors (dual sensor feedback, or dual loop control): one position sensor connected to the motor and the other position sensor connected to the load. The motor-connected sensor is primarily used in the velocity control loop to maintain closed loop stability, whereas the load-connected sensor is primarily used for accurate position sensing and control. Without the motor-connected sensor, the closed loop system may be unstable. Without the load-connected sensor, desired positioning accuracy cannot be achieved. In systems where the backlash is much smaller than the positioning accuracy required, the backlash can be ignored.

4. *Stiffness:* the transmission components are not perfectly rigid. They have finite stiffness. The stiffness of the transmission box between input and output shaft is rated with a *torsional or translational stiffness* parameter.

5. *Break-away friction:* This friction torque (or force) is an estimated value and highly dependent function of the lubrication condition of the moving components. This is the minimum torque or force needed at the input shaft to move the mechanism.

6. *Back driveability:* Motion conversion mechanisms involve two shafts, the input shaft and output shaft. In the normal mode of operation, the motion and torque in the input shaft are transmitted to the output shaft with a finite efficiency. Back driveablity refers to the transmission of power in the opposite direction, that is the motion source is provided at the output shaft and transmitted to the input shaft. Most spur gear, belt and pulley type mechanisms are back driveable with the same efficiency in both directions. Rotary-to-linear motion conversion mechanisms such as lead-screw and ball-screws have different efficiencies and are not necessarily back driveable. Ball-screws are considered back driveable for all cases. The back driveability of a lead-screw depends on the lead angle. If the lead angle is below a certain value (i.e., $30°$), the back driveability may be in question. Furthermore, it also highly depends on the lubrication condition of the mechanism since it affects the friction force that must be overcome in order to back drive the mechanism. Worm-gear mechanisms are not back driveable. There are applications which benefit from that, such as raising a very heavy load and in the case of power failure in the input shaft motor, the mechanism is not supposed to back drive under gravitational force and is suppose to hold the load in position. In short, when back driveability is required, two variables are of interest: (i) η_b efficiency in the back drive direction, and (ii) F_{fric} friction force to overcome in order to initiate motion, which is highly dependent on the lubrication conditions.

3.2 ROTARY TO ROTARY MOTION TRANSMISSION MECHANISMS

3.2.1 Gears

Gears are used to increase or decrease the speed ratio between the input and output shaft. The effective gear ratio is obvious (Figure 3.1). Assuming that the gears do not slip, the linear distance traveled by each gear at the contact point is same,

$$s_1 = s_2 \tag{3.18}$$

$$\Delta\theta_1 \cdot r_1 = \Delta\theta_2 \cdot r_2 \tag{3.19}$$

$$N = \frac{\Delta\theta_1}{\Delta\theta_2} \tag{3.20}$$

$$N = \frac{r_2}{r_1} \tag{3.21}$$

Since the pitch of each gear must be same, the number of teeth on each gear is proportional to their radius,

$$N = \frac{\dot{\theta}_1}{\dot{\theta}_2} = \frac{N_2}{N_1} = \frac{r_2}{r_1} \tag{3.22}$$

where N_1 and N_2 are the number of gear teeth on each gear. It can be shown that for an ideal gear box (100% power transmission efficiency),

$$P_{out} = P_{in} \tag{3.23}$$

$$T_{out} \cdot \dot{\theta}_{out} = T_{in} \cdot \dot{\theta}_{in} \tag{3.24}$$

Hence,

$$N = \frac{N_2}{N_1} = \frac{\dot{\theta}_{in}}{\dot{\theta}_{out}} = \frac{T_{out}}{T_{in}} \tag{3.25}$$

The reflection of inertia and torque from the output shaft to the input shaft can be determined by using the energy and work relationships. Let the rotary inertia of the load on the output shaft be J_1 and the load torque be T_1. Let us express the kinetic energy of the load

$$KE = \frac{1}{2} \cdot J_1 \cdot \dot{\theta}_{out}^2 \tag{3.26}$$

$$= \frac{1}{2} \cdot J_1 \cdot (\dot{\theta}_{in}/N)^2 \tag{3.27}$$

$$= \frac{1}{2} \cdot J_1 \cdot \frac{1}{N^2} \cdot (\dot{\theta}_{in})^2 \tag{3.28}$$

$$= \frac{1}{2} \cdot J_{in,eff} \cdot (\dot{\theta}_{in})^2 \tag{3.29}$$

where the reflected inertia (inertia of the load seen by the input shaft) is

$$J_{in,eff} = \frac{J_1}{N^2} \tag{3.30}$$

Similarly, let us determine the effective load torque seen by the input shaft. The work done by a load torque T_1 over an output shaft displacement $\Delta\theta_{out}$ is

$$W = T_1 \cdot \Delta\theta_{out} \tag{3.31}$$

$$= T_1 \cdot \frac{\Delta\theta_{in}}{N} \tag{3.32}$$

$$= T_{in,eff} \cdot \Delta\theta_{in} \tag{3.33}$$

The effective reflective torque on the input shaft as a result of the load torque on the output shaft is

$$T_{in,eff} = \frac{T_1}{N} \tag{3.34}$$

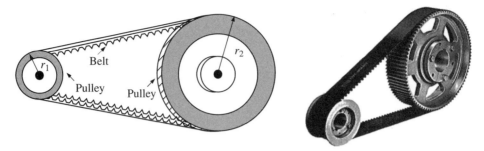

FIGURE 3.2: Rotary to rotary motion conversion mechanisms: timing belt and toothed pulley.

The same concept of kinetic energy and work of the tool is used in all of the other mechanisms to determine the reflected inertia and torque between output and input shafts.

3.2.2 Belt and Pulley

The gear ratio of a belt-pulley mechanism is the ratio between the input and output diameters. Assuming no slip between the belt and pulleys on both shafts, the linear displacement along the belt and both pulleys should be equal (Figure 3.2),

$$x = \Delta\theta_1 \cdot r_1 = \Delta\theta_2 \cdot r_2 \tag{3.35}$$

The effective gear ratio is by definition,

$$N = \frac{\Delta\theta_1}{\Delta\theta_2} \tag{3.36}$$

$$= \frac{r_2}{r_1} \tag{3.37}$$

$$= \frac{d_2}{d_1} \tag{3.38}$$

The inertia and torque reflection between the input and output shaft has the same relationship as the gear mechanisms.

Example Consider a spur-gear mechanism with a gear ratio of $N = 10$. Assume that the load inertia connected to the output shaft is a solid steel material with diameter $d = 3.0$ in, length $l = 2.0$ in. The friction related torque at the load is $T_1 = 200$ lb $\cdot$ in. The desired speed of the load is 300 rev/min. Determine the necessary speed at the input shaft as well as reflected inertia and torque.

The necessary speed at the input shaft is related to the output shaft speed by a kinematic relationship defined by the gear ratio,

$$\dot{\theta}_{\text{in}} = N \cdot \dot{\theta}_{\text{out}} = 10 \cdot 300 \, \text{rpm} = 3000 \, \text{rpm} \tag{3.39}$$

The inertia and torque experienced at the input shaft due to the load alone (which we call the reflected inertia and the reflected torque) are

$$J_{\text{in,eff}} = \frac{1}{N^2} \cdot J_1 \tag{3.40}$$

$$T_{\text{in,eff}} = \frac{1}{N} \cdot T_1 \tag{3.41}$$

The mass moment of inertia of the cylindrical load,

$$J_1 = \frac{1}{2} \cdot m \cdot (d/2)^2 \tag{3.42}$$

$$= \frac{1}{2} \cdot \rho \cdot \pi \cdot (d/2)^2 \cdot l \cdot (d/2)^2 \tag{3.43}$$

$$= \frac{1}{2} \cdot \rho \cdot \pi \cdot l \cdot (d/2)^4 \tag{3.44}$$

$$= \frac{1}{2} \cdot (0.286/386) \cdot \pi \cdot 2.0 \cdot (3/2)^4 \, \text{lb} \cdot \text{in} \cdot \text{s}^2 \tag{3.45}$$

$$= 0.0118 \, \text{lb} \cdot \text{in} \cdot \text{s}^2 \tag{3.46}$$

where $g = 386 \, \text{in/s}^2$, the gravitational acceleration, is used to convert the weight density to mass density. Hence, the reflected inertia and torque are

$$J_{\text{in,eff}} = \frac{1}{10^2} \cdot 0.0118 = 0.118 \times 10^{-3} \, \text{lb} \cdot \text{in} \cdot \text{s}^2 \tag{3.47}$$

$$T_{\text{in,eff}} = \frac{1}{10} \cdot 200 = 20.0 \, \text{lb} \cdot \text{in} \tag{3.48}$$

3.3 ROTARY TO TRANSLATIONAL MOTION TRANSMISSION MECHANISMS

The rotary to translational motion transmission mechanisms convert rotary motion to linear translational motion. Translational motion is also refered as linear motion. Both terms will be used interchangeably in the following discussions. In addition, torque input is converted to force at the output. It should be noted that all of the rotary to translational motion transmission mechanisms discussed here are back drivable, meaning that they also make the conversion in the reverse direction.

3.3.1 Lead-Screw and Ball-Screw Mechanisms

Lead-screw and ball-screw mechanisms are the most widely used precision motion conversion mechanisms which transfer rotary motion to linear motion (Figure 3.3). The lead-screw

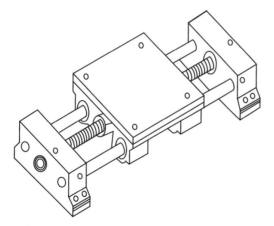

FIGURE 3.3: Rotary to translational motion conversion mechanism: lead-screw or ball-screw with linear guide bearings.

is basically a precision threaded screw and nut. In a screw and nut pair, typically we turn the nut and it advances linearly on the stationary screw. In the lead-screw, used as a motion conversion mechanism, the nut is not allowed to make rotation around the screw, but is supported by linear (translational) bearings to move. The screw is rotated (i.e., by an electric motor). Since the screw is not allowed to travel, but only rotate, the nut moves translationally along the screw. The tool is connected to the nut. Hence, the rotary motion of the motor connected to the screw is converted to the translational motion of the nut and the tool.

Ball-screw design uses precision ground spherical balls in the groove between the screw and nut threads to reduce backlash and friction in the motion transmission mechanism. Any lead-screw has a finite backlash typically in the order of micrometer range. By using preloaded springs, a set of spherical bearing-type balls are used to reduce the backlash. Hence, they are called ball-screws. Most XYZ table-type positioning devices use ball-screws instead of lead-screws. However, the load carrying capacity of ball-screws is less than that of the lead-screws since the contact between the moving parts (lead and nut) is provided by point contacts of balls. On the other hand, a ball-screw has less friction than lead-screw.

The kinematic motion conversion factor, or effective gear ratio, of a lead-screw is characterized by its pitch, p rev/in or rev/mm. Therefore, for a lead-screw having a pitch of p, which is the inverse of the distance traveled for one turn of the thread called lead, $l = 1/p$ in/rev or rev/mm, the rotational displacement ($\Delta\theta$ in units of rad) at the lead shaft and the translational displacement (Δx) of the nut is related by

$$\Delta\theta = 2\pi \cdot p \cdot \Delta x \tag{3.49}$$

$$\Delta x = \frac{1}{2\pi \cdot p} \cdot \Delta\theta \tag{3.50}$$

The effective gear ratio may be stated as

$$N_{ls} = 2\pi \cdot p \tag{3.51}$$

Let us determine the inertia and torque seen by the input end of the lead-screw due to a load mass and load force on the nut. We follow the same method as before and use energy–work relations. The kinetic energy of the mass m_1 at a certain speed $\dot{x}$ is

$$KE = \frac{1}{2}m_1 \cdot \dot{x}^2 \tag{3.52}$$

Noting the above motion conversion relationship,

$$\dot{x} = \frac{1}{2\pi \cdot p} \cdot \dot{\theta} \tag{3.53}$$

Then

$$KE = \frac{1}{2}m_1 \cdot \frac{1}{(2\pi \cdot p)^2} \cdot \dot{\theta}^2 \tag{3.54}$$

$$= \frac{1}{2}J_{eff} \cdot \dot{\theta}^2 \tag{3.55}$$

Then, the effective rotary inertia seen at the input shaft (J_{eff}) due to a translational mass on the nut (m_1) is

$$J_{eff} = \frac{1}{(2\pi \cdot p)^2} \cdot m_1 \tag{3.56}$$

$$= \frac{1}{N_{ls}^2} \cdot m_1 \tag{3.57}$$

It should be noted that m_1 is in units of mass (not weight, weight = mass $\cdot$ g), and the J_{eff} is the mass moment of inertia. Therefore if the weight of the load is given, W_1,

$$J_{\text{eff}} = \frac{1}{(2\pi \cdot p)^2} \cdot (W_1/g) \tag{3.58}$$

$$= \frac{1}{N_{1s}^2} \cdot (W_1/g) \tag{3.59}$$

The lead-screw has a very large effective gear ratio effect. Notice that for $p = 2, 5, 10$, the net gear ratio is $N = 4\pi, 10\pi, 20\pi$, respectively, and the inertial reflection of a translational mass is a factor determined by the square of the effective gear ratio.

Let us determine the reflected torque at the input shaft due to a load force, F_1. The work done by a load force during a incremental displacement is

$$Work = F_1 \cdot \Delta x \tag{3.60}$$

The corresponding rotational displacement is

$$\Delta x = \frac{1}{2\pi \cdot p} \cdot \Delta\theta \tag{3.61}$$

Hence,

$$Work = F_1 \cdot \Delta x \tag{3.62}$$

$$= F_1 \cdot \frac{1}{2\pi \cdot p} \cdot \Delta\theta \tag{3.63}$$

$$= T_{\text{eff}} \cdot \Delta\theta \tag{3.64}$$

The equivalent torque seen at the input shaft (T_{eff}) of the lead-screw due to the load force F_1 at the nut is

$$T_{\text{eff}} = \frac{1}{2\pi \cdot p} \cdot F_1 \tag{3.65}$$

$$= \frac{1}{N} \cdot F_1 \tag{3.66}$$

Example   Consider a ball-screw motion conversion mechanism with a pitch of $p = 10$ rev/in. The mass of the table and workpiece is $W_1 = 1000$ lb and the resistance force of the load is $F_1 = 1000$ lb. Determine the reflected rotary inertia and torque seen by a motor at the input shaft of the ball-screw.

$$J_{\text{eff}} = \frac{1}{(2\pi \cdot 10)^2 \text{ rad}^2/\text{in}^2} \cdot (1000/386) \text{ lb}/(\text{in/s}^2) \tag{3.67}$$

$$= 6.56 \times 10^{-3} \text{ lb} \cdot \text{in} \cdot \text{s}^2 \tag{3.68}$$

The torque that is reflected on the input shaft due to the load force F_1 is

$$T_{\text{eff}} = \frac{1}{2\pi \cdot p} \cdot F_1 \tag{3.69}$$

$$= \frac{1}{2\pi \text{ rad/rev} \cdot 10 \text{ rev/in}} \cdot 1000 \text{ lb} \tag{3.70}$$

$$= \frac{100}{2\pi} \text{ lb} \cdot \text{in} \tag{3.71}$$

$$= 15.91 \text{ lb} \cdot \text{in} \tag{3.72}$$

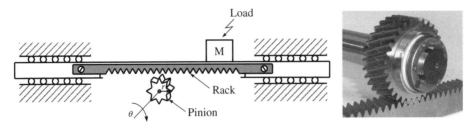

FIGURE 3.4: Rotary to translational motion conversion mechanism: rack and pinion mechanism. The advantage of the rack and pinion mechanism over the lead-screw mechanism is that the translational motion range can be very long. The lead-screw length is limited by the torsional stiffness. In rack and pinion mechanisms, since the translational part does not rotate, it does not have the reduced torsional stiffness problem due to the long length.

3.3.2 Rack and Pinion Mechanism

The rack and pinion mechanism is an alternative rotary to linear conversion mechanism (Figure 3.4). The pinion is the small gear. The rack is the translational (linear) component. It is similar to a gear mechanism where one of the gears is a linear gear. The effective gear ratio is calculated or measured from the assumption that there is no slip between the gears,

$$\Delta x = r \cdot \Delta \theta \tag{3.73}$$

$$\Delta \dot{x} = r \cdot \Delta \dot{\theta} \tag{3.74}$$

$$V = r \cdot w \tag{3.75}$$

where $\Delta \theta$ is in radian units. Hence, the effective gear ratio is

$$N = \frac{1}{r}; \quad \text{if} \quad \Delta\theta \text{ is in rad units or w is in rad/s units} \tag{3.76}$$

$$= \frac{1}{2\pi \cdot r}; \quad \text{if} \quad \Delta\theta \text{ is in rev units or w is in rev/s units} \tag{3.77}$$

The same mass and force reflection relations we developed for lead-screws apply for the rack and pinion mechanisms. The only difference is the effective gear ratio.

3.3.3 Belt and Pulley

Belt and pulley mechanisms are used both as rotary to rotary and rotary to linear motion conversion mechanisms depending on the output point of interest. If the load (tool) is connected to the belt and used to obtain linear motion, then it acts as a rotary to linear motion conversion device (Figure 3.5). The relations we developed for the rack and pinion

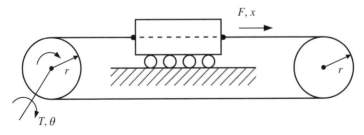

FIGURE 3.5: Rotary to translational motion conversion mechanism: Belt and pulley mechanism where both pulleys have the same diameter. The output motion is taken from the belt as the translational motion.

mechanism identically apply here. This mechanism is widely used in low inertia, low load force, and high bandwidth applications such as coil winding machines.

3.4 CYCLIC MOTION TRANSMISSION MECHANISMS

3.4.1 Linkages

One degree of freedom linkages define a nonlinear relationship between input and output motion variables as

$$\theta_{\text{out}} = f(\theta_{\text{in}}) \qquad (3.78)$$

where θ_{in} is the input motion variable, θ_{out} output motion variable, and $f(\cdot)$ is the nonlinear geometric function. The position, velocity, and acceleration of the output variable $(\theta_{\text{out}}(t), \dot{\theta}_{\text{out}}(t), \ddot{\theta}_{\text{out}}(t))$ are directly determined by the input variable $(\theta_{\text{in}}(t))$ and its time derivatives $(\dot{\theta}_{\text{in}}(t), \ddot{\theta}_{\text{in}}(t))$ and the partial derivatives of the nonlinear function with respect to the input variable $(\partial f(\theta_{\text{in}})/\partial \theta_{\text{in}}, \partial^2 f(\theta_{\text{in}})/\partial \theta_{\text{in}}^2)$.

Linkages are generally one degree of freedom, kinematically closed chain, robotic manipulators. The motion of one member (output link) of the linkage is a periodic function of the motion of another linkage member (input link). The most common linkages include

1. slider-crank mechanism (i.e., used in internal combustion engines (Figure 3.6)),

2. four-bar mechanism (Figure 3.7).

The slider-crank mechanism is the most widely recognized since it is used in every internal combustion engine. It can be used to convert the translational displacement of the slider to the rotary motion of the crank or vice versa. The length of the crank arm and the connecting link determine the geometric relationship between the slider translational displacement and crank angular rotation.

Four-bar linkage can convert the cyclic motion of the input arm (i.e., rotation of the input arm shaft) to the limited range cyclic rotation of the output shaft. By selectively choosing the input and output shaft, as well as the lengths of the linkages, different motion conversion functions are obtained. The motion control of such linkages is rather simple, since they are mostly used with a constant speed input shaft motion, and the resultant motion is obtained at the output shaft.

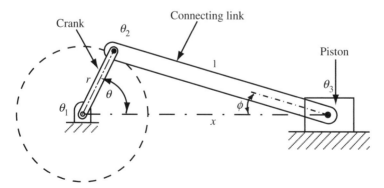

FIGURE 3.6: Rotary to translational motion conversion mechanism: slider-crank mechanism. In an internal combustion engine, the mechanism is used as translational (piston motion is input) to rotary motion (crank shaft rotation) conversion mechanism.

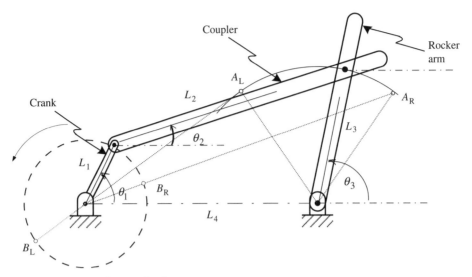

FIGURE 3.7: Four bar mechanism.

From the kinematics of the slider-crank mechanism, the following relations can be derived,

$$x = r\cos\theta + l\cos\phi \tag{3.79}$$

$$l\sin\phi = r\sin\theta \tag{3.80}$$

$$\cos\phi = [1 - \sin^2\phi]^{1/2} \tag{3.81}$$

$$= \left[1 - \left(\frac{r}{l}\sin\theta\right)^2\right]^{1/2} \tag{3.82}$$

$$x = r\cos\theta + l\left[1 - \left(\frac{r}{l}\sin\theta\right)^2\right]^{1/2} \tag{3.83}$$

$$\dot{x} = -r\dot{\theta}\left[\sin\theta + \frac{r}{2l}\frac{\sin(2\theta)}{\left[1 - \left(\frac{r}{l}\sin\theta\right)^2\right]^{1/2}}\right] \tag{3.84}$$

where r is the radius of the crank (length of the crank link), l is the length of the connecting arm, x is the displacement of the slider, θ is the angular displacement of the crank, ϕ is the angle between the connecting arm and displacement axis. The position and speed of piston motion and crank motion are related by the above geometric relations. The acceleration relation can be obtained by taking the time derivative of the speed relation [7].

Example Consider a crank-slider mechanism with the following geometric parameters: $r = 0.30$ m, $l = 1.0$ m. Consider the simulation of a condition that the crank shaft rotates at a constant speed $\dot{\theta}(t) = 1200$ rpm. Plot the displacement of the slider as a function of crank shaft angle from 0 to 360 degrees of rotation (for one revolution) and the linear speed of the slider.

Since we have the geometric relationship between the crank shaft angle and speed versus the slider position and speed (Eqn. 3.83, 3.84), we simply substitute for r and l the above values, and calculate the x and $\dot{x}$ for $\theta = 0$ *to* 2π rad and $\dot{\theta} = 1200 \cdot (1/60) \cdot 2\pi$ rad/s.

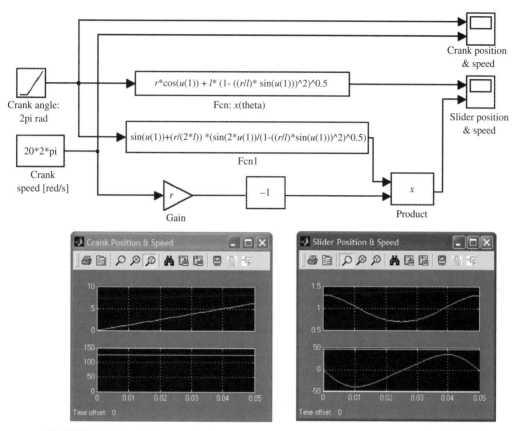

FIGURE 3.8: Simulation result of slider crank mechanisms: $r = 0.3\,\mathrm{m}$, $l = 1.0\,\mathrm{m}$, speed of crank shaft is constant at $\dot{\theta} = 1200\,\mathrm{rpm}$. The resulting slider position and speed functions are shown in the figure.

The MATLAB® or Simulink® environment can be conveniently used to generate the results (Figure 3.8).

3.4.2 Cams

Cams convert the rotary motion of a shaft into translational motion of a follower (Figure 3.9). The relationship between the translational motion and rotary motion is not a fixed gear ratio, but a nonlinear function. The nonlinear cam function is determined by the machined cam profile. A cam mechanism has three major components:

1. the input shaft,
2. the cam,
3. the follower.

If the input shaft axis of the cam is parallel to the follower axis of motion, such cams are called *axial* cams. In this case, cam function is machined into the cylindrical surface along the axis of input shaft rotation. If they are perpendicular to each other, such cams are called *radial* cams and the cam function may be machined either along the outside diameter or face diameter. All cam profiles can be divided into periods called *rise, dwell, and fall* in various combinations. For instance, a cam profile can be designed such that for one revolution of the input shaft, the follower makes a cyclic motion that is various combinations of *rise, fall,*

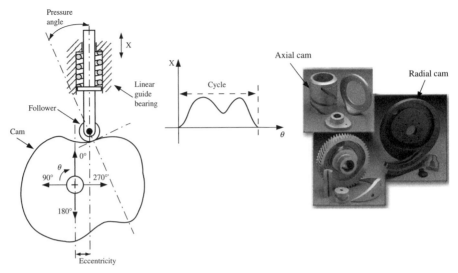

FIGURE 3.9: Rotary to translational motion conversion mechanism: cam mechanism.

dwell; that is, *rise, dwell, fall, dwell*, or *rise, fall, dwell*. In addition, a cam function may be designed such that the follower makes single or multiple cycles per revolution of the input shaft, that is one, two, three, or four follower cycles per input shaft revolution. It is common to design rise and fall periods of the cam as symmetric. During the dwell periods, the follower is stationary. Therefore, during dwell period the follower position is constant, and speed and acceleration are zero. If symmetric cam functions are used for rise and fall periods, then we are only concerned with the cam function design for the rise period.

A cam profile is defined by the following relationship,

$$x = f(\theta); \quad 0 < \theta < 2\pi \tag{3.85}$$

where x is the displacement of the follower, and θ is the rotation of the input shaft. The cam function is periodic. The cam follower motion repeats for every revolution of the input shaft.

In addition to selecting an appropriate cam function, there are three other important parameters to consider in cam design (Figure 3.7):

1. *Pressure angle:* measured as the angle between the follower motion axis and the axis perpendicular to the common tangent line at the contact point between cam and follower (Figure 3.7). A cam should be machined such that the pressure angle stays less than about 30° in order to make sure the side loading force on the follower is not too high.

2. *Eccentricity:* the offset distance between the follower axis and cam rotation axis in the direction perpendicular to the cam motion. By increasing eccentricity, we can reduce the effective pressure angle, and hence the side loading forces on the follower. However, as the eccentricity increases, the cam gets larger and less compact.

3. *Radius of curvature:* the radius of the cam function curvature along its periphery. The radius of curvature should be a continuous function of the angular position of the cam input shaft. Any discontinuity in the radius of curvature is essentially reflected as a non-smooth cam surface. In general, the radius of curvature should be at least 2 to 3 times larger than the radius of the follower. The main considerations are the continuity and ability of the follower to maintain contact on the cam at all times.

The displacement, velocity, acceleration, and jerk (time derivative of acceleration) are important since they directly affect the forces experienced in the mechanism. In cam design, quite often we focus on the first, second, and third derivatives of the cam function with respect to θ instead of t, time. In the final analysis, we are interested in the time derivative values, since they determine the actual speed, acceleration, and jerk. The relationships between the derivatives of cam function with respect to θ and t are as follows.

$$\frac{dx}{d\theta} = \frac{df(\theta)}{d\theta}; \qquad \frac{dx}{dt} = \frac{dx}{d\theta}\dot{\theta} \tag{3.86}$$

$$\frac{d^2x}{d\theta^2} = \frac{d^2f(\theta)}{d\theta^2}; \qquad \frac{d^2x}{dt^2} = \frac{d^2x}{d\theta^2}\dot{\theta}^2 + \frac{dx}{d\theta}\ddot{\theta} \tag{3.87}$$

$$\frac{d^3x}{d\theta^3} = \frac{d^3f(\theta)}{d\theta^3}; \qquad \frac{d^3x}{dt^3} = \frac{d^3x}{d\theta^3}\dot{\theta}^3 + \frac{d^2x}{d\theta^2}(3\cdot\dot{\theta}\cdot\ddot{\theta}) + \frac{dx}{d\theta}\frac{d\ddot{\theta}}{dt} \tag{3.88}$$

Notice that when $\dot{\theta} = constant$, then $\ddot{\theta} = 0$, and $d\ddot{\theta}/dt = 0$. When the input shaft of the cam speed is constant, which is the case in most applications, then the relationship between the time derivatives and derivatives with respect to θ of the cam function simplify as follows,

$$\frac{dx}{d\theta} = \frac{df(\theta)}{d\theta}; \qquad \frac{dx}{dt} = \frac{dx}{d\theta}\dot{\theta} \tag{3.89}$$

$$\frac{d^2x}{d\theta^2} = \frac{d^2f(\theta)}{d\theta^2}; \qquad \frac{d^2x}{dt^2} = \frac{d^2x}{d\theta^2}\dot{\theta}^2 \tag{3.90}$$

$$\frac{d^3x}{d\theta^3} = \frac{d^3f(\theta)}{d\theta^3}; \qquad \frac{d^3x}{dt^3} = \frac{d^3x}{d\theta^3}\dot{\theta}^3 \tag{3.91}$$

Significant effort is made in selecting the $f(\theta)$ cam function in order to shape the first, second, and even the third derivative of it so that desired results (i.e., minimize vibrations) are obtained from the cam motion. *As a general rule in cam design, the cam function should be chosen so that the cam function, first, and second derivatives (displacement, speed, and acceleration functions) are continuous and the third derivative (jerk function) discontinuities (if any) are finite.* In general these continuity conditions are applied to cam function derivatives with respect to θ. If the input cam speed is constant, the same continuity conditions are then satisfied by the time derivatives as well. For instance, a cam function with trapezoidal shape results in discontinuous velocity and infinite accelerations. Therefore, pure trapezoidal cam profiles are almost never used. The alternatives for the selection of functions that meet the continuity requirements in the cam function are many. Some functions are defined in analytical form (parts of the cam function are portions of the sinusoidal functions joined to form a continuous function) and some are custom developed by experimentation and stored in numerical form, that is two column data of θ_i versus x_i. The four types of basic cam functions are discussed below (Figure 3.10).

1. *Cycloidal displacement* cam function has an acceleration curve for the rise portion that is sinusoidal with single frequency,

$$\frac{d^2x(\theta)}{d\theta^2} = C_1 \cdot \sin(f_1\theta) \tag{3.92}$$

where C_1 is constant related to the displacement range of x and f_1 is determined by the portion of input shaft rotation to complete the *rise portion of cam motion*. For

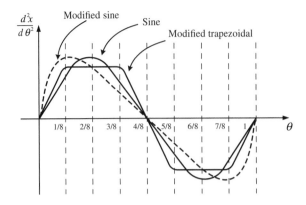

FIGURE 3.10: Commonly used cam profiles. The acceleration function is shown as a function of the driving axis. Sinusoidal, modified sin, and modified trapezoidal functions are common cam profiles.

instance, if rise motion is completed in $\theta_{\text{rise}} = 1/2\,\text{rev} = \pi\,\text{rad}$,

$$f_1\theta_{\text{rise}} = 2\pi \tag{3.93}$$

$$f_1 = \frac{2\pi}{\theta_{\text{rise}}} \tag{3.94}$$

$$= \frac{2\pi}{\pi} \tag{3.95}$$

$$= 2 \tag{3.96}$$

Similarly, if rise motion is completed in $1/3\,\text{rev} = 2\pi/3\,\text{rad}$, then $f_1 = 3$. Both displacement and velocity functions are also sinusoidal functions. The speed and displacement functions can be readily obtained by integrating the acceleration curve,

$$\frac{dx(\theta)}{d\theta} = -C_1\frac{1}{f_1}\cdot\cos(f_1\theta)|_0^\theta \tag{3.97}$$

$$= \frac{C_1}{f_1} - C_1\frac{1}{f_1}\cdot\cos(f_1\theta) \tag{3.98}$$

$$x(\theta) = \left[\frac{C_1}{f_1}\theta - C_1\frac{1}{f_1^2}\cdot\sin(f_1\theta)\right]\Bigg|_0^\theta \tag{3.99}$$

$$= \frac{C_1}{f_1}\theta - C_1\frac{1}{f_1^2}\cdot\sin(f_1\theta); \qquad for \quad 0 \le \theta \le \theta_{\text{rise}} \tag{3.100}$$

Let us assume that the rise portion is to occur in $1/2$ revolution of the cam input shaft. Then, $f_1 = 2$. Let the rise distance be $x_{\text{rise}} = 0.2\,\text{m}$. Then the constant C_1 can be determined by evaluating the displacement cam function at the end of the rise cycle for $\theta = \pi$,

$$0.2 = \frac{C_1}{2}\pi \tag{3.101}$$

$$C_1 = \frac{2\cdot0.2}{\pi} \tag{3.102}$$

Assuming that the cam has symmetric rise and down portions without any dwell portion, the complete period of motion for the cam is $\theta = 0$ to $2 \times \theta_{\text{rise}}$. Acceleration, speed, and displacement curves for the *down portion* of the cam are all mirror images of the rise portion. Mirror image and original function relationship is as follows. The original function for rise motion is x_{rise},

$$x_{\text{rise}}(\theta) = f(\theta); \quad 0 \le \theta \le \theta_{\text{rise}} \tag{3.103}$$

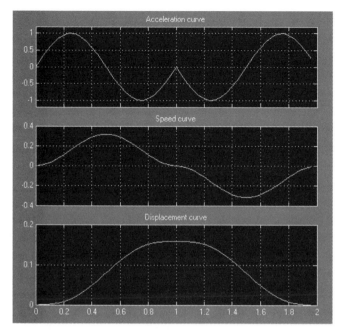

FIGURE 3.11: Sinusoidal acceleration cam profile: acceleration, speed, and displacement function for a cam design where half of the shaft revolution is used for the rise motion and the other half is used for the down motion symmetrically.

then, the mirror image (let us call that x_{down})

$$x_{\text{down}}(\theta) = f(2\theta_{\text{rise}} - \theta); \quad \theta_{\text{rise}} \leq \theta \leq 2\theta_{\text{rise}} \tag{3.104}$$

Hence the mirror image of the displacement cam function $x(\theta)$ is $x(2\theta_{\text{rise}} - \theta)$,

$$x(\theta) = \frac{C_1}{f_1}(2\theta_{\text{rise}} - \theta) - C_1\frac{1}{f_1^2} \cdot \sin(f_1(2\theta_{\text{rise}} - \theta)); \quad \theta_{\text{rise}} \leq \theta \leq 2\theta_{\text{rise}} \tag{3.105}$$

Figure 3.11 shows the acceleration, speed, and displacement curves for one cycle of the cam motion. If there is a dwell portion for the input shaft rotation range of $\theta_{\text{rise}}, \theta_{\text{rise}} + \theta_{\text{dwell}}$, then the dwell portion of the cam function following the rise portion would be as follows,

$$x(\theta) = x_{\text{rise}}(\theta_{\text{rise}}) \ ; \ \theta_{\text{rise}} \leq \theta \leq \theta_{\text{rise}} + \theta_{\text{dwell}} \tag{3.106}$$

$$\frac{dx(\theta)}{d\theta} = 0 \tag{3.107}$$

$$\frac{d^2x(\theta)}{d\theta^2} = 0 \tag{3.108}$$

and for the down portion of the cam,

$$x(\theta) = \frac{C_1}{f_1}(2\theta_{\text{rise}} + \theta_{\text{dwell}} - \theta) - C_1\frac{1}{f_1^2} \cdot \sin(f_1(2\theta_{\text{rise}} + \theta_{\text{dwell}} - \theta)); \tag{3.109}$$

$$\theta_{\text{rise}} + \theta_{\text{dwell}} \leq \theta \leq 2\theta_{\text{rise}} + \theta_{\text{dwell}} \tag{3.110}$$

$$\frac{dx(\theta)}{d\theta} = -\frac{C_1}{f_1} + C_1 \frac{1}{f_1} \cdot \cos(f_1(2\theta_{\text{rise}} + \theta_{\text{dwell}} - \theta)) \tag{3.111}$$

$$\frac{d^2x(\theta)}{d\theta^2} = C_1 \cdot \sin(f_1(2\theta_{\text{rise}} + \theta_{\text{dwell}} - \theta)) \tag{3.112}$$

2. *Modified sine* function is a modified version of the first function, cycloidal displacement function. Compared to the original version, this version results in lower peak acceleration and speed values while maintaining a similar cam displacement function. This cam function uses at least two frequencies of the sinusoidal profile. Pieces of the two sine functions are combined to form a smooth cam function.

$$\frac{d^2x(\theta)}{d\theta^2} = C_1 \cdot \sin(f_1\theta) + C_2 \cdot \sin(f_2\theta) \tag{3.113}$$

where C_1, C_2 constants would have non-zero values in some segments and zero values in other segments of the cam. In other words, in some segments of the cam function one of the sections of the sinusoidal function is used, in other segments the other sinusoidal function is used with its appropriate segment connecting to the preceeding and the following curve. Let θ_{rise} be the period of input shaft rotation for the rise period. f_1, f_2 are two constant frequencies.

$$f_1 = \frac{2\pi}{\theta_{\text{rise}}/2} = \frac{4\pi}{\theta_{\text{rise}}} \tag{3.114}$$

$$f_2 = \frac{2\pi}{3\theta_{\text{rise}}/2} = \frac{4\pi}{3\theta_{\text{rise}}} \tag{3.115}$$

The acceleration function is constructed from the segments of two sinusoidal functions as follows,

$$\frac{d^2x(\theta)}{d\theta^2} = A_{\text{o}} \cdot \sin\left(\frac{2\pi}{(\theta_{\text{rise}}/2)}\theta\right); \qquad for\ 0 \le \theta \le \frac{1}{8}\theta_{\text{rise}} \tag{3.116}$$

$$\frac{d^2x(\theta)}{d\theta^2} = A_{\text{o}} \cdot \sin\left(\frac{2\pi}{(3\theta_{\text{rise}}/2)}\theta + (\pi/3)\right); \ for\ \frac{1}{8}\theta_{\text{rise}} \le \theta \le \frac{7}{8}\theta_{\text{rise}} \tag{3.117}$$

$$\frac{d^2x(\theta)}{d\theta^2} = A_{\text{o}} \cdot \sin\left(\frac{2\pi}{(\theta_{\text{rise}}/2)}\theta - 2\pi\right); \qquad for\ \frac{7}{8}\theta_{\text{rise}} \le \theta \le \theta_{\text{rise}} \tag{3.118}$$

The acceleration function like this describes the movement of the follower for the rise period. If the follower cycle is made of *rise and fall* without any dwell periods, the complete displacement cycle is performed over an input shaft rotation of $\theta = 2\theta_{\text{rise}}$. Displacement and speed curves are determined by directly integrating the acceleration curve with zero initial condition for speed and zero initial condition for displacement at the beginning of integration.

The fall motion curves are simply obtained using a mirror image of the rise motion curves. During the dwell portion (if used in the cam design), the cam follower displacement stays constant, and speed and acceleration are zero.

3. *Modified trapezoidal* acceleration function modifies the standard trapezoidal acceleration function around the points where the acceleration changes slope. The acceleration function is smoothed out with a smooth function, such as a sinusoidal function, hence eliminating the jerk discontinuity. If purely trapezoidal acceleration profiles are used, there would be discontinuity in the jerk function when the acceleration function slope changes from a finite value to zero value (constant acceleration). In order to reduce the effect of this in terms of vibrations in the mechanism, the

acceleration profile function includes sinusoidal functions at the corners of the curve and constant acceleration functions in other segments of the cam. Let θ_{rise} be the period of input shaft rotation for the rise period. The modified trapezoidal acceleration profile is formed by constructing a function where the pieces of it are obtained from a sinusoidal function with period $\theta_{\text{rise}}/2$ and constant acceleration function.

Let us consider a modified trapezoidal acceleration function. The acceleration function has positive and negative (symmetric to the positive) portions. The variable portions of the acceleration function are implemented with portions of the sinusoidal, and constant acceleration portions are implemented with a constant value (Figure 3.16),

$$\frac{d^2x}{d\theta^2} = A_{\text{o}} \cdot \sin(2\pi\theta/(\theta_{\text{rise}}/2)); \ \ for \ 0 \le \theta \le \theta_{\text{rise}}/8 \tag{3.119}$$

$$\frac{d^2x}{d\theta^2} = A_{\text{o}}; \ \ for \ \frac{1}{8}\theta_{\text{rise}} \le \theta \le \frac{3}{8}\theta_{\text{rise}} \tag{3.120}$$

$$\frac{d^2x}{d\theta^2} = A_{\text{o}} \cdot \sin(2\pi\theta/(\theta_{\text{rise}}/2) - \pi); \ \ for \ \frac{3}{8}\theta_{\text{rise}} \le \theta \le \frac{5}{8}\theta_{\text{rise}} \tag{3.121}$$

$$\frac{d^2x}{d\theta^2} = -A_{\text{o}}; \ \ for \ \frac{5}{8}\theta_{\text{rise}} \le \theta \le \frac{7}{8}\theta_{\text{rise}} \tag{3.122}$$

$$\frac{d^2x}{d\theta^2} = A_{\text{o}} \cdot \sin(2\pi\theta/(\theta_{\text{rise}}/2) - 2\pi); \ \ for \ \frac{7}{8}\theta_{\text{rise}} \le \theta \le \theta_{\text{rise}} \tag{3.123}$$

Again, the acceleration functions describe half of the displacement cycle of the follower. For a symmetric return function, the mirror image of the same acceleration function is implemented in the cam profile.

4. *Polynominal functions* of many different types, that is including third, fourth, fifth, sixth, or seventh order polynomials, are used for cam functions.

$$x(\theta) = C_0 + C_1\theta + C_2\theta^2 + C_3\theta^3 + ... + C_{\text{n}}\theta^{\text{n}} \tag{3.124}$$

Typically, the input shaft is driven at a constant speed, and output follower motion is obtained as a cyclic function of input shaft rotation. Before the development of computer controlled programmable machines, almost all of the automated machines had one large power source driving a shaft. Many different cams generated desired motion profiles at individual stations from the same shaft. Hence, all the stations are synchronized to a master shaft. The synchronization is fixed as a result of the machined profiles of cams. If a different automated control is needed, cam profiles had to be physically changed. As programmable computer controlled machines replace the fixed automation machines, the use of mechanical cams has decreased. Instead, cam functions are implemented in software to synchronize independently controlled motion axes. If a different type of synchronization is required, all we need to do is to change the cam software. In mechanical cam synchronized systems, we would have to physically change the cam.

Example   Consider a cam with its follower as shown in Figure 3.9. Let us consider the displacement, speed, acceleration, and jerk profiles for modified sine cam functions. Assume that the input shaft of the cam runs at a constant speed, hence $\ddot{\theta} = 0$, $\frac{d}{dt}\ddot{\theta} = 0.0$. Assume that nominal operating speed of input shaft of the cam is $\dot{\theta} = 2\pi \cdot 10 \ \text{rad/s} \ (\dot{\theta} = 10 \ \text{rev/s} = 600 \ \text{rev/min})$, the total displacement of the cam follower is 2.0 in, and the period of cam motion for a complete up–down cycle is once per revolution of the input shaft. Since the input shaft speed is constant, the time derivatives and derivatives with respect to input shaft angle are related to each other as

$$\dot{x} = \frac{dx}{d\theta}\dot{\theta}; \quad \ddot{x} = \frac{d^2x}{d\theta^2}(\dot{\theta})^2; \quad \frac{d}{dt}\ddot{x} = \frac{d^3x}{d\theta^3}(\dot{\theta})^3 \tag{3.125}$$

We will assume that the cam has rise and fall periods without any dwell period in between them. For one half of a cycle (rise period), the cam profile is defined in three sections as a function of input shaft displacement which are regions $0 \le \theta \le \frac{1}{8}\theta_{\text{rise}}$, $\frac{1}{8}\theta_{\text{rise}} \le \theta \le \frac{7}{8}\theta_{\text{rise}}$, $\frac{7}{8}\theta_{\text{rise}} \le \theta \le \theta_{\text{rise}}$.

It can be shown that the *modified sine cam function* in the rise region is as follows ($Cam_1(\theta)$),

$$\frac{d^2x(\theta)}{d\theta^2} = A_o \cdot \sin\left(\frac{4\pi}{\theta_{\text{rise}}}\theta\right); \quad 0 \le \theta \le \frac{1}{8}\theta_{\text{rise}} \tag{3.126}$$

$$\frac{dx(\theta)}{d\theta} = \left(\frac{A_o\theta_{\text{rise}}}{4\pi}\right) \cdot \left(1 - \cos\left(\frac{4\pi}{\theta_{\text{rise}}}\theta\right)\right) \tag{3.127}$$

$$x(\theta) = \frac{A_o\theta_{\text{rise}}}{4\pi} \cdot \left(\theta - \frac{\theta_{\text{rise}}}{4\pi}\sin\left(\frac{4\pi}{\theta_{\text{rise}}}\theta\right)\right) \tag{3.128}$$

Likewise, the cam functions for the other period of motion are defined as follows ($Cam_2(\theta)$),

$$\frac{d^2x(\theta)}{d\theta^2} = A_o \cdot \sin\left(\frac{4\pi}{3\theta_{\text{rise}}}\theta + \frac{\pi}{3}\right); \quad \frac{1}{8}\theta_{\text{rise}} \le \theta \le \frac{7}{8}\theta_{\text{rise}} \tag{3.129}$$

$$\frac{dx(\theta)}{d\theta} = V_{s1} - A_o \cdot \left(\frac{3\theta_{\text{rise}}}{4\pi}\right)\left(\cos\left(\frac{4\pi}{3\theta_{\text{rise}}}\theta + \frac{\pi}{3}\right)\right) \tag{3.130}$$

$$x(\theta) = X_{s1} + V_{s1} \cdot \theta - A_o \cdot (3\theta_{\text{rise}})^2 \cdot \left(\sin\left(\frac{4\pi}{3\theta_{\text{rise}}}\theta + \frac{\pi}{3}\right)\right) \tag{3.131}$$

and similarly, for the period of $\frac{7}{8}\theta_{\text{rise}} \le \theta \le \theta_{\text{rise}}$, ($Cam_3(\theta)$)

$$\frac{d^2x(\theta)}{d\theta^2} = A_o \cdot \sin(4\pi\theta/\theta_{\text{rise}} - 2\pi); \quad \frac{7}{8}\theta_{\text{rise}} \le \theta < \theta_{\text{rise}} \tag{3.132}$$

$$\frac{dx(\theta)}{d\theta} = V_{s2} - A_o \cdot \left(\frac{\theta_{\text{rise}}}{4\pi}\right)\left(\cos\left(\frac{4\pi}{\theta_{\text{rise}}}\theta - 2\pi\right)\right) \tag{3.133}$$

$$x(\theta) = X_{s2} + V_{s2} \cdot \theta - A_o \cdot \left(\frac{\theta_{\text{rise}}}{4\pi}\right)^2 \cdot \left(\sin\left(\frac{4\pi}{\theta_{\text{rise}}}\theta - 2\pi\right)\right) \tag{3.134}$$

where the constants $X_{s1}, X_{s2}, V_{s1}, V_{s2}$ are determined to meet the continuity requirements at the boundaries of the cam function sections. Using the total travel range of the follower, we can determine the constant A_o as a function of the total travel range specified. Below are the five equations from which the above five constants can be calculated.

$$Cam_1\left(\frac{1}{8}\theta_{\text{rise}}\right) = Cam_2\left(\frac{1}{8}\theta_{\text{rise}}\right) \tag{3.135}$$

$$Cam_2\left(\frac{7}{8}\theta_{\text{rise}}\right) = Cam_3\left(\frac{7}{8}\theta_{\text{rise}}\right) \tag{3.136}$$

$$\frac{d}{dt}\left(Cam_1\left(\frac{1}{8}\theta_{\text{rise}}\right)\right) = \frac{d}{dt}\left(Cam_2\left(\frac{1}{8}\theta_{\text{rise}}\right)\right) \tag{3.137}$$

$$\frac{d}{dt}\left(Cam_2\left(\frac{7}{8}\theta_{\text{rise}}\right)\right) = \frac{d}{dt}\left(Cam_3\left(\frac{7}{8}\theta_{\text{rise}}\right)\right) \tag{3.138}$$

$$Cam_3(\theta_{\text{rise}}) = x_{\text{rise}} \tag{3.139}$$

The cam function is the combination of Equations 3.128, 3.131, 3.134 and their mirror images, ordered in reverse, and θ substituted by $2\theta_{rise} - \theta$ as an independent variable in the cam functions. The cam function then,

$$Cam_1(\theta) = \frac{A_o\theta_{rise}}{4\pi} \cdot \left(\theta - \frac{\theta_{rise}}{4\pi}\sin\left(\frac{4\pi}{\theta_{rise}}\theta\right)\right); \qquad 0 \leq \theta \leq \frac{1}{8}\theta_{rise} \tag{3.140}$$

$$Cam_2(\theta) = X_{s1} + V_{s1} \cdot \theta - A_o\left(\frac{3\theta_{rise}}{4\pi}\right)^2 \cdot \left(\sin\left(\frac{4\pi}{3\theta_{rise}}\theta + \frac{\pi}{3}\right)\right);$$
$$\frac{1}{8}\theta_{rise} \leq \theta \leq \frac{7}{8}\theta_{rise} \tag{3.141}$$

$$Cam_3(\theta) = X_{s2} + V_{s2} \cdot \theta - A_o\left(\frac{\theta_{rise}}{4\pi}\right)^2 \cdot \left(\sin\left(\frac{4\pi}{\theta_{rise}}\theta - 2\pi\right)\right);$$
$$\frac{7}{8}\theta_{rise} \leq \theta \leq \theta_{rise} \tag{3.142}$$

$$Cam_4(\theta) = Cam_3(2\theta_{rise} - \theta); \quad \theta_{rise} \leq \theta \leq \frac{9}{8}\theta_{rise} \tag{3.143}$$

$$Cam_5(\theta) = Cam_2(2\theta_{rise} - \theta); \quad \frac{9}{8}\theta_{rise} \leq \theta \leq \frac{15}{8}\theta_{rise} \tag{3.144}$$

$$Cam_6(\theta) = Cam_1(2\theta_{rise} - \theta); \quad \frac{15}{8}\theta_{rise} \leq \theta \leq 2\theta_{rise} \tag{3.145}$$

For the purpose of manufacturing a cam, we only need the displacement functions for a complete revolution of the input shaft.

3.5   SHAFT MISALIGNMENTS AND FLEXIBLE COUPLINGS

Mechanical systems always involve two or more shafts to transfer motion. There is always a finite accuracy with which the two shafts can be aligned in the axial direction. Any shaft misalignments will result in loads on the bearings and cause vibration, and hence reduce the life of the machinery. In order to reduce the vibration and life reducing effects of shaft misalignment, flexible couplings are used between shafts (Figure 3.12).

There are two main categories of flexible shaft couplings:

1. couplings for large power transfer between shafts and motors,
2. couplings for precision motion transfer at low powers between shafts and motors.

High precision motion systems include motors with very low friction and yet very delicate bearings. Such motors are very sensitive to shaft misalignments. The bearing failure of the motor as a result of excessive shaft misignment is a very common reliablity problem. Therefore, in most high performance servo motor applications, the motor shaft is coupled to the load via a flexible coupling. The flexible couplings provide the ability to make the system more tolerant to shaft misalignments. However, it comes at the cost of reduced stiffness of the mechanical system. Therefore, designers must make sure that the stiffness of the coupling does not interfere with the desired motion bandwidth (especially in variable speed and cyclic positioning applications).

Couplings are rated by the following parameters:

1. maximum and rated torque capacity,
2. torsional stiffness,
3. maximum allowed axial misalignment,
4. rotary inertia and mass of the coupling,

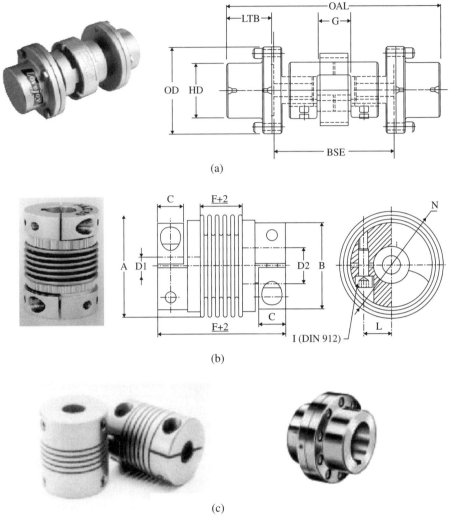

FIGURE 3.12: Flexible couplings used between connecting shafts in motion transmission mechanisms to compensate for the shaft misalignments and to protect bearings.

5. input and output shaft diameters,
6. input/output shaft connection method (set screw, clamped with keyway),
7. design type (bellow or helical coupling).

3.6 ACTUATOR SIZING

Every motion axis is powered by an actuator. The actuator may be electric, hydraulic or pneumatic power based. The size of the actuator refers to its power capacity and must be large enough to be able to move the axis under the given inertial and load force/torque conditions. If the actuator is undersized, the axis will not be able to deliver the desired motion, that is cannot deliver desired acceleration or speed levels. If the actuator is oversized, it will cost more and the motion axis will have slower bandwidth since, as the actuator size gets larger (larger power levels), the bandwidth gets slower as a general rule. Therefore, it is important to properly size the actuators for a motion axis with a reasonable margin of safety. Along with determining the proper actuator size for an application, the size of the gear mechanism needs to be determined unless the actuator is directly coupled to the load (Figure 3.14). The focus of this section is on sizing a rotary electric motor type actuator. The same concepts can be used for other types of actuators.

The question of actuator sizing is a question of determining the following requirements for an axis under worst operating conditions (i.e., largest expected inertia and resistive load),

1. maximum torque (also called peak torque) required, T_{max},
2. rated (continuous or root mean squared, RMS) torque required, T_r,
3. maximum speed required, $\dot{\theta}_{max}$,
4. positioning accuracy required, $(\Delta\theta)$,
5. gear mechanism parameters: gear ratio, its inertial and resistive load (force/torque), stiffness, backlash characteristics.

Once the torque requirements are determined, then the amplifier current and power supply requirements are directly determined from them.

In general, accuracy and maximum speed requirements of the load dictate the gear ratio. Below, we will assume that a gear mechanism with an appropriate gear ratio is selected and focus on determining the actuator size. For a given application, the load motion requirements specify the desired positioning accuracy and maximum speed. Let us call that Δx and $\dot{x}_{max}$. The desired positioning accuracy and maximum speed at the actuator (i.e., rotary motor) are determined by,

$$\Delta\theta = N \cdot \Delta x \tag{3.146}$$

$$\dot{\theta} = N \cdot \dot{x}_{max} \tag{3.147}$$

A gear ratio range is defined by the minimum accuracy and the maximum speed requirement, that is in order to provide the desired accuracy (Δx) for a motor with a given actuator positioning accuracy $(\Delta\theta)$, the gear ratio must be

$$N \geq \frac{\Delta\theta}{\Delta x} \tag{3.148}$$

In order to provide the desired maximum speed for a given maximum speed capacity of the motor (not to exceed the maximum speed capability of the motor), the gear ratio must be smaller than or equal to

$$N \leq \frac{\dot{\theta}_{max}}{\dot{x}_{max}} \tag{3.149}$$

Hence, the acceptable gear ratio range is defined by the accuracy and maximum speed requirement,

$$\frac{\Delta\theta}{\Delta x} \leq N \leq \frac{\dot{\theta}_{max}}{\dot{x}_{max}} \tag{3.150}$$

Some of the most commonly used motion conversion mechanisms (also called gear mechanisms) are shown in Figure 3.15. Notice that the gear mechanism adds inertia and possible load torque to the motion axis in addition to performing the gear reduction role and coupling the actuator to the load. In precision positioning applications, the first requirement that must be satisfied is accuracy.

The actuator needs to generate torque/force in order to move two different categories of inertia and load (Figure 3.14):

1. load inertia and force/torque (including the gear mechanism),
2. inertia (and any resistive force) of the actuator itself. For instance, an electric motor has a rotor with finite inertia and that inertia is important for how fast the motor can accelerate and decelerate in high cycle rate automated machine applications. Similarly, a hydraulic cylinder has a piston and large rod which has mass.

The torque/force and motion relationship for each axis is determined by Newton's Second Law. Let us consider it for a rotary actuator. The same relationships follow for translational actuators by replacing the rotary inertia with mass, torque with force, and angular acceleration with translational acceleration ($\{J_T, \ddot{\theta}, T_T\}$ replace with $\{m_T, \ddot{x}, F_T\}$),

$$J_T \cdot \ddot{\theta} = T_T \tag{3.151}$$

where J_T is the total inertia reflected on the motor axis, T_T is the total net torque acting on the motor axis, and $\ddot{\theta}$ is angular acceleration. The *reflected* inertia or torque means the equivalent inertia or torque seen at the motor shaft after the gear reduction effect is taken into account.

There are three issues to determine for the actuator sizing (Figure 3.14),

1. the net inertia (it may be a function of the position of the motion conversion mechanism, Figure 3.15),
2. determine the net load torque (it may be a function of the position of the motion conversion mechanism and speed),
3. the desired motion profile (Figure 3.10).

Let us discuss the first item, detemination of inertia. Total inertia is the inertia of the rotary actuator and the reflected inertia,

$$J_T = J_m + J_{l,eff} \tag{3.152}$$

where $J_{l,eff}$ includes all the load inertias reflected on the motor shaft. For instance, in the case of a ball-screw mechanism, this includes the inertia of the flexible coupling (J_c) between the motor shaft and ball-screw, ball-screw inertia (J_{bs}), and load mass inertia (due to W_l),

$$J_{l,eff} = J_c + J_{bs} + \frac{1}{(2\pi p)^2}(W_l/g) \tag{3.153}$$

Notice that the total inertia that the actuator has to move is the sum of the load (including motion transmission mechanism) and the inertia of the moving part of the actuator itself.

The total torque is the difference between the torque generated by the motor (T_m) minus the resistive load torques on the axis (T_l),

$$T_T = T_m - T_{l,eff} \tag{3.154}$$

where T_l represents the sum of all external torques. If the load torque is in the direction of assisting the motion, it will be negative, and the net result will be the addition of two torques. The T_l may include friction (T_f), gravity (T_g), and process related torque and forces (i.e., an assembly application may require the mechanism to provide a desired force pressure, T_a), nonlinear motion related forces/torques if any (i.e., Corriolis forces and torques, T_{nl}),

$$T_l = T_f + T_g + T_a + T_{nl} \tag{3.155}$$

$$T_{l,eff} = \frac{1}{N} \cdot T_l \tag{3.156}$$

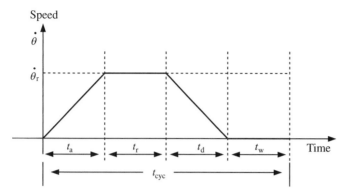

FIGURE 3.13: A typical desired velocity profile of a motion axis in programmable motion control applications such as automated assembly machines, robotic manipulators.

Notice that the friction torque may have a constant and speed dependent component to represent the Coulomb and viscous friction, $T_f(\dot{\theta})$.

For actuator sizing purposes, these torques should be considered for the worst possible case. However, care should be exercised as too much safety margin in the worst case assumptions can lead to unnecessarily large actuator sizing. Once the friction, gravitational loading, task related forces, and other nonlinear force coupling effects in articulated mechanisms are estimated, the mechanism kinematics is used to determine the reflected forces on the actuator axis. This reflection is a constant ratio for simple motion conversion mechanisms such as gear reducers, belt-pulleys, and lead-screws. For more complicated mechanisms such as linkages, cams, and multi degrees of freedom mechanisms, the kinematic reflection relations for the inertias and forces are not constant. Again, these relationships can be handled using worst case assumptions in simpler forms, or using more detailed nonlinear kinematic model of the mechanism.

Finally, we need to know the desired motion profile of the axis as a function of time. Generally, we assume a worst case cyclic motion. The most common motion profile used is a trapezoidal velocity profile as a function of time (Figure 3.13). The typical motion includes a constant acceleration period, then a constant speed period, then a constant deceleration period, and a dwell (zero speed) period.

$$\dot{\theta} = \dot{\theta}(t); \ \ 0 \leq t \leq t_{cyc} \tag{3.157}$$

Once the inertias, load torques, and desired motion profile are known, the required torque as a function of time during a cycle of the motion can be determined from

$$T_m(t) = J_T \cdot \ddot{\theta}(t) + T_{l,eff} \tag{3.158}$$

$$= J_T(\theta) \cdot \ddot{\theta}(t) + T_{f,eff}(\dot{\theta}) + T_{g,eff}(\theta) + T_{a,eff}(t) + T_{nl,eff}(\theta, \dot{\theta}) \tag{3.159}$$

Notice that before we calculate the torque requirements, we need to guess the inertia of the actuator itself, which is not known yet. Therefore, this calculation may need to be iterated few times.

Once the required torque profile is known as a function of time, two sizing values are determined from it: the maximum and root-mean square (RMS) value of the torque,

$$T_{max} = max(T_m(t)); \tag{3.160}$$

$$T_r = T_{rms} = \left(\frac{1}{t_{cyc}} \int_0^{t_{cyc}} T_m(t)^2 \, dt \right)^{1/2} \tag{3.161}$$

From the desired motion profile specification, we determine the maximum speed the actuator must deliver using the kinematic relations. In order to design an optimal motion control axis, the actuator sizing and the motion conversion mechanism (effective gear ratio) should be considered together. It is possible that a very small gear ratio may require a motor with very large torque requirement, and yet run at very low speeds, hence making use of a small part of the power capacity of the motor. An increased gear ratio would then require a smaller torque motor and that the motor would operate at higher speeds on average, hence making more use of the available power of the motor.

Once the torque requirements are known, the drive current and power supply voltage requirements can be directly determined for a given electric motor-drive system. Similarly, for hydraulic actuators, once the force requirements are determined, we would pick the supply pressure and determine the diameter of the linear cylinder. The speed requirement would determine the flow rate. Once these are known, then the size of the valve and pump can be determined.

Actuator Sizing Algorithm (Figure 3.14, 3.15):

1. Define the geometric relationship between the actuator and load. In other words, select the type of motion transmission mechanism between the motor and the load (N).

2. Define the inertia and torque/force characteristics of the load and the transmission mechanism, that is define the inertia of the tool as well as the inertia of the gear reducer mechanisms (J_1, T_1).

3. Define the desired cyclic motion profile in the form of load speed versus time ($\dot{\theta}_1(t)$).

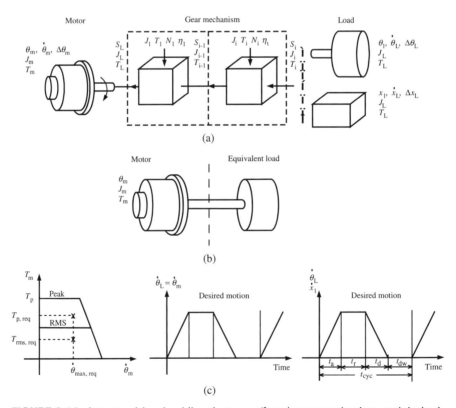

FIGURE 3.14: Actuator sizing: load (inertia, torque/force), gear mechanism, and desired motion must be specified.

Mechanism type	Mechanism characteristics				Input characteristics			Output characteristics		
	n_1	η_1	J_1	T_1	S_2	J_2	T_2	S_1	J_1	T_1
Gear 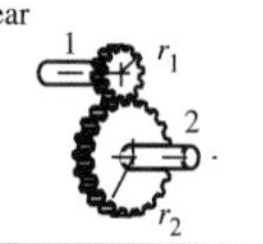	$\dfrac{r_2}{r_1}$	≤ 1.0	J_{r1} J_{r2}	$T_c.\mathrm{sgn}(\dot\theta_1)$ $+c\,\dot\theta_1$	θ_2	J_2	T_2	$\theta_1 = n\,\theta_2$ $=(\dfrac{r_2}{r_1})\theta_2$	$J_{r1} + \left(\dfrac{J_{r2}}{n^2\eta}\right)$	$T_i + \dfrac{1}{n\eta}T_2$
Belt-Pulley 	$\dfrac{r_2}{r_1}$	≤ 1.0	J_{r1} J_{r2} W_{belt}	$T_c.\mathrm{sgn}(\dot\theta_1)$ $+c\,\dot\theta_1$	θ_2	J_2	T_2	$\theta_1 = n\,\theta_2$ $=(\dfrac{r_2}{r_1})\theta_2$	$J_{r1} + \dfrac{J_{r2}+J^2}{n^2\eta}$ $+ \dfrac{1}{2}\left(\dfrac{W_{belt}}{g}\right)(r_1^2+r_2^2)$	$T_i + \dfrac{1}{n\eta}T_2$
Ball screw or cam 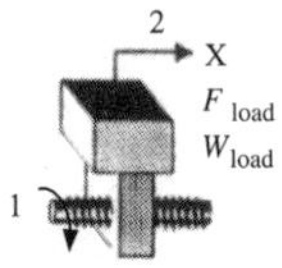 ($X=\theta/2\pi p$)	$2\pi p$	≤ 1.0	J_{lead} W_{load}	$T_c.\mathrm{sgn}(\dot\theta_1)$ $+c\,\dot\theta_1$	X	W_{load}	F_{load}	$\theta_1 = n\,X$ $=(2\pi p)\,X$	$J_{load} + \left(\dfrac{W_{load}}{g}\right)\left(\dfrac{1}{n^2\eta}\right)$	$T_i + \dfrac{1}{n\eta}F_{load}$
Rack pinion 	$\dfrac{1}{r_p}$	≤ 1.0	J_{pinion} W_{rack}	$T_c.\mathrm{sgn}(\dot\theta_1)$ $+c\,\dot\theta_1$	X	W_{load}	F_{load}	$\theta_1 = n\,X$ $=(\dfrac{1}{r_p})\,X$	J_{pinion} $+ \left(\dfrac{W_{rack}}{g}\right)\left(\dfrac{1}{n^2\eta}\right)$ $+ \left(\dfrac{W_{load}}{g}\right)\left(\dfrac{1}{n^2\eta}\right)$	$T_i + \dfrac{1}{n^2\eta}F_{load}$
Conveyor belt 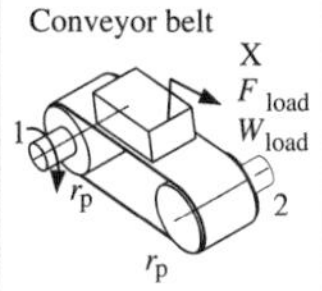	$\dfrac{1}{r_p}$	≤ 1.0	J_{p1} J_{p2} W_{belt}	$T_c.\mathrm{sgn}(\dot\theta_1)$ $+c\,\dot\theta_1$	X	W_{load}	F_{load}	$\theta_1 = n\,X$ $=(\dfrac{1}{r_p})\,X$	$J_{p1} + J_{p2}$ $+\left(\dfrac{W_{belt}+W_{load}}{g}\right)\left(\dfrac{1}{n^2\eta}\right)$	$T_i + \dfrac{1}{n^2\eta}F_{load}$

FIGURE 3.15: Commonly used motion conversion (gear) mechanisms and their input–output relationships.

4. Using the reflection equations developed above, calculate the reflected load inertia and torque/forces ($J_{1,\text{eff}}, T_{1,\text{eff}}$) that will effectively act on the actuator shaft as well as the desired motion at the actuator shaft ($\dot\theta_{\text{m}}(t) = \dot\theta_{\text{in}}(t)$).

5. Guess an actuator/motor inertia from an available list (or make the first calculation with zero motor inertia assumption), and calculate the torque history, $T_{\text{m}}(t)$, for the desired motion cycle. Then calculate the peak torque and RMS torque from $T_{\text{m}}(t)$.

6. Check if the actuator meets the required performance in terms of peak and RMS torque, and maximum speed capacity ($T_{\text{p}}, T_{\text{rms}}, \dot\theta_{\text{max}}$). If the above selected actuator/motor from the available list does not meet the requirements (i.e., too small or too large), repeat the previous step by selecting a different motor.

7. The continuous torque rating of most electric servo motors is given for 25 °C ambient temperature and an aluminum face mount for heat dissipation considerations. If the nominal ambient temperature is different form 25 °C, the continuous (RMS) torque capacity of the electric motor should be derated using the following equation for a temperature,

$$T_{\text{rms}} = T_{\text{rms}}(25\,^\circ\text{C})\sqrt{(155 - \text{Temp}^\circ\text{C})/130} \qquad (3.162)$$

If the T_{rms} rating is exceeded, the temperature of the motor winding will increase proportionally. If the temperature rise is above the rated temperature for the winding insulation of the motor, the motor will be damaged permanently.

3.6.1 Inertia Match Between Motor and Load

The ratio of the motor's rotor inertia and the reflected load inertia is always a concern in high performance motion control applications. It is a rule of thumb that the ratio of motor inertia to load inertia should be between one-to-one and up to one-to-ten,

$$\frac{J_m}{J_1/N^2} = \frac{1}{1} \sim \frac{1}{10} \tag{3.163}$$

The one-to-one match is considered the optimal match. Below, we show that the one-to-one inertial match is *optimal* only in the ideal case where the motor drives a purely inertial load and that this inertia ratio results in *minimum heating of the motor.*

Let us consider the case that the motor is coupled to a purely inertial load through an effective gear ratio. The torque and motion relationship is

$$T_m(t) = \left(J_m + \frac{1}{N^2}J_1 \right) \cdot \ddot{\theta}_m \tag{3.164}$$

$$= \left(J_m + \frac{1}{N^2}J_1 \right) \cdot N \cdot \ddot{\theta}_1 \tag{3.165}$$

Minimal heating occurs when the required torque is minimized, since torque is proportional to current and heat generation is related to current ($P_{elec} = R \cdot i^2$, where P_{elec} is the electric power dissipated at the motor windings due to its electrical resistance R and current i). The minimum torque occurs at the gear ratio where the derivative of T_m with respect to N is equal to zero,

$$\frac{d}{dN}(T_m) = \left(J_m + \frac{1}{N^2}J_1 \right) \ddot{\theta}_1 + \left(\frac{-2N}{N^4}J_1 \right) \cdot N\ddot{\theta}_1 \tag{3.166}$$

$$= \ddot{\theta}_1 \cdot \left(J_m - \frac{1}{N^2}J_1 \right) \tag{3.167}$$

$$= 0 \tag{3.168}$$

Therefore, the optimal gear ratio between the motor and a purely inertial load which minimizes the torque requirements (hence, the heating), is

$$J_m = \frac{1}{N^2}J_1 \tag{3.169}$$

$$= J_{1,reflected} \tag{3.170}$$

It is important to note that this ideal inertia match (1:1) between the motor's rotor inertia and reflected load inertia is optimal only for purely inertial loads. In applications where the load may be dominated by friction or other application related load torque or forces, the ideal inertia match may not be a good design.

Example   Consider a rotary motion axis driven by an electric servo motor. The rotary load is directly connected to the motor shaft without any gear reducer. The rotary load is a solid cylindrical shape made of steel material, $d = 3.0$ in, $l = 2.0$ in, $\rho = 0.286$ lb/in^3. The desired motion of the load is a periodic motion (Figure 3.14). The total distance to be traveled is $1/4$ of a revolution. The period of motion is $t_{cyc} = 250$ ms, and the dwell portion of it is $t_{dw} = 100$ ms, and the remaining part of the cycle time is equally divided between acceleration, constant speed, and deceleration periods, $t_a = t_r = t_d = 50$ ms. Determine the required motor size for this application.

This example matches the ideal model shown in Figure 3.14. The load inertia is

$$J_1 = \frac{1}{2} \cdot m \cdot (d/2)^2 \tag{3.171}$$

$$= \frac{1}{2} \cdot \rho \cdot \pi \cdot (d/2)^2 \cdot l \cdot (d/2)^2 \tag{3.172}$$

$$= \frac{1}{2} \cdot \rho \cdot \pi \cdot l \cdot (d/2)^4 \tag{3.173}$$

$$= \frac{1}{2} \cdot (0.286/386) \cdot \pi \cdot 2.0 \cdot (3/2)^4 \, \text{lb} \cdot \text{in} \cdot \text{s}^2 \tag{3.174}$$

$$= 0.0118 \, \text{lb} \cdot \text{in} \cdot \text{s}^2 \tag{3.175}$$

Let us assume that we will pick a motor which has a rotor inertia the same as the load so that there is an ideal load and motor inertia match, $J_m = 0.0118 \, \text{lb} \cdot \text{in} \cdot \text{s}^2$. The acceleration, top speed, and deceleration rates are calculated from the kinematic relationships,

$$\theta_a = \frac{1}{2} \dot{\theta} \cdot t_a = \frac{1}{4} \cdot \frac{\pi}{2} \tag{3.176}$$

$$\dot{\theta} = 2 \cdot \theta_a/t_a = 2 \cdot (1/4) \cdot (\pi/2) \, \text{rad}/(0.05 \, \text{s}) \tag{3.177}$$

$$\quad = 80\pi/16 \, \text{rad/s} = 40/16 \, \text{rev/s} = 2400/16 \, \text{rev/min} = 150 \, \text{rev/min} \tag{3.178}$$

$$\ddot{\theta}_a = \dot{\theta}_a/t_a = (80\pi/16)(1/0.05) = 1600\pi/16 \, \text{rad/s}^2 = 100 \, \pi \, \text{rad/s}^2 \tag{3.179}$$

$$\ddot{\theta}_r = 0.0 \tag{3.180}$$

$$\ddot{\theta}_d = -100 \, \pi \, \text{rad/s}^2 \tag{3.181}$$

$$\ddot{\theta}_{dw} = 0.0 \tag{3.182}$$

The required torque to move the load through the desired cyclic motion can be calculated as follows,

$$T_a = (J_m + J_1) \cdot \ddot{\theta} = (0.0118 + 0.0118) \cdot (100\pi) = 7.414 \, \text{lb} \cdot \text{in} \tag{3.183}$$

$$T_r = 0.0 \tag{3.184}$$

$$T_d = (J_m + J_1) \cdot \ddot{\theta} = (0.0118 + 0.0118) \cdot (-100\pi) = -7.414 \, \text{lb} \cdot \text{in} \tag{3.185}$$

$$T_{dw} = 0.0 \tag{3.186}$$

Hence, the peak torque requirement is

$$T_{max} = 7.414 \, \text{lb} \cdot \text{in} \tag{3.187}$$

and the RMS torque requirement is

$$T_{rms} = \left(\frac{1}{0.250} \left(T_a^2 \cdot t_a + T_r^2 \cdot t_r + T_d^2 \cdot t_d + T_{dw}^2 \cdot t_{dw} \right) \right)^{1/2} \tag{3.188}$$

$$= \left(\frac{1}{0.250} (7.414^2 \cdot 0.05 + 0.0 \cdot 0.05 + (-7.414)^2 \cdot 0.05 + 0.0 \cdot 0.1) \right)^{1/2} \tag{3.189}$$

$$= 4.689 \, \text{lb} \cdot \text{in} \tag{3.190}$$

Therefore, a motor which has rotor inertia of about $0.0118 \, \text{lb} \cdot \text{in} \cdot \text{s}^2$, maximum speed capability of 150 rev/min or better, peak and RMS torque rating in the range of 8.0 lb·in and 5.0 lb·in range would be sufficient for the task.

This design may be improved further by the following consideration. The top speed of the motor is only 150 rpm. Most electric servo motors run in the 1500 rpm to 5000 rpm range where they deliver most of their power capacity, while maintaining a constant torque capacity up to these speeds. As a result, it is reasonable to consider a gear reducer between the motor shaft and the load in the range of $10:1$ to $20:1$ and repeat the sizing calculations. This will result in a motor that will run at a higher speed and will have lower torque requirements.

3.7 HOMOGENEOUS TRANSFORMATION MATRICES

The geometric relationships in simple one degree of freedom mechanisms can be derived using basic vector algebra. The derivation of geometric relations for multi degrees of freedom mechanisms, such as robotic mechanisms, is rather difficult using three-dimensional vector algebra. The so called 4×4 homogeneous transformation matrices are very powerful matrix methods to describe the geometric relations [8]. They are used to describe the geometric relations of a mechanism between the absolute values of

1. displacement variables,
2. the relations between the incremental changes in displacements, and
3. force and torque transmission through the mechanism.

The position and orientation of a three-dimensional object with respect to a reference frame can be uniquely described by the position coordinates of a point on it (three-components of position information in three-dimensional space) and the orientation of it (described by three angles). The position coordinates are associated with a point and are unique for a given point with respect to a reference coordinate frame. Orientation is associated with an object, not a point. The best way to describe the position and orientation of an object is to attach a coordinate frame to the object, and describe the orientation and the origin coordinates of the attached coordinate frame with respect to a reference frame. For instance, the position and orientation of a tool held by the gripper of a robotic manipulator can be described by a coordinate frame attached to the tool (Figure 3.16). The position coordinates of the origin of the attached coordinate frame and its orientation with respect to another reference frame also describe the position and orientation of the tool on which the frame is attached.

The transformation of an object between any two different orientations can be accomplished by a sequence of three independent rotations. However, the number and sequence of rotation angles to go from one orientation to another is not unique. There are many possible rotation combinations to make a desired orientation change. For instance, an orientation change between any two different orientation of two coordinate frames can be accomplished by a sequence of three angles such as

1. roll, pitch, and yaw angles,
2. Z,Y,X Euler angles,
3. X,Y,Z Euler angles.

There are 24 different possible combinations of a sequence of three angles to go from one orientation to another. *Finite rotations are not commutative. Infinitesimal rotations are commutative.* That means, the order of a sequence of finite rotations makes a difference in the final orientation. For instance, making a 90° rotation about the x-axis followed by another 90° rotation about the y-axis results in a different orientation than that of making a 90° rotation about the y-axis followed by another 90° rotation about the x-axis. However, if the rotations are infinitesimal, the order does not matter.

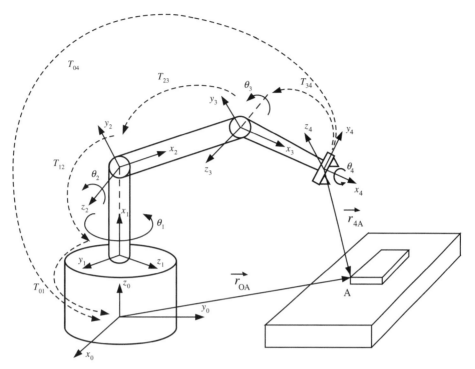

FIGURE 3.16: Multi degree of freedom mechanisms: a robotic manipulator with three joints. We use this example to illustrate how to describe the position and orientation of one coordinate frame with respect to another. If we attach a coordinate frame to a workpiece, we can describe the position and orientation of it with respect to other coordinate frames through the coordinate frame attached to it.

The 4x4 homogeneous transformation matrices describe the position of a point on an object and the orientation of the object in three-dimensional space using a 4x4 matrix. The first 3x3 portion of the matrix is used to define the orientation of a coordinate frame fixed to the object with respect to another reference coordinate frame. The last column of the matrix is used to describe the position of the origin of the coordinate frame fixed to the object with respect to the origin of the reference coordinate frame. The last row of the matrix is [0 0 0 1]. A general 4x4 homogeneous transformation matrix T has the following form (Figure 3.17),

$$T = \begin{bmatrix} e_{11} & e_{12} & e_{13} & x_A \\ e_{21} & e_{22} & e_{23} & y_A \\ e_{31} & e_{32} & e_{33} & z_A \\ 0 & 0 & 0 & 1 \end{bmatrix} \tag{3.191}$$

For instance, the second column $[e_{12} \ e_{22} \ e_{32}]^T$ is the cosines of the angles between the unit vector $\vec{e}_2$ of the coordinate frame A and the unit vectors $\vec{i}, \vec{j}, \vec{k}$ of the reference frame O (Figure 3.17). Column vectors are the cosine angles for the second coordinate frame unit vectors when they are expressed in terms of the unit vectors of the first coordinate frame. Row vectors are the cosine angles for the first coordinate frame unit vectors when they are expressed in terms of the unit vectors of the second coordinate frame.

They describe the position and orientation of the coordinate frame A with respect to the coordinate frame 0. The columns of the (3x3) portion of the matrix which contain the

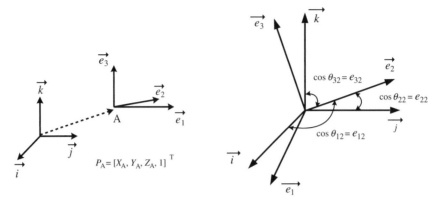

FIGURE 3.17: Use of (4x4) coordinate transformation matrices to describe the kinematic (geometric) relationships between different objects in three-dimensional space. A point is described with respect to a reference coordinate frame by its three position coordinates. An object is described by a coordinate frame fixed to it: the position of its origin and its orientation with respect to the reference coordinate frame.

orientation information are the cosine angles between the unit vectors of the coordinate frames. Notice that, even though we know that the orientation of one coordinate frame with respect to another can be described by three angles, the general form of the rotation portion of the (4x4) transformation matrix (the (3x3) portion) requires nine parameters. However, they are not all independent. There are six constraints between them, leaving three independent variables (Figure 3.17). The six constraints are

$$\vec{e}_1 \cdot \vec{e}_1 = 1.0 \tag{3.192}$$
$$\vec{e}_2 \cdot \vec{e}_2 = 1.0 \tag{3.193}$$
$$\vec{e}_3 \cdot \vec{e}_3 = 1.0 \tag{3.194}$$
$$\vec{e}_1 \cdot \vec{e}_2 = 0 \tag{3.195}$$
$$\vec{e}_1 \cdot \vec{e}_3 = 0 \tag{3.196}$$
$$\vec{e}_2 \cdot \vec{e}_3 = 0 \tag{3.197}$$

where

$$\vec{e}_i = e_{1i}\vec{i} + e_{2i}\vec{j} + e_{3i}\vec{k}; \quad i = 1, 2, 3 \tag{3.198}$$

are the unit vectors along each of the axes of the attached coordinate frame expressed in terms of its components in the unit vectors of the other coordinate frame. The use of cosine angles in describing the orientation of one coordinate frame with respect to another one is a very convenient way to determine the elements of the matrix.

The 4x4 homogeneous transformation matrices are the most widely accepted and powerful (if not the computationally most efficient) method to describe kinematic relations. The algebra of transformation matrices follows basic matrix algebra. Let us consider the coordinate frames numbered 1, 2, 3, 4 and a point A on the object where the position coordinates of the point with respect to the third coordinate frame are described by r_{4A} (Figure 3.17). The description of the coordinate frame 4 (position of its origin and orientation) with respect to coordinate frame 0 can be expressed as,

$$T_{04} = T_{01} \cdot T_{12} \cdot T_{23} \cdot T_{34} \tag{3.199}$$

where T_{01}, T_{12}, T_{23}, and T_{34} are the description of the origin position coordinates and orientations of the axes of coordinate frames 1 with respect to 0, 2 with respect to 1, and that of 3 with respect to 2. The position coordinate vector of point A can be expressed with respect to coordinates 2 and 3 as follows (Figure 3.17),

$$r_{3A} = T_{34} \cdot r_{4A} \tag{3.200}$$

$$r_{2A} = T_{23} \cdot r_{3A} = T_{23} \cdot T_{34} \cdot r_{4A} \tag{3.201}$$

$$r_{1A} = T_{12} \cdot r_{2A} = T_{12} \cdot T_{23} \cdot T_{34} \cdot r_{4A} \tag{3.202}$$

$$r_{0A} = T_{01} \cdot r_{1A} = T_{01} \cdot T_{12} \cdot T_{23} \cdot T_{34} \cdot r_{4A} \tag{3.203}$$

where, $r_{4A} = [\,x_{4A} \ \ y_{4A} \ \ z_{4A} \ \ 1\,]^{\mathrm{T}}$. The $r_{3A}, r_{2A}, \ r_{1A}, \ r_{0A}$ are similarly defined. Notice that T_{12} is the description (position coordinates of the origin and the orientation of the axes of coordinate system 2 with respect to coordinate system 1) of coordinate system 2 with respect to coordinate system 1. Then the reverse description, that is the description of coordinate system 1 with respect to coordinate system 2, is the inverse of the previous transformation matrix,

$$T_{21} = T_{12}^{-1} \tag{3.204}$$

The (4x4) transformation matrix has a special form and the inversion of the matrix also has a special result. Let

$$T_{12} = \begin{bmatrix} R_{12} & p_A \\ - - - - - - & \\ 0 \ \ 0 \ \ 0 & 1 \end{bmatrix} \tag{3.205}$$

Then, the inverse of this matrix can be shown as

$$T_{12}^{-1} = \begin{bmatrix} R_{12}^{\mathrm{T}} & -R_{12}^{\mathrm{T}} \cdot p_A \\ - - - - - - & \\ 0 \ \ 0 \ \ 0 & 1 \end{bmatrix} \tag{3.206}$$

Also notice that the order of transformations is important (multiplication of matrices are dependent on the order),

$$T_{12} \cdot T_{23} \neq T_{23} \cdot T_{12} \tag{3.207}$$

For a general purpose multi degrees of freedom mechanism, such as a robotic manipulator, the relationship between a coordinate frame attached to the tool (the coordinates of its origin and orientation) with respect to a fixed reference frame at the base can be expressed as a sequence of transformation matrices where each transformation matrix is a function of one of the axis position variables. For instance, for a four degrees of freedom robotic manipulator, the coordinate system at the wrist joint can be described with respect to base as (Figure 3.17)

$$T_{04} = T_{01}(\theta_1) \cdot T_{12}(\theta_2) \cdot T_{23}(\theta_3) \cdot T_{34}(\theta_4) \tag{3.208}$$

where $\theta_1, \theta_2, \theta_3, \theta_4$ are the positions of axes driven by motors. Any given position vector relative to the fourth coordinate frame (r_{4A}) can be expressed with respect to the base as follows,

$$r_{0A} = T_{04} \cdot r_{4A} \tag{3.209}$$

$$= T_{01}(\theta_1) \cdot T_{12}(\theta_2) \cdot T_{23}(\theta_3) \cdot T_{34}(\theta_4) \cdot r_{4A} \tag{3.210}$$

where $r_{4A} = [\,x_{4A}, \ y_{4A}, \ z_{4A}, \ 1\,]^{\mathrm{T}}$ the three coordinates of the point A with respect to the coordinate frame 4.

The Denavit-Hartenberg method [7] defines a standard way of attaching coordinate frames to a robotic manipulator such that only four numbers (one variable, three constant parameters) are needed per one degree of freedom joint to represent the kinematic relationships.

Given the 4x4 transformation matrix description of a coordinate frame attached to the tool, T_{0T}, with respect to a reference coordinate frame 22, we can determine the position coordinates of the tool-coordinate frame (x_T, y_T, z_T) as well as orientation angles (i.e., roll, yaw, pitch) of the tool-coordinate frame with respect to the reference coordinate frame and form the corresponding $\underline{x}$ vector.

In generic terms, the relationship between the coordinates of the tool and the joint displacement variables can be expressed as,

$$\underline{x} = \underline{f}(\underline{\theta}) \tag{3.211}$$

The vector variable $\underline{x}$ represents the Cartesian coordinates of the tool (i.e., position coordinates x_P, y_P, z_P in a given coordinate frame and oriention angles where three angles can be used to describe the orientation). The description of the position coordinates of a point with respect to a given reference frame is unique. However, the orientation of an object with respect to a reference coordinate frame can be described by many different possible combinations of angles. Hence, the orientation description is not unique. The vector variable $\underline{\theta}$ represents the joint variables of the robotic manipulator, that is for a six joint robot $\underline{\theta} = [\theta_1, \theta_2, \theta_3, \theta_4, \theta_5, \theta_6]^T$.

The $\underline{f}(\underline{\theta})$ is called the *forward kinematics* of the mechanism, which is a vector nonlinear function of joint variables,

$$\underline{f}(\underline{\theta}) = [f_1(\underline{\theta}), f_2(\underline{\theta}), f_3(\underline{\theta}), f_4(\underline{\theta}), f_5(\underline{\theta}), f_6(\underline{\theta})]^T \tag{3.212}$$

The inverse relationship, that is the geometric function which defines the axis positions as functions of tool position and orientation, is called the *inverse kinematics* of the mechanism,

$$\underline{\theta} = \underline{f}^{-1}(\underline{x}) \tag{3.213}$$

The inverse kinematics function may not be possible to express in one analytical closed form for every mechanism. It must be determined for each special mechanism on a case by case basis. For a six revolution joint manipulator, a sufficient condition for the existence of the inverse kinematic solution in analytical form is that three consecutive joint axes must intersect at a point. Forward and inverse kinematic functions of a mechanism relate the joint positions to the tool positions.

If a robotic manipulator is always taught (i.e., by using a "teach mode") the points the tool tip must go through in three-dimensional space, the controller can record the corresponding joint angles at each point. Then, some form of interpolation can be used to define intermediate points to generate the desired motion path. In this type of robotic manipulator, where the desired motion of the manipulator is defined by a teach mode, inverse kinematics are never needed. However, in applications where not all of the target points are taught, inverse kinematics are needed. For instance, target tool tip position may need to be determined in real-time based on a vision sensor measurement of the target workpiece position and orientation. Then, the corresponding joint angles must be calculated using inverse kinematic relations.

The *differential* relationships between joint axis variables and tool variables are obtained by taking the differential of the forward kinematic function. The resultant matrix that relates the differential values of joint and tool position variables (in other words, it relates the velocities of joint axes and tool velocity) is called the *Jacobian matrix* of the mechanism (Figure 3.17).

$$\underline{\dot{x}} = \frac{df(\underline{\theta})}{d\underline{\theta}} \cdot \underline{\dot{\theta}} \tag{3.214}$$

$$= J\underline{\dot{\theta}} \tag{3.215}$$

where the J matrix is called the Jacobian of the mechanism. Each element of the Jacobian matrix is defined as

$$J_{ij} = \frac{\partial f_i(\theta_1, \theta_2, \theta_3, \theta_4, \theta_5, \theta_6)}{\partial \theta_j} \tag{3.216}$$

where $i = 1, 2, ..., m$ and $j = 1, 2, .., n$, where n is the number of joint variables. For a six degrees of freedom mechanism, $n = m = 6$. If the mechanism has less than six degrees of freedom, the Jacobian matrix is not a square matrix. Likewise, the inverse of the Jacobian, J^{-1} relates the changes in the tool position to the changes in the axis displacements,

$$\underline{\dot{\theta}} = J^{-1} \cdot \underline{\dot{x}} \tag{3.217}$$

If the Jacobian is not invertible in certain positions, these positions are called the *geometric singularities* of the mechanism. It means that at these locations, there are some directions that the tool cannot move no matter what the change is in the joint variables. In other words, no joint axis variable combination can generate a motion in certain directions at a singularity point. A robotic manipulator may have many singularity points in its workspace. The geometric singularity is directly a function of the mechanical configuration of the manipulator. There are two groups of singularities,

1. Workspace boundary: a given manipulator has a finite span in three-dimensional space. The locations that the manipulator can reach are called the *workspace*. At the boundary of workspace, the manipulator tip cannot move out, because it has reached its limits of reach. Hence, all points in the workspace boundary are singularity points since at these points there are directions along which the manipulator tip cannot move.

2. Workspace interior points: these singularity points are inside the workspace of the manipulator. Such singularity points depend on the manipulator geometry and generally occur when two or more joints line up.

The same Jacobian matrix also describes the relationship between the torques/forces at the controlled axes and the force/torque experienced at the tool. Let the tool force be *Force* and the corresponding tool position differential displacement be $\delta \underline{x}$. The differential work done is

$$Work = \underline{\delta x}^T \cdot \underline{Force} \tag{3.218}$$

Note that the Jacobian relationship

$$\underline{\delta x} = J \cdot \underline{\delta \theta} \tag{3.219}$$

and the equivalent work done by the corresponding torques at the controlled axes can be expressed as

$$Work = \underline{\delta \theta}^T \cdot \underline{Torque} \tag{3.220}$$

$$= \delta x^T \cdot Force \tag{3.221}$$

$$= (J \cdot \delta \theta)^T \cdot Force \tag{3.222}$$

Hence, the force–torque relationship between the tool and joint variables is

$$\underline{Torque} = J^T \cdot \underline{Force} \tag{3.223}$$

and the inverse relationship is

$$\underline{Force} = (J^T)^{-1} \cdot \underline{Torque} \tag{3.224}$$

Notice that at the *singular configurations* (geometric singularities) of the mechanism, those configuration of the mechanism at which the inverse of the Jacobian matrix does not exist, there are some force directions at the tool which do not result in any change in the axes, torques. They only result in reaction forces in the linkage structure, but not in the actuation axes. Another interpretation of this result is that there are some directions of tool motion where we cannot generate force no matter what combination of torques are applied at the joints.

There are different methods for the calculation of the Jacobian matrix for a mechanism [9,10]. The inverse Jacobian matrix can be either obtained analytically in symbolic form or calculated numerically off-line or on-line (in real-time). However, real-time numerical inverse calculations present a problem both in terms of the computational load and also the possible numerical stability problems around the singularities of the mechanism. The decision regarding the Jacobian matrix and its inverse computations in real-time should be made on a mechanism by mechanism basis.

Example Consider two coordinate frames numbered 1 and 2 as shown in Figure 3.18. Let the origin coordinates of the second coordinate frame have the following coordinates, $r_{1A} = [x_{1A} \ y_{1A} \ z_{1A} \ 1]^T = [-0.5 \ 0.5 \ 0.0 \ 1]^T$. Orientations of axes are such that X_2 is parallel to Y_1, Y_2 is parallel but in opposite direction to X_1, and Z_2 has the same direction as Z_1. Determine the vector description of point P whose coordinates are given in second frame as $r_{2P} = [x_{2P} \ y_{2P} \ z_{2P} \ 1]^T = [1.0 \ 1.0 \ 0.0 \ 1]^T$.

Using the homogeneous transformation matrix relationship between coordinate frames 1 and 2,

$$r_{1P} = T_{12} \cdot r_{2P} \tag{3.225}$$

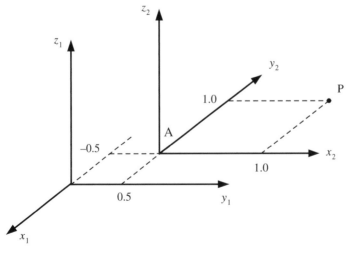

FIGURE 3.18: Describing the position and orientation of one coordinate frame with respect to an other.

where T_{12} is described by the orientation of the second coordinate frame and position coordinates of its origin with respect to the first coordinate frame.

$$T_{12} = \begin{bmatrix} e_{11} & e_{12} & e_{13} & x_{1A} \\ e_{21} & e_{22} & e_{23} & y_{1A} \\ e_{31} & e_{32} & e_{33} & z_{1A} \\ 0 & 0 & 0 & 1 \end{bmatrix} \tag{3.226}$$

$$= \begin{bmatrix} 0.0 & -1.0 & 0.0 & -0.5 \\ 1.0 & 0.0 & 0.0 & 0.5 \\ 0.0 & 0.0 & 1.0 & 0.0 \\ 0 & 0 & 0 & 1 \end{bmatrix} \tag{3.227}$$

Notice that the orientation portion of the matrix is the coefficients of the relationships between the unit vectors,

$$\vec{i} = \cos\theta_{11} \cdot \vec{e}_1 + \cos\theta_{12} \cdot \vec{e}_2 + \cos\theta_{13} \cdot \vec{e}_3 \tag{3.228}$$

$$= e_{11} \cdot \vec{e}_1 + e_{12} \cdot \vec{e}_2 + e_{13} \cdot \vec{e}_3 \tag{3.229}$$

$$= 0.0 \cdot \vec{e}_1 + (-1.0) \cdot \vec{e}_2 + 0.0 \cdot \vec{e}_3 \tag{3.230}$$

$$\vec{j} = \cos\theta_{21} \cdot \vec{e}_1 + \cos\theta_{22} \cdot \vec{e}_2 + \cos\theta_{23} \cdot \vec{e}_3 \tag{3.231}$$

$$= e_{21} \cdot \vec{e}_1 + e_{22} \cdot \vec{e}_2 + e_{23} \cdot \vec{e}_3 \tag{3.232}$$

$$= 1.0 \cdot \vec{e}_1 + (0.0) \cdot \vec{e}_2 + (0.0) \cdot \vec{e}_3 \tag{3.233}$$

$$\vec{k} = \cos\theta_{31} \cdot \vec{e}_1 + \cos\theta_{32} \cdot \vec{e}_2 + \cos\theta_{33} \cdot \vec{e}_3 \tag{3.234}$$

$$= e_{31} \cdot \vec{e}_1 + e_{32} \cdot \vec{e}_2 + e_{33} \cdot \vec{e}_3 \tag{3.235}$$

$$= (0.0) \cdot \vec{e}_1 + (0.0) \cdot \vec{e}_2 + (1.0) \cdot \vec{e}_3 \tag{3.236}$$

Hence,

$$r_{1P} = T_{12} \cdot r_{2P} \tag{3.237}$$

$$= [-1.5 \ 1.5 \ 0.0 \ 1]^T \tag{3.238}$$

Example   The purpose of this example is to illustrate graphically that the order of a sequence of finite rotations is important. If we change the order of rotations, the final orientation is different (Figure 3.19). In other words, $T_1 \cdot T_2 \neq T_2 \cdot T_1$. Let T_1 represent a rotation about the x-axis by $90°$, and T_2 represent a rotation about the y-axis by $90°$. Figure 3.19 shows the sequence of both T_1 followed by T_2 and T_2 followed by T_1. The resulting final orientations are different since the order of rotations are different. Finite rotations are not commutative. We can show this algebraically for this case.

$$T_1 = \begin{bmatrix} 1.0 & 0.0 & 0.0 & 0.0 \\ 0.0 & 0.0 & 1.0 & 0.0 \\ 0.0 & -1.0 & 0.0 & 0.0 \\ 0 & 0 & 0 & 1 \end{bmatrix} \tag{3.239}$$

$$T_2 = \begin{bmatrix} 0.0 & 0.0 & -1.0 & 0.0 \\ 0.0 & 1.0 & 0.0 & 0.0 \\ 1.0 & 0.0 & 0.0 & 0.0 \\ 0 & 0 & 0 & 1 \end{bmatrix} \tag{3.240}$$

Clearly, $T_1 \cdot T_2 \neq T_2 \cdot T_1$.

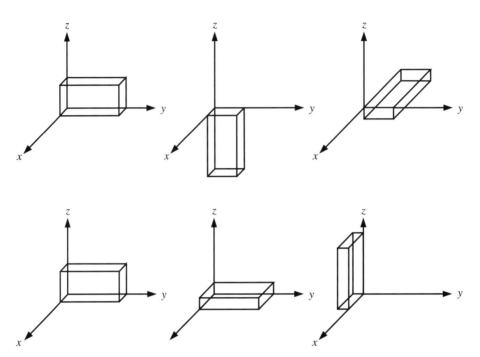

FIGURE 3.19: Two finite rotation sequence performed in reverse order. The final orientations are different. Finite rotations are not commutative.

Example   Consider the geometry of a two-link robotic manipulator (Figure 3.20). The geometric parameters (link lengths l_1 and l_2) and joint variables are shown in the figure.

1. Derive the tip position P coordinates as a function of joint variables (forward kinematic relations).
2. Derive the Jacobian matrix that relates the joint velocities to the tip position coordinate velocities.

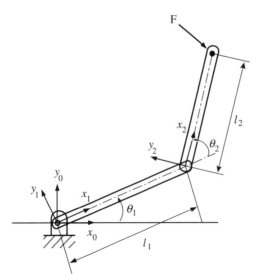

FIGURE 3.20: Kinematic description of a two joint planar robotic manipulator.

3. If there is a load with weight W at the tip, determine the necessary torques at joints 1 and 2 in order to balance the load.

In this example, we are asked to determine the following relations,

$$\underline{x} = \underline{f}(\underline{\theta}) \tag{3.241}$$

$$\underline{\dot{x}} = J(\theta)(\underline{\dot{\theta}}) \tag{3.242}$$

$$\underline{Torque} = J^{\mathrm{T}}(\theta)(\underline{Force}) \tag{3.243}$$

Let us attach three coordinate frames to the manipulator. Coordinate frame 0 is attached to the base and fixed, coordinate frame 1 is attached at joint 1 to link 1 (moves with link 1), coordinate frame 2 is attached at joint 2 to link 2 (moves with link 2). The position vector of the tip with respect to coordinate frame 2 is simple, $r_{2\mathrm{P}} = [\, l_2 \;\; 0 \;\; 0 \;\; 1 \,]^{\mathrm{T}}$. The transformation matrices between the three coordinate frames are functions of θ_1 and θ_2 as follows,

$$T_{01} = \begin{bmatrix} \cos(\theta_1) & -\sin(\theta_1) & 0.0 & 0.0 \\ \sin(\theta_1) & \cos(\theta_1) & 0.0 & 0.0 \\ 0.0 & 0.0 & 1.0 & 0.0 \\ 0 & 0 & 0 & 1 \end{bmatrix} \tag{3.244}$$

$$T_{12} = \begin{bmatrix} \cos(\theta_2) & -\sin(\theta_2) & 0.0 & l_1 \\ \sin(\theta_2) & \cos(\theta_2) & 0.0 & 0.0 \\ 0.0 & 0.0 & 1.0 & 0.0 \\ 0 & 0 & 0 & 1 \end{bmatrix} \tag{3.245}$$

Hence, the tip vector description in base coordinates,

$$r_{0\mathrm{P}} = [x_{0\mathrm{P}} \;\; y_{0\mathrm{P}} \;\; 0 \;\; 1]^{\mathrm{T}} = T_{01} \cdot T_{12} \cdot r_{2\mathrm{P}} \tag{3.246}$$

The vector components of $r_{0\mathrm{P}}$ can be expressed as individual functions for more clarity,

$$x_{0\mathrm{P}} = x_{0\mathrm{P}}(\theta_1, \theta_2; l_1, l_2) = l_1 \cdot \cos(\theta_1) + l_2 \cdot \cos(\theta_1 + \theta_2) \tag{3.247}$$

$$y_{0\mathrm{P}} = y_{0\mathrm{P}}(\theta_1, \theta_2; l_1, l_2) = l_1 \cdot \sin(\theta_1) + l_2 \cdot \sin(\theta_1 + \theta_2) \tag{3.248}$$

$$z_{0\mathrm{P}} = 0.0 \tag{3.249}$$

The Jacobian matrix for this case can simply be determined by taking the derivative of the forward kinematic relations with respect to time and expressing the equation in the matrix form to find the Jacobian matrix.

$$\dot{x}_{0\mathrm{P}} = \frac{d}{dt}(x_{0\mathrm{P}}(\theta_1, \theta_2; l_1, l_2)) = J_{11}\dot{\theta}_1 + J_{12}\dot{\theta}_2 \tag{3.250}$$

$$\dot{y}_{0\mathrm{P}} = \frac{d}{dt}(y_{0\mathrm{P}}(\theta_1, \theta_2; l_1, l_2)) = J_{21}\dot{\theta}_1 + J_{22}\dot{\theta}_2 \tag{3.251}$$

where it is easy to show that the elements of the Jacobian matrix J

$$J = \begin{bmatrix} J_{11} & J_{12} \\ J_{21} & J_{22} \end{bmatrix} \tag{3.252}$$

$$= \begin{bmatrix} -l_1 \cdot \sin(\theta_1) - l_2 \cdot \sin(\theta_1 + \theta_2) & -l_2 \cdot \sin(\theta_1 + \theta_2) \\ l_1 \cdot \cos(\theta_1) + l_2 \cdot \cos(\theta_1 + \theta_2) & l_2 \cdot \cos(\theta_1 + \theta_2) \end{bmatrix} \tag{3.253}$$

The torque needed to balance a weight load, $F = [F_x \ F_y]^T = [0 \ -W]^T$, at the tip is determined by

$$\begin{bmatrix} Torque_1 \\ Torque_2 \end{bmatrix} = \begin{bmatrix} J_{11} & J_{21} \\ J_{12} & J_{22} \end{bmatrix} \begin{bmatrix} 0.0 \\ -W \end{bmatrix} \tag{3.254}$$

which shows the necessary static torque at each joint to balance a weight at the tip for different positions of the manipulator. Notice that since the Jacobian matrix is 2x2, it is relatively simple to obtain the inverse Jacobian in analytical form.

$$J^{-1} = \frac{1}{l_1 \cdot l_2 \cdot \sin(\theta_2)} \begin{bmatrix} l_2 \cos(\theta_1 + \theta_2) & l_2 \cdot \sin(\theta_1 + \theta_2) \\ -l_1 \cos(\theta_1) - l_2 \cos(\theta_1 + \theta_2) & l_1 \sin(\theta_1) - l_2 \cdot \sin(\theta_1 + \theta_2) \end{bmatrix}$$
$$\tag{3.255}$$

Notice that when $\theta_2 = 0.0$, the mechanism is at a singular point, which is indicated by the $\sin(\theta_2)$ term in the denominator of the inverse Jacobian equation above.

3.8   A CASE STUDY: AUTOMOTIVE TRANSMISSION AS A "GEAR REDUCER"

An automotive transmission is a gear reducer. It changes the speed ratio (gear ratio) between the input and output shaft. The input shaft is connected to the crank shaft of the engine, the output shaft is connected to the lower powertrain which is connected to the differential (or differentials in four wheel drive vehicles) to drive the wheels. The gear ratio is not constant, but can be one of a finite number of gear ratios (i.e., in five speed transmission, five different the gear ratios can be engaged). In continuously variable transmissions, the gear ratio can be changed to any value between a minimum and a maximum value, instead of one of a finite number of selections. The mechanism by which the gear ratio is changed from one ratio to another determines the type of transmission. In manual shift transmissions, the gear is changed by the operator by moving a lever. In automatic transmissions, the gear is changed by an electronic control module (ECM) which controls a set of electrically actuated clutch/brakes or valves. In continuously variable transmissions (CVT), i.e. toroidal CVT, the gear ratio is changed by continuously varying a drive diameter, hence the gear ratio can be any value between a minimum and maximum value (hence the name "continuously varying"), instead of having a finite number of discrete gear ratio numbers, that is, 5-speed (or 6-speed) transmission meaning there are 5 (or 6) gear ratios to select from.

3.8.1   The Need for a Gearbox "Transmission" in Automotive Applications

A gearbox is used to change the speed and torque ratio between the input and output shafts. If we assume that the gearbox operates with 100% efficiency, the input power and output power are equal to each other. As the gearbox changes the speed ratio between input–output shafts, it also changes the torque ratio in the inverse ratio.

In automotive and mobile equipment motion, the power output from engine is defined as a function of speed using the so called "lug-curve" (see Section 1.2). At a given speed, the maximum power capacity of an engine is known. Hence, the maximum available torque output from the engine is known as a function of engine speed. The gearbox (better known as the transmission in automotive applications) essentially scales that lug-curve at different gear ratios as shown in Figure 3.24, excluding the effect of the torque converter. A typical

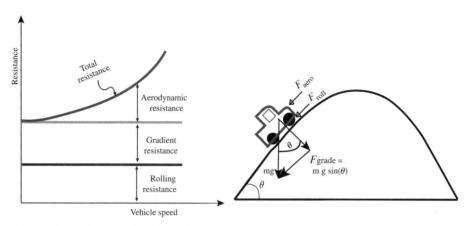

FIGURE 3.21: Resistive load force (or equivalent torque) in mobile equipment and automotive applications: aerodynamic, gradient, and rolling resistance forces.

transmission has five forward gears. Hence, there are five "the transmission" output speed–torque curves which are scaled versions of the engine lug curve. A continuously variable transmission (CVT) can provide any gear ratio value between a minimum and maximum gear ratio value it is designed for, not just a finite number of discrete gear ratios. Therefore, a CVT can achieve an infinite number of different scaled versions of the lug-curve between the minimum and maximum scaled curves (i.e., between gear ratio 1 and gear ratio 5). The set of lug-curves at different gear ratios shows the maximum torque the engine–transmission combination can generate at various speed ranges at the transmission output shaft. The gear ratio allows us to trade torque for speed, that is reduce speed and increase torque between input and output shafts or vice versa. If the demand for torque changes due to the operating conditions of the vehicle, we can change the gear ratio to select a different speed–torque curve capability in order to better meet the demand.

In mobile equipment applications, there are three major categories of resistance to motion that must be overcome by the engine–transmission output shaft. The resistance categories are:

- *Aerodynamic resistance force* due to the friction between the mobile equipment body and air. This force, and the equivalent torque that the engine must overcome, is a function of the equipment shape and its aerodynamic properties. In general, the aerodynamic resistance force grows almost exponentially with the speed of the equipment (Figure 3.21),

$$F_{\text{aero}} = F_{\text{aero}}(V_{\text{vehicle}}) \qquad (3.256)$$

- *Gravity resistance force* (also called *gradient force*) is the force acting on the vehicle due to the gravity force component in non-zero gradients. This force is equal to the weight of the vehicle times the sinus of the gradient angle,

$$F_{\text{gradient}} = W_{\text{vehicle}} \cdot \sin(\theta) \qquad (3.257)$$

where θ is the gradient angle. If the vehicle is moving up in the gradient, it is a resisting force, if it is moving down, it is an assisting force.

- *Rolling resistance force* is due to the continuous deformation of the tires as the vehicle moves. In general this force is a function of the tire pressure, vehicle mass, and ground traction conditions. It does not vary much as a function of the vehicle speed.

$$F_{\text{roll}} \approx Constant \qquad (3.258)$$

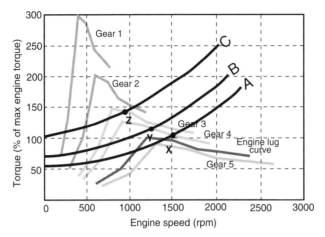

FIGURE 3.22: Engine lug curve between torque–speed, and scaled versions of it at different gear ratios of the transmission (assuming 100% transmission efficiency), which shows the speed and torque profile that is delivered at the transmission output shaft, over laid with resistance curves to determine steady-state operating conditions.

Total resistance force for a given vehicle mass, tire pressure, ground traction condition, a gradient, and vehicle speed can be expressed as,

$$F_{resist} = F_{aero}(V_{vehicle}) + F_{gradient} + F_{roll} \tag{3.259}$$

The total resistance force is determined by the three components as shown in Figure 3.21. For different gradients and traction conditions, we would have different curves (i.e. A, B, C) as shown in Figure 3.22. As a vehicle changes gradient and ground conditions while it is traveling, the total resistance to it changes between these curves. If the vehicle is desired to maintain a certain speed despite these variations, the output torque (or equivalent force delivered to the tires) must be large enough to overcome the total resistance.

Therefore, as the operating conditions change, the gear ratio must be changed in order to meet the demand. The gear ratio decision can be made either by the driver or by a control computer. The main principle is to overlay the torque–speed curves for different gear ratios over the actual resistance load curves. Let us assume that the vehicle is operating in gear 4 at full throttle along the load curve A, at steady-state speed corresponding to about 1500 rpm at the engine output shaft (point X), where the torque output from engine–transmission is equal to the load. Consider that the road gradient changes and that now the load curve is the curve B. At that speed, the load is larger than the torque capacity of the engine–transmission at the current gear. If no gear change is made, the vehicle speed (and engine speed) will decrease until the load curve and torque curve intersects at a point (point Y). At that point engine speed is reduced to about 1250 rpm. If the gradient further increases to load curve C, the engine–transmission cannot maintain the vehicle speed at the current gear ratio, because the maximum output torque capacity at the current gear (gear 4) is always smaller than the load. If the gear is not reduced, the vehicle stops and the engine stalls. The only way to keep the vehicle moving is to reduce the gear ratio to gear 3 (or a lower gear) so that the torque output capacity of the engine can meet the load, hence the engine speed stabilizes at the steady-state speed of about 950 rpm (point Z).

3.8.2 Automotive Transmission: Manual Shift Type

A typical automotive manual transmission has five gear ratios. Figure 3.23 shows a view of a manual transmission concept. The input shaft is the crank shaft from the engine. The

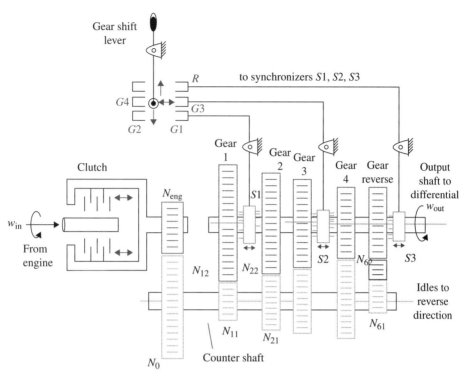

FIGURE 3.23: Manual transmission used in automotive applications: operating principle of a manual transmission (five forward gears, one reverse gear), (Source: How Things Work). Automatic shift version of the manual transmission, the difference between "manual shift" and "automatic shift" versions is in the way the synchronizers are moved. In the manual shift version, the synchronizers are moved manually via a mechanical linkage mechanism. In the automatic shift version, the synchronizers are moved by an electronically controlled actuation mechanism (i.e., an electro-hydraulic circuit which controls the flow of a pressurized fluid to move the synchronizers as well as engage/disengage clutches).

counter-shaft has seven gears: five gears for five forward gear ratios, one gear for reverse gear, and one gear on the left to couple the counter shaft to the input shaft. The output shaft similarly has six gears: five gears for five forward gears and one for reverse gear. The output shaft is connected to the differential which drives the wheels. There is also a clutch between the engine crank shaft and the input shaft to connect and disconnect the two shafts temporarily from each other during gear shifts. The gears on the counter-shaft are rigidly connected to it and rotate at the same speed with the counter-shaft. Whereas the gears on the output shaft are free-spining. A gear can be rigidly coupled to the output shaft using the "synchronizers." The synchronizers rotate at the same speed with the output shaft, but they are grooved on the shaft so that they can slide along the shaft while rotating. The idler gear is used to reverse the direction of rotation in the connection between the counter-shaft gear and the reverse gear.

The gear selector fork is operated by the driver. It has three fingers (arms) to push. Each finger operates a fork which is connected to a synchronizer. Each synchronizer can be connected to one of two adjacent gears, and hence can be used to select one of two gears. As a result, this design can support six different gear ratios (five forward and one reverse) using three synchronizers. When none of the synchronizers are engaged, the *neutral gear* is selected, and no power is transmitted from counter-shaft to the output shaft.

When the driver wants to change gear,

1. First, the clutch is disengaged so the engine does not power the counter-shaft.

2. Then the gear selector lever is moved to the desired position which then moves the synchronizer into contact with the selected gear on the output shaft. Hence, that gear is now connected to the output shaft rigidly, instead of free-spinning. The process of moving the synchronizer to engage a selected gear takes some time. Before the clutch is re-engaged, the synchronizer couples with the selected gear, and then brings its speed to the same speed of the output shaft, and locks it to the output shaft.

3. After that, the clutch should be re-engaged. If this speed synchronization does not happen well with proper sequence and timing (i.e., the clutch is re-engaged too early, before the synchronizer brings the selected gear to output shaft speed and locks it in), then there will be some slip and audible noise in the gear shift until the selected gear is fully coupled with the output shaft. On the other hand, if the clutch is re-engaged too late, the vehicle will lose speed due to too operating without a power source for too long.

The *automatic shift* versions of this type of transmission are commonly used in agricultural equipment and automotive applications. The difference in the design is in the mechanism by which the gears are engaged and disengaged. In the manual version, this is done by the driver using a shift lever mechanism which physically moves the synchronizers with a fork mechanism to engage/disengage desired gears. In the automatic version, there are two main components,

1. An electronic control module (ECM) makes the decisions on which gear to shift, based on various sensor signals (i.e., operator pedal position, engine speed, transmission output speed), and sends the corresponding command signals (i.e., current) to the actuation devices (i.e., valve solenoids).

2. An actuation mechanism that would include an electro-hydraulic (or electro-pneumatic or electric) solenoid actuated valves, cylinders, and hydraulic supply pressure in order to move the selected synchronizer to the desired gear engagement position. Cylinders in this type of applications are so small and have a very short stroke, they are often referred as "pistons."

In other words, the decision to shift the gear is made by the ECM, the actual motion to make the shift is accomplished by a motion control system (which is controlled by the ECM) which may be an electro-hydraulic, electro-pneumatic, or all electric circuit. A common feature of any automatic shift transmission is that the components of the shift motion mechanism are typically an electro-hydraulic system which has:

- hydraulic supply/return: hydraulic supply pressure line from a pump and return line to oil reservoir,
- valves: solenoid actuated valves to control the flow or pressurized fluid into the "actuators,"
- actuators: small pistons/cylinders which are moved by the flow from the valves and then the pistons are used to engage/disengage the selected clutch or brakes or move synchronizers,
- clutch/brakes or synchronizers: highly integrated compact disc clutches and brakes are built into the transmission frame.

At any given time, the transmission is in either neutral position (none of the gears are engaged) or in one of the six gear positions (5 forward, 1 reverse). The gear ratio at any gear

$$N_i = \frac{w_{out}}{w_{in}} \tag{3.260}$$

$$N_i = \frac{N_{eng}}{N_0} \cdot \frac{N_{i1}}{N_{i2}}; \quad i = 1, 2, 3, 4, 5 \tag{3.261}$$

$$N_r = -\frac{N_{eng}}{N_0} \cdot \frac{N_{61}}{N_{62}}; \quad \textit{reverse gear ratio} \tag{3.262}$$

where N_i's represent the number of teeth of each gear. Similarly, the torque output delivered to the output shaft at any gear is

$$P_{out} = \eta_{eff} \cdot P_{in} \tag{3.263}$$

$$T_{out} \cdot w_{out} = \eta_{eff} \cdot T_{eng} \cdot w_{eng} \tag{3.264}$$

$$w_{out} = N_i \cdot w_{eng} \tag{3.265}$$

$$T_{out} = \eta_{eff} \cdot \frac{1}{N_i} \cdot T_{eng} \tag{3.266}$$

where η_{eff} is the efficiency of the transmission, which is in the range of 85 to 97% (0.85 to 0.97) range, $w_{eng} = w_{in}$ input speed is same as the engine crank shaft speed. Assuming 100% efficiency for the transmission,

$$T_{out} = \frac{1}{N_i} \cdot T_{eng} \tag{3.267}$$

Notice that as the gear number in automotive applications increases (i.e., gear 1, gear 2, gear 3, gear 4, gear 5), the gear ratio value N_i increases, that is $N_1 = 0.3$, $N_2 = 0.5$, $N_3 = 0.66$, $N_4 = 0.75$, $N_5 = 1.0$. Hence, for a given engine speed, the vehicle speed gets larger. However, in turn the torque delivered to the output shaft gets smaller as N_i increases. For that reason, when a vehicle is climbing a hill, we need more torque, hence gear should be shifted down to a lower gear.

The steady-state engine power is characterized by the so-called "lug curve," which is torque versus speed at full throttle for a given engine as discussed in Section 1.2. Each gear of the transmission effectively scales this curve, assuming 100% transmission efficiency at all gears for the sake of simplicity. Plotting the scaled lug curves (that is the output torque and speed at the transmission output shaft) on the same figure for five different gears shows us the speed–torque ranges each gear should be used in (Figure 3.24).

As an example, consider the following simple lug curve data for an engine under full throttle conditions, and the gear ratios in gears 1 through 5,

$$w_{eng} = [\,600 \quad 900 \quad 1200 \quad 1500 \quad 1800 \quad 2100 \quad 2400\,]\,(rev/min) \tag{3.268}$$

$$T_{eng} = [\,25 \quad 50 \quad 100 \quad 95 \quad 80 \quad 75 \quad 70]\,(N \cdot m) \tag{3.269}$$

$$N_i = [\,1/3.0 \quad 1/2.0 \quad 1./1.5 \quad 1/1.33 \quad 1/0.9]; \quad i = 1, 2, 3, 4, 5 \tag{3.270}$$

Figure 3.24 shows the lug curve and the scaled versions of the lug curve to indicate the torque at the transmission output shaft at different transmission output shaft speeds.

For instance, if the load torque is very large (i.e., climbing a very steep hill), the transmission should operate in gear 1 and the maximum achievable speed is shown in the figure. Likewise, if the load torque is very small and top speed is of importance, then the transmission should be operated at a high gear (i.e., gear 5). Figure 3.24 shows the maximum load torque and speed that can be supported at each gear.

Let us assume that the engine is operating at full throttle, and the transmission is in gear 3. Then the operating curve of the vehicle is the gear 3 speed–torque curve shown. If

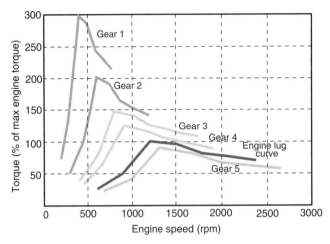

FIGURE 3.24: Engine lug curve between torque–speed, and scaled versions of it at different gear ratios of the transmission (assuming 100% transmission efficiency) which shows the speed and torque profile that is delivered at the transmission output shaft.

the vehicle speed corresponds to 1000 rpm output shaft speed at the transmission output shaft connecting to the differential, and the load torque is constant at about 120% of the maximum engine torque, then the available torque delivered at that condition is larger than the load torque (torque delivered to output shaft is about 140% of the maximum engine torque at gear 3 at that speed), then the vehicle will accelerate until the output torque drops and equals load torque on the curve for gear 3. Again, we assume the engine is at full throttle, which means that the engine is operating at full capacity (maximum fuel rate, i.e., gas pedal is pressed to its maximum displacement). In order to find the steady-state operating condition under a constant load at a given gear, draw a horizontal line from the current operating load condition. The point where this line crosses the lug-curve at the given gear ratio gives us the steady-state operating condition, and hence the steady-state operating speed where the output torque matches the load torque. In this example case, the vehicle speed will stabilize in steady state at a speed corresponding to about 1500 rpm output shaft speed. Similarly, if the vehicle is running at this speed and load torque increases to 140%, perhaps due to an increase in road slope, then the vehicle speed would slow down until the engine (and hence vehicle speed) drops to a value on gear 3 curve where the produced torque is equal to the load torque. In this case this occurs around 1000 rpm. As a further discussion, if the load torque is 160% of the maximum torque, the gear has to be shifted to the lower gear, that is gear 2, in order to meet the load demand. If the gear is maintained at gear 3 under this condition, the engine will stall (stop), since at gear 3 there is no way the engine can output as large a torque as demanded. The only way to meet the load demand is to reduce the gear ratio, hence increase the torque multiplication factor to deliver more torque to the output shaft. In this case, if the gear is shifted to gear 2, the vehicle speed would stabilize at a speed that corresponds to the output shaft speed of the transmission at about 800 rpm under full throttle condition.

3.8.3 Planetary Gears

Figure 3.25 shows two examples of *planetary gear* sets. Figure 3.25a shows the most commonly used form of planetary gears with an internal ring gear. Figure 3.25b differs only in the fact that the ring gear is not an internal gear, but an external gear. Planetary

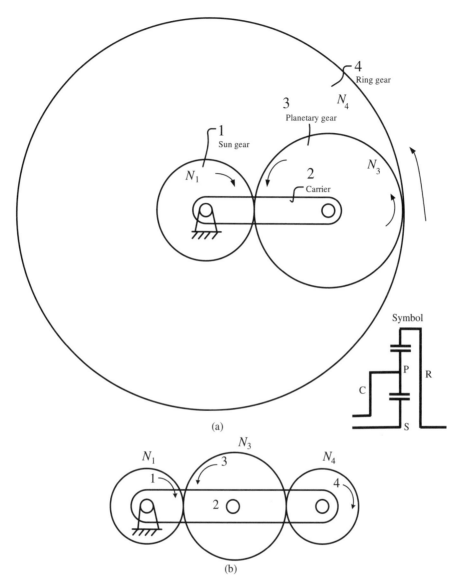

FIGURE 3.25: Planetary gears: (a) ring gear is an internal gear, (b) ring gear is an external gear.

gears are also called *epicyclic gears*. There are four main components of a planetary gear:

1. sun gear,
2. carrier (also called planetary carrier or arm),
3. planetary gear,
4. ring gear.

There are five possible conditions of planetary gear applications (Figure 3.26): in four of the conditions one of the components is fixed.

1. Sun gear is fixed, ($w_1 = 0.0$): ring gear, planetary gear, and planetary carrier all rotate in the same direction. This is called the *walking planetary gear condition* (Figure 3.26a).

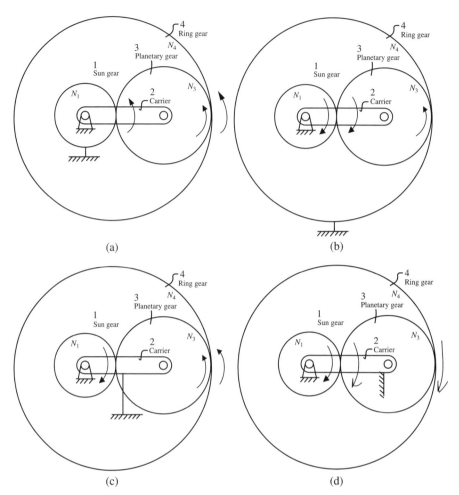

FIGURE 3.26: Four common configurations of planetary gears: (a) sun gear is fixed: directions of rotation of planetary gear and carrier are same, (b) ring gear is fixed: directions of rotation of planetary gear and carrier are opposite, (c) planetary carrier is fixed, non-planetary operation (idler). (d) planetary gear is fixed to the carrier (cannot rotate about the carrier axis). Locked condition. All components rotate as one unit, gear ratio is 1:1.

2. Ring gear is fixed ($w_4 = 0.0$): sun gear and planetary carrier rotate in the same directions, planetary gear rotates about its own shaft in the opposite direction (Figure 3.26b). This is also called the *walking planetary gear condition*.

3. Planetary carrier is fixed ($w_2 = 0.0$): functions as a non-planetary gear. The planetary gear acts as an idler gear. This is called the *idling planetary gear condition* (Figure 3.26c).

4. The planetary gear is fixed, while the sun, ring, and carrier are allowed to move (Figure 3.26d). In this condition, all components move as one unit, as if they are connected to each other with a one-to-one (1 : 1) gear ratio.

5. None of the planetary gear components are fixed (Figure 3.25), two of the components are used as the input shaft and one of the component outputs is used as the output. This results in a continuously variable transmission (CVT or also sometimes called "split torque transmission" (STT), because the torque delivered to the output shaft is "split" between two input shaft sources).

If the *carrier* is fixed (not moving), the planetary gear reduces to a standard gear, with the gear ratio being the ratios of each gear (Figure 3.26c),

$$N_{31} = \frac{w_3}{w_1} = sgn\left(\frac{w_3}{w_1}\right)\frac{N_1}{N_3} = -\frac{N_1}{N_3} \tag{3.271}$$

$$N_{41} = \frac{w_4}{w_1} = sgn\left(\frac{w_4}{w_1}\right)\frac{N_1}{N_3} \cdot \frac{N_3}{N_4} = sgn\left(\frac{w_4}{w_1}\right)\frac{N_1}{N_4} = -\frac{N_1}{N_4} \tag{3.272}$$

where the gear that is connected to both input shaft and output shaft (in this case the planetary gear) acts as an *idler gear*. This is called the *"idler"* or *"idling"* mode of operation for a planetary gear. The idler gear does not change the gear ratio, but changes the direction of rotation only. The contribution of the idle gear to the gear ratio is multiplication by -1. The sign of rotation in the gear ratio can also be incorporated into the definition of the gear ratio. It is customary to consider the direction of rotation in counter-clockwise as positive, and the rotation in clockwise direction as negative. The $sgn(\cdot)$ is negative for a planetary gear with internal ring gear (Figure 3.25a) for both equations. The $sgn(\cdot)$ is negative for the first, and positive for the second, equation for the planetary gear type with external ring gear shown in (Figure 3.25b).

If the *planetary gear* is fixed about its local axis of rotation on the carrier, the relative rotational speed of the planetary gear with respect to the carrier is zero, ($w_{32} = 0.0$), then the gear mechanism is in *locked condition*, where the angular speed of the sun, carrier and the ring gear are all same, in other words, have all gear ratio of 1:1 with respect to each other (Figure 3.26d).

$$w_1 = w_2 = w_4 \tag{3.273}$$

$$N_{2,1} = N_{4,1} = 1 \tag{3.274}$$

$$w_3 = w_2 + w_{32} \tag{3.275}$$

$$w_3 = w_2 + 0 \tag{3.276}$$

$$w_3 = w_2 \tag{3.277}$$

In the planetary gear mode, the carrier is not fixed, but moving. This mode is also referred to as the "walking" mode. In walking mode, either the sun gear or the ring gear is held stationary. In the case of the stationary sun gear, the planetary gear always turns in the same direction on its pin as the planet carrier rotation (Figure 3.26a). In the case of the stationary ring gear, the planetary gear turns in the opposite direction on its pin as the planetary carrier (Figure 3.26b).

In planetary mode, the gear ratio relationships are applicable in terms of the *relative velocities* of the components with respect to the carrier. *The relative gear ratio definition with respect to the planetary carrier is the key principle in defining the gear ratio relationship of planetary gears,* analogous to the non-planetary case where the carrier is fixed. Below are the two fundamental kinematic equations for planetary gear ratio calculations

$$N_{r31} = \frac{w_{32}}{w_{12}} = \frac{w_3 - w_2}{w_1 - w_2} = sgn\left(\frac{w_{32}}{w_{12}}\right)\frac{N_1}{N_3} = -\frac{N_1}{N_3} \rightarrow \frac{w_3 - w_2}{w_1 - w_2} = -\frac{N_1}{N_3} \tag{3.278}$$

$$N_{r41} = \frac{w_{42}}{w_{12}} = \frac{w_4 - w_2}{w_1 - w_2}$$

$$= sgn\left(\frac{w_{42}}{w_{12}}\right)\frac{N_1}{N_3} \cdot \frac{N_3}{N_4} = sgn\left(\frac{w_{42}}{w_{12}}\right)\frac{N_1}{N_4} = -\frac{N_1}{N_4} \rightarrow \frac{w_4 - w_2}{w_1 - w_2} = -\frac{N_1}{N_4} \tag{3.279}$$

where we define the above gear ratios as *relative*, as indicated by the subscript "r."

Of the four elements of a planetary gear, two of them can be specified as inputs. In other words, planetary gears have two degrees of freedom that can be specified by the user, that is, the speed (absolute angular velocity) of the any two components. Then, the absolute speeds of the other two components can be found from the above equations. Once that is known, the gear ratios can be expressed between any two absolute angular speeds, that is

$$N_{31} = \frac{w_3}{w_1}, N_{41} = \frac{w_4}{w_1}.$$

If we take one of the components as the output shaft of interest, and consider two input shafts as input shafts, this configuration can be used as a "continuously variable gear ratio" or a "continuously variable transmission" (CVT). The output shaft speed is a function of two input shaft speeds. If we consider one of the input shafts as the primary input shaft which has a gear ratio to the output shaft, then we can consider the second input shaft to effectively change the gear ratio between the primary input shaft speed and output shaft speed since we know how the second input shaft speed contributes to the output shaft speed.

For instance, let us consider that the input shafts are the ring and sun gears, and the output shaft is the carrier. The carrier speed can be calculated as (Figure 3.25a).

$$\frac{w_4 - w_2}{w_1 - w_2} = -\frac{N_1}{N_4} \tag{3.280}$$

$$w_2 = \frac{N_1}{N_1 + N_4} \cdot w_1 + \frac{N_4}{N_1 + N_4} \cdot w_4 \tag{3.281}$$

where the gear ratio between w_2 and w_1 is shown. But we can consider the contribution of w_4 as effectively changing that gear ratio. This method is used in *continuously variable transmissions* (CVT). Similarly, for the same condition, we can derive the gear ratio relationship between the two input shafts and the planetary gear (component number 3) from the above equations. It can be shown that the gear ratio relationship is

$$\frac{w_3 - w_2}{w_1 - w_2} = -\frac{N_1}{N_3} \tag{3.282}$$

$$w_3 = \frac{N_1}{N_3} \left(\frac{N_1 + N_3}{N_1 + N_4} - 1 \right) \cdot w_1 + \frac{N_4}{N_3} \left(\frac{N_1 + N_3}{N_1 + N_4} \right) \cdot w_4 \tag{3.283}$$

In most planetary gear applications, one of the three components other than the carrier (typically the sun gear, planetary gear, or ring gear) is fixed, one of the remaining two components acts as the input shaft, and the remaining two components, rotation is kinematically determined by the appropriate gear ratio where one of the two remaining components is the output shaft. If the carrier is fixed, the gear mechanism is in non-planetary mode.

For the planetary gear type shown in Figure 3.26b, and for this special configuration (ring gear is fixed) and the sun gear is input, the absolute gear ratios can be shown from above equations as follows (set $w_4 = 0.0$),

$$N_{21} = \frac{w_2}{w_1} = \frac{N_1}{N_1 + N_4} \tag{3.284}$$

$$N_{31} = \frac{w_3}{w_1} = \frac{N_1}{N_3} \cdot \left(\frac{N_1 + N_3}{(N_1 + N_4)} - 1 \right) \tag{3.285}$$

It is trivial to show that $N_{32} = N_{31}/N_{21}, \ N_{41} = N_{42} = N_{43} = 0.0$.

This configuration (ring gear fixed) is used in *final drive* gear reducers inside the wheel in large construction equipment applications. There is a planetary gear reducer between the output shafts of the differential and the tire-wheel assembly. The sun gear is connected to the differential output shafts (input to the planetary gear reducer of the final drive), the

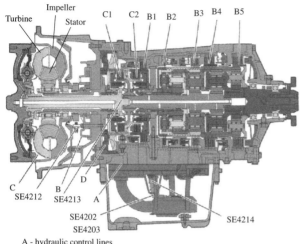

	C- Clutch		B - Brake				
	C1	C2	B1	B2	B3	B4	B5
Reverse 1		(a)					(a)
Reverse 2		(a)				(a)	
Neutral N (L)						(b)	
Neutral N (H)							(b)
Forward 1	(a)						(a)
Forward 2	(a)					(a)	
Forward 3	(a)				(a)		
Forward 4	(a)		(a)				
Forward 5	(a)			(a)			
Forward 6	(a)	(a)					

A - hydraulic control lines
B - pump, C - torque convertor,
D - retarder (hydraulic brake)

(a) - Activated, (b) Activated with low pressure

FIGURE 3.27: Automatic transmission used in heavy equipment applications showing the use of planetary gears, controlled clutches, and brakes to select gear ratios. The torque converter provides a variable hydrodynamic coupling (variable gear ratio) between the engine input shaft and gear reducer section of the transmission. The table on the right shown which clutch/brake combination must be engaged to select a particular gear (Reproduced with permission from Volvo Trucks).

ring gear is fixed to the vehicle frame, and the output shaft is the planetary carrier which is connected to the tire-wheel assembly (Figure 3.34).

In transmission applications, multiple stages of such planetary gears are stacked in series (Figure 3.27). Then, using electrically (or electro-hydraulically) controlled multi disc clutches and brakes, we can decide in real-time which component to fix and which component to rotate in each planetary stage. Through controlled clutches and brakes, we can change from one gear ratio to another, for example from N_{21} to N_{31} by disconnecting the motion of the gear 2, and then connecting the motion of gear 3 to the output shaft by disengaging and engaging the clutches and brakes for the respective gears. For a good smooth transition, the transient dynamics of this clutch/brake disengagement/engagement motion must be carefully controlled. For instance, simple ON/OFF control of the clutch/brake disengagement/engagement would result in very rough and unacceptable response (i.e., similar to a beginner driver's gear shift on a manual shift transmission).

The multi stage planetary gear mechanism, shown in Figure 3.27, provides ten different gear ratios, as shown on the table on the right: six gear ratios for forward motion, two gear ratios for reverse motion, and two gear ratios for the neutral condition. This type of automatic transmission is used in articulated trucks and similar mobile and construction equipment applications. A common design characteristic of multi stage planetary gears is that adjacent stages share a component with each other. For instance, the stage 1 ring gear, stage 2 carrier (Figure 3.28), and the stage 3 ring gear are all rigidly connected to each other as one component. Similarly, the stage 3 planetary carrier is connected to stage 4 ring gear, which is then rigidly connected to the stage 5 ring gear. The stage 4 carrier is connected to the sun gear of stage 5. Notice that the sun gear of stage 5 is the output shaft of this multi stage planetary gear mechanism (Figure 3.28).

The electro-hydraulically controlled multi disc brakes (B1 through B5) and clutches (C1 and C2) are engaged or disengaged, as shown in the table, to select the desired gear ratio. It is important to note that the transition of the state of a clutch or brake from engaged

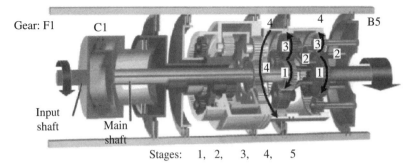

Gear: F1   C1   Input shaft   Main shaft   Stages: 1, 2, 3, 4, 5

FIGURE 3.28: Automatic transmission used in heavy equipment applications showing the use of planetary gears, controlled clutches, and brakes to select gear ratios: condition for gear Forward 1 (F1) (Reproduced with permission from Volvo Trucks).

to disengaged or from disengaged to engaged state is important for smoothness of the shift quality. To this end, the transition of the clutch/brake engagement/disengagement is controlled by pressure regulating valves that control the fluid flow to the clutches/brakes.

Consider the clutch and brake engagement conditions for forward gear 1: F1 (Figure 3.28).

1. Clutch 1 (C1) locks the input shaft to the main shaft.

2. Brake 5 (B5) locks the planetary carrier of stage 5 to the transmission housing. As a result, the planetary stage 5 is in the *idling* condition, which means that it is acting as a regular gear since its planetary carrier is fixed.

3. This forces the ring gear of stage 4 (item 4,4), which is shared by stages 4 and 5, to rotate in the opposite direction relative to the sun gear numbered (4,1) in the figure. The output shaft (connected to sun gear of stage 5, number (5,1) in the figure) is then driven in the opposite direction to the ring gear. The net result is that output shaft and input shaft are in the same direction, hence the forward gear.

The gear ratio between input and output shaft speeds can be found as follows. Note that the gear numbers of each stage are defined as follows, $N_{i,j}$, where i is the planetary stage number (1, 2, 3, 4, 5), and j is the gear number in that stage: 1 for sun gear, 2 for planetary carrier, 3 for planetary gear, 4 for ring gear. The speeds are also defined with the same notation, $w_{i,j}$, where i is the planetary stage number (i = 1, 2, 3, 4, 5), and j = 1, 2, 3, 4 the component numbers of that stage, sun, carrier, planet, and ring gears, respectively. The output shaft speed (the shaft speed of the sun gear of planetary stage 5) is determined by the ring gear speed of stage 4 as follows. Stage 5 operates in idle condition because its planetary carrier is fixed, and the ring gear of stage 4 and stage 5 are the same, and the speed of the output shaft is the same as the speed of the planetary carrier of stage 4,

$$w_{\text{out}} = w_{5,1} \tag{3.286}$$

$$= w_{4,2} \tag{3.287}$$

$$\frac{w_{5,1}}{w_{5,4}} = -\frac{N_{5,4}}{N_{5,1}} \tag{3.288}$$

$$w_{5,4} = w_{4,4} \quad (\textit{same component}) \tag{3.289}$$

$$w_{\text{out}} = -\frac{N_{5,4}}{N_{5,1}} \cdot w_{4,4} \tag{3.290}$$

From the basic planetary gear relationship derived above, we can write the following for planetary stage 4,

$$w_{4,2} = \frac{N_{4,1}}{N_{4,1} + N_{4,4}} \cdot w_{4,1} + \frac{N_{4,4}}{N_{4,1} + N_{4,4}} \cdot w_{4,4} \tag{3.291}$$

$$w_{\text{out}} = w_{5,1} \tag{3.292}$$

$$= w_{4,2} \tag{3.293}$$

where equations are simply modified by their subscripts to indicate the planetary stages. Given the three relationships above, the output speed $w_{\text{out}} = w_{5,1} = w_{4,2}$ can be calculated as function of input speed $w_{4,1}$. In other words, we can eliminate $w_{4,4}$ from the above equations, resulting in a gear ratio relationship between $w_{4,1}$ (the input shaft speed) and $w_{4,2}$ which is the same as the output shaft speed of the transmission. It can be shown that the resuting gear ratio relationship is

$$w_{4,2} = \frac{N_{4,1}}{N_{4,1} + N_{4,4}} \cdot w_{4,1} + \frac{N_{4,4}}{N_{4,1} + N_{4,4}} \cdot w_{4,4} \tag{3.294}$$

$$= \frac{N_{4,1}}{N_{4,1} + N_{4,4}} \cdot w_{4,1} - \frac{N_{4,4}}{N_{4,1} + N_{4,4}} \frac{N_{5,1}}{N_{4,4}} \cdot w_{4,2} \tag{3.295}$$

$$= \left[1 + \frac{N_{4,4}}{N_{4,1} + N_{4,4}} \frac{N_{5,1}}{N_{4,4}}\right]^{-1} \frac{N_{4,1}}{N_{4,1} + N_{4,4}} \cdot w_{4,1} \tag{3.296}$$

where the gear ratio at this gear is

$$GR_1 = \left[1 + \frac{N_{4,4}}{N_{4,1} + N_{4,4}} \frac{N_{5,1}}{N_{4,4}}\right]^{-1} \frac{N_{4,1}}{N_{4,1} + N_{4,4}} \tag{3.297}$$

$$= \frac{N_{4,1}}{N_{4,1} + N_{4,4} + N_{5,1}} \tag{3.298}$$

Example: Planetary Gears   Consider Figure 3.25a with the following gear parameters: $N_1 = 25$, $N_3 = 50$, $N_4 = 100$ gears. Inputs are specified as: sun gear speed $w_1 = 100$ rev/min and the ring gear is stationary, $w_4 = 0.0$ rev/min. Find the speed of the carrier and the planetary gear and absolute gear ratios between the sun gear and the carrier, and the sun gear and the planetary gear.

For the direct application of the two equations we have above, first we will find the absolute velocity of the two components, then calculate the absolute gear ratios.

$$N_{r41} = \frac{w_{42}}{w_{12}} = \frac{w_4 - w_2}{w_1 - w_2} = sgn(\cdot)\frac{N_1}{N_3} \cdot \frac{N_3}{N_4} = sgn(\cdot)\frac{N_1}{N_4} \tag{3.299}$$

$$= \frac{0 - w_2}{100 - w_2} = -\frac{25}{100} \tag{3.300}$$

$$w_2 = 20 \text{ rev/min} \quad \textit{same direction as } w_1 \tag{3.301}$$

$$N_{r31} = \frac{w_{32}}{w_{12}} = \frac{w_3 - w_2}{w_1 - w_2} = sgn(\cdot)\frac{N_1}{N_3} \tag{3.302}$$

$$= \frac{w_3 - 20}{100 - 20} = -\frac{25}{50} \tag{3.303}$$

$$w_3 = -20 \text{ rev/min} \quad \textit{opposite direction to } w_1 \tag{3.304}$$

Notice that w_3 is the absolute angular velocity of the planetary gear. It has two components contributing to its absolute angular velocity: angular velocity of the carrier (w_2, component 2) and angular velocity of the planetary gear with respect to the carrier about it own rotation axis, w_{32}. As indicated above,

$$w_{32} = w_3 - w_2 \tag{3.305}$$

$$w_{32} = -20 - 20 \, \text{rpm} \tag{3.306}$$

$$= -40 \, \text{rpm} \tag{3.307}$$

which is the angular velocity of the planetary gear about its own rotation axis.

Then the net gear ratios between sun gear and carrier (N_{21}), and sun gear and planetary gear (N_{31})

$$N_{21} = \frac{w_2}{w_1} = \frac{20}{100} = \frac{1}{5} \tag{3.308}$$

$$N_{31} = \frac{w_3}{w_1} = \frac{-20}{100} = -\frac{1}{5} \tag{3.309}$$

The absolute gear ratios could also be calculated for this case using the equations

$$N_{21} = \frac{N_1}{N_1 + N_4} = \frac{25}{25 + 100} = \frac{1}{5} \tag{3.310}$$

$$N_{31} = \frac{N_1}{N_3} \left(\frac{N_1 + N_3}{N_1 + N_4} - 1 \right) = \frac{25}{50} \left(\frac{25 + 50}{25 + 100} - 1 \right) = -\frac{1}{5} \tag{3.311}$$

In transmission applications, we could obtain two different gear ratios from this single planetary gear stage: use a brake (B1) to lock the ring gear, and clutch (C1) to engage the carrier to the output shaft, hence obtain a gear ratio of $\frac{1}{5}$ between the input and output shafts. Alternatively, use a brake (B1) to lock the ring gear, and a second clutch (C2) to engage the planetary gear to the output shaft to obtain a gear ratio of $-\frac{1}{5}$ between input and output shafts.

3.8.4 Torque Converter

Figure 3.27 shows an automatic transmission based on a planetary gear mechanism. The input section of an automatic transmission has a *torque converter* which functions as a *hydrodynamic flexible coupling* between the engine and the transmission (Figure 3.29a and b)[1]. It can also be viewed as an automatic clutch with slip capability. As a result, it provides a damping to the power transmission line. Torque converter provides a variable gear ratio between its input–output shafts. The main disadvantage of a torque converter is its lower power conversion efficiency compared to direct drive. Because the power is transmitted by the moving fluid, it creates heat, hence the losses. In addition, a cooling system may be needed to remove the heat.

The torque converter has three main components:

1. *The impeller*, which is connected to the input shaft of the torque converter, which is also the engine output shaft. The impeller converts the mechanical energy of its input shaft to the hydrodynamic energy of the coupling fluid.

2. *The turbine*, which is connected to the output shaft of the torque converter, which is connected to the input shaft of the transmission's planetary gear set, and converts the hydrodynamic power of the fluid back into mechanical rotational motion of its shaft.

3. *The stator* is a component that redirects the coupling fluid between the turbine and the impeller, typically stationary, however in some cases it can be free-wheeling or

---

[1] This sub-section is based on Lectures by Dr. Richard Ingram.

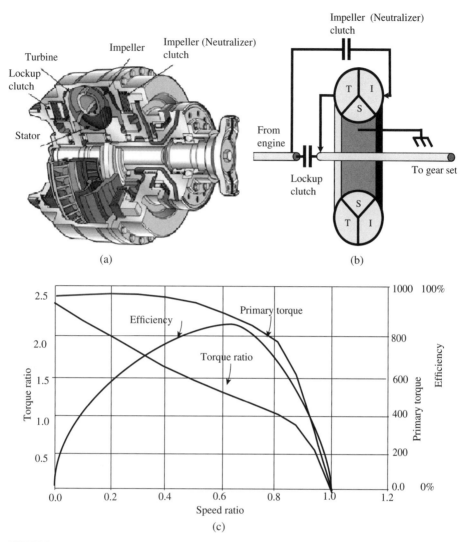

FIGURE 3.29: Torque converter: (a) components: impeller, turbine, stator. There are also impeller (neutralizer) clutch and lock-up clutch in heavy equipment applications. (b) Functional relationship torque converter components. (c) Steady-state input–output relationship for torque converter: torque ratio and primary torque as a function of speed ratio, and efficiency of torque converter as a function of speed ratio.

connected to the impeller shaft with a one-way clutch which allows it to turn only in the same direction as the impeller, not reverse.

There are two common modifications to the torque converter in heavy equipment applications (Figure 3.29a and b):

- *Neutralizer clutch* (also called *impeller clutch*) which is used to disconnect the input shaft of the torque converter (shaft of the impeller) from the engine shaft so that partial power or no power is transmitted from the engine to the torque converter. It is typically a proportionally controlled, not ON/OFF, type clutch. When it is fully engaged (ON), the full power transmission from engine to torque converter occurs. When it is fully disengaged (OFF, in neutralized state), there is no power transfer from the engine to the torque converter's impeller shaft, hence the engine and the rest of the powertrain is effectively disconnected. In a partially engaged state via a proportional pressure control circuit on the neutralizer clutch, partial power is transmitted between the engine and the torque converter. This proportional control capability of the neutralizer clutch can be used to limit the torque transmitted to the transmission, that is to limit the rimpull force (traction torque) for traction control, or to manage how much power is allowed to go to the transmission in order to allocate the rest of the available engine power to the other hydraulic systems. This is sometimes referred as *the power management algorithm* or *power management strategy* in vehicle control terminology. Power management refers to the controlled distribution of available engine power among two or more subsystems, that is transmission, steering, and implement hydraulic systems.

- *Lock-up clutch* which is used to mechanically lock the impeller shaft and turbine shaft without slip, effectively providing a rigid coupling and eliminating the torque converter's hydrodynamic coupling function. This also referred to as the *direct drive mode* of the torque converter, where the torque converter is effectively non-functional. This is an ON/OFF type of clutch with a controlled dynamic transition between ON and OFF states. It is used in heavy equipment applications after the first gear to eliminate the inefficiency of the torque converter. In such applications, the torque converter is used (lock-up clutch is not engaged) only on the first gear. When the lock-up clutch is engaged, the torque converter section is not transmitting power through the hydrodynamic coupling. In order to minimize the power loss in the circulated fluid between impeller-turbine-stator, the stator is generally of the free-wheeling type instead of being fixed to the housing when the lock-up is engaged.

The only components of a torque converter that are subjected to wear, are the friction discs of the neutralizer and lock-up clutches and the seals. The rest of the components are rather highly reliable, rigid components that rarely need servicing.

The torque converter behavior is modeled by two steady-state algebraic functions (Figure 3.29c): torque ratio and primary torque. The torque ratio (also called torque multiplication) is higher when there is a larger relative speed difference between the input and output shafts. When the impeller and turbine shaft speeds are the same, there is no torque transmitted between the two shafts. The two characteristic functions of the torque converter are defined as follows, as a function of speed ratio $N_w = w_{turb}/w_{imp}$

1. torque ratio, $N_T(N_w)$, between impeller and turbine shafts as a function of the speed ratio, and
2. primary torque, $T_p(N_w)$, as a function of the speed ratio.

For a given torque converter, these two characteristic functions can be measured as follows (Table 3.1):

1. connect the torque converter between an engine and a dynamometer (also called *a dyno*),
2. set the engine speed to the rated speed, that is $w_{rated} = 1800$ rpm, via a closed loop engine speed controller so that the engine maintains that speed,
3. control the dyno to maintain desired speeds from zero to rated speed at selected intervals in steady state, that is again using a closed loop dyno speed controller, $w_{dyno} = 0$ rpm, 100 rpm, ..., 1800 rpm,

TABLE 3.1: Table for measurement and calculations to characterize the steady-state input–output behavior of a torque converter: determining torque gain ($N_T(N_w)$) and primary torque ($T_p(N_w)$) as a function of speed ratio ($N_w = w_{turb}/w_{imp}$).

Controlled	Controlled	Measure	Measure	Calculate	Calculate
$w_{eng} = w_{rated}$	w_{dyno}	$T_{eng} = T_p$	T_{dyno}	$N_w = \dfrac{w_{dyno}}{w_{eng}}$	$N_T = \dfrac{T_{dyno}}{T_{eng}}$
1800 rpm	0 rpm	x	x	...	...
1800 rpm	100 rpm	x	x	...	...
...	...	x	x	...	...
...	...	x	x	...	...
1800 rpm	1800 rpm	x	x	...	...

4. then measure the following variables in steady state: engine output shaft (impeller) and dyno shaft (turbine) speeds as well as torques,

$$w_{imp} = w_{eng} = w_{rated} \tag{3.312}$$

$$w_{turb} = w_{dyno} \tag{3.313}$$

$$T_p = T_{imp} = T_{eng} \tag{3.314}$$

$$T_{turb} = T_{dyno} \tag{3.315}$$

5. then calculate and plot the following data (Figure 3.29c):

$$N_w = w_{dyno}/w_{eng} = w_{dyno}/w_{rated} \tag{3.316}$$

$$N_T = T_{turb}/T_{imp} \tag{3.317}$$

Plot the two variables

- $T_p(N_w)$ versus speed ratio N_w
- $N_T(N_w) = T_{dyno}(N_w)/T_{eng}(N_w)$ versus speed ratio N_w.

These two functions characterize the steady-state input–output relation of a torque converter.

For a given torque converter, we would have the $T_p(N_w), N_T(N_w)$ functions. In specific steady-state operating conditions, if the input and output shaft speeds of the torque converter are known (given: w_{imp}, w_{turb}), then the input and output torques can be calculated as

$$T_{imp}(N_w, w_{imp}) = T_p(N_w) \cdot \left(\frac{w_{eng}}{w_{rated}}\right)^2 \tag{3.318}$$

$$T_{turb}(N_w, w_{imp}) = N_T(N_w) \cdot T_{imp}(N_w, w_{imp}) \tag{3.319}$$

where $w_{eng} = w_{imp}, w_{turb}$ are impeller and turbine speeds, T_{imp}, T_{turb} are torques at impeller and turbine shafts. Notice that, in the above measurements to characterize a torque converter, the impeller (engine output shaft) speed is maintained at a constant rated speed and $T_p(N_w)$ is obtained for a constant impeller speed as turbine speed varied from zero to the rated speed. To obtain the impeller torque at a general operating speed of the engine, we use the parabolic relationship as shown above. In the most general form $T_{imp} = T_{imp}(N_w, w_{eng})$. When $w_{eng} = w_{rated}$, then this relationship reduces to the $T_{imp}(N_w) = T_p(N_w)$.

The magnitude of these two functions depends on the physical size of the torque converter. The exact shape of these two functions is determined by the shape of the blades in the impeller, turbine, and stator, which are customized differently for different machine applications. For instance, the blade shapes in a torque converter in a truck would be different from those in a passenger car, because these two machines require different torque converter behavior.

In order to understand the role and dynamics of the torque converter in the vehicle drive train dynamic simulation, consider the torque converter as a nonlinear coupling device that is flexible or that can have different slip between input and output speeds. Consider the beginning of a simulation, where the engine has a speed and the vehicle has a speed.

1. The impeller speed of the torque converter is determined by the engine speed,

$$w_{imp} = w_{eng} \tag{3.320}$$

2. Turbine speed is determined by the speed of the vehicle which is reflected to the turbine by the lower powertrain gear ratio,

$$w_{turb} = \frac{1}{N_{lp}} \cdot w_{tire} \tag{3.321}$$

where w_{tire} is the vehicle tire speed, and N_{lp} is the total gear ratio of the lower powertrain (between turbine speed and tire speed).

3. Then, T_{imp} and T_{turb} torques are determined from the steady-state characteristics of the torque converter.

The efficiency of torque converter is defined as

$$\eta_{tc} = \frac{P_{out}}{P_{in}} \tag{3.322}$$

$$= \frac{T_{out} \cdot w_{out}}{T_{in} \cdot w_{in}} \tag{3.323}$$

$$= \frac{T_{turb} \cdot w_{turb}}{T_{imp} \cdot w_{imp}} \tag{3.324}$$

$$= N_w \cdot N_T \tag{3.325}$$

The maximum efficiency of a torque converter occurs around 0.7 to 0.85 speed ratio range and is around the 85–90% range. When the speed ratio of the torque converter is zero, the power conversion efficiency is zero because output speed, and hence the output power, is zero. When the speed ratio is one (both impeller and turbine shaft speeds are the same), the power conversion efficiency is also zero because the torque output (and torque ratio) is zero. Between these extremes, the efficiency curve is similar to a parabolic function of speed ratio, where the maximum efficiency is achieved around the 0.7 to 0.85 speed ratio range (Figure 3.29c).

For a specific machine application, the engine and torque converter should be sized properly. This is called *engine–torque converter matching*. The basic objective is to match the power level of the engine and the power transfer capability of the torque converter. Different machine applications require different engine–torque converter matching in that the maximum efficiency of the torque converter occurs at the rated speed of the engine (full-matched case), at above the rated speed of the engine (light-matched case), and at below the rated spedd of the engine (heavy-matched case). Roading machines typically have heavy to full match, whereas non-roading machines (such as wheel type loaders) typically have light to full match.

For a given torque converter, we have the torque gain ($N_T(N_w)$) and primary torque ($T_p(N_w)$) functions as functions of the speed ratio. Let us consider a load torque and its reflection on the impeller. In other words, the impeller torque is the reflected load torque acting on the engine output shaft. The intersection of the engine lug curve and this

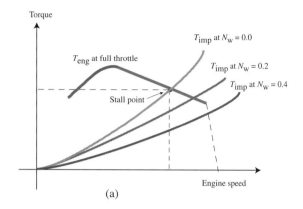

(a)

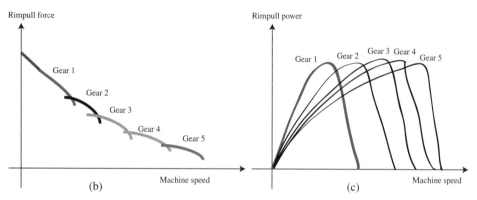

(b)   (c)

FIGURE 3.30: Steady-state operating condition of a machine with an engine and torque converter based transmission: (a) At full throttle, the steady-state engine operating condition is determined by the intersection of the engine lug curve and impeller torque curve for a given torque converter speed ratio. For different speed ratios, different impeller torque curves give us different steady-state operating conditions for the engine. (b) The calculated rimpull force as a function of machine speed: given a gear setting, calculate equivalent force delivered at the tire–ground interaction as a result of the torque converter output torque as a function of different speed ratios and engine speed in steady state. (c) Rimpull power: rimpull force times the machine travel speed. Notice that the maximum power delivered at the rimpull is about the same for all gears, where the shape of the power curve as a function of machine speed is simply stretched out as a function of machine speed for different gears.

$T_{\text{imp}}(w_{\text{eng}})$ due to load for different speed ratio $N_{\text{w}} = 0.0, 0.1, ..., 1.0$ curves defines the steady-state operating speed and torque of the engine when the engine is at full throttle (Figure 3.30a). Then, at a given torque converter speed ratio (i.e., $N_{\text{w}} = 0.0, 0.1, ..., 1.0$), we can find the *stall point*, hence the engine torque and engine speed ($T_{\text{imp}} = T_{\text{eng}}$ and w_{eng}, on Figure 3.30a).

Given the current gear ratio of the planetary gear set and the rest of the gear ratios (differential and final drive) are known for a given machine, we can calculate the *rimpull force* at this condition,

$$V_{\text{machine}} = N_{\text{total}} \cdot N_{\text{w}} \cdot w_{\text{eng}} \qquad (3.326)$$

$$F_{\text{rimpull}} = \frac{1}{N_{\text{total}}} \cdot \eta_{\text{lp}} \cdot T_{\text{turb}} \qquad (3.327)$$

where N_{total} is the total gear ratio from the torque converter output shaft to the tire–ground contact point (it includes the average radius of the tire, R_{ave}, in the gear ratio between the shaft rotation and linear displacement of the tire),

$$N_{total} = N_{planetary} \cdot N_{diff} \cdot N_{final\ drive} \cdot R_{ave} \qquad (3.328)$$

where $N_{planetary}$ is the gear ratio at the currently selected gear (ratio between the speed of the torque converter output shaft which is turbine and the speed of the planetary gear set output shaft), N_{diff} is the gear ratio at the differential, $N_{final\ drive}$ is the gear ratio of the final drive, η_{lp} is the power transmission efficiency of the lower powertrain from turbine shaft to the tires.

For a given gear condition (i.e., gear 1, gear 2, gear 3), if we repeat this calculation for different speed ratios of the torque converter, $N_w = 0.0, 0.1, \ldots, 1.0$, we can plot the rimpull force developed at the tire–ground interaction at that gear as a function of machine translational speed. If we repeat this for different gears (different N_{total} values), then we obtain different rimpull force versus machine speed curves for different gears (Figure 3.30b). This set of rimpull force–machine speed curves for different gear ratios defines the tractive (rimpull) force capability of a machine. In steady state, for a given gear ratio of the transmission, each curve shows the maximum rimpull force the machine can generate at different machine speeds. Actual *measured rimpull force curves* versus machine speed would be a little different than the *calculated rimpull force curves* due to slip between tires and ground as well as friction losses, hence there is some error in calculating the machine speed based on the engine speed.

Rimpull force versus machine speed curves for a machine which has a drive train without torque converter (a direct coupling between engine and gear reducer set, i.e., planetary gear set) would look like the curves shown in Figure 3.24. When a torque converter is included in the coupling between the engine and gear set, the rimpull force versus machine speed curves are a little smoother, and the rimpull force is a little smaller compared to the direct drive case. This is expected since the torque converter is less efficient than the direct coupling.

The power delivered to the rim of the machine (rimpull power) is the engine power minus the transmission losses. Notice that the maximum rimpull power is about same for all gears. The difference is simply due to efficiency differences in the transmission at different gear ratios. The shape of the rimpull power curve is simply stretched for different gears (Figure 3.30c).

One simple measurement to confirm the engine and drive-train capabilities of a machine is to measure the *stall point*. That is, for a given gear ratio (i.e., gear 1), set the engine to full throttle, and load the machine such that the machine speed is zero ($N_w = 0.0$), and measure the rimpull force and the engine speed at that point (maximum rimpull at machine stall speed). This measured data then can be compared with the specifications (i.e., maximum rimpull force on Figure 3.30b) to verify the accuracy of how well the machine at hand meets the specified performance. About 3–5% variation between specifications and actual measured values is normal as a result of manufacturing variations and measurement errors.

3.8.5 Clutches and Brakes: Multi Disc Type

Clutches and brakes are very common components in motion transmission mechanisms such as transmissions. They both involve two shafts. A clutch transmit torque from one moving shaft to another moving shaft. In the case of brakes, the second shaft is stationary. By controlling a combination of clutches and brakes, different gear ratios are obtained from the transmission.

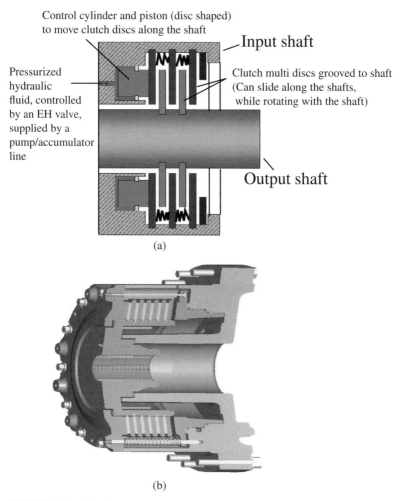

Control cylinder and piston (disc shaped)
to move clutch discs along the shaft

Input shaft

Pressurized
hydraulic
fluid, controlled
by an EH valve,
supplied by a
pump/accumulator
line

Clutch multi discs grooved to shaft
(Can slide along the shafts,
  while rotating with the shaft)

Output shaft

(a)

(b)

FIGURE 3.31: Multi disc clutch and brake concept. (a) Two-dimensional cross-sectional view of a multi disc clutch, (b) three-dimensional cut-away view of a multi disc brake.

Multi disc clutch and brakes are perhaps the most versatile types among others. The basic principle of torque transmission, via clutch and brakes, is the friction between two discs. If we press two discs against each other, the friction between them will transmit torque from one to another. Now, we extend this idea to a set of discs, where one set is connected to shaft one, the other set connected to shaft two. In addition they are connected to their respective shafts through axial grooves so that they can move translationally (slide along the shaft) along the rotation axis. The two sets of discs would be placed in alternating order in order to maximize the friction surface. The actuation mechanism for the clutch and brakes can be mechanical, air pressure, hydraulic, electro-hydraulic, or electromagnetic. The role of the control mechanism is to provide the control power to simply force or release (engage or disengage) the friction discs against each other. Figure 3.31 shows the concept of a multi disc clutch. If either one of the shafts is fixed, it functions as a brake. If a hydraulic mechanism controls the fluid flow to the piston chamber to regulate the pressure, then we control the friction between the discs that connect the shafts to each other. The fluid flow, which is controlled in such a way to as regulate the pressure, is controlled by an electro-hydraulic proportional valve where the current applied to the valve solenoid will result in a proportional pressure. In clutch/brakes, where the transmitted torque is proportionally controlled, the pressure at the piston control chamber is regulated proportionally to the desired torque transmission. Such multi disc clutch/brake components are commonly used in automotive transmissions to engage or disengage desired gear ratios.

The actuation mechanism may also be based on a direct electrical coil actuation. When the current is applied to the coil, it acts as an electro-magnet, and is used to engage/disengage the multi disc brake/clutch. If the coil current is proportionally controlled, then the multi disc brake is capable of proportional operation.

The mechanical dimensions and the number of disc-plates used determines the maximum torque capacity of the multi disc clutch/brake. The dynamic response depends on the type of actuation method, that is mechanical linkage, pneumatic, hydraulic, electro-hydraulic, or direct electro-magnetic. Typical response times vary from 15 ms to 500 ms for electro-magnetically controlled types.

3.8.6 Example: An Automatic Transmission Control Algorithm

Figure 3.32a shows the basic logical blocks of an automatic transmission control algorithm. In automatic transmission control, the real-time control algorithm in the electronic control module (ECM) essentially tries to do what a good professional driver tries to do with a manual shift transmission; accomplish a "smooth and yet fast enough" gear shift using various sensory information about the condition of the vehicle.

The algorithm monitors three main sensory signals:

1. transmission gear selection lever, operated by the driver (i.e., P, R, N, D for park, reverse, neutral, and direct automatic mode),

2. engine speed sensor (input speed to transmission),

3. transmission output speed sensor.

Based on the operator gear selection, the algorithm selects a desired gear ratio. If the operator selects P or N, the corresponding gear ratio selection is directly made based on that selection of the operator. If the operator selects R or D, the gear ratio is selected based on the engine speed and transmission output speed sensor. The algorithm estimates the vehicle speed based on the transmission output speed (neglecting the effects of wheel slip), and decides on the appropriate gear. This decision is based on the principle illustrated in Figure 3.32.

Given current gear and current engine speed, from the gear shift table we can determine *up-shift speed* and *down-shift speed* for the transmission output speed (Figure 3.32b). If the current transmission output speed is between these two values, no gear change is made. If it is below the down-shift speed value, the gear is shifted down one gear. If it is larger than the up-shift speed value, the gear is shifted up one gear. In order to make sure the gear shift does not occur due to noise in the signal, the measured transmission output speed is verified to be out of the range defined by up-shift and down-shift speed values for a period of time, that is 100 ms, before an actual up-shift or down-shift decision is made.

Next, using a look-up table, the algorithm determines which clutch/brake combination should be engaged and disengaged (Figure 3.32a). An example of such a table is shown in Figure 3.27. Once the decision of which clutch/brake combination needs to be engaged or disengaged is made, the next block on the control algorithm implements a pressure control algorithm to achieve the desired pressures in those clutch/brakes as a function of time as shown in Figure 3.32c. This is to control the transient response of the gear shift. The figure shows the typical desired pressure profile for smooth gear shifting in automatic transmissions. In large equipment applications, typical clutch/brake engagement

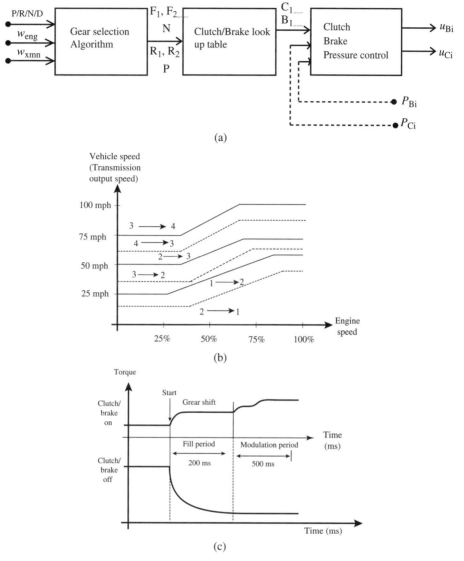

FIGURE 3.32: Automatic transmission control algorithm: (a) different logical components of an automatic transmission control algorithm, (b) gear selection logic, (c) controlled pressure as a function of time in engaged and disengaged clutch/brakes.

and disengagement takes between 300–700 ms. Notice that a clutch/brake engagement is performed in two phases:

1. fill phase (up to 200 ms), and
2. modulation phase (up to 500 ms).

During part of the gear shift process, engine torque is not transmitted from the transmission input shaft to the output shaft. As a result, the lower powertrain (components between the transmission output shaft and wheels) does not get power from the engine during part of the gear shift time period. That is why it is important that the gear shift happens smoothly without too much time delay so that the vehicle does not lose noticeable speed. On the other hand, if the shift happens too fast there will be too much variation in the speed

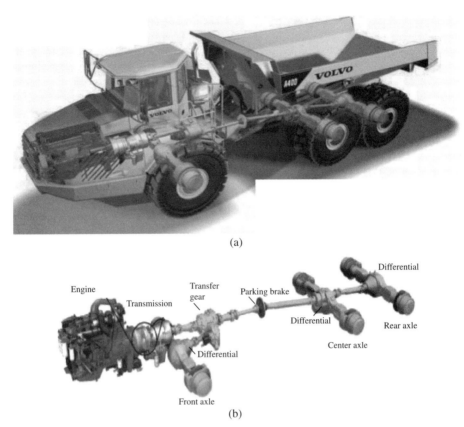

(a)

(b)

FIGURE 3.33: Powertrain of an articulated truck: Volvo Model A series, (a) picture of the articulated truck showing the powertrain, (b) powertrain components: engine, transmission with torque converter, differentials, (front, center, rear, inter axle (two of them)) and multi disc brakes. Each differential has locking mechanism under operator (or ECM) control for traction control purposes (Reproduced with permission from Volvo Trucks).

of the vehicle (too much accelleration or deceleration). Therefore, a good balance must be found between a "smooth yet fast enough gear shift." The clutch/brake modulation is accomplished by controlling the solenoid operated valves that control the pressure in the respective control cylinders of the clutches/brakes Figure 3.31. Each valve used to control one of the clutch/brake is proportionally operated. It has at least three ports: the pump pressure supply port, tank pressure supply port, output pressure port. By controlling the position of the valve spool, the output pressure of the valve is proportionally controlled between the pump supply pressure port and tank pressure port. Such valves are referred to as *electronically controlled pressure control* (ECPC) valves. The logic of pressure modulation may be open loop based on a time profile or closed loop based on measured pressure feedback. Automatic transmission control algorithms have different names based on the type of control logic they implement in the "gear selection algorithm" block (Figure 3.32a) such as *full throttle shift*, *part throttle shift*, *controlled throttle shift*.

3.8.7 Example: Powertrain of Articulated Trucks

An articulated truck powertrain is shown in Figures 3.33 and 3.34 where a picture of an articulated truck is shown along with the components of powertrain. The powertrain components are

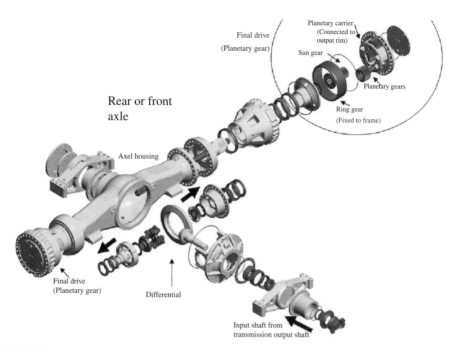

Planetary carrier
(Connected to
output rim)

Final drive

(Planetary gear)

Sun gear

Planetary gears

Rear or front
axle

Ring gear
(Fixed to frame)

Axel housing

Final drive
(Planetary gear)

Differential

Input shaft from
transmission output shaft

FIGURE 3.34: Axle (front or rear) of a vehicle: differential, final drive based on a single stage planetary gear (used in heavy equipment applications), and brakes in the final drive housing. The sun gear is driven by the shaft coming from the differential, the ring gear is fixed to the frame, the planetary carrier is connected to the output shaft (to the tire-wheel). Reprinted Courtesy of Caterpillar Inc.

1. The diesel engine which is the mechanical power source for the machine.

2. The automatic transmission with torque converter. The word "automatic" transmission means that the gear shifting at the planetary stage is handled automatically by the ECM. The torque converter provides the "flexible coupling" function between the diesel engine and planetary gear stage. In this particular case, there is also a *hydraulic retarder* in the transmission (also see Figure 3.27, component D). The role of the hydraulic retarder is to provide the braking role by simply spinning the hydraulic fluid in the transmission and transforming the mechanical energy into heat. As a result, the hydraulic retarder is used as part of the braking system to decelerate the machine. In many operating conditions it can meet the deceleration requirement of the machine such that the disc-brakes do not need to be applied, hence improving the brake life due to reduced usage of brakes.

3. A transfer gear is used to transmit power from transmission output to both front axle and center/rear axels.

4. The front/center/rear axels each have a differential (Figure 3.35), left and right axle shafts. On each axel shaft there are disc-brakes and a final gear reduction mechanism. In addition, the differentials have locking mechanisms which are controlled by operator input (or automatically by the ECM) for the purpose of traction control.

The *differential gear*, shown in Figure 3.35, has three shafts: one input shaft and two output shafts. The relative motion of the planetary gear (component number 3) adds in the opposite direction to the left and the right output shafts. The purpose of the differential mechanism is to allow different output speeds between two output shafts driven by the

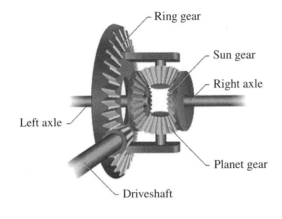

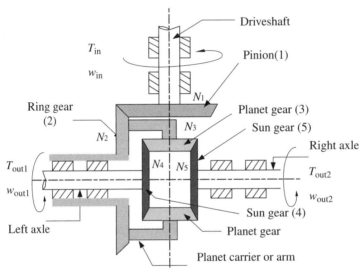

FIGURE 3.35: Automotive differential gear as an example.

same input shaft. In mobile and automotive applications this is needed when the vehicle travels over a path that has curvature, not a straight line. In that case, in order to avoid tire slip, the outside wheels must rotate at a faster rate than the inside wheels.

The fundamental relationship for differential operation can be expressed under two conditions:

1. *unlocked differential*,
2. *locked differential*.

The *unlocked differential* mode is the normal mode of operation of the differential. In this case the output torques transmitted to each output shaft are the same. Output shaft torques cannot be different, because the reaction forces on the planetary gear from the output shafts must be the same, but the output shaft speeds can be different which is determined by the dynamics of the whole system.

In the *locked differential* mode, that is the planetary gear is not allowed to rotate about its own axis, the output speeds of the two shafts must be the same, but the transmitted torques to output shafts can be different. In this case the differential becomes a regular gear reducer, and its "differential" function is cancelled. In unlocked and locked differential conditions, the following fundamental relations hold,

$$T_{out1} = T_{out2}; \quad unlocked \ differential \ mode \tag{3.329}$$

$$w_{out1} = w_{out2}; \quad locked \ differential \ mode \tag{3.330}$$

where T_{out1}, T_{out2} are output torques, and w_{out1}, w_{out2} are output shaft speeds of the differential.

When in locked condition (or unlocked condition and the planetary gear speed around its carrier axis is zero, hence both output shaft speeds are same), the differential's gear ratio is (Figure 3.35)

$$N_{diff} = \frac{N_1}{N_2} \tag{3.331}$$

$$w_{out1} = w_{out2} = N_{diff} \cdot w_{in} \tag{3.332}$$

When it is in unlocked condition, output speeds can be different, which is determined by the dynamic interactions of the load and traction conditions and the type motion path the vehicle is following (i.e., a curved path),

$$w_{out1} = N_{diff} \cdot w_{in} + \Delta w \tag{3.333}$$

$$w_{out2} = N_{diff} \cdot w_{in} - \Delta w \tag{3.334}$$

where Δw speed is added and subtracted from two output shafts due to the rotation of the planetary gear about its carrier axis, which results in "differential" speeds between the two output shafts.

Normally, a differential operates in the unlocked case. If the vehicle travels on curved paths, the differential allows the outside wheel to rotate faster than the inside wheel in order to accomodate the speed differential needed. This is provided by the rotation of the planetary gear about its own axis. This is the main function of an unlocked differential operation, reducing tire slip and wear in curved paths.

The drawback of the unlocked differential condition is that the torques transmitted to each output shaft are the same. The implication of this is that if the two wheels have different traction conditions on the ground, then one of the wheels which has less traction will spin much faster than the other wheel. Eventually, the maximum torque that can be delivered to each wheel is limited by the smallest of the traction torques available between the tires and ground. Consider a case where a vehicle is on straight line path, but the left wheel has almost no traction due to ice, while the right wheel has good ground traction. If we start from a stopped condition, the left wheel spins at a high speed and provides not much traction to the vehicle, and the right wheel will move very slowly or will not move at all (or even rotate in reverse direction) depending on how small the traction torque available between the tire and icy surfacis. The differential will not be able to direct more torque to the wheel that has more traction. The maximum torque that the differential can transmit is limited by the smaller of the two traction torques that can be supported by each tire–ground interaction. Under these conditions, the differential is better to be locked to provide better traction at the expense of giving up the differential function (different output shaft speeds) temporarily. When the differential is locked, both output shafts must then rotate at the same speed. This eliminates the condition that the maximum torque transmitted is limited by the minimum traction. As a result, traction is improved. However, the penalty paid is that if the vehicle is moving on a curved path, there will be tire slip and more wear on tires, and steering quality will be poorer, since the wheel speeds are forced to be same by locking the differential.

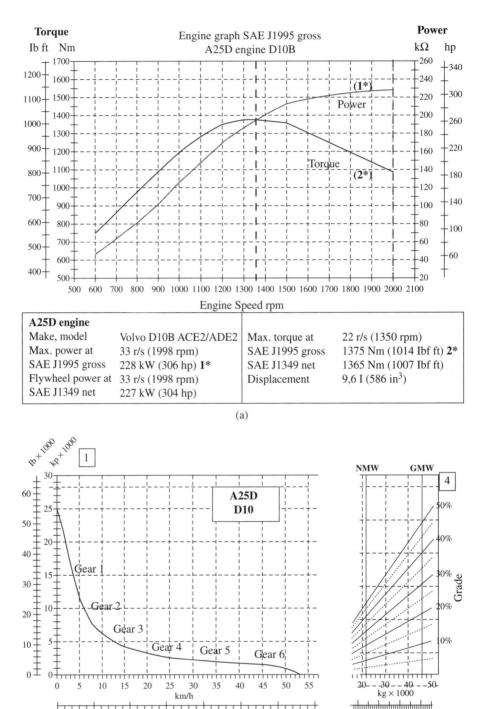

FIGURE 3.36: Engine lug curve, machine speed versus rim pull force capacity: (a) engine lug curve for torque and power output as a function of engine speed at full throttle, (b) rim pull force at different gear ratios and machine speeds. In order to determine the maximum uphill grade the machine can climb at a given speed and machine weight, draw a vertical line from the desired machine speed to the rim pull curve, from there draw a horizontal line to the grade curve. Read the grade value which this horizontal line crosses for the machine weight (Reproduced with permission from Volvo Trucks).

The traction force (rim pull force) capability and speed of the truck directly determine its productivity capacity. The performance curve that defines the maximum traction force (rimpull force) as a function of machine speed is determined by the engine torque–speed curve (lug curve), transmission (torque converter and planetary gear reducer set), and the lower powertrain gear ratios (differential, final drive). Figure 3.36 shows a manufacturer specified performance curve for rimpull versus machine speed. A commonly used performance criteria for such machines is their grading capability. That is the maximum up-hill grade it can climb at a given machine speed and machine weight. In order to determine the maximum uphill grade capability the machine can climb at a *given machine speed and machine weight* (Figure 3.36),

1. draw a vertical line from the given machine speed to rim pull curve,
2. from there draw a horizontal line to the grade curve,
3. then, read the grade value which this horizontal line crosses for the machine weight.

This is the maximum uphill grade the machine can climb at the given machine speed and machine weight. For instance, at about 15 km/hr machine speed, and machine weight of 40 000 kg, the maximum uphill grade this machine can climb is about 10%. If the machine weight is 20 000 kg, the machine can climb about 20% grade at the same speed.

3.9 PROBLEMS

1. Consider a gear reducer as shown in Figure 3.1. Let the diameter of the gear on the input shaft be equal to $d_1 = 2.0$ in and width $w_1 = 0.5$ in. Assume that the gear is made of steel and that it is a solid frame without any holes. The ouput gear has the same width and material, and the gear reduction from input to output is $N = 5$, $(d_2 = 10.0$ in). The length and diameters of the shafts that extend to the sides of the gears are $d_{s1} = 1.0$ in, $d_{s2} = 1.0$ in and $l_{s1} = 1.0$ in, $l_{s2} = 1.0$ in. Let us consider that there is a net load torque of $T_{L2} = 50$ lb in at the output shaft?

   1. Determine the net rotary inertia reflected on the input shaft due to two gears and two shafts.
   2. Determine the necessary torque at the input shaft to balance the load torque.
   3. If the input shaft is actuated by a motor that is controlled with $1/10$ degree accuracy, what is the angular positioning accuracy that can be provided at the output shaft.

2. Repeat the same analysis and calculations for a belt and toothed pulley mechanism. The gear ratios and the shaft sizes are the same. The load torque in the output shaft is the same. Neglect the inertia of the belt. Comment on the functional similarities. Also discuss practical differences between the two mechanisms.

3. Consider a linear positioning system using a ball-screw mechanism. The ball-screw is driven by an electric servo motor. The ball screw is made of steel, has length of $l_{ls} = 40$ in, and diameter of $d_{ls} = 2.5$ in. The pitch of the lead is $p = 4$ rev/in (or the lead is 0.25 in/rev). Assume that the lead-screw mechanism is in vertical direction and moving a load of 100 *lbs* against the gravity up and down in z-direction.

   1. Determine the net rotary inertia reflected on the input shaft of the motor.
   2. Determine the necessary torque at the input shaft to balance the weight of the load due to gravity.
   3. If the input shaft is actuated by a motor that is controlled with $1/10$ degree accuracy, what is the translational positioning accuracy that can be provided at the output shaft?

4. For problem 3, consider that the typical cyclic motion that the workpiece is to make is defined by a trapezoidal velocity profile. The load is to be moved a distance of 1.0 in in 300 ms, to wait there for 200 ms and then move in the reverse direction. This motion is repeated continuously.

Assume the 300 ms motion time is equally divided between acceleration, run, and deceleration times, $t_a = t_r = t_d = 100$ ms (Figure 3.13).

1. Calculate the necessary torque (maximum and continuous rated) and maximum speed required at the motor shaft. Select an appropriate servo motor for this application.
2. If an incremental position encoder is used on the motor shaft for control purposes, what is the required minimum resolution in order to provide a tool positioning accuracy of 40 μin.
3. Select a proper flexible coupling for this application to be used between the motor shaft and the lead-screw. Include the inertia of the flexible coupling in the inertia calculations and motor sizing calculations. Repeat step 1.

5. Repeat the same analysis and calculations of Problem 3 and 4, if a rack and pinion mechanism was used to convert the rotary motion of motor to a translational motion of the tool.

1. First, determine the rack and pinion gear that gives the same gear ratio (from rotary motion to translational motion) as the ball-screw mechanism.
2. Discuss the differences between the ball-screw, rack and pinion, and belt and pulley (translational version) mechanisms.

6. Given a four-bar linkage (Figure 3.7), derive the geometric relationship between the following motion variables of the mechanism. Let the link lengths be $l_1 = 1.0\,\text{m}, l_2 = 2.0, l_3 = 1.5\,\text{m}, l_4 = 1.0\,\text{m}$.

1. Input is the angular position $(\theta_1(t))$ of link 1, output is the angular position of link 3, $(\theta_3(t))$. Find $\theta_3 = f(\theta_1)$. Plot θ_3 for one revolution of link 1 as function of θ_1.
2. Determine the x and y coordinates of the tip of the link 3 during the same motion cycle. Plot the results on the x-y plane (path of the tip of link 3 during one revolution of link 1).

7. Consider the cam and follower mechanism shown in Figure 3.9. The follower arm is connected to a spring. The follower is to make an up–down motion once per revolution of the cam. The travel range of the follower is to be 2.0 in in total.

1. Select a modified trapezoidal cam profile for this task.
2. Assume the input shaft to the cam is driven at 1200 rpm constant speed. Calculate the maximum linear speed and linear accelerations experienced at the tool tip.
3. Let the stiffness of the spring be $k = 100$ lb/in and the mass of the follower and the tool it is connected to $m_f = 10$ lb. Assume the input shaft motion is not affected by the dynamics of the follower and tool. The input shaft rotates at constant speed at 1200 rev/min. Determine the net force function at the follower and tool assembly during one cycle of the motion and plot the result. Notice that

$$F(t) = m_f \ddot{x}(t) + k \cdot x(t) \tag{3.335}$$

and $x(t), \ddot{x}(t)$ are determined by the input shaft motion and cam function. What happens if the net force $F(t)$ becomes negative? One way to assume that $F(t)$ does not become negative is to use a preloaded spring. What is the preloading requirement to ensure $F(t)$ is always positive during the planned motion cycle? The preload spring force can be taken into account in the above equation as follows,

$$F(t) = m_f \ddot{x}(t) + k \cdot x(t) + F_{pre} \tag{3.336}$$

where $F_{pre} = k \cdot x_o$ is a constant force due to the preloading of the spring. This force can be set to a constant value by selection of spring constant and initial compression.

8. Consider an electric servo motor and a load it drives through a gear reducer (Figure 3.14).

1. What is the generally recommended relationship between the motor inertia and reflected load inertia?
2. What is the optimal relationship in terms of minimizing the heating of the motor?
3. Derive the relationship for the optimal relationship.

9. Consider the coordinate frames $x_2 y_2 z_2$ attached to link 2 and $x_3 y_3 z_3$ attached to link 3 in Figure 3.16. Let the length of link 2 be l_2.

1. Given that joint 2 and joint 3 axes are parallel to each other (z_2 is parallel to z_3), write the 4x4 transformation matrix that describes the coordinate frame 3 with respect to coordinate frame 2; that is, $T_{23} = ?$
2. Determine T_{32}, the description of the coordinate frame 2 with respect to coordinate frame 3 (Hint: $T_{32} = T_{23}^{-1}$).

10. Consider the problem in Figure 3.20. Let $l_1 = 0.5$ m, $l_2 0.25$ m, and F be the weight of a payload mass of $m = 100$ kg, then $F = 9.81 \cdot 100\,N$, in the vertical negative direction of Y_0. Determine the joint torques necessary to hold the load in position at the following conditions. Neglect the lengths of the links.

1. $\theta_1 = 0°$, $\theta_2 = 0°$
2. $\theta_1 = 30°$, $\theta_2 = 30°$
3. $\theta_1 = 90°$, $\theta_2 = 0°$

11. The objective of this problem is to illustrate the problem of *backlash* in motion control systems and how to deal with it. Consider the closed loop position control system shown in Figure 10.4. Assume that the gear motion transmission mechanism is a lead-screw type (Figure 3.3). Further, let us model the components as follows:

- the motor dynamics as inertia only (J_m) with no damping, and current to torque gain is K_T,
- a position sensor connected to the motor that gives N_{s1} count/rev number of counts per revolution,
- amplifier as a voltage to current gain (K_a) with all filtering effects neglected,
- lead-screw and the load it carries is modeled with its effective gear ratio ($N = 1/(2\pi p)$) where p(rev/mm) is the pitch and mass m (rotary inertia of the lead screw is neglected). Assume there is load force acting on the inertia (F_l). In addition consider that the lead screw has a backlash of x_b, which we will assume is a constant value.
- control algorithm is implemented with an analog op-amp as a form of PID controller.

Assume the following numerical values for the system components: $J_m = 10^{-5}$ kg $\cdot$ m^2, $K_T = 2.0$ Nm/A, $N_{s1} = 2000$ count/rev, $K_a = 2A/V$, $p = 0.5$ rev/mm, $m = 100$ kg, $F_l = 0.0N$, $x_b = 0.1$ mm. For simplicity, use the following relationship for the total inertia acting on the motor (although during the period of motion when backlash is in effect and the lead-screw is not moving the nut, the load inertia is not coupled to the motor; but we will neglect this) as well as motor torque and transmitted force to the moving mass,

$$J_t = J_m + \frac{1}{(2\pi p)^2} \cdot m \tag{3.337}$$

$$T_l = \frac{1}{2\pi p} \cdot F_l \tag{3.338}$$

(a) If the desired positioning accuracy of the load is 0.001 mm, draw a control system block diagram and sensors to achieve this (Hint: due to backlash, we must have a load-coupled position sensor. Let the resolution of that sensor be called N_{s2} counts/mm. Further, we should have a measurement accuracy that is 2 to 5 times better than desired positioning accuracy).
(b) Develop a dynamic model of the closed loop control system (i.e., using Simulink®). Simulate the motion in response to a rectangular pulse, that is initial position and commanded position are at zero until $t = 1.0$ s, then a step position command of 1.0 mm and back to zero position command at $t = 3.0$ s, and continue simulations until $t = 5.0$ s. Use the motor-coupled position sensor for velocity loop with a P-only gain, and a load-coupled position sensor in the position loop with a PD-type control. Adjust the gains in order to achieve a good response.
(c) What happens if you use only the motor-coupled position sensor, not the load-coupled position sensor? Show your claim with simulation results. Modify component parameters if necessary to illustrate your point.

(d) What happens if you use only the load-coupled position sensor, not the motor-coupled position sensor? Show your claim with simulation results. Modify component parameters if necessary to illustrate your point.

Simulate a condition as follows: the engine runs at a constant speed. The friction coefficient at each tire–ground contact is constant and the same for all tires, and the vehicle weight is equally distributed in all tires. The initial direction of motion and steering is a straight line motion. The steering angle is zero at all times, which means all the wheels are aligned to move in a straight line. Based on a different desired gear selection as a function of time (i.e., step changes in gears at specific points in time), assume instantaneous gear shift and clutch engagement/disengagement. Assume the clutch stays engaged at all times except when the neutral gear is selected.

12. (This problem can be assigned as a small research project for students).

Develop a mathematical model of a four wheel drive vehicle including the following components of the powertrain (Figure 3.37):

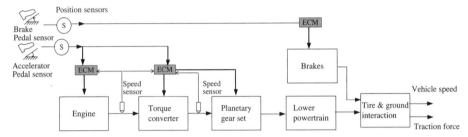

FIGURE 3.37: Automotive powertrain block diagram: engine, transmission (torque converter and planetary gear set), lower powertrain, brake, and tire–ground interaction.

1. Engine is to be modeled as a lug curve for different throttle values (without any transient dynamics), $T_{eng}(\theta_{throttle}, w_{end}$,

$$J_{eng} \cdot \dot{w}_{eng}(t) = T_{eng}(\theta_{throttle}, w_{eng}) - T_{imp}(t) \qquad (3.339)$$

$T_{eng}(\theta_{throttle}, w_{end})$. This function is commonly referred as the "engine map." The maximum torque of the engine at full throttle is 1000 N m. Define an reasonable engine map based on this information, that is, $T_{eng}(\theta_{throttle}, w_{end})$, look up table.

2. Transmission is to be modeled on a torque converter and a planetary gear set. We model the torque converter with its steady-state torque ratio and primary torque functions, whereas the planetary gear set is assumed to shift to the desired gear instantaneously. The transient time in the gear shifting is neglected, as well as the inertial and stiffness characteristics of the transmission. Simply model the planetary gear set as an ideal gear ratio device.

$$T_{imp}(t) = T_p(w_{turb}/w_{eng}) \cdot \frac{1}{w_r^2} \cdot w_{eng}^2 \qquad (3.340)$$

$$T_{turb} = N_t(w_{turb}/w_{eng}) \cdot T_{imp}(t) \qquad (3.341)$$

$$N_p = N_p(gear) \qquad (3.342)$$

$$T_{out}(t) = N_p \cdot T_{turb}(t) \qquad (3.343)$$

$$w_{out}(t) = \frac{1}{N_p} \cdot w_{turb}(t) \qquad (3.344)$$

The torque converter is to be modeled as two steady-state functions: primary torque and torque ratio, which are defined as a function of the speed ratio. The efficiency of the torque converter is assumed to be 100%. Define the two steady-state curves of the torque converter

$$T_p(w_{eng}, w_{turb}) = 1000 - 4000 \cdot \left(\left(\frac{w_{turb}}{w_{eng}} \right) - 0.5 \right)^2 \tag{3.345}$$

$$N_r(w_{eng}, w_{turb}) = 2.5 \cdot \left(1 - \frac{w_{turb}}{w_{eng}} \right) \tag{3.346}$$

where w_r is the rated speed where torque converter performance is given, and N_p is the gear ratio of the planetary gear set at the selected $gear = \{-1, 0, 1, 2, 3, 4, 5\}$, $N_p = \{-2.5, 0, 2.5, 1.5, 1.0, 0.75, 0.5\}$.

3. There are two identical differentials: one for the front wheel pair, and one for the rear wheel pair. Let us assume they are all in locked condition. The final drive gear ratio does not exist. In other words the output shafts of the differentials drive the wheels directly.

$$J_{lp} \cdot \dot{w}_{wheel}(t) = T_{out}(t) - T_{brake}(t) - \frac{1}{R_{wheel}} \cdot F_{traction}(t) \tag{3.347}$$

4. Friction between the wheel tires and ground is based on a constant friction coefficient and assumed to be the same for all.

$$m_{vehicle} \cdot \ddot{x}(t) = F_{traction}(t) - F_{load}(t) \tag{3.348}$$

where

$$F_{traction} = \mu_r \cdot \frac{1}{R_{wheel}} \cdot T_{out}(t) \tag{3.349}$$

$$F_{load} = F_{friction} + F_{drag} + F_{gravity} \tag{3.350}$$

The whole vehicle weight, $W_v = m_v \cdot g$, is equally distributed among the four tires, where m_v is the total mass of the vehicle, $g = 9.81 \mathrm{m/s^2}$, μ_t is the traction coefficient. Assume identical traction conditions at each tire. The student is asked to specify reasonably realistic parameters and performance for the components as follows: Given the performance curves and parameters,

$$T_{eng}(\theta_{throttle}, w_{eng}) \quad \textit{engine map} \tag{3.351}$$

$$w_r = 2400 \, \mathrm{rpm} \tag{3.352}$$

$$N_p(-1, 0, 1, 2, 3, 4, 5) = \{-2.5, 0, 2.5, 1.5, 1.0, 0.75, 0.5\} \tag{3.353}$$

$$J_{lp} = 1.0 \, \mathrm{kg \, m^2} \tag{3.354}$$

$$R_{wheel} = 0.5 \, \mathrm{m} \tag{3.355}$$

$$m_{vehicle} = 1\,000 \, \mathrm{kg} \tag{3.356}$$

$$\mu_r = 0.0 \quad \text{for case 1} \tag{3.357}$$

$$= 0.9 \quad \text{for case 2} \tag{3.358}$$

$$F_{friction}(t) = 0.0 \tag{3.359}$$

$$F_{drag}(t) = 0.0 \tag{3.360}$$

$$F_{gravity}(t) = m_{vehicle} \cdot 9.81 \cdot \sin(15°) \tag{3.361}$$

Simulate a condition as follows. The vehicle speed is zero initially at time $t = 0.0 \, \mathrm{s}$. The engine runs at a constant speed and at a constant throttle, 100%. The initial direction of motion and steering is a straight line motion. The slope of the ground is 15° and the vehicle is climbing. The gear is shifted from neutral to first gear at time $t = 1.0 \, \mathrm{s}$ and the throttle is at 100%. Until that time, the engine is simply running at a steady-state speed of 2200 rpm. The friction coefficient at each tire–ground contact is constant and the same for all tires, and the vehicle weight is equally distributed in all tires.

Note: This model can be made more accurate by including dynamics for the engine response and its controls, dynamics of the clutch and its transient engagement/disengagement, inertia-stiffness and gear-change transient dynamics of the transmission, inertia of the axles and differential, and a more accurate model of friction at each tire and ground contact. Furthermore, the steering motion can be included along with the differential dynamics in unlocked condition.

CHAPTER 4

MICROCONTROLLERS

Most of the discussions in this chapter are based on the PIC 18F452 microcontroller (or PIC 18F4431 version which has a quadrature encoder interface). The following manuals can be downloaded from **http://www.microchip.com** and should be used as a reference and as part of this chapter:

1. PIC 18FXX2 Data Sheet or PIC 18F4431 Data Sheet
2. MPLAB IDE V6.xx Quick Start Guide
3. MPLAB C18 C Compiler Getting Started
4. MPLAB C18 C Compiler Users' Guide
5. MPLAB C18 C Compiler Libraries

4.1 EMBEDDED COMPUTERS VERSUS NON-EMBEDDED COMPUTERS

The digital computer is the brain of a mechatronic system. As such, it is called the *controller* when used for the control function of an electro-mechanical system. Any computer with proper I/O interface devices (digital and analog I/O) and software tools can be used as a controller. For instance, a desktop PC can be used as a process controller by adding an I/O expansion board and control software. Clearly, there are many hardware components on a desktop PC (a non-embedded computer) that are not needed for process control functions. An embedded computer uses only the necessary hardware and software components and is much smaller than a non-embedded computer, such as a desktop PC. An embedded computer used as the controller of a mechatronic system is referred to as the *embedded controller*. A microcontroller is the main building block of an embedded computer.

Figure 4.1a shows a comparison between an *embedded* and *non-embedded* (i.e., desktop) computer. Figure 4.1b shows a picture of a commercially available embedded controller used in mobile equipment applications. The main differences between embedded and non-embedded computers are as follows:

1. Embedded computers are generally used in real-time applications. Therefore, they have hard real-time requirements. *Hard* real-time requirement means that certain tasks must be completed within a certain amount of time, or the computer must react to an external event within a certain time. Otherwise the consequences may be very serious. The consequences of not meeting the real-time response requirements in a desktop application are not as serious.

2. Embedded computers are not general purpose computing machines, but have more specialized architectures and resources. For instance, a desktop computer would have

Mechatronics with Experiments, Second Edition. Sabri Cetinkunt.
© 2015 John Wiley & Sons, Ltd. Published 2015 by John Wiley & Sons, Ltd.
Companion Website: www.wiley.com/go/cetinkunt/mechatronics

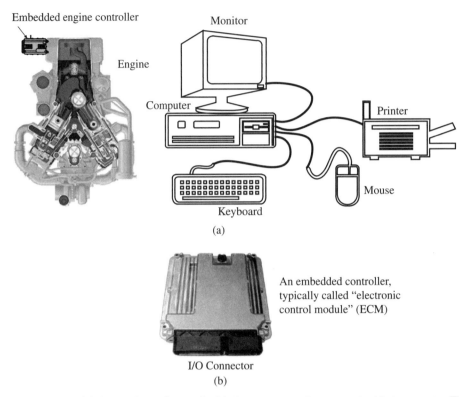

FIGURE 4.1: (a) Comparison of an embedded computer and a non-embedded computer. The embedded computer has just enough resources for the application, must operate in a harsh environment, has a smaller physical size and has hard real-time requirements. (b) Picture of a rugged "embedded controller" used in off-highway equipment applications.

a hard disk drive, floppy disk drive, CD/DVD drive, and tape drive for permanent data storage, whereas an embedded computer may have battery backed RAM or flash or ROM memory to store just the application software. Embedded microcontrollers have limited resources in terms of power (i.e., they may be powered by a battery), memory, and CPU speed. Embedded computers are dedicated to specific tasks. They do not store general purpose programs such as word processors, graphics programs, and so on.

3. Embedded controllers and I/O interfaces are more integrated at the chip design level than general purpose computers, that is I/O interface channels such as analog to digital (ADC) and digital to analog (DAC) converters are integrated into the microchip hardware.

4. The physical size of the embedded computer is typically required to be very small, which may be dictated by the application.

5. Embedded controllers operate in extreme environmental conditions (i.e., an embedded controller for a diesel engine must operate under conditions of large temperature and vibration variations).

6. Embedded computers invariably incorporate a *watchdog timer* circuit to reset the system in case of a failure.

7. Embedded microcontrollers may have a dedicated debugging circuit on the chip so that the timing of all the I/O signals can be checked and the application program

debugged on the target hardware. Debugging tools (hardware and software) are a very important part of an embedded system development suite. Unlike desktop applications, real-time embedded applications must be debugged and I/O signals must be checked for worst case conditions.

8. The developer must know the details of the embedded system hardware (bus architecture, registers, memory map, interrupt system), since the application software development is influenced by the hardware resources of the embedded computer.

9. Interrupts play a very important role in almost all embedded controller applications. It is through the interrupts that the embedded controller interacts with the controlled process and reacts to events in real-time.

10. As the complexity increases, embedded systems require a real-time operating system (RTOS). RTOS provides already tested I/O and resource management software tools. For instance, Ethernet communication may be provided by RTOS functions instead of being written by ourselves. Furthermore, RTOS can provide task scheduling, guaranteed interrupt latency, and resource availability.

Perhaps the most significant factors that differentiate embedded computers from desktop computers are the *real-time requirement*, *limited resources*, and *smaller physical size*. The real-time performance of an embedded system is defined by how fast it responds to interrupts and how fast it can switch tasks.

Programming for a real-time application using a microcontroller/DSP, versus programming for a non-real time application such as using a desktop PC, has the following differences:

1. Memory resources in a desktop PC are very large and conserving memory is not a concern for the programmer. Whereas memory resources in microcontrollers are limited and memory space should be used carefully in order to not exceed that available. In real-time programming, depending on the microcontroller/DSP type and the development environment we use, the development environment at the compilation and link time may allow us to decide how to utilize the available memory for variables and constants, in order to best fit the application program into the available memory.

2. Computational time requirements are also not much of a concern in non-real time PC applications. Whereas, in real-time applications, everything must run in real-time and our program must be able to react to events in-time and in real-time. Therefore, long delays or long loops can be a problem in real-time applications. In real-time programming, we make use of interrupt-driven event handling to make sure the microcontroller responds to external "interrupt" events fast.

3. In real-time applications, safety is of great importance. If something unexpected happens, we need to define an ordered way of ending current program execution and be able to restart in a timely and defined order. In desktop applications, the so-called "blue screen" condition (a system failure) may simply be solved by rebooting the system and waiting for a few minutes. In real-time systems, waiting for a few minutes, even for a few seconds, may not be acceptable.

4. In real-time programming involving real-world hardware I/Os, the application software deals with reading and writing of bits, that is the status of a switch or turn ON/OFF a light. Data and registers are accessed in such a way that we refer to specific bits for read and write operations. We can think of I/O operations as read/write operations for a table of bits which corresponds to the I/O. In other words, all of the I/O data is simply the values of a set of registers and their bits. For instance, configuring and reading an ADC involves writing to a set of configuration registers

(writing their appropriate bits to "1" or "0"), and then when the conversion data is ready (determined by reading a specific bit from a status register), reading the data register's first 10 bits for a 10-bit ADC converter. For example, PIC 18F452 has about 100 special function registers (SFR) each of which is 8-bits long (B0-B7). This means that a total of 800 bits are used to handle (configure, read, write) all of the I/O. In non-real-time applications, we tend to work with data structures that are long and organized in some hierarchy as opposed to dealing with specific bits.

Design Steps of an Embedded Microcontroller-Based Mechatronic System

Design of a microcontroller-based mechatronic system includes the following steps. These steps encompass the microcontroller and its interface to the electromechanical system. We assume the electromechanical system is already designed.

Step 1: Specifications – define the purpose and function of the device. What are the required inputs (sensors and communication signals) and required outputs (actuators and communication signals)? This is the hardware requirement of microcontroller. What are the logical functions required? This is the software requirement of microcontroller.

Step 2: Selection – select a proper microcontroller that has the necessary I/O capabilities to meet the hardware requirements as well as the CPU and memory capabilities to meet the software requirements.

Step 3: Selection – select a proper development tool set for the microcontroller, that is PC-based development software tools and hardware tools for debugging.

Step 4: Selection – identify electronic components and sub-systems necessary to interface the microcontroller to the mechatronic system.

Step 5: Design – complete schematics for the hardware interface circuit, including each component identification and its connections in the interface circuit.

Step 6: Design – write the pseudo-code of the application software structure, its modules and flow chart.

Step 7: Implement – build and test the hardware.

Step 8: Implement – write the software.

Step 9: Implement – test and debug the hardware and the software.

Microcontroller Development Tools

Typical development tools used in an embedded controller development environment are grouped into two categories: hardware tools and software tools (Figure 4.2). The hardware tools include:

1. a desktop/laptop PC to host most of the development software tools,
2. the target processor (i.e., an evaluation board or the final target microcontroller hardware),
3. debugging tools such as a ROM emulator, logic analyzer,
4. EEPROM/EPROM/Flash writer tools (also called the "programmer", because it is used to download the program to the target microcontroller).

The software tools include the target processor specific

1. assembler, compiler, linker, and debugger,
2. real-time operating system (not required).

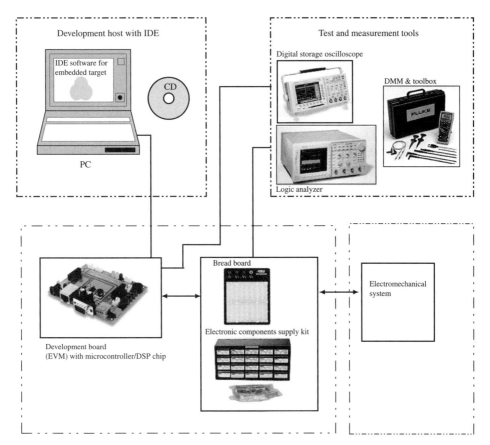

FIGURE 4.2: The components of a development setup for a microcontroller-based control system: PC as host development environment including the development software tools for the microcontroller, communication cable, microcontroller board, breadboard, test and measurement tools, and electronic components supply kit.

The compiler includes a *start-up code* for the embedded system to boot-up the system on RESET, check resource integrity, and load the application program from a known location and start its execution. The assembly code can be mixed with the C-code. However, we will not discuss the assembly code here.

Furthermore, the compiler generates *relocatable code*, which is then used by the "linker" to specifically locate it in the physical memory based on the selections made in the application file for the linker. This file is prepared by the developer to properly use the target system memory resources. The linker decides where in the memory to place the program code and the data.

In particular, the debugging tools deserve special attention. The debugging requirements for embedded applications are significantly more stringent that those of desktop applications. The reason is that the consequences of a failure are a lot more serious in real-time systems. The simplest form of debugging tool includes a debug kernel on the target system which communicates with a more comprehensive debugging software on the host computer. More recent embedded controllers include on-chip debugging circuitry which improves the debugging capability. A ROM emulator allows us to use RAM in place of ROM during the development phase so that the host can change the target code and write to the target hardware quickly.

Microcontroller Development Tools for PIC 18F452
The development board for PIC 18F452 (and PIC 18F4331) is the PICDEM 2 Plus Demo board plus various peripheral hardware devices that allow the PIC 18F452 microcontroller to effectively interface with electro-mechanical devices. These include a proper power supply through a 9V AC/DC adapter and regulator, a 4 MHZ clock, RS-232 interface, LCD, four LEDs, and a prototyping area.

The MPLAB ICD 3 module is a low cost debug and development tool that connects between the PC and the PICDEM 2 Plus Demo board (or the designer's target board) allowing direct in-circuit debugging of PIC 18F452 microcontroller (Figure 4.2). It is also used as the communication and programmer tool between PC and the PIC Development Board. Using the MPLAB ICD 3, PC communicates with PIC Development Board, downloads programs to it, and debugs programs while running it on the PIC microcontroller.

Programs can be executed in real-time or single step mode, watch variables established, break points set, memory read/writes accomplished, and more. It is also used as a development programmer for the microcontrollers to download and save the code in the microcontroller memory.

Hardware development tools are:

1. The PC as the host for development tools.
2. The development board (PICDEM 2 Plus Demo board), which has the microcontroller (PIC 18F452), its power supply, support circuits, and space for breadboard for custom application specific hardware interface development.
3. The communication cable and in-circuit debugger hardware (MPLAB ICD 3) for downloading and debugging programs.

Software development tools that run on a PC and are interfaced to the board via the communication cable are:

1. MPLAB IDE V.6xx integrated development environment (IDE) which includes an editor, assembler (MPASM), linker (MPLINK), debugger, and software simulator (MPLAB SIM) for the PIC chip.
2. MPLAB C18 C-compiler (works under MPLAB IDE V.6xx).

There are two debugger tools:

1. A debugger in software MPLAB SIM, which simulates the target PIC microcontroller on the PC, and allows debugging of the logic on the PC. This of course not a real-time debugger.
2. A real-time hardware debugger using the ICD3. ICD3 is also used at the communication device between the PC and PIC board, and as a programmer to download the executable code from the PC to PIC Board.

It is recommended that basic logic functionality be debugged using the MPLAB SIM simulator (software debugger). Then, when the I/O verifications is needed, the application program should be debugged on the target PIC board using the ICD3 hardware. Using ICD3, we can run the application program on the target PIC microcontroller and run it in real-time. However, during debugging we are most likely to run the program in sections, put "break points" to stop the program, "Step by Step" execute the code, and examine ("Watch") variables.

The PIC microcontroller can be programmed in a number of ways, using both C-language and assembly language. Our lab experiments require the use of the MPLAB IDE together with the MPLAB C18 ANSI-compliant C compiler installed on a PC. The language tool suite also consists of the MPASM assembly language interpreter, the MPLINK object linker, and debugger. The C18 compiler supports mixing assembly language and C language instructions in the same file. A block of assembly code included in a C source file must be labeled by

```
_asm

....

_endasm
```

The basic sequences for programming the PIC, using a PC as the development tool, are as follows:

1. On the PC, create a new project in the MPLAB IDE environment. Setup the project environment by selecting the target PIC microcontroller, specify directories to access libraries, add source codes to the project, and specify the project name. For each project, the following configurations must be made in the MPLAB IDE environment:

   (a) Create a project: `MPLAB IDE > Project> New`.
   (b) Select the target PIC microcontroller: `MPLAB IDE > Configure > Select Device: PIC18F452`.
   (c) Configure project options: `MPLAB IDE > Project > Select Language Toolsuite: Microchip C18 Toolsuite`.
   (d) Configure project options: `MPLAB IDE > Project > Set Language Tool Locations: MPLAB C18 C Compiler` and,
   `....... >Set Language Tool Locations: MPLINK Object Linker`.
   (e) Configure project build options: `MPLAB IDE > Project > Build Options...>Project`, then setup the options in the window under different tabbed pages (options: General, Assembler, Compiler, Linker options).
   (f) Add source files to the project: in the project window, right click on "Source Files" and select "Add Files."
   (g) Select a script file for the linker: in the project window, right click on "Linker Scripts," and select "Add Files," and then select file "18f452.lkr" from the "lkr" directory.

2. Write the program source code in C language, using the built-in editor or any ASCII text editor, and save it as *filename.c*. Add other relevant files to your project. `MPLAB IDE > File > New` and `MPLAB IDE > File > Save`.

3. Build (compile and link) the project in the MPLAB IDE environment. This converts the high-level C code to the corresponding hex files which contain the binary coded machine instructions: `MPLAB IDE > Project > Build All`.

4. First debug the program using the software simulator (SIM, provided as part of MPLAB IDE) for the PIC chip on the PC. This is a non-real-time simulation of the PIC chip. Once the program is debugged using the non-real-time simulator (MPLAB SIM), then it can be transferred to the PIC microcontroller. If programming in C language, the program must be re-built using a different linker script file (18*f*452*i.lkr* file for building a program to run on the actual PIC 18F452 chip, instead of 18*f*452.*lkr* which is used for debugging the program on the PC). Then, transfer the program from the PC to the PIC board through the MPLAB IDE environment and communication cable (MPLAB ICD 3):
   `MPLAB IDE > Programmer > Select Programmer` and
   `...........>Settings` and
   `MPLAB > Debugger > Connect`.

`MPLAB > Debugger > Program` to download the application program to the PIC microcontroller.

The debugging commands for non-real-time software debugger (MPLAB SIM) and hardware debugger MPLC ICD3 are largely the same. Typical debugging commands to find errors in the program are similar to debugging commands for other high level programming languages. Typically we need to be able to

1. **Run, Halt, Continue** the program execution.

2. **Single step, Step into** the function, **Step over** functions.

3. Setup "**Breakpoints**" at various lines in the program. When the program reaches that line, the program execution will halt. To setup a "Breakpoint", open the source file, place the cursor on the desired line, right-click on the line, and then select "Set Breakpoint" from the menu. A red icon should appear on that line to indicate that this is a breakpoint line. When the program execution reaches this line, the program will halt before executing this line.

4. Setup a "**Watch window**" to view the values of selected variables and registers, that is when the program stops at a breakpoint, values of various variables can be examined to check for errors.

   `MPLAB IDE > Wiew > Watch`

5. Give the run command to the PIC chip from the PC. Debug the code on the PIC chip with software and hardware debugging tools (i.e., MPLAB ICD3). When debugging a program, disable the watchdog timer (WDT). The watchdog timer can be enabled/disabled from the IDE menus. Otherwise, while the program is paused for debugging, the WDT will reset the processor.

4.2 BASIC COMPUTER MODEL

Let us consider the operation of a basic computer using a human analogy (Figure 4.3). As shown in the Figure 4.3, the human has a brain to process information, eyes to read, hands to reach various components, and fingers to write. There is also a clock. On the desk, there is a deck of cards which has the instructions to follow, a chalk, an eraser, a black board, input–output trays, and two pockets with one card each for quick access to read/write things on.

The analogy between this human model and a computer is as follows:

```
brain                     --- CPU
wall clock                --- clock
deck of instruction cards --- read only memory (ROM)
chalk-eraser-black-board  --- random access memory (RAM)
pocket cards              --- accumulators (also called registers)
input-output tray         --- I/O devices
eyes, hands and arms      --- bus to access resources (read/write)
```

There are seven basic components of a computer:

1. The CPU which is the brain of the computer. The CPU is made of a collection of arithmetic/logic units and registers. For instance, every CPU has
   (a) a program counter (PC) register which holds the address of the next instruction to be fetched from the memory,

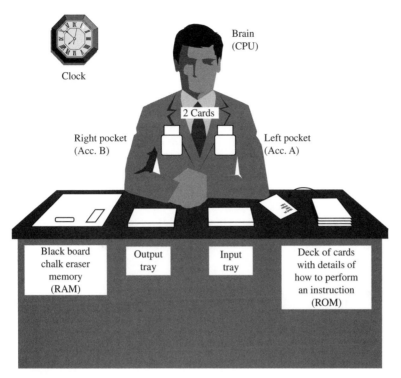

FIGURE 4.3: Basic computer and human analogy.

    (b) an instruction decoder register, which interprets the fetched instruction and passes on the appropriate data to other registers,

    (c) an arithmetic logic unit (ALU), which is the "brain inside the brain," executes the mathematical and logical operations.

2. A clock – a computer executes even the simplest instruction at the tick of a new clock cycle. The clock is like the heart of a human body. Nothing happens in the computer without the clock.

3. A ROM – read only memory which contains information on how to execute the basic instruction set. It can only be read from and cannot be written to. It maintains data when power is lost. EPROM is an erasable programmable ROM. An EPROM chip has a window where the memory can be erased by exposing it to ultraviolet light and then be reprogrammed. EEPROM is an electrically erasable programmable ROM where the memory can be re-written by electrical signals in the communication interface without any ultraviolet light.

4. RAM memory – random access memory serves as an erasable blackboard where the information can be read from and written to. Data is lost if power is lost. *Static* RAM and *dynamic* RAM are two common RAM types of memory. Static RAM stores data in flip-flop circuits and does not require a refresh–write cycles to hold the data as long as power is not lost. Dynamic RAM requires periodic refresh–write cycles to hold the data even when power is maintained.

5. Registers are a few specific memory locations which can be accessed faster than the other RAM memory locations.

6. I/O devices – every computer must interact with a user and external devices to perform a useful function. It must be able to read in information from the outside world, process it, then output information to the outside world. An I/O device interface chip is also called *peripheral interface adaptor* (PIA).

7. A bus to allow communication between the CPU and devices (memory, I/O devices). The bus includes a set of lines that includes those to provide power, address, data, and control signals.

Let us assume that the human is supposed to pick a new card from the deck (ROM) at the tick of every minute of the clock, read it, execute the instruction, and continue this process until the instruction cards say to stop. Here is a specific example. The program is to read numbers from the input tray. It reads until five odd numbers are read. It adds the five odd numbers and writes the result to the output tray at the end. Then, the program execution stops.

Minute: 1, 2, 3, 4,...- pick a card at the tick of every minute from the deck of cards (ROM). Assume that the first four cards in the deck include the following instructions.

Card 1: Read a number from the input tray, write it on the card on the left pocket (Accumulator A).

Card 2: Is the number odd? If yes, write it on the blackboard.

Card 3: Are there five numbers on the black board? If yes – go to Card 4, if no – go to Card 1.

Card 4: Add all numbers, write the total number to the output tray and stop.

Notice the following characteristics in a computer program:

- normally the program instructions are executed sequentially,
- the order of execution can be changed using the conditional statements,
- the CPU, clock, ROM, RAM and accumulators, and I/O are the key components of a basic computer operation.

The digital computer is a collection of many ON/OFF switches. The transistor switches are so small that a 1000×1000 array of them (1 000 000 transistor switches) can be built on a single chip. A combination of transistor switches can be used to realize various logic functions (i.e., AND, OR, XOR) as well as mathematical operations (+, − , *, /).

Every CPU has an instruction set that it can understand. This is called the *basic instruction set* or the *machine instructions*. Each instruction has a unique binary code that tells the CPU what operation to perform and programmable operands for the source of data. Some microprocessors are designed to have a smaller instruction set. They have fewer instructions but execute them faster than general purpose microprocessors. Such microprocessors are called *reduced instruction set computers* (RISC).

Assembly language is a set of mnemonic commands corresponding to the basic instruction set. The mnemonic commands make it easier for programmers to remember the instructions instead of trying to remember their binary code (also called *hex code*, because it is easier to code each of four binary digits with one hex character 0–9 and A–F). A program written in assembly language must be converted to machine instructions before it can be run on a computer. This is done by the *assembler*.

All of the instructions in a program written in a high level language must first be reduced to the combinations of the basic instruction set the CPU understands. This is done by the compiler and the linker, which translates the high level instructions to low level machine instructions. The basic instruction set is microprocessor specific. The build process (compile and link) of the C18 compiler generates various files including the executable file with extension "*.hex". In addition, the filename extensions "*.map" contains the list of variable names and the allocated memory address for them. The filename extension "*.lst" contains the machine code (disassembled) generated for each line of the assembly and C-code in a program. These files can be useful in the debugging process.

High level programming languages try to hide the processor specific details from the programmer. This allows the programmer to program different microprocessors with a single high level language. In microcontroller applications, it is generally necessary to understand the hardware and assembly language capabilities of a particular microcontroller in order to fully utilize its capabilities.

A computer program minimally needs the following machine instructions,

- Access memory (and I/O devices), and perform read/write operations between the CPU registers and the memory (and I/O devices). These operations are typically called

  LOAD address (read from address) and

  STORE address (write to address).

- Mathematical (add, subtract, multiply) and logical operations, such as

  ADD address (implied to be added to the content of one the accumulators),

  SUB address (implied to be subtracted from the content of one the accumulators),

  AND, OR, NOT, JUMP, CALL, RETURN.

In high level languages, data structures (variables, structures, classes, etc.) are used to manage the information. Assembly language does not provide data structures to manage data in a program. It is up to the programmer to define variables and allocate proper memory address space for them. Assembly language provides access instructions to the memory and I/O devices as the source and destination for data. In addition it provides operators (mathametical, logical, etc.) as well as various decision making instructions such as GOTO, JUMP, CALL, RETURN. Using these instructions, we operate on the data in memory and I/O devices.

The computer CPU accesses the resources (ROM, RAM, I/O devices) through a *bus* which is the hardware connection between the components of a computer. The bus has four basic groups of lines Figure 4.4:

1. power bus
2. control bus
3. address bus
4. data bus.

The power bus provides the power for the components to operate. Each line carries a TTL level signal: 0 VDC or 5 VDC (OFF or ON). The control bus indicates whether the CPU wants to read or write. The address bus selects a particular device or memory location by specifying its address. The data bus carries the data between the devices. Notice that if address bus is 16 lines (16-bit), there can be $2^{16} = 65\,536$ distinct addresses the CPU can access.

An I/O operation between the CPU and the memory (or I/O devices) involves the following steps at the bus interface level:

- the CPU places the address of the memory or I/O device on the address bus,
- the CPU sets the appropriate control lines on the control bus to indicate whether it is an input or output operation,

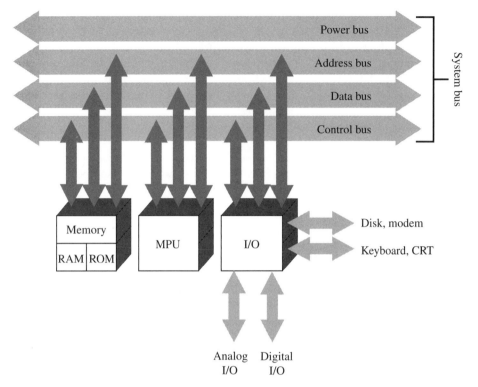

FIGURE 4.4: General functional grouping of lines in a computer bus: power bus, address bus, data bus, control bus.

- the CPU reads the data bus if it is an input operation, or write, the data to the data bus if it is a write operation.

In short, performing I/O between the CPU and the other computer devices involves controlling all three components of the bus: address bus, control bus, data bus. In a high level language such as C, when we have a line of code as follows,

```
x = 5.0 ;
```

the compiler generates the necessary code to assign a memory address for x, places the address of it in the address bus, turns on the write control lines on the control bus, and writes the data "5.0" on the data bus.

4.3 MICROCONTROLLER HARDWARE AND SOFTWARE: PIC 18F452

The term *microprocessor* refers to the central processing unit (CPU) and its on-chip memory. The term *microcontroller* (μC) refers to a chip that integrates the microprocessor and many I/O device interfaces such as ADC, DAC, PWM, digital I/O and communication bus (Figure 4.5a). A microcontroller is an integrated microprocessor chip with many I/O interfaces. It results in a small footprint (physical size) and low cost. In microcontroller applications, the amount of memory typically needed is in the order of tens

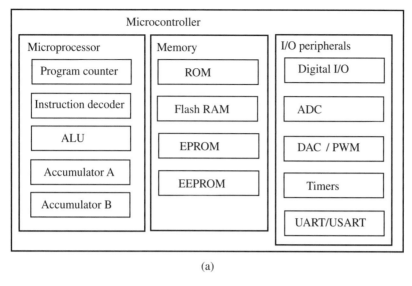

(a)

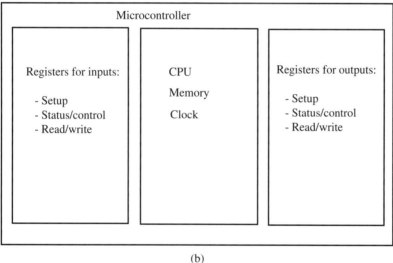

(b)

FIGURE 4.5: Microprocessor and microcontroller comparison: (a) microcontroller includes a microprocessor and I/O peripherals on the same chip. (b) High level programming view of a microcontroller.

of kilobytes, as opposed to the hundreds of megabytes of memory commonly available in desktop applications. As a result, the cost of microcontrollers is less compared to general purpose computers, which makes them good candidates for embedded controller applications.

From a high level programming point of view, that is using C language or a MATLAB®/Simulink® type graphical development environment, programming an embedded controller for a real-time control application is a matter of defining the logic between inputs and outputs (Figure 4.5b).

The logic is application dependent and is implemented through basic constructs of a high level programming language such as *if…, for… , while(…)* , basic AND, OR type logic operators as well as mathematical operators and functions (+, - , * , / , sin(..), cos(..) *etc.*), data structures to manage the data, and a hierarchy of functions for modular programming.

All of the input and output access is simply a matter of accessing a set of registers corresponding to the inputs and outputs. Generally, there are three kinds of register for any I/O:

1. Setup registers used to "setup" (configure) the operation of the I/O, that is, if a pin is programmed to be a digital input, digital output, or analog I/O. This is typically done once at the beginning of the application program.

2. Status/Control registers, used to determine the status of the I/O at any time, as well as initiate actions, that is to tell an ADC converter to start a conversion or to determine if the current ADC conversion process is completed.

3. Read/write registers, which are the "data" register for the I/O. For input, it is to be read, for output, it is to be written. Once the I/Os are setup, reading or writing an I/O is simply a matter of reading/writing the corresponding data register, just like reading and writing to a variable in memory.

In embedded programming, the I/O control is a matter of writing to/reading from these three groups of registers. This read/write operation may quite often read/write individual bits of registers (i.e., writing 0 or 1 to a particular bit in a register, or reading to determine if a particular bit in a register is 0 or 1), as well as the whole register byte-by-byte or word-by-word. Therefore, we can say that at the lowest level (register level) programming of an embedded controller, the programming involves the read/write operation of bits. High level programming language support with processor specific library functions often are used in order to make the program development generic and easily portable to different microcontrollers by simply configuring the compiler libraries for different processors. For real-time programming, understanding how I/O works and how real-time interrupts are handled are key to successful real-time application software development.

4.3.1 Microcontroller Hardware

As users of microcontrollers in mechatronic systems, we need to understand the hardware of microcontrollers. We will study this from the inside out. In the discussion, we are interested in the functionality of the microcontroller components as opposed to how they are designed or manufactured.

The main hardware features that one needs to understand for any microcontroller or digital signal processor (DSP) are as follows:

1. the pin-out on the chip that identifies the role of each pin,

2. the registers in the CPU that define the "brain" structure and internal workings of the microprocessor, memory, the bus structure, which defines how the CPU communicates with the rest of the memory and I/O resources,

3. the support chips such as the real-time clock, watch-dog timer, interrupt controller, programmable timers/counters, analog to digital converter (ADC), and pulse width modulation (PWM) module.

Most of the output pins of the microcontroller are implemented as *three-state* devices (the state of the line can be one of three states: LOW (OFF), HIGH (ON), High Impedance states), which allows them to be configured as either input or output under software control.

The application program is stored in the ROM type memory so that the program is not lost when the power is turned OFF to the computer. As a result, whenever the power is turned ON, the computer knows what to do (that is the program and necessary data in the ROM). The data generated after the program starts to run and which is not needed when the program starts is called the temporary data, and is stored in the RAM. When power is turned OFF, information in the RAM is lost. The exception to this is battery backed-RAM, where the data is not lost when power is turned OFF to the computer because the battery still keeps the RAM powered. Flash memory has the characteristics of both ROM and RAM.

Flash memory is a type of EEPROM memory which is packaged as "flash-disk" or "flash-stick" as a permanent data storage device, that is a solid-state permanent data store device without any mechanically moving parts. It retains its content when power is turned OFF, like a ROM, and it can be re-written, like a RAM. Read access time is in the order of that of dynamic RAM. Erase and re-write time is significantly slower compared to RAM. It offers random access read, but not random access write operations because it can be erased and re-written in "blocks." Typically, flash memory can be used for about 100 000 re-write cycles.

A microcontroller is a single-chip integrated circuit (IC) computer. PIC stands for peripheral interface controller. PIC is a trade name given to a series of microcontrollers manufactured by Microchip Technology Inc. We will be using the *PIC 18F452* chip for the laboratory experiments. The PIC 18F452 has five bidirectional input–output ports, named Ports A to E. Port A is 7-bit wide port while Port E is 3-bits wide (Figure 4.6). The other ports, B to D, are all 8-bit ports. Port pins are labeled RA0 through RA6, RB0 through RB7, RC0 through RC7, RD0 through RD7, and RE0 through RE2. Hence, the total number of pins at ports A through E is $7 + 8 + 8 + 8 + 3 = 34$ out of a 40-pin DIP package of PIC 18F452 chip (Figure 4.6). The remaining pins are used for V_{DD} (two pins), V_{SS} (two pins), $OSC1/CLK1, MCLR/V_{PP}$. Most of the pins are software configurable for one of multiple functions between general purpose I/O and peripheral I/O. Using registers in the chip, one function is selected for each pin under software control.

The PIC 18F452 is an 8-bit microcontroller based on the Harvard architecture. It is available in 40-pin DIP, 44-pin PLCC, and 44-pin TQFP packages. In the 44-pin packages, 4 of the pins are not used and labeled as NC. The Harvard architecture separates the program memory and the data memory. During a single instruction cycle, both program instructions

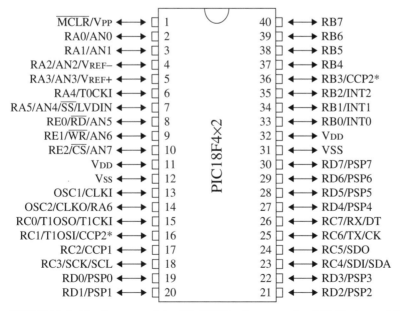

FIGURE 4.6: Pin diagrams of the PIC 18F452 microcontroller (DIP 40-pin model shown).

and data can be accessed simultaneously. The PIC 18F452 contains a RISC processor. This particular RISC processor has 75-word basic instruction set.

PIC 18F452 chip is compatible with clock speeds between 4 MHz and 40 MHz. The clock speed is determined by two factors:

1. hardware: the external crystal oscillator or ceramic resonators along with a few capacitors and resistors,

2. software: configuration register settings to select the operating mode of the processor.

PIC 18F452 supports up to 31 levels of stack, which means up to 31 subroutine calls and interrupts can be nested (Figure 4.8) for the low priority interrupts. Stack space in the memory holds the return address from the function calls and from interrupt service routines. Stack space is neither part of the program memory nor data memory.

The program counter (PC) register is 21-bits long (Figure 4.7). Hence, the address space for program memory can be up to 2 MB. PIC 18F4431 is flash memory based and has 16 KB of FLASH program memory (addresses 0000h through 6FFFh) which is 8 K of two-byte (single-word) instruction space (Figure 4.7). Data memory is implemented in RAM and EEPROM. Data RAM (static RAM, (SRAM)) has 4096 (4 KB) bytes of space via the 12-bit data bus address, of which 768 bytes is implemented as SRAM, and 256 bytes implemented as EEPROM on PIC 18F4431. The address space F60h through FFFh (160 bytes) is reserved for Special Function Registers (SFR) each of which is an 8-bit register. SFRs are used to configure and control the operation of the core CPU and I/O peripherals. It is through accessing these registers (160 8-bit SFRs), that practically all of the I/O operation of the microcontroller is accomplished. The rest of the data memory can be considered as General Purpose Registers, which are used as a scratch pad for data storage.

In C-language using a C18 compiler, using the "rom" directive, data variables can be explicitly directed to be allocated in program memory space if data memory space is not sufficient for a given application. Similarly, the "ram" directive in data declarations allocates memory in data memory.

```
rom char c ;
rom int n  ;

ram float x ;
```

Some of the important locations in the program memory map are as follows:

1. The RESET vector is at address 0x0000h.

2. The high priority interrupt vector is at address 0x0008h.

3. The low priority interrupt vector is at address 0x0018h.

The RESET condition is the startup condition of the processor. When the processor is RESET, the program counter content is set to 0x0000h, hence the program branches to this address to get the address of the instruction to execute. The C-compiler places the beginning address of the **main()** function at this location. Hence, on RESET, the **main()** function of a C-program is where the program execution starts. The source of RESET can be:

1. Power-on-reset (POR): when an on-chip V_{DD} rise is detected, a POR pulse is generated. On power-up, the internal power-up timer (PWRT) provides a fixed time-out delay to keep the processor in RESET state so that V_{DD} rises to an acceptable and stable level. After the PWRT time-out, the oscillator startup timer (OST) provides another time delay of 1024 cycles of oscillator period.

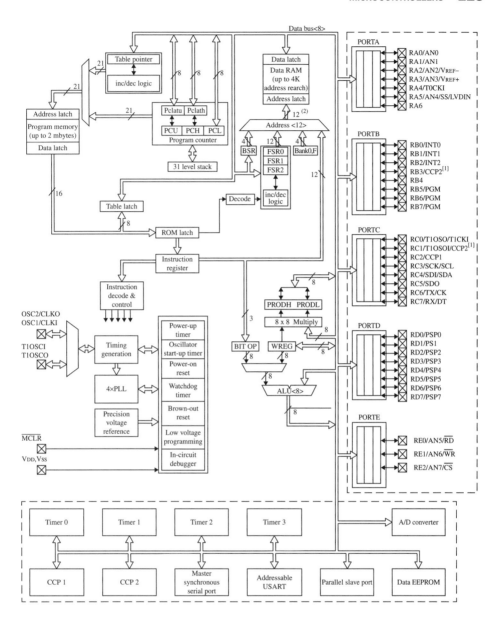

FIGURE 4.7: Block diagram of the PIC 18F452: registers, bus, peripheral devices, and ports (Fig.1-2 from PIC 18Fxx2 Data Sheet). Reproduced with permission from Microchip Technology, Inc.

2. MCLR reset during normal operation or during SLEEP: the MCLR input pin can be used to RESET the processor on demand.

3. Brown-out reset (BOR): when the V_{DD} voltage goes below a specified voltage level for more than a certain amount of time (both parameters are programmable), the BOR reset is activated automatically.

4. Watch dog timer (WDT) reset: WDT can be used to reset the processor if it times-out during normal operation in order to clear the state of the processor, or it can also be used to wake-up the processor from SLEEP mode. In WDT reset during normal operation, the program counter is initialized to 0x0000h, whereas in WDT wake-up mode, the program counter is incremented by 2 to continue to the instructions where it was left.

5. RESET instruction, stack full and stack underflow conditions result in software initiated reset conditions.

When PIC 18F452 executes a SLEEP instruction, the processor is held at the beginning of an instruction cycle. The processor can be woken-up from SLEEP mode by either external RESET, Watchdog Timer Reset or an external interrupt.

4.3.2 Microprocessor Software

Addressing Modes   Data memory space in PIC 18F452 is accessed by a 12-bit long address space, which means that it is 4096 bytes long. This space is organized as 16 banks of 256 bytes of memory (Figure 4.8). Banks 0–14 are used as General Purpose Registers (GPRs). GPRs can be used for general data storage. The upper half of bank 15 (F80h-FFFh) is used as Special Function Registers (SFR). Lists of all registers of PIC 18F452 are provided in the PIC 18F452 Users' Manual. SFRs are used for configuration, control, and status information of the microcontroller. Most features of the microcontroller are configured by properly setting the SFRs. EEPROM data memory space is 256 bytes and is accessed indirectly through SFRs. There are four SFRs involved in accessing EEPROM: EECON1, EECON2, EEDATA, and EEADR. EEDATA and EEADR registers are used to

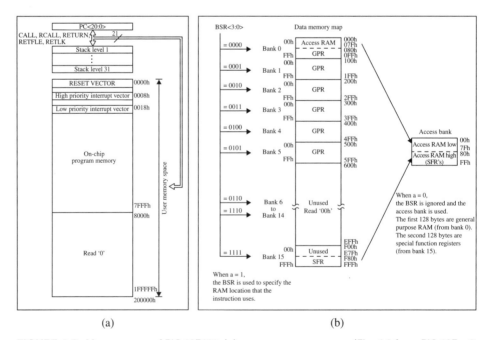

(a)   (b)

FIGURE 4.8: Memory map of PIC 18F452: (a) program memory map (Fig. 4.2 from PIC 18Fxx2 Data Sheet), (b) data memory map (Fig. 4.7 from PIC 18Fxx2 Data Sheet). Reproduced with permission from Microchip Technology, Inc.

hold the data and address for EEPROM. EECON1 and EECON2 registers control the access to EEPROM.

There are two addressing modes: *direct addressing* and *indirect addressing*. In *direct addressing mode*, Bank Select Register (BSR which is at the FE0h address in the SFR section of Bank 15 in data RAM) bits 3:0 (4-bits) are used to select one of 16 banks, and the bits 11:4 (8-bits) from the opcode are used to select one of 256 memory locations in the selected bank.

Instruction Set   The PIC 18Fxxx chip instruction set has 75 *instructions* (see PIC 18F452 or 18F4431 Users' Manual for a list and description of the basic instruction set). All, except three instructions, are 16-bit single word instructions. Three instructions are double words (32-bit). Each instruction has a unique binary code. When the microcontroller decodes the instruction code, the dedicated hardware circuit performs a function, such as increment the value in accumulator, add two numbers, compare two numbers. An instruction cycle is equal to four oscillator cycles. These four oscillator cycles per instruction cycle are called Q1, Q2, Q3, and Q4 cycles. Therefore, if a 4 MHz oscillator is used, an instruction cycle is 1 µs.

Obtaining an instruction from the memory is called *fetch*. Decoding it to determine what to do is called *decode*. And performing the operation of the instruction is called *execute*. Each one of these fetch, decode, and execute phases of the instruction takes one instruction cycle. The fetch–decode–execute cyles are pipelined, meaning executed in parallel, so that each instruction executes in one instruction cycle. In other words, while one instruction is decoding and executing, another instruction is fetched from the memory at the same time. Single word instructions typically take one instruction cycle, and two-word instructions take two instruction cycles to execute.

The CPU understands only the binary codes of its instruction set. All programs must ultimately be reduced to a collection of binary codes supported by its instruction set, specific to that microcontroller. However, it is clear that writing programs with binary codes is very tedious. *Assembly language* is a collection of three to five characters that are used as *mnemonics* corresponding to each instruction. Using assembly language instructions is much easier than using binary code for instructions. The *assembler* is a program that converts the assembly language code to the equivalent binary code (machine code) instructions that can be understood by the CPU. At the assembly language level, the instructions can be grouped into the following categories:

1. Data access and movement instructions used to read, write, and copy data between locations in the memory space. The memory space locations can be registers, RAM, or peripheral device registers. Examples of instructions in this category are: MOV, PUSH, POP.

2. Mathematical operations: add, subtract, multiply, increment, decrement, shift left, and shift right. Example instructions include ADDWF, SUBWF, MULWF, INCF, DECF, RLCF, RRCF.

3. Logic operations: such as AND, OR, inclusive OR. Example PIC 18F452 instructions include ANDWF, IORWF, XORWF.

4. Comparison operations: example instructions include CPFSEQ (=), CPFSGT (>), CPFSLF (<).

5. Program flow control instructions to change the order of program execution based on some logical conditions: GOTO, CALL, JUMP, RETURN.

In order to effectively use assembly language, one must very closely understand the hardware and software architecture (pinout, bus, registers, addressing modes) of a particular microprocessor. High level languages, such as C, attempt to relieve the programmer from the details of tedious programming in assembly language. For instance, in order to perform a long mathematical expression in assembly language, we would have to write a sequence of add, subtract, and multiply operations. Whereas in C we can directly type the expression in a single statement. Although the basic mathematical and logical operations are expressed the same way for different processors in C, the specific hardware capabilities of a processor cannot be made totally generic. For instance, one must understand the interrupt structure in a particular microcontroller in order to be able to make use of it even with C programming language. Quite often, I/O related operations require access to certain registers that are specific to each microcontroller design. Therefore, a good programmer needs to understand the architecture of a particular microcontroller even if a high level language such as C is used as the programming language.

4.3.3 I/O Peripherals of PIC 18F452

The input/output (I/O) hardware connection to the outside world is provided by the pins of the microcontroller. Figure 4.6 shows the pins of PIC 18F452 and their functions.

I/O Ports   There are five ports on PIC 18F452: PORTA (7-pin), PORTB (8-pin), PORTC (8-pin), PORTD (8-pin), and PORTE (3-pin). Each port has three registers for its operation. They are:

- TRISx register, where "x" is the port name A, B, C, D, or E. This is the *data direction* or *setup register*. The value of the TRISx register for any port determines whether that port acts an input (i.e., reads the value present on the port pins to the data register of the port) or an output (writes contents of the port data register to the port pins). Setting a TRISx register bit (= 1) makes the corresponding PORTx pin an input. Clearing a TRISx register bit (= 0) makes the corresponding PORTx pin an output.
- PORTx register. This is the data register for the port. When we want to read the status of the input pins, this register is read. When we write to this register, its content is written to the output latch register.
- LATx register (output latch). The content of the latch register is written to the port.

As an example, the code below sets all pins of PORT C as output, and all pins of PORT B as input.

```
TRISC = 0; /* Binary 00000000 */
PORTC = value ; /* write value variable to port C */

TRISB = 255; /* Binary 11111111 */
value = PORTB ; /* read the data in port B to variable value. */
```

The MPLAB C18 compiler provides C-library functions to setup and use I/O ports. For example, Port B interrupt-on-change and pull-up resistor functions can be enabled/disabled using these functions.

```
#include <portb.b>

OpenPORTB (PORTB_CHANGE_INT_ON & PORTB_PULLUPS_ON  ) ;
                 /* Configure interrupts and internal pull-up resistors on PORTB */
ClosePORTB ();     /* Disable  ................................................ */
```

```
OpenRBxINT (PORTB_CHANGE_INT_ON & RISING_EDGE_INT &  PORTB_PULLUPS_ON );
                                        /* Enable interrupts for PORTB pin x */
CloseRBxINT () ;                        /* Disable ....................... */


EnablePullups() ;  /* Enable the internal pull-up resistors on PORTB */

DisablePullups(); /* Disable .................................... */
```

PWM Input and PWM Output Signal An analog value, that is a value other than two state (ON or OFF), can be coded to a pulse-width value signal which is called pulse-width-modulation (PWM). The advantage of a PWM signal over an analog voltage level signal is its immunity against noise. For instance, if we have an analog voltage level signal in the 0 VDC to 10 VDC range, such as 4 VDC level, a noise voltage level of 0.1 V results in 2.5% error in the signal value. However, if the same signal is coded as a PWM signal as 40% duty cycle (that is the signal is at ON state (i.e., 5 VDC) for 40% of the time and OFF state (i.e., 0 VDC) the other 60% of the time, the 0.1 VDC noise is not large enough to confuse or change the ON/OFF level of the signal, and hence results in no error in the signal. In short, PWM signals are more robust and insensitive to noise compared to analog voltage level signals.

There are two variables that define a PWM signal (Figure 4.9):

1. PWM frequency,
2. duty cycle.

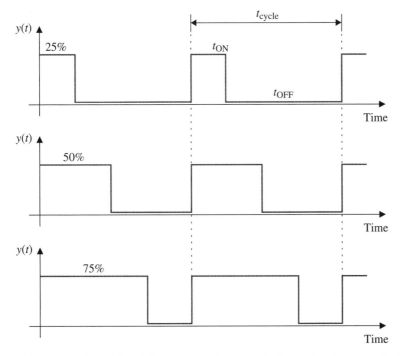

FIGURE 4.9: PWM signal: frequency and duty cycle. Examples shown are for 25, 50, 75% duty cycle.

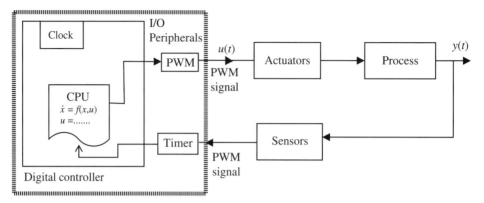

FIGURE 4.10: PWM signal input and output interface in embedded controllers.

The PWM frequency is the carrier frequency of the signal. For output signals, that is the PWM output signal from an embedded controller to a current amplifier. The PWM frequency should be at least 10 times (or even 100 times) larger than the bandwidth of the device (i.e., current aplifier) that the PWM signal is sent to. Similarly, if the PWM signal is an input signal to an embedded controller from a sensor, the PWM frequency of the signal should be at least 10 or more times faster than the highest frequency content expected in the measured variable.

The *duty cycle* is the value of the signal. For instance, 25% duty cycle means the signal is ON 25% of one period of the signal and 75% of the period is OFF. Resolution of the PWM output signal is directly a function of the number of bits the PWM function. For instance, an 8-bit resoultion PWM output function has a resolution of 1 part in $2^8 = 256$ for the full range of the signal. A 25% duty cycle is indicated by writing a number $duty\_cycle = 256/4 = 64$ to the register corresponding to the duty cycle of the PWM output peripheral (Figure 4.10).

A PWM input signal interface to an embedded controller can be accomplished by a timer (Figure 4.10), where timer values are captured at the rising and falling edges of the PWM signal to calculate the ON and OFF periods of the signal. By capturing the ON and OFF durations, we can calculate the PWM signal frequency and duty cycle,

$$w_{PWM} = \frac{1}{t_{ON} + t_{OFF}} \tag{4.1}$$

$$duty\_cycle = \frac{t_{ON}}{t_{ON} + t_{OFF}} \tag{4.2}$$

where t_{ON}, t_{OFF} are the ON time period and OFF time period captured by the timer where the PWM input signal is connected.

Similarly, there are amplifiers which require frequency as their command signal. And there are sensors whose output signal for the measured variable is a variable frequency, such as engine speed sensors. Frequency input can be best handled either by a counter or timer module. For instance, a counter can count the number of pulses per unit time period from which we can estimate the frequency. Alternatively, the time period of each pulse can be measured by a timer from which we can calculate the frequency. Digital output ports can be used to generate variable frequency output signals, however there is no dedicated frequency output peripheral integral to the PIC microcontroller. A DAC converter output (analog voltage) can be converted to a proportional frequency with an operational amplifier circuit. Likewise, a variable frequency signal can be converted to an analog voltage signal with an operational amplifier circuit. Such circuits are standard in op-amp cookbook reference books.

Capture/Compare/Pulse Width Modulation (PWM)
The capture/compare/ PWM (CCP) module is used to perform one of three functions: capture, compare, or generate a PWM signal. PIC 18F452 has two CCP modules: CCP1 and CCP2. The operation of both these modules is identical, with the exception of a special event trigger. The operation of CCP involves the following:

1. Input/Output pin: RC2/CCP1 and RC1/CCP2 pins for CCP1 and CCP2, respectively. For the capture function, this pin must be configured as input, for compare and PWM functions, it must be configured as output.

2. Timer source: in capture and compare mode, TIMER1 or TIMER3, and in PWM mode TIMER2 can be selected as the timer source.

3. Registers: used to configure and operate the CCP1/CCP2 modules.
   (a) CCP1CON (or CCP2CON) register: 8-bit register used to configure the CCP1 (CCP2) module and select which operation is desired (capture, compare, or PWM).
   (b) CCPR1 (CCPR2) register: 16-bit register used as the data register. In capture mode, it holds the captured value of TIMER1 or TIMER3 when the defined event occurs in RC2/CCP1 pin. In compare mode, it holds the data value being constantly compared to the TIMER1 and TIMER3 value and when they match, the output pin status is changed or interrupt is generated (the action taken is programmable via the CCP1CON or CCP2CON register settings). In PWM mode, CCP1 (CCP2) pin provides the 10-bit (1 part in $2^{10} = 1024$ resolution) PWM output signal. PWM signal period information is encoded in the PR2, and duty cycle in CCP1CON bits 5:4 (2-bit) and CCPR1 (8-bit) registers which make up the 10-bit PWM duty cycle resolution. TIMER2 and register settings are used to generate the desired PWM signal.
   (c) PR2 register: used to set the PWM period in microseconds.
   (d) T2CON register: used to select the prescale value of TIMER2 for the PWM function.

In *capture mode*, the CCPR1 (CCPR2) register captures the 16-bit value of TIMER1 or TIMER3 (depending on which one of them has been selected in the setup) when an external event occurs on pin RC2/CCP1 (RC1/CCP2). These pins must be configured as input in capture mode. CCP1CON (CCP2CON) register bits are set to select the capture mode to be one of the following (Figures 4.11, 4.12, 4.13): every falling edge, every rising

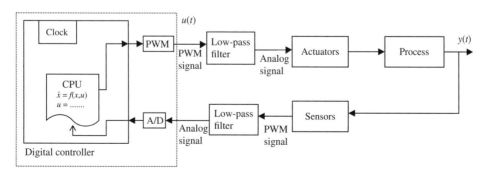

FIGURE 4.11: PWM signal to analog voltage conversion in an embedded control system.

Capture mode operation:

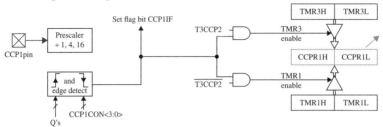

Compare mode operation:

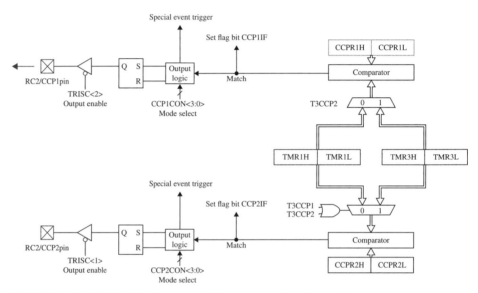

FIGURE 4.12: CCP Module: Capture and Compare modes of operation.

PWM mode operation:

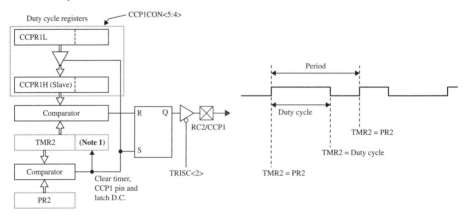

FIGURE 4.13: CCP Module: PWM mode of operation.

edge, every fourth rising edge, every 16th rising edge. When a capture occurs, the interrupt request flag bit CCP1IF (PIR register, bit 2) is set and must be cleared in software. If another capture interrupt occurs before the CCPR1 register value is read, it will be overwritten by the new captured value. The C18 compiler library provides the following functions to implement capture.

```
#include <capture.h>
#include <timers.h>

OpenCapture1(C1_EVERY_4_RISE_EDGE & CAPTURE_INT_OFF) ;
  /* Configure capture 1 module: capture at every 4th rising edge of capture 1
     pin signal, no interrupt on capture. */

OpenTimer3(TIMER_INT_OFF & T3_SOURCE_INT) ;
  /* Timer3 is the source clock to capture on trigger. */

while (!PIR1bits.CCP1IF) ; /* Wait until Capture module 1 has captured */
result = ReadCapture1() ;   /* Read the captured value */

....                        /* Process captured data */

if(!CapStatus.Cap1OVF)  /* Check (if needed) if there was any overflow condition*/
{
 ...                        /* Further processing of captured data if needed */
}
```

In *compare mode*, the 16-bit CCPR1 (CCPR2) register value is constantly compared to the TIMER1 or TIMER3 register value. When they match, RC2/CCP1 pin (RC1/CCP2 pin) is controlled to either high, low, toggle, or unchanged state. The choice of action on the compare match event is made by proper settings of CCP1CON (CCP2CON) registers. In compare mode, the RC2/CCP1 (RC1/CCP2) pin must be configured for output. It is possible to select to generate interrupt on the "compare match" event. In this case, the output pin status is not changed. That action is left to the interrupt service routine (ISR).

In *PWM mode*, the CCP1 pin produces PWM output with 10-bit resolution. The CCP1 pin is multiplexed with the PORTC data latch, and the output is obtained from the PORTC-2 (RC2) pin. The TRISC-2 bit must be cleared to make the CCP1 pin an output. PWM output has two parameters: *period* which defines the frequency of PWM signal and *duty cycle* which defines the percentage of ON-time of the signal within each period. The PWM period obtained is a function of the frequency of the oscillator used to drive the PIC 18F452.

Example register values are given in the code below. The code configures the CCP2 port for PWM of 0.256 ms period (frequency 3.9 KHz) and a duty cycle of 25%.

```
PR2 = 255;     /* Sets the period for the PWM1 output channel in
                  microseconds*/
CCP1CON = 12;  /* Activates the PWM mode in the CCP register*/
CCPR1L = 63;   /* Sets the duty cycle for the PWM output */
TRISC = 0;     /* Configures PortC for output */
T2CON = 62;    /* Configures Timer 2 for prescale value of 1:1 and postscale 1:8*/
```

The PR2 register sets the PWM period, while the duty cycle is set by CCPR1L and CCP1CON< 5 : 4 > bits. PWM period is given by:

$$PWM_{period} = (PR2 + 1) \times 4 \times (Clock\ Period) \times (Timer\ Prescale\ Value) \quad (4.3)$$

$$= 256 \times 4 \times \frac{1}{4\ MHz} \times 1 = 0.256\ ms \quad (4.4)$$

$$PWM_{\text{Dutycycle}} = CCPR1L : CCP1CON < 5 : 4 > \times \tag{4.5}$$

$$(\textit{Clock Period}) \times (\textit{Timer Prescale Value}) \tag{4.6}$$

$$= 252 \times \frac{1}{4 \text{ MHz}} \times 1 = 0.063 \text{ ms} \tag{4.7}$$

C library functions used to setup and operate the PWM outputs (PWM1 or PWM2)are shown below,

```
#include <pwm.h>
#include <timers.h>

OpenTimer2(TIMER_INT_OFF & T2_PS_1_4 & P2_POST_1_8) ;
        /* Setup TIMER2: disable interrupt, set pre and post scalers to 4 and 8 */

OpenPWM1(char period ) ;
        /* Enable and setup PWM1 module output signal "period" (8-bit value). */
   ...  /* PWM period=(period+1)*4*Tosc*(TIMER2 Prescaler)*/

SetDCPWM1(unsigned int duty_cycle) ;
        /* Set "duty cycle" of PWM output signal (10-bit value) */
        /*  High Time of PWM  =  ( duty_cycle *Tosc)  */
...
ClosePWM1() ; /* Disable PWM1 output module */
```

Analog to Digital Converter (ADC) The analog to digital converter (A/D or ADC) allows conversion of an analog input signal to a digital number. The ADC on the PIC 18F452 has a 10-bit range and eight multiplexed input channels. Unused analog input channel pins can be configured as digital I/O pins. The analog voltage reference is software selected to be either the chip positive and negative supply (V_{DD} and V_{SS}) or the voltage levels at RA3/V_{REF+} and RA2/V_{REF-} pins.

There are two main registers to control the operation of ADC:

1. ADCON0 and ADCON1 registers: to configure the ADC and control its operation, that is:
   - enable ADC,
   - select the method to trigger the ADC conversion "START" (timer source or polled-programming by directly writing to a bit of the control register),
   - select which channel to sample,
   - select the conversion rate,
   - format the conversion result.

2. ADRESH and ADRESL registers: to hold the ADC converted data in 10-bit format in the two 8-bit registers.

A 10-bit ADC conversion takes $12 \times T_{AD}$ time period. The T_{AD} time can be software selected to be one of seven possible values: $2 \times T_{OSC}$, $4 \times T_{OSC}$, $8 \times T_{OSC}$, $16 \times T_{OSC}$, $32 \times T_{OSC}$, $64 \times T_{OSC}$, or internal A/D module RC oscillator.

ADC must be setup with proper register value settings. Then, when ADCON0 bit:2 is set, the ADC conversion process starts. At that point, sampling of the signal is stopped, the charge capacitor holds the sampled voltage, and the ADC conversion process converts the analog signal to a digital number using the successive approximation method. The conversion process takes about $12 \times T_{AD}$ time periods. When an ADC conversion is complete

1. the result is stored in ADRESH and ADRESL registers,

2. ADCON0 bit:2 is cleared, indicating ADC conversion is done,

3. the ADC interrupt flag bit is set.

The program can determine whether ADC conversion is complete or not by either checking the ADCON bit:2 (cleared) or waiting for ADC conversion complete interrupt if it is enabled. The sampling circuit is automatically connected to the input signal again, and the charge capacitor tracks the analog input signal. The next conversion is started again by setting the ADCON0 bit:2.

The ADC can be configured for use either by directly writing to the registers, or by using C library functions present in the C-18 compiler. The C functions OpenADC(..), ConvertADC (), BusyADC() and ReadADC(), CloseADC() are used to configure the ADC module, begin the conversion process, and read the result of conversion, and disable the ADC converter. Examples of their use are given below:

```
#include <p18f452.h>
#include <adc.h>
#include <stdlib.h>
#include <delays.h>

OpenADC(ADC_FOSC_32 & ADC_RIGHT_JUST & ADC_8ANA_0REF, ADC_CH0 & ADC_INT_OFF);
    /* Enable ADC in specified configuration. */
    /* Select: clock source, format (right justified), reference voltage source,
         channel 0 , enable/disable interrupt on ADC-conversion completion */
while(1)
{
  ConvertADC();          /* Start the ADC conversion process*/
  while ( BusyADC() ) ; /* Wait until conversion is complete:
  return 1 if busy, 0 if not.*/
  result = ReadADC();   /* Read converted voltage; store in variable 'result'.
                           10-bit conversion result will be stored in the
                           least or most significant 10-bit portion of
                           result depending on how ADC was configured
                           in OpenADC(...) call.    */

  Delay1KTCYx(1)  ;   /* Delay for 1000 TCY cycle */

}

CloseADC();    /* Disable the ADC converter */
```

Here, the arguments in the OpenADC(..) function are used to configure the ADC module to use the specified clock sources, reference voltages and select the ADC channel. A detailed description of the arguments is provided in the C-18 users guide.

Timers and Counters   In real-time applications, there are many cases when different tasks need to be performed at different periodic intervals. For instance, in an industrial control application, the status of doors in a building may need to be checked every minute, the parking lot gate status may need to be checked every hour. The best way to generate such periodic interrupts with different frequencies is to use a programmable timer/counter chip. A timer/counter chip operates as a timer or as a counter. In the timer mode of operation, it counts the number of clock cycles, hence it can be used as a time measurement peripheral. The clock source can be an internal clock or external clock. In counter mode, it counts the number of signal state transitions at a defined pin, that is, at rising edge or falling edge.

The PIC 18F452 chip supports four timers/counters: TIMER0, TIMER1, TIMER2, and TIMER3. TIMER0 is an 8-bit or 16-bit software selectable, TIMER1 is a 16-bit, TIMER2 is an 8-bit timer, and TIMER3 is a 16-bit timer/counter. Timer/counter operation is controlled by setting up appropriate register bits, reading from and writing to certain registers. Timer operation setup requires enable/disable, source of signal, type of operation (timer or counter), and so on. When a timer overflows, interrupt is generated. The timer operation is controlled by a set of registers:

1. The INTCON register is used to configure interrupts in the microcontroller.

2. The TxCON register is used to configure TIMERx, where "x" is 0, 1, 2, or 3 for the corresponding timer. Configuration of a timer involves the selection of the timer source signal (internal or external), pre- or post-scaling of the source signal, source signal edge selection, enable/disable timer.

3. The TRISA register port may be used to select an external source for the TIMER at port A pin 4 (RA4).

4. The TMRxL and TMRxH are 8-bit register pairs that make up the 16-bit timer/counter data register.

TIMER2 sets the interrupt status flag when its data register is equal to the PR2 register, whereas TIMER0, TIMER1 and TIMER3 set the interrupt status flag when there is overflow from maximum count to zero (from FFh to 00h in 8-bit mode, and from FFFFh to 0000h in 16-bit mode). The registers used to indicate the interrupt status are INTCON, PIR1, PIR2. The interrupt can be masked by clearing the appropriate timer interrupt mask bit of the timer interrupt control registers.

Counter mode is selected by setting the T0CON register bit 5 to 1. In counter mode, TIMER0 increments on every state transition (on either falling or rising edge) at pin RA4/T0CKI.

C library functions used to setup and operate timers are (the following functions are available for all timers on the PIC microcontroller, TIMER0, TIMER1, TIMER2, TIMER3),

```
#include <timers.h>

OpenTimer0(char config)  ; /* Open and configure timer 0: enable interrupt,
                             select 8/16-bit mode, clock source, prescale value */
...
result = ReadTimer0()  ;   /* read the timer 0*/
...
WriteTimer0(data) ;        /* write to timer register 0 */
...
CloseTimer0() ;            /* close (disable) timer 0 and its interrupts */
```

Often, a certain amount of time delay is needed in the program logic. C-library functions provide programmable time delay where the minimum delay unit is one instruction cycle. Therefore, the actual real-time delay accomplished by these functions depends on the processor operating speed.

```
#include <delays.h>
....
Delay1TCY(); /* Delay 1 instruction cycle  */
Delay10TCYx(unsigned char unit);
        /* unit=[1, 255], Delay period = 10*unit instruction cycle */
Delay100TCYx(unsigned char unit);
        /* unit=[1, 255], Delay period = 100*unit instruction cycle */
Delay1KTCY(unsigned char unit);
        /* unit=[1, 255], Delay period = 1000*unit instruction cycle */
Delay10KTCY(unsigned char unit);
        /* unit=[1, 255], Delay period = 10000*unit instruction cycle */
....
```

Watch Dog Timers A watch dog timer (WDT) is a hardware timer, which is used to reboot or take a predefined action if it expires. The watch dog timer keeps a "watch eye" on the system performance. If something gets stuck, the watch dog timer can be used to reset everything. A watch dog timer is essential in embedded controllers. The WDT counts down from a programmable preset value. If it reaches zero before the software resets the counter to preset value, it is assumed that something is stuck. Then the processor's reset line is asserted. The time-out period can be programmed to a value between 4 ms to 131.072 s.

The watchdog timer (WDT) on the PIC chip is an on-chip RC oscillator. It works even if the main clock of the CPU is not working. WDT can be enabled/disabled, and the time-out period can be changed under software control. When the WDT times-out, it generates a RESET signal which can be used to restart the CPU or wake-up the CPU from SLEEP mode. There are three registers involved in configuration and use of WDT on PIC 18F452 (and PIC 18F4431): CONFIG2H, RCON, WDTCON. If CONFIG2H register bit 0 (WDTEN bit) is 1, WDT cannot be disabled by other software. If this bit is cleared, then WDT can be enabled/disabled by WDTCON register, bit 0 (1 for enable, 0 for disable). When WDT times-out, RCON register bit 3 (TO bit) is cleared. The time-out period is determined in hardware, and extended by post-scaler in software (CONFIG2H register, bits 3:1).

4.4 INTERRUPTS

4.4.1 General Features of Interrupts

An interrupt is an event which stops the current task the microprocessor is executing, and directs it to do something else. And when that task is done, the microprocessor resumes the original task. An interrupt can be generated by two different sources: 1. hardware interrupts (external), 2. software (internal) generated interrupt with an instruction in assembly language, such as **INT n**.

When an interrupt occurs, the CPU does the following:

1. Finishes currently executing statement.
2. Saves the status, flags, and registers in the stack so that it can resume its current task later.
3. Checks the interrupt code, looks at the interrupt service table (vector) to determine the location of the *interrupt service routine* (ISR), a function executed when the interrupt occurs.
4. Branches to the ISR and executes it.
5. When ISR is done, restores the original task from the stack and continues.

When an interrupt is generated, the main task stops after some house-keeping tasks, and the address location of the ISR for this interrupt needs to be determined. This information is stored in the interrupt service vector. The table has default ISR addresses. If you want to assign a different ISR address (or name) for an interrupt number, the old address should be saved, then a new ISR address should be written. Later, when the application terminates, the old ISR address should be restored.

In a given computer control system, there can be more than one interrupt source and they may happen at the same time. Therefore, different interrupts need to be assigned different priority levels to determine which one is more important. Higher priority interrupts can suspend lower priority interrupts. Therefore, nested interrupts can be generated (Figure 4.14).

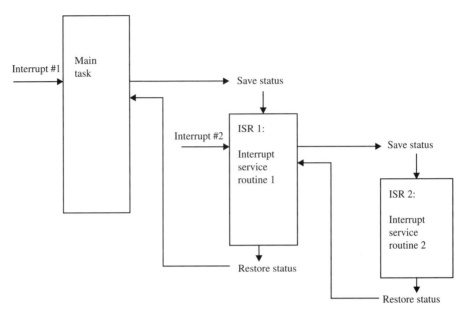

FIGURE 4.14: Nesting of interrupts is possible through a priority level.

The time between the generation of interrupt to the time the program branches to the ISR for the interrupt (after saving the current state of microprocessor and house keeping operations, taking care of other higher priority interrupts) is called the *interrupt latency time*. Smaller interrupt latency time is generally desired. Due to multiple interrupt sources and priorities, it may be impossible to calculate the worst possible interrupt latency time since it may depend on the number of higher priority interrupts generated at the same time.

In critical applications, if interrupting the current process is not tolerable, interrupts can be disabled temporarily with appropriate assembly instructions or C functions.

Let us assume that there are two interrupt sources: interrupt 1 and interrupt 2 (Figure 4.14). Let us assume that interrupt 2 has higher priority than interrupt 1. While the main program is executing, let us assume that interrupt 1 occured. The processor will save the status of microprocessor, determine the interrupt number, look in the *interrupt service vector table* for the address of the *interrupt service routine (ISR)* for this particular interrupt, and jump to that ISR-1. Let us further assume that while it is executing ISR-1, interrupt 2 occurs. The processor will save the status, and jump to ISR-2. When ISR-2 is done, the processor returns to ISR-1, restores the previous state, and continues the ISR-1 from where it was interrupted. When ISR-1 is finished, it will restore the state of the main task before the interrupt 1 occured, and continue the main task.

4.4.2 Interrupts on PIC 18F452

PIC 18F452 supports external and internal interrupts (Figure 4.15). There are 17 interrupt sources. In addition, external interrupts can be defined to be active in rising or falling edges of input signals.

It supports two levels of interrupt priority: high priority and low priority. Each interrupt source is assigned one of the two priority levels by setting appropriate register bits. A high priority interrupt vector is at 000008h, and a low priority interrupt vector is at 000018h. The address of ISR routine for each interrupt category must be stored at these addresses. High priority interrupts override any low priority interrupt in progress. Only

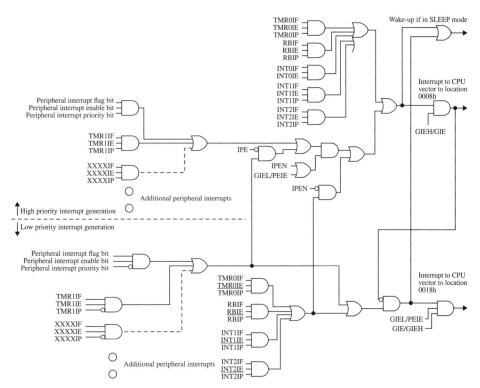

FIGURE 4.15: Interrupts and their logic in PIC 18F452 (from PIC 18Fxx2 Data Sheet). Reproduced with permission from Microchip Technology, Inc.

one high priority interrupt can be active at a given time, and they cannot be nested. Low priority interrupts can be nested by setting the INTCON register bit 6 (GIEL bit = 1) at the beginning of the ISR.

There are ten registers used to control the interrupts in PIC 18F452: RCON, INTCON, INTCON2, INTCON3, PIR1, PIR2, PIE1, PIE2, IPR1, IPR2. Each interrupt source, except INT0 which is always a high priority interrupt, has three bits for enable/disable and priority setup, and to determine interrupt status (also called *flag* or *request*) register space:

1. one bit to enable/disable the interrupt (enable bit),
2. one bit to set the priority level (priority bit), and
3. one bit to indicate the current status of interrupt (flag bit).

These bits are distributed over the ten 8-bit registers referenced above.
Interrupt sources include:

1. Three external interrupt pins: RB0/INT0, RB1/INT1, RB2/INT2 can be programmed to be edge triggered interrupts on rising or falling edges by setting appropriate bits in the above 10 registers. Both global interrupt enable and individual interrupt enable for a specific interrupt source must be enabled for the interrupt to be enabled. When interrupt occurs, the corresponding interrupt status bit (also called interrupt flag) is set (INTxF), which can be used to determine the source of the interrupt in ISR. INT0 is always high priority, whereas the priority of INT1 and INT2 can be programmed as high or low. These interrupts can wake-up the processor from SLEEP state.

2. PORTB pins 4, 5, 6, 7 can be programmed to generate interrupt on change-of-state (on rising edge or falling edge). PORTB pins 4–7 change-of-state interrupts can be enabled or disabled as well as priority set as a group under program control.

3. "ADC conversion done" is another interrupt, which is generated when an ADC conversion is completed. When ADC conversion complete interrupt is generated (assuming it is enabled), PIR1-bit 6 is set. The logic to trigger the start of the ADC conversion must be handled separately (i.e., after reading the current ADC conversion at the ISR routine, or periodically performing ADC conversion based on a TIMER generated signal).

4. TIMER interrupts (interrupt-on-overflow generated by TIMER0, TIMER1, TIMER3, and when TIMER2 data register = PR2 register). When timer interrupts are enabled and timer overflow occurs, resulting in interrupt generation, interrupt status flag bits in corresponding registers associated with the interrupt source are set that can be used in the ISR routine to determine the source of the interrupt (INTCON bit 1 for TIMER0, PIR1-bit 0 for TIMER1, PIR1-bit 1 for TIMER2, and PIR2-bit 1 for TIMER3).

5. CCP1, CCP2 interrupts (once enabled and priorities are selected to be either as high or low) are generated as folows: in capture mode, a TIMER1 register capture occurred, and in compare mode a TIMER1 register match occurred.

6. Communication peripheral interrupts (master synchronous serial port, addressable USART, parallel slave port, EEPROM write interrupt).

For external interrupts at INT0-INT2 (RB0-RB2) pins or PORTB pins 4–7 (RB3-RB7) input change interrupt, expected interrupt latency time is three to four instruction cycles. Notice that PIC 18F452 groups all interrupts into two categories and provides two ISR addresses: a low priority group with ISR address stored at 000018h, and a high priority group with ISR address stored at 000008h. Within each group, application software must be implemented by testing interrupt flag bits (bits contained in the 10 registered listed above) to determine which one of the interrupt sources in a high or low priority group has triggered the interrupt.

The RESET signal can be considered as a non-maskable interrupt. The address for the program code to execute on the RESET signal is stored at the vector address 0000h.

The following example C-code shows the use of an external interrupt and its software handling in a PIC 18F452 microcontroller. There are two parts to setting up a function as an interrupt service routine (ISR) with high or low priority.

1. First: interrupt service vector locations for high priority interrupt (address 000008h) and low priority (address 000018h), must be setup to branch to the desired ISR routine.

2. Second: the desired ISR function logic must be defined. Then each ISR function should be defined. The ISR routine should be proceeded by a `#pragma interrupt` ..... to indicate that the function is an ISR for high priority interrupt or a `#pragma interruptlow` ..... to indicate that the function is an ISR for low priority interrupt, not just a regular C-function. Furthermore, the ISR cannot have any argument or return type.

```
/* ......High Priority Interrupt Service Routine (ISR) Setup */

#pragma code HIGH_INTERRUPT_VECTOR = 0x8
                        /* Place the following code starting at
                           address 0x08, which is the location for
                           high priority interrupt vector */
```

```
void at_high_interrupt_vector(void)
{
    _asm
  goto High_ISR
    _endasm
}
#pragma code   /* Restore the default program memory allocation. */


#pragma interrupt High_ISR
void High_ISR(void)
{

  ....
  ....
  ....

}


/* ...... Low Priority Interrupt Service Routine (ISR) Setup */

#pragma code LOW_INTERRUPT_VECTOR = 0x18
                  /*  Place the following code starting at
                      address 0x18, which is the location for
                      low priority interrupt vector */
void  at_low_interrupt_vector(void)
  {
    _asm
  goto Low_ISR
    _endasm
  }
#pragma code
      /* Restore the default program memory allocation. */


#pragma interruptlow  Low_ISR
void Low_ISR(void)
{

  ....
  ....
  ....

}
#pragma code   /* Restore the default program memory allocation. */
```

Example: Interrupt Handling with PIC 18F452 on PICDEM 2 Plus Demo Board   This example requires:

1. hardware: the PICDEM 2 Plus Demo board with PIC 18F452 microcontroller chip and MPLAB ICD 3 (in-circuit debugger) hardware,

2. software: MPLAB IDE v6.xx and MPLAB C18 development tools.

The INT0 pin is connected to a switch which is an external interrupt. It is a high priority interrupt. Therefore, when this interrupt occurs, the program will branch to the interrupt vector table location 0x0008h and get the address of the interrupt service routine to branch to. By default, this address contains the address of the interrupt service routine called high_ISR. Hence, when INT0 occurs, the program will branch to the high_ISR function, after saving the contents of certain registers and return address in the stack. We can type in the code for the interrupt service routine in high_ISR or call another function from there and just have the "call" or "goto" statement in the high_ISR function. In addition, we need to setup certain registers for the interrupt to function properly, that is, enable interrupt, define the active state of interrupt. Interrupt service routine (ISR) must have no input arguments and no return data.

```c
#include <p18f452.h>  /* Header file for PIC 18F452 register declarations */
#include <portb.h>    /* For RB0/INT0 interrupt */

/* Interrupt Service Routine (ISR) logic */

void toggle_buzzer(void) ;

#pragma code HIGH_INTERRUPT_VECTOR = 0x8
                    /* Specify where the program address to be stored */
void  at_high_interrupt_vector(void)
{
  _asm
      goto High_ISR
  _endasm
}
#pragma code    /* Restore the default program memory allocation. */

#pragma interrupt  High_ISR
void High_ISR(void)
{
   CCP1CON = ~CCP1CON &  0x0F
                    /* Toggle state of buzzer: OFF if ON, ON if OFF */
   INTCONbits.INT0IF = 0
                    /* Clear flag to avoid another inter-
rupt due to same event*/
}

/* Setup of the ISR */

void EnableInterrupt(void)
{
   RCONbits.IPEN = 1   ; /* Enable interrupt priority levels */
   INTCONbits.GIEH = 1 ; /* Enable high priority interrupts */
  }

void InitializeBuzzer(void)
{
  T2CON = 0x05 ;          /* postscale 1:1, Timer2 ON, prescaler 4 */
  TRISCbits.TRISC2 = 0 ; /* configure CCP1 module for buzzer operation */
  PR2 = 0x80  ;          /* initialize PWM period */
  CPPR1l = 0x80 ;        /*  ........ PWM duty cycle */
}

void main (void)
{

  EnableInterrupts() ;
  InitializeBuzzer() ;

  OpenRB0INT(PORTB_CHANGE_INT_ON & PORTB_PULLUPS_ON & FALLING_EDGE_INT) ;
                    /* Enable RB0/INT0 interrupt, configure RB0 pin for
                        interrupt, trigger interrupt on falling edge */

  CCP1CON = 0x0F  ; /* Turn ON buzzer */

  while (1) ;   /* Wait indefinitely.   */
                /* when the interrupt occurs,
                    the corresponding ISR will be executed */
}
```

Example: Timer0 Interrupt Below is a sample code which sets up Timer0 to generate interrupt. Timer0 is defined as a low priority interrupt. Interrupt vector table location for low priority interrupts is 000018h . As in the previous example, a shell form of low priority ISR for a low priority interrupt address is placed at the 0x18 vector address, and "goto" statement is used to direct the program execution to the Timer0 ISR.

```c
#include <p18f452.h> /* Header file for PIC 18F452 register declarations */
#include <timer.h>   /* For timer interrupt  */

/* Interrupt Service Routine (ISR) logic */

void Timer0_ISR(void) ;

#pragma code LOW_INTERRUPT_VECTOR = 0x18
         /* Specify where the program address to be stored */

void Low_ISR(void)
{
  _asm
      goto Timer0_ISR
  _endasm
}

#pragma code  /* Restore the default program memory allocation. */



#pragma interruptlow Timer0_ISR
void Timer0_ISR(void)
{
   static unsigned char led_display = 0 ;

   INTCONbits.TMR0IF = 0 ;
   led_display = ~led_display & 0x0F  /* toggle LED display */
   PORTB = led_display ;

}

/* Setup of the ISR */

void EnableInterruptTimer0(void)
{
   TRISB = 0 ;
   PORTB = 0 ;
   OpenTimer0 (TIMER_INT_ON & TO_SOURCE_INT & TO_16BIT) ;
   INTCONbits.GIE = 1 ;
}


void main (void)
{

   EnableInterruptTimer0() ;

   while (1) ;   /* Wait indefinitely.   */
                 /* when the interrupt occurs,
                    the corresponding ISR will be executed */
}
```

PIC 18F4431 Microcontroller   The PIC 18F4431 version of the PIC 18Fxxxx family microcontrollers has a quadrature encoder interface at the following pins for an incremental encoder interface:

- RA2, RA3, RA4 for Index (INDX), Channel A (QEA), Channel B (QEB), respectively.

It also supports 8 channels of PWM output via the following pins:

- RB3:0, pins 0 through 3, can be configured for PWM0, PWM1, PWM2, PWM3, respectively.
  RB4 and RB5 pin can be configured for PWM5 and PWM4 output signals, respectively.
  RD6 and RD7 pins can be configured for PWM6 and PWM7 output.

Interrupts on-change-of-state for pins RB pins 4 through 7 (RB7:4) are effective only if these pins are configured as input. If any of these pins is configured for output, the interrupt for that pin is disabled.

Timers on PIC 18F4431 are TIMER0, TIMER1, TIMER2, and TIMER5. The timer control register for each timer (T0CON, T1CON, T2CON, T5CON) is an 8-bit register which is used to configure (control) the operation of the timer. The bits allows us to: enable/disable timer mode or counter mode, 8-bit or 16-bit mode, clock source (internal or external), pre-scaler/post-scaler values. Timer 2 and Timer 5 have PR2 and PR5 re-registers which define the maximum value to which the timer is to count, then roll-over and generate interrupt. Timer 1 and CCP1 compare configurations that can be used to trigger a periodic "ADC conversion start" trigger to the ADC module. Timer 2 can be used as the source to define the period (base frequency) of the PWM output pin. Other timers generate interrupt when the counting from 00h to FFh is complete and overflow to 00h occurs. For counter operation, an external source should be selected. In this mode, pre and post-scaling values are used. In timer mode, pre and post-scaler values are ignored and timers are incremented every instruction cycle. For each timer, if the interrupt is enabled via the corresponding interrupt enabled bit on the corresponding timer control register, when a timer interrupt occurs, the corresponding timer interrupt flag is set.

Motion feedback module (MFM) has a quadrature encoder interface (EQI), which is used to interface to an incremental encoder for position feedback. Once the position feedback is available, the microcontroller can perform closed loop position control in any actuated motion control system, that is electric or electro-hydraulic or pneumatic. The QEI interface has three pins: QEA, QEB, INDX which are to be connected to the A, B, and Index channels of an incremental encoder (the Index channel of an incremental encoder is sometimes referred as Index C or Z). Using the pulses on Channel A and B, and the phase relationships between them, the module keeps track of position information using a 16-bit up/down counter register. An INDX pin can also be connected to another external digital input other than the Encoder's Index channel, that is a registration sensor (see Chapter 10 for registration applications and possible use of these).

The following registers are used in the configuration and use of the MFM (QEI) module:

```
QEICON    8-bit  configuration register

POSCNT    16-bit up/down counter: position register

MAXCNT    16-bit maximum count register
```

```
VELR      16-bit  velocity estimation register

DFLTCON   8-bit  four noise filters on the three
          channels of the QEI and T5CKI pin: enable/disable.

PIR3      8-bit interrupt flag register;
             bit OC1IF for velocity update,
                 IC2QEIF for position update,
                 IC3QEIF for direction change.

INTCON    8-bit interrupt control register
```

The QEI module is configured using the QEICON register (8-bit register). Bit settings determine if x2 or x4 (quadrature) decoding is used. Using the phase angle relationship between Channel A and B, the MFM module determines the direction of motion (forward + or reverse -), and hence increments the position register (POSCNT) up or down. the POSCNT register is reset to zero when it either overflows or reaches MAXCNT register value ((QEICON<4:2>= 110 or 010), or on INDX signal if it was configured to do so using QEICON register (QEICON<4:2>= 101 or 001).

If the position counter reset interrupt is enabled, interrupt is generated (IC2QEIF flag is set) on a reset of the POSCNT register when

1. the POSCNT register rolls over FFFFh to 0000h in period mode (QEICON<4:2>= 110 or 010),
2. the MAXCNT register value is reached by the POSCNT register (QEICON<4:2>= 110 or 010),
3. the INDX pulse is detected (QEICON<4:2>= 101 or 001).

The POSCNT register resets to 00h if it was moving in a forward (+) direction from which it continues to count up, and to FFh if it was moving in a reverse (-) direction from which it continues to count down. In addition, the module can accurately estimate velocity both at high speed and low speed and writes the velocity estimation to the VELR register which the application program can access.

The same pins are software configurable for input capture function using TIMER5: CAP1/QEA, CAP2/EQB, CAP3/INDX.

PIC 18F4431 supports 4-pairs (8 channels, 4 pairs where in each pair one channel is a complement of the other channel) of PWM channels using a PWM module, at pins PWM0-7. Each pair have independent PWM generators (independent PWM frequency and pulse width). The PWM module can be configured for automatic dead-time insertion, edge and center-aligned output modes, up to 14-bit resolution. PWM frequency can be changed on-the-fly under software control.

4.5 PROBLEMS

1. In the PIC 18F452 microcontroller, how is the RESET condition generated and what sequence of events happens under that condition?

2. What is the role of "watch dog timer"? How does it work in the PIC 18F452?

3. How many hardware interrupt priorities are supported in the PIC 18F452? Describe how these interrupts are enabled/disabled, how an interrupt is handled, and the sequence of events that occurs in the microcontroller when an interrupt event occurs.

4. List the input/output (I/O) ports available on the PIC 18F452 microcontroller. Give examples of C-code to show how they are setup and used in a real-time application software.

5. What is the role of capture/compare/PWM port on the PIC 18F452 microcontroller? Describe how this port is setup and used in different applications.

6. Discuss the differences between "polling driven" versus "interrupt driven" programming. Give examples for polling driven and interrupt driven applications. What are the internal and external interrupt sources to the microcontroller?

7. Discuss analog and digital converter (ADC) inputs available on the PIC 18F452 chip: how many channels, resolution of each channel, multiplexed or not, how ADC operation is controlled in software.

8. PWM input and output signals are often converted to analog voltage signals using low-pass filters. Design and simulate the following in Simulink®: a PWM signal source with 1 KHz base frequency and adjustable pulse-width modulation (PWM source), an analog filter with time constant of 0.01 s. Simulate the input and output of the low pass filter for a PWM duty cycle of 25, 50, and 100% the low pass filter transfer function is

$$D(s) = \frac{1}{0.01\,s + 1} \tag{4.8}$$

CHAPTER 5

ELECTRONIC COMPONENTS FOR MECHATRONIC SYSTEMS

5.1 INTRODUCTION

Analog and digital electronic components are an integral part of mechatronic devices. Most commonly used analog and digital electronic devices are discussed in this chapter. We start with linear circuits, semiconductors, and discuss electronic switching devices (diodes and transistors). This is followed by a discussion of analog operational amplifiers (op-amps). Examples of proportional, derivative, integral, low pass, and high pass filters using op-amps are discussed. Digital electronic circuit components are discussed last. The focus of the discussions is on the input–output functionality of the devices, not the detailed modeling of the input–output dynamics nor the details of their construction and manufacturing.

5.2 BASICS OF LINEAR CIRCUITS

Basic *passive components* of electrical circuits are (Figure 5.1) the resistor, capacitor, and inductor. By passive components, we mean that the component can be connected to a circuit with two terminal points without requiring its own power supply. Basic *variables* in an electrical circuit are

1. voltage (potential difference analogous to pressure in hydraulic systems), and
2. current (flow of electrons, analogous to fluid flow).

The basic *parameters* of a typical electrical component involve three properties,

1. resistance, R,
2. capacitance, C,
3. inductance, L.

Current is the rate of charge flow through a conductor in a circuit,

$$i(t) = \frac{dQ(t)}{dt} \tag{5.1}$$

where $Q(t)$ is the amount of electrical charge, "electrons," that passes through the conductor. The smallest unit of electrical charge is the electrical charge of an electron or of a proton. The amount of electrical charge of an electron (e^-) and that of proton (p^+) is the same

Mechatronics with Experiments, Second Edition. Sabri Cetinkunt.
© 2015 John Wiley & Sons, Ltd. Published 2015 by John Wiley & Sons, Ltd.
Companion Website: www.wiley.com/go/cetinkunt/mechatronics

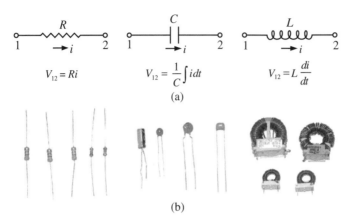

FIGURE 5.1: Basic components of electrical circuits: (a) symbols of resistance (R), capacitance (C), inductance (L). (b) Pictures of resistors, capacitors, and inductors.

except that they are opposite signs, in units of Coulomb,

$$|e^-| = |p^+| = 1.60219 \times 10^{-19} \text{ C} \tag{5.2}$$

The passive components define relationships between current and voltage potential difference between their two nodes. An ideal *resistor* has a potential difference between its two ends proportional to the current passing through it. The proportionality constant is called the resistance, R

$$V_{12}(t) = R \cdot i(t) \tag{5.3}$$

The resistance of a material is a function of the material property, that is the *resistivity* (ρ) and its geometry. As the cross-section of the conductor increases, resistance to the flow of charges reduces, as they have more room to move through. As the length of the conductor increases, resistance increases. Hence, the resistance of a conductor is

$$R = \rho \frac{l}{A} \tag{5.4}$$

where ρ is the *resistivity* of the material ($\rho = 1.7 \times 10^{-8} \, \Omega \cdot$ m for copper, $\rho = 2.82 \times 10^{-8} \, \Omega \cdot$ m for aluminum, $\rho = 0.46 \, \Omega \cdot$ m for germanium, $\rho = 640 \, \Omega \cdot$ m for silicon, $\rho = 10^{13} \, \Omega \cdot$ m for hard rubber, $\rho = 10^{10}$ to $10^{14} \, \Omega \cdot$ m for glass), l is the length, A is the cross-sectional area of the conductor. It is also important to point out that the resistivity of materials, and hence the resistance, is a function of temperature. The variation of resistance as a function of temperature varies from material to material. In general, many materials, but not all, have the following resistance and temperature relationship,

$$R(T) = R_0[1 + \alpha(T - T_0)] \tag{5.5}$$

where $R(T)$ is the resistance at temperature T, R_0 is the resistance at temperature T_0, and α is the rate of change in resistance as a function of temperature. *Superconductors* are materials that have almost zero resistance below a certain critical temperature, T_c. For each superconductor, there is a critical temperature T_c below which the above relationship does not hold. The resistance suddenly drops to almost zero at $T \leq T_c$. This critical temperature for mercury is about $T_c = 4.2 \, ^\circ$K.

There are two major types of resistors: wire wound and carbon types. Wire wound resistors are used for high current applications such as regenerative power dumping as heat in motor control. Carbon type resistors are used in low current signal processing applications where the resistance value is color coded on the component using a four color code. The maximum power rating of a resistor defines the maximum power it can dissipate without damage,

$$P = V_{12} \cdot i = V_{12} \cdot (V_{12}/R) = R \cdot i^2 = \frac{V_{12}^2}{R} < P_{max} \tag{5.6}$$

which defines the maximum voltage drop or maximum current that can be present across the resistor.

A *capacitor* stores electric charges, and hence creates an electric field (electric voltage) across its terminals. It is made of two conducting materials separated by an insulating material such as air, vacuum, glass, rubber, or paper. The two conductor surfaces store equal but opposite charges. As a result, there is a voltage potential across the capacitor. The voltage potential is proportional to the amount of charge stored and the characteristics of the capacitor. There are four major capacitor types: mica, ceramic, paper/plastic film, and electrolytic. Electrolytic type capacitors are polarized. Therefore, the positive terminal must be connected to the positive side of the circuit.

An ideal capacitor generates a voltage potential difference between its two nodes proportional to the stored electrical charge (integral of current conducted)

$$V_{12}(t) = \frac{Q}{C} = \frac{1}{C} \int_0^t i(\tau) \cdot d\tau \tag{5.7}$$

where C is called the capacitance. An ideal capacitor induces a 90° phase angle between its voltage and current across its terminals due to the integral function. When a capacitor is fully charged, it stores a finite amount of charge,

$$Q = \int_0^t i(\tau) \cdot d\tau = C \cdot V_{12} \tag{5.8}$$

The size of a capacitor is indicated by the amount of charge it can store (capacitance, C). A given capacitor can hold voltage up to a certain amount, V_{max} (called the rated voltage). Capacitance (C) is a measure of how much charge (Q) the capacitor can hold for a given voltage potential, (V). A capacitor cannot hold a voltage potential above a maximum value, V_{max}. If the rated voltage is exceeded, the capacitor breaks down and cannot hold the charges. The construction principle of a capacitor includes two conductors separated by insulating material. The capacitance is proportional to the surface area of the capacitor (conductors inside the capacitor) which holds the charge and inversely proportional to the distance between them. The proportionality constant is the permittivity of the medium that separates the conductors, ϵ,

$$C = \frac{\epsilon \cdot A}{l} = \frac{\kappa \epsilon_0 \cdot A}{l} \tag{5.9}$$

where κ is the *dielectric constant*, ϵ_0 is the *permittivity* of free space. The dielectric constants for some materials are as follows: $\kappa = 1.00$ for vacuum, $\kappa = 3.4$ for nylon, $\kappa = 5.6$ for glass, $\kappa = 3.7$ for paper. Clearly, as the dielectric constant of the insulating material increases, the capacitance of the capacitor increases. The capacitance is proportional to the surface area of the capacitor plates and inversely proportional to the distance between them. The maximum voltage that the capacitor can handle before break-down, called *break-down voltage* or *rated voltage*, increases with the increase in dielectric constant of the insulator. The break-down voltage of the capacitor also increases with the distance parameter l.

Energy stored in a capacitor can be calculated as follows,

$$dW = V \cdot dQ = \frac{Q}{C} dQ \tag{5.10}$$

$$W = \frac{1}{2} \frac{Q^2}{C} \tag{5.11}$$

and the $W \leq W_{max}$, where W_{max} is the maximum energy storage capacity of the capacitor,

$$W_{max} = \frac{1}{2} C \cdot V_{max}^2 \tag{5.12}$$

Notice that capacitors block the DC voltage and pass the AC voltage. In other words, the DC voltage will build a potential difference in the capacitor until they are equal, provided that the DC source voltage is below the break-down voltage of the capacitor. The AC voltage simply alternates the charge and discharge of the capacitor. Unlike the DC component, which is blocked by the capacitor, the AC component of the voltage is passed.

An ideal *inductor* generates a potential difference proportional to the rate of change of current passing through it

$$V_{12} = L \cdot \frac{di(t)}{dt} \tag{5.13}$$

where L is called the inductance. Notice that, regardless of the direction of current ($i > 0$ or $i < 0$), the voltage across the inductor is proportional to the rate of change of the current. Let V_{10} and V_{20} be the voltages at point 1 and 2 across the inductor with reference to ground. Then,

$$V_{10} > V_{20} \quad when \quad \frac{di}{dt} > 0 \tag{5.14}$$

$$V_{10} < V_{20} \quad when \quad \frac{di}{dt} < 0 \tag{5.15}$$

The inductor is made with a coil of a conductor around a core, like a solenoid. The core can be a magnetic or an insulating material. The inductance value, L, is a function of permeability of the core material, number of turns in the coil, cross-sectional area, and length. Permeability of a material is a measure of its ability to conduct electromagnetic fields. It is analogous to the electrical conductivity. If the material composition of the space around an inductor changes, that is due to motion of device components in a solenoid, the inductance changes.

In deriving the equation for electric circuits, two main relationships are used (Kirchhoff's Laws);

1. current law,

2. voltage law (Figure 5.2).

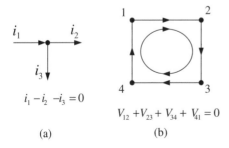

FIGURE 5.2: Kirchhoff's electric circuit laws: (a) current law: algebraic sum of currents at any node is zero. (b) voltage law: algebraic sum of voltages in a loop is zero.

Current law states that the algebraic sum of currents at a node is equal to zero, which is a statement of the "conservation of charge." Voltage law states that the sum of voltages in a loop is equal to zero, which is a statement of the "conservation of potential." For instance in Figure 5.2, Kirchoff's current law states that

$$i_1 - i_2 - i_3 = 0 \qquad (5.16)$$

and Kirchoff's voltage law states that

$$V_{12} + V_{23} + V_{34} + V_{41} = 0 \qquad (5.17)$$

5.3 EQUIVALENT ELECTRICAL CIRCUIT METHODS

Quite often, we need to reduce a two-terminal circuit with multiple components into an equivalent simpler circuit with a voltage source and an impedance or a current source and an impedance. For now, assume that *impedance* is a generalized form of *resistance*. Two of the most well-known equivalent circuit analysis methods are discussed below. These methods are useful in determining input and output loading errors in coupled electrical circuits, that is measurement errors introduced due to the effect of a measuring device in an electrical circuit.

5.3.1 Thevenin's Equivalent Circuit

Thevenin's equivalent circuit consists of *a voltage source in series with an equivalent resistor*. Any section of a linear circuit with multiple resistors (plus voltage and current sources) components can be replaced with a Thevenin's equivalent circuit.

Consider the circuit shown in Figure 5.3. Our objective is to determine the equivalent voltage source and series resistance value for the circuit shown in the dotted line. Hence, we can examine the interaction between the load resistance R_L and the rest of the circuit.

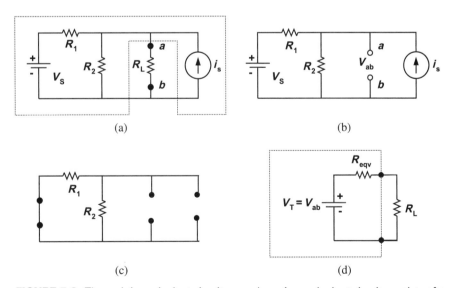

FIGURE 5.3: Thevenin's equivalent circuit procedure. An equivalent circuit consists of an open circuit voltage and an equivalent resistor in series.

Thevenin's equivalent circuit analysis requires the calculation of two parameters: V_T and R_{eqv}. The following are the standard steps to do that.

1. Identify the sub-circuit whose Thevenin's equivalent circuit is to be determined. Identify the two point terminals (a and b) out of the sub-circuit. Remove the load resistor between the two points (Figure 5.3a).
2. Calculate the open circuit voltage between a and b, $V_T = V_{ab}$, (Figure 5.3b).
3. Short-circuit voltage sources (i.e., zero source) and disconnect all current sources (open circuit the current sources) (Figure 5.3), and
4. calculate the equivalent series resistor (R_{eqv}) in the remaining circuit (excluding the R_L), R_{eqv} (Figure 5.3c). The equivalent resistance of the circuit inside the dotted box (as viewed from V_{ab}) is represented by R_{eqv} and also called the *output resistance* of the sub-circuit. The circuit inside the dotted box can be viewed as a voltage source plus a resistance the source has in addition to the load (Figure 5.3d).

As far as the two points a and b are concerned, the circuit inside the dotted block is equivalent to a voltage source $V_T = V_{ab}$ and series resistor R_{eqv}.

5.3.2 Norton's Equivalent Circuit

Norton's equivalent circuit consists of *a current source in parallel with an equivalent resistor* (Figure 5.4). The equivalent resistor is the same as the resistor calculated in Thevenin's equivalent circuit. The value of the current source (i_N) is same as the value of the current that would flow in a short-circuit of the output terminals, that is to replace the output resistance between points a and b with a short-circuit connection. The procedure for obtaining Norton's equivalent circuit is similar to the procedure for Thevenin's equivalent circuit. The conversions can be made between the equivalent Thevenin's circuit and Norton's circuit, that is between two points in a circuit any voltage source with series resistance can be replaced by an equivalent current source and parallel resistor, and visa versa. It can be shown that

$$V_T = i_N \cdot R_{eqv} \tag{5.18}$$

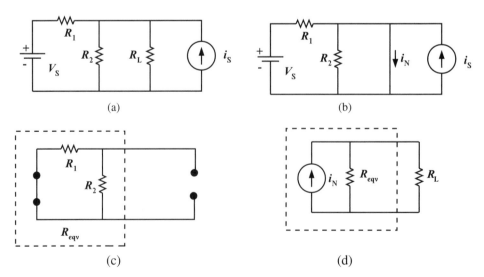

(a)   (b)

(c)   (d)

FIGURE 5.4: Norton's equivalent circuit procedure. The equivalent circuit consists of a current-source and an equivalent resistor in parallel.

The steps to follow to find the Norton's equivalent circuit are as follows (Figure 5.4),

1. Short-circuit the load resistor, and then calculate the Norton source current (i_N) through the short wire jumping across the open connection points where the load resistor used to be (Figure 5.4b).

2. Short-circuit voltage sources and open circuit current sources.

3. Calculate the total resistance between the open connection points (excluding the load resistance, Figure 5.4c).

4. Draw the Norton equivalent circuit with the Norton equivalent current source in parallel with the Norton equivalent resistance (Figure 5.4d). The load resistor re-attaches between the two open points of the equivalent circuit.

Example   In the example shown in Figure 5.3, let the following values for the parameters be

$$V_s = 10\ V, \quad R_1 = 7\,\Omega, \quad R_2 = 3\,\Omega, \quad i_s = 5\ \text{Amp} \tag{5.19}$$

Then, it is easy to show that (Figure 5.3b),

$$V_s = R_1 \cdot i_1 + R_2 \cdot (i_1 + i_s) \tag{5.20}$$

$$10\ V = 7 \cdot i_1 + 3 \cdot (i_1 + 5) \tag{5.21}$$

$$= 10 \cdot i_1 + 15 \tag{5.22}$$

$$i_1 = -0.5\ A \tag{5.23}$$

The voltage potential V_{ab} is

$$V_{ab} = 3 \cdot (-0.5 + 5.0) = 13.5\ V \tag{5.24}$$

The equivalent resistance of the circuit after the voltage sources are short-circuited, and current sources are open-circuited (Figure 5.3c),

$$\frac{1}{R_{eqv}} = \frac{1}{7} + \frac{1}{3} \tag{5.25}$$

$$R_{eqv} = \frac{7 \cdot 3}{7 + 3} \tag{5.26}$$

$$= 2.1\,\Omega \tag{5.27}$$

Thevenin's equivalent circuit of this example is represented by a voltage source $V_T = 13.5\ V$ and equivalent resistor $R_{eqv} = 2.1\,\Omega$ (Figure 5.3d).

For instance, the current on the load resistor R_L can be calculated as a function of the load resistance,

$$i_R = \frac{V_T}{R_L + R_{eqv}} \tag{5.28}$$

$$= \frac{13.5}{R_L + 2.1} \tag{5.29}$$

$$i_R = i_R(R_L; V_T, R_{eqv}) \tag{5.30}$$

It is instructive to obtain the Norton equivalent of the same circuit and confirm that $V_T = R_{eqv} \cdot i_N$. Referring to Figure 5.4b, the current i_N is sum of the current due to current source and voltage source,

$$i_N = i_s + i_1 \tag{5.31}$$

Notice that no current will pass through R_2 since there is zero resistance path parallel to it, therefore,

$$i_1 = \frac{V_s}{R_1} \tag{5.32}$$

Hence,

$$i_N = 5A + \frac{10}{7}A = \frac{45}{7}A \tag{5.33}$$

Referring to Figure 5.4c, the equivalent resistance in the dotted block as seen by the load is again,

$$\frac{1}{R_{eqv}} = \frac{1}{R_1} + \frac{1}{R_2} \tag{5.34}$$

$$= \frac{1}{7} + \frac{1}{3} \tag{5.35}$$

$$R_{eqv} = \frac{7 \cdot 3}{7 + 3} = 2.1\,\Omega \tag{5.36}$$

And it is confirmed that the relationship between Thevenin's and Norton's equivalent circuit,

$$V_T = R_{eqv} \cdot i_N \tag{5.37}$$

$$= 2.1 \cdot \frac{45}{7} = 13.5\,V \tag{5.38}$$

5.4 IMPEDANCE

5.4.1 Concept of Impedance

The term *impedance* is a generalization of the *resistance*. If a circuit consists of a potential source and a resistor, the impedance and the resistance are the same thing. Impedance is a generalized version of resistance when a circuit contains other components such as capacitive and inductive components. The input–output relationship between voltage and current in a resistor circuit is

$$\frac{V(t)}{i(t)} = R \tag{5.39}$$

Let us take the Fourier transform of this,

$$\frac{V(jw)}{i(jw)} = R \tag{5.40}$$

where in the frequency domain the input–output ratio is constant, R, not a function of frequency, w, and $j = \sqrt{-1}$. The *impedance* of a resistor is a real constant.

Let us examine an RL circuit (Figure 5.5a), and consider the input (current)–output (voltage) relationship in the frequency domain. In Figures 5.5c and d, if the switch is disconnected from the voltage supply lead (A) and connected to the other lead (B), that is putting the voltage supply out of the circuit, the $V(t)$ in the equations below would be set to zero.

$$V(t) = R \cdot i(t) + L\frac{di(t)}{dt} \tag{5.41}$$

$$\frac{V(jw)}{i(jw)} = (R + jwL) \tag{5.42}$$

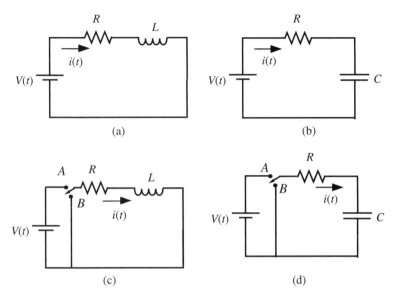

FIGURE 5.5: (a) RL circuit, (b) RC circuit, (c) RL circuit with a switch, (d) RC circuit with a switch.

where $Z(jw)$ is defined as the *impedance* as follows,

$$Z(jw) = \frac{V(jw)}{i(jw)} = (R + jwL) \tag{5.43}$$

which can be viewed as generalized resistance that is a complex function of frequency. The complex function of frequency has a magnitude and phase for each frequency. Impedance is a transfer function between current (input) and voltage (output).

Admittance, $Y(jw)$, is defined as the inverse of impedance,

$$Y(jw) = \frac{1}{Z(jw)} = Re(Y(jw)) + jIm(Y(jw)) \tag{5.44}$$

where the real part of admittance ($Re(Y(jw))$) is called the *conductive part* or the *conductance* and the imaginary part ($Im(Y(jw))$) is called the *susceptive part* or *susceptance*.

Similarly, let us consider the RC circuit (Figure 5.5b). The voltage and current relationship of the circuit is

$$V(t) = R \cdot i(t) + \frac{1}{C} \int_0^t i(\tau)d\tau \tag{5.45}$$

$$V(jw) = \left(R + \frac{1}{Cjw}\right) \cdot i(jw) \tag{5.46}$$

$$\frac{V(jw)}{i(jw)} = \frac{1 + RC\,jw}{Cjw} \tag{5.47}$$

The impedance of the RC circuit is defined as

$$Z(jw) = \frac{V(jw)}{i(jw)} = \frac{1 + RC\,jw}{Cjw} \tag{5.48}$$

Notice that if we were interested in the relationship between the input voltage and the voltage across the capacitor ($V_C(t)$) as the output voltage (i.e., the RC circuit is used as a

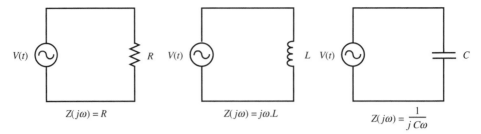

FIGURE 5.6: Concept of impedance: generalized resistance as impedance.

filter between input and output voltages), the transfer function between them is

$$V_C(t) = \frac{1}{C} \int_0^t i(\tau)d\tau \tag{5.49}$$

$$V_C(jw) = \frac{1}{C \cdot jw} \cdot i(jw) \tag{5.50}$$

$$i(jw) = C \cdot jw \cdot V_C(jw) \tag{5.51}$$

and

$$\frac{V_C(jw)}{V(jw)} = \frac{1}{1 + RC\,jw} \tag{5.52}$$

which defines a low pass filter between the input voltage and output voltage.

It is rather easy to show that the impedance of a resistor, capacitor, and inductor are as follows (Figure 5.6),

$$Z_R(jw) = R \tag{5.53}$$

$$Z_C(jw) = \frac{1}{C\,jw}; \ \textit{also called capacitive reactance} \tag{5.54}$$

$$Z_L(jw) = jw\,L; \ \textit{also called inductive reactance} \tag{5.55}$$

Impedance between two points in a circuit is the Fourier transform of the voltage (output) and current (input) ratio. It is also called the transfer function between voltage (output) and current (input). Impedance is a complex quantity. It has magnitude and phase, and is a function of frequency. When we refer to large or small impedance, we generally refer to its magnitude.

Example Consider the RL and RC circuits shown in Figure 5.5c, d, respectively. A switch in each circuit is connected to the supply voltage (A side) and B side at specified time instants. Assume the following circuit parameters; $L = 1000\,\text{mH} = 1000 \times 10^{-3}\,\text{H} = 1\,\text{H}$, $C = 0.01\,\mu\text{F} = 0.01 \times 10^{-6}\,\text{F}$, $R = 10\,\text{k}\Omega$. Let the supply voltage be $V_s(t) = 24\,\text{VDC}$. Assume in each circuit, initial conditions on the current is zero and initial charge in the capacitor is zero. Starting time is $t_o = 0.0\,s$. At time $t_1 = 100\,\mu\text{s}$ the switch is connected to the supply, and at time $t_2 = 500\,\mu\text{s}$, the switch is disconnected from the supply and connected to the B side of the circuit. Plot the voltage across each component and current as function of time for the time period of $t_0 = 0.0\,s$ to $t_f = 1000\,\mu\text{s}$.

For the RL circuit, when it is connected to the supply,

$$V_s(t) = L \cdot \frac{di(t)}{dt} + R \cdot i(t); \ t_1 \leq t \leq t_2 \tag{5.56}$$

and when it is disconnected from the supply and connected to the B side,

$$0 = L \cdot \frac{di(t)}{dt} + R \cdot i(t); \ t_2 \leq t \leq t_f \tag{5.57}$$

with the initial conditions of current being the last value during the previous state. Another way of treating this problem is to consider the first equation with a pulse input voltage,

$$V_s(t) = L \cdot \frac{di(t)}{dt} + R \cdot i(t); \ 0 \leq t \leq t_f \tag{5.58}$$

where $V_s(t)$ is a pulse signal, that is the circuit is connected to 24 VDC (switch is on A side) and 0.0 VDC (switch is on B side),

$$V_s(t) = 24 \cdot (1(t - t_1) - 1(t - t_2)) \tag{5.59}$$

where $1(t)$ represent the unit step function, $1(t - t_1)$ represent the unit step function shifted in time by t_1.

The solution of the above differential equations can be shown to be as follows, or more easily with Simulink® (Figure 5.7).

$$i(t) = 0.0; \ t_0 \leq t \leq t_1 \tag{5.60}$$
$$i(t) = 24/(10 \times 10^3) \cdot (1 - e^{-(t-t_1)/(L/R)}) \, \text{A} \tag{5.61}$$
$$= 2.4 \times (1 - e^{-(t-0.0001)/0.0001}) \, \text{mA}; \quad t_1 \leq t \leq t_2 \tag{5.62}$$
$$i(t) = i(t_2) \cdot (e^{-(t-t_2)/(L/R)}) \, \text{A} = 2.356 \cdot e^{-(t-0.0005)/0.0001} \, \text{mA}; \ t_2 \leq t \leq t_f \tag{5.63}$$

Then the current across the resistor and inductor in each time period can be found by

$$V_R(t) = R \cdot i(t) \tag{5.64}$$
$$V_L(t) = L \cdot \frac{di(t)}{dt} = V_s(t) - R \cdot i(t) \tag{5.65}$$

We can confirm our physical expectation by the plots of these results. Initially, the inductor will oppose, by generating an opposing voltage, the large rate of change in current and the opposition will get smaller as the rate of change of current gets smaller, and as the current reaches steady-state. When the voltage source switches suddenly to zero volts, the current will start to decrease and the inductor will generate voltage in opposition to that change. In other words, the inductive voltage will be negative since $di(t)/dt$ is negative.

For an RC circuit, the circuit relationships are, when connected to supply,

$$V_s(t) = R \cdot i(t) + V_c(t) \tag{5.66}$$
$$= R \cdot i(t) + \frac{1}{C} \left(Q(t_1) + \int_{t_1}^{t} i(\tau)d\tau \right); \quad t_1 \leq t \leq t_2 \tag{5.67}$$

and when it is disconnected from the supply and connected to the B side,

$$0.0 = R \cdot i(t) + \frac{1}{C} \left(Q(t_2) + \int_{t_2}^{t} i(\tau)d\tau \right); \quad t_2 \leq t \leq t_f \tag{5.68}$$

The voltage across the capacitor at any given time (t) is the initial voltage due to the initial charge at a given time (t_0) plus the net change in the charge (integral of the current from the initial time (t_0) to present time (t))

$$V_c(t) = \frac{1}{C} \left(Q(t_0) + \int_{t_0}^{t} i(\tau)d\tau \right) \tag{5.69}$$

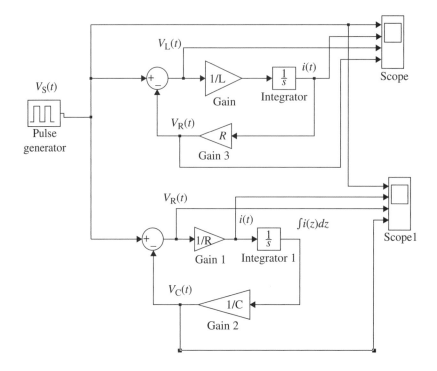

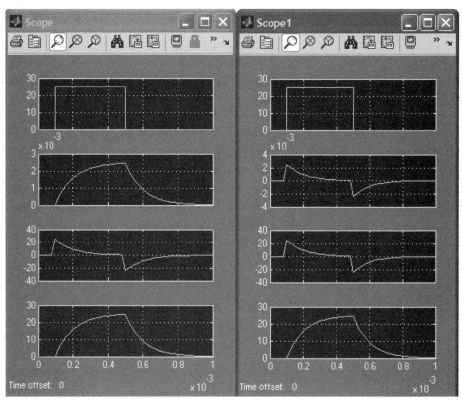

FIGURE 5.7: Simulink® model of the RL and RC circuits with a switch and simulation results.

Again, the above two different states of the RC circuit can be represented with the first equation with a pulse input voltage representation,

$$V_s(t) = R \cdot i(t) + \frac{1}{C} \left(\int_{t_1}^{t} i(\tau) d\tau \right); \quad t_1 \leq t \leq t_f \tag{5.70}$$

where we used the fact that $Q(t_1) = 0.0$ since initially the capacitor is assumed to be uncharged and

$$V_s(t) = 24 \cdot (1(t - t_1) - 1(t - t_2)) \tag{5.71}$$

The above differential equation can be solved using Laplace transforms method for $i(t)$, then $V_c(t)$ and $V_R(t)$ can be obtained from the current–voltage relationship across the capacitor and resistor components. Using Simulink® for numerical solution is also an easier option (Figure 5.7).

$$V_c(t) = 0.0; \quad t_0 \leq t \leq t_1 \tag{5.72}$$
$$V_c(t) = 24 \cdot (1 - e^{-(t-t_1)/(RC)}) \, V = 24 \cdot (1 - e^{(t-0.0001)/0.0001}) \, V; \, t_1 \leq t \leq t_2 \tag{5.73}$$
$$V_c(t) = V_c(t_2) \cdot (e^{-(t-t_2)/(RC)}) \, V = 23.56 \cdot e^{-(t-0.0005)/0.0001} \, V; \, t_2 \leq t \leq t_f \tag{5.74}$$

It is important to recognize that both circuits are first-order filters and that the response speed is dominated by the time constant of the circuit, $\tau = L/R = 100 \, \mu s$ for the RL circuit and $\tau = RC = 100 \, \mu s$ for the RC circuit. Similarly, we can confirm our physical expectations by comparing these results. Initially, the capacitor is uncharged and there is no voltage across it. Hence the current will be developed as if there is only a resistor on the circuit. Then, as a result of the current, the capacitor will start to charge and develop voltage potential across it. As a result, the available voltage across the resistor will decrease. Eventually, when the capacitor charge is so large that the voltage across t is equal to the supply voltage, the voltage across the resistor will be zero, hence the current will be zero. When the circuit is switched to zero supply voltage, the capacitor voltage will drive the current in the opposite direction until the charge it holds is fully drained. The charge and discharge of the capacitor is an exponential function, whose time constant is determined by $R \cdot C$.

5.4.2 Amplifier: Gain, Input Impedance, and Output Impedance

Electronic circuits consist of connections between components where the output of one component is connected to the input of another component. The input and output impedances of connected components are important and can significantly affect the transmitted signal. For instance, op-amps are used to amplify and filter its input signal before passing it to the next component.

An ideal amplifier, in its simplest form, amplifies its input signal (Figure 5.8) and presents the result as its output signal. It does not change the original signal shape (frequency content) due to the fact that the signal is connected to the amplifier. Consider an operational amplifier (op-amp) connected to an input source (i.e., a sensor signal) and an output load. Ideally, the op-amp has a gain (K_{amp}), input impedance (Z_{in} or R_{in}), and output impedance (Z_{out} or R_{out}). An ideal op-amp has infinite input impedance and zero output impedance. In reality, it has a very large input impedance and very small output impedance. The Z_{in} is the generalized version of R_{in}, and Z_{out} is the generalized version of R_{out}.

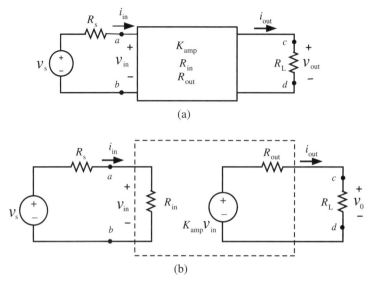

FIGURE 5.8: Input and output loading effects as a result of the connection of an op-amp to an input signal source and output load. (a) input and output connection of the op-amp device, (b) model of the op-amp with input impedance, output impedance, and open circuit voltage gain.

The open circuit voltage gain, K_{amp}, is assumed constant for low frequency signals relative to the bandwidth of the amplifier,

$$V_{out,open} = K_{amp} \cdot V_{in} \qquad (5.75)$$

where the V_{in} is the voltage across the input terminals of the amplifier, and $V_{out,open}$ is the voltage across the output terminals of the amplifier when it is in open loop condition.

Let us recall the definition of the *impedance* and the fact that it is a generalization of resistance. For the amplifier model shown in Figure 5.8, the input and output impedances are defined as,

$$Z_{in}(jw) = \frac{V_{in}(jw)}{i_{in}(jw)} \qquad (5.76)$$

$$Z_{out}(jw) = \frac{V_{out}(jw)}{i_{out}(jw)} \qquad (5.77)$$

The input impedance is the generalized resistance seen by the input signal source. The output impedance is the generalized resistance of a voltage source which is the same as the output resistance of its equivalent Thevenin's circuit. The gain, input, and output impendances are three parameters that effectively define the performance characteristics of an amplifier.

5.4.3 Input and Output Loading Errors

Let us consider an amplifier connected between a voltage source and a load (Figure 5.8). The input voltage is affected by the input impedance of the amplifier,

$$V_{in} = \frac{R_{in}}{R_{in} + R_s} v_s \qquad (5.78)$$

and the output voltage is affected by the output impedance,

$$V_{out} = \frac{R_L}{R_L + R_{out}} K_{amp} V_{in} \tag{5.79}$$

Notice that the ideally, when $R_{in} \to \infty$ and $R_{out} \to 0$, $V_{in} = v_s$ and $V_{out} = K_{amp} v_s$, where the input and output loading effects are zero. In reality, the finite input and output impedance introduces so-called *loading errors* to the signal amplification. This is minimized when the input impedance is very large relative to the source impedance, and output impedance is very small relative to the load impedance. The net gain of the amplifier in a real amplifier is

$$v_o = \left(\frac{R_L}{R_L + R_{out}} \right) \left(\frac{R_{in}}{R_{in} + R_s} \right) K_{amp} v_s \tag{5.80}$$

where the portion of the amplifier gain

$$\frac{R_L}{R_L + R_{out}} \frac{R_{in}}{R_{in} + R_s} \tag{5.81}$$

is due to the input and output loading effects. In order to minimize the effect of input loading errors,

$$R_{in} \gg R_s \tag{5.82}$$

Similarly, in order to minimize the output loading errors,

$$R_{out} \ll R_L \tag{5.83}$$

Furthermore, more accurate analysis of the amplifier would take the frequency dependance of the input and output impedance into account ($Z_{in}(jw)$ instead of R_{in}, $Z_{out}(jw)$ instead of R_{out}).

Example Let us consider a sensor, a signal amplifier, and a measurement device, all connected in series as shown in Figure 5.8. The sensor provides a voltage proportional to the measured physical variable. Let us call that voltage $v_s(t) = 10$ V at steady-state condition. For simplicity let us assume that the gain of the amplifier is unity, $K_{amp} = 1.0$. Determine the voltage measured at the measuring device (i.e., digital voltmeter or oscilloscope) for the following two conditions,

1. $R_s = 100\,\Omega$, $R_L = 100\,\Omega$, $R_{in} = 100\,\Omega$, $R_{out} = 100\,\Omega$
2. $R_s = 100\,\Omega$, $R_L = 100\,\Omega$, $R_{in} = 1\,000\,000\,\Omega$, $R_{out} = 1\,\Omega$

For the first case, the measured voltage is

$$v_{out} = \left(\frac{R_L}{R_L + R_o} \right) \left(\frac{R_{in}}{R_{in} + R_s} \right) K_{amp} v_s \tag{5.84}$$

$$= \left(\frac{100}{100 + 100} \right) \left(\frac{100}{100 + 100} \right) 1 \, v_s \tag{5.85}$$

$$= 0.25 \cdot v_s \tag{5.86}$$

$$= 2.5 \text{ V} \tag{5.87}$$

which shows an error of 75% of the correct value of the voltage.

For the second case the measured voltage is

$$v_{out} = \left(\frac{R_L}{R_L + R_{out}}\right)\left(\frac{R_{in}}{R_{in} + R_s}\right) K_{amp} \, v_s \tag{5.88}$$

$$= \left(\frac{100}{100 + 1}\right)\left(\frac{1\,000\,000}{1\,000\,000 + 100}\right) \cdot 1 \cdot v_s \tag{5.89}$$

$$= \left(\frac{100}{101}\right)\left(\frac{1\,000\,000}{1\,000\,100}\right) \cdot v_s \tag{5.90}$$

$$= 0.990 \cdot v_s \tag{5.91}$$

$$= 9.90 \text{ V} \tag{5.92}$$

which shows an error of 1% of the correct value of the voltage. Notice that there are two components to the error in the gain due to input and output loading, one due to the input loading (K_1) and and the other due to the output loading (K_2),

$$v_{out} = K_2 \cdot K_1 \cdot K_{amp} \, v_s \tag{5.93}$$

where,

$$K_1 = \left(\frac{R_{in}}{R_{in} + R_s}\right) \tag{5.94}$$

$$= \frac{1\,000\,000}{1\,000\,100} \tag{5.95}$$

$$K_2 = \left(\frac{R_L}{R_L + R_{out}}\right) \tag{5.96}$$

$$= \frac{100}{101} \tag{5.97}$$

This example illustrates that in order to minimize the effect of loading errors, components should have very large input impedance (ideally infinite), and very small output impendace (ideally zero).

5.5 SEMICONDUCTOR ELECTRONIC DEVICES

Electronic systems include components made using semiconductor materials, such as diodes, SCRs, transistors, and integrated circuits (ICs). Pictures of some common semiconductor devices are shown in Figure 5.9. Because current flow is accomplished through the flow of electrons in the solid crystal structure of semiconductor materials, components made from semiconductor materials are also called solid-state devices. The difference between an electrical and electronic system lies in whether solid-state components (diodes, transistors, ICs) are used in the circuit or not.

5.5.1 Semiconductor Materials

Semiconductor materials are the group IV elements in the periodic table. The most commonly used semiconductor materials are silicon (Si) and germanium (Ge). Semiconductor materials have an electrical conductivity property that is somewhere in between the conductors and insulators (non-conductors), hence the name *semiconductors*. A silicon atom has 14 electrons, and four of these electrons revolve in the outermost orbit around the nucleus (Figure 5.10a). In its pure form, silicon is not much use as a semiconductor. The crystalline

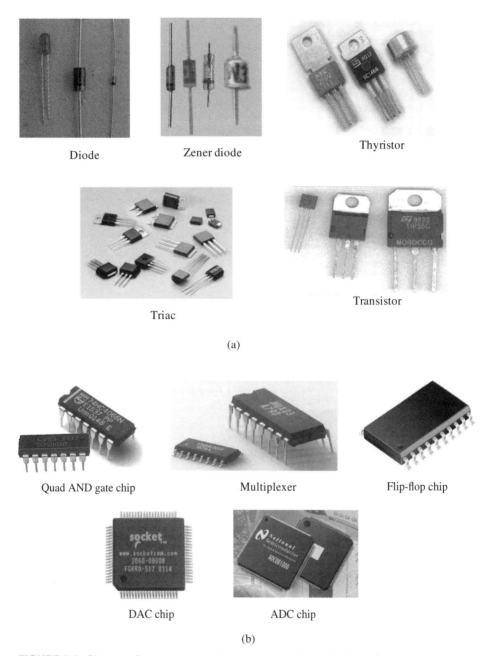

Diode   Zener diode   Thyristor

Triac   Transistor

(a)

Quad AND gate chip   Multiplexer   Flip-flop chip

DAC chip   ADC chip

(b)

FIGURE 5.9: Pictures of some commonly used semiconductor devices: discrete devices (diode, transistor) and integrated circuit chips (ICs).

structure of silicon is similar to the diamond structure of carbon which is a very stable structure (Figure 5.10b). A three-dimensional view of the crystal structure shows that each silicon atom is surrounded by four other atoms and they each share one electron in each connection (Figure 5.10a,b,c).

Impurities from group III and group V elements are added to pure silicon. The resulting silicon is also called *doped silicon*. When a material from group III of the periodic

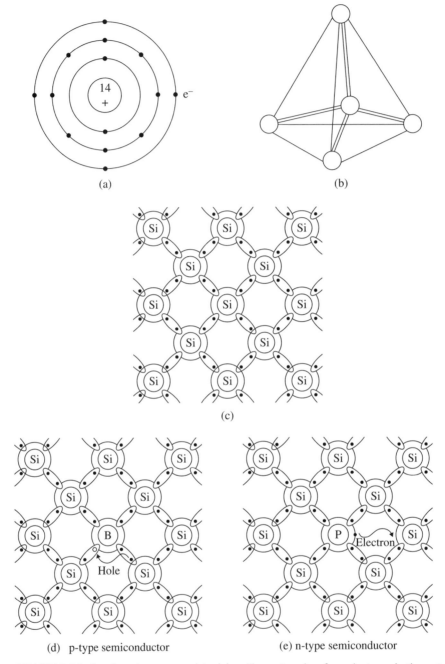

(a)

(b)

(c)

(d) p-type semiconductor

(e) n-type semiconductor

FIGURE 5.10: Semiconductor materials: (a) a silicon atom has four electrons in the outermost oribit. (b and c) each silicon atom is surrounded by four other atoms in the crystal structure where it is shown both in three-dimensional and two-dimensional representation. (d) p-type, and (e) n-type semiconductor material (doped silicon) crystal structure.

table (i.e., boron (B)) is added to the silicon, there will be one missing electron for each added impurity atom in the crystal structure. One of the four connections around the group III atom will be missing an electron (Figure 5.10d). This effective positive charge, also called a *hole*, is loosely connected to the atom. Its ability to move around the crystal structure gives the material the ability to conduct electricity. Since the net added electrical charge is positive, such semiconductors are called *p-type semiconductors* (Figure 5.10d). Similarly, if the added material is from group V of the periodic table (i.e., phospore (P), arsenic (As)), there will be an extra electron around the added impurity atom. This provides the conductor property to the semiconductor material. Since the net added electrical charge to the crystal structure is a negative charge due to the extra electron, this type is called *n-type semiconductor* (Figure 5.10e). All semiconductor materials conduct electricity through the holes and electrons introduced by the doping material into its crystal structure.

Semiconductor materials are the basis of electronic switches which contain various junctions of n-type and p-type semiconductor materials. The conductivity property of these junctions are controlled by a base current under our control or a function of the voltage bias between the two terminals. The device acts as a conductor (closed switch) and as a non-conductor (insulator, open switch) under different operating conditions.

5.5.2 Diodes

A diode is a two-terminal device made of a p-n junction (Figure 5.11a). It works like an electronic switch that allows current flow only in one direction. It functions like a one-directional check valve. It lets the flow in one direction when the pressure is above a threshold value and it blocks the flow in the reverse direction. It is a conductor in the positive direction, and an insulator in the negative direction. Once it starts conducting, the resistance of it is negligable. It is like a closed switch. The voltage and current relationship across a diode has three regions: forward bias region, reverse bias region, and reverse break-down region. When the voltage is larger than the forward bias voltage value ($V_F = 0.3$ V for germanium diode, $V_F = 0.6$ V to 0.7 V for silicon diode), $V_D > V_F$, the diode becomes a conductor and acts as a closed switch. Hence, the current increases very fast indicating very low resistance at this device. The actual value of the current is determined by the rest of the circuit. In this region, the diode can be considered as a closed switch with negligable resistance and a small voltage drop of V_F. When the diode is in the reverse bias region, $V_Z < V_D < V_F$, then the diode acts as an insulator or open switch. In other words, it acts like a very large resistor. When the $V_D < V_Z$, the diode operates in the brea-kdown region, and acts as a closed switch. This is referred to as the *avalanche effect* and a large current flows in the reverse direction. If this reverse current is larger than a rated value for the diode, the component will fail. The design parameters of a diode include:

1. V_F, forward bias voltage.
2. V_Z, reverse break-down voltage.
3. f_{sw}, maximum switching frequency the diode can respond to.
4. $i_{FWD,ave}$, the forward average (continuous) current the diode can conduct without damage at a given temperature. As temperature increases, this current rating decreases.
5. $i_{FWD,peak}$, the forward peak (short period, i.e., 10 ms) current the diode can conduct without damage. More specifically, this rating is a function of the repetetion rate (pulse width), and as the repetetion (pulse rate) increases, this rated current decreases.
6. $i_{REV,max}$, the maximum reverse current the diode will conduct while the reverse bias is voltage below the break-down voltage.
7. P_{nom} nominal power dissipation at the diode, which is approximately the voltage drop across it times the current it is conducting. Generally this is a small amount since at all operating modes of a diode, either the voltage drop across it or the current through it is very small.
8. Temperature range the diode is rated to operate, i.e. $-65\,°C$ to $200\,°C$.

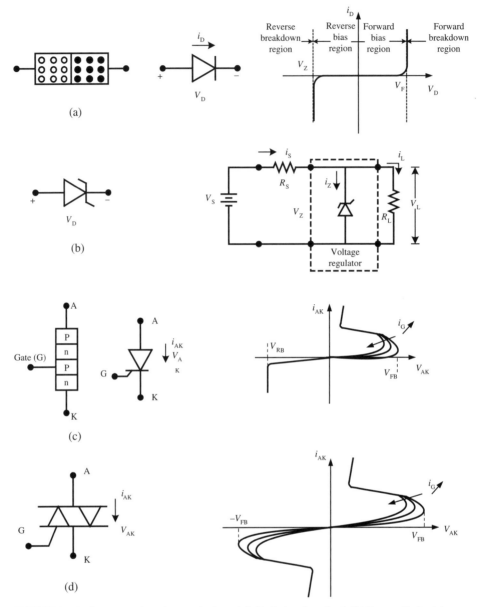

FIGURE 5.11: Some semiconductor devices: (a) diode (pn-junction), (b) Zener diode, (c) thysristor (silicon controller rectifier or SCR), (d) Triac.

The most common applications of diodes are half-wave and full-wave rectification of AC signals to DC signals.

Zener Diode A Zener diode is a special type of diode which makes use of the reverse break-down voltage (V_Z). It operates in the break-down region without destroying itself. Given a well-defined reverse break-down voltage, a Zener diode is generally used as a voltage regulator under varying supply and load voltage conditions (Figure 5.11b). Any voltage above the V_Z value across the Zener diode, and hence across the load, will be reduced to V_Z due to very high conductance across the diode. Therefore, when the Zener diode is conducting, the current across the load and diode are

$$i_L = \frac{V_Z}{R_L}; \quad V_S \geq V_Z \tag{5.98}$$

$$i_S = \frac{V_s - V_Z}{R_S}; \quad V_s \geq V_Z \tag{5.99}$$

$$i_Z = i_S - i_L; \quad V_S \geq V_Z \tag{5.100}$$

When the Zener diode is not conducting,

$$V_L < V_Z \tag{5.101}$$

$$i_Z = 0.0 \tag{5.102}$$

$$i_L = i_S \tag{5.103}$$

$$R_1 \cdot i_L = V_L < V_Z \tag{5.104}$$

The power dissipated on a Zener diode is

$$P_Z = i_Z \cdot V_Z = \left[\frac{(V_S - V_Z)}{R_S} - \frac{V_Z}{R_L} \right] \cdot V_Z \tag{5.105}$$

which must not exceed the power rating of the diode. The main design parameters of a diode are the maximum power dissipation capacity ($P_{Z,max}$, i.e., 0.25 W to 50 W) and the reverse breakdown voltage (V_Z, i.e., 5 V to 100 V range).

Thyristors Thyristors (also called *silicon controller rectifiers* (SCR)) are a three-terminal device made of three junctions (junctions of PNPN) using two p-type and two n-type semiconductor materials (Figure 5.11c). The three terminals are anode (A), cathode (K), and gate (G). The input–output relationship in steady-state is the voltage (v_{AK}) and current (i_{AK}) between A and K where this relationship is a function of the gate current (i_G). SCR is a controllable diode. The gate is used to turn ON the SCR, but it cannot be used to turn it OFF. SCR turns OFF when the i_{AK} goes to zero, then the gate can be used to turn it ON again. In order to turn ON the SCR, it must be forward biased ($V_{AK} \geq V_{FB}(i_G)$) plus the gate current must be above the *latching current* specification ($i_G > i_{G,min}$), the minimum amount of current required to turn ON the SCR. The forward voltage drop V_{FB} is in the range of 0.5 V to 1.0 V. The current conducted by the SCR can range between 100 mA to 100 A. The power dissipated across the SCR is the voltage drop across it times the conducting current,

$$P_{SCR} = V_{drop} \cdot i_{AK} < P_{max} \tag{5.106}$$

By controlling the timing of the gate current while the SCR is forward biased, a controlled portion of a forward alternating voltage can be conducted. This is used in speed control of DC motors from an AC source. If SCR is forward biased suddenly (i.e., $dV_{AK}/dt > 50 \text{ V}/\mu s$), it may turn ON even though base current is not applied. To reduce such undesirable effects, a low pass passive RC-circuit (also called a *snubber circuit*) is used in parallel with the SCR. In addition, when a SCR is turned ON, the rate of current change, $di(t)/dt$, may be very large. In order to reduce that, the load should have inductance that is larger than a minimum required value. If the inductance is very low, the rate of current change can be too large.

Triacs Triac is equivalent to two SCRs connected in reverse-parallel configuration (Figure 5.11d). While an SCR can be turned ON only in the forward biased direction, a triac can be turned ON in both directions. Typical voltage drop across a triac is about 2 V while it is conducting. This voltage times the load current determines the energy dissipation in the form of heat by a triac. SCR and triac components are generally used to switch AC loads. The amount of power that is allowed to conduct is controlled by the timing gate voltage of SCR and triac relative to the AC voltage applied between terminals A and K.

Example Consider the Zener diode application circuit shown in Figure 5.11b. Let us assume that $V_s = 24\,\text{V}$, $R_s = 1000\,\Omega$, $V_Z = 12\,\text{V}$, and R_L varies between $1000\,\Omega$ and $2000\,\Omega$. Let us determine the currents in each branch of the circuit, voltage across the load, and power dissipated across the Zener diode. The Kirchoff's voltage law in the loop including the power supply and the zener diode gives,

$$V_s = R_s \cdot i_s + V_Z \tag{5.107}$$

$$24 = 1000 \cdot i_s + 12 \tag{5.108}$$

$$i_s = \frac{24 - 12}{1000} = 0.012\,\text{A} = 12\,\text{mA} \tag{5.109}$$

Since the voltage drop across the Zener diode is limited to $V_Z = 12\,\text{V}$, the same voltage potential exists across the parallel load resistor,

$$V_{R_L} = V_Z = R_L \cdot i_L \tag{5.110}$$

$$12 = R_L \cdot i_L \tag{5.111}$$

$$i_L = \frac{12}{R_L} \tag{5.112}$$

when $R_L = 1000\,\Omega$, $i_L = 12.0\,\text{mA}$, and when $R_L = 2000\,\Omega$, $i_L = 6.0\,\text{mA}$. Since

$$i_s = i_L + i_Z \tag{5.113}$$

The current across the diode is

$$i_Z = i_s - i_L \tag{5.114}$$

which varies as the load resistance varies. In other words, the current across the zener diode is $i_Z = 0.0\,\text{mA}$ when $R_L = 1000\,\Omega$ and $i_Z = 6.0\,\text{mA}$ when $R_L = 2000\,\Omega$. The Zener diode provides a constant 12 V voltage across the load resistor. As the load resistor varies, the Zener diode dumps the excess current while providing the $V_Z = 12\,\text{V}$ constant voltage across the two terminals. If $R_L < 1000\,\Omega$, that is $R_L = 500\,\Omega$, the diode does not conduct. Assume that the diode is not conducting, then, $i_Z = 0$,

$$i_s = i_L = \frac{V_s}{R_L + R_s} = \frac{24}{1500} = 0.016\,\text{A} = 16\,\text{mA} \tag{5.115}$$

$$V_L = R_L \cdot i_L = 500\,\Omega \cdot 16\,\text{mA} = 8\,\text{V} < V_Z \tag{5.116}$$

So, the Zener diode acts as a component to dump the excess voltage, but does not make up for lower voltages. The maximum power dissipated across the Zener diode in this example is

$$P_Z = V_Z \cdot i_Z = 12\,\text{V} \cdot 6\,\text{mA} = 72\,\text{mW} \tag{5.117}$$

Therefore, a Zener diode with $1/4\,\text{W}$ power rating and 12 V breakdown voltage rating would be sufficient for this circuit.

Figure 5.12 shows the application of diodes for *voltage surge protection* due to inductive loads. Such a use of diodes is called *free-wheeling diodes*. When a diode is placed parallel to the transistor in motor control applications, it is called a *by-pass diode*. When there is an inductive load (a coil winding of the conductor, i.e., in relays, solenoids, motors), sudden switching of the current either by mechanical switches or electronic switches

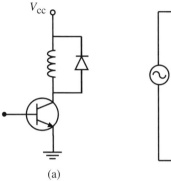 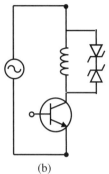

V_{cc}

(a)   (b)

FIGURE 5.12: Diode application: voltage surge protection when inductive load is present in the circuit. (a) For DC circuits, (b) for AC circuits.

(transistors) results in large voltage surges due to the inductive load. Recall that for an inductor,

$$V(t) = L \cdot \frac{di(t)}{dt} \qquad (5.118)$$

When the current is switched suddenly there can be large voltage surges since $di(t)/dt$ can be very large. Especially when the current is switched OFF, the surge in current does not have path to travel but shows up across the switch, which can be damaged. In addition, the voltage surge maybe so large as to damage the insulation of the conductor. In order to reduce the effects of the surge voltages, a diode is used across each inductive load. If the voltage supplied to the inductive load is DC, a single diode is sufficient. If it is AC, then two opposite Zener diodes are used to handle the voltage surges in both directions (Figure 5.12). In order to quickly dissipate the power trapped in the inductive load after the switch-OFF, some designs include a small resistor in series with the diodes.

 Light emitting diodes (LEDs) and light sensitive diodes (LSDs) are diodes that give out light with an intensity that is proportional to the current and pass currents proportional to the received light intensity, respectively. Furthermore, the light frequency can be modulated or pulsed (i.e., up to 10 MHz range) where the LEDs and LSDs respond only to the selected frequency range of the modulated light, hence they would not be affected by the ambient light. LEDs and LSDs are used as *opto-couplers* in electrical circuits in order to electrically isolate two circuits and couple them optically (Figure 5.13).

Example   Consider the circuits shown in Figure 5.14. Part (a) of the figure shows a half wave rectifier. The output voltage across the load resistor is simply the positive portions of the input voltage minus the bias voltage of the diode. Part (b) of the figure shows the same circuit with a low pass filter inserted between the diode and load resistor. This method is used in DC power supplies in order to smooth out the pulsation of AC input voltage and generate a DC output voltage. Let us examine the behavior of each circuit. For both cases,

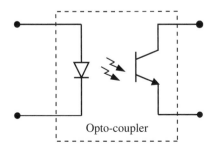

Opto-coupler

FIGURE 5.13: Opto-coupler symbol: light emitting diode (LED) and light sensitive transistor (LST) pair used to electrically isolate two sides of the circuit using optical light as the coupling medium.

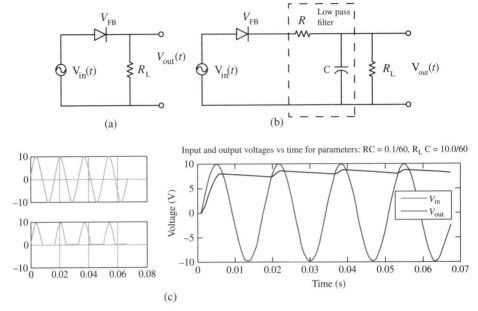

FIGURE 5.14: Simple DC power supply using a diode: (a) using a single diode for half wave rectification function, (b) using a diode and a low pass filter. (c) Simulated results for each circuit. For good DC power supply behavior, we need RC to be small compared to input signal frequency (that is $1/RC$ larger than input signal frequency), and $R_L C$ to be large (that is $1/R_L C$ smaller than input signal frequency).

input voltage is an AC voltage with 60 Hz frequency and 10 V magnitude, and the forward bias voltage of the diode is $V_{FB} = 0.7$ V.

For case (a), the output voltage is simply

$$V_{out}(t) = V_{in}(t) - V_{FB} : \text{ when } V_{in}(t) \geq V_{FB} \tag{5.119}$$

$$= 0; \text{ when } V_{in}(t) < V_{FB} \tag{5.120}$$

Since $V_{in}(t) = 10.0 \cdot \sin(2\pi \, 60 \, t)$,

$$V_{out}(t) = 9.3 \cdot \sin(2\pi \, 60 \, t) : \text{ when } V_{in}(t) \geq 0.7 \text{ V} \tag{5.121}$$

$$= 0; \text{ when } V_{in}(t) < 0.7 \text{ V} \tag{5.122}$$

The results are plotted in the same figure.

For case (b), we need to consider two cases:

1. when the diode is conducting, and
2. when the diode is not conducting.

Unlike case (a), in case (b) when the diode is not conducting, the current (and voltage) at the output circuit is not zero due to the capacitor. When the diode is conducting,

$$(V_{in}(t) - V_{FB}) \geq V_{out}(t) \tag{5.123}$$

the full circuit is active. When the diode is not conducting,

$$(V_{in}(t) - V_{FB}) < V_{out}(t) \tag{5.124}$$

the capacitor C and load resistor R_L form a complete circuit and the current flows due to the non-zero voltage in the capacitor. Assume that the charge in the capacitor is zero when the circuit is first turned on.

(i) When the diode is conducting; $(V_{in}(t) - V_{FB}) \geq V_{out}(t)$

$$V_{in}(t) - V_{FB} - R \cdot i(t) = V_{out}(t) \tag{5.125}$$

$$i(t) = i_C(t) + i_{R_L}(t) \tag{5.126}$$

$$i_{R_L}(t) = \frac{1}{R_L} \cdot V_{out}(t) \tag{5.127}$$

$$i_C(t) = C \cdot \frac{dV_{out}(t)}{dt} \tag{5.128}$$

Let $V_{in}^*(t) = V_{in}(t) - V_{FB}$, and substitute the $i(t)$, $i_C(t)$, $i_{R_L}(t)$ in the first equation in order to obtain the dynamic relationship between the input voltage and output voltage. The result can be shown to be

$$RC \frac{dV_{out}(t)}{dt} + \left(1 + \frac{R}{R_L}\right) V_{out}(t) = V_{in}^*(t) \tag{5.129}$$

where the initial value of output voltage at the begining of each cycle (when the diode first starts to conduct) is the voltage across the capacitor, $V_{out}(t_i) = V_C(t_i)$. During the time that the diode conducts, we want the output voltage to quickly track the input voltage. In order to accomplish this we need a small time constant, which means RC should be small compared to the input frequency, or equivalently $1/RC$ should be larger than the input frequency, that is ten times larger $1/RC = 10 \cdot 60$ or $RC = 1/(10 \cdot 60)$.

(ii) When the diode is not conducting, $(V_{in}(t) - V_{FB}) < V_{out}(t)$, we have a capacitor C (with initial charge and voltage from the last instant when diode was conducting) and load resistor R_L forming a closed electrical circuit. Hence, the dynamic behavior of the voltage and current relationship during that period is described by

$$V_C(t_i) \; ; \quad \textit{given from previous phase or initial condition} \tag{5.130}$$

$$V_{out}(t) = V_C(t) = V_C(t_i) + \frac{1}{C} \int_{t_i}^t i_C(\tau)d\tau = R_L \cdot i_{R_L}(t) \tag{5.131}$$

$$i_C(t) = -i_{R_L} = -\frac{V_C(t)}{R_L} \tag{5.132}$$

Evaluating this equation, it can be shown that the output voltage dynamics during the time when the diode is not conducting is defined by

$$\frac{1}{R_L C} \int_{t_i}^t V_{out}(\tau)d\tau + V_{out}(t) = V_C(t_i) \tag{5.133}$$

or

$$\frac{dV_{out}(t)}{dt} + \frac{1}{R_L C} V_{out}(t) = 0.0; \tag{5.134}$$

with the initial condition of

$$V_{out}(t_i) = V_C(t_i) \tag{5.135}$$

where $V_C(t_i)$ is the voltage across the capacitor (which is same as the output voltage) the last time the diode was conducting. Notice that the time domain analytical solution of the above equation is (which can be numerically confirmed by MATLAB®/Simulink® simulation)

$$V_{out}(t) = V_C(t_i) \cdot e^{-(t-t_i)/(R_L C)} \tag{5.136}$$

which shows that in order to make the decay rather slow and obtain a DC output voltage, $R_L C$ should be as large as possible. R_L will be function of the load connected to the output, therefore we have more control over the value of capacitor, C, to make sure that $R_L C$ is large enough to provide a smooth output DC voltage relative to the oscillation frequency of the input voltage. Results are shown for 60 Hz input voltage and $RC = 0.1/60$, $R_L C = 10.0/60$.

In summary, the diode alone only passes the positive half of the input signal and acts as a half wave rectifier. The diode plus the low pass filter circuit smooth out the oscillations of the AC input and provide almost constant DC output voltage. In order to get a smooth DC output voltage, the $1/(RC)$ should be much larger than the input signal frequency and $1/(R_L C)$ should be much smaller than the input signal frequency. The performance of the simple DC power supply shown in Figure 5.14b can be simulated for different values of $R \cdot C$ and $R_L \cdot C$ relative to the frequency of the input signal which is 60 Hz in this example.

```
% Simulation of the diode and low pass filter circuit for
DC power supply operation.

    t0 = 0.0 ;
    tf = 4.0/60.0 ;
    t_sample = 0.001 ;
    x_out =  0 ;
    u_out = 0 ;

% Start the simulation loop...

    u = 0.0 ;
    x = 0.0 ;

    for (t = t0 : t_sample : tf )
      u1 = 10.0 * sin(2*pi*60*t) ;
      Vfb = 0.7 ;
      u = u1 - Vfb ;
      x = rk4('process_dynamics',t,t+t_sample, x, u) ;
            x_out=[x_out ; x' ] ;
            u_out=[u_out ; u1 ] ;
    end


% ..Plot results....
    t_out=t0:t_sample:tf  ;
    t_out = [t_out' ; tf+t_sample ] ;
    subplot(111)
       plot(t_out,u_out(:,1),'r',t_out,x_out(:,1),'b') ;
       title('Input and output voltages vs time for
       parameters: RC=0.1/60, R_LC = 10.0/60') ;

%

function xdot= process_dynamics(t,x,u)
%
%   describes the dynamic model: o.d.e's
%   returns  xdot vector.
%
      R= 0.1/60. ;
      C= 1.0 ;
      R_L = 10./60. ;
```

```
if (u >= x )
    xdot = (1/(R*C))*(u - (1+(R/R_L)) * x) ;
else
    xdot = -(1/(R_L*C))*x ;
end
```

5.5.3 Transistors

A transistor can function in two ways:

1. as an ON/OFF switch,
2. as a proportional amplifier.

It is an electronic switch (solid-state switch) that can be opened or closed completely or partially. Another way of looking at a transistor is that it is a variable resistor where the resistor value is controlled by the gate current. It is an active element which is used to modulate (control) the flow of power from source to load. A transistor is a three-terminal device made of two junctions of p-type and n-type semiconductors. If the junctions are made of n-p-n order, it is called npn-type transistor. If the junctions are made of p-n-p order, it is called pnp-type transistor. The difference between them in a circuit application is that the supply connection polarity is opposite, hence the direction of current is opposite. The three terminals are collector (C), emitter (E), and base (B) (Figures 5.13 and 5.15). One of the important features of input and output relations of a transistor is that the input strongly affects the output. However, the output variations (i.e., due to load changes) do not affect the input. Therefore, the base of a transistor is always part of the input circuit which controls the output circuit. Unlike passive components such as resistors, capacitors, and inductors, the transistor is an active device. It requires a power supply to operate and can increase the power between input and output. The input circuit acts as a control circuit which controls a much larger power level in the output circuit using a much smaller power level at the input circuit. The transistor is the electronic analogy of a hydraulic valve.

The main types of transistors are discussed below. Bipolar junction transistors (BJTs) are the most common transistors. Metal oxide semiconductor field effect transistors (MOSFETs) require smaller gate current, have better efficiency, and have higher switching frequency. Insulated gate bipolar transistors (IGBTs) are a more recent transistor type which attempts to combine the advantages of BJTs and MOSFETs. A transistor performance is characterized by the maximum base current, current between the two main terminals (collector-emitter, drain-source), forward bias voltage, reverse breakdown voltage, maximum switching frequency, forward current gain, reverse voltage gain, input impedance, and output impedance.

Bipolar Junction Transistors (BJTs)   There are two types of BJTs: npn-type and pnp-type, each type being a three-terminal device (Figure 5.13a). The input–output relationship between collector (C) and emmitor (E) terminals is controlled by the base (B) signal. A BJT can be used as an electronic switch (ON/OFF) or a proportional amplifier (output current i_C is proportional to input current i_B). The voltage and current relationship across the C and E terminals of the transistor (V_{CE}, i_{CE}) is a function of the base current. There are three main design parameters of interest: current gain (*beta*), voltage drop between base and emitter when the transistor is conducting ($V_{BE} = V_{FB}$), and the minimum voltage drop across the collector and emitter when the transistor is saturated or fully ON ($V_{CE} = V_{SAT}$). Note that $i_E = i_B + i_C$ always.

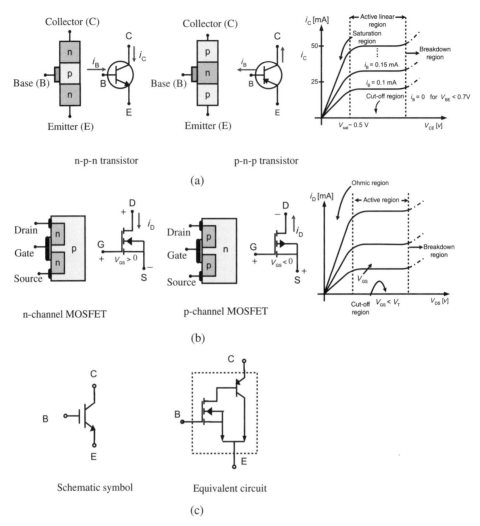

FIGURE 5.15: Some semiconductor devices: transistors. (a) Bipolar junction transistors (BJT), (b) metal oxide semiconductor field effect transistor (MOSFET), and (c) insulated gate bipolar transistor (IGBT).

There are three main modes of operation of the BJT in steady-state,

1. *Cut-off Region:* $V_{BE} < V_{FB}, i_B = 0$, hence, $i_C = 0, V_{CE} > V_{supply}$. The transistor acts like an OFF-state switch. There is no current flow between the C and E. V_{FB} can vary between 0.6 V to 0.8 V as a result of manufacturing variations. In valve analogy, the valve is closed.

2. *Active Linear Region:* $V_{BE} = V_{FB}, i_B \neq 0; i_C = \beta\, i_B, V_{SAT} < V_{CE} < V_{supply}$. The typical value of $V_{SAT} \approx 0.2$ V to 0.5 V. Transistor acts like a current amplifier. The output current, i_C, is proportional to the base current, i_B, where the proportionality constant (gain) is a design parameter of the transistor, which is typically around 100 and can range from 50 to 200. Good circuit designs should not rely on this open-loop transistor gain for it can vary significantly from one copy to another of the same transistor as well as being a function of temperature. The current i_C is also function of the V_{CE} slightly. Under constant base current conditions, the collector current

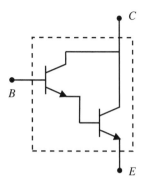

C

B

E **FIGURE 5.16:** Darlington transistor.

increases slightly as the collector to emitter voltage increases. In valve analogy, valve is partially open.

3. *Saturation Region:* $V_{BE} = V_{FB}$, $i_B > i_{C,max}/\beta$, $V_{CE} \leq V_{SAT} \approx 0.2\,V - 0.5\,V$. In this mode the transistor operates like a closed (ON) switch between C and E terminals. The actual value of i_C is determined by the circuit preceeding the collector, which is analogous to a completely open valve where the flow rate is determined by the supply and load pressures.

In approximate calculations, it is customary to assume $V_{FB} = V_{SAT} = 0.0$ and $i_E = i_C$ in transistor circuits. In valve analogy, the valve is fully open.

BJT type transistors can support collector currents in the range of 100 mA to 10 A. The BJTs rated for above 500 mA current are called *power transistors* and must be mounted on a heat sink.

In order to obtain higher gain, β, from BJTs, a commonly used transistor is the Darlington transistor which gives a gain that is the multiplication of the two stages of the transistors (Figure 5.16)

$$\beta = \beta_1 \cdot \beta_2 \tag{5.137}$$

where the gain β of a Darlington transistor can be in the order of 500 to 20 000.

Note that the power dissipation across a BJT is

$$P_{BJT} = V_{CE} \cdot i_{CE} \tag{5.138}$$

which should be below the rated power of the transistor in a given design. The power dissipation across the transistor is a key factor to consider for two different reasons:

1. failure of the component due to excessive heating and the associated heat dissipation issues,

2. efficiency of the component by reducing the wasted heat energy.

When a transistor operates in fully ON (in the saturated region), the voltage drop across it is very small, $V_{CE} \approx 0.4\,V$. Hence the power dissipation is small. Similarly, when the transistor is fully OFF (in the cutoff region), even though the voltage drop across it is large, the current conducted is zero, $i_C \approx 0.0$. Hence power dissipation is again small. The observation we make here is that when the transistor is operated either in fully ON or fully OFF mode, the power dissipation is minimized and its operational efficiency is improved. Let us consider that we control the gate of the transistor with a high frequency signal (i.e., $w_{sw} = 10\,KHz$). One period of the gate signal is $t_{sw} = 100\,\mu s$ long. If we control the width (portion) of the signal within each period that will saturate the transistor (fully ON) and the width of the signal that will turn OFF the transistor, we can control the average gain

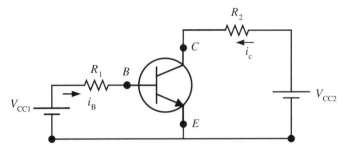

FIGURE 5.17: Common emitter configuration and voltage amplifier usage of a transistor.

of the transistor by the average of the ON–OFF widths of the gate signal. By increasing and decreasing the ON–OFF width periods within each t_{sw} period, we can vary the net average gain. Since only two states of transistor control are needed, the gate signal only has to have two voltage levels: a high level to saturate the transistor and a low level to cutoff the transistor. Then the only control problem is controlling the ON–OFF widths of the switching pulses. This is called the *pulse width modulated* (PWM) control of transistors. The PWM control method results in more efficient operation of the transistors. It operates a transistor in one of two regions: cutoff region (OFF state) and saturation region (ON state).

If the transistor is operated in the *active region*, which is the region where the output current is linearly proportional to the gate current, we use the transistor in linear (or proportional) mode. In this case both the i_C and V_{CE} can have reasonably finite values. Hence, in linear gate signal control mode, the power dissipation is larger. In linear operating mode, the efficiency of the transistor is lower compared to the efficiency in PWM mode.

In the *common-emitter* configuration (input and output have the emitter as common as shown in Figure 5.17) of a transistor, the input circuit voltage–resistor combination determines the base current. Base current times the transistor gain determines the collector current. Then, the voltage drop across the C and E terminals is determined by the voltage balance in the output circuit loop given the known value of collector current, voltage and resistor. If the calculated i_C requires a V_{CE} value that is less than $V_{CE,sat} = 0.2$ V, then the transistor saturates. Then the actual i_C value is determined by the output circuit.

Metal Oxide Semiconductor Field Effect Transistors (MOSFETs)   A MOSFETs are also a three-terminal transistors with terminals called drain (D), source (S), and gate (G) (Figure 5.13b). Like BJTs, there are two types of MOSFETs: npn-type and pnp-type. The input (control) variable is the gate voltage, V_G. The output is the drain current, i_D. There are three regions of operation of a MOSFET,

1. *Cutoff Region:* $V_{GS} < V_T$ and $i_G = 0$, hence, $i_D = 0$, where V_T is the gate-source threshold voltage. The threshold voltage takes on different values for different types of FETs (i.e., $V_T \approx -4$ V for junction FETs, $V_T \approx -5$ V for MOSFETs in depletion mode, $V_T \approx 4$ V for MOSFETs in enhancement mode). There is negligible current flow through the D terminal and the connection between D and S terminals is in an open switch state.

2. *Active Region:* $V_{GS} > V_T$, hence, $i_D \propto (V_{GS} - V_T)^2$, $V_{DS} > (V_{GS} - V_T)$. The transistor functions as a voltage controlled current amplifier, where the output current is proportional to the square of the net GS voltage. Notice that for the MOSFET to operate in this region, V_{DS} must be above a certain threshold. Otherwise, the MOSFET operates in the *ohmic region*.

3. *Ohmic Region:* When V_{GS} is large enough ($V_{GS} > V_T$) and $V_{DS} < V_{GS} - V_T$, the i_D is determined by the source circuit connected to the D terminal and the transistor behaves like a closed switch between D and S terminals. $V_{GS} \gg V_T$, then and $i_D = V_{DS}/R_{ON}(V_{GS})$. The MOSFET acts like a nonlinear resistor where the value of the resistance is nonlinear function of the gate voltage.

It should be noted that the advantages of MOSFETs over BJTs are higher efficiency, higher switching frequency, and better thermal stability. On the other hand, MOSFETs are more sensitive to static voltage. In general, MOSFETs have been replacing BJTs in low voltage (<500 V) applications.

Insulated Gate Bipolar Transistors (IGBT)   IGBTs are the new alternative to BJTs in high voltage (>500 V) applications where they combine the advantages of BJTs and MOSFETs. IGBT is a four layer device with three terminals similar to BJTs (Figure 5.15c). IGBTs are widely used in motor drive applications and operate in PWM mode for high efficiency.

Example   In Figure 5.17, let $V_{CC1} = 5$ V, $V_{CC2} = 25$ V, $R_1 = 100$ KΩ, $R_2 = 1$ KΩ. The base current is (assuming $V_{BE} = 0.0$ for simplicity)

$$i_B = \frac{5\,V}{100\,K\Omega} = 0.05\,mA \tag{5.139}$$

Let the gain of the transistor be $\beta = 100$. The collector current is

$$i_C = i_C(i_B, V_{CE}) \tag{5.140}$$
$$i_C \approx \beta \cdot i_B = 5\,mA \tag{5.141}$$

Hence, the voltage balance in the output circuit dictates that

$$25\,V = 1000 \cdot 5 \cdot 10^{-3} + V_{CE} \tag{5.142}$$
$$V_{CE} = 25 - 5 = 20\,VDC \tag{5.143}$$

In general, the output voltage across the transistor for the common-emitter configuration shown in Figure 5.17 is (let $V_{out} = V_{CE}, V_{in} = V_{CC1}, V_s = V_{CC2}$)

$$V_{out} = V_s - \frac{R_2}{R_1} \cdot \beta \cdot V_{in} \tag{5.144}$$

and the voltage potential across the output circuit resistor (V_{R2}) is

$$V_{R2} = \frac{R_2}{R_1} \cdot \beta \cdot V_{in} \tag{5.145}$$

The common-emmiter configuration of the transistor as shown in Figure 5.17 can be used as a voltage amplifier. If the output voltage is taken as the voltage across the transistor,

$$V_{out} = V_s - K_1 \cdot V_{in} \tag{5.146}$$

or if the output voltage is taken as the voltage across the resistor preceeding the collector,

$$V_{out} = K_1 \cdot V_{in} \tag{5.147}$$

Example   In order to illustrate the fact that a transistor can be operated like a switch (ON or OFF) as well as a proportional amplifier, let us consider the following cases of input voltage in the same Figure 5.17. Furthermore, let us more accurately assume that the maximum voltage drop across the base and emitter is $V_{BE} = 0.7$ V, and the minimum voltage drop across the collector and emitter is $V_{CE} = 0.2$ V.

1. Case 1: $V_{CC1} = 0.0$ V. Then $i_B = 0$, hence, $i_C = 0$. Therefore, the voltage measured between the collector and the emitter (let us call it the output voltage, V_{out}) is $V_{out} = V_{CC2} = 25$ VDC. We can call this the OFF state of the transistor.

2. Case 2: $V_{CC1} = 30.7$ V (more precisely, the input voltage at the gate is large enough to generate a large base current that will saturate the transistor output circuit). Then $i_B = (30.7 - 0.7)/100$ K $= 0.3$ mA, hence, $i_C = \beta i_B = 30$ mA. Therefore, the voltage measured between the collector and the emitter (let us call it the output voltage, V_{out}) is $V_{out} = V_{CC2} - R_2 \cdot i_C = 25 - 1000 \cdot 30$ mA $= -5$ V. Since the output voltage at the collector cannot be less than the emitter voltage plus the 0.2 V minimum voltage drop across the collector and emitter, it must be, $V_{out} = 0.2$ V. Hence the actual collector current saturates at $i_C = (V_{CC2} - V_{CE})/R_2 = (25 - 0.2)/1000 = 24.8$ mA and the output voltage at the collector is $V_{out} = 0.2$ V. We can call this the ON state of the transistor.

3. Case 3: When the input voltage is above cutoff voltage ($V_{BE} = 0.7$ V) but below the saturation voltage, the transistor operates as a proportional voltage amplifier. The saturation voltage can be calculated as follows. The minimum voltage drop across C and E when the transistor is saturated is

$$V_{CE} = 0.2 \text{ V}; \text{ when saturated} \tag{5.148}$$

At this point, the current is

$$i_C = \frac{V_{CC2} - V_{CE}}{R_2} = \frac{25 - 0.2}{1000} = 24.8 \text{ mA} \tag{5.149}$$

The base current must be

$$i_B = \frac{1}{\beta} i_C = \frac{1}{100} i_C \tag{5.150}$$

$$= \frac{V_{in,sat} - V_{BE}}{R_1} \tag{5.151}$$

$$V_{in,sat} = 0.248 \text{ mA} \cdot 100 \text{ K}\Omega - 0.7 = 24.1 \text{ V} \tag{5.152}$$

Base voltage value larger than that operates the transistor in the saturation region.

$$V_{in} < V_{BE}; \text{ transistor is in fully OFF state} \tag{5.153}$$

$$V_{BE} < V_{in} < V_{in,sat}; \text{ linear region} \tag{5.154}$$

$$V_{in} > V_{in,sat}; \text{ saturated region (fully ON state)} \tag{5.155}$$

Case 1 and case 2 are shown as a mechanical switch analogy in Figure 5.18. Figure 5.18a shows the so-called sinking connection and Figure 5.18b shows the so-called sourcing connection of the transistor to the load. The names *sinking* and *sourcing* are given from the point of view of the transistor in that it either *sinks* the current from the load or *sources* current to the load.

Example A transistor circuit shown in Figure 5.19 can be used to switch the power on a load which is controlled by a mechanical switch. Let's assume that $V_{BE} = 0.7$ V and $V_{CE} = 0.2$ V. If the load is of inductive type, a suppression diode should be used in parallel with the load. When the transistor goes from the ON to OFF stage very fast, large voltage transients develop due to inductance. These large voltage transients can be easily large enough to damage the transistor if it is larger than the forward break-down voltage of the

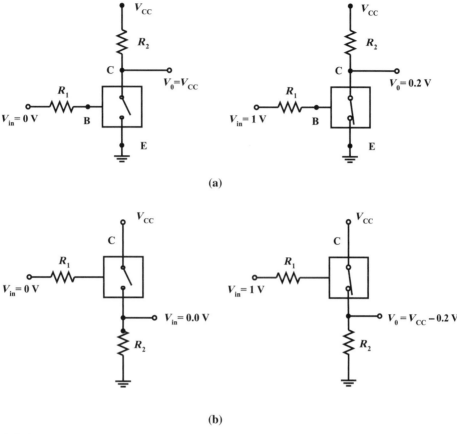

(a)

(b)

FIGURE 5.18: Mechanical switch analogy for a transistor: (a) Current sinking connection (sinking current from the load). Also called a low-side (output) switch. (b) Current sourcing connection (sourcing current to the load). Also called a high side (output) switch.

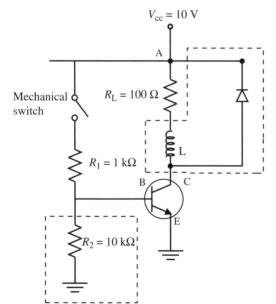

FIGURE 5.19: A transistor circuit used as an electronic switch.

transistor. The diode limits the voltage to the supply voltage by providing a current flow path during that transient period. The current flow due to inductive voltage is then dissipated in the diode-load loop within a short period of time. The resistor (R_2) which connects the base to the ground is not necessary, but makes the circuit a better one by providing a ground to the base when the transistor is not turned ON (when the mechanical switch is open). In the following discussion, assume the components in the dotted blocks do not exist for simplicity. Let $L = 0.0\,\text{H}$ and $R_2 = 0.0\,\Omega$. When the mechanical switch is OFF, the base voltage and current are zero, the transistor is OFF, no current flows through the load. When the mechanical switch is ON, the base current and collector current are,

$$i_B = \frac{V_{AB}}{R_1} = \frac{(10 - 0.7)V}{1000\,\Omega} = 9.3\,\text{mA} \tag{5.156}$$
$$i_C = \beta \cdot i_B = 0.93\,\text{A} \tag{5.157}$$

where it is assumed that $\beta = 100$. But the maximum current that the supply can support is

$$i_{C,\text{max}} = \frac{V_{CC} - V_{CE}}{R_L} = \frac{10 - 0.2}{100} = 0.098\,\text{A} \tag{5.158}$$

Hence, the transistor saturates with the maximum voltage drop across the load, $V_L = 10 - 0.2$, and the minimum voltage drop across the transistor, $V_{CE} = 0.2\,\text{V}$. The transistor acts like a very low resistance switch when it is saturated. Voltage drop across it is $V_{CE} = 0.2\,\text{V}$. Providing excess base current to saturate the transistor is a good idea and provides a safety margin to make sure the transistor is fully turned ON (in its saturation region) and hence providing the maximum voltage drop across the load. Notice that when the transistor is ON, the current drain through the resistor R_2 is

$$i_2 = \frac{V_{BE}}{R_2} = \frac{0.7}{10\,000} = 0.07\,\text{mA} \tag{5.159}$$

which is a negligable load.

Example Another common use of a transistor is in a current-source circuit (Figure 5.20). Notice that this circuit does not have any voltage amplification, but it has current amplification. Output voltage tracks the base voltage as shown below with the exception of the small voltage drop across the base and emitter. Let us assume voltage drop across the base-emitter is $V_{BE} = 0.6\,\text{V}$.

$$V_{out} = V_{in} - 0.6; \; when \; V_{in} \geq 0.6\,\text{V} \tag{5.160}$$
$$V_{out} = 0; \; when \; V_{in} < 0.6\,\text{V} \tag{5.161}$$

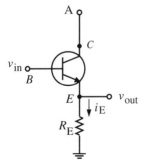

FIGURE 5.20: A transistor circuit as a current-source.

The current through the R_E,

$$i_E = \frac{V_{out}}{R_E} = \frac{V_{in} - 0.6}{R_E} \tag{5.162}$$

and the current through the collector,

$$i_C = \beta \cdot i_B \tag{5.163}$$

$$i_E = i_C + i_B = (\beta + 1) \cdot i_B \tag{5.164}$$

$$= \frac{(\beta + 1)}{\beta} i_C \tag{5.165}$$

and

$$i_C = \frac{\beta}{\beta + 1} i_E = \frac{\beta}{\beta + 1} \frac{V_{in} - 0.6}{R_E} \tag{5.166}$$

as long as the transistor is not saturated, $(V_C > V_E + 0.2)$, the current output (i_E) is proportional to the base voltage. This circuit does not have any voltage gain, but it does have current gain, and hence power gain, between input and output signals.

Note that

$$V_{in} = V_{BE} + V_E \tag{5.167}$$

$$V_E = V_{in} - V_{BE} \geq 0.0 \tag{5.168}$$

$$V_C = V_{CE} + V_E \tag{5.169}$$

The voltage at the emitter follows the base voltage. Current through the collector and emitter is proportional to the base voltage.

Example   There are many applications where a load resistor varies and despite these variations it is desired to provide a constant current through the load.

Let us modify the circuit of the previous example as shown in Figure 5.21. The main additions are an input circuit supply voltage V_{CC1}, a resistor R_1, and a Zener diode with breakdown voltage of V_Z. In the output circuit, we have a load resistor R_L connected between the output supply V_{CC2} and the collector terminal of the transistor.

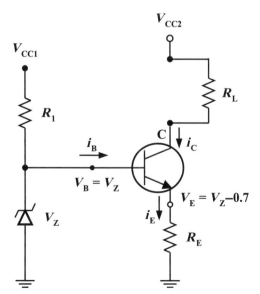

FIGURE 5.21: A transistor circuit as a constant current-source.

Notice that as long as the Zener diode is conducting, the voltage across the base and ground will be $V_B = V_Z$, and the maximum base current is

$$i_B = (V_{CC1} - V_Z)/R_1 \qquad (5.170)$$

The voltage between the emitter and ground is

$$V_E = V_Z - V_{BE} = V_Z - 0.7 \qquad (5.171)$$

Hence, current across the resistor R_E is

$$i_E = V_E/R_E = (V_Z - 0.7)/R_E \qquad (5.172)$$

The collector current which also is the load current is

$$i_C = i_E - i_B = \frac{\beta}{\beta + 1} i_E \qquad (5.173)$$

Notice that i_B is very small compared to i_E. The i_E is constant as long as the $V_{CC1} > V_Z$ and the R_E is constant. Therefore, the i_C will remain a constant current despite the variations in the load resistance R_L.

Bias Voltage   So far we have considered transistor amplifiers that amplify current or voltage and that when the input signal is zero, the output signal is zero as well, that is

$$V_{out} = K_v \cdot V_{in} \qquad (5.174)$$

Many applications require offset voltage in the input and output voltage relationship as follows,

$$V_{out} = V_0 + K_v \cdot V_{in} \qquad (5.175)$$

where V_0 is the bias voltage. There are many different versions of circuit designs to introduce bias voltage to the amplifier, one of which is shown in Figure 5.22. When the input voltage (V_{in}) is zero, there is still a base current due to the power supply voltage (V_{CC}) and the resistor R_B. Given the supply voltage, load resistance, amplifier current gain, and the desired bias voltage: V_{CC}, R_L, β, V_0, determine the bias resistance (R_B) needed.

The output circuit relation of the transistor is

$$V_{CC} = R_L \cdot i_C + V_{CE} = R_L \cdot i_C + V_{out} \qquad (5.176)$$

$$i_C = \frac{V_{CC} - V_{out}}{R_L} \qquad (5.177)$$

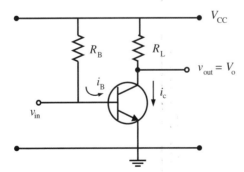

FIGURE 5.22: Bias voltage circuit for an amplifier.

The input circuit of the transistor has the relationship (assuming the transistor is operating in the linear active region), and $V_{\text{in}} = 0.0$,

$$i_B = \frac{1}{\beta} \cdot i_C \tag{5.178}$$

$$i_B = \frac{V_{R_B}}{R_B} \tag{5.179}$$

$$= \frac{V_{CC} - V_{BE}}{R_B} \tag{5.180}$$

$$\approx \frac{V_{CC}}{R_B} \tag{5.181}$$

Let use denote $V_{\text{out}} = V_0$ to indicate the bias voltage, since $V_{\text{in}} = 0$. From the above relations, we can determine the bias resistor needed to provide the desired bias voltage for the given supply, load, and transistor,

$$V_0 = V_{CC} - R_L \cdot i_C \tag{5.182}$$

$$= V_{CC} - R_L \cdot \beta \cdot i_B \tag{5.183}$$

$$= V_{CC} - R_L \cdot \beta \cdot \frac{V_{CC}}{R_B} \tag{5.184}$$

$$V_0 = V_{CC} \left(1 - \frac{\beta \cdot R_L}{R_B} \right) \tag{5.185}$$

The bias resistor is then

$$R_B = \beta \cdot \frac{V_{CC}}{V_{CC} - V_0} \cdot R_L \tag{5.186}$$

It should be noted again that this is only one of many possible and simple ways of introducing bias voltage to a transistor amplifier. In this design, the bias voltage is a linear function of the amplifier current gain, β, which is known to vary significantly due to manufacturing variations and temperature. In a given initial design, if it turns out that the bias voltage (V_0) is larger than anticipated as a result of variations in β, then the designer may need increase or reduce R_B.

Example   Consider the bias voltage circuit for a transistor as shown in Figure 5.22. Let the parameters of this circuit be given as follows,

$$V_{CC} = 12 \, \text{V}, \quad \beta = 100, \quad R_L = 10 \, \text{K}\Omega \tag{5.187}$$

Determine the value of the bias resistor, R_B, so that the bias voltage is $V_0 = 6 \, \text{V}$.
    From the above equation, the bias resistor is

$$R_B = \beta \cdot \frac{V_{CC}}{V_{CC} - V_0} \cdot R_L \tag{5.188}$$

$$= 100 \cdot 2 \cdot 10 \, \text{K}\Omega \tag{5.189}$$

$$= 2 \, \text{M}\Omega \tag{5.190}$$

After this initial design, if the circuit indicates that the V_0 is different than the desired value (due to the variation in the transistor gain β), then the value of the resistor R_B should be modified according to the above relation until the desired bias voltage is obtained.

5.6 OPERATIONAL AMPLIFIERS

Operational amplifiers (op-amps) were first developed in the late 1940s. Today, op-amps are linear integrated circuits which are low cost, reliable, and hundreds of millions of units are made per year. As the name implies, an op-amp performs an *operational amplification* function where the *operation* may be add, subtract, multiply, filter, compare, convert, and so on. Op-amp is a device with a very high open loop gain (ideally infinity, in reality finite but very large). The main function of an op-amp is often defined by external feedback components. There are a number of operational amplifiers manufactured by multiple vendors under licence. For instance, op-amp 741 is available under the trade names LM741, NE741, μA741 by different manufacturers. LM117/LM217/LM317 is a higher performance version of 741. LM117, LM217, LM317 are rated for military, industrial and commercial applications, respectively. The 301, LM339, LM311 (comparator IC chip), LM317 (voltage regulator), 555 timer IC chip, XR2240 counter/timer are widely used op-amps. The integrated circuit design incorporates multiple discrete devices (transistors, resistors, capacitors, diodes) into one compact chip which is typically in dual-inline-package (DIP) form.

The 741 op-amp has 17 BJT transistors, 12 resistors, 1 capacitor, and 4 diodes. Note that the actual internal component count and design may vary from manufacturer to manufacturer. The coding standard used to identify an op-amp includes the manufacturer code, op-amp functional code, rating for commercial or military use, and mechanical package form (i.e., DIP). In order to use an op-amp in a design, the designer needs the physical size, functionality described in the data sheet, and pinout information of the op-amp. With the wide availability and proven designs of integrated circuit (IC) op-amps, it is very rare that one designs circuits from discrete components (using transistors, resistors, capacitors, etc.) for analog signal processing purposes. For almost every analog signal processing need, there is an IC op-amp available in the market. Most often, the designer does not need to know all the internal circuit details of the IC op-amp. The characteristics of the IC op-amp that are needed are its input–output function (i.e., low pass filter), gain, bandwidth and input/output impedances. Most IC op-amp packages include multiple stages in the IC design, that is input stage generally includes a differential amplifier for its high input impedance, followed by an intermediate gain stage and finally an output stage.

5.6.1 Basic Op-Amp

Figure 5.23 shows the DIP package and symbol for a basic op-amp. Notice that in an 8-pin DIP package, the integrated circuit chip is actually a small portion of the DIP package. A

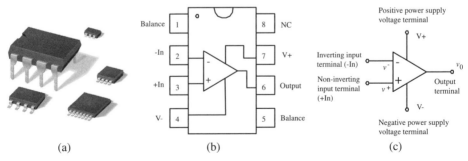

FIGURE 5.23: Operation amplifier: (a) op-amp DIP-packages, (b) op-amp pins, (c) op-amp symbol.

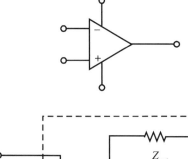

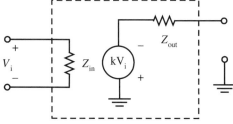

FIGURE 5.24: Model of an op-amp: single-ended output op-amp. Idealized model assumes infinite input impedance and open loop gain, and zero output impendance, $Z_{in} = \infty$, $Z_{out} = 0$, and $K = \infty$.

triangle is always used as a symbol of amplification in electrical circuits. The symbol of a basic op-amp shows the following five terminal, connections,

1. Power supply (bipolar) terminals (V^+, V^-, i.e., ±15 VDC, ±12 VDC, ±9 VDC, ±6 VDC).

2. Inverting ($-$) and noninverting ($+$) input terminals, each referenced to ground, with voltages v^- and v^+.

3. Output terminal referenced to ground, where output voltage is designated as V_o.

This is an open loop op-amp. Most uses of op-amps involve adding external components to it between its terminals to implement the desired functionalities. However, key to understanding the applications of op-amps (open loop or closed loop) is its open loop properties. Here are the idealized assumptions of an open loop op-amp. Notice that in practice, actual parameters are very close to the idealized assumptions that the performance difference between the two is negligible in most cases.

Idealized assumptions on op-amps are as follows (Figure 5.24):

1. Input impedance of the op-amp is infinite. In reality it is a very large number compared to source impedance.

2. Output impedance of an op-amp is zero. In reality it is a small number compared to load impedance.

3. Open loop gain of the op-amp (K_{OL}) is infinite. In reality, it is a very large number, such as 10^5 to 10^6.

4. Bandwidth of dynamic response is assumed to be infinite, but in reality it is a finite large number.

The implications of these assumptions are as follows, which are very useful in deriving the input–output relationship of any op-amp configuration.

1. Because input impedance is infinite, current flow into the op-amp from either input terminal is zero, $i^- = i^+ = 0$.

2. In addition, the differential voltage between the two input terminals is zero, $v^+ = v^-$ or $E_d = v^+ - v^- = 0$.

Note that these are approximate conclusions based on idealized assumptions. In reality, they are not exactly zero, but close. The output voltage $V_o = K_{OL} \cdot E_d$, finite since K_{OL} is a very large number and E_d is a very small number. When the op-amp output is saturated, the output voltage is equal to the supply voltage,

$$V_o = V_{sat} = V^+ \quad or \quad V_o = -V_{sat} = V^- \tag{5.191}$$

The implication of the fact that a practical op-amp has large input impedance and very small output impedance makes it ideal as a buffer between a signal source (input signal) and a load (output signal). Using an op-amp (i.e., unity gain op-amp) between a signal source and a load isolates the input signal from the effects of the load. This is a fundamental benefit and is widely used in processing sensor signals.

Noise in Electrical Circuits   Any unwanted component of a signal is broadly categorized as *noise*. Noise can be induced on a signal by electric and magnetic field interferences, as well as ground loops. *Electrical field interference* occurs due to very large voltage sources in the vicinity of the signal carrying conductors. Large electric potentials can add unwanted voltages ("noise") on the signal. Noise due to *magnetic field interference* occurs due to large magnetic fields in the vicinity of the signal. Magnetic fields (also called *electromagnetic fields*) are created by large currents, strong permanent magnets, radio and TV stations. Variations in a magnetic field induce a voltage on the conductors around it. As the magnetic field changes so does the induced voltage, which is considered a noise. In general, the closer the signal carrying conductor to these large electrical voltage source or electromagnetic fields, the larger the induced voltage noise is. The induced noise is minimized by minimizing the coupling between the two: our signal and electric field or magnetic field. This is best done by minimizing the *permitivity* ("electrical conductivity") of the medium for electric field coupling, and *permeability* ("electromagnetic conductivity") of the medium for electromagnetic field coupling.

In order to minimize the interference related noise (either as a result of electric fields or magnetic fields, which is called electromagnetic interference (EMI)), designers practice the following general guides:

1. Keep the low power signals as far away from high voltage and high current signals as possible. In electric power panels, the high power signals (amplifier and actuation signals) and low power signals (sensor and controller signals) should always be grouped separately and physically routed as far away from each other as possible in order to minimize the noise induced on the low power signals due to high power signals.

2. *Twisting* the two wires of a cable has the effect of cancelling out the interference noise that may be induced on each of the wires. This works assuming that both wires have the same induced noise voltage, which is a fairly good assumption in practice. *Coaxial* cables provide even better protection against noise.

3. *Shielding* of the twisted cable pair and grounding it provides further protection against noise. The shield is typically a metal foil that wraps the twisted cable and "shields" them from external noise sources. The shield should be connected to ground only at one end.

Grounding related noise is another commonly encountered noise category. The term *ground* refers to providing a *zero voltage reference* for electrical circuits. There are three

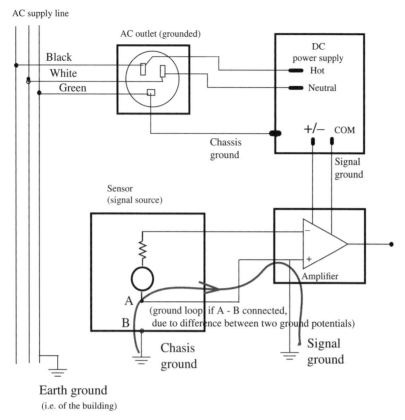

AC supply line

FIGURE 5.25: Grounding in an electrical circuit: earth ground, chassis ground, signal ground.

different kinds of grounds (Figure 5.25):

1. earth ground,
2. chassis ground,
3. signal (COMMON) ground.

Consider a single phase AC line in a residential building. The power plug to the supply line has three leads: the hot wire (usually black color) which carries the AC power, the neutral wire (usually white color) which is the reference, and the ground wire (usually green color) which is physically connected to the earth via a large copper conductor. For instance, the earth ground would be common to all outlets in a residential home, that is to say that all of the ground wires from different outlets in the home would terminate at the same point for the earth ground connection. Since earth is assumed to have infinite capacitance, as various circuits dump current to it, its voltage potential does not change.

In general, chassis ground (grounding the metallic enclosure) is connected to the earth ground as a safety measure. That way, if there is any voltage potential that may develop on the chassis due to unintended electrical leakage, it would be "dumped" quickly to the earth, hence not allowing a dangerous voltage level buildup on the chassis. Connecting the chassis ground to the earth ground is an important safety measure in electrical devices (Figure 5.25, see how the DC power supply connector connects the chasis ground to the earth ground through the AC outlet). A device's chassis is grounded only if it is connected to the earth ground. For instance, in home appliance applications, sometimes there is only a two-prong electrical outlet in a room, and we have an electrical device with a three-prong power cord. If a three-prong connector is connected to another adaptor that has three prongs on one side and two prongs on the other side (hot and neutral), the third prong is not connected to the electrical circuit, and the net result is that the chassis of the device is not grounded.

If there is a short from power lines to the enclosure or some charge has accumulated on the enclosure, having it (the enclosure) grounded provides a conductive path to the ground so that there will not be a high voltage built up on the enclosure. If a person touches such an enclosure, there will not be an electric shock. If the enclosure is not grounded, and there is a leak to the enclosure, it will result in a high potential build-up. When a person touches it, the person's body will provide a conductive path which is dangerous. For example, the frame of an electric motor should be grounded. In case the windings make contact with the frame due to some damage, the ground connection provides a path for the current to ground. If a person touches the motor frame under that condition, there would not be any danger. If the motor frame was not grounded, then there would be a serious danger of electric shock. The same applies for home appliances such as electric heaters, electric ovens, washers, and so on. The so-called *ground fault interrupter (GFI)* type receptacle works on the principle that it monitors any voltage leakage and voltage build-up on the connected equipment enclosure. If there is voltage build-up, the GFI receptacle disconnects the power from the line to the device, effectively behaving like a circuit breaker. Hence, it provides a safety measure against such voltage leakage and potentially harmful conditions.

Signal ground (also referred as COMMON ground or just COM in circuits) is used as the zero voltage reference for sensor signals.

Voltage is a relative quantity measured always between two points. *Ground* in any electrical circuit provides the local zero voltage level against which other voltages are referenced. Ideally, there would be one common ground point in all circuits and all voltages would be referenced to it. However, practical circuits often have more than one ground point. When the actual voltage level of these multiple ground points are different from each other, two different kinds of errors occur:

1. The voltages measured against different grounds will not be compared correctly, since their values are referenced against different "zero" levels.

2. If the different ground points are electrically connected to each other via a conductive path and they have different voltage levels, there will be a current flow between the grounds. This is called the *ground loop*, and results in error (noise) on the signal. Therefore, in circuits with multiple grounds, there should not be a conductive path between the grounds. The key to avoiding ground loops is to make sure that there is no conductive path between different grounds in a circuit. For instance, as shown in Figures 5.25 and 5.26, the shield should be grounded only at one end, not both. Similarly, signal common wire in single-ended signals and ground wire in differential-ended signals should be grounded at one end only.

Single-Ended Voltage Signal and Differential-Ended Voltage Signal

When a signal is presented and measured as a voltage difference between a conductor and a ground, it is called a *single-ended signal*. When a signal is presented and measured as the difference in voltage between two conductors both of which are referenced to the same ground, it is called a *differential-ended signal*. In order to conduct a single-ended signal over a conductor, one signal wire and a ground wire are needed. One ground wire may be shared by multiple single-ended signals. In order to conduct a differential-ended signal over a conductor, two signal wires and a ground wire are needed. Again, the ground wire may be shared by multiple differential-ended signals (Figure 5.26a and b). For example,

Single-ended signal: $V_{in} = V_2 - V_1 = V_2 - V_{GND} = V_2$

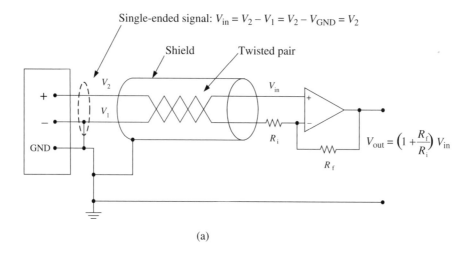

(a)

Differential-ended signal: $V_{in} = (V_2 - V_{GND}) - (V_1 - V_{GND}) = V_2 - V_1$

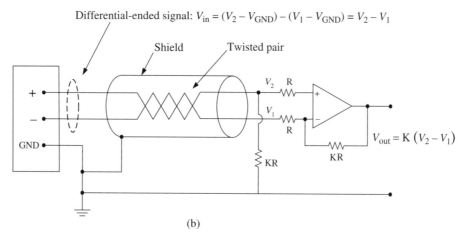

(b)

FIGURE 5.26: (a) Single-ended signal, (b) differential-ended signal and operational amplifier.

a 16-channel ADC (plus a ground wire for zero reference voltage, hence a total of 17 wires) may be configured to handle 16 single-ended (SE) signals or 8 differential-ended (DE) signals (Figure 5.27a and b). Both single-ended and differential-ended signals should always be grounded only at one end (source or destination), not at both ends, in order to avoid ground loop related noise.

In Figure 5.26a, a *single-ended signal* is connected to an operational amplifier. If there is external noise, it will be additive to the signal line (V_2) that is not grounded, while the grounded line (V_1) will dump the noise quickly. As a result, external noise will be amplified with the same gain as the desired signal component. Figure 5.26b shows a differential-ended signal connected to a differential op-amp. Differential amplifier inputs are not grounded. It amplifies the difference between the two inputs. None of the inputs are grounded, rather each carries a voltage level relative to the ground (GND). The actual signal information is the difference between the two lines. When unwanted noise is induced, it is fairly accurate to assume that it will be additive to both of the lines by the same amount. Hence, the difference taken at the amplifier will cancel out this so called *common-mode signal*. As long as the noise signal is induced equally on both conductors, to differential amplifier can effectively cancel the noise.

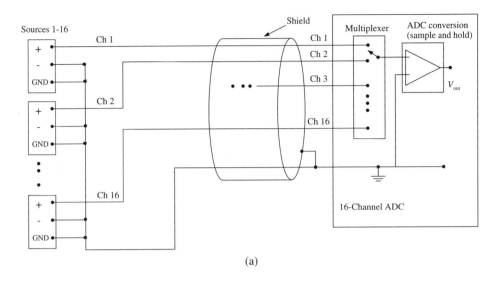

(a)

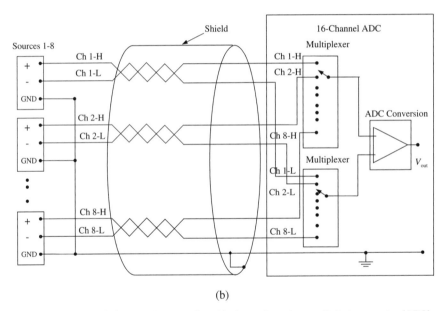

(b)

FIGURE 5.27: (a) and (b) show the use of an 16-channel analog to digital converter (ADC) on a data acquisition system for sampling a set of 16-channel single-ended signals and 8-channel differential-ended signals.

The performance characteristics of a differential amplifier are defined by the differential gain, common mode gain, and common-mode rejection ratio (CMMR),

$$K_{\text{diff}} = \frac{V_{\text{out1}} - V_{\text{out2}}}{V_{\text{in1}} - V_{\text{in2}}} \tag{5.192}$$

$$K_{\text{cm}} = \frac{V_{\text{out1}} - V_{\text{out2}}}{((V_{\text{in1}} + V_{\text{in2}})/2)} \tag{5.193}$$

$$CMRR = \frac{K_{\text{diff}}}{K_{\text{cm}}} \tag{5.194}$$

Notice that we only need to know two of these parameters. The most commonly referenced ones are K_{diff} and *CMRR*. Ideally, the differential gain is the desired gain of the amplifier (i.e., 1.0 or 10.0) and the common mode gain is zero. That is, the noise induced on both lines that are common is rejected totally by the differential amplifier. Hence, an ideal differential amplifier would have infinite CMRR. In reality, the common-mode gain is small but finite, hence CMRR is large, but finite. Typical values of CMRR are in the order of 80 dB to 120 dB[1], which means that the amplifier amplifies the differential signal 10^4 to 10^6 times more than the common-mode signal (noise).

The output voltage and input voltage relationship of a differential amplifier can be expressed as

$$V_{out} = K_{diff} \cdot (V_{in1} - V_{in2}) + K_{cm} \cdot (V_{in1} + V_{in2})/2 \qquad (5.195)$$

$$= K_{diff} \cdot (V_{in1} - V_{in2}) + \left(\frac{K_{diff}}{CMMR}\right) \cdot (V_{in1} + V_{in2})/2 \qquad (5.196)$$

In an ideal differential amplifier, CMMR is infinite, hence $K_{diff}/CMMR = K_{cm} = 0.0$. In reality it is of course a very small, but finite gain.

The parameters of a real op-amp that are of interest for designers are as follows:

1. open loop gain (i.e., 10^4 to 10^7 range),
2. bandwidth (i.e., 1 MHz range),
3. input impedance (i.e., $Z_{in} = R_{in} = 10^6 \, \Omega$ range),
4. output impedance (i.e., $Z_{out} = R_{out} = 10^2 \, \Omega$ range),
5. common mode rejection ratio (CMRR) for differential amplifiers, (i.e., *CMRR* = 60 dB *to* 120 dB range)
6. power supply required (i.e. 1.5 VDC to 30 VDC range, unipolar or bipolar),
7. maximum power dissipated internally by the op-amp,
8. maximum input voltage and differential input voltage levels,
9. operating temperature range (commercial rating: 0 to 70 °C, industrial rating: −25 to 85 °C, military rating: −55 to 125 °C).

Example Consider a sensory signal in a medical application, that is an EKG signal, where the nominal value of the signal is about 10 mV, yet the expected noise in the signal can vary from 0 V to 100 mV. Let us consider two different ways of transmitting and amplifying this signal: (1) single-ended signal, and (2) differential-ended signal. In both cases let us assume that the gain of the op-amp is $K_{diff} = 40$ dB $= 100$ V/V, and the common-mode rejection ratio of the differential op-amp is $CMMR = 100$ dB $= 10^5$ V/V. We assume that the noise is induced equally in both conductors when differential-ended signal is used.

If single-ended signal is used, output signal voltage level would vary between

$$V_o = K_{diff} \cdot V_i \qquad (5.197)$$

$$V_o = 100 \cdot 10\,\text{mV} = 1\,\text{V}; \; when \; V_i = 10\,\text{mV} \qquad (5.198)$$

$$V_o = 100 \cdot 110\,\text{mV} = 11\,\text{V}; \; when \; V_i = 10\,\text{mV} + 100\,\text{mV} \; (noise) \qquad (5.199)$$

Clearly, the single-ended signal and op-amp cannot differentiate the difference between the actual signal and noise. If a differential-ended signal is used, the output signal voltage level

---

[1] $K_{dB} = 20 \log_{10} K$ or $K = 10^{\frac{K_{dB}}{20}}$, i.e., $K = 0.01, 0.1, 1, 10, 100$ is same as $K_{dB} = -40, -20, 0, 20, 40$, respectively.

would vary between

$$V_o = K_{diff} \left(V_{diff} + \frac{1}{CMMR} \cdot V_{common} \right) \tag{5.200}$$

$$V_o = 100 \cdot (10\,mV + (1/10^5) \cdot 0.0) = 1\,V; \quad when \; V_i = 10\,mV \tag{5.201}$$

$$V_o = 100 \cdot (10\,mV + (1/10^5) \cdot 100\,mV) = 1.0001\,V \tag{5.202}$$

$$; \quad when \; V_i = (10 + 100)\,mV \tag{5.203}$$

Clearly, the differential-ended signal and op-amp are able to amplify the desired portion of the signal and attenuate very effectively the noise portion of the signal. The key assumption for this to work is that the same noise signal is induced on both conductors (both inputs of the differential amplifier).

5.6.2 Common Op-Amp Circuits

Op-amps are used in both open loop and closed loop configurations. Various op-amp circuits for specific functions are discussed below: comparator, inverting and non-inverting amplifier, sum, derivative, integral, and various filters.

Comparator Op-Amp   The comparator functionality is to compare two signals, and turn ON (i.e., $V_o = V_{sat}$) or OFF ($V_o = -V_{sat}$) the output of the op-amp based on relative values of the two input signals (Figure 5.28). A reference signal V_{ref} is connected to the inverting ($-$) input terminal. The other input signal (V_i) is connected to the noninverting ($+$) input terminal. The output of the op-amp will be either $V_o = V_{sat}$ or $V_o = -V_{sat}$. The

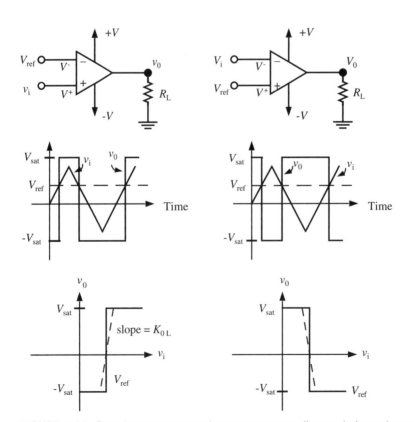

FIGURE 5.28: Open loop op-amp used as a comparator: direct polarity and reverse polarity.

op-amp circuit determines whether input signal V_i is above or below the reference signal, V_{ref}. By changing the connection of V_{ref} and V_i into $(-)$ and $(+)$ terminals, the output polarity of the op-amp is changed. The open loop op-amp circuit input–output relationship as a comparator can be summarized as

$$V_o = K_{OL} \cdot (V_i - V_{ref}); \tag{5.204}$$

$$if \ - V_{sat} < K_{OL} \cdot (V_i - V_{ref}) < V_{sat} \tag{5.205}$$

$$= V_{sat}; \ if \ K_{OL} \cdot (V_i - V_{ref}) > V_{sat} \tag{5.206}$$

$$= -V_{sat}; \ if \ K_{OL} \cdot (V_i - V_{ref}) < -V_{sat} \tag{5.207}$$

Let us consider an op-amp with the following parameters,

$$K_{OL} = 10^6, \ V_{sat} = 12 \, \text{VDC} \tag{5.208}$$

Then, when the difference between the two input terminal voltages is within $\pm 12 \, \mu V$, the output is proportional to the difference with a voltage output between $\pm 12 \, V$. Otherwise, the output is either saturated at $12 \, V$ or at $-12 \, V$.

The same circuit is also used as a PWM modulator which converts the reference V_i analog voltage level signal to a pulse width modulated (PWM) signal when V_{ref} is a fixed frequency periodic signal (i.e., triangular or sinusoidal signal). Similarly, comparator op-amp configurations are also used as signal generator circuits. The LM311 (and LM111/LM211) family of integrated circuit op-amps are commonly used as high-speed comparators.

Example Consider the op-amp comparator circuits shown in Figure 5.29. Assume that the resistances at the input terminals are

$$R_1 = 4 \, k\Omega \quad R_2 = 6 \, k\Omega \tag{5.209}$$

and the supply voltage $V_{C1} = 10 \, \text{VDC}$. Let the saturation output voltage be $V_{sat} = 13 \, \text{VDC}$. The input voltage $v_{in} = 9 \sin(2\pi t)$. Notice that the difference between the two circuits is the connection of the reference voltage and input voltage to the op-amp terminals. Draw the output voltage as function of time for both circuits. Notice that the reference voltage is

$$V_{ref} = \frac{R_2}{R_1 + R_2} V_{C1} = \frac{6}{4 + 6} \cdot 10 = 6 \, V \tag{5.210}$$

Notice that this is a comparator op-amp circuit. Let us neglect the raise time transient of the output voltage response at this timescale range. Hence, we can show the change of output state like a step function. When the $(v^+ - v^-) > V_{sat}/K_{OL} \approx 0.0$, the output voltage is $V_o = V_{sat}$ and otherwise, when the $(v^+ - v^-) < -V_{sat}/K_{OL} \approx 0.0$, the output voltage is $V_o = -V_{sat}$. For all practical purposes, for the first connection, when $V_{in} > 6 \, V$, the $V_{out} = V_{sat} = 13 \, V$, and when $V_{in} < 6 \, V$, the $V_{out} = -V_{sat} = -13 \, V$. For the second configuration, the output polarity would be opposite to the first configuration. The input–output voltage is shown in Figure 5.29.

Example The circuit shown in Figure 5.30 determines whether input voltage V_{in} is inside the window defined by two reference voltages, $V_{ref,l}, V_{ref,h}$. The output voltage will be V_{sat} if the input voltage is outside the window. It will be $-V_{sat}$ if the input is inside the voltage window defined by $V_{ref,l}, V_{ref,h}$. The LED will be ON when the input voltage is outside the window.

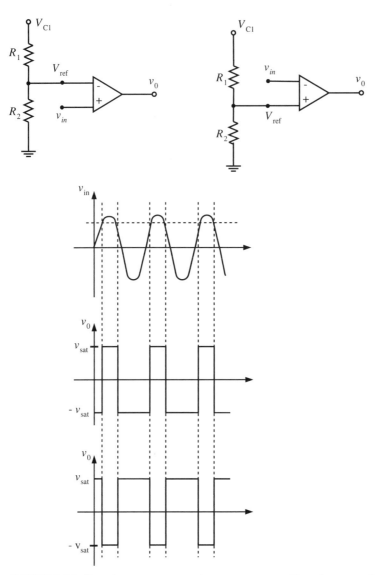

FIGURE 5.29: Open loop op-amp used as a comparator: V_{ref} reference voltage and V_{in} input voltage (i.e., voltage from a sensor) are switched at the op-amp terminals between the two cases. The net result is the output polarity change.

Op-Amps with Positive Feedback   Modifications to the op-amp feedback loop on the non-inverting terminal can provide *hysteresis* in the comparator to implement ON/OFF relay control with hysteresis, such as is used in home temperature control. A well-known circuit with this function is the *Schmitt trigger* (inverting and noninverting types, Figure 5.31). The op-amp uses positive feedback which results in a hysteresis relationship between input and output. *Positive feedback is the key to the hysteresis function realization.* Such circuits are not only used in ON/OFF control systems with desired hysteresis, but also reject undesirable noise in the signal provided the noise magnitude is smaller than the hysteresis magnitude (Figure 5.31). A more general form of the Schmitt trigger op-amp circuit allows adjustment of both the hysteresis magnitude and center input voltage value about which the hysteresis is designed. Schmitt trigger circuits are also used

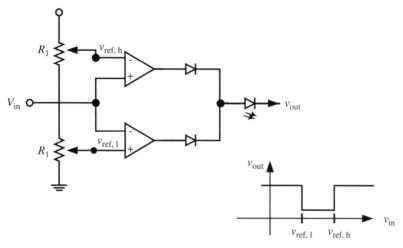

FIGURE 5.30: Two open loop op-amps used as a window comparator: $V_{ref,l} \leq V_{in} \leq V_{ref,h}$.

in digital circuits to eliminate the noise, for example the LM7414 integrated circuit package incorporates six Schmitt triggers in one IC package for digital circuit applications. Eliminating the switch bouncing from the electrical signal is called *de-bouncing*, which is a typical application of the Schmitt trigger in digital circuits.

Consider an inverting Schmitt trigger op-amp (Figure 5.31b). The voltage at the (+) terminal is same as the voltage across R_2,

$$V_{R_2} = \frac{R_2}{R_1 + R_2} \cdot V_0 = \frac{R_2}{R_1 + R_2} \cdot V_{sat} \tag{5.211}$$

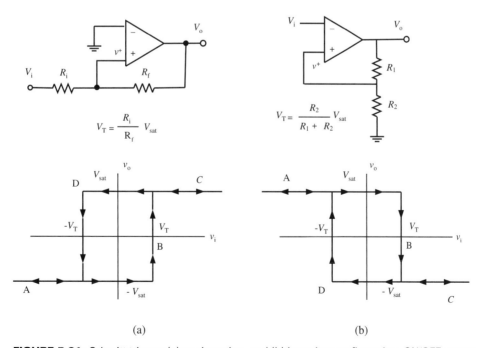

(a) (b)

FIGURE 5.31: Schmitt trigger: (a) noninverting, and (b) inverting configuration. ON/OFF output with hysteresis function.

where $V_o = V_{sat}$ or $V_o = -V_{sat}$, the output of the amplifier is essentially always saturated depending on the signals on its input terminals (+) and (−).

$$v^+ = \frac{R_2}{R_1 + R_2} \cdot V_{sat} \ during \ V_o = V_{sat} \tag{5.212}$$

$$v^+ = -\frac{R_2}{R_1 + R_2} \cdot V_{sat} \ during \ V_o = -V_{sat} \tag{5.213}$$

Let us trace the operation on the op-amp along the input–output relationship curve that has the hysteresis loop. Assume that initially we start on the curve from the left hand side,

$$V_o = V_{sat}, \ v^+ = \frac{R_2}{R_1 + R_2} \cdot V_{sat}, \ v_i < 0 \tag{5.214}$$

For the op-amp output to change state, $V_o = -V_{sat}$ (move to the C part of the curve), the inverting input must slightly exceed $(v_i = v^-) > (v^+)$. In other words, we move along B on the input–output curve, where v^- must be larger than $v^+ = \frac{R_2}{R_1+R_2} V_{sat} > 0$. Here, we neglect the transient response details of the change in the output state of op-amp. At that point, $v^+ = \frac{R_2}{R_1+R_2} V_{sat}$ and $V_o = -V_{sat}$. In order to return to the previous state, now the inverting input must be slightly more negative than the noninverting input which is at the value of $v^+ = -\frac{R_2}{R_1+R_2} V_{sat}$. This can happen on the curve to the left side of region D. Then, the state of the op-amp output is switched back to ON (D line along the hysteresis loop). The noninverting Schmitt trigger circuit works with the same principle except the output polarity is different.

Let us consider the noninverting Schmitt trigger circuit shown in Figure 5.28a. The voltage on inverting terminal v^- is connected to ground zero. The voltage on the noninverting terminal, v^+, is a value between V_i and V_o as follows (assuming that there is no current flow through the op-amp). The current through the circuit between V_i and V_o, through resistors R_i and R_f is as follows,

$$i = \frac{V_i - V_o}{R_i + R_f} \tag{5.215}$$

where a positive direction of the current is assumed as the direction from V_i to V_o. Then, the voltage at v^+, which is the point between the resistors is

$$v^+ = V_i - R_i \cdot i \tag{5.216}$$

$$= V_i - R_i \cdot \frac{V_i - V_o}{R_i + R_f} \tag{5.217}$$

$$= \frac{1}{R_i + R_f}(R_f V_i + R_i V_o) \tag{5.218}$$

Now, let us consider the case starting with $V_i \ll 0.0$, then $V_o = -V_{sat}$. Then, $v^+ < 0.0 = V^-$, and then $V_o = -V_{sat}$ will continue to be in negative saturation output values. The only way the output will switch state, that is $V_o = V_{sat}$, is when $v^+ > v^- = 0.0$, which can happen (see above equation for v^+) if

$$R_f V_i + R_i V_o > 0.0 \tag{5.219}$$

$$R_f V_i - R_i V_{sat} > 0.0 \tag{5.220}$$

$$V_i > \frac{R_i}{R_f} V_{sat} \tag{5.221}$$

At this point and above values of V_i, the output V_o will switch to a $V_o = V_{sat}$ value. In order for the op-amp output to return to the previous state, $V_o = -V_{sat}$, the following condition must be met: $v^+ < v^- = 0.0$. Then, again referring to the equation describing v^+, this can happen if

$$R_f V_i + R_i V_o < 0.0 \qquad (5.222)$$

$$R_f V_i + R_i V_{sat} < 0.0 \qquad (5.223)$$

$$V_i < -\frac{R_i}{R_f} V_{sat} \qquad (5.224)$$

Hence, the op-amp output will switch from $V_o = V_{sat}$ state to $V_o = -V_{sat}$ state when $V_i \le -\frac{R_i}{R_f} V_{sat}$. Similarly, the op-amp output will switch from $V_o = -V_{sat}$ state to $V_o = V_{sat}$ state when $V_i \ge \frac{R_i}{R_f} V_{sat}$.

Notice that the nonlinear hysteresis function is accomplished by the feedback of the output signal to the noninverting (positive) input of the op-amp, not the negative input. Feedback to the negative input is used for implementing linear functions with an op-amp.

Schmitt trigger functionality is also used in digital circuits. For instance, the desired output of a device is ON (for 5 VDC) or OFF (for 0 VDC). The input voltage might not be exactly 0 VDC or 5 VDC but somewhere in between (Figure 5.32). Instead of defining a single transition voltage, a hysteresis region can be defined at 3.0 VDC with a 2.0 (+1 and −1) VDC hysteresis. This would result in an ON output state when voltage transitions from 0 VDC to 5 VDC at 4 VDC and above, and back to the OFF state when the input signal transitions from 5 VDC region to 0 VDC region at 2 VDC and below. Schmitt triggers are used in *de-bouncing* of input signals, and avoiding unintended ON/OFF state changes due to noise. For instance, when a mechanical or electromechanical input switch is closed or opened, the transition from ON to OFF state, or OFF to ON state, does not happen in one single clean switching but by bouncing of the contact. This results in a noisy input signal as shown in Figure 5.32c. Schmitt triggers can be used to "clean-out" or "de-bounce" the switch signal. Digital IC packages typically have two, four, six or more Schmitt triggers in one DIP package (i.e., 7413, 7414, 7418, 7419, 74310, 4093, 40106, 4024, 4040, 4022).

Example   Consider the op-amp circuit shown in Figure 5.31b. Let us assume that the saturation voltage is $V_{sat} = \pm 13$ V and that the input signal $V_i(t)$ is a sinusoidal signal with magnitude 10 V and frequency 10 Hz,

$$V_i(t) = 10\sin(2\pi t) \qquad (5.225)$$

Draw the output voltage of the op-amp for the following values of the feedback resistors,

1. $R_1 = 100\,\text{k}\Omega$, $R_2 = 100\,\Omega$
2. $R_1 = 100\,\text{k}\Omega$, $R_2 = 100\,\text{k}\Omega$

For case 1, the width of the hysteresis voltage, V_T is

$$V_T = \frac{R_2}{R_1 + R_2} \cdot V_{sat} = \frac{100}{100\,000 + 100} \cdot 13 \qquad (5.226)$$

$$= 13/1001 \text{ V} \approx 13\,\text{mV} \qquad (5.227)$$

The output voltage will switch between $V_{sat} = +13$ V and $-V_{sat} = -13$ V when the input signal passes the band of -13 mV and 13 mV around zero voltage. Due to the hysteresis, the output state switching is dependent upon the previous direction of the input voltage crossing the hysteresis band. Notice that due to the scale, the hysteresis behavior is not visually detectable in the figure.

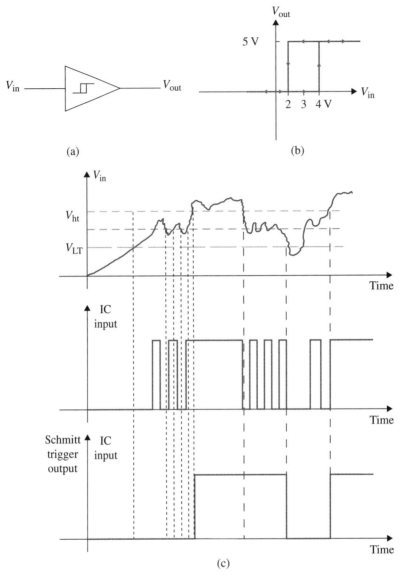

FIGURE 5.32: Schmitt trigger circuit in digital circuits: (a) symbol, (b) input–output relationship, (c) a noisy input signal, ON/OFF signal at the input port of an IC without Schmitt trigger, and ON/OFF signal after it passes through the Schmitt trigger, which is also referred to as *switch-debouncing*.

For case 2, the width of the hysteresis voltage, V_T is

$$V_T = \frac{R_2}{R_1 + R_2} \cdot V_{sat} = \frac{100\,000}{100\,000 + 100\,000} \cdot 13 \tag{5.228}$$

$$= 6.5\,V \tag{5.229}$$

In this case, the output voltage switches between the $V_{sat} = +13\,V$ and $-V_{sat} = -13\,V$ when the input signal goes through the hysteresis band of $+6.5\,V$ and $-6.5\,V$. The hysteresis magnitude is large and visually observed on the figure. The input signal and output signal for both of the cases above are shown in Figure 5.33.

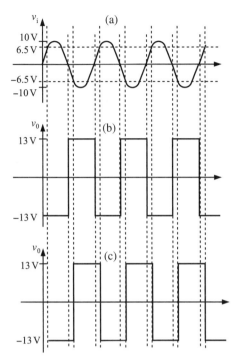

FIGURE 5.33: Schmitt trigger (inverting configuration) op-amp circuit used to convert a periodic input signal to a square wave output signal.

Inverting Op-Amp The functionality is to amplify the input voltage to output voltage with a negative gain. Neglecting the transient delay of response between input and output voltages,

$$V_o(t) = K_{CL} \cdot V_i(t) \tag{5.230}$$

Inverting the op-amp (Figure 5.34a) connects the (+) input terminal to ground, and the input signal is connected to the (−) input terminal. There are two resistors around the op-amp: R_i and R_f. Let us show the relationship between V_i and V_o using the ideal op-amp assumptions. Recall that ideal op-amp assumptions state $i^+ = i^- = 0$, $E_d = v^+ - v^- = 0$, $i_f = i_{in}$. Notice that since the noninverting terminal of the op-amp is grounded,

$$v^+ = v^- = 0 \tag{5.231}$$

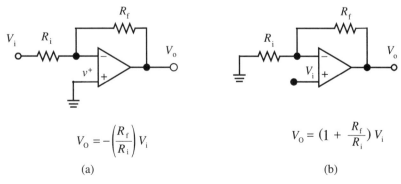

$$V_O = -\left(\frac{R_f}{R_i}\right) V_i$$

(a)

$$V_O = \left(1 + \frac{R_f}{R_i}\right) V_i$$

(b)

FIGURE 5.34: Basic feedback (closed loop) configuration of op-amps: (a) inverting, and (b) noninverting configuration as a gain amplifier.

Then,

$$i_{in} = V_i/R_i \tag{5.232}$$

$$i_f = i_{in} \tag{5.233}$$

$$V_f = R_f \cdot i_f = R_f \cdot V_i/R_i \tag{5.234}$$

Hence, since the output voltage will have opposite polarity to V_f,

$$V_o = -V_f = -\frac{R_f}{R_i} \cdot V_i \tag{5.235}$$

$$V_o = K_{CL} \cdot V_i \tag{5.236}$$

where the gain of the inverting op-amp

$$K_{CL} = -\frac{R_f}{R_i} \tag{5.237}$$

Noninverting Op-Amp   A noninverting amplifier simply amplifies an input voltage to output voltage with a positive gain. This is accomplished by the feedback connections shown in Figure 5.34b. Following the same ideal op-amp assumptions ($v+ = v^-, i^+ = i^- = 0$), the input–output relationship (neglecting transient response differences) can be derived as follows,

$$v^+ = v^- = V_i \tag{5.238}$$

$$i_{in} = V_i/R_i \tag{5.239}$$

$$i_f = i_{in} \tag{5.240}$$

$$V_0 = (R_i + R_f) \cdot i_f \tag{5.241}$$

Since this is a noninverting amplifier,

$$V_o = \frac{R_i + R_f}{R_i} \cdot V_i \tag{5.242}$$

$$= K_{CL} \cdot V_i \tag{5.243}$$

where the gain of the noninverting op-amp is

$$K_{CL} = 1 + \frac{R_f}{R_i} \tag{5.244}$$

which is always larger than one.

Example   A special case of the noninverting op-amp is obtained when there are no resistors in the configuration. Effectively, this is same as $R_f = 0$, and $R_i = \infty$. Hence, the gain of the amplifier is unity. Such an op-amp configuration is called the *voltage follower op-amp or buffer op-amp* and used to isolate the source and load (Figure 5.27b). The voltage gain is unity, but the current gain is larger than unity in order to isolate the source from the load.

Example   Let us consider a noninverting op-amp with a saturation output voltage of $V_{sat} = 13\,V$, $R_i = 10\,k\Omega$, and $R_f = 10\,k\Omega$. The input voltage and output voltage relationship can be easily determined by

$$V_o = \frac{R_i + R_f}{R_i} \cdot V_i \tag{5.245}$$

$$= 2.0 \cdot V_i \tag{5.246}$$

As a result, the input voltage range is limited to the ± 6.5 V range. Beyond that, the output voltage saturates at 13 V.

Notice that a typical op-amp can handle input voltages up to the supply voltage values or a few volts less. However, the feedback resistor values can be such that the output of the linear amplifier saturates much earlier than the maximum input voltage the op-amp can accept. For further discussion, consider the same example with $R_i = 1$ kΩ and $R_f = 99$ kΩ. Then, the nominal gain of the noninverting op-amp is 100.0. The output voltage would saturate when the input voltage is outside the range of ± 0.13 V.

Differential Input Op-Amp The desired function is to determine the difference between two signals and possibly multiply the difference with a gain,

$$V_o = K \cdot (V_1 - V_2) \tag{5.247}$$

which is used in closed loop control circuits as the summing junction that is find the difference between a command signal and sensor signal. (Figure 5.35a) shows a differential input op-amp circuit. In its general form, the input–output relationship can be obtained using the superposition principle. The output is the sum of the outputs due to the inverting input and the noninverting input. The superposition principle can be used in the derivation: (i) connect V_2 to ground and solve for $v'_o = K_1 \cdot V_1$, and (ii) connect V_1 to ground and solve for $v''_o = K_2 \cdot V_2$. Then, add them together to get $V_o = v'_o + v''_o$. The output due to input at its noninverting terminal is (Figure 5.34b)

$$v^+ = \frac{R_2}{R_1 + R_2} V_1 \tag{5.248}$$

$$v'_o = \frac{R_3 + R_4}{R_3} v^+ \tag{5.249}$$

$$= \frac{R_3 + R_4}{R_3} \frac{R_2}{R_1 + R_2} \cdot V_1 \tag{5.250}$$

And the output due to input at its inverting terminal is (Figure 5.34a)

$$v''_o = -\frac{R_4}{R_3} \cdot V_2 \tag{5.251}$$

The total output is

$$V_o = v'_o + v''_o \tag{5.252}$$

$$V_o = \left(\frac{R_2}{R_1 + R_2}\right)\left(\frac{R_3 + R_4}{R_3}\right) \cdot V_1 - \left(\frac{R_4}{R_3}\right) \cdot V_2 \tag{5.253}$$

Note that when $R_1 = R_2 = R_3 = R_4$, the input–output relationship is

$$V_o = V_1 - V_2 \tag{5.254}$$

Similarly, when $R_1 = R_3 = R$ and $R_2 = R_4 = K \cdot R$,

$$V_o = K \cdot (V_1 - V_2) \tag{5.255}$$

One of the main usages of differential op-amps is in amplifying noise sensitive signals. As discussed in Figure 5.25, single-ended signals are referenced with respect to ground. Any noise induced on the signal wire coming into the op-amp would be amplified. This is particularly problematic when the noise signal is comparable to the actual signal magnitude. In such cases, it is best to transmit the signal voltage in differential-ended format. That is using two wires and the signal information is the voltage difference between the two wires. If any noise is induced during the transmission, it would be induced on both lines and the

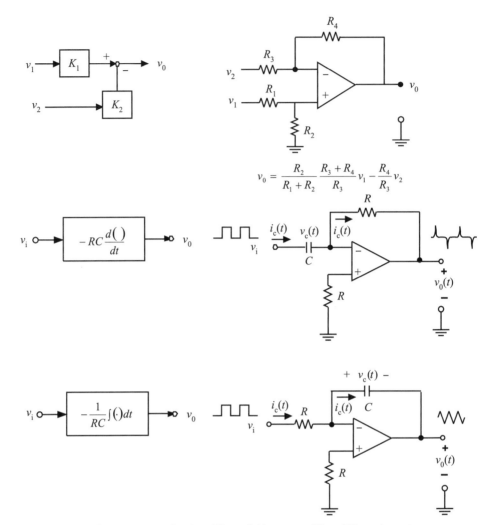

FIGURE 5.35: Some op-amp circuits: differential input amplifier, differentiator, integrator.

difference between them would still be unaffected by noise. Amplification of differential-ended signals is one of the most common applications of differential op-amps. Later we will review an improved version of this op-amp, that is the instrumentation amplifier.

Example Consider the differential op-amp circuit shown in Figure 5.35. Calculate the values of the resistors in order to obtain the following input–output voltage relationship, $v_0 = v_1 - 2v_2$.

The output voltage and input voltage relationship for the differential op-amp is

$$v_0 = \frac{R_2}{R_1 + R_2} \frac{R_3 + R_4}{R_3} \cdot v_1 - \frac{R_4}{R_3} v_2 \qquad (5.256)$$

Since we want v_2 to have a gain of 2, then

$$\frac{R_4}{R_3} = 2 \qquad (5.257)$$

$$R_4 = 2 \cdot R_3 \qquad (5.258)$$

Since we want v_1 to have a gain of 1, then

$$\frac{R_2}{R_1 + R_2} \frac{R_3 + R_4}{R_3} = 1 \tag{5.259}$$

$$R_1 = 2 \cdot R_2 \tag{5.260}$$

Let $R_2 = R_3 = 10\,\text{k}\Omega$, then $R_1 = R_4 = 20\,\text{k}\Omega$.

Derivative Op-Amp   The desired function is to take the derivative of the input voltage signal and provide that as an output voltage signal,

$$V_o(t) = K \frac{d}{dt}(V_i(t)) \tag{5.261}$$

Figure 5.35 shows an op-amp circuit for differentiation. Using the ideal op-amp assumptions, the input–output relationship is derived as follows,

$$i_c = C \cdot \frac{dV_i(t)}{dt} \tag{5.262}$$

$$i_f = i_c \tag{5.263}$$

$$V_f = R \cdot i_f \tag{5.264}$$

$$V_o = -V_f \tag{5.265}$$

Hence,

$$V_o = (-RC) \cdot \frac{dV_i(t)}{dt} \tag{5.266}$$

Integrating Op-Amp   If we change the locations of the resistor and capacitor in the derivative op-amp, we obtain an integrating op-amp circuit (Figure 5.35). The desired function is

$$V_o(t) = K \int (V_i(\tau)d\tau) + V_o(0) \tag{5.267}$$

where $V_o(0)$ is the initial voltage. The derivation of the I/O relationship is straightforward,

$$i_c = V_i(t)/R \tag{5.268}$$

$$i_f = i_c \tag{5.269}$$

$$V_f(t) = \frac{1}{C} \int_0^t i_f(\tau)d\tau \tag{5.270}$$

$$V_o(t) = -V_f(t) \tag{5.271}$$

$$= -\frac{1}{RC} \int_0^t V_i(\tau)d\tau \tag{5.272}$$

where the initial voltage values in the integrations have been neglected.

Next we present the op-amp circuits and the input–output relation for filtering operations used in signal processing and control systems. We provide op-amp circuits for the low pass, high pass, band pass, and band reject (notch) filters. It should be noted that digital implementations of filters in software provide more flexibility than the analog op-amp implementations. However, op-amp implementation is simpler as it does not require a real-time software.

Low Pass Filter Op-Amp   The low pass filter passes the low frequency content of a signal and suppresses the high frequency content (Figure 5.36). The break frequency at

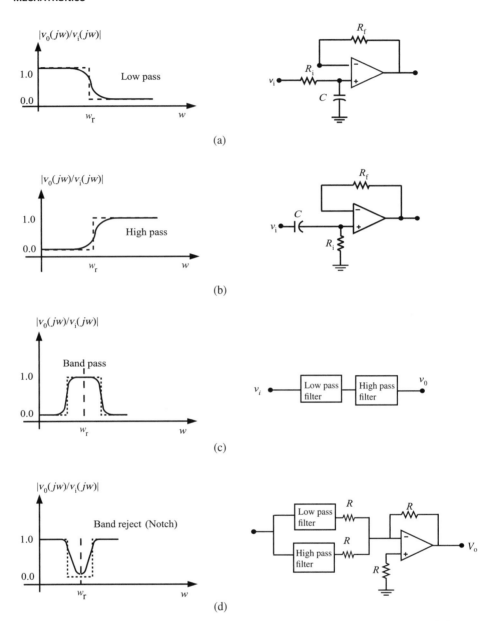

FIGURE 5.36: Some op-amp filter circuits: (a) low pass, (b) high pass, (c) band pass, (d) band reject (notch) filters.

which the transition from low to high frequency occurs is defined by the filter parameters. In addition, the rate of transition and the phase lag are determined by the filter order. The frequency domain input–output voltage relationship of a low pass op-amp filter is

$$\frac{V_0(jw)}{V_{in}(jw)} = \frac{1}{1 + \tau_1 \cdot jw} \tag{5.273}$$

where the time constant of the first-order filter $\tau_1 = RC$.

The input–output voltage relationship for the low pass filter can be derived as follows. The current into the op-amp is zero. Hence, the voltage at the (−) and (+) terminals of the op-amp is the same. Since current into the op-amp is zero, voltage at the (−) terminal is the same as the output voltage. That is,

$$i^- = i^+ = 0.0 \tag{5.274}$$

$$V^- = V^+ \tag{5.275}$$

$$= V_o \tag{5.276}$$

Notice that the output voltage, $V_o = V^+$ is the voltage across the capacitor C. The voltage across the capacitor is the output voltage and is related to the current and capacitance value as (assuming zero initial voltage at $t = 0$, $V_o(0) = 0.0$),

$$V_o(t) = \frac{1}{C} \int_0^t i(\tau)d\tau \tag{5.277}$$

$$V_o(s) = \frac{1}{Cs}i(s) \tag{5.278}$$

The current in the circuit is

$$i(t) = \frac{V_i(t) - V_o(t)}{R} \tag{5.279}$$

$$i(s) = \frac{V_i(s) - V_o(s)}{R} \tag{5.280}$$

$$R \cdot i(s) = V_i(s) - \frac{1}{Cs}i(s) \tag{5.281}$$

$$\left(R + \frac{1}{Cs}\right) \cdot i(s) = V_i(s) \tag{5.282}$$

$$\frac{i(s)}{V_i(s)} = \frac{Cs}{RCs + 1} \tag{5.283}$$

Hence the transfer function between the output voltage and input voltage is

$$\frac{V_o(s)}{V_i(s)} = \frac{1}{RCs + 1} \tag{5.284}$$

$$\frac{V_o(jw)}{V_i(jw)} = \frac{1}{1 + jRCw} \tag{5.285}$$

Notice that when $w = \frac{1}{RC}$ rad/s, the magnitude ratio of the output voltage to input voltage is

$$\left|\frac{V_o(jw)}{V_i(jw)}\right| = \frac{1}{\sqrt{2}} = 0.707 \tag{5.286}$$

where the value $w_c = \frac{1}{RC}$ rad/s or $f_c = \frac{1}{2\pi RC}$ Hz is called the cutoff frequency of the filter, that is the frequency at which the output signal magnitude is 0.707 times the input signal magnitude in steady-state. In other words, the output signal is attenuated by 3 dB in comparison to the input signal.

High Pass Filter Op-Amp   The high pass filter suppresses the low frequency content of the input signal, and passes the high frequency content. An op-amp implementation of a high pass filter is shown in Figure 5.36. The frequency domain input–output voltage relationship of a high pass op-amp filter is

$$\frac{V_o(jw)}{V_{in}(jw)} = \frac{j\tau_1 \cdot w}{1 + \tau_1 \cdot jw} \tag{5.287}$$

where the time constant of the first-order filter $\tau_1 = RC$. Notice that the only difference between the low pass and high pass filter is the placement of the resistor and capacitor in the (+) input terminal.

Similarly, the input–output voltage relationship can be derived for the high pass filter. Notice that the location of the resistor and capacitor on the circuit is swapped compared to the low pass filter. Following a similar derivation process, it is straightforward to derive the input–output voltage relationship.

$$V_o(t) = R \cdot i(t) \tag{5.288}$$

$$V_o(s) = R \cdot i(s) \tag{5.289}$$

The voltage across the capacitor,

$$V_i(t) - V_o(t) = \frac{1}{C} \int_o^t i(\tau) d\tau \tag{5.290}$$

$$V_i(s) - V_o(s) = \frac{1}{Cs} i(s) \tag{5.291}$$

By substituting the relationship for $i(s) = \frac{1}{R} V_o(s)$, it can be shown that

$$\frac{V_o(s)}{V_i(s)} = \frac{RCs}{1 + RCs} \tag{5.292}$$

$$\frac{V_o(jw)}{V_i(jw)} = \frac{jRCw}{1 + jRCw} \tag{5.293}$$

which represents the high pass filter transfer function. Again, notice that at $w = w_c$ where $w_c = \frac{1}{RC}$ rad/s or $f_c = \frac{1}{2\pi RC}$ Hz, the magnitude ratio is 0.707. Except that in the high pass filter, the filter passes the frequency content above this frequency and attenuates the frequency content below that frequency. The low pass filter does the opposite.

Band Pass Filter Op-Amp   A band pass filter passes a selected narrow band of frequencies, and suppresses the rest (Figure 5.36). The design parameters of the filter are to select the frequency band to pass: center frequency and the width around that frequency, w_r and Δw_B. It can be realized by a low pass and high pass filter in series.

Band Reject Filter Op-Amp   A band reject filter passes all frequencies except a selected narrow band of frequencies (Figure 5.36). This is also called a *notch filter*, and basically does the opposite of the band pass filter. The design parameters of a notch filter are to select the center frequency and the width of the frequency around the center frequency that will be suppressed, w_r and Δw_B. It can be realized by a low pass and high pass filter in parallel.

Instrumentation Op-Amp   An instrumentation op-amp is used to amplify small sensor signals in noisy environments (Figure 5.37). It is a modified version of the differential op-amp with improved performance characteristics (i.e., higher input impedance, easy adjustment of gain). The instrumentation amplifier has a higher CMRR (common mode

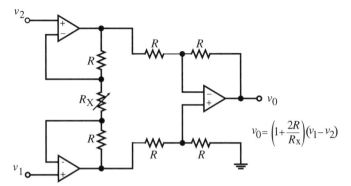

FIGURE 5.37: Instrumentation op-amp: modified version of a differential op-amp with improved characteristics to amplify sensor signals in noisy environments.

rejection ratio) which is an advantage for noisy environments, and the gain of the op-amp is adjustable with a single resistor (R_x) for both input terminals. Notice that the op-amp at the output side of Figure 5.37 is the same as the difference amplifier shown in Figure 5.31a with unity gain. The voltages at the inverting terminals of the top and bottom op-amp are v_1 and v_2 based on the ideal op-amp assumptions. Then the current through R_x can be calculated as,

$$i = \frac{v_2 - v_1}{R_x} \tag{5.294}$$

where we took the positive direction of the current from top to bottom. If v_1 is larger than v_2, the direction of the current would be opposite (negative value). Since current flow into the op-amps must be zero, this current i must also pass through resistors R above and below the resistor R_x. Then the voltage difference between the outputs of top and bottom op-amps is

$$v_2' - v_1' = i \cdot R + i \cdot R_x + i \cdot R = (v_2 - v_1) \cdot \left(1 + \frac{2R}{R_x}\right) \tag{5.295}$$

Finally the last difference op-amp performs the amplification of -1

$$v_o = (v_1' - v_2') = \left(1 + \frac{2R}{R_x}\right) \cdot (v_1 - v_2) \tag{5.296}$$

Current-to-Voltage Converter and Voltage-to-Current Converter In electronics circuits, we often need to convert a current signal to a proportional voltage and convert a voltage signal to a proportional current signal. The first case is used typically in sensor signal transmission where a current source (sensor) indicates the value of a measured variable. At the controller end, we may need to convert the current to a proportional voltage signal. Figure 5.38a shows a modified version of an inverting op-amp used as a current-to-voltage converter. Notice that since the current flow between the two terminals of the op-amp are zero, the source current must flow through the resistor R. Furthermore, since the terminals of the op-amp is connected to the ground, the output voltage must be negative relative to the ground when the current is positive. Hence, the input current to output voltage conversion relationship is

$$V_{out} = -R \cdot i_{in} \tag{5.297}$$

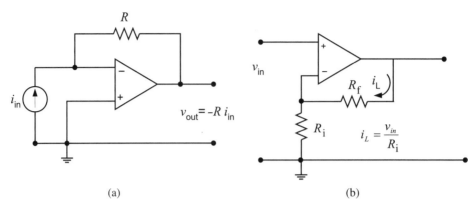

FIGURE 5.38: (a) Current to voltage converter op-amp circuit. (b) Voltage to current converter op-amp circuit.

Figure 5.38b shows a circuit which performs the opposite function: converts voltage to current. An op-amp is used as a current source. This is a noninverting op-amp. The only difference here is that the circuit output that we are interested in is the current through the load resistor, R_L. Such a function is used to drive small DC motors and solenoids. For instance, the input voltage is the command signal from a controller which is proportional to the desired torque. In DC motors, torque is proportional to the current passing through its windings. Therefore, in order to obtain the desired torque output from the motor, we must provide a current value proportional to the commanded voltage signal. In the figure, the load (i.e., motor winding) is shown as having resistance R_f (or can be impedance Z_f for more general case of electrical load) and is placed in the feedback loop of the op-amp. Since this is a noninverting op-amp,

$$V_{out} = \frac{R_i + R_f}{R_i} \cdot V_{in} \tag{5.298}$$

and the output current over the load is

$$i_L = \frac{V_{out}}{R_i + R_f} = \frac{1}{R_i} V_{in} \tag{5.299}$$

Notice that the load current is independent of the load resistance (R_f) and its variations. For linear operating regions without saturation,

$$V_{out} = (R_i + R_f) \cdot i_L < V_{sat} \tag{5.300}$$

When the output load requires current levels above 0.5 mA, the output current should be amplified using power transistors (i.e., BJTs, MOSFETs or IGBTs).

Figure 5.39 shows simple examples of the use of these two op-amp configurations. In the first case, a solar cell is used to generate a current proportional to the light it receives. The input current is converted to a proportional output voltage by the op-amp. Hence, the voltage output is proportional to the light received. This circuit can be used as an analog light intensity sensor. The other circuit input voltage is manually adjusted, and the output current is proportional to the voltage presented by the resistor R_3. The output current is proportional to this voltage, where the proportionality constant is R_2. Hence, the intensity of light emitted by the LED is proportional to the input voltage. The 741 op-amp can be replaced by other similar op-amps.

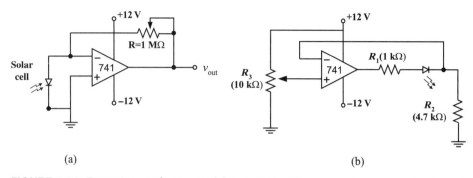

FIGURE 5.39: Example op-amp circuits: (a) current-to-voltage converter op-amp circuit used to provide output voltage that is proportional to the current generated by the solar cell, (b) voltage-to-current converter op-amp circuit where the intensity of LED light is proportional to the current, which is proportional to the input voltage pickup by the adjustable resistor R_3.

Example Show that the following Figures 5.40a and b represent a low pass and a high pass filter, like those shown in Figure 5.36, except that the sign of the input–output gain is negative.

$$\frac{V_{out}(s)}{V_{in}(s)} = -\frac{1}{RCs+1}; \quad low\ pass \tag{5.301}$$

$$\frac{V_{out}(s)}{V_{in}(s)} = -\frac{RCs}{RCs+1}; \quad high\ pass \tag{5.302}$$

where $R_1 = R_2 = R$ is assumed.

The transfer function in the frequency domain between input and output voltages can be derived by following the op-amp idealized assumptions and Kirchoff's current and voltage laws. Consider the low pass filter. The voltage at the positive terminal is grounded, hence the voltage potential at the negative terminal is also grounded since $v^+ = v^-$,

$$v^+ = 0 \tag{5.303}$$

$$v^- = 0 \tag{5.304}$$

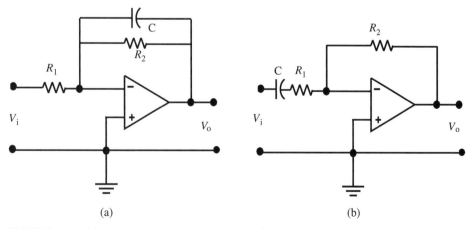

FIGURE 5.40: (a) Low pass, inverting, active filter, (b) high pass, inverting, active filter.

Then we can calculate the current over the resistor R_1.

$$i_1 = \frac{V_i}{R_1} \tag{5.305}$$

Also note that no current would flow into the op-amp, then the same current must pass through the R_2 and C combination (noting Kirchoff's current law),

$$i_1(t) = \frac{V_i(t)}{R_1} = i_{R2}(t) + i_c(t) \tag{5.306}$$

$$= \frac{V_o(t)}{R_2} + i_c(t) \tag{5.307}$$

$$V_o(t) = \frac{1}{C} \int_0^t i_c(\tau)d\tau \tag{5.308}$$

where we use the fact that the current flow over the R_2 and C is determined by the output voltage and the ground voltage at the negative input (inverting) terminal. We will add the negative sign to the input–output voltage relation at the end of the derivation. If we take the Laplace transform of the above equations, we can easily find the transfer function betwen input voltage and output voltage,

$$i_1(s) = \frac{V_i(s)}{R_1} = \frac{V_o(s)}{R_2} + Cs \cdot V_o(s) \tag{5.309}$$

$$\frac{V_o(s)}{V_i(s)} = -\frac{R_2}{R_1} \cdot \frac{1}{1 + R_2 Cs} \tag{5.310}$$

where we added the negative sign to indicate that the sign relationship between input voltage and output voltage is opposite.

$$\frac{V_o(s)}{V_i(s)} = -\frac{R_2}{R_1} \cdot \frac{1}{R_2 Cs + 1} \tag{5.311}$$

For the case of $R_1 = R_2 = R$ we have

$$\frac{V_o(s)}{V_i(s)} = -\frac{1}{RCs + 1} \tag{5.312}$$

The same reasoning applies for the derivation of the transfer function between input voltage and output voltage for the high pass filter. The only difference is the definition of the output voltage. In the above derivation, we would use

$$V_o(t) = R \cdot i(t) \tag{5.313}$$

and it can be shown, by following the same procedure for the low pass filter, that the transfer function for the high pass filter is

$$\frac{V_o(s)}{V_i(s)} = -\frac{RCs}{RCs + 1} \tag{5.314}$$

5.7 DIGITAL ELECTRONIC DEVICES

The logic ON state is represented by 1 and the logic OFF state is represented by 0. Two of the most popular digital device types are *transistor–transistor logic* (TTL) and *complementary metal oxide silicon* (CMOS). Logic device families differ from each other in terms of their power consumption and the speed of operation. When mixed series of logic ICs are used in a circuit, the most important factors to consider are the current loading and current driving capacity of the device. The design should make sure that each device can drive the gates it is connected to and does not present an overload current to the other devices.

The TTL nominal voltage level for logic 1 is 5 VDC, and for logic 0 is 0 VDC. The supply voltage must be in the range of 4.75 V to 5.25 V. Per gate power consumption of a TTL device is in the order of few miliamps. The output current sinking capacity is about 30 mA. A CMOS device supply voltage can be in the 3 to 18 VDC range. Per gate power consumption of a CMOS device is 80% less than that of a TTL device. The CMOS logic family requires a less stringent power supply voltage, consumes less power but is sensitive to static voltages. Digital logic devices are realized by complex networks of transistors on an integrated circuit (IC). The basic device number for the TTL NAND gate is 7400 (the military version is 5400 for extended temperature range) and has four NAND gates. The variations of the 7400 series take the following forms: 74L00 for low power but slower devices, 74H00 for high power and high speed, and a more recent version is the 74LS00 series for low power but high speed performance. The TTL series IC chips are numbered in 74LSxxx form where the xxx code is assigned based on the chronological introduction of the device into the market.

When an IC chip is used in a circuit and some of the pins are not used, they should be connected to common or pulled up to high voltage state. Open pins are not a recommended design practice since they tend to fluctuate and give a bad logic state.

5.7.1 Logic Devices

AND, OR, XOR, NOT (inverter), NAND, and NOR are the most common logic gates in digital electronics. The AND gate has two inputs and one output. The output logic is 1 if both inputs are 1, otherwise it is logic 0 state. The AND logic among more than two inputs can be implemented by cascading many AND gates in one chip. The NOT gate inverts the logic state of the input. The output is always at the opposite logic state to that of the input. Combining NOT gates with AND and OR, we form NAND and NOR gates. The gate symbols and logic diagrams are illustrated in Figure 5.41 and Figure 5.42.

5.7.2 Decoders

Decoders are used as device selection components in a computer bus system. When a computer places the address of a device on the bus, the decoder attached to each device checks that and gives either an ON or OFF output signal if it is the addressed device. So, only one decoder should give logic 1 output in response to a unique address in the bus. By combining AND and NOT gates, it is straightforward to design address decoders. In general, an 8-bit general purpose decoder is set to respond to a specific address by setting 8-dip switches ON/OFF which define the address of the device the decoder is connected to (Figure 5.43).

5.7.3 Multiplexer

A multiplexer is used to connect one of multiple input lines to an output line. A typical application of a multiplexer is with analog to digital converters (ADC). For instance, there can be four or eight analog signal channels connected to one ADC. Under program control, each channel is connected to the ADC for conversion. Such ADC converters are referred to as 4-channel multiplexed ADC or 8-channel multiplexed ADC. The desired channel is

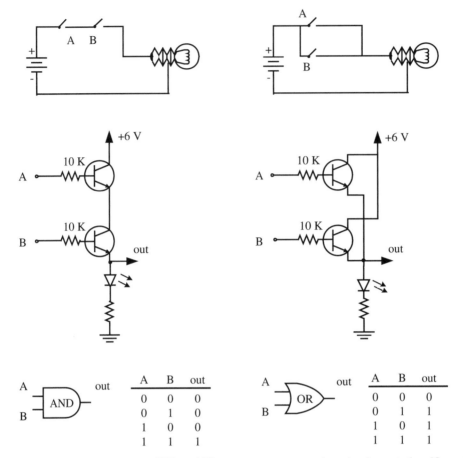

FIGURE 5.41: Logic gates: AND and OR gates - concept, transistor implementation, IC symbol.

selected by providing the binary code on the multiplexer control lines, that is two control lines are needed for a 4-channel multiplexer, three control lines are needed for an 8-channel multiplexer (Figure 5.44). A 4-channel multiplexer circuit is shown in Figure 5.45. Depending on the code presented in channel select lines A and B (00, 01, 10, 11) one of the four channels (Ch1, Ch2, Ch3, Ch4) is connected to the output under program control.

5.7.4 Flip-Flops

Flip-flop circuits are implemented using a combination of AND, OR, and NOT gates. The most common flip-flop types are D, RS, and JK flip-flops (Figure 5.46). RS flip-flop is commonly used as a debouncer for single-pole double-throw (SPDT) mechanical switches. When a SPDT mechanical switch is closed and opened (ON/OFF), it makes multiple ON-OFF contacts in the order of a millisecond time period. In human terms, this may look like a single OFF/ON transition, but in the digital electronics time scale, the many transitions of OFF/ON are detected. RS flip-flop is used in *debouncing* the mechanical switch input.

D-type Flip-Flop   The logic level present in D-input is transferred to output Q and latched when E-input is ON or transition of state on the E-channel occurs. When the E-input is low, the D-input is ignored and Q maintains its previous state. The D-latch is used to strobe and buffer (latch) an input signal state. D-type flip flops are used in groups to

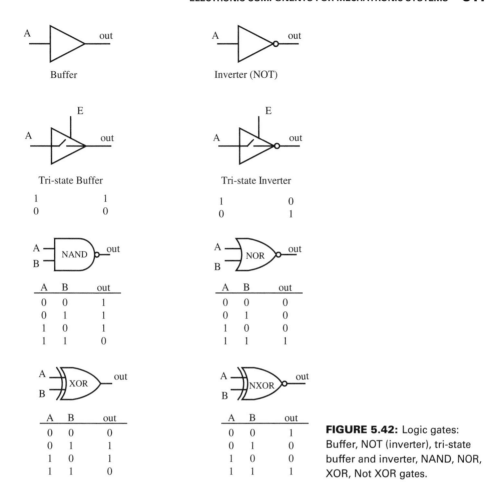

FIGURE 5.42: Logic gates: Buffer, NOT (inverter), tri-state buffer and inverter, NAND, NOR, XOR, Not XOR gates.

form n-bit data buffers (Figure 5.47). For output operations from a digital computer to D/A converter or discrete output lines, the data from the computer bus is latched and maintained (until updated by the computer) using D-flip flops.

RS-type Flip-Flop A reset-set flip-flop has two inputs (R and S) and two outputs (Q, $\bar{Q}$) where the two outputs are the opposite of each other. The logic between the input and

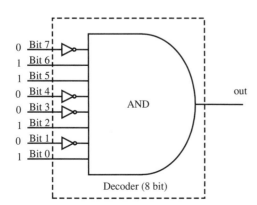

FIGURE 5.43: An 8-bit address decoder circuit implemented with NOT and AND gates.

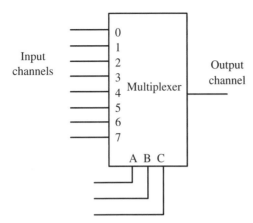

FIGURE 5.44: Multiplexer circuit and its function.

output terminals is as follows:

- when S = 1 and R = 0, then Q = 1 (is "set" to the 1),
- when S = 0 and R = 1, Q = 0 (is "reset" to 0).

When both S and R are 0, then the outputs do not change. The state of R and S both being 1 is not allowed. If that occurs, outputs both go to 0. Basically, the output Q is set (Q = 1) and reset (Q = 0) by S and R pulses.

JK-type Flip-Flop   JK flip-flop is similar to RS flip-flop with only one difference: if both inputs are 1 simultaneously, the outputs of JK reverse their states. The I/O transfer is

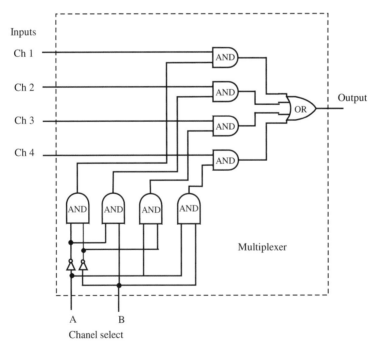

FIGURE 5.45: A multiplexer circuit for selecting one of four digital input channels. For instance, digital code on channel select lines AB, selects the following channels: 00 select channel 1, 01 selects channel 2, 10 selects channel 3, 11 selects channel 4.

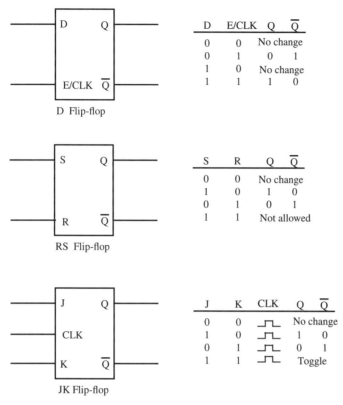

D	E/CLK	Q	$\overline{Q}$
0	0	No change	
0	1	0	1
1	0	No change	
1	1	1	0

D Flip-flop

S	R	Q	$\overline{Q}$
0	0	No change	
1	0	1	0
0	1	0	1
1	1	Not allowed	

RS Flip-flop

J	K	CLK	Q	$\overline{Q}$
0	0	⊓	No change	
1	0	⊓	1	0
0	1	⊓	0	1
1	1	⊓	Toggle	

JK Flip-flop

FIGURE 5.46: Flip-flops: D-type, RS-type, JK-type.

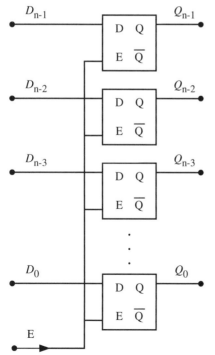

FIGURE 5.47: n-bit data latch buffer using n-set of D-type flip-flops for data input and output between a digital computer bus and I/O devices. Data is placed in the data bus ($D_0 \ldots D_{n-1}$), and then the E-line is pulsed to latch the data in the D-flip flops. After that, the changes in the data bus do not affect the output of the D-flip flops (latched data) until a new pulse in the E-line.

triggered either by a predefined edge transition signal of a clock or input sampled in a high state of clock signal and transferred to the output at the trailing edge of the clock signal. JK flip-flops are used in counters. For instance, four JK-flip flops can be used to count up to sixteen.

5.8 DIGITAL AND ANALOG I/O AND THEIR COMPUTER INTERFACE

Figure 5.48 shows the interface between a generic computer bus and a parallel input and a parallel output device. For the purpose of the present discussion, let us consider the

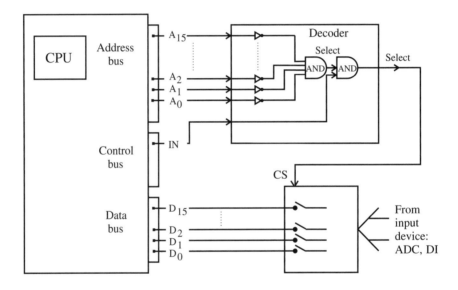

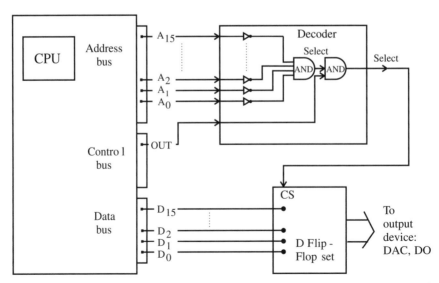

FIGURE 5.48: Interface circuit between a digital computer and parallel data input and output device. For instance, parallel data input may be the register of an ADC, parallel output may be the register of a DAC.

digital computer as having a CPU and a bus. Here we assume that the CPU includes the microprocessor's clock, the CPU, the memory (random access memory (RAM) and read-only memory (ROM)). The bus is made of three main groups of digital lines: *address bus* lines used to address devices and memory by the CPU, *control bus* lines used to indicate whether the operation is a read or a write as well as used for interface handshaking signals between CPU and I/O devices, and a *data bus* which carries the data between the CPU and I/O devices (or memory). Let us assume that the CPU executes programmed instructions to perform logic and I/O with the external devices.

The address decoder circuit of each I/O device uniquely specifies the address of the I/O device on the bus. The data lines of the device are connected to the data bus of the computer. Every device that the CPU communicates with (reads or writes) must be addressed by the CPU first. A particular device is selected on the address bus via the decoder of the device whose address has been placed on the address bus by the CPU. The address number to which the decoder responds by turning its output ON is set by jumpers or DIP switches which define whether or not each address bus line has an inverter or direct connection.

The control bus of the computer is used to strobe the I/O device to process the data bus information. When a read (input) or write (output) operation is performed by the CPU, the following sequence of events are generated by the CPU at the machine instruction level:

1. The CPU places the address of the I/O device on the address bus. Only one device decoder will provide an active output in response to a unique address in the address bus.
2. The CPU places the data on the data bus for the write operation, and
3. The CPU turns on the OUT signal of the control bus to tell the I/O device that the data is ready.
4. When the I/O device is given enough time to read (or signals the CPU that it is done via a handshake line in the control bus), the OUT signal is dropped, and CPU goes on with other operations.

For read (input) operations, the sequence of steps 2 and 3 are changed:

1. The CPU places the address of the I/O device on the address bus. Only one device decoder will provide an active output in response to a unique address in the address bus.
2. The CPU turns on the IN signal of the control bus to tell the I/O device it is ready to read the data.
3. The CPU reads the data on the data bus, and
4. When the CPU is done (transfer the data to its registers from the data bus), it drops the IN signal, and CPU goes on with other operations.

In process control applications, the output device is a set of D-type flip-flops connected to discrete output lines or a D/A converter, and the input device is a set of R-S flip-flops connected to discrete input lines or an A/D converter (Figure 5.49).

Example Figure 5.50 shows the interface between a digital computer data output line and a relay. The digital data line is set high or low under software control. This controls the opto-coupler, which then turns the transistor ON/OFF. The transistor powers the control circuit to the relay coil. Once the relay coil is energized, its contact conducts the current in the output circuit which may turn ON/OFF a device, such as a light or a motor. Notice that

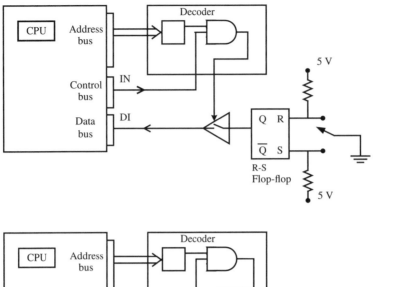

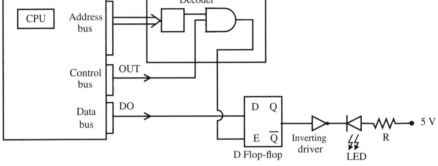

FIGURE 5.49: Interface circuit between a digital computer and discrete input and discrete output devices. A R-S flip-flop is used to eliminate the debouncing problem of the switch. The D flip-flop is so that the data is maintained at the output of the flip-flop even when this device is no longer addressed.

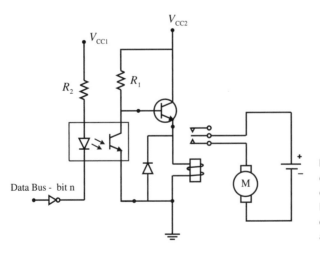

FIGURE 5.50: ON/OFF control of a relay through an opto-coupler and transistor. Data bus-bit-n output in this circuit can come from a circuit as shown in Figure 5.49.

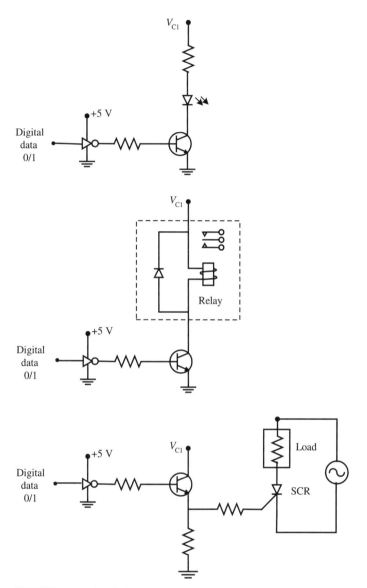

FIGURE 5.51: ON/OFF control using a digital computer data bus line: (a) LED (ON/OFF) light output, (b) relay ON/OFF control, (c) SCR control in an AC circuit. In all cases, a transistor is used as an electronic switch between the computer and the output circuit.

in this example, since no flip-flop is used, the data line must be dedicated for the control of this relay.

Example Figure 5.51 shows various interfaces between a digital computer data output line and an ON/OFF output device through a transistor. In case 1, an output LED light is turned ON/OFF by the computer data bus via a transistor. In case 2 and 3, a relay and a SCR are controlled (turned ON/OFF) respectively. By controlling the timing of the SCR gate signal, a proportional control is approximately achieved.

5.9 D/A AND A/D CONVERTERS AND THEIR COMPUTER INTERFACE

A D/A converter converts a digital number to an analog signal, that is generally a voltage level. An A/D converter does the opposite. It converts an analog signal to a digital number. D/A and A/D converters are essential components in interfacing the digital world to the analog world.

The process of converting signals from analog form to digital form involves two operations:

1. sample the signal,
2. quantize the signal to the resolution the A/D can represent with n-bits.

Sampling is the process where a finite number samples of a continuous signal is taken and converted into a discrete number sequence. The samples are the only information available regarding the signal in the computer.

Let us consider how a D/A converter converts a digital number to an analog signal (Figure 5.52). The voltage potential across the resistor bank at the output of the D flip-flop memory device is V_h and held constant, that is 10 VDC. The resistor value R_f is selectable to usually one of four different values. This is used to change the output range of the D/A converter. If a bit is ON (D0-D7, which is same as Q0-Q7), there will be a current flowing through the corresponding resistor. If the bit is OFF, there will not be any current flowing through that resistor. Therefore, since the sum of current at point A of the op-amp should be zero, the output voltage for a given digital number N sent to the D/A converter can be calculated as follows:

$$V_0/R_f = V_h(b_0/R + b_1/(R/2) + \dots + b_7/(R/2^7))$$
$$V_0 = V_h(R_f/R)N$$

where, b_i is zero (0) or one (1) representing the value of bit i, N is the corresponding number sent to the D/A converter. The range it can cover is zero to V_{range}, where

$$V_{range} = V_h(R_f/R) * (2^n - 1) \qquad (5.315)$$

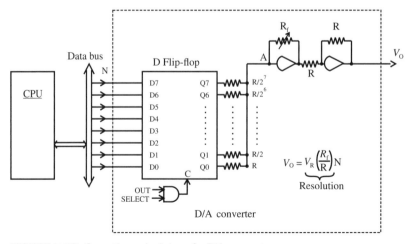

FIGURE 5.52: Operating principles of a D/A converter.

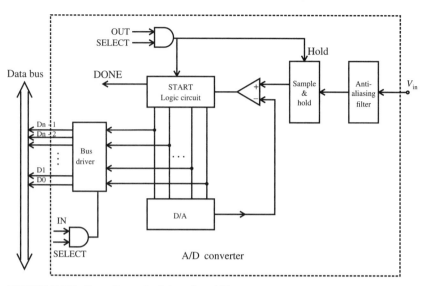

FIGURE 5.53: Operating principles of an A/D converter.

The resolution of a D/A converter (the smallest voltage change that can be made at the output of the D/A converter) is

$$\Delta V = V_h(R_f/R) = V_{range}/2^n - 1 \qquad (5.316)$$

In that range, the D/A can provide $(2^n - 1)$ different levels of voltage.

The basic A/D converter operation uses a D/A converter plus an additional circuit (Figure 5.53). The analog signal is passed through an anti-aliasing filter. This filter may or may not be an integral part of the A/D converter. Then the signal is sampled and held at the sampled level (Figure 5.54). While it is being held, a D/A converter cycles through a sequence of numbers based on a search algorithm to match the D/A output signal to the sampled signal. When the two are equal, the comparator will indicate that the correct signal level is determined by the D/A. The digital representation of the sampled analog signal is the same as the digital number used by the D/A to generate the equal signal. The conversion is accurate to the resolution of the A/D converter. Then, the CPU will be signalled about the fact that the A/D conversion is complete.

This signaling is typically done in two ways:

1. Set a bit in the ADC status register to indicate the "end-of-conversion ADC." By reading the value of this bit, the program can determine whether the current ADC conversion process is completed or not. This method is used in the "polled" (programmed) method of handling the ADC conversion.

2. Set the interrupt bit in the ADC control register (or status register, depending on where the interrupt bit is implemented for the particular microcontroller) to generate an interrupt to the CPU, indicating that the ADC conversion process is complete. This is used in the interrupt driven ADC conversion handling method. The application program must be setup to service this interrupt, which is simply providing a subroutine (called "interrupt service routine" (ISR)) to read the ADC data register.

The CPU would then read the data on the D/A portion of the A/D through the data bus. Notice that this type of A/D conversion is an interative process and takes a longer time

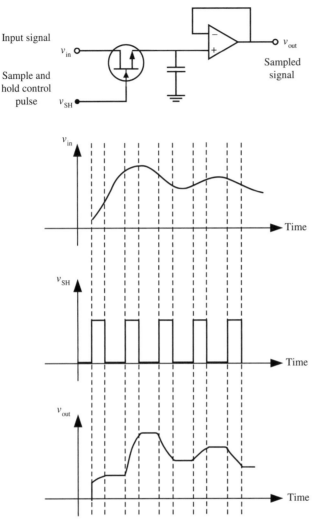

FIGURE 5.54: Sample and hold circuit using a buffer op-amp, JFET type transistor switch, and a capacitor.

to complete than the D/A conversion process. There are other types of A/D converters, such as the flash A/D converter which makes the conversion faster without using an D/A converter in the circuit, but these are also more expensive.

The sample and hold circuit is a fundamental circuit for analog to digital signal conversion. For the conversion, we must sample the signal in periodic intervals (which must be fast enough to capture the frequency details of the signal) and hold the sampled value constant while the A/D converter tries to determine its digital equivalent. A basic sample and hold circuit includes a voltage follower op-amp (buffer), a JFET type transistor switch to turn ON and OFF the connection of the input voltage, and a capacitor at the op-amp input terminal to charge and hold the signal constant. The sampling process is done typically in fixed frequency, that is $f_s = 1$ kHz. The input transistor switch is turned ON every $T_s = 1/f_s$ s, for a very short period of time that is long enough to charge the capacitor to the input voltage level present, that is $T_{on} = 0.05 \cdot T_s$. Then, the transistor switch is turned OFF and the capacitor holds the last (sampled) value of the input voltage constant. The T_{on}

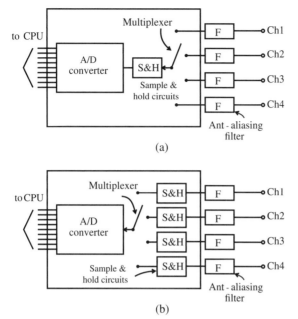

FIGURE 5.55: Multiplexed multi channel analog signal and sampling via a single A/D converter: (a) multiplexed sample-and-hold circuit, (b) simultaneous sample-and-hold circuit.

time must be long enough so that the circuit output voltage reaches the input voltage value within a desired accuracy. Typical values of T_{on} are in the order of microseconds. It should be pointed out that the sample and hold circuit shown in Figure 5.54 is only a basic concept circuit. Actual sample and hold circuits involve two or more op-amps and are a bit more complicated. One op-amp is used to buffer the input and the other op-amp is used to buffer the sampled output.

Quite often an A/D converter is used to sample and convert signals from multiple channels. A multiplexer has many input channels and one output channel. It selectively connects one of the input channels to the output channel. The selection is made by a digital code in a set of digital lines in the interface. Figure 5.55 shows two different implementations. In case (a), there is only one sample and hold circuit. Therefore, the signals converted from different channels are not taken at the same time instant. For instance, the A/D converter selects channel 1, samples-hold-converts. This takes finite time, T_{conv}. Then it selects-samples-holds-converts channel 2. This takes another T_{conv} time. Hence, the channel 3 will be sampled 2 T_{conv} time later than channel 1. This is called the *sequential sample-and-hold*.

In case (b), there are individual sample and hold circuits for each channel. All the channels are sampled at the same time and signals are held at that value. Then the A/D conversion goes through each channel using the multiplexer. This is called the *simultaneous sample-and-hold circuit*.

Quantization Error of A/D and D/A Converters: Resolution

An n-bit device can represent 2^n different states. The resolution of the device is one part in $2^n - 1$. In physical signal terms, the resolution is 1 part in the whole range divided by $2^n - 1$. Notice that we need to mark a range into four discrete levels (Figure 5.56). Similarly, if we have the ability to represent 2^n states, we can divide a given range into 2^n levels. Therefore, the resolution of an n-bit A/D or D/A converter is

$$V_{range}/(2^n - 1) \qquad (5.317)$$

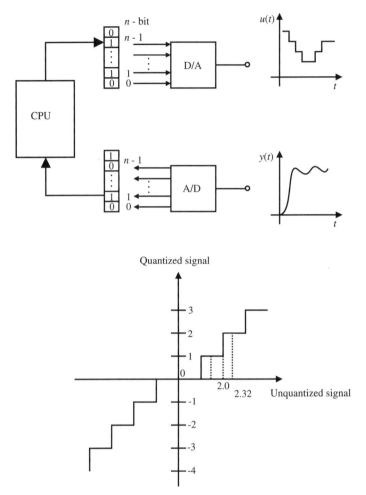

FIGURE 5.56: Resolution and range of A/D and D/A converters.

Quantization is the result of the fact that the A/D, D/A, and also the CPU are finite word and finite resolution devices (Figure 5.56). Let us consider that we are sampling a signal in the 0 to 7 V DC range. Assume that we have a 3-bit A/D converter. The A/D converter can represent only 0, 1, 2, 3, 4, 5, 6, 7 VDC signals exactly. If the analog signal was 4.6 V, it can be represented as either 4 or 5 by the A/D conversion. If the truncation method is used in the quantization, it will be converted to 4. If the round-off method is used, it will be converted to 5. The maximum value of the quantization error is the resolution of the A/D or D/A converter.

For instance, an 8-bit A/D converter can represent 256 different states. If an analog sensor with a signal range of 0 to 10 VDC is connected to it, the smallest change the A/D can detect in the signal is 10/255 VDC. Any change in the signal less than 10/255 VDC will not be detected by the computer. Similarly, if the A/D converter is 12-bit, the smallest change detectable in the signal is equal to the signal range divided by $2^{12} - 1 = 4095$. This is called the resolution of the A/D converter. If it is 16-bit A/D converter, its resolution is one part in $2^{16} - 1 = 65\,535$.

The D/A converter converts the digital numbers into analog signals. This is also referred to as the signal reconstruction stage. Discrete values of control action are presented

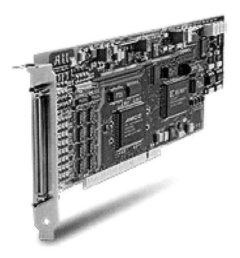

FIGURE 5.57: A commercially available data acquisition board for a PC bus (Model-KPCI-1801HC by Keithley). I/O capabilities are 12-bit 32-Ch differential ended or 64-Ch single ended analog input (multiplexed A/D converter), 12-bit 2-Ch D/A converter, 4-Ch digital input, 8-Ch digital output, maximum sample rate is 333 kHz. Reproduced with permission from Keithley Instruments, Inc.

to the D/A converter once per sampling period. The interpolation performed by the D/A converter for the values of the signal during the period between the updates is called the reconstruction approximation. The most commonly used form of D/A converter is the zero-order-hold type where the current value of the signal is held constant until a new value is sent. The resolution of the D/A converter is the smallest change it can send out. This is one part in the whole range it can cover, 1 part in $2^n - 1$, where n is the number of bits the D/A converter has. If the D/A converter has an output range of -10 VDC to $+10$ VDC, lets us call $R = 10 - (-10) = 20$ VDC. If the D/A converter is an 8-bit converter, it can have $N = 2^8$ different states. Therefore, the smallest change it can make in the output is $R/N = R/(2^n) = 20/(255)$. Clearly, as the number of bits of the A/D and D/A gets larger (i.e., 16-bit), the resolution and the resultant quantization error becomes less significant.

Figure 5.57 shows a commercially available data acquisition card for PCs (PCI or ISA bus). The board has 12-bit resolution 32-channel differential ended or 64-channel single ended multiplexed A/D converter, 12-bit 2-channel D/A converter, 4-channel digital input, and 8-channel digital output lines. The maximum sampling rate that is supported by the board is 333 kHz (333 samples/s).

Example Consider the adjustable gain operational amplifier shown in Figure 5.58. Such an amplifier may be used between a sensory voltage signal and input channel of an ADC where ADC channel may have a fixed range, that is ± 10 VDC, yet the input voltage range may vary from one application to another. Hence a programmable gain is needed in order to best utilize the available resolution and range of the ADC. Assume that the ADC has ± 10 VDC input voltage range. Determine the resistor values at the feedback path of the op-amp so that the circuit has the gain of $0.1, 1.0, 10.0$, in order to scale a sensory input voltage range of 0–100 V, 0–10 V, and 0–0.1 V, respectively, to 0–10 V range of ADC.

The gain of the first op-amp is $K_1 = -R_f/R_i$, where $R_i = 1.0\,\text{M}\Omega$. The second op-amp has simply a negative one gain $K_2 = -1$, hence the overall gain of the two op-amps is $K = K_1 \cdot K_2 = R_f/R_i$. Let $R_i = 1.0\,\text{M}\Omega$. Then it is straightforward to determine the

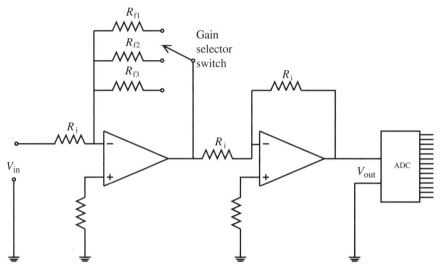

FIGURE 5.58: A circuit with two operational amplifiers and selectable gain: by selecting which switch to connect on the feedback resistor path, the gain of the op-amp circuit is adjusted. Such circuits are used as the voltage range scaler in ADC input as well as DAC output circuits.

necessary gains,

$$K = \frac{R_f}{R_i} \tag{5.318}$$

$$R_f = K \cdot R_i \tag{5.319}$$

$$R_{f1} = 0.1 \cdot 1\,M\Omega = 100\,K\Omega\ for\ K = 0.1 \tag{5.320}$$

$$R_{f2} = 1.0 \cdot 1\,M\Omega = 1.0\,M\Omega\ for\ K = 1.0 \tag{5.321}$$

$$R_{f3} = 10.0 \cdot 1\,M\Omega = 10\,M\Omega\ for\ K = 10.0 \tag{5.322}$$

The same circuit may also be used at the output side of a DAC where the standard output range of a DAC may need to be scaled for the next device.

5.10 PROBLEMS

1. Consider the RC and RL circuits shown in Figure 5.5c–d. Let $V_s(t) = 12$ VDC, $R = 100\,k\Omega$, $L = 100\,mH$, $C = 0.1\,\mu F$. Let us assume that initially the current in the circuit and charge capacitor is zero. Simulate the current and voltage across each component for the case that the switch is connected to the supply voltage at time zero, and flipped over instantly to B position at time $t = 250\,\mu s$. Solve this problem using Simulink® and present the results in five plots including the state of the switch as a function of time, plus $i(t)$, $V_L(t)$, $V_R(t)$, $V_C(t)$. Experiment with the system response by varying the circuit component parameters R, C, L. How does increasing R affect the time constant of the system in RC and RL circuits?

2. Consider a voltage amplifier using a bipolar junction transistor as shown in Figure 5.17. Let the resistances $R_1 = R_2 = 10\,k\Omega$, and supply voltage $V_{cc2} = 24$ VDC. Assume that the current gain of the transistor is 100, and the base to emitter voltage drop when it is conducting is 0.7 VDC. Calculate and tabulate the following results for the following cases of input voltage, $V_{in} = V_{cc1} = 0.0, 0.5, 0.7, 0.75, 0.80, 0.85, 0.9, 1.0$. The results to be calculated are base current (i_b), collector current (i_c), emitter current (i_e), and output voltage (V_o) measured between the collector and emitter.

3. Design an op-amp circuit that will change the offset and slope of an input voltage and provide an output voltage. Mathematically, the desired relationship between the input and output voltages of the op-amp circuit is

$$V_{out} = K_1 \cdot (V_{in} - V_{offset})$$ (5.323)

For numerical calculations, assume the V_{in} range is 2.0–3.0 V. The desired output voltage range is 0.0–10 V. An example application where such a circuit may be useful is shown in Figure 5.59.

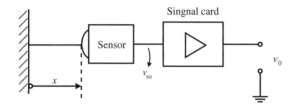

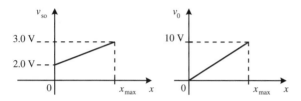

FIGURE 5.59: Sensor head (transducer) and signal conditioner to adjust the offset voltage and the gain on the signal generated by the sensor.

4. Consider the op-amp circuit shown in Figure 5.60. Derive the relationship between the input voltages V_{in1}, V_{in2}, and the output voltage V_{out} as function of R_1, R_2, R_3, R_4, R_5. Select values for the resistors so that the following relationship is obtained,

$$V_{out} = 5.0 \cdot (V_{in1} - 3.0 \cdot V_{in2})$$ (5.324)

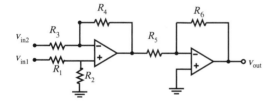

FIGURE 5.60: Op-amp circuit for problem 4.

5. Determine the minimum voltage at the base of the transistor (V_{in}) that will saturate the transistor and hence provide the maximum glow from the LED (Figure 5.61). Assume the following for the

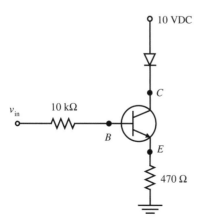

FIGURE 5.61: Figure for problem 5.

LED and transistor: the forward bias voltage of the LED is $V_f = 2.5$ V, and the minimum voltage drop between collector and emitter of transistor is $V_{CE} = 0.3$ V (i.e., when the transistor is fully saturated), and the gain of the transistor $\beta = 100$.

6. Design an op-amp circuit to implement a PD (proportional plus derivative) control where the proportional gain and the derivative gain are both adjustable. The circuit should have two input voltages ($V_{i1}(t)$, $V_{i2}(t)$) and one output voltage ($V_o(t)$). The desired mathematical function to be realized by the op-amp circuit is

$$V_o(t) = K_p \cdot (V_{i1} - V_{i2}(t)) + K_d \cdot d/dt((V_{i1} - V_{i2}(t))) \qquad (5.325)$$

7. Consider the op-amp circuits shown in Figure 5.62. Derive the input–output voltage relationships.

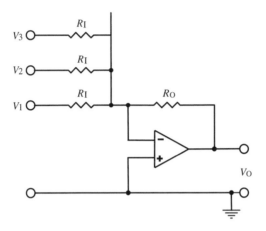

Noninverting

Inverting

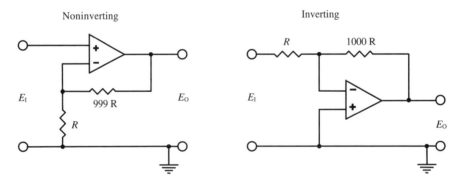

FIGURE 5.62: Figure for problem 7.

8. Consider the op-amp circuits shown in Figure 5.63. Determine the output voltage when the input voltage is $V_i = 0.1$ V, and the current that flows to the inverting input.

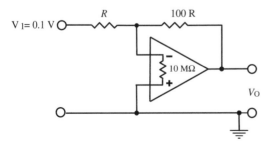

FIGURE 5.63: Figure for problem 8.

9. Consider a data acquisition system that samples signals from various sensors. Let us assume the following: there are four sensors to be sampled. The voltage outputs from each sensor are in the following ranges for sensors 1 through 4: ± 10 VDC, ± 1 VDC, 0 to 5 VDC, and 0 to 2 VDC. Expected maximum frequency content in each signal is 1 kHZ, 100 Hz, 20 Hz and 5 Hz for sensor 1 through 4, respectively. The error introduced due to sampling should not be more than $\pm 0.01\%$ (1 part in 10 000) of the maximum value of the signal. Determine the specifications for a 4-channel ADC (analog to digital converter) for this application that will meet the requirements. Specifiy the minimum sampling rate for each channel (according to the sampling theorem) and recommend a practical sampling rate for each channel.

10. Consider the RL and RC circuits shown in Figure 5.5c, d, respectively. The switch in each circuit is connected to the supply voltage (A side) and B side at specified time instants. Assume the following circuit parameters; $L = 1000$ mH $= 1000 \times 10^{-3}$ H $= 1$ H, $C = 0.01$ μF $= 0.01 \times 10^{-6}$ F, $R = 10$ kΩ. Let the supply voltage be $V_s(t) = 24$ VDC. Assume in each circuit, the initial condition on the current is zero and initial charge in the capacitor is zero. Starting time is $t_o = 0.0$ s. At time $t_1 = 100$ μs the switch is connected to the supply, and at time $t_2 = 500$ μs, the switch is disconnected from the supply and connected to the B side of the circuit. Plot the voltage across each component and current as a function of time for the time period of $t_0 = 0.0$ s to $t_f = 1000$ μs.

CHAPTER 6

SENSORS

6.1 INTRODUCTION TO MEASUREMENT DEVICES

Measurements of variables are needed for monitoring and control purposes. Typical variables that need to be measured in a data acquisition and control system are:

1. position, velocity, acceleration,
2. force, torque, strain, pressure,
3. temperature,
4. flow rate,
5. humidity.

Figure 6.1 shows the basic concept of a measurement device. The measurement device is called the *sensor*. We will discuss different types of sensors to measure the above listed variables. Figure 6.2 shows an example of sensors used on a construction equipment. A sensor is placed in the environment where a variable is to be measured. The sensor is exposed to the effect of the measured variable. There are three basic phenomenon in effect in any sensor operation:

1. The change in the measured physical variable (i.e., pressure, temperature, displacement) is translated into a change in the property (resistance, capacitance, magnetic coupling) of the sensor. This is called the *transduction*. The change of the measured variable is converted to an equivalent property change in the sensor. The transduction relationship, that is the relationship between the measured variable and the change in the sensor material property, is the fundamental physical principle of the sensor operation. It is desirable that this relationship is repeatable and does not vary with other environmental variables. For instance, a pressure sensor output voltage as a function of pressure should not change much due to changes in the ambient temperature.
2. The change in the property of the sensor is translated into a low power level electrical signal in the form of voltage or current.
3. This low power sensor signal is amplified, conditioned (filtered), and transmitted to an intelligent device for processing, for example to a display for monitoring purposes or use in a closed loop control algorithm.

Sensor types vary in the transduction stage in measuring a physical variable. In response to the physical variable, a sensor may be designed to change its resistances, capacitance, inductance, induced current, or induced voltage.

In any measurement system, accuracy is a major specification. Let us clarify the terminology used regarding accuracy. Figure 6.3 shows the meaning of *accuracy,*

Mechatronics with Experiments, Second Edition. Sabri Cetinkunt.
© 2015 John Wiley & Sons, Ltd. Published 2015 by John Wiley & Sons, Ltd.
Companion Website: www.wiley.com/go/cetinkunt/mechatronics

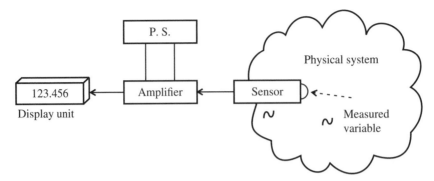

FIGURE 6.1: The components of a sensor: sensor head, amplifier, power supply, display or processing unit.

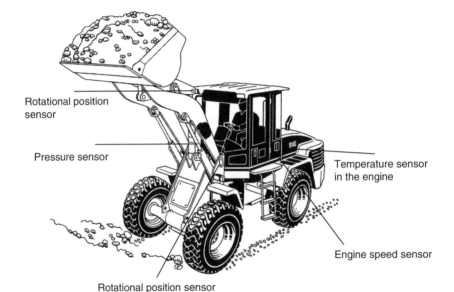

FIGURE 6.2: Examples of various types of sensors used on construction equipment. There are more sensors on modern construction equipment. Only a small set of examples are shown in this figure. Reprinted Courtesy of Caterpillar Inc.

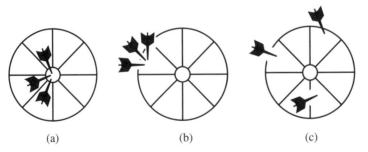

FIGURE 6.3: The definitions of accuracy and repeatability: (a) accurate, (b) repeatable, but not accurate, (c) not repeatable, not accurate. Resolution is the smallest positional change the arrow can be placed on the target (imagine that the target has many small closely spaced holes and the arrow can only go into one of these holes).

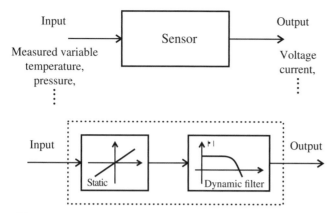

FIGURE 6.4: Input–output model of a sensor: steady-state (static) input and output relationship plus the dynamic filtering effect.

repeatablity, and resolution. Resolution refers to the smallest change in the measured variable that can be detected by the sensor. Accuracy refers to the difference between the actual value and the measured value. Accuracy of a measurement can be determined only if there is another way of more accurately measuring the variable so that the sensor measurement can be compared with it. In other words, accuracy of a measurement can be determined only if we know the true value of the variable or a more accurate measurement of the variable. Repeatability refers to the average error in between consecutive measurements of the same value. The same definitions apply to the accuracy of a control system as well. In a measurement system, repeatability can be at best as good as the resolution. Resolution (the smallest change the sensor can detect on the measured variable) is the property of the sensor. Repeatability (the variation in the measurement of the same variable value among different measurement samples) is the property of the sensor in a particular application environment. Hence, the repeatability is determined both by the sensor and the way it is integrated into a measurement application.

Let us focus on the input–output behavior of a generic sensor as shown in Figure 6.4. A sensor has a dynamic response bandwidth as well as steady-state (static) input–output characteristic. The dynamic response of a sensor can be represented by its frequency response or by its bandwidth specification. The bandwidth of the sensor determines the maximum frequency of the physical signal that the sensor can measure. For accurate dynamic signal measurements, the sensor bandwidth must be at least one order of magnitude (x10) larger than the maximum frequency content of the measured variable.

A sensor can be considered as a filter with a certain bandwidth and static input–output characteristics. Let us focus on the static input–output relation of a generic sensor. An ideal sensor would have a linear relationship between the sensed physical variable (input) and the output signal. This linear relationship is a function of the *transduction* and amplification stage. Typical non-ideal characteristics of a sensor include (Figure 6.5):

1. gain changes,
2. offset (bias or zero-shift) changes,
3. saturation,
4. hysteresis,
5. deadband,
6. drift in time.

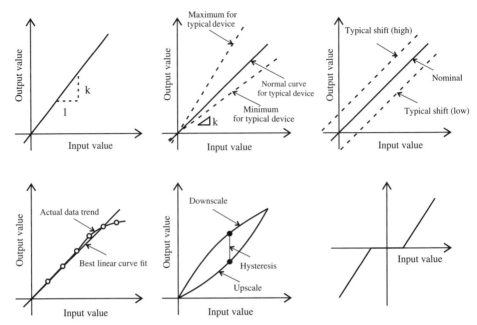

FIGURE 6.5: Typical nonlinear variations of static input–output relationship from the ideal behavior of a sensor.

The static input–output relationship of a sensor can be identified by changing the physical variable in known increments, then waiting long enough for the sensor output to reach its steady-state value before recording it, and repeating this process until the whole measurement range is covered. The result can be plotted to represent the static input–output characteristic of the particular sensor. If these non-ideal input–output behaviors are repeatable, then a digital signal processor can incorporate the information into the sensor signal processing algorithm in order to extract the correct measurement despite the nonlinear behavior. Repeatability of the nonlinearities is the key requirement for accurate signal processing of the sensor signals. If the nonlinearities are known to be repeatable, then they can be compensated for in software in order to obtain accurate measurement.

In general, a sensor needs to be *calibrated* to customize it for an application. In any control system application such as an automated machine in an assembly line or a mobile equipment, which may involve hundreds of sensors, one of the first steps in implementing a control system is the *sensor calibration*. That is to establish the desired input–output relationship for each sensor such as the relationship shown in Figure 6.5a. If the sensor exhibits drift in time, then it must also be calibrated periodically. Sensor calibration refers to adjustments in the sensor amplifier to compensate for the above variations so that the input (measured physical variable) and output (sensor output signal) relationship stays the same. The sensor calibration process involves adjustments to compensate for variations in gain, offset, saturation, hysteresis, deadband, and drift in time.

Figure 6.6 shows a circuit for a resistance type sensor and its signal amplification using an op-amp. The sensor transduction is based on the change of resistance as a function of the measured variable (i.e., temperature, strain, pressure). The resistance change is converted to voltage change which is typically a small value. Then it is amplified by an inverting type op-amp to bring the sensor signal (voltage) to a practical level. Notice that the resistor R_1 is used to calibrate the sensor for offset (bias) adjustments. The R_1 and R_s act as the voltage divider. The op-amp is in inverting configuration where the resistors R_2

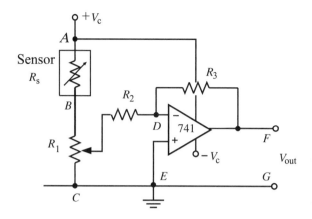

FIGURE 6.6: Resistive type sensor signal amplification using an op-amp. The sensor transduction principle is based on the physical relationship that the resistance of the sensor varies as a function of the measured variable.

and R_3 determine the gain of the amplifier. It can be easily shown that the output voltage from the circuit as a function of the sensor resistance is,

$$V_{out} = -\frac{R_3}{R_2} \cdot \frac{R_1(y)}{R_s(x) + R_1} \cdot V_c \tag{6.1}$$

where x represents the measured variable, $R_s(x)$ is the resistance of the sensor which varies as a function of the measured variable, $R_1(y)$ represents the potentiometer adjustment of the R_1 as it is connected to the op-amp input to provide the reference voltage. $R_1(y)$ is the resistance amount from the point it is connected to the R_2 to the ground connection at point C. We can use $R_1(y)$ to adjust the offset voltage. The sensor gain can further be adjusted by making R_2 or R_3 adjustable resistors (potentiometers).

6.2 MEASUREMENT DEVICE LOADING ERRORS

Loading errors in measurement systems are errors introduced to the measurement of the variable due to the sensor and its associated signal processing circuit. There are two types of loading errors:

1. mechanical loading error,
2. electrical loading error.

An example of mechanical loading error is as follows. Consider that we want to measure the temperature of a liquid in a container. If we insert a mercury-in-glass thermometer in the container, there will be a finite amount of heat transfer between the liquid and the thermometer. The measurement will stabilize when both temperatures of the liquid and thermometer are the same. Therefore, the fact that the thermometer is inserted into the liquid and there is a heat transfer between them has changed the original temperature of the liquid. Clearly, this mechanical loading error would be very large if the volume of the liquid were small compared to the size of the thermometer, and would be neglegable if the liquid volume were very large compared to that of the thermometer. Strictly speaking, it is not possible to perfectly measure a physical quantity since the very act of introducing a sensor changes the original physical environment. Therefore, every measurement system has some mechanical loading error. The design question is how to make that error as small as possible.

The electrical loading error issue exists in electrical circuits used in measurement systems. Consider the voltage measurement across the resistor in the following figure

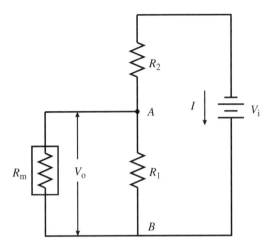

FIGURE 6.7: Electrical loading error in measurement systems and sensors.

(Figure 6.7). Assume that the measurement device has an internal resistance of R_m. Without the measurement device, the ideal value of the voltage we want to measure is

$$V_o^* = \frac{R_1}{R_1 + R_2} \cdot V_i \tag{6.2}$$

Once the measurement device is connected to the circuit at points A and B, it changes the electrical circuit. The equivalent resistance in between point A and B is

$$R_1^* = \frac{R_1 R_m}{R_1 + R_m} \tag{6.3}$$

and the measured voltage across points A and B is

$$V_o = \frac{R_1^*}{R_1^* + R_2} \cdot V_i \tag{6.4}$$

which is

$$V_o = V_i \frac{R_1 R_m / (R_1 + R_m)}{R_1 R_m / (R_1 + R_m) + R_2} \tag{6.5}$$

Notice that as $R_m \to \infty$, $V_o = V_o^*$. However, if R_m is close to the value of R_1, the V_o/V_o^* deviates from unity, that is $R_m = R_1$

$$V_o = V_i \frac{R_1}{R_1 + 2R_2} \tag{6.6}$$

In most measurement systems, the relationship between the R_m and R_1 is such that

$$R_m = 10^3 \cdot R_1 \tag{6.7}$$

Consider for simplicity that $R_1 = R_2$. The voltage measured (V_o) and the ideal voltage should have been measured if R_m was infinity (V_o^*),

$$V_o = V_i \frac{1000}{2001} \tag{6.8}$$

and the ideal voltage

$$V_o^* = V_i \frac{1000}{2000} \tag{6.9}$$

The voltage measurement error percentage due to the electrical loading in this case is

$$e_v = \frac{(1000/2001) - (1/2)}{(1/2)} \cdot 100 = 0.0499\% \qquad (6.10)$$

For cases where the measurement device input impedance is about 10^3 times the resistance of the equivalent two port device whose voltage is measured, the voltage measurement error introduced due to the electrical loading effect is negligible.

Let us consider a poor case of loading error condition: $R_1 = R_2 = R_m$. Then the mesasured voltage,

$$V_o = V_i \frac{R_1 R_m/(R_1 + R_m)}{R_1 R_m/(R_1 + R_m) + R_2} \qquad (6.11)$$

$$= V_i \frac{R_1^2/(2R_1)}{R_1^2/(2R_1) + R_1} \qquad (6.12)$$

$$= V_i \frac{R_1/2}{R_1/2 + R_1} \qquad (6.13)$$

$$= V_i \frac{1/2}{1/2 + 1} \qquad (6.14)$$

$$= V_i \cdot \frac{1}{3} \qquad (6.15)$$

The voltage measurement error due to electrical loading error in this case V_0 is and the ideal voltage is V_o^*

$$V_o^* = V_i \frac{1000}{2000} \qquad (6.16)$$

$$V_0 = V_i \frac{1}{3} \qquad (6.17)$$

$$e_v = \frac{\left(\frac{1000}{2000} - \frac{1}{3}\right)}{1/2} \cdot 100 \qquad (6.18)$$

$$= \frac{1}{3} \cdot 100 \qquad (6.19)$$

$$= 33.33 \qquad (6.20)$$

Therefore, in order to minimize the effect of electrical loading errors due to the circuits used for the measurement, the measurement device should have a large input resistance (input impedance). The larger the input impedance, the smaller the electrical loading error in the voltage measurement.

6.3 WHEATSTONE BRIDGE CIRCUIT

A Wheatstone bridge circuit is used to convert the change in resistance into voltage output. It is a standard circuit used as part of sensor signal conditioners (Figure 6.8). A Wheatstone bridge has

1. a power supply voltage, V_i, and
2. four resistances arranged in a bridge circuit, R_1, R_2, R_3, R_4.

One of the resistance branches is the resistance of the sensor. The sensor resistance changes as a function of the measured variable, i.e. RTD resistance as a function of temperature or strain-gauge resistance as a function of strain.

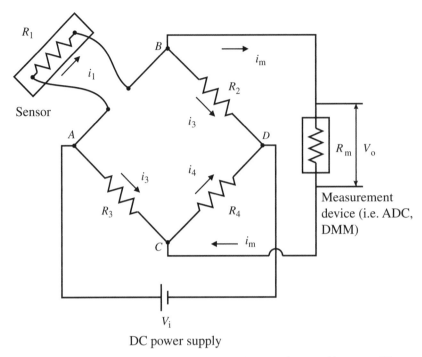

FIGURE 6.8: Wheatstone bridge circuit, which is very often used in many different types of sensor applications as part of the signal conditioner and amplifier circuit of the sensor, to convert variation in the resistor to voltage.

Let us consider the case where the bridge is balanced and the voltage differential between points B and C is zero ($V_{BC} = 0$) and that the current passing through the measurement device (i.e., galvonometer, digital voltmeter, or ADC circuit for a data acquisition system) is zero ($i_m = 0$). Since $V_{BC} = 0$,

$$i_1 R_1 - i_3 R_3 = 0 \qquad (6.21)$$
$$i_2 R_2 - i_4 R_4 = 0 \qquad (6.22)$$

and $i_m = 0$,

$$i_1 = i_2 \qquad (6.23)$$
$$i_3 = i_4 \qquad (6.24)$$

Solving these equations, the following relations must hold for the bridge to be balanced ($V_{BC} = 0$ and $i_m = 0$),

$$\frac{R_1}{R_2} = \frac{R_3}{R_4} \qquad (6.25)$$

Now, let us examine two different methods for using the Wheatstone bridge circuit in order to measure the variation in resistance in the form of proportional output voltage.

6.3.1 Null Method

R_1 represents the sensor whose resistance changes as a function of the measured variable. R_2 is a calibrated adjustable resistor (either manually or automatically adjusted). R_3 and R_4 are resistances fixed and known for the circuit. Initially the circuit is calibrated so that the balanced bridge condition is maintained and no voltage is observed across points B and C. Assume that the sensor resistance R_1 changes as a function of the measured variable. Then, we can adjust R_2 in order to maintain the balanced bridge condition, $V_{BC} = 0$,

$$\frac{R_1}{R_2} = \frac{R_3}{R_4} \tag{6.26}$$

Then, when the $V_{BC} = 0$, the resistance of the sensor (R_1) can be determined since R_3, R_4 are fixed and known, and R_2 can be read from the adjustable resistor,

$$R_1(x) = \frac{R_3}{R_4} \cdot R_2(adjusted) \tag{6.27}$$

Notice that this method of determining the sensor resistance is not sensitive to the changes in the supply voltage, V_i. But it is suitable only for measuring steady-state or slowly varying resistance changes, hence slowly varying changes in the measured variable (i.e., temperature, strain, pressure). Once $R_1(x)$ is known, from the transduction relationship of the sensor, the measured variable x can be determined (i.e., from the input–output graph or map of the transduction principle).

6.3.2 Deflection Method

In order to measure time varying and transient signals, the deflection method should be used. In this case, three legs of the resistor bridge have fixed resistances, R_2, R_3, R_4. The R_1 is the resistance of the sensor. As the sensor resistance changes, non-zero output voltage V_{BC} is measured. For our first derivation, let us assume that the voltage measurement device has infinite input resistance, $R_m \rightarrow \infty$, so that no current flows through it despite a finite voltage potential across points B and C, $V_{BC} \neq 0, i_m = 0$. For the following derivation, let $V_o = -V_{BC}$ for output voltage polarity;

$$i_1 R_1 + V_{BC} - i_3 R_3 = 0 \tag{6.28}$$
$$V_{BC} = i_3 R_3 - i_1 R_1 \tag{6.29}$$
$$V_o = -i_3 R_3 + i_1 R_1 \tag{6.30}$$

Since $i_m = 0$, $i_1 = i_2$ and $i_3 = i_4$,

$$i_1 = V_i/(R_1 + R_2) \tag{6.31}$$
$$i_3 = V_i/(R_3 + R_4) \tag{6.32}$$

Then,

$$V_o = V_i \left(\frac{R_1}{R_1 + R_2} - \frac{R_3}{R_3 + R_4} \right) \tag{6.33}$$

In most Wheatstone bridge circuit applications with sensors, the bridge is balanced at a reference condition, $V_o = 0$ and the initial values of the resistance arms are the same, $R_1 = R_2 = R_3 = R_4 = R_o$. Let

$$R_1 = R_o + \Delta R \tag{6.34}$$

where the ΔR is the variation from the calibrated nominal resistance. If we substitute these relations, it can be shown that

$$V_o/V_i = \frac{\Delta R/R_o}{4 + 2\Delta R/R_o} \tag{6.35}$$

In general, $\Delta R/R_o \ll 1$, and the above relation can be approximated,

$$V_o/V_i = \frac{\Delta R/R_o}{4} \tag{6.36}$$

Another convenient way of expressing this relationship is

$$V_o = \frac{V_i}{4R_o} \Delta R \tag{6.37}$$

In Equation 6.20, if $R_1 = R_3 = R_4 = R_0$ is constant, but $R_2 = R_0 + \Delta R$, the same output voltage and resistance variation relationship holds except the polarity of the output voltage would be opposite (negative). Similarly, in Equation 6.20, if $R_1 = R_2 = R_4 = R_0$ is constant, but $R_3 = R_0 + \Delta R$, the same output voltage and resistance variation relationship holds except the polarity of the output voltage would be opposite (negative). If $R_1 = R_2 = R_3 = R_0$ is constant, but $R_4 = R_0 + \Delta R$, the same output voltage and resistance variation relationship holds except the polarity of the output voltage would be positive. This can be easily shown by making these substitutions in Equation (6.33).

Notice that if the sensor signal conditioner has an ADC converter and an embedded digital processor, the above approximation is not necessary. The more complicated relationship can be used in the software for a more accurate estimation of the measured variable using Equation (6.33 or 6.35).

Now, let us relax the assumption on the measurement device resistance. Let us assume that it is not infinite, but a large finite value. The result is same as the electrical loading error condition. If the R_m is much larger than R_0, that is 10^3 times, the error introduced to the measurement due to non-infinite R_m is negligible.

Example Consider that a RTD type temperature sensor is used to measure the temperature of a location. The two terminals of the sensor are connected to the R_1 position of a Wheatstone bridge circuit. The sensor temperature–resistance relationship is as follows

$$R = R_0[1 + \alpha(T - T_o)] \tag{6.38}$$

where from the sensor calibration data it is known that $\alpha = 0.004\,°C^{-1}$, $T_o = 0\,°C$ reference temperature, and $R_o = 200\,\Omega$ at temperature T_o. Assume that $V_i = 10\,VDC$, and $R_2 = R_3 = R_4 = 200\,\Omega$. What is the temperature when the $V_o = 0.5\,VDC$?

Let us assume that the input resistance of the voltage measurement device is infinity,

$$V_o = V_i \cdot \frac{\Delta R/R_o}{4} \tag{6.39}$$

Find ΔR, then $R = R_o + \Delta R$, calculate T from

$$R = R_0(1 + \alpha(T - T_o)) \tag{6.40}$$

The resulting numbers are $\Delta R = 40\,\Omega$, $T = 50\,°C$.

Let us consider that the input resistance of the output voltage measuring device is $R_m = 1\,M\Omega$, instead of infinite. The resulting measurement would indicate the following temperature,

$$V_o = V_i \cdot \frac{\Delta R/R_o}{4(1 + R_o/R_m)} \tag{6.41}$$

which gives

$$\Delta R = 40 \cdot 1.0002\,\Omega \tag{6.42}$$

and the more accurate tempareture measurement is

$$T = 40 \cdot 1.0002/0.8 \tag{6.43}$$
$$= 50 \cdot 1.0002 \tag{6.44}$$
$$= 50.01\,°\text{C} \tag{6.45}$$

Note that if the input resistance of the measurement device is small compared to the nominal resistance in the Wheatstone bridge, the measurement error will be much larger. For instance, if $R_\text{m} = 1000\,\Omega$, the resulting temperature measurement would be

$$T = 50 \cdot 1.2 = 60\,°\text{C} \tag{6.46}$$

which has a 20% error in measurement due to the low input impedance of the measurement device relative to the impedance of the circuit.

6.4 POSITION SENSORS

There are two kinds of *length* measurements of interest: (i) absolute position (the distance between two points), (ii) incremental position (the change in the position). If a sensor can measure the position of an object on power-up relative to a reference (the distance of the object from a reference point on power-up), we call it an *absolute position sensor*. If the sensor cannot tell the distance of the object from a reference on power-up, but can keep track of the change in position from that point on, we call it an *incremental position sensor*. Examples of absolute position sensors include a calibrated potentiometer, absolute optical encoder, linear variable differential transformer, resolver, and capacitive gap sensor. Examples of incremental position sensors include incremental optical encoder and laser interferometer. Most of the position sensors have rotary and translational (linear) position sensor versions.

6.4.1 Potentiometer

A potentiometer relates the absolute position (linear or rotary) into the resistance (Figures 6.9 and 6.10). The resistance change is converted to a proportional voltage change in the electrical circuit portion of the sensor. Hence, the relationship between the measured

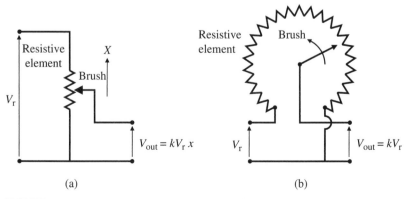

(a)   (b)

FIGURE 6.9: Linear and a rotary potentiometer for position measurement.

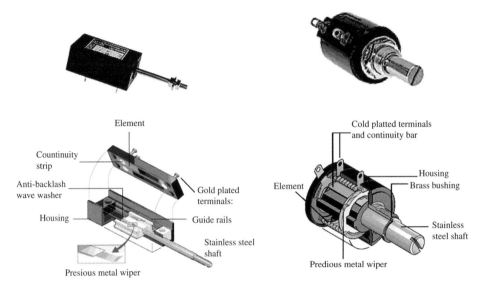

FIGURE 6.10: Pictures of linear and a rotary potentiometers.

physical variable, translational displacement x or rotary displacement θ, and the output voltage for an ideal potentiometer is

$$V_{\text{out}} = k \cdot V_{\text{r}} \cdot x \qquad (6.47)$$

or

$$V_{\text{out}} = k \cdot V_{\text{r}} \cdot \theta \qquad (6.48)$$

The sensitivity, $k \cdot V_{\text{r}}$ of the potentiometer in Equations 6.47 and 6.48 is a function of the winding resistance and the physical shape of the winding. The range and resolution of the potentiometer are designed into the sensor as a balanced compromise: the higher the resolution, the smaller the range of the potentiometer. Due to the brush-resistor contact, the accuracy is limited. As the contact arm moves over the resistor winding, the output voltage changes in small discrete steps, which define the resolution of the potentiometer. For very long length measurements where the distance may not be a straight line, that is 5.0 m curve, a spring loaded multi turn rotary potentiometer arm is connected to a string. Then the string moves with the measured curve distance. As the string is pulled, the potentiometer arm moves around the multi turn resistor. The output voltage is then proportional to the length of the pulled string. The total distance measured can be any shape. Potentiometers are considered low cost, low accuracy, limited range, simple, reliable, absolute position sensors. The typical resistance of the potentiometer is around 1 KΩ per inch. Since there is a supply voltage, there will be a finite power dissipation over the potentiometer. However, it is a small amount of power and less than 1 W/in.

6.4.2 LVDT, Resolver, and Syncro

The linear variable differential transformer (LVDT), resolver, and syncro are sensors which operate based on the *transformer principle* of electromagnetism. The key to their operating principle is that the change in the position of the rotor element changes the electromagnetic coupling (magnetic flux linkage) between the two windings (Figure 6.11, and Figure 6.12), primary and secondary windings. One of these windings (typically the primary winding) is excited with an external AC voltage. As a result, the induced voltage between the two

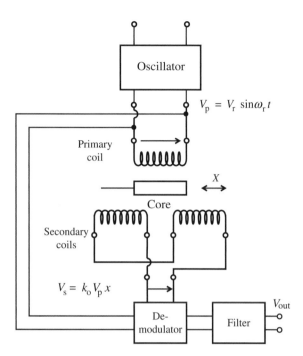

FIGURE 6.11: Linear variable differential transformer (LVDT) and its operating principles. An oscillator circuit generates the excitation signal for the primary winding. The demodulator circuit removes the high frequency signal content and obtains the magnitude of the induced voltage, which is related to the core position.

windings changes in relation to the position. Hence, we have a well-defined relationship between the induced voltage and the position. In LVDTs, both windings are stationary, and a rotor core made of a material with high magnetic permeability couples the two windings electromagnetically. In rotary LVDTs (resolvers and syncros), the primary winding is located on the rotor, and the secondary winding on the stator. Either the rotor winding or the stator windings can be excited externally by a known voltage, and the induced voltage on the other winding is measured which is related to the position. The operating principles of a LVDT, resolver, and syncro are shown in Figures 6.11–6.16. Notice that the syncro is just a three-phase stator version of the resolver.

The LVDT is an absolute position sensor. On power-up, the sensor can tell the position of the magnetic core relative to the neutral position. The LVDT's primary winding is excited by a sinusoidal voltage signal. The induced voltage on the secondary windings has the same frequency except that the magnitude of the voltage is a function of the position of the magnetic core. In other words, the displacement modulates the magnitude of the induced voltage. As the core displacement increases from the center, the magnitude of the voltage differential between the two stator windings increases. The core material must have a large magnetic permeability compared to air, such as iron–nickel alloy. A non-magnetic stainless

FIGURE 6.12: Pictures of LVDTs (left) and a resolver (right).

steel rod is used to connect the core to the part whose displacement is to be measured. The sign (direction) of the magnitude of the voltage differential is determined by relating the induced voltage phase to the reference voltage phase. It is a function of the direction of the magnetic core displacement from neutral position. The primary winding is excited by

$$V_p(t) = V_r \cdot \sin(w_r t) \tag{6.49}$$

and the voltage differential between the secondary windings is

$$V_s(t) = k_0 \cdot V_p(t) \cdot x(t) \tag{6.50}$$
$$= k_0 \cdot V_r \cdot \sin(w_r t) \cdot x(t) \tag{6.51}$$

Once the the $V_s(t)$ is demodulated in frequency, the output signal is presented as a DC voltage,

$$V_{out}(t) = k_1 \cdot x(t) \tag{6.52}$$

which is proportional to the core displacement.

LVDTs can be used for high resolution position measurement (i.e., $1/10\,000$ in resolution) but with a relatively small range (up to 10 in range). Excitation frequency of the primary coil is in the range of 50 Hz to 25 kHz. The bandwidth of the sensor is about $1/10$ of the excitation frequency (Figure 6.12).

The resolver (Figure 6.13) and syncro (Figure 6.16) sensors operate on the same principle as the LVDT. Resolvers compete with encoders in position measurement applications. In general, resolvers have better mechanical ruggedness, but lower bandwidth than the encoders. Resolvers have two stator windings (90 degrees out of phase), whereas syncros have three stator windings (120 degrees out of phase).

Let us examine the operation of the resolver where we will consider that the rotor winding is excited externally by a sinusoidal voltage signal. The induced voltage in the stator windings is measured. The magnitude of induced voltages is a function of the angular position of the rotor. The induced voltages at the two stator windings are 90 degrees out of phase with each other since they are mechanically placed at a 90 degree phase angle. Therefore, the resolver is an absolute rotary position sensor with one revolution range. The

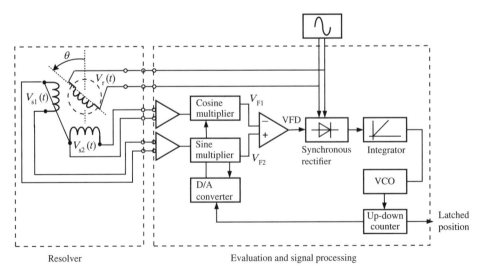

FIGURE 6.13: Resolver and its operating principle: resolver sensor head and the signal processing circuit (also called the resolver to digital converter, RTDC).

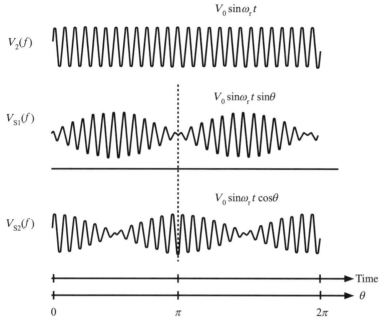

FIGURE 6.14: Excitation voltage for the rotor winding and the induced stator voltages. The induced voltages are shown as a function of time for a constant rotor speed over a period of one revolution.

angular position changes beyond one revolution are kept track of through digital counting circuits. The rotor excitation voltage is (generated by an oscillator circuit),

$$V_r(t) = V_{ref} \cdot \sin(w_r t) \tag{6.53}$$

The induced voltage on the stator windings are (Figure 6.14)

$$V_{s1}(t) = k_0 \cdot V_r(t) \cdot \sin(\theta(t)) = k_0 \cdot V_{ref} \cdot \sin(w_r t) \cdot \sin(\theta(t)) \tag{6.54}$$

$$V_{s2}(t) = k_0 \cdot V_r(t) \cdot \cos(\theta(t)) = k_0 \cdot V_{ref} \cdot \sin(w_r t) \cdot \cos(\theta(t)) \tag{6.55}$$

Next, the V_{s1}, V_{s2} are multiplied by sin and cos functions in the RTDC circuit,

$$V_{f1}(t) = k_0 \cdot V_r(t) \cdot \sin(\theta(t)) \cdot \cos(\alpha(t)) \tag{6.56}$$

$$V_{f2}(t) = k_0 \cdot V_r(t) \cdot \cos(\theta(t)) \cdot \sin(\alpha(t)) \tag{6.57}$$

Then the output of the error amplifier is

$$\Delta V_f(t) = k_0 \cdot V_{ref} \cdot \sin(w_r t) \cdot \sin(\theta(t) - \alpha(t)) \tag{6.58}$$

Then, the signal is demodulated to remove the $\sin(w_r t)$ component by a rectifier and low pass filter. Then, the signal is fed to an integrator which provides input to the voltage controlled oscillator (VCO). The output of the VCO is fed to an up-down counter to keep track of the position. This results in making the up-down counter increase or decrease α in a direction so that $\sin(\theta - \alpha)$ approaches zero. The value of the up-down counter is converted to angle α by a DAC as an analog signal to feed the sin and cos multiplication circuit. The iteration on α continues until $\alpha = \theta$, at which time the rotor position information is latched from the up-down counter as digital data. The algorithm finds the angle θ by iteratively changing α. The iteration stops when ΔV_f is equal to zero, at which time it means that

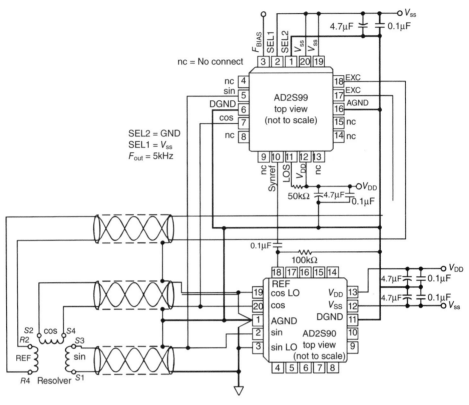

FIGURE 6.15: An example circuit for resolver signal processing. It can be used for a resolver which has one rotor and two stator windings. AD2S99 chip provides the reference oscillator signal, and an AD2S90 chip processes the resolver signals for conversion to digital data.

$\theta = \alpha$. Functionally, the end result is that the output position value presented is proportional to the rotor angle

$$V_{out}(t) = k_1 \cdot \theta(t) \tag{6.59}$$

This circuit can be implemented with integrated circuits AD2S99 programmable oscillator chip and AD2S90 resolver to digital converter (RTDC) chip by analog devices (Figure 6.15). The AD2S99 chip is a programmable sinusoidal oscillator. The input supply voltage is ± 5 VDC. It provides the sinusoidal excitation voltage for the primary winding. The excitation frequency is programmable to be one 2 kHz, 5 kHz, 10 kHz, or 20 kHz. It accepts the induced voltage signals from the secondary winding pair at the sin and cos pins. In addition, it provides a synchronous square wave reference output signal that is phase locked with the sin and cos signal. This signal is used by the AD2S90 chip for converting the resolver signal to a digital signal. AD2S90 is the main chip that performs the conversion of the resolver analog signals to digital form. It provides the resulting digital position signal in two different format: a 12-bit serial digital code and an incremental encoder equivalent A and B channel signal. In motor control applications, many amplifiers accept the resolver signal and use it for current commutation. In addition, they output an encoder equivalent (encoder emulation) signal which can be used by a closed loop position controller. It emulates a 1024-line per revolution incremental encoder A and B channels. When the A and B channel signals are decoded with ×4, it results in 4096 counts/rev resolution which is equivalent to a 12-bit resolution.

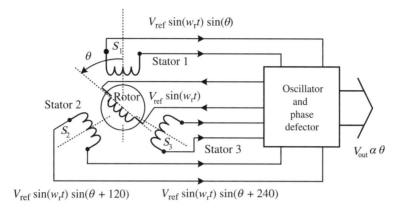

$V_{\text{ref}} \sin(w_r t) \sin(\theta)$

Stator 1

θ

S_1

Rotor

Stator 2

$V_{\text{ref}} \sin(w_r t)$

S_2

S_3

Stator 3

Oscillator
and
phase
defector

$V_{\text{out}} \, \alpha \, \theta$

$V_{\text{ref}} \sin(w_r t) \sin(\theta + 120)$ $V_{\text{ref}} \sin(w_r t) \sin(\theta + 240)$

FIGURE 6.16: Syncro and its operating principle.

There are other types of signal processing circuits to extract the θ angle information from the resolver phase voltages. For instance, both stator voltages can be sampled at the same frequency ($w_r/(2\pi)$ Hz) where the start of the sampling is synchronized to the $V_r(t)$ by a 90 degree phase angle in order to sample the maximum magnitudes. In other words, by sampling V_{s1}, V_{s2} at the same frequency as the w_r, we achieve demodulation by sampling. The sampled signals are

$$V_{s1}^{\text{adc}} = V_{\text{mag}} \cdot \sin(\theta) \tag{6.60}$$

$$V_{s2}^{\text{adc}} = V_{\text{mag}} \cdot \cos(\theta) \tag{6.61}$$

where V_{mag} is the sampled value of the $k_0 \cdot V_{\text{ref}} \cdot \sin(w_r t)$ portion of the signal. Then, we can compute the arctan of the two signals to obtain the angle information,

$$\theta = \arctan\left(\frac{V_{s1}^{\text{adc}}}{V_{s2}^{\text{adc}}}\right) \tag{6.62}$$

This method uses two channels of an A/D converter and a digital computational algorithm for the arctan($\cdot$) function. Such a circuit for an RTDC (resolver to digital converter) can be implemented using ADMC401 chip (AD converter) and a AD2S99 oscillator chip (both by analog devices).

The operating principle of the syncro is almost identical to that resolved. The only difference is that there are three stator phases with 120 degrees of mechanical phase angle. Hence, the induced voltages in the three stator phases will be 120 degrees apart,

$$V_{s1} = k_0 \cdot V_{\text{ref}} \cdot \sin(w_r t) \cdot \sin(\theta) \tag{6.63}$$

$$V_{s2} = k_0 \cdot V_{\text{ref}} \cdot \sin(w_r t) \cdot \sin(\theta + 120) \tag{6.64}$$

$$V_{s3} = k_0 \cdot V_{\text{ref}} \cdot \sin(w_r t) \cdot \sin(\theta + 240) \tag{6.65}$$

and these signals can be processed to extract the angular position information with circuits similar to the ones used for resolvers.

Commercial LVDT and resolver sensors are packaged such that the input and output voltages to the sensor are DC voltages. The input circuit includes a modulator to generate an AC excitation signal from the DC input. The output circuit includes a demodulator which generates a DC output voltage from the AC voltage output of the secondary windings. For instance, LVDT Model -240 (by Trans-Tek Inc.) has a position measurement range in 0.05–3.0 in, 24 VDC input voltage, 300 Hz bandwidth, and an internally generated carrier frequency of 13 kHz.

FIGURE 6.17: Pictures of a rotary (a) and a linear encoder (b).

6.4.3 Encoders

There are two main groups of encoders: absolute encoders and incremental encoders. Absolute encoders can measure the position of an object relative to a reference position at any time. The output signal of the absolute encoder presents the absolute position in a digital code format. An incremental encoder can measure the change in position, not the absolute position. Therefore, the incremental encoder cannot tell the position relative to a known reference. If absolute position information is needed from incremental encoder measurement, the device must perform a so called "home-ing" motion sequence in order to establish its reference position after the power up. From that point on, the absolute position can be kept track of by digital counting. An absolute encoder does not need a counting circuit if the total position change is within the range covered by the absolute encoder.

Encoders can also be classified based on the type of position they measure: translational or rotary. Figure 6.17 shows pictures of linear (translational) and rotary encoders. Encoders are the most widely used position sensors in electric motor control applications such as brush-type DC motors, brushless DC motors, stepper motors, and induction motors. It is estimated that over 70% of all motor control applications with position sensor use encoders as position sensors.

The operating principles of linear and rotary encoders are identical. In one case, there is a rotary disk, in the other case there is a linear scale. Figure 6.18 shows the components of a rotary and a linear incremental and absolute encoder. The main difference between the rotary and linear encoders is the glass scale with the printed pattern to interrupt the light as it moves.

An encoder has the following components (Figures 6.19, 6.20):

1. a disk or linear scale with light and dark patterns printed on it,
2. a light source (LED possibly with a focusing lens),
3. two or more photodetectors,
4. a stationary mask.

In the case of an incremental encoder, the disk pattern is a uniform black and opaque printed pattern around the disk (Figure 6.19). As the disk rotates and the angle changes, the disk pattern interrupts and passes the light. If the disk is metal (which may be necessary in applications where the environmental conditions are extreme in terms of vibrations, shock forces, and temperature), the same principle is used as reflected light instead of pass-through light in counting the number of incremental position changes. The photodetector output turns ON and OFF everytime the disk pattern passes over the LED light (Figure 6.19). Therefore, the change in angular position can be measured by counting the number of state changes of the photodetector output. Let us assume that there are 1000 lines over the

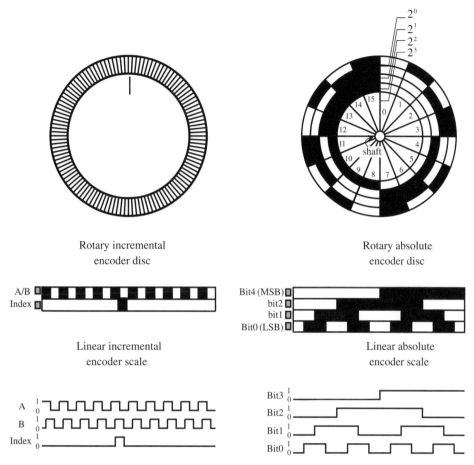

A/B
Index

Bit4 (MSB)
bit2
bit1
Bit0 (LSB)

Rotary incremental
encoder disc

Rotary absolute
encoder disc

Linear incremental
encoder scale

Linear absolute
encoder scale

A
B
Index

Bit3
Bit2
Bit1
Bit0

FIGURE 6.18: Disks of incremental (rotary and linear) and absolute encoders and the typical encoder output signals.

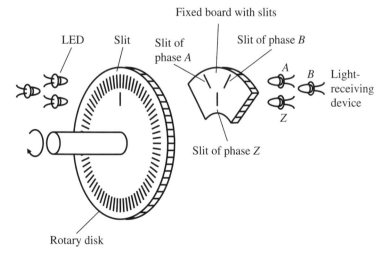

FIGURE 6.19: Components and operating principle of a rotary incremental encoder: rotary disk with slits, LEDs, phototransistors (light receiving devices) mask.

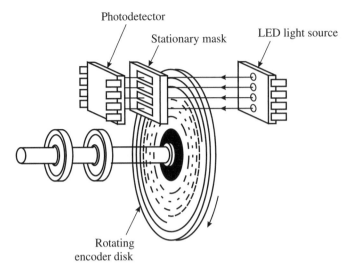

Photodetector

Stationary mask

LED light source

Rotating encoder disk

FIGURE 6.20: Components and operating principle of a rotary absolute encoder: rotary disk with absolute position coding in gray scale, LED set, phototransistor set, mask.

disk. Therefore, the photodetector output will change state 1000 times per revolution of the disk. Each pulse of the photodetector means $360°/1000°$ of angular position change in the shaft. If only one photodetector is used, the change in position can be detected, but not the direction of the change. By using two photodetectors (usually called A and B channels of the encoder output), which are displaced from each other by 1/2 of the size of a single grading on the disk (or an integer number plus 1/2 of the size of a single grading), the direction of the motion can be determined. If the disk direction of rotation changes, the phase between the A and B channels changes from $+90°$ to $-90°$. In addition to the dual photodetectors A and B, an incremental encoder also has a third photodetector which turns ON (or OFF) for one pulse period per revolution. This is accomplished by a single slit on the disk. Using this channel (usually called the C or Z channel), the absolute position of the disk can be established by a home motion sequence. On power-up, the angular position of the shaft relative to a zero reference position is unknown. The encoder can be rotated until the C channel is turned on. This position can be used as the zero reference position for keeping track of the absolute position.

Finally, each output channel of the encoder (A,B,C) may have a complementary channel ($\bar{A}, \bar{B}, \bar{C}$) which is used as protection against noise. Figure 6.21 illustrates how the complementary encoder output channels can be used to eliminate position measurement errors due to noise. Notice that for this approach to work, the same noise signal is assumed to be present on each channel. In short, the complementary channels improve the noise immunity of the encoder, but this is not a solution for all possible noise conditions.

The linear incremental encoder works in the same principle, except that instead of a rotary disk, it has a linear scale. Furthermore, the linear scale is stationary and the light assembly moves.

An absolute encoder differs from the incremental encoder in the printed pattern on the disk and the light assembly (LEDs and photodetectors). Figure 6.20 shows the components of an absolute encoder. Notice the coded pattern on the disk. Each discrete position of the disk corresponds to a unique state of photo detectors. Therefore, at any given time (i.e., after power-up), the absolute position on the shaft can be determined uniquely. The encoder can tell us the absolute position within one revolution. Multi turn absolute

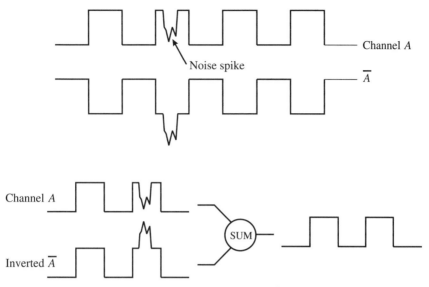

FIGURE 6.21: Use of complementary channel signals $\bar{A}$, $\bar{B}$ in conjuction with A and B channels to cancel noise, hence increasing the noise immunity of the sensor.

encoders are also available. The resolution of the absolute encoder is determined by the number of photodetectors. Each photodetector output represents a bit on the digitally coded position information. If the absolute encoder has 8 photodetectors (8-bit), the smallest position change that can be detected is $360°/(2^8) = 360/256°$. If the encoder is 12-bit, the resolution is $360°/(2^{12}) = 360/4096°$. For each position, the absolute encoder outputs a *unique code*. The coding of an absolute encoder is not necessarily a binary code. Gray codes are known to have better noise immunity and are less likely to provide wrong reading compared with binary code.

An encoder performance is specified by:

1. Resolution: number of counts per revolution. This is the number of lines on the disk or linear scale for incremental encoder. For an absolute encoder with N-bit resolution (N set of LED and photodetector pairs), there are 2^N counts per revoution.

2. Maximum speed: the maximum speed of the encoder can be limited by electrical and mechanical reasons. The maximum state change capacity of the photodetectors determines the maximum speed limit set by electrical capability of the encoder. The frequency output of the encoder (resolution times the maximum speed) must be below that value. The maximum speed is also limited by the mechanical bearings of the encoder.

3. Encoder output channels available: A, B, C, $\bar{A}, \bar{B}, \bar{C}$ for incremental encoders.

4. Electrical output signal type: TTL, open collector or differential line driver type. The differential line driver type is used for long cables and noisy environments.

5. Mechanical limits: maximum radial and axial loads, sealing for dusk and fluid, vibration and temperature limits.

Count Multiplication and Interpolation   By detecting the phase between A and B channels of an incremental encoder, the number of counts per line can be effectively multiplied by 1, 2, or 4. Since A and B channels are 90° out of phase, with one cycle of

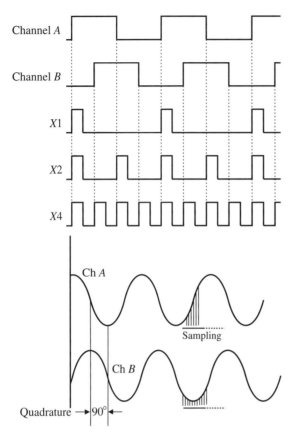

FIGURE 6.22: Digital processing of incremental encoder signal channels A and B in order to increase resolution: quadrature decoding to increase resolution by x4, and digital oversampling of sinusoidal output signal from the encoder.

line change, we can actually obtain up to four counts by counting the transition in each channels (Figure 6.22). The quadrature count multiplication (×4) is standard in industry. For a quadrature incremental encoder the number of counts per revolution is four times the number of lines per revolution.

If further position resolution improvement is needed beyond the ×4 quadrature decoding of what is physically coded to disk, then sinusiodal output photodetectors must be used. The output of the so called sinusodial encoders are not discrete ON/OFF levels, but sinusoidal signals as a function of the angular position within one cycle of disk lines (Figure 6.22). Typical signal voltage level is about 1.0 VDC peak to peak. By sampling the level of the signal finely, the effective position resolution of the encoder can be increased, that is by ×1024 using a 10-bit sampling circuit. This type of interpolation requires a specifically designed interpolation circuit. Notice that the repeatability of the mechanical dimension of the printed pattern on the disk is ultimately the smallest resolution that can be achieved. The electronic interpolation improves the resolution by predicting the intermediate positions through oversampling of a sinusoidal signal. The interpolation based resolution improvement can be easily lost by noise in the sinusoidal output signal. An example of sinusoidal output linear incremental encoder is Series LR/LS (by Dynapar Corp). The sinusoidal output of the linear encoder can be sampled and interpolated to 250 nm resolution using a 10-bit sampling and interpolation circuit.

Example Consider an incremental encoder with 2500 lines/rev resolution and a decoder which uses ×4 decoding logic. Then the resolution of the encoder is

10 000 counts/rev. Assume that the photodetectors in the decoder circuit can handle A and B channel signals up to 1 MHz frequency.

1. Determine the maximum speed the encoder and decoder circuit can handle.
2. If the encoder resolution was 25 000 lines/rev and the same decoder was used, what is the maximum speed that could be handled by the decoder circuit?

Since the decoder can handle 1 MHz A and B channel signals, the maximum speed that can be measured without data over run is

$$w_{max} = \frac{1 \cdot 10^6 \, \text{pulse/s}}{2500 \, \text{pulse/rev}} = 400 \, \text{rev/s} = 24\,000 \, \text{rpm} \tag{6.66}$$

The angular position measurement resolution is 2500 lines/rev $\times$ 4 = 10 000 count/rev, then 360/10 000 degrees/count, which is the smallest change in angular position that can be detected.

If the encoder resolution is increased by a factor of 10, then the position measurement resolution is increased by a factor of 10, 360/100 000 degrees/count. However, the maximum speed the decoder can handle is reduced by the same factor,

$$w_{max} = \frac{1 \cdot 10^6 \, \text{pulse/s}}{25\,000 \, \text{pulse/rev}} = 40 \, \text{rev/s} = 2400 \, \text{rpm} \tag{6.67}$$

While the position measurement resolution is improved by a factor of 10, from 360/10 000 degrees/count to 360/100 000 degrees/count, maximum speed capability is reduced also by the same factor of 10, from 24 000 rpm to 2400 rpm.

Therefore, if we need to increase the position measurement resolution by increasing encoder resolution, we reduce the maximum speed measurement capacity if we use the same decoder circuit. If we want to increase position measurement resolution without losing the maximum speed measurement capacity, the maximum frequency capacity of the decoder circuit must be increased by the same factor. Decoder circuits which can handle encoder signal frequency of up to 2 MHz are common. Decoders up to 50 MHz input frequency capacity are also available.

6.4.4 Hall Effect Sensors

The Hall effect (named after Edward Hall, 1879) is the phenomenon that semiconductor and conductor materials develop an induced voltage potential when the material is in a magnetic field and has a current passing through it. The relationship between the induced voltage, the current, and the magnetic field strength is a vector relationship (Figure 6.23). When a sheet of semiconductor (i.e., gallium-arsenide, GaAs) or conductor material has a current passing through it, and is placed in a magnetic field, a voltage is induced in the

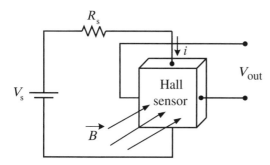

FIGURE 6.23: Principle of Hall effect and its usage in sensor design.

direction perpendicular to both the current and the magnetic field vector. If the direction of the magnetic field or the current changes, so does the direction of the induced voltage. The induced voltage magnitude is also function of the material type. It is very small for conductors, but large enough for semiconductors such that this effect is widely used in sensor designs. The sensor output voltage is function of the magnetic field strength (B), the current (i) and the material of the sensor,

$$V_{out} = V_{out}(B, i, material) \tag{6.68}$$

In a sensor application, the current (i) is fixed by the sensor power supply voltage (V_s) and resistor (R_s). A Hall effect sensor requires an external power supply for the current i and magnetic field density (B) for it to work. The material of the sensor is also fixed. Then, the output voltage varies as a function of the magnetic field strength for a given Hall sensor. In the sensor by design, the measured variable must be arranged such that it changes the magnetic field strength over the Hall sensor. For instance, the measured variable may be the position of a magnet relative to the sensor. Similarly, the magnet may be part of the sensor, and a ferrous metal object entering the field changes the magnetic permeability (magnetic conductivity) around the sensor, and hence changes the magnetic field density (B). As a result the output voltage is changed (i.e., Figure 6.24). This principle is also used in presence sensors which have only two state ON/OFF outputs. Typical voltage levels produced by the Hall transducer are at the millivolts level, hence must be amplified by an op-amp for processing in a measurement and control system. The op-amp may amplifiy the Hall effect sensor voltage linearly to measure analog distance or can convert it to an ON/OFF output signal for presence sensing. Since the sensing element of the Hall sensor can be a semiconductor, both the sensing element and digital signal processing circuit can be integrated into a single silicon chip. For instance AD22151 by Analog Devices is one such integrated Hall effect sensor. It operates from a 5 VDC power supply and can operate in an environment with a temperature range of −40 to +150°C. Such integrated

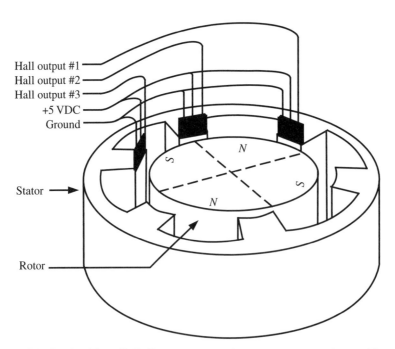

FIGURE 6.24: Three Hall effect sensors used to sense commutation positions in a brushless DC motor.

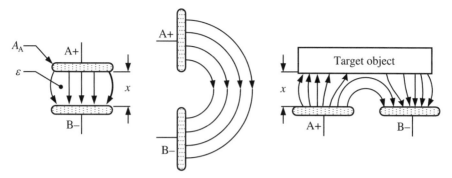

FIGURE 6.25: Operating principle of capacitive gap sensor: the presence of the target object changes the effective capacitance.

Hall effect sensors also include a built-in temperature sensor so that the variations in the gain of the sensing head as temperature varies can be compensated for. Figure 6.24 shows an application of Hall effect sensors in position sensing of the rotor of a brushless DC motor. Brushless DC motors need current commutation, that is the shaping of the desired current in each phase as a function of rotor position, for proper operation. In trapezoidal current commutated brushless DC motors, we need to know six ranges of rotor position in order to properly commutate the current. Three Hall effect sensors, each operating in ON/OFF modes are used to provide rotor position information for current commutation. The sensor heads and permanent magnets are arranged such that at any given position, one or two of the sensors are ON.

6.4.5 Capacitive Gap Sensors

A capacitive gap sensor measures the distance between the front face of the sensor and a target object. It is a non-contact distance sensor. The target material should have high relative permittivity[1]. Metals, plastics with carbon are good target materials for sensing. The capacitance between the sensor and the target material is related approximately (Figure 6.25),

$$C = \frac{\epsilon \cdot A}{x} \qquad (6.70)$$

where $\epsilon = \kappa \cdot \epsilon_0$, ϵ is the *permittivity of the medium*, κ is the *dielectric constant* of the medium separating the sensor and target material, ϵ_0 is the *permittivity of free space*, A is the area of the plates, x is the distance to be measured. Therefore, there is a well-defined relationship between the distance (gap, x) and the capacitance, C. The effective dielectric constant κ between the sensor head and target material is a function of the target material type (conductive, i.e., iron, aluminum or non-conductive non metal, i.e., plastic). Notice that since the net capacitance is a function of the target material, the effective sensing distance varies with different target material types. The excitation circuit for the sensor works to maintain a constant electric field magnitude between the sensor head and the target object. As the capacitance changes, the required current to do so changes proportionally.

---

[1] The force, F, between two charged particles, q_1 and q_2, which are separated from each other by distance r in free space, is

$$F = k_e \frac{q_1 \cdot q_2}{r^2} \qquad (6.69)$$

where $k_e = 1/(4\pi \cdot \epsilon_0) = 8.9875 \cdot 10^9 \; \mathrm{N \, m^2/C^2}$, is called the *Coulomb constant* and ϵ_0 is the *permittivity of free space*.

Hence, there is a relationship between the measurable current effort and capacitance. The excitation circuit which maintains the constant capacitance by controlling the current uses a high frequency modulation signal (i.e., 20 KHz modulation frequency).

The resolution of the capacitive gap sensor is typically in the micrometer range ($1 \ \mu m = 10^{-6} \ m$), but it can be made as low as a few nanometers ($1 \ nm == 10^{-9} \ m$) for special applications. The range is limited to about 10 mm. The frequency response of specific capacitive sensors can vary as a function of the sensed gap distance. The bandwidth of a given sensor varies as a function of the gap measured as a percentage of its total range. The smaller the gap distance measured relative to the sensor range, the higher the bandwidth of the sensor. For instance, a non-contact capacitive gap sensor can have 1000 Hz bandwidth for measuring gap distances within 10% of its range, while the same sensor may have about 100 Hz bandwidth while measuring gap distances around 80% of its range.

Capacitive presence sensors provide only two state ON/OFF output, and sense the change in the oscillator circuit signal amplitude. When a target object enters the field sensing distance of the sensor, the capacitance increases and the magnitude of oscillations increases. A detection and output circuit then controls the ON/OFF state of a transistor.

The capacitive gap sensor can also be used to sense the presence, density, and thickness of non-conducting objects. The non-conductor materials (such as epoxy, PVC, glass) which have a different dielectric constant than air can be detected since the presence of a such material in front of the probe instead of air results in a change in the capacitance.

6.4.6 Magnetostriction Position Sensors

Magnetostriction linear position sensors are widely used in hydraulic cylinders. Figure 6.26 shows the basic operating principle of the sensor. A permanent magnet moves with the object whose position is to be measured. The sensor head sends a current pulse along a wire which is housed inside a waveguide. The interaction between the two magnetic fields: the magnetic field of the permanent magnet and the electromagnetic field of the current pulse, produces a torsional strain pulse on the waveguide. The torsional strain pulse travels at about 9000 ft/s speed. The time it takes for the strain pulse to arrive at the sensor head is proportional to the distance of the permanent magnet from the sensor head. Therefore, by measuring the time period between the current pulse sent out and the strain pulse reflected back, the distance can be measured,

$$x = V \cdot \Delta t \tag{6.71}$$

where V is the travel speed of the torsional strain pulse which is known, Δt is measured, and hence the position, x can be determined as the measured distance.

Typical resolution is in the range of $2 \ \mu m$, and the range can be in the order of 10.0 m. The bandwidth of the sensor is typically in the 50–200 Hz range and is limited by the torsional strain pulse travel speed and the range (length) of the particular sensor.

Maximum possible sensor bandwidth, that is the maximum frequency the position signal is obtained, is determined by the maximum position the sensor can measure and the travel speed of the wave (about 9000 ft/s). For instance for a sensor with $x_{max} = 4 \ ft$ measurement range, the maximum bandwidth is

$$w_{max} = \frac{1}{\Delta t} = \frac{V}{x_{max}} \tag{6.72}$$

$$= \frac{9000 \ ft/s}{4 \ ft} \tag{6.73}$$

$$= 2.25 \ kHz \tag{6.74}$$

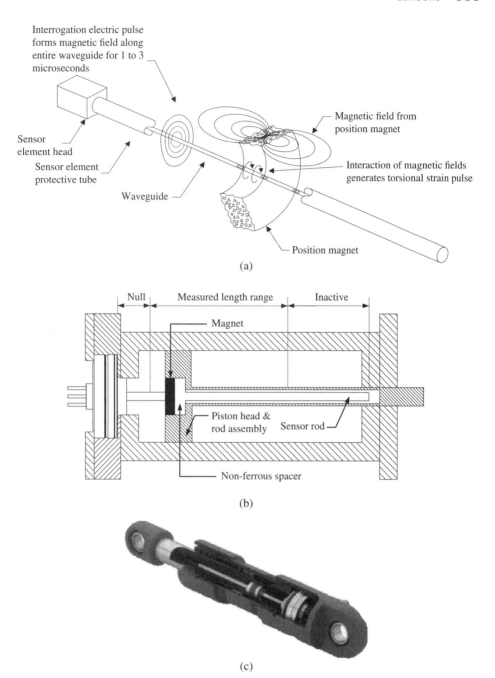

Interrogation electric pulse
forms magnetic field along
entire waveguide for 1 to 3
microseconds

Sensor
element head

Sensor element
protective tube

Waveguide

Magnetic field from
position magnet

Interaction of magnetic fields
generates torsional strain pulse

Position magnet

(a)

Null   Measured length range   Inactive

Magnet

Piston head &
rod assembly   Sensor rod

Non-ferrous spacer

(b)

(c)

FIGURE 6.26: Operating principle of magnetostriction absolute position sensor: an industry standard sensor used for cylinder position measurement in electro-hydraulic motion control systems. (a) Operating principle of the sensor; (b) in-cylinder mechanical placement of the sensor; (c) cut-away picture of a hydraulic cylinder with a magnetostriction sensor.

However, in practice averaging is used in the position signal processing. In other words, multiple wave signals are obtained in a row, such as ten samples in a row, and then the average value of that measurement is presented as one position measurement information. If averaging is used, the actual bandwidth will be smaller by a proportional amount. In this case, if five sample averaging is used, the sensor's maximum bandwidth would be $2.25\,\text{kHz}/5 = 450\,\text{Hz}$ or if ten sample averaging is used, the sensor's maximum bandwidth would be $2.25\,\text{kHz}/10 = 225\,\text{Hz}$. The magnetostrictive absolute position sensor is one of the most rugged position sensing technologies currently available and is widely used in harsh environment applications, especially with hydraulic cylinders.

6.4.7 Sonic Distance Sensors

Sound is transmitted through the propagation of pressure in the air. The speed of sound in the air is nominally 331 m/s at 0 °C dry air (343 m/s at 20 °C dry air). Two of the important characteristics of sound waves are *frequency* and *intensity* (amplitude). The human ear can hear the frequencies in the range of 20 Hz to 20 kHz. Frequencies above this range are called ultrasonic frequencies.

Sonic distance sensors measure the distance of an object by measuring the time period between the sent ultrasonic pulse and the echoed pulse (Figure 6.27). The sensor head sends ultrasonic sound pulses at high frequency (i.e., 200 kHz), and measures the time period between the sent pulses and the echoed pulses. For instance, it may send a short pulse of 200 kHz frequency and fixed intensity every 10 ms, where the time period of the sent pulse is only a few milliseconds. Before a new pulse is sent out, the time instant of the reflected sound pulse is measured. This process continues periodically, every 10 ms in this example. Knowing the speed of sound, a digital signal processor embedded to the sensor can calculate the distance of the object,

$$x = V_{\text{sound}} \cdot \Delta t \qquad (6.75)$$

where V_{sound} is known, Δt is measured, x is calculated. The range of the sensor can be up to 20 m with a minimum range of 5 cm. Notice that the sonic distance sensor is not appropriate for very short distance measurements, that is under 5 cm. The frequency response of the sensor (distance measurement update rate) varies with the distance measured, in general it can be around 100 Hz. Clearly, sonic sensors cannot be used in high bandwidth servo positioning applications which require a 1.0 ms or faster position loop update period. Typical applications of the sonic distance sensor include the roll diameter measurement, web loop length measurement, liquid level measurement, and box presence measurement on conveyors (Figure 6.28).

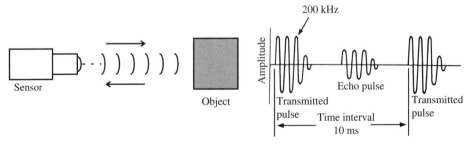

FIGURE 6.27: Operating principle of a sonic distance sensor.

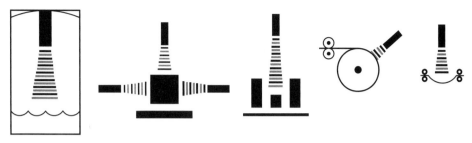

FIGURE 6.28: Applications of of a sonic distance sensor in various manufacturing processes: (a) liquid level sensing, (b) two-dimension sensing (i.e., box height and width), (c) one-dimension sensing (i.e., box height), (d) roll diameter sensing, (e) web slack sensing.

6.4.8 Photoelectic Distance and Presence Sensors

Sensors which measure the distance or presence of an object from a reference point using "light" as the transduction mechanism are collectively called "photoelectric sensors" or "light sensors." The light sensors which provide only two discrete outputs (ON/OFF) are called *photoelectric presence sensors* (Figure 6.29). The sensor emitter sends a light beam. The receiver (phototransistor) turns its output either ON or OFF depending on the received light. The light threshold that separates the ON and OFF states is adjustable by the op-amps in the electronic circuit of the sensor. The light sensors are immune from the ambient light variations due to the modulated nature of the emitted light. The light emitting diodes (LED) are tuned to emit light at high frequency. Figure 6.30 shows the frequency spectrum of light waves. The phototransistors (receivers) are also tuned to respond to that light frequency and reject the other frequencies, much like tuning a radio channel. As a result, the sensor output is very robust against ambient light conditions. If the receiver is across from the emitter, it is called a *through light sensor* (or through beam photoelectric sensor). If the receiver and emitter are on the same sensor head, and the receiver turns ON/OFF its output based on the amount of reflected light, it is called a *reflective light sensor* (or reflective beam photoelectric sensor).

A category of light-based distance and presence sensors uses reflected light intensity for measurement. The presence sensors have a two state output: ON or OFF, representing the presence or absence of an object at the target location. Such reflected light based presence sensors (sensors with ON or OFF output) work on the principle of reflected light strength. An emitter sends a focused light. The object reflects the light. The reflected light

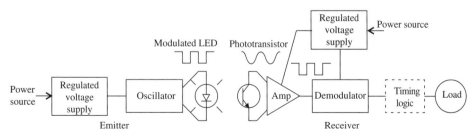

FIGURE 6.29: Components and operating principle of a through-beam light presence sensor. Reflective and diffusive versions of this type of sensor can also be used for distance sensing. (a) Liquid level sensing, (b) part dimension sensing (height and width), (c) part height sensing, (d) roll diameter sensing, (e) web slack level sensing.

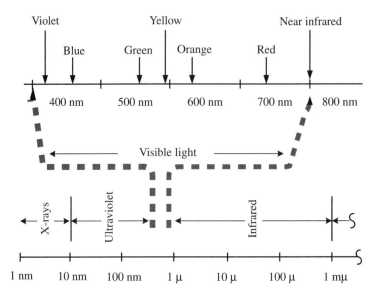

FIGURE 6.30: Frequency spectrum of light.

is captured on a detector. If the object is within a range of expected distance, the reflected light strength will be such that the output of the detector is ON. If the object is outside that range (or the reflectivity of the object is weaker even though it is in the expected distance range of the sensor), the output of the sensor will be OFF.

Figure 6.31a shows various applications of light-based sensors in presence and distance sensing applications. The same type of sensor, using the so-called *reflected light triangularization* principle, is used in measuring the distance between the sensor head and the object reflecting the light (Figure 6.31b). The sensor head has a laser emitter and charge coupled device (CCD) receiver. The emitter sends a laser light, and the reflected light is captured on the CCD receiver. The location on the CDD array where the reflected light arrives is a function of the distance of the object from the sensor head. An on-board microcontroller or DSP makes the necessary geometric calculations to relate the CCD position information to the object position information based on a simple trigonometric relation. In order to deal with noise in the signal, the microcontroller typically uses an averaging method in its calculations. It is common to have a laser based distance sensor with 1.0 m range, 10 μ resolution, and 4 kHz update rate.

The light sensors have the largest range and resolution ratio. The bandwidth of the sensor is very high (in the order of a few kHz) and it is not sensitive to the distance measured due to the high speed of light. They are also very small in physical size and easy to install.

Photoelectric sensors are widely used in industry in the form of "safety light curtains" (Figure 6.32). A set of emitter–receiver pairs forms a linear line of light beams. When any portion of the light beam is interrupted by an object passing through, the sensor output turns ON. Typically, the sensor output is connected to drive a circuit breaker to shutdown the power to the machine. Notice that the light curtain is formed by a finite number (i.e., 10 to 100 range) of emitter–receiver pairs. Therefore, the object must be at least as thick as the light curtain resolution to be detectable. Typical resolution of the light curtain is around 25 mm to 30 mm. Small objects can potentially pass through the light curtain without interrupting the light.

Detecting the presence/absence of
a seal (rubber piece) on a electrolytic capacitor

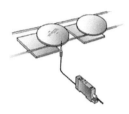

Detecting the wafers in drying furnace

Detecting the passing of PCBs in a reflow furnace

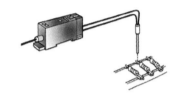

Detecting the lead-wires of resistors

(a)

Sensor head

CCD
Receiver array

Laser emitter

Reflected light

A

B

Target object
at position 1

Target object
at position 2

(b)

FIGURE 6.31: (a) Applications of reflective types of light sensors: presence sensors with
ON/OFF output, and (b) distance sensor using triangularization in a laser sensor.

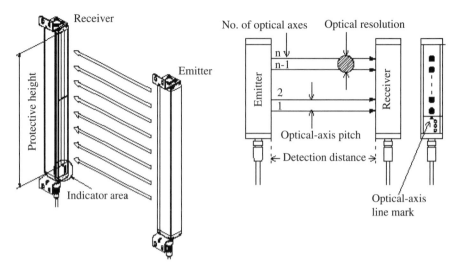

FIGURE 6.32: A through-beam emitter–receiver array light sensor used as a safety light curtain.

6.4.9 Presence Sensors: ON/OFF Sensors

A special class of the position related sensors are those which sense the presence of an object with a sensing range and provide one of two discrete outputs: ON or OFF. Such sensors are collectively called *presence sensors* or *ON/OFF sensors*.

We have previously discussed the light based sensors which can be used to detect the presence of an object within the viewing field of the sensor. There are three different types of light based (photoelectric) presence sensors: through-beam, reflective, diffusive (Figure 6.33). The sensor has a frequency tuned emitter and a receiver head. Depending on the receiver light and the adjustable threshold on the receiver head, the output of the sensor is turned ON or OFF.

Inductive and capacitive proximity sensors are two of the most common non-contact presence sensors used in industry. Their operating principles are shown in Figure 6.34. The sensing range of these sensors is in the 1 cm to 10 cm range. Typical maximum switching frequency of an inductive presence sensor is about 1 ms (1 kHz) and the switching frequency of a capacitive presence sensor is about 10 ms (100 Hz). Notice that while the inductive proximity sensors sense only the metal targets, capacitive proximity sensors can sense non-metallic targets as well.

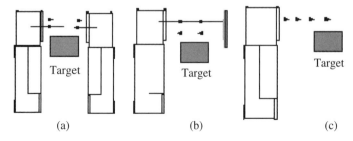

FIGURE 6.33: Operating principles of photoelectric presence sensors (a) through-beam type, (b) reflective type, (c) diffusive type.

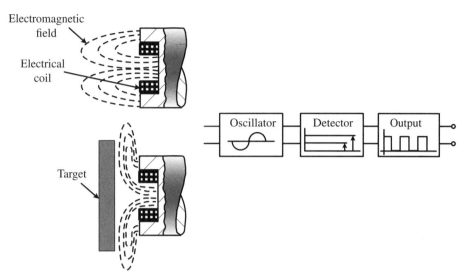

FIGURE 6.34: Operating principles of a proximity sensor.

An inductive sensor has the following main components: sensor head (ferrous core and conductor coil winding around it), oscillating current supply circuit, detection and output circuit. The oscillating supply circuit establishes an oscillating current, hence an oscillating electromagnetic field, around the sensor head. When a metallic object enters the field of the sensor, it changes the electromagnetic field density around the sensor since the effective magnetic permeability of the surrounding environment changes. The oscillating electromagnetic field induces eddy currents on the target metallic object. The eddy current losses draw energy from the supply circuit of the sensor, and hence reduce the magnitude of oscillations. The detection circuit measures the drop in the oscillation current magnitude, and switches (turn a ON/OFF) the output circuit transistor. Inductive sensors operate with electromagnetic fields. Inductive sensors can detect metal objects. For a given inductive sensor, the detection range is higher for ferrous metals (i.e., iron, stainless steel) than the detection range for non-ferrous metals (i.e., aluminum, copper). Capacitive sensors operate with electrostatic fields. The objects sensed by capacitive sensors must have an effect in changing the electrostatic field through the effective change in capacitance. This is typically accomplished by changing the dielectric constant around the sensor, hence changing the effective capacitance. Notice that, as a sensed part enters the field of a proximity sensor (inductive or capacitive type), the electric field around the sensor head is gradually changed. The change in the field (electromagnetic field for inductive type, electrostatic field for capacitive type) strength occurs in proportion to the position of the sensed object. As a result a threshold value in the sensed change is used to decide present (ON) or not-present (OFF) decision of the sensor.

An inductive proximity sensor and a gear set is often used as a position and velocity sensor for applications which require low resolution but very rugged position and speed sensors (Figure 6.35). As each gear tooth passes by the proximity sensor, the output of the sensor changes state between ON and OFF. This basically is equivalent to a single channel encoder. Using a single channel proximity sensor, the change of direction can not be detected. Therefore, this type of sensor is appropriate for applications where the rotational direction of the shaft is in only one direction (i.e., engine output shaft rotation direction is always in the same direction). The speed is determined by the frequency of the pulses from the proximity sensor.

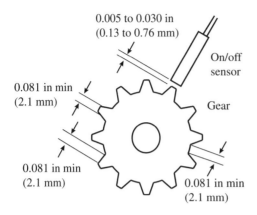

0.005 to 0.030 in
(0.13 to 0.76 mm)

On/off
sensor

0.081 in min
(2.1 mm)

Gear

0.081 in min
(2.1 mm)

0.081 in min
(2.1 mm)

FIGURE 6.35: Operating principles of a proximity sensor (inductive, capacitive, Hall effect types) and a gear on a shaft to measure rotary position and speed.

6.5 VELOCITY SENSORS

6.5.1 Tachometers

Construction of a tachometer is identical to the construction of a brush-type DC motor, except that it is smaller in size since the tachometer is used for measurement purposes, not for the purpose of converting electrical power to mechanical power like an electric motor actuator. A tachometer involves a rotor winding, a permanent magnet stator and commutator-brush assembly (Figure 6.36).

A tachometer is a passive analog sensor which provides an output voltage proportional to the velocity of a shaft. There is no need for external reference or excitation voltages. Let us consider the dynamic model of the electrical behavior of a brush type DC motor. Let the resistance and inductance of the rotor winding be R and L, respectively. The back EMF constant of the motor (which will be used as tachometer) is K_{vw}. The dynamic relationship between the terminal voltage, V_t, current, i in the rotor winding, and the angular speed of the rotor, w is

$$0 = L \cdot \frac{d}{dt} i(t) + R \cdot i(t) + K_{\text{vw}} \cdot \dot{\theta}(t) \qquad (6.76)$$

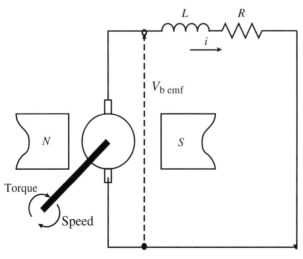

FIGURE 6.36: Operating principle of a tachometer for speed sensing.

where the voltage due to back electromotive force is

$$V_{bemf} = K_{vw} \cdot \dot{\theta}(t) \tag{6.77}$$

If we neglect the effect of inductance, $L = 0$, then the tachometer relationship can be approximated

$$R \cdot i(t) = -K_{vw} \cdot \dot{\theta}(t) \tag{6.78}$$

$$V_R(t) = -K_{vw}\dot{\theta}(t) \tag{6.79}$$

$$\dot{\theta}(t) = -\frac{1}{K_{vw}} \cdot V_R(t) \tag{6.80}$$

where $V_R(t)$ is the voltage across the resistor which can be easily measured (but note that resistance R varies with temperature), and K_{vw} is known by design of the tachometer. Then the speed can be calculated.

By design, tachometers have L and R parameters that result in a small current (i). In steady-state, the output voltage is proportional to the shaft velocity. The proportionality constant is a parameter determined by the size of the tachometer, type of winding, and permanent magnets used in its construction. Ideally, the gain of the tachometer should be constant, but in reality it varies with temperature (T) and rotor position (θ) due to a finite number of commutators. Hence, the sensor output voltage and angular speed relation can be expressed as,

$$V_{out}(t) = K_{vw} \cdot \dot{\theta}(t) \tag{6.81}$$

where

$$K_{vw} = K_{vw}(T, \theta) \tag{6.82}$$

The value of the gain at rated temperature and nominal rotor position is called the K_{vwo}. The gain changes as a function of temperature. In addition, due to the finite number of commutators, the gain has a periodic ripple as a function of rotor position. The frequency of the ripple is equal to the number of commutators. The ripple due to commutation is reflected on the output voltage which is generally less than 0.1% of the maximum output voltage.

The parameters that specify the performance of a tachometer are

1. speed to voltage gain, $K_{vw}(V/rpm)$,
2. maximum speed, w_{max}, limited either by the bearings or the magnetic field saturation,
3. inertia of the rotor,
4. maximum expected ripple voltage and frequency.

A typical dynamic time constant of a tachometer is in the range of $10–100\,\mu s$.

For very high speed applications, the winding and magnets are designed such that $K_{vw}(V/krpm)$ is a low value, whereas for very sensitive low speed applications, it is designed such that the K_{vw} is a high value.

Example Consider a tachometer with a gain of $2\,V/1000\,rpm = 2\,V/krpm$. It is interfaced to a data acquisition system through an analog to digital converter (ADC) which has 12-bit resolution and $\pm 10\,V$ input range. The sensor specifications state that the ripple voltage due to commutators on the tachometer is 0.25% of the maximum voltage output.

1. Determine the maximum speed that the sensor and data acquisition system can measure.
2. What are the measurement errors due to the ripple voltage and due to ADC resolution?
3. If the ADC was 8-bit, which one of the error sources is more significant: ripple voltage or ADC resolution?

Since the input to ADC saturates at 10 V, the maximum output from tachometer should be 10 V. Hence, the maximum speed that can be measured is

$$w_{max} = \frac{10\,V}{2\,V/krpm} = 5\,krpm = 5000\,rpm \qquad (6.83)$$

The measurement error due to ripple voltage is

$$E_r = \frac{0.25}{100} \cdot 10\,V = 0.025\,V \quad or \quad \frac{0.25}{100} \cdot 5000\,rpm = 12.5\,rpm \qquad (6.84)$$

The measurement error due to the ADC resolution is one part in 2^{12} since ADC has 12-bit resolution over the full range of measurement ($\pm 10\,V = 20\,V$ range),

$$E_{ADC} = \frac{20\,V}{2^{12}} = \frac{20}{4096} = 0.00488\,V \quad or \quad \frac{1}{4096} \cdot 10\,000\,rpm = 2.44\,rpm \quad (6.85)$$

If the ADC is 12-bit resolution, the measurement error is dominated by the ripple voltage. Measurement error due to ADC resolution is negligable. If the ADC is 8-bit, then the measurement error due to the ADC resolution is

$$E_{ADC} = \frac{20\,V}{2^{8}} = \frac{20}{256} = 0.078\,V \; or \; \frac{1}{256} \cdot 10\,000\,rpm = 39.0625\,rpm \qquad (6.86)$$

In this case, the measurement error due to ADC resolution is larger than the error due to the ripple voltage of the sensor.

6.5.2 Digital Derivation of Velocity from Position Signal

In most motion control systems, velocity is a derived information from position measurements. Accuracy of the derived (estimated) velocity depends on the resolution of the position sensor. The velocity can be derived from the position measurement signal either by analog differentiation using an op-amp circuit or digital differentiation using the sampled values of the position signal.

The ideal op-amp differentiator has the following input signal to output signal relationship,

$$V_{out}(t) \approx -K\frac{d}{dt}(V_{in}(t)) \qquad (6.87)$$

where $V_{in}(t)$ represents the position sensor signal, $V_{out}(t)$ represents the estimated velocity signal.

Clearly, digital derivation is more flexible since it allows more specific filtering and differentiation approximations,

$$V = \frac{\Delta X}{T_{sampling}} \qquad (6.88)$$

Velocity Estimation using Encoders at Very Low Speed Motion Applications   An interesting method of velocity estimation in low speed applications using encoders is to determine the time period between two consecutive pulse transitions in order to increase velocity estimation accuracy. In very low velocity applications, the number of pulse changes in an encoder count within a given sampling period may be as low as one or two counts. If one of the count changes occurs at the edge of the sampling period, the velocity estimation can be different by 50% whether that count occured right before or right after the sampling period ended. Notice that typically, $T_{sampling} = 1$ ms, and ΔX is 1 or 2 counts in low speed applications. One count variation in the position change, depending on when that happens (right before the $T_{sampling}$ period expired or right after it), can vary estimated velocity by 100%.

Instead of keeping a fixed time period and counting the number of pulses during that period to estimate the velocity, we measure *the time period between two consecutive pulses using a high resolution timer*, that is using a µs resolution timer/counter. Then, the estimated velocity is

$$V = 1[count]/T_{period} \tag{6.89}$$

Since the time period measurement is very accurate, the velocity estimation is more accurate than the first case. Let us assume that we have an encoder with a resolution of 10 000 cnt/rev (including a ×4 quadrature count multiplication), and that it is connected to a shaft which rotates at 6.0 rev/min. Assume that the sampling rate is $T_{sampling} = 1.0$ ms. This results in 60 000 counts/min or equivalently 1 cnt/ms.

Let us assume that due to small variations in the velocity, there were two count changes within one sampling period instead of the normal one pulse, one at the beginning of the sampling period, and one at the 900 µs mark. The velocity estimation using the fixed sampling period approach would give

$$V = 2\,cnt/1.0\,ms \tag{6.90}$$

$$= 2\,cnt/1.0\,ms(60\,000\,ms/1\ min)(1\,rev/10\,000\,cnt) \tag{6.91}$$

$$= 12\,rev/min. \tag{6.92}$$

If the second pulse transition did not happen at the 900 µs mark, but beyond the 1.0 ms sampling period, that is at 1.001 ms, the estimated velocity would have been,

$$V = 1\,cnt/1.0\,ms \tag{6.93}$$

$$= 1\,cnt/1.0\,ms(60\,000\,ms/1\ min)(1\,rev/10\,000\,cnt) \tag{6.94}$$

$$= 6\,rev/min. \tag{6.95}$$

If the velocity was estimated by *the time period measurement method*, then the instantenous velocity would be accurately estimated. Let us assume that we measure the time period between two consecutive pulses using a 1 µs resolution timer/counter. Let us assume that in one case the measured time period between two pulses is 900 µs and in another case it is 901 µs due to timer/counter resolution. The estimated velocity from both of these cases is

$$V_1 = 1\,cnt/0.9\,ms = 6/0.9\,rpm = 6.6667\,rpm \tag{6.96}$$

and the second estimated velocity is

$$V_2 = 1\,cnt/0.901\,ms = 6/0.901\,rpm = 6.6593\,rpm \tag{6.97}$$

Clearly, using the time period measurement between two consecutive pulses using a 1 µs resolution timer/counter results in 0.11% error in velocity measurement. Whereas, counting the number of pulses during a fixed sampling period can result in velocity measurement errors as high as 100%.

6.6 ACCELERATION SENSORS

Three different types of acceleration sensors, also called *accelerometers*, each based on a different *transduction principle*, are discussed as follows.

1. An inertial motion based accelerometer, where the sensor consists of a small mass-damper-spring in an enclosure and mounted on the surface of an object whose acceleration is to be measured. The displacement (x) of the sensor inertia is proportional

to the magnitude of acceleration ($\ddot{X}$) of the object provided that the frequency of acceleration is well within the natural frequency (bandwidth) of the sensor.

$$\ddot{X} \rightarrow x \rightarrow V_{\text{out}} \tag{6.98}$$

2. Piezoelectric based accelerometers provide a charge (q) proportional to the inertial force as a result of the acceleration. Piezoelectric materials provide a charge proportional to the strain which is proportional to the inertial force.

$$\ddot{X} \rightarrow F \rightarrow q \rightarrow V_{\text{out}} \tag{6.99}$$

3. Strain-gauges can be used for acceleration measurement if the sensor can transduct a strain (ϵ) proportional to the acceleration. Once that is accomplished, the change in the strain is measured by the change in the strain-gauge resistance and hence as an output voltage from a Wheatstone bridge and op-amp circuit.

$$\ddot{X} \rightarrow F \rightarrow \epsilon \rightarrow R \rightarrow V_{\text{out}} \tag{6.100}$$

6.6.1 Inertial Accelerometers

An inertial accelerometer is basically a small mass-spring-damper system with high natural frequency. Consider the figure shown in Figure 6.37, which is the concept of an inertial accelerometer connected to a body whose acceleration is to be measured. Figure 6.38 shows pictures of a number of accelerometers. The dynamic relations between the relative displacement of the sensor inertia with respect to its enclosure, x, and the acceleration of the body, $\ddot{x}_{\text{base}}$, are

$$m \cdot (\ddot{x}(t) + \ddot{x}_{\text{base}}(t)) + c \cdot \dot{x}(t) + k \cdot x(t) = 0 \tag{6.101}$$

$$\ddot{x}(t) + (c/m) \cdot \dot{x}(t) + (k/m) \cdot x(t) = -\ddot{x}_{\text{base}}(t) \tag{6.102}$$

Notice that if the accelerometer parameters m, c, k are chosen such that the motion of the accelerometer is critically damped, then the displacement of the accelerometer relative to its enclosure, $x(t)$, is proportional to the acceleration of the base in steady-state. The speed of response is determined by the c/m and k/m ratios. Let

$$c/m = 2\xi w_{\text{n}} \tag{6.103}$$

$$k/m = w_{\text{n}}^2 \tag{6.104}$$

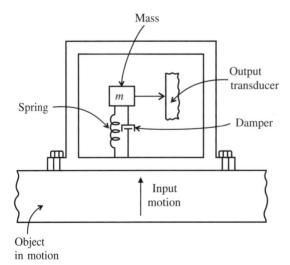

FIGURE 6.37: Operating principle of an inertial accelerometer.

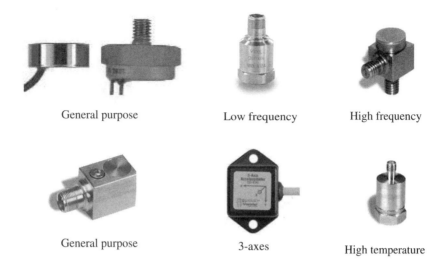

General purpose   Low frequency   High frequency

General purpose   3-axes   High temperature

FIGURE 6.38: Pictures of various accelerometers.

The values of m, c, k are chosen such that $\xi = 0.7$ to 1.0 range, and w_n can be chosen up to a few hundred Hz. The smaller the mass and the stiffer the spring constant of the sensor, the higher will be its bandwidth.

Let us consider a sinusoidal base displacement and acceleration as a function of time,

$$x_{base}(t) = A \, \sin(w \, t) \tag{6.105}$$

The resulting acceleration of the base as a result of constant magnitude sinusoidal displacement of the base is,

$$\ddot{x}_{base}(t) = -A \, w^2 \, \sin(w \, t) \tag{6.106}$$

The steady-state response of the displacement of the sensor inertia with respect to its housing, $x(t)$, is the steady-state solution for

$$\ddot{x}(t) + (c/m) \cdot \dot{x}(t) + (k/m) \cdot x(t) = -\ddot{x}_{base}(t) \tag{6.107}$$

$$\ddot{x}(t) + (c/m) \cdot \dot{x}(t) + (k/m) \cdot x(t) = A \, w^2 \, \sin(w \, t) \tag{6.108}$$

In steady-state,

$$x_{ss}(t) = \frac{A \, (w/w_n)^2 \, \sin(wt - \phi)}{\{[1 - (w/w_n)^2]^2 + [2 \, \xi(w/w_n)]^2\}^{1/2}} \tag{6.109}$$

where

$$\phi = \tan^{-1} \frac{2\xi(w/w_n)}{1 - (w/w_n)^2} \tag{6.110}$$

The steady-state displacement of the sensor has the following properties:

1. the sensor displacement has the same frequency as the acceleration of the base,

2. there is a phase shift between the sensor displacement and base acceleration and the phase angle is a function of acceleration frequency (w) as well as the sensor parameters (ξ, w_n),

3. the magnitude of the sensor displacement is proportional to the acceleration magnitude. However, the proportionality constant is a function of acceleration frequency (w) as well as the sensor parameters (ξ, w_n).

The magnitude of the sensor displacement, which is a sinusoidal function, is

$$|x_{ss}(t)| = \frac{A\,(w/w_n)^2}{\{[1-(w/w_n)^2]^2 + [2\,\xi(w/w_n)]^2\}^{1/2}} \qquad (6.111)$$

There are two different purposes for this type of sensor,

1. If the sensor is intended to be used to measure the acceleration of the base, then we are interested in the ratio of the magnitude of the response to the magnitude of excitation,

$$\left|\frac{x_{ss}(t)}{A\,w^2}\right| = \frac{(1/w_n)^2}{\{[1-(w/w_n)^2]^2 + [2\,\xi(w/w_n)]^2\}^{1/2}} \qquad (6.112)$$

2. If the sensor is intended to be used as a seismic instrument (i.e., for earthquake measurements) and measure the displacement of the base, then we are interested in the ratio of

$$\left|\frac{x_{ss}(t)}{A}\right| = \frac{(w/w_n)^2}{\{[1-(w/w_n)^2]^2 + [2\,\xi(w/w_n)]^2\}^{1/2}} \qquad (6.113)$$

The magnitude ratio of the sensor displacement to the magnitude of the acceleration for case (1) as a function of frequency is shown in Figure 6.39. In order to have a proportional relationship (constant magnitude ratio) between the sensor output and the measured quantity (acceleration in this case) within the frequency range of interest, it is necessary that (Figure 6.39a)

$$w \ll w_n \; ; \; w\ in\ [0, w_{max}] \qquad (6.114)$$

where w_{max} is the maximum frequency content of the acceleration which we expect to measure.

On the other hand, if the sensor is used to measure the displacement of the base, as in seismic applications, in order to have a proportional relationship (constant magnitude

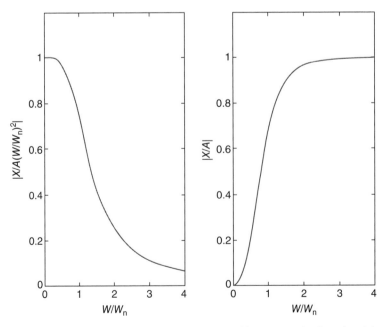

FIGURE 6.39: Inertial accelerometer output and input magnitude ratios: (a) sensor output to base acceleration magnitude ratio, (b) sensor output to base displacement ratio.

ratio) between sensor output and measured quantity (base displacement in this case), it is necessary that (Figure 6.39b)

$$w \gg w_n \;\; ; \;\; w_{min} < w < w_{max} \tag{6.115}$$

This relationship states that an inertial accelerometer used to measure the seismic displacement should be used in a frequency range that is larger than a certain minimum frequency, w_{min}, and that frequency should be at least as large or more than the natural frequency of the sensor,

$$w_n < w_{min} < w \tag{6.116}$$

In other words, the natural frequency of an accelerometer used for seismic displacement measurement should be very small. Since the $w_n = \sqrt{k/m}$, the mass of the seismic accelerometer should be very large and the spring constant should be small. This explains the reason for the large size of seismic accelerometers compared to normal accelerometers.

A given inertial acceleration sensor can measure accelerations with a frequency content below its natural frequency, and displacements with a frequency content above its natural frequency.

An inertial accelerometer cannot accurately measure seismic displacements that have a frequency lower than the natural frequency of the sensor. Likewise, an inertial accelerometer cannot measure accelerations that have a frequency content closer to or larger than the natural frequency of the sensor. Inertial accelerometers used for seismic displacement measurement must have as small a natural frequency as possible. Inertial accelerometers used for acceleration measurement must have as large a natural frequency as possible.

Example: Gyroscope Used as an Orientation Angle Sensor Gyroscopes (also called *inertial gyros*) are used in ships, airplanes, and space vehicles for measuring rotational speed and orientation angles of the vehicle. The gyroscope shown in Figure 6.40 is used to measure one orientation angle of the device. The gyro has a rotating rotor which

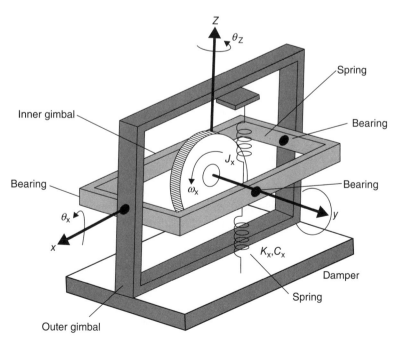

FIGURE 6.40: Inertial gyro used to measure the rate of rotation: Inertial gyro construction and operating principle.

is typically driven at a constant speed by a DC electric motor. The rotor is supported by two bearings on the inner gimbal, which in turn is supported by two bearings on the outer gimbal. The inner gimbal is further connected to the outer gimbal via a pair of springs. The basic dynamic principle of operation of the gyro is that a rotational rate (angular velocity) about the z-axis, that is $\dot{\theta}_z(t)$ results in a proportional angular displacement about the x-axis in steady-state, that is

$$\theta_x(t) = K \cdot \frac{d\theta_z(t)}{dt} \tag{6.117}$$

As a result, when this gyro is mounted on a vehicle by securing its outer gimbal to the vehicle, the angular displacement about the x-axis is a measure of the rotational velocity about the z-axis. This rotational speed about the z-axis can be integrated over time, given the initial condition of the angle about the z-axis, to keep track of the orientation of the vehicle about the z-axis. The dynamic relationship for this gyro is

$$J_x \frac{d^2\theta_x(t)}{dt^2} + C_x \frac{d\theta_x(t)}{dt} + K_x\theta_x(t) = J_y \cdot w_y \cdot \frac{d\theta_z(t)}{dt} \quad ; \; transient \tag{6.118}$$

$$K_x\theta_x(t) = J_y \cdot w_y \cdot \frac{d\theta_z(t)}{dt} \quad ; \; steady\text{-}state \tag{6.119}$$

$$\theta_x(t) = \frac{J_y \cdot w_y}{K_x} \cdot \frac{d\theta_z(t)}{dt} \tag{6.120}$$

$$\theta_x(t) = K\frac{d\theta_z(t)}{dt} \tag{6.121}$$

where J_x, C_x, K_x are the rotational moment of inertia of the gyro rotor and its inner gimbal about the x-axis, the damping coefficient, and the spring constant. J_y is the rotary inertia of the gyro rotor about the y-axis, w_y is the speed of the gyro rotor, which is typically kept constant by a controlled electric motor.

6.6.2 Piezoelectric Accelerometers

Some materials (such as natural quartz crystal, silicon dioxide, barrium titanite, lead zirconate titanate (PZT)), called *piezo crystals*, produce a charge in response to a force (or deformation) applied to them. This is called the *direct piezoelectric* effect. The same materials also have the reverse phenomenon, that is, they produce force in response to an applied charge. This is called the *reverse piezoelectric* effect. The Greek word "piezo" means "to squeeze" or "pressure." Quartz has excellent temperature stability and shows almost no decay in its piezoelectric properties over time. PZTs are polarized by applying very high DC voltages at high temperatures. PZTs show a natural decay in their piezoelectric properties over time, hence may need to be periodically calibrated or re-polarized.

Piezoelectric accelerometers work on the principle that the acceleration times the sensor inertia will apply a force on the sensor. Mechanically, the piezo element acts as a very precise and stiff spring in the sensor design. As a result of the piezoelectric phenomena, the charge output, hence the output voltage, from the sensor is proportional to the inertial force

$$\ddot{x} \rightarrow F \rightarrow q \rightarrow V_{out} \tag{6.122}$$

$$q = C \cdot V_{out} \tag{6.123}$$

where q is the charge produced by the piezoelectric material, C is the effective capacitance, and V_{out} is the produced output voltage.

A calibrated piezoelectric acceleration sensor has the following input–output relationship,

$$V_{out} = K \cdot \ddot{x} \tag{6.124}$$

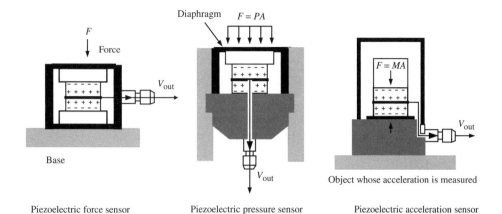

Piezoelectric force sensor   Piezoelectric pressure sensor   Piezoelectric acceleration sensor

FIGURE 6.41: Piezoelectric principle and its usage in force, pressure, and acceleration sensors.

Notice that although the piezo element has a finite stiffness and acts as a stiff spring, the actual deformation of the piezo element is very small due to its large stiffness. Therefore, it can be considered almost as if there is no deformation. Piezoelectric based acceleration sensors have a range as high as 1000 g, with sensor bandwidth up to 100 kHz. The limiting factor in the frequency response of the sensor is the charge amplifier bandwidth, which is much slower than the sensor element. The bandwidth of the sensor element plus the charge amplifier can be up to a few kHz range. Piezoelectric based sensors are most appropriate for measuring signals that are time varying. They are not appropriate for measuring static or low frequency signals since the charge due to the load will slowly discharge.

The design principles of piezoelectric based sensors for pressure, force, strain, and acceleration measurements are very similar. The external force (in the case of a force sensor it is the force sensed, in case of a pressure sensor it is the pressure times the surface area of the diaphragm, in the case of acceleration sensor it is the inertial force, $m_{sensor} \cdot \ddot{x}$) induces a strain on the piezoelement of the sensor. The output charge is proportional to the strain (Figure 6.41).

As a example, the accelerometer Model-339B01 by PCB Piezotronics has the following characteristics: voltage sensitivity $K = 100\,mV/g$, frequency range up to 2 kHz, amplitude range up to 50 g with a resolution of 0.002 g.

6.6.3 Strain-gauge Based Accelerometers

The operating principle of a strain-gauge based accelerometer is very similar to an inertial accelerometer. The only difference is that the spring function is provided by a cantilever flexible beam (Figures 6.42–6.45a). In addition, the strain in the cantilever beam is measured, instead of the displacement. The strain is proportional to the inertial force, hence the acceleration. The voltage output from the sensor, proportional to strain, is obtained from the standard Wheatstone bridge circuit.

$$\ddot{x} \rightarrow F \rightarrow \sigma \rightarrow \epsilon \rightarrow \Delta R \rightarrow V_{out} \qquad (6.125)$$

where $F = m\ddot{x}$, $\sigma = F/A$, $\epsilon = (1/E)\sigma$, $\Delta R = G\epsilon$, m is inertia, A is the cross-sectional area of deformation, E Youngs' Modulus of elasticity constant, G the strain-gauge factor.

$$V_{out}(t) = K \cdot \ddot{x}(t) \qquad (6.126)$$

The range of the accelerations this type of sensor can measure is very similar to those of piezoelectric type accelerometers, that is up to 1000 g. The bandwidth of the sensor can be as high as a few kHz.

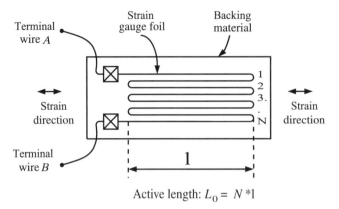

Active length: $L_0 = N * l$

FIGURE 6.42: A typical strain gauge for strain measurement.

6.7 STRAIN, FORCE, AND TORQUE SENSORS

6.7.1 Strain Gauges

The most common strain measurement sensor is a *strain gauge*. The strain gauge transduction principle is based on the relationship between the change in length and its resulting change in the resistance of a conductor. Typical strain-gauge material used is constantan (55% copper and 45% nickel). A fine wire of strain-gauge material is given a directional shape and bonded to the part surface using adhesive bonding materials. The resistance and strain relationship is

$$\frac{\Delta R}{R_0} = G \frac{\Delta L}{L_0} \tag{6.127}$$

where G is called the *gauge factor* of the sensor, R_0 is the nominal resistance, and L_0 nominal length under no strain conditions. In order to increase the resulting resistance change under given strain conditions, we need to pack more length, L, into a sensor size. That is the reason for the many rounds of conductor wire in one direction in the construction of a strain gauge (Figures 6.42 and 6.43). Ideally, strain is the property at a point on a material and is

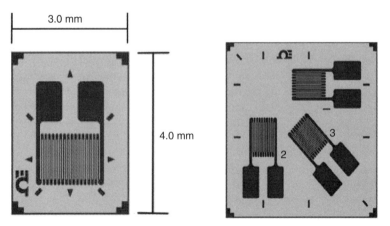

FIGURE 6.43: Pictures of typical strain gauges for strain measurement. Adapted from Omega.com.

the ratio of change in the length to the total length in that direction,

$$\epsilon = \frac{\Delta L}{L} \tag{6.128}$$

The change in the strain-gauge resistance is converted to a proportional voltage using a Wheatstone brige,

$$V_{\text{out}} = K_1 \cdot \frac{\Delta R}{R} \tag{6.129}$$

$$= K_1 \cdot G \cdot \frac{\Delta L}{L} \tag{6.130}$$

$$= K_2 \cdot \frac{\Delta L}{L} \tag{6.131}$$

$$= K_2 \cdot \epsilon \tag{6.132}$$

Note that a strain gauge has a finite dimension. It is bounded to the surface of the part over a finite area. Therefore, the measured strain is the average strain over the area occupied by the strain gauge. The bonding of the strain gauge to the workpiece is very important for two reasons: (i) it needs to provide a uniform mechanical linkage between the surface and sensor in order to make the measurement correctly, (ii) the bonding material electrically isolates the sensor and the part. The strain gauges themselves have a very high bandwidth. It is possible to build strain gauge sensors that have almost 1 MHz bandwidth. Using silicon crystal materials, strain gauge size can be minaturized while having a very large sensor bandwidth.

6.7.2 Force and Torque Sensors

Force and torque sensors operate on the same principle. There are three main types of force and torque sensors,

1. spring displacement based force/torque sensors,
2. strain-gauge based force/torque sensors,
3. piezoelectric based force sensors.

Let us consider a weighing station as a force sensor application. If we have a calibrated spring with a known spring coefficient, K_{spring}, the load force (for torque measurement, we would use an equivalent torsional spring) can be measured as displacement,

$$F = K_{\text{spring}} \cdot x \tag{6.133}$$

Using any of the position sensors, the displacement x can be converted to a proportional voltage, and hence we obtain the force (or torque) measurement,

$$V_{\text{out}} = K_1 \cdot x \tag{6.134}$$

$$= K_2 \cdot F \tag{6.135}$$

The strain-gauge based force and torque sensors measure the force (or torque) based on the measured strain. Again, the sensor relies on an elastic sensing component. Typical force/torque sensors are called *load cells*. A load cell has built in elastic mechanical components on which strain gauges are mounted (Figure 6.44). As the load cell experiences the force (or torque), it deforms a small amount which induces a strain on the sensing element. The strain is measured by the strain gauge. As a result, since the force/torque to strain relationship is linear by the design of the load cell, the measurement is proportional to the force/torque.

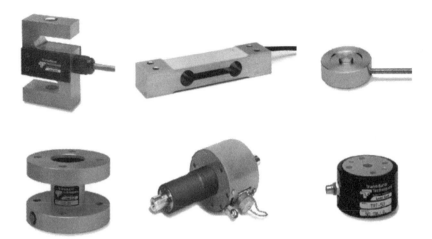

FIGURE 6.44: Various load cells using strain gauges for force and torque measurement.

The other alternative for strain-gauge based force/torque sensor is to mount the strain gauges directly over the shaft on which we want to measure the force/torque. In some applications, it is not possible to install a load cell. Force or torque on a shaft must be measured without significantly changing the mechanical design. Figure 6.45 shows such force and torque sensing using four strain gauge pairs on a shaft. Notice that most force and torque sensors use symmetrically bonded strain gauges to reduce the effect of temperature variations and drift of the strain gauge output.

There are two inherent assumptions in the stain-gauge based force/torque sensing:

1. The strain in the material is small enough such that the deformation is in the elastic range, and that

$$\epsilon = \frac{1}{E}\sigma = \frac{1}{E}\frac{F}{A} \tag{6.136}$$

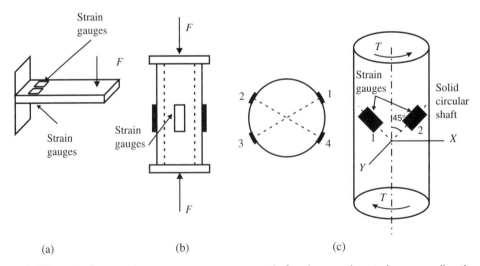

FIGURE 6.45: Force and torque measurement on a shaft or beam using strain gauges directly mounted on the part.

2. The strain gauge is subjected to the same strain. The strain–resistance change relationship is

$$\epsilon = \frac{1}{G} \frac{\Delta R}{R} \tag{6.137}$$

where the change in the resistance is converted to an output voltage by a Wheatstone bridge circuit,

$$\frac{\Delta R}{R} = \frac{4}{V_i} \cdot V_o \tag{6.138}$$

As a result, the relationship between the strain gauge output voltage and force is

$$F = \left(\frac{4 \cdot E \cdot A}{V_i \cdot G}\right) \cdot V_o \tag{6.139}$$

Notice that the stain gauge output voltage to force calibration requires information about the material (E) on which the strain gauge is bonded, the cross-section area of the part (A), sensor gauge factor (G), and Wheatstone bridge circuit reference voltage (V_i).

Finally, force can be measured using piezoelectric sensors, similar to piezoelectric pressure transducers (Figure 6.41). The piezoelectric sensor creates a charge proportional to the force acting on it. The advantage of this is that it does not introduce an additional flexibility into the system as part of the sensor. The only elasticity it introduces is the elasticity of piezoelectric quartz cystal which has a modulus of elasticity in the range of $100\,\mathrm{GPa} = 100 \cdot 10^9\,\mathrm{N/m^2}$. The typical bandwidth of a piezo based force sensor is about $10\,\mathrm{kHz}$.

Example Consider the force measurement using a strain gauge on a shaft under compression (Figure 6.45b). Let us consider that the shaft material is steel. The elastic Young's modulus $E = 2 \cdot 10^8\,\mathrm{kN/m^2}$, and cross-sectional area of the shaft is $A = 10.0\,\mathrm{cm^2}$. We have a strain gauge bonded on the shaft in the direction of the tension. The nominal resistance of the strain gauge is $R_0 = 600\,\Omega$, the gauge factor is $G = 2.0$. The other three legs of the Wheatstone bridge also have constant resistances of $R_2 = R_3 = R_4 = 600\,\Omega$. The reference voltage for the Wheatstone bridge is $10.0\,\mathrm{VDC}$. If the output voltage measured $V_{out} = 2.0\,\mathrm{mV}$, what is the force?

Notice that the stress–strain relationship, assuming the deformation is within the elastic range,

$$\sigma = \frac{F}{A} \tag{6.140}$$

$$\epsilon = \frac{1}{E}\sigma \tag{6.141}$$

$$\frac{\Delta R}{R} = G \cdot \epsilon \tag{6.142}$$

$$V_{out} = \frac{V_i}{4 \cdot R_0}\Delta R \tag{6.143}$$

Hence,

$$V_{out} = \frac{V_i \cdot G}{4 \cdot E \cdot A} \cdot F \tag{6.144}$$

The force corresponding to the $V_{out} = 2.0\,mV\,DC$ is $F = 80\,000\,N$. It is also interesting to determine the change in the strain-gauge resistance for this force measurement,

$$V_{out} = \frac{V_i}{4 \cdot R_o} \Delta R \qquad (6.145)$$

$$2 \cdot 10^{-3} = \frac{10}{4 \cdot 600} \Delta R \qquad (6.146)$$

$$\Delta R = 0.480\,\Omega \qquad (6.147)$$

Notice that the percentage change in the resistance of the strain gauge is

$$\frac{\Delta R}{R} \cdot 100 = \frac{0.480}{600} \cdot 100 = 0.08\% \qquad (6.148)$$

Since the gauge factor $G = 2$, the strain (change in the length of the shaft is)

$$\epsilon = \frac{\Delta l}{l} \qquad (6.149)$$

$$= \frac{1}{G} \frac{\Delta R}{R} \qquad (6.150)$$

$$= \frac{1}{2} \cdot 0.08\% \qquad (6.151)$$

$$\frac{\Delta l}{l} = 0.04\% \qquad (6.152)$$

$$\Delta l = \frac{0.04}{100} \cdot l \qquad (6.153)$$

6.8 PRESSURE SENSORS

Absolute pressure is measured relative to a perfect vacuum where the pressure is zero. The local atmospheric pressure is the pressure due to the weight of the air of the atmosphere at that particular location (Figure 6.46). Therefore, the local pressure varies from one location

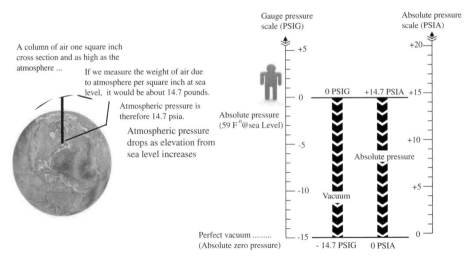

FIGURE 6.46: Absolute pressure, gauge pressure, atmospheric pressure definitions. Adapted from Bosch-Rexroth.

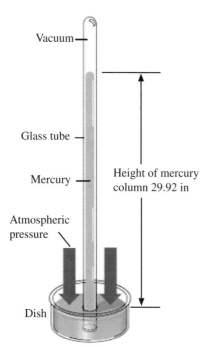

Vacuum

Glass tube

Mercury

Atmospheric
pressure

Height of mercury
column 29.92 in

Dish

FIGURE 6.47: Barometer to measure the local
atmospheric absolute pressure.

to another as a function of the height of the location from sea level. Average atmosheric absolute pressure due to the weight of the air of the atmosphere is $14.7\,lb/in^2 = 14.7\,psi$. This means that the weight acting over an area of $1\,in^2$ due to the weight of the air in the atmosphere is 14.7 lb. Therefore, the absolute atmospheric pressure at sea level is nominally $14.7\,lb/in^2(14.7\,psia)$.

Pascal's law states that pressure in a contained fluid is transmitted equally in all directions. Using this physical principle, a *barometer* is used to measure absolute pressure. The pressure acting on the fluid due to the atmosphere at the surface is balanced by the pressure due to the fluid (i.e., mercury) weight in the tube (Figure 6.47). Using this measurement method it can be observed that the atmospheric pressure is equivalent to the pressure applied by a 29.92 in(760 mm) column height of mercury. A column of mercury with 29.92 in height and $1\,in^2$ cross-section has 14.7 lbs of weight. If water is used in the barometer instead of mercury, the height of the water in the tube to balance the atmospheric pressure would be 33.95 in, which produces the same $14.7\,lb/in^2$ pressure since the density of water is lower than mercury. Notice that the top of the tube in the barometer must be a vacuum. In order to establish a vacuum at the top of the tube, the tube filled with mercury or water is placed upside down in the container filled with mercury. The height of the mercury will drop until its height is 29.92 in if filled with mercury, or 33.95 in if filled with water, which generates a pressure of 14.7 psi at the surface level of the container.

The pressure sensed by most pressure sensors is the *relative pressure* with respect to the local atmospheric pressure. However, a sensor can be calibrated to measure the absolute pressure as well (Figure 6.46). If a sensor measures pressure relative to the vacuum pressure, it is referred to as the absolute pressure and the units denoted as psia. If a sensor measures pressure relative to the local atmospheric pressure, it is referred to as the *relative* or *gauge pressure*, and the unit used to indicate that is psig. The notation psi refers to psig by standard convention in notation.

Notice that the atmospheric absolute pressure varies from the nominal value $(14.7\,\mathrm{lb/in}^2(14.7\,\mathrm{psia}))$ due to variations in

1. height from sea level,
2. temperature and the resulting variations in air density

$$p_{\mathrm{atm}} = p_{\mathrm{atm},0} + \Delta p(h, T) \qquad (6.154)$$
$$= 14.7 + \Delta p(h, T)\,\mathrm{psia} \qquad (6.155)$$

6.8.1 Displacement Based Pressure Sensors

The basic *transduction* principle in this type of pressure sensor is to convert the pressure into a proportional displacement, and then convert this displacement to a proportional electrical voltage. Figure 6.48 shows various concepts of pressure sensors where the pressure is proportional to the displacement of the sensing element (Bourdon C-tube, bellows, diaphragm). The motion of the flexible sensing element can be translated into a proportional voltage by various methods including position sensing, capacitance change, strain change, and piezoelectric effects. For instance, the Bourdon tube based pressure sensor can be

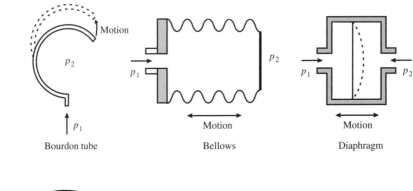

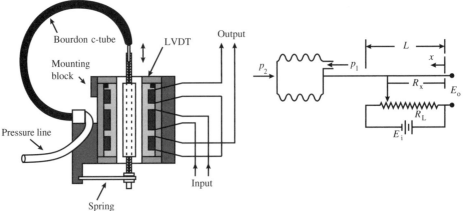

FIGURE 6.48: Various pressure sensor concepts: pressure to displacement transduction and then displacement measurement in order to obtain pressure measurement.

connected to an LVDT or a linear potentiometer to get a voltage signal proportional to the pressure.

$$\Delta x = k_p \cdot \Delta P \tag{6.156}$$

$$V_{out} = k_{vx} \cdot \Delta x \tag{6.157}$$

$$= (k_p \cdot k_{vx}) \cdot \Delta P \tag{6.158}$$

$$= k_{pv} \cdot \Delta P \tag{6.159}$$

The pressure sensors shown in Figure 6.48 measure the relative pressure between p_1 and p_2, that is the pressure difference between them, $\Delta p = p_1 - p_2$. If either p_1 or p_2 is the vacuum pressure (absolute zero pressure), then the sensor measures the absolute pressure.

6.8.2 Strain-Gauge Based Pressure Sensor

The pressure induced deformation of the diaphragm is measured by strain gauges on it. The strain on the diaphragm is proportional to the pressure. The resistance of the strain gauge changes in proportion to its strain. Using a Wheatstone bridge circuit, a proportional output voltage is obtained from the strain gauge. The relationship between pressure and strain gauge voltage output is (Figure 6.49),

$$\Delta x = k_p \cdot \Delta P \tag{6.160}$$

$$\epsilon = k_1 \cdot \Delta x \tag{6.161}$$

$$V_{out} = k_2 \cdot \epsilon \tag{6.162}$$

$$= k \cdot \Delta P \tag{6.163}$$

An example of a strain-gauge based differential pressure sensor is the Model P2100 series by Schaevitz Inc. The sensor has two pressure input ports and can measure the pressure differential between the two ports up to 5000 psi. The natural frequency of the sensor is 4 kHz for pressures up to 75 psi and 15 kHz for pressures upto 1000 psi. For instance, such a sensor is appropriate to measure the pressure differential between two sides of a hydraulic cylinder.

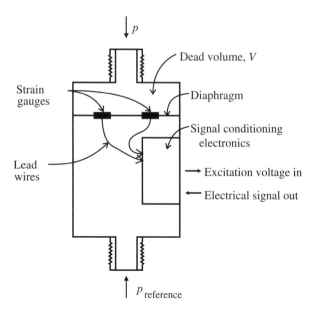

FIGURE 6.49: Pressure sensor using a diaphragm and strain gauge. Pressure is proportional to the strain induced on the diaphragm. The strain is measured as a proportional voltage output by a Wheatstone bridge circuit.

6.8.3 Piezoelectric Based Pressure Sensor

Piezoelectric based pressure sensors are the most versatile pressure sensor types. The pressure of the diaphragm is converted to a force acting on the piezoelectric element (Figure 6.41). The piezoelectric element will generate a voltage proportional to the force acting on it which is proportional to the pressure. The piezo based pressure sensors can have bandwidth in the order of kHz range.

The charge produced by the piezoelectric effect as a result of pressure is

$$q = K_{qp} \cdot p \tag{6.164}$$

and the output voltage is

$$V_{out} = \frac{q}{C} \tag{6.165}$$

Then,

$$V_{out} = \frac{K_{qp}}{C} \cdot p \tag{6.166}$$
$$= K_1 \cdot p \tag{6.167}$$

6.8.4 Capacitance Based Pressure Sensor

The diaphragm pressure sensing concept can also be used to change the capacitance between two charged plates inside the sensor. The displacement of the diaphragm results in a proportional change in capacitance. Using an operational amplifier, reference voltage, and reference capacitor, the change in the capacitance of the sensor can be converted to a voltage output signal proportionally (Figure 6.50).

$$x = K_1 \cdot \Delta P \tag{6.168}$$
$$C = \frac{K_2 \cdot A}{x} \tag{6.169}$$

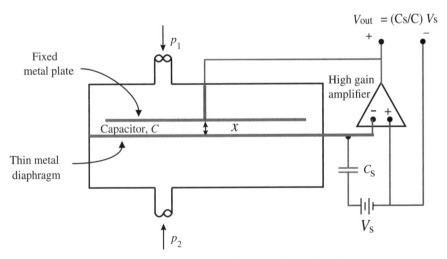

FIGURE 6.50: Capacitive pressure sensor and its operating principle.

Using the operational amplifier,

$$V_{out} = V_s \cdot (C_s/C) \tag{6.170}$$

$$= \frac{V_s \cdot C_s}{K_2 \cdot A} \cdot x \tag{6.171}$$

$$= \frac{V_s \cdot C_s \cdot K_1}{K_2 \cdot A} \cdot \Delta P \tag{6.172}$$

The signal flow relationship for the sensor operation is as follows,

$$\Delta P \rightarrow x \rightarrow C \rightarrow V_{out} \tag{6.173}$$

where the pressure differential results in a change in the distance between two plates of the capacitive sensor which in turn changes the capacitance of the sensor.

6.9 TEMPERATURE SENSORS

Three classes of temperature sensors are discussed below:

1. sensors which change physical dimension as a function of temperature,
2. sensors which change resistance as a function of temperature (RTD and thermistors), and
3. sensors which work based on thermoelectric phenomena (thermocouples).

Pictures of various RTD and thermocouple type temperature sensors are shown in Figure 6.51.

6.9.1 Temperature Sensors Based on Dimensional Change

Temperature is an indicator of the molecular motion of matter. Most metals and liquids change their dimension as a function of temperature. In particular, mercury is used in glass thermometers to measure temperature since its volume increases proportionally with the temperature. Then the glass tube can be scaled to indicate the measured temperature (Figure 6.52). It has a typical accuracy of about $\pm 0.5\,°C$. Similarly, bimetallic solid materials change their dimension as a function of temperature. As a result, they can be used as a temperature sensor by converting the change in the dimension of the bimetallic component into a voltage. However, mercury is rarely used as a temperature sensor element due to its highly toxic nature and environmental concerns if it is spilled.

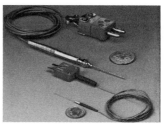

FIGURE 6.51: Pictures of various temperature sensors: thermocouples and RTDs.

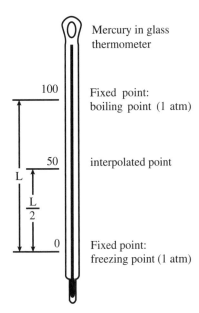

Mercury in glass
thermometer

Fixed point:
boiling point (1 atm)

interpolated point

Fixed point:
freezing point (1 atm)

FIGURE 6.52: Mercury-in-glass thermometer for
manual temperature measurement. It uses the
transduction principle that the volume of the mercury
expands as a linear function of temperature.

6.9.2 Temperature Sensors Based on Resistance

RTD Temperature Sensors   A RTD (resistance temperature detector) temperature
sensor operates on the transduction principle that the resistance of the RTD material changes
with the temperature. Then the resistance change can be converted to a proportional voltage
using a Wheatstone bridge circuit.

A good approximation to the resistance and temperature relationship for most RTD
materials is

$$R = R_0(1 + \alpha(T - T_0)) \tag{6.174}$$

where α is the sensitivity of the material resistance to temperature variation. The construction of a typical RTD sensor is shown in Figure 6.53. The sensitivity constant α for various
materials is shown in Table 6.1.

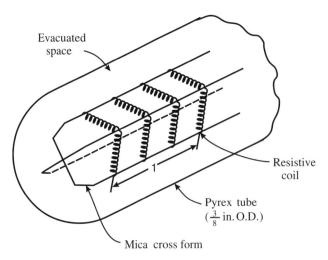

Evacuated
space

Resistive
coil

Pyrex tube
($\frac{3}{8}$ in. O.D.)

Mica cross form

FIGURE 6.53: Resistance temperature detector (RTD) sensor for temperature measurement.

TABLE 6.1: RTD temperature sensor materials and the resistance–temperature sensitivity coefficient.

Material	α
Aluminum	0.00429
Copper	0.0043
Gold	0.004
Platinum	0.003927
Tungsten	0.0048

RTDs may be used to measure the cryogenic temperature to approximately 700 °C temperature range. Platinum is one of the most common materials used in RTD sensors. The main advantages of RTD sensors are that the resistance–temperature relationship is fairly linear over a wide temperature range and the measurement accuracy can be as small as ±0.005 °C. Furthermore, the drift of the sensor over time is very small, typically in the range of less than 0.1 °C/year. As a result, RTDs do not require frequent calibration. A RTD is a passive device. It has a resistance where the resistance changes linearly with temperature. One good way of converting the change in resistance to voltage is to use the RTD in a Wheatstone bridge circuit. The dynamic response of the RTD sensor is relatively slow compared to other temperature sensors. RTDs can not be used to measure high frequency transient temperature variations.

Thermistor Temperature Sensors Thermistor sensors are based on semiconductor materials where the resistance of the sensing element reduces exponentially with the temperature. The typical resistance and temperature relationship for a thermistor is approximately,

$$R = R_o \cdot e^{\beta \left(\frac{1}{T} - \frac{1}{T_o} \right)} \tag{6.175}$$

where β is also a function of temperature and a property of the semiconductor material. The variation in the resistance of a thermistor for a given temperature change is much larger than the variation in resistance of a RTD sensor. This type of sensor is used for their high sensitivity, high bandwidth, and ruggedness compared to RTDs. However, the manufacturing variations in thermistors can be large from one sensor to another. Therefore, they cannot be used as direct replacements to one another. As a result, each sensor must be properly calibrated before replacement.

6.9.3 Thermocouples

Thermocouples are perhaps the most popular, easy to use, and inexpensive temperature sensors. A thermocouple has two electrical conductors made of different metals. The two conductors are connected as shown in Figure 6.54. The key requirement is that the

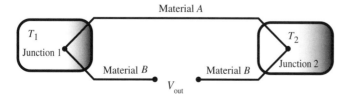

FIGURE 6.54: Thermocouple temperature sensor and its operating principle.

connections between the two conductors at both ends must form a good electrical connection. The fundamental *thermoelectric phenomenon* is that there is a voltage differential developed between the open circuit end of the conductor proportional to the temperature of the one of the junctions relative to the temperature of the other junction. The thermoelectric phenomena is a result of the flow of both heat and electricity over a conductor. This is called the *Seebeck effect*, named after Thomas J. Seebeck who first observed this phenomenon in 1821. The voltage differential measured at the output of the thermocouple is approximately proportional to the temperature differential between the two points (V_{out} in Figure 6.54),

$$V_{out} \approx K \cdot (T_1 - T_2) \tag{6.176}$$

Notice that the proportionality constant is a function of the thermocouple materials. The thermocouple materials refer to the material types used for conductors A and B. Furthermore, it is not exactly constant, but varies with temperature by a small amount.

The proportionality constant is a very good approximation for many types of thermocouples over large temperature ranges. This makes the thermocouples very attractive sensors due to their linearity over large temperature ranges. The voltage output of the thermocouple is in the milli Volt (mV) range and must be amplified by an op-amp circuit before it is used by a data acquisition system.

A thermocouple measures the temperature difference between its two junctions. In order to measure the temperature of one of the junctions, the temperature of the other junction must be known. Therefore, a reference temperature is required for the operation of the thermocouple. This reference can be provided by either *ice-water* or by a built-in electronic reference temperature. The measurement error in most thermocouples is around ± 1 to $2\,°C$. Different thermocouple material pairs are designated with a standard letter to simplify references to them (Table 6.2).

In most cases, the output of the thermocouple is processed by a digital computer system. The reference temperature is provided by a thermistor based sensor as part of the thermocouple interface circuit of the data aquisition board (DAQ). Multiple thermocouples can be connected in series to sum the sensor generated signal or in parallel to measure the average temperature over a finite area. Computer interface cards for thermocouple signal processing make use of the standard thermocouple tables for the voltage to temperature conversion for each specific type of thermocouple, instead of using linear approximation to the voltage–temperature relationship. Such standard reference tables are generated by organizations such as the National Institute of Standards and Technology (NIST) for different types of thermocouples.

TABLE 6.2: Thermocouple types and their applications.

Type	Material A	Material B	Applications
E	Chromel (90% nickel, 10% chromium)	Constantan (55% copper, 45% nickel)	Highest sensitivity, $< 1000\,°C$
J	Iron	Constantan	Non-oxidizing environment, $< 700\,°C$
K	Chromel (90% nickel, 10% chromium)	Alumel (94% nickel, 3% manganese, 2% aluminum, 1% silicon)	$< 1400\,°C$
S	Platinum and 10% rhodium	Platinum	Long term stability, $< 1700\,°C$
T	Copper	Constantan	Vacuum environment, $< 400\,°C$

6.10 FLOW RATE SENSORS

There are four main groups of sensors to measure the flow rate of a fluid (liquid or gas) passing through a cross-sectional area:

1. mechanical flow rate sensors,
2. differential pressure measurement based flow rate sensors,
3. thermal flow rate sensors,
4. mass flow rate sensors.

6.10.1 Mechanical Flow Rate Sensors

There are three major types of mechanical flow rate sensors: positive displacement flow rate sensors, turbine flow meters, and drag flow meters. Their operating principle is based on the volume displaced by the fluid flow and drag between the fluid and the sensing element.

Positive Displacement Flow Meters Positive displacement flow meters work on the same principle as the positive displacement hydraulic pumps and motors (see Chapter 7). The positive displacement pumps (and motors) are so named because they displace a well-defined fluid volume per revolution. For instance, a gear pump or piston pump sweeps a fixed amount of volume per revolution. This is called the *displacement* of the pump in units of $D = \text{Volume}/\text{Revolution}$. If the rotational speed of the pump is known, w_{shaft}, then the amount of fluid flow rate (Q)that passes through the pump or motor is determined by

$$Q = D \cdot w_{\text{shaft}} \tag{6.177}$$

The same principle can be used as a sensor in *positive displacement flow meters (PDFM)*. The most popular PDFM is the gear type (Figure 6.55). The flow meter usage has more similarity to hydraulic motor usage than hydraulic pump usage. The flow force drives the flow meter. The flow meter is designed such that the hydraulic energy spent in driving the meter is minimal. Since the *displacement* of the flow meter (D) is known for a given pump, if we measure the angular speed of the shaft, then the flow rate can be calculated using the same equation given above.

FIGURE 6.55: Positive displacement type flow meters: gear and lobe type positive displacement flow meters are shown.

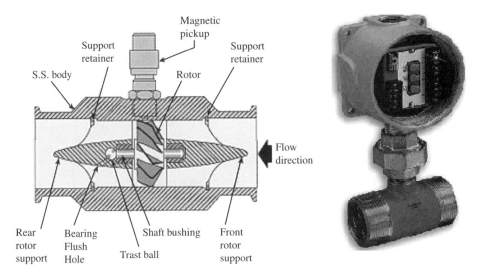

FIGURE 6.56: Turbine type flow rate sensor.

The analogy between the tachometer (speed sensor) and DC brush-type electric motor is similar to the analogy between the positive displacement type flow meters and hydraulic motors. Both motor and sensor version work on the same principle, except that in sensor applications the objective of the design is not to convert energy from one from to another (electric to mechanical or hydraulic to mechanical), but rather to measure speed or flow rate.

Turbine Flow Meters   The turbine flow meter (Figure 6.56) has a turbine on a shaft placed inline with the direction of the flow. As a result of the drag between the fluid flow and the turbine, the turbine rotates about its shaft.

$$Flow\ rate \rightarrow Turbine\ speed \rightarrow Speed\ sensor \rightarrow V_{out} \qquad (6.178)$$

The speed of the turbine is proportional to the flow speed (hence the flow rate) of the fluid. The linear proportionality constant is an approximate relationship and holds well at high flow rates. Such flow meters are not suitable for measuring low flow rates. The speed of the turbine is measured and converted to output voltage using any one of the speed sensors.

Drag Flow Meters and Vortex Flow Meters   The drag based flow meter inserts a sensing object into the flow so as to pickup drag force from the flow. The drag force is measured by a strain-gauge based force sensor (Figure 6.57). It turns out that the drag force is proportional to the square of the speed.

$$F_{drag} = \frac{C_d\ A\ \rho\ u^2}{2} \qquad (6.179)$$

where C_d is the drag coefficient calibrated for the specific sensor, A is the drag surface of the sensing object, ρ is fluid density, u is the speed of flow. The transduction principle and sensor output relationship is as follows,

$$Flow\ rate \rightarrow Drag\ Force \rightarrow Strain \rightarrow V_{out} \qquad (6.180)$$

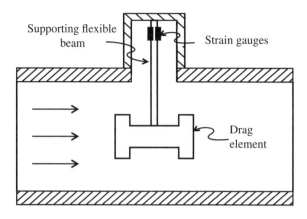

FIGURE 6.57: Strain-gauge based drag measurement type flow rate sensor.

A vortex flow meter uses an object inserted into the flow field where the object sheds vortices as the flow passes over its surface (Figure 6.58). It turns out that the frequency of the vortex shedding is proportional to the speed of the flow, hence flow rate. Then a transducer that counts the vortex frequency can provide a proportional output voltage. As a result, the sensor output voltage can be calibrated to represent the flow rate.

6.10.2 Differential Pressure Flow Rate Sensors

Flow rate sensors based on differential pressure measurements make use of the Bernoulli's equation, which is a relationship between the pressure and speed of fluid flow at two different points. It is estimated that over 50% of all flow rate measuring devices are based on differential pressure type sensors. Assume that the height of the fluid does not change relative to a reference plane between the two points. Then the pressure and speed of the fluid at two separate cross-sections are related by

$$p_1 + \frac{\rho\, u_1^2}{2} = p_2 + \frac{\rho\, u_2^2}{2} \tag{6.181}$$

Pitot Tube A pitot tube is a differential pressure measurement sensor which makes use of the Bernoulli equation for a special case (Figures 6.59, 6.60). The pressure is measured at two points. At one of the points, the speed of the fluid is zero $u_2 = 0$.

Then, measuring the differential pressure $p_2 - p_1$ allows us to calculate the fluid velocity along its flow stream. Once the fluid velocity is known, assuming that it is the average fluid speed, the flow rate can be calculated using the cross-sectional area information.

$$u_1 = \frac{2}{\rho}(p_2 - p_1)^{1/2} \tag{6.182}$$

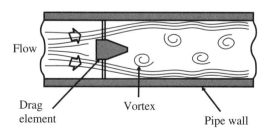

Flow

Drag element   Vortex   Pipe wall

FIGURE 6.58: Vortex frequency counting type flow rate sensor.

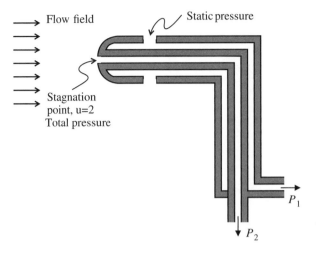

FIGURE 6.59: Pitot tube for flow rate sensing via diffential pressure measurement.

Note that Bernoulli's equation is valid for a compressible fluid flow up to a Mach number of about 0.2. For higher Mach numbers, the relationship can be modified and the sensor calibrated to accurately obtain the flow speed as a function of the differential pressure and Mach number.

As the flow passes around the Pitot tube, vortices are shed as a result of the tube surface and fluid interaction. The frequency of the vortices is a function of the flow rate and the pitot tube diameter. A given Pitot tube can experience a certain vortex shedding frequency around a certain flow rate which may result in exciting the natural frequency of the Pitot tube. Each Pitot tube sensor specification includes the range of flow rate to avoid in order to make sure that it is not excited around its natural frequency.

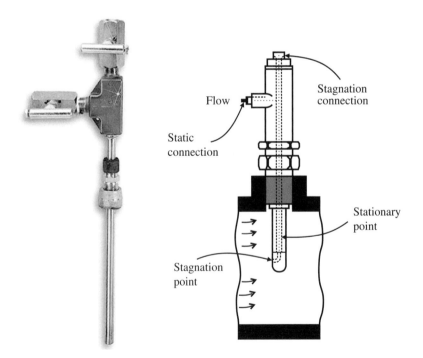

FIGURE 6.60: Pitot tube sensor picture and typical application.

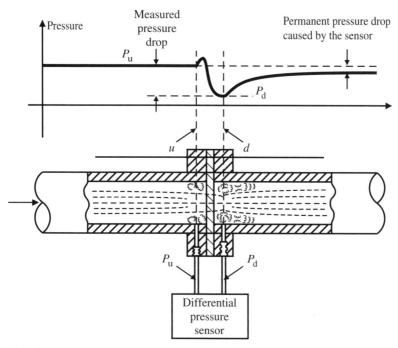

FIGURE 6.61: Standard obstruction orifices to measure flow rate via differential pressure measurements.

Obstruction Orifices Another method of flow rate measurement is to insert a standard profile obstruction orifice on the pipe where the flow rate measurement is desired (Figure 6.61). The pressure differential at the input and output side of the standard obstruction profile is measured and related to the flow rate. There are many different types of standardized obstruction orifices. The flow rate is related to the differential pressure, cross-sectional area, and the geometeric shape of the the standard obstruction orifice,

$$Q = f(p_1, p_2, A, \textit{Geometry of Obstruction}) \qquad (6.183)$$

where A is the cross-sectional area. Different obstruction orifice shapes and sizes are calibrated using a higher accuracy flow rate sensor in order to define the above relationship for each specific orifice only as a function of the pressure differential. Hence, for a given obstruction (geometric shape and size is defined), the flow rate as a function of pressure differential is a calibrated data table,

$$Q = f(\Delta p) \qquad (6.184)$$

6.10.3 Flow Rate Sensor Based on Faraday's Induction Principle

Faraday's electromagnetic induction principle is used in some flow rate sensors to measure the flow rate of an electrically conductive liquid. The fluid must be electrically conductive for this sensing technique to work.

The basic principle is that if a conductor moves in a magnetic field, there is a voltage developed across the conductor. This is referred to as the *generator principle* in electric motors. The same principle is used to measure the flow rate of a fluid. Given a fixed electromagnetic field (generated by a permanent magnet or electro-magnet (coil with

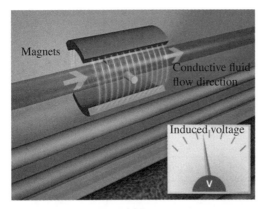

(a) Principle of operation

(b) An application example

FIGURE 6.62: Flow rate sensor based on Faraday's electromagnetic induction principle. B is the magnetic field strength, V is the velocity of the fluid flow (flow rate), E is the voltage developed as a result of B and V. Let B be constant, then E (induced voltage) is proportional to V (fluid flow speed).

current passing through it)), and a conductive fluid passing through it at a certain speed, there will be voltage potential developed in the direction perpendicular to both the magnetic field and flow vector and the magnitude of the voltage is proportional to the speed (flow rate) of the fluid (Figure 6.62).

6.10.4 Thermal Flow Rate Sensors: Hot Wire Anemometer

The most well-known thermal measurement based flow rate sensor is the *hot wire anemometer*. The basic transduction principle is as follows: there is a heat transfer between any two objects with different temperatures. The rate of heat transfer is proportional to the temperature difference between them. In the case of a flow rate sensor, the two objects are the sensor head and the fluid around it (Figure 6.63). The effective heat transfer coefficient between the sensor and the fluid is dependent upon the speed of the flow. This relationship is

$$\dot{H} = (T_{\mathrm{w}} - T_{\mathrm{f}}) \, (K_{\mathrm{o}} + K_1 u^{1/2}) \tag{6.185}$$

where $\dot{H}$ is the heat transfer rate, T_{w} the temperature of the tungsten wire used by the sensor, T_{f} the temperature of the fluid, u is the fluid flow speed, K_{o} and K_1 are sensor calibration constants.

Tungsten wire

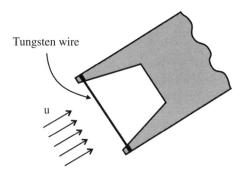

FIGURE 6.63: Hot wire anemometer for flow rate measurement using the thermal heat transfer principle.

This relationship is used in the hot wire anemometer. A tiny probe with a tunsgten wire (length in the range of 1 to 10 mm and diameter in the range of 1 to 15 μm) is placed in the flow field. The resistance of the tungsten probe is proportional to its temperature.

$$R_{\mathrm{w}} = R_{\mathrm{w}}(T_{\mathrm{w}}) \tag{6.186}$$

As a current is passed through the tungsten wire, heat is transferred from the wire to the fluid.

$$\dot{H} = R_{\mathrm{w}} \cdot i^2 \tag{6.187}$$

The heat transfer rate depends both on the temperature difference and fluid speed. The tungsten wire current is controlled in such a way that its temperature (and hence resistance) is held constant. The amount of heat transferred can be estimated from the current and resistance measurements on the sensor. Assuming that the fluid temperature is also constant (or measured separately by a temperature sensor), we can calculate the flow speed. In other words, K_0, K_1 is known in advance. We assume that the fluid temperate is also known, T_{f}. Then, we control the current through the tungsten wire in order to control its temperature, T_{w}, and while achieving that we measure $\dot{H} = R_{\mathrm{w}} \cdot i^2$, the heat rate it generates. Hence, we can calculate the fluid speed u, from which the flow rate can be estimated.

Example applications of a *thermal flow rate sensor* are shown in Figure 6.64, which operate based on the "hot film anomometer" principle. The principle is that a flowing gas transfers heat from a heated probe and the heat transfer rate is proportional to the flow rate of gas. Then, if a sensor head is heated in a controlled way, we can determine the gas flow rate based on how much heat is lost.

6.10.5 Mass Flow Rate Sensors: Coriolis Flow Meters

The coriolis flow meter measures the mass flow rate as opposed to the volume flow rate. Therefore, it is not sensitive to temperature, pressure, or viscosity variation in the fluid flow. The sensor includes a U-shaped tube, a magnetically excited base which excites the U-shaped tube at a fixed frequency, that is around 80 Hz (Figure 6.65). The interaction between the inertial force of the incoming fluid in one arm of the U-shaped tube and the vibration of the tube creates a force in perpendicular direction to the direction of flow and the direction of the tube vibration. The forces acting on the two sides of the U-shaped tube are in opposite directions, which creates a twist torque around the tube. The twist torque, hence the twist angle of the tube, is proportional to the mass flow rate of the fluid.

The frequency of the twist is the same as the frequency of the tube's base oscillation. The output of the twisting motion is measured as an oscillating angular displacement. The magnitude of the twisting oscillations is proportional to the flow rate.

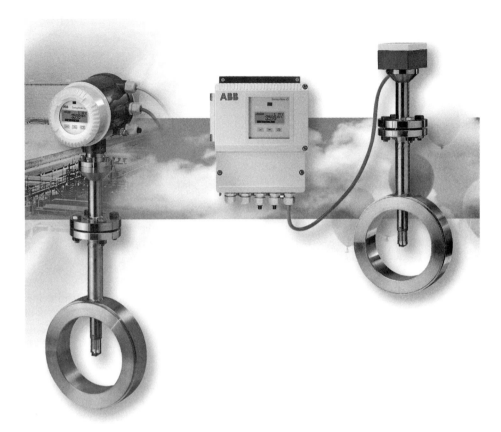

FIGURE 6.64: Thermal flow rate sensor head (transducer, or also called a "probe") and the signal conditioning/processing unit (which may be packaged together or separately and connected by a cable). Reproduced with permission from ABB Ltd.

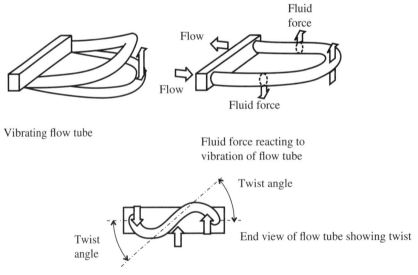

Vibrating flow tube

Fluid force reacting to vibration of flow tube

Twist angle

Twist angle

End view of flow tube showing twist

FIGURE 6.65: Coriolis mass flow rate sensor.

6.11 HUMIDITY SENSORS

Relative humidity is defined as the percentage ratio of the amount of water vapor in moist air, versus the amount of water vapor in saturated air at a given temperature and pressure. Relative humidity is strongly affected by temperature.

The main types of humidity sensors use the *capacitance, resistive, and optical reflection* principles in the transduction stage.

Capacitive humidity sensors use a polymer material which changes capacitance as a function of humidity. The relationship is fairly linear. The sensor is designed as parallel plates with porous electrodes on a substrate. The electrodes are coated with a dielectric polymer material that absorbs water vapor from the environment with changes in humidity. The resulting change in dielectric constant causes a variation in capacitance. The variation in capacitance is converted to voltage by an appropriate op-amp circuit to provide a proportional voltage output.

Resistance based humidity sensors use materials on an electrode whose resistance change as a function of humidity. In general, the resistance–humidity relationship is an exponential relationship, and hence requires digital signal processing so that the voltage output is proportional to the measured humidity. Capacitive humidity sensors are more rugged and have less dependency on the temperature than the resistive humidity sensors.

The chilled mirror hygrometer (CMH) is one of the most accurate humidity measurement sensors. It measures humidity by the dew point method (Figure 6.66). The operating principle is based on the measurement of the reflected light from a condensation layer which forms over a cooled mirror. A metallic mirror with good thermal conductivity is chilled by a thermoelectric cooler to a temperature such that the water on the mirror surface is in equilibrium with the water vapor pressure in the gas sample above the mirror surface. When the mirror is chilled to the point that dew begins to form and the equilibrium is maintained, a beam of light is directed at the mirror surface, and the photodetectors measure the reflected light. The reflected light is scattered as a result of the dew droplets on the mirror surface. In order to maintain a constant reflected light, the photodetector output is used to control the thermoelectric heat pump to maintain the mirror at dew point temperature. Then the measured temperature is related to the humidity of the gas sample. The resolution of the CMH humidity sensor is about one part in 100 of its measurement range.

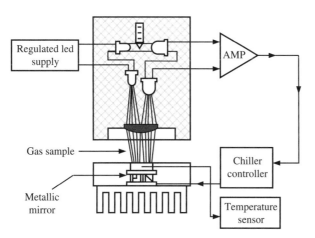

FIGURE 6.66: Humidity sensor: chilled mirror hygrometer.

6.12 VISION SYSTEMS

Vision systems, also called computer vision or machine vision, are general purpose sensors. They are called the "smart sensors" in industry since what is sensed by a vision system totally depends on the image processing software. A typical sensor is used to measure a variable, that is temperature, pressure, length, and so on. A vision system can be used to measure shape, orientation, area, defects, differences between parts, and so forth. Vision technology has improved significantly over the past 20 years, to the extent that they are rather standard "smart sensing components" in most factory automation systems for part inspection and location detection. Their lower cost makes them increasingly attractive for use in automated processes. Furthermore, vision systems are now standard in mobile equipment safety systems to detect obstacles and avoid collisions, especially when used with radar based obstacle detection systems.

There are three main components of a vision system (Figures 6.67, 6.68),

1. vision camera: this is the sensor head, made of a photosensitive device array such as charge coupled device (CCD) and optical lens,

2. image processing computer hardware (converts the CCD voltage to digital data) and software to process the image,

3. lighting system.

The basic principle of operation of a vision system is shown in Figure 6.67. The vision system forms an image by measuring the reflected light from objects in its field of view. The rays of light from a source (i.e., ambient light or structured light) strike the objects in the field of view of the camera. The part of the reflected light from the objects reaches the sensor head. The reflected light may be passed through an optical lens then to the CCD. The sensor head is made of an array of photosensitive solid-state devices such as photodiodes or charge coupled devices (CCD) where the output voltage at each element is proportional to the time integral of the light intensity received. The sensor array is a finite

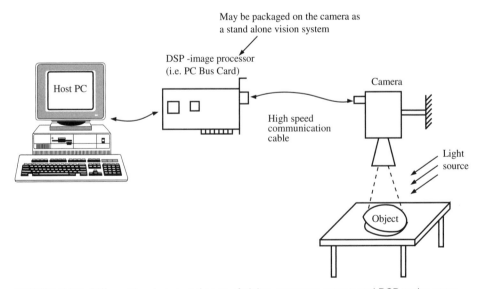

FIGURE 6.67: Different hardware packages of vision systems: sensor and DSP at the same physical location, or sensor head and DSP are at different physical location and digital data is transferred from sensor head to the DSP using a high speed communication interface.

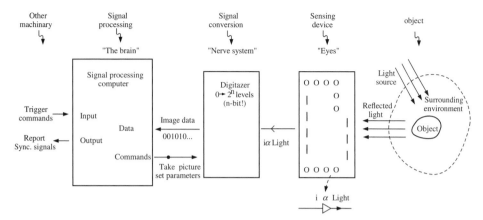

FIGURE 6.68: Components and functions of a vision system.

number of CCD elements in a line (i.e., 512 elements, 1024 elements, etc.) for the so-called *line-scan cameras* or a finite array of two-dimensional distribution (i.e., 512 × 512, 640 × 640, 1024 × 1024) as shown in Figure 6.69. A field of view in the real world coordinates with dimensions $[x_f, y_f]$ is mapped to the $[n_x, n_y]$ discrete sensor elements. Each sensor element is called *a pixel*. The spatial resolution of the camera, that is the smallest length dimension it can sense in x and y directions, is determined by the number of pixels in the sensor array and the field of view that the camera is focused on,

$$\Delta x_f = \frac{x_f}{n_x} \tag{6.188}$$

$$\Delta y_f = \frac{y_f}{n_y} \tag{6.189}$$

where $\Delta x_f, \Delta y_f$ are the smallest dimensions in x and y directions the vision system can measure. Clearly, the larger the number of pixels, the better the resolution of the vision system. A camera with a variable focus lens can be focused to different field of views by adjusting the lens focus without changing the distance between the camera and field of view, and hence changing the spatial resolution and range of the vision system.

The light source is a very important, but often neglected, part of successful vision system design. The vision system gathers images using the reflected light from its field of

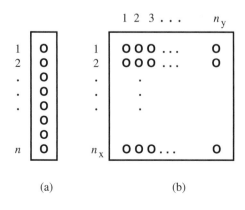

(a)   (b)

FIGURE 6.69: Vision sensor head types: (a) line-scan camera where the sensor array is arranged along a line, and (b) two-dimensional camera where the sensor array is arranged over a rectangular area.

view. The reflected light is highly dependent on the source of the light. There are four major lighting methods used in vision systems:

1. back lighting, which is very suitable in edge and boundary detection applications,

2. camera mounted lighting, which is uniformly directed on the field of view and used in surface inspection applications,

3. oblique lighting, which is used in inspection of the surface gloss type applications,

4. co-axial lighting which is used to inspect relatively small objects, such as threads in holes on small objects.

An image at each individual pixel is sampled by an analog to digital converter (ADC). The smallest resolution the ADC can have is 1-bit. That is the image at the pixel would be considered either white or black. This is called a *binary image*. If the ADC converter has 2-bits per pixel, then the image in each pixel can be represented in one of four different levels of gray or color. Similarly, an 8-bit sampling of pixel signal results in $2^8 = 256$ different levels of gray (*gray scale image* or colors in the image). As the sampling resolution of pixel data increases, the gray scale or color resolution of the vision system increases. In gray scale cameras, each pixel has one CCD element whose analog voltage output is proportional to the gray scale level. In color sensors, each pixel has three different CCD element for three main colors (red, blue, green). By combining different ratios of the three major colors, different colors are obtained.

Unlike a digital camera used to take pictures where the images are viewed later on, the images acquired by a computer vision system must be processed at periodic intervals in an automation environment. For instance, a robotic controller needs to know whether a part has a defect or not before it passes away from its reach on a conveyor. The available processing time is in the order of milliseconds and even shorter in some applications such as visual servo applications. Therefore, the amount of processing necessary to evaluate an image should be minimized.

Let us consider the events involved in an image acquisition and processing.

1. A control signal initiates the exposure of the sensor head array (camera) for a period of time called *exposure time*. During this time, each array collects the reflected light and generates an output voltage. This time depends on the available external light, and camera settings such as aperture.

2. Then the image in the sensor array is locked and converted to digital signal (A to D conversion).

3. The digital data is transfered from the sensor head to the signal processing computer.

4. Image processing software evaluates the data and extracts measurement information.

Notice that as the number of pixels in the camera increases, the computational load, and the processing time, increases since the A/D conversion, data transfer, and processing all increase with the increasing number of pixels and the resolution of each pixel (i.e., 4-bit, 8-bit, 12-bit, 16-bit). The typical frame update rate in commercial two-dimensional vision systems is at least 30 frames/s. Line-scan cameras can easily have frame update rate around 1000 frames/s.

The effectiveness of a vision system is largely determined by its software capabilities. That is, what kind of information it can extract from the image, how reliably can it do it, and how fast can it do it. Standard image processing software functions include the following capabilities.

1. Thresholding an image: once an image is aquired in digital form, a threshold value of color or gray scale can be selected, and all pixel values below that value (white value)

are set to one fixed low value, and all pixel values above that value are set to a high value (i.e., black value). This turns the image into a binary image. Various detection algorithms can be run quickly on such an image, such as the edge detection algorith.

2. Edge detection of an object: an edge is detected when a sharp change occurs from one pixel to another in the gray scale value of the image. Once such a transition is detected between two pixels in a search direction on the image array, then all the pixels around the transition pixel are searched to determine the edge boundary.

3. Color or gray scale distribution (also called the histogram of image): this is a plot of the gray scale distribution of the image, that is how many pixels (y-axis) has a given gray scale (x-axis).

4. Connectivity of the object (detect discontinuities).

5. Image comparison with a reference image in memory (also called template matching), that is image system may store a set of "good part" images. A real-time image is compared to the stored images to determine whether it matches one of the template images or not.

6. Position and orientation of an object relative to another reference.

7. Dimensions (length, area) of an image: once the boundaries of a part or parts in an image are detemined, the dimensions and area information can be easily calculated.

8. Character recognition, that is recognizing letters and numerals.

9. Geometric image transformations, that is mathematical operations on the matrix data of the image to move, rotate, stretch the image.

Example Consider a vision system with 1024×1024 pixel resolution. The camera processes 60 frames per second. The resolution of the the the A/D converter system is 8-bit. What is the amount of data (bytes) processed per second? If the camera is focused into an surface with $10\,\text{cm} \times 10\,\text{cm}$ dimensions, what is the spatial resolution in measurement?

Each pixel holds one byte of data since the ADC converter has 8-bit resolution. The number of bytes processed per second is equal to the number of data bytes per frame times the number of frames per second,

$$N = 1024 \times 1024 \times 60 \tag{6.190}$$
$$= 62\,914\,560\,\text{bytes/s} \tag{6.191}$$
$$\approx 63\,\text{MB/s} \tag{6.192}$$

Clearly, the amount of data processed per second is very large in a high resolution camera. The smallest distances in x and y directions the system can measure are

$$\Delta x_\text{f} = \frac{10\,\text{cm}}{1024} = 0.00976\,\text{cm} = 0.0976\,\text{mm} \tag{6.193}$$
$$\Delta y_\text{f} = \frac{10\,\text{cm}}{1024} = 0.00976\,\text{cm} = 0.0976\,\text{mm} \tag{6.194}$$

which indicates that the vision system can measure dimensions in x and y coordinates with about 0.1 mm accuracy.

6.13 GPS: GLOBAL POSITIONING SYSTEM

The global positioning system (GPS) is a location measurement system on Earth, based on signals received from a group of satellites (about 24 satellites) orbiting the Earth at an

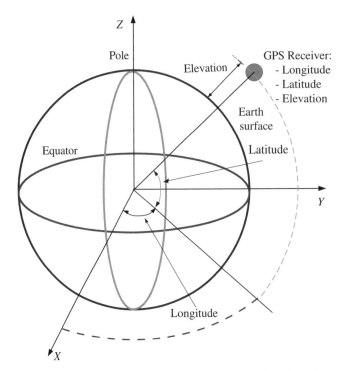

FIGURE 6.70: GPS data: longitude, latitude, and elevation of a receiver on earth, used to measure the position of the receiver on Earth.

altitude of approximately 20 000 km [11]. GPS provides us with the *longitude, latitude,* and *elevation* data for the GPS receiver placed at any point on Earth as shown in Figure 6.70.[2]

The trajectories of the locations of the satellites around the Earth are very predictable and can be determined accurately by the Earth-based control stations. The position of the satellites is known relative to the fixed locations of the control stations on Earth and is updated periodically by the control stations, which are operated by the US Department of Defense (US-DoD). The location of each GPS satellite is transmitted from the control stations to the satellites, then from the satellites to Earth as a broadcast signal. The GPS satellites broadcast their position information. An individual GPS receiver on Earth determines its global position coordinates by processing the received GPS broadcast signals. GPS receivers are used for *positioning, locating, navigating flights and cars, warehouse management, fleet tracking, mining site management, and surveying* applications.

There are three major components to a GPS system:

1. space based orbiting group of GPS satellites,
2. Earth based operational control stations,
3. GPS receiver.

A GPS user needs to only setup the GPS receiver. As GPS receiver units become smaller and less expensive, there is a growing number of applications for GPS and it has become accessible to virtually everyone. In transportation applications, GPS assists pilots and drivers in pin-pointing their locations and avoiding collisions. Farmers can use GPS to guide equipment and control accurate distribution of fertilizers and other chemicals. To facilitate

---

[2] This section is based on the tutorial information provided on GPS technology at the web page www.Trimble.com.

grading of large areas, GPS controlled earthmoving equipment, including bulldozers, back-hoes, road graders, excavators, and compactors, can greatly increase productivity during the land development process.

Applications of GPS can be divided into two major categories:

1. **Location and Tracking**: The most common and basic use of GPS is the determination of a location of a receiver on Earth. The GPS signal is used to determine the "where" ("locating where it is now") an object is and is it moving ("tracking"). GPS offers the location data (latitude, longitude, and elevation) for any receiver point on Earth. Tracking is the process of monitoring the movement of things. The tracked object must have a GPS receiver to detect its position information, and the ability to transmit that data to the tracking station. By processing the GPS broadcast signal, a GPS receiver determines its location on Earth. By transmitting this location information to a remote control station (i.e., by wireless radio signal transmission), it enables a tracking capability at a central control location. As a result, a central control station can received the reported location information from multiple GPS receivers on a fleet, and keep track of their locations.

2. **Navigation and Geospatial Information System (GIS)**: GIS is a collection of computer software and geographic data for processing all forms of geographically referenced information in a database format. GPS is the "sensor" used to measure and provide the location data to all this geographically referenced information. In other words, for any point of interest, the "where" information is provided by GPS. The "what and how" information is provided by the GIS software system. GIS systems are widely used in surveying (obtaining a digital terrain map of a section of Earth's surface) for road building, mining, and farming applications. Apart from locating a point on the planet, GPS signal processing tools also provide information as to how to travel from one location to some other location on the planet. GPS provides navigation information for ships and planes. Given a desired destination, if the current position of an object can be determined via GPS, then a motion path can be planned. This planned path can be update in real-time as the object moves closer to the desired destination. Obstacles along the way or other constraints can be imposed to modify the planned path.

6.13.1 Operating Principles of GPS

GPS employs about 24 satellites in 20 000 km circular orbits [12]. These satellites are placed in six orbit planes with four operational satellites in each plane. The satellites orbit the Earth every 11 hours and 58 minutes. A GPS receiver receives the GPS broadcast signal and and uses the "*trilateration*" method to calculate the receiver's location. A GPS receiver must be locked on to signals from at least three satellites to determine its location (latitude, longitude, and altitude). Using a fourth satellite signal allows the GPS receiver to determine its position more accurately. Due to the synchronization error between the receiver clock and the atomic clocks on the GPS satellites, the signal from the fourth satellite will not intersect at an exact point in the trilateration. Then, the GPS receiver firmware searches for a single correction that can be applied to all four satellite signals that would make the trilateration algorithm work.

The basic setup of the GPS receiver is shown in Figure 6.71. The signals transmitted from the satellites are received by the antenna. This signal is then filtered and amplified using a signal conditioning unit. The conditioned signal is later digitized by an ADC (analog-digital converter). This digitized signal is then sent to the software block for post processing. The user position (the GPS receiver position) obtained can be presented in terms

FIGURE 6.71: GPS receiver product examples. By Stepshep (Own work) [CC-BY-SA-3.0 (http://creativecommons.org/licenses/by-sa/3.0)], via Wikimedia Commons.

of longitude, latitude, and elevation. This data can also be used by other electronic control modules on board the vehicle for computing the actuator commands and control signals.

A list of major GPS received suppliers are given in the references (www.trimble.com, www.garmin.com, www.novatel.ca, www.magelllangps.com, www.sokkia.com).

Trilateration is the method of determining the relative position of an object using basic geometry. To understand the concept of trilateration, let us assume that we need to determine the location of an object on a 2D plane (Figure 6.72).

Let us assume that we have three source (broadcast) signals A, B, C. The location of these three signal sources A, B, C are known relative to a fixed reference point on Earth. This information is determined by the Earth based operational control stations and sent to satellites periodically. Let us assume that the receiver is 10 miles away from source A. This puts the object on the perimeter of a circle with a center as point A and a radius of 10 miles. Next, we also know that the object is 20 miles from source B. This puts the object on the perimeter of circle B, with a center as point B and a radius of 20 miles (Figure 6.72a). Now, we know that the object is located at one of the two points of the intersection of the two circles as shown in Figure 6.72a. To decide which of the two points is the true location, we can define a third circle, for example as being 15 miles away from source C. This defines the exact two-dimensional (2D) coordinates of the object to be located [X,Y].

This concept of trilateration is extended to determine the location of any object in three-dimensional space (Figure 6.72b). A GPS receiver measures its distance from three (3) satellites. If we know that the object is 10 miles from satellite A, we can place the object on a sphere of 10 miles radius with center as satellite A. If the object also lies 20 miles from satellite B, we know that the object also lies on a second sphere with a 20 mile radius centered on satellite B. This puts the object anywhere on the circumference of the circle formed by the intersection of the two spheres. Then a measurement from a third satellite C, we get a third sphere, which intersects the circle on two points. The Earth itself acts as the fourth sphere. Only one of the two possible points will be actually on the surface of the Earth. Receivers generally look to four or more satellites to improve accuracy and provide precise altitude (*z*-coordinate) information.

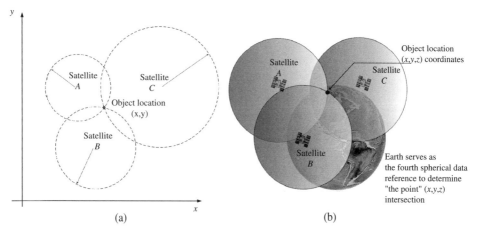

FIGURE 6.72: (a) Concept of two-dimensional trilateration method, (b) concept of three-dimensional trilateration method (Source: Trimble.com).

The distance between a GPS satellite and the GPS receiver is determined by measuring the amount of time it takes for the radio signal (the GPS signal) to travel from the satellite to the receiver. In other words it amounts to measuring the time taken for signal transmission from the GPS satellite to the receiver and calculating the distance by multiplying it with the signal travel speed. Radio signals travel at the speed of light which is approximately 300 000 km/s.

$$S = v \cdot t$$

where S = distance from satellite to receiver, to be determined

v = velocity of radio signal, which is known, 300 000 km/s

t = travel time (measured variable) of the GPS signal (6.195)

from the satellite to the receiver (6.196)

The travel time of a GPS signal from a satellite to a receiver on Earth is in the range of 60–100 ms. This is an inherent time delay in position measurement and its effect on closed loop control should be considered. Assume that the clocks on the GPS satellites and the ground receivers are perfectly synchronized. To understand the concept, let us assume that when a signal is transmitted from the satellite, the "time of transmission" is encoded in the signal. The receiver records the "receive time" of the signal. Then the travel time is calculated at the receiver as the difference.

In reality, both the satellite and receiver generate a pseudo-random code. Each satellite has a unique pseudo-random code. The receiver knows this code for each satellite. The GPS receiver manufacturers program these known codes into the firmware (embedded software) on each receiver. By comparing the time difference (phase) between the satellite's pseudo-random code received compared to the receiver's code, we determine the time period it took to reach us. This gives us the signal travel time from the GPS satellite. However, it should be noted that all this is possible only if the clock on the receiver and the clock on the satellite are accurately synchronized.

If the receiver's clock was perfect, then perfect measurements from three satellites would intersect at a single point, which is the receiver position. However, with imperfect measurements, because of imperfect clocks and signal transmission medium (atmospheric) distortions, a fourth satellite measurement will not intersect with the first three. The computer on board the receiver looks at the offset from universal time and looks for a single correction factor for the receiver's clock that it can make to all the timing values to make all four measurements intersect at a single point. The measured values from the three satellites do not intersect at one single point for 2D case (four satellites for 3D case). However, when the clock correction value is applied to all the three measurements, the measurements from the satellites intersect to locate the object. This correction brings the clock into "synchronization" with the atomic accuracy clock of the satellites.

6.13.2 Sources of Error in GPS

There are five groups of major sources of error in GPS, briefly discussed below:

1. **Clock Errors:** Any difference between the GPS satellite based clock and receiver based clock leads to direct position measurement error.

2. **Errors due to Atmospheric Conditions:** Earth's atmosphere refracts the GPS signals and this causes an error in measurement of distance. Furthermore, the induced error will be different between different satellites and the receiver.

3. **Errors due to Relative Location of GPS Satellites:** The relative location of the satellites with respect to each other and the receiver also plays an important role in determining the position measurement accuracy. Good satellite distribution geometry is obtained when the satellites used for measurement are spread out in space.

4. **Multipath Error:** The basic concept of GPS assumes that a GPS signal travels straight from the satellite to the receiver. Unfortunately, in the real world the signal will also bounce around on the local environment and get to the receiver. The result is many signals arriving at the receiver, first the direct one, then a bunch of delayed reflected ones. If the bounced signals are strong enough they can confuse the receiver and cause erroneous measurements.

6.13.3 Differential GPS

Standard GPS signal accuracy is negatively affected by various atmospheric distortions, hence reducing the GPS positioning accuracy. The GPS accuracy can be improved by the so called "differential GPS" (DGPS).

The underlying principle of differential GPS (DGPS) is that any two receivers that are relatively close together on Earth will experience the same atmospheric errors. Let us assume that one of the receivers is stationary, and the other receiver may be moving. DGPS requires that one of the GPS receivers be set up on a precisely known location. This GPS receiver is the base (reference) station whose location on Earth is precisely known relative to a reference location on Earth, that is control station. The base station receiver calculates its position based on satellite signals and compares it to the known location. The difference is applied to the GPS data recorded by the second GPS receiver. The basic assumption is that since the base station and receiver are closely located on Earth, any GPS error induced (i.e., due to atmospheric conditions, multi path, or relative location errors) will be induced by almost the same amount on both signals received at the base station and the receiver. If we can determine that induced error using the known location of the base station, then we can cancel the same amount of error from the signal received by the receiver. The corrected information can be applied to data on the second receiver and transmitted in real-time in the field using radio transmitters (Figure 6.73).

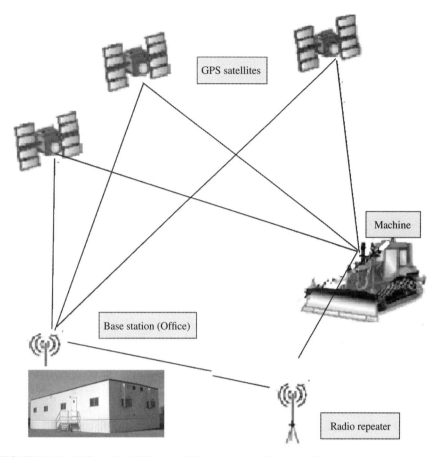

FIGURE 6.73: Differential GPS in real-time operation (Source: Trimble.com).

6.14 PROBLEMS

1. Describe the operating principles and derive the excitation voltage, induced voltage, and output voltage relationship for the type of rotary variable differential transformer called a syncro (Figure 6.14). Suggest a signal conditioner circuit design using integrated circuits (ICs) to support the excitation voltage requirement and process the output signal of the syncro to provide an analog voltage or digital output signal that is proportional to the angular position of the rotor.

2. Consider a translational (linear) positioning stage driven by a servo motor (Figure 3.3). The travel range of the table is 50 cm. The required positioning accuracy of the stage is 0.1 μm. a) Provide a list of possible position sensors for this application, discuss their relative advantages and disadvantages. What type of sensor is most appropriate for this application? b) Assume that a motor based rotary incremental encoder is used for position sensing. What should be the resolution of the encoder, if the ball-screw pitch is 0.5 rev/mm? If the decoder circuit for the encoder can handle encoder signal frequency upto 1 MHz, what is the implication of this on the motion limits? c) Assume that the stage has a tool and that we need to detect the contact between the tool and a workpiece. After detecting the contact, we need to further move the stage in the same direction by 0.1 mm. What kind of sensor(s) would be approporiate for this purpose? d) Discuss the advantages and disadvantages of placing the position sensor on the motor shaft (rotary displacement, angle, sensor attached to motor) or on the tool (linear displacement sensor attached to tool).

3. Consider a rotary motion axis and its speed control (i.e., speed control of a spindle of a CNC machine tool or speed control of a precision conveyor). Assume that both high speed (w_{max} = 3600 rpm) and low speed w_{min} = 1 rpm speed control is required with 0.01 rpm regulation accuracy. a) Determine specifications for a tachometer to meet the speed control requirements. b) Select an optical incremental encoder and derive speed information from the position pulses digitally. What is the speed estimation accuracy at minimum speed if the sampling period is 1.0 ms? Show how the speed estimation can be improved with the *time period* measurement technique. What is the maximum pulse frequency of the selected encoder at the maximum speed?

4. Consider the following measurement problems: i) seismic displacement measurement for earthquake monitoring, ii) three dimensional (horizontal, vertical, and lateral) acceleration of a car body during travel, iii) and measurement of the same during an crash test. Discuss the expected range of frequency content of the signals in each application and select appropriate sensors for the measurement.

5. Consider the stress and force measurement on a rectangular beam. Assume that a horizontal force is applied at the tip of the beam, and the other end is clamped to a stationary base (Figure 6.74). A strain gauge is bonded on the surface at the mid point along the length of the beam in the direction of the deformation. a) What other information is needed in order to measure the strain, stress, and external force applied on the beam? b) If the maximum force expected is 1.0 kNt, select a strain gauge for the measurement with a Wheatstone bridge circuit. Assume the material of the beam is steel and cross-sectional dimensions (in the direction perpendicular to the force) are 10 cm × 5 cm. c) Can the same measurement system be used to measure a vertical force? If so, what other information is needed about the beam?

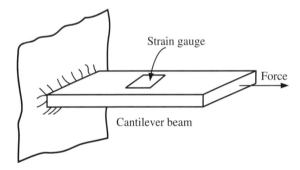

Strain gauge

Force

Cantilever beam

FIGURE 6.74: Strain, stress, and force measurement on a deformable beam under external load.

6. **(a)** Consider a strain gauge with nominal resistance of 120 Ω. It is glued on the surface of a structure. The structure deforms nominally 0.001 strain level ($\epsilon = \Delta l / l = 0.001$). Assume that the gauge factor of the strain gauge is 2.0. Determine the change in the nominal resistance of the strain gauge under this condition.

(b) Design a circuit to measure the strain as proportional voltage output to a data acquisition board which has 0–5 V input range. When the strain level is 0.001, the output voltage should be 5.0 V.

7. Consider the hydraulic motion control system shown in Figure 7.3. It is necessary to measure the pressure difference between pump output and the valve output. In other words,

$$\Delta p = \max(p_P - p_A, p_P - p_B) \tag{6.197}$$

which is typically used to control the pump displacement to provide a constant pressure drop across the valve (called the *load sensing* pump control method). It is anticipated that the maximum pump output pressure is 15 MPa and the pressures at point A and B will be lower than that. Select a set of pressure transducers which will provide the desired measurement range and 1% accuracy over a frequency range of 0 Hz to 1 KHz.

8. **(a)** What are the basic temperature measurement principles and main differences between the following temperature sensors: 1) RTD, 2) thermistor, 3) thermocouple?

(b) What type of temperature sensor would you recommend for the following cases and why? 1) Temperature measurement of a small volume electronic part with highly transient temperature condition. 2) Temperature measurement of a liquid in a large container.

9. Consider a PC based temperature measurement system. The PC has a data acquisition card that has 12-bit resolution, and can multiplex 16 channels of analog signals (16-channel, multiplexed ADC with 12 bit resolution). The ADC card has a gain of 10 which amplifies the thermocouple voltage before the ADC sampling. The range of input voltage to the board is 0 to 100 mV, which is amplified to the 0–1.0 V range before the ADC sampling and conversion. Assume that all 16 channels use J-type thermocouples and that the PC card has a reference junction compensator which provides a reference temperature equivalent to 0 °C. Draw a functional block diagram of this data acquisition system. What is the voltage input if the nominal temperature is 100 °C? What is the voltage input if the nominal temperature is 200 °C? What is the estimated measurement error in units of °C, assuming that the error in reference junction compensator is 0.5 °C and the accuracy of the thermocouples is ±0.25 °C. In calculating the measurement error, use the worst case scenario where the measurement errors are additive ($e_{sum} = \sum_{i=0}^{n} e_i$). Given the digital value of the voltage conversion by ADC, what is needed in software to relate the voltage to temperature data? Discuss if this would be different for different types of thermocouples (i.e., each channel may have a different thermocouple type, J-type, R-type, B-type, T-type).

10. Faraday's law of induction states that an electromotive force (EMF) voltage is induced on a conductor due to changing magnetic flux and that the induced voltage opposes the change in the magnetic flux. Application of this forms the basis of many electric motors and electric sensors (see Section 8.2). A back EMF is induced in the conductors moving in a magnetic field established by the field magnets (permanent magnets or electromagnets (Figure 8.9(d))). As the conductor moves in the magnetic field, there will be force acting on the charges in it just as there is a force acting on a charge moving in a magnetic field. Let us consider a constant magnetic field B, a conductor with length l moving in perpendicular direction to magnetic field vector, and its velocity $\dot{x}$ (Figure 8.9(d)). The induced voltage because of this motion is,

$$V_{emf} = B \cdot \dot{x} \cdot l \qquad (6.198)$$

This same principle can be used in a flow rate sensor, where the magnetic field B is induced by a permanent or electromagnet of the sensor, $\dot{x}$ is the speed of the flow across the cross-sectional area A. The induced voltage across the moving fluid in the direction perpendicular to both directions of B and $\dot{x}$ is proportional to $B \cdot \dot{x}$ for a given sensor size. Then, the flow rate can be calculated from $Q = A \cdot \dot{x}$. The fluid must be electrically conductive for this principle to work. a) Sketch the design of a magnetic flow rate sensor based on this principle. b) Find a magnetic flow rate sensor on the web and identify its operating characteristics (i.e., minimum conductivity of the fluid, maximum flow rate it can measure, accuracy and bandwidth).

CHAPTER 7

ELECTROHYDRAULIC MOTION CONTROL SYSTEMS

7.1 INTRODUCTION

In hydraulic systems, the power transmission medium is pressurized hydraulic fluid. If the control of the pressurized fluid flow is done by electrical means, then they are called electrohydraulic (EH) systems. If the control of the fluid is done by a combination of mechanical and hydraulic mechanisms, then they are called hydro-mechanical systems.

Figure 7.1 shows the essential components used in an electrohydraulic system. In electronically controlled hydraulic systems, the operator input devices are connected to sensors which generate the command signal to the electronic control module (EMC), which then controls the valves and pumps/motors. The main components of a EH system are:

1. engine or electric motor (mechanical power source),
2. pump (mechanical to hydraulic power conversion), along with oil reservoir, filter, heat exchanger, relief valve),
3. main valve (for controlling, "metering," the flow into the actuator),
4. hydraulic cylinder or motor (actuator),
5. sensors (position, velocity, pressure, flow rate),
6. electronic control module (ECM),
7. operator input devices (connected to integrated sensors which are then connected to the ECM),
8. other items, such as accumulators, safety or special function valves, hoses.

The concept of a hydraulic system is shown in Figure 7.2. The basic components of a electrohydraulic motion control system are shown in Figure 7.3. A closed loop controlled version of a single axis EH control system is shown in Figure 7.4 where sensors at the valve, cylinder, and application tool may be included as part of a closed loop digital control system. Hydraulic systems deal with the *supply and control* of fluid *pressure, flow rate and flow direction*.

The mechanical power source (usually an internal combustion engine in mobile equipment or an electrical induction motor in industrial applications), hydraulic pump, and reservoir form the so called *hydraulic power supply unit*. The tank (reservoir) acts as the hydraulic fluid supply and reservoir for the return lines. The volumetric size of the reservoir is typically 3 to 5 times the gallons/minute (gpm) flow rate of the pump, that is if the pump is rated at 10 gpm, the reservoir size can be in the 30–50 gallons range. The "pump" converts the mechanical power received from the mechanical power source into hydraulic

Mechatronics with Experiments, Second Edition. Sabri Cetinkunt.
© 2015 John Wiley & Sons, Ltd. Published 2015 by John Wiley & Sons, Ltd.
Companion Website: www.wiley.com/go/cetinkunt/mechatronics

FIGURE 7.1: Components of an electrohydraulic control system: input devices (joysticks), electronic control module (ECM), hydraulic tank, accumulators, pumps, motors, cylinders, valves, hoses and filters. Reproduced with permission from Parker Hannifin.

power in the form of pressurized fluid at its outlet port. The pump is the heart of a hydraulic system. If the pump fails, no hydraulic power can be transmitted to the final actuators. In industrial applications, an electric motor is typically used to provide mechanical power to the pump, whereas the internal combustion engine is the mechanical power source to drive the pump in mobile equipment applications. Figure 7.5 shows two common examples of hydraulic power units for industrial applications. The JIC style is one of the most common hydraulic power units. The main benefit of the design is its simplicity and easy access to main components for servicing. The main drawback of it is the fact the pump inlet port is at a higher elevation than the fluid in reservoir. This may lead to a cavitation problem at the pump inlet port. The overhead design eliminated the cavitation problem to a large extent. In addition, the pump may be submerged in the reservoir so that the circulation of fluid in the reservoir also helps cool the pump. However, the submerged style makes servicing of

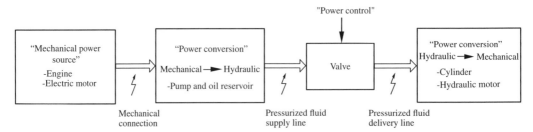

FIGURE 7.2: Concept of a hydraulic power system: the main functional components needed are 1. mechanical power source device, 2. mechanical to hydraulic power conversion device, 3. hydraulic power control device, 4. hydraulic to mechanical power conversion device.

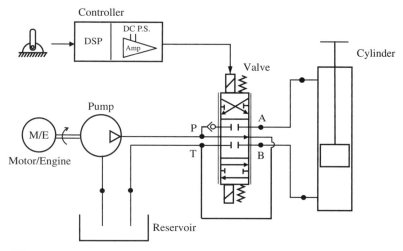

FIGURE 7.3: Symbolic representation of the components of an electrohydraulic motion system for one axis motion: hydraulic power unit (pump and reservoir), valve, actuator, controller. Since there are no sensors used for the control decision, this is an open loop control system. It is also called an *operator in the loop* control system.

the pump more difficult. There are other types of hydraulic power units, such as L-shaped, vertically mounted motor on reservoir, and T-shaped reservoir.

The *valve* is the key control element. A valve is used to *control* (also called *modulate* or *throttle* or *meter*) the fluid flow. The valve is the metering device. The flow can be regulated in terms of its *rate*, *direction*, and *pressure*. The type of the valve used and the "control objective" that drives the valve vary from application to application. The most common form is a proportional flow control valve. The valve is controlled by a solenoid or torque motor where the current in the windings generates a proportional spool displacement. The spool displacement opens the metering ports, and hence controls the flow rate. The valve amplifier (considered as part of the controller block in Figure 7.3) provides the amplified version of the low power command signal from the controller. A *proportional valve* (or a proportional hydraulic control component) is a device which provides an output that is proportional to the input. For instance, an electrohydraulic proportional flow control valve

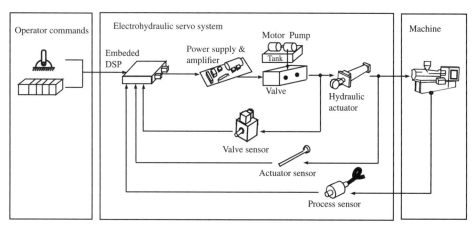

FIGURE 7.4: Basic components of a closed loop controlled electrohydraulic motion system for one axis motion (pump, valve, cylinder, controller, sensors).

(a)   (b)

FIGURE 7.5: Hydraulic power units for industrial applications which include electric motor, pump, reservoir, filtration, heat exchanger. a) JIC style hydraulic power unit where the pump is on top of the reservoir and must have good suction in the inlet pipe in order to avoid cavitation in the pump input port. b) Overhead reservoir where the lowest level of oil is at a higher elevation than the pump. This design is less likely to have a cavitation problem. The pump can also be submerged in the reservoir which helps its cooling, but it is more difficult to service.

provides a flow rate (output) which is proportional to the current command signal to its solenoid driver. An electrohydraulic proportional pressure control valve provides a pressure at the output port (output) which is proportional to the electrical command signal (input) to its solenoid drivers. Such a valve, including the solenoid and driver circuit, is considered a proportional device. In practice, the input–output relationship of devices are hardly ever a perfectly proportional linear relationship, but are close to it. The actuator, power delivery component, of the EH motion system can be either a linear cylinder or a rotary hydraulic motor. For steady-state considerations, the speed of the actuator is proportional to the flow rate.

Hydraulic circuits can be represented using one of three forms (Figure 7.6). Pictorial or cut-away representations of components and their interconnections are useful in physically understanding how each component works and visualizing the flow path. However, the engineering standard is to use the symbolic representation of the components as shown in Figure 7.6c. With symbols, we simplify the process of drawing and representing information about components even though they do not graphically display the physical construction or size of the component. This is generally the case in all engineering circuit drawings including electrical circuits, where symbols are used to represent components and their input–output relationship as opposed to conveying their physical construction information. Therefore, it is very important for an engineer to understand the physical construction and the operating principles of hydraulic components, as well as their symbolic representations in order to be able draw, read, and properly interpret circuit diagrams.

Applications of hydraulic motion control systems include:

1. Mobile equipment, such as construction equipment, that generates its power from an internal combustion engine, and delivers power to work tools via pressurized hydraulic fluid using pump, valve, and cylinder/motor components.

2. Industrial factory automation applications:
   (a) presses (punch presses, transfer presses),
   (b) injection molding machines,

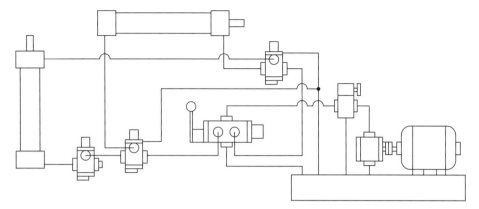

(a) - Picture representation of hydraulic circuit components

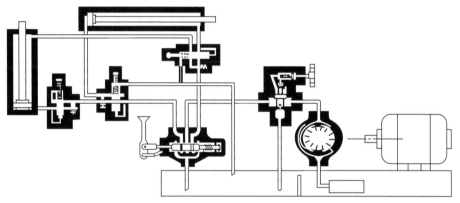

(b) - Cut-away representation of hydraulic circuit components

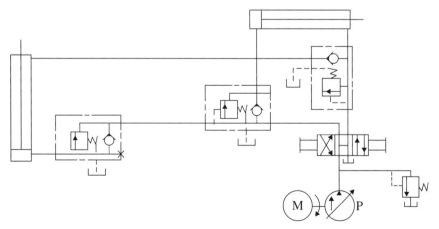

(c) Symbolic representation of hydraulic circuit components

FIGURE 7.6: Hydraulic circuit diagrams can be represented in three different forms for communicating information: (a) pictorial representation of the components in the circuit and their connections to each other, (b) cut-away view of the components and their connections to each other, (c) symbolic representation (this is the standard engineering representation) of the components and their connections to each other.

    (c) sheet metal thickness control drives in steel mills,

    (d) winches and hoists.

3. Civil and military aircraft flight control systems and naval ships:
    (a) primary flight control systems that involve the motion of wing surfaces, rudder, and elevator,
    (b) secondary flight control systems that involve spoilers and and trim surfaces,
    (c) engine fuel rate delivery control systems.

4. Military ground vehicles:
    (a) turret motion control,
    (b) hydrostatic traction drive systems.

Guidelines for Understanding a Hydraulic Circuit   Hydraulic circuit diagrams are the standard engineering tools used to describe a hydraulic circuit. A hydraulic circuit diagram shows the connections between various hydraulic, mechanical, and electrical components using lines for connections and standard symbols for components such as pumps, valves, and cylinders. At first glance, a real hydraulic circuit diagram may look very complicated. However, a systematic approach to reading it is useful. It is not unusual to spend multiple days studying a given hydraulic circuit in order to fully understand it. The following general guidelines (ten point guidelines) are recommended in interpreting a given hydraulic circuit diagram:

1. Identify the main hydraulic power supply: hydraulic oil reservoir along with its filters and heat exchangers, pump (or pump group), possibly line relief valve, and power source (i.e., electric motor or engine).

2. Identify the main hydraulic actuators: cylinders and motors.

3. Identify the main flow control valves that connect the pump supply to the actuators; the type of valve (ON/OFF, proportional, servo), open center/closed center, pressure compensated or not and so on.

4. Identify the safety valves (relief valves) in the supply side and on the actuator (line) side (i.e., line relief valves, cross-over relief valves).

5. Determine the main hydraulic power flow paths from pumps to valves, then to the actuators, and return lines from actuators to valves and back to tanks.

6. If a pilot control circuit exists, identify the pilot control pump supply and return lines, its hydraulic oil reservoir and oil conditioning components, as well as the pilot control valves.

7. Then identify the pilot control valves which take either a mechanical motion signal from an operator input device or from a digital computer as the command signal, and convert it to an amplified pilot pressure to actuate the main valve. If the main valve is single stage (i.e., low flow rate applications), then the operator control signal may be connected directly to the main flow control valve without a pilot stage. The "pilot" signals represent the "control" signals to the main power stage. The word "piloting" is used to describe the low power control signal that goes into the main power stage control elements, that is main valves. In large power applications, the electrical control signals (first stage of control) are amplified to pilot hydraulic control signals (second stage of control) which are then used to actuate (control) the main stage components.

8. Identify the operator generated (or automatically generated or remote by an electronic control module (ECM)) command signal sources, that is levers, joysticks, steering wheels, pedals.

9. Identify the sensors, and the logic by which they are used in the control decisions: electrical sensors are interfaced to an electronic control module. Mechanical sensing mechanisms and hydraulic sensing lines are directly used in a hydro-mechanical control logic in the circuit. For instance, the load sensing control of a pump can be implemented either as a hydro-mechanical control or a digital control.

10. Determine any hydro-mechanically built-in *priority* and *over-ride type logic* in the circuit. Then, study the control logic software used to digitally control the hydraulic system.

The control of an EH motion system may involve the control of the valves, the pump and the motor. Furthermore, the control logic may be based on "open loop" or "closed loop" concepts. If a human operator is in the loop, a fixed displacement pump with variable input speed (i.e., variable engine speed) controlled by an operator pedal, and a valve actuated proportional to the operator lever input (open loop control) is sufficient. The operator would adjust the pump input speed with pedal, and hence controls the hydraulic supply power. At the same time, the operator controls the valve shifting with a lever based on his/her observations, and hence modulates the delivery to the actuator. In automated systems, closed loop control of the valve and pump may be necessary. The design of a electro-hydraulic motion control system involves the following steps,

1. *Specifications:* the first step in any design is to specify the requirements that the system must meet, that is performance, operating modes, fail-safe operation.

2. *System concept design:* given the specifications, an appropriate system design concept must be developed, that is it is an open circuit hydraulic or closed circuit hydraulic system, open loop or closed loop controlled.

3. *Component sizing and selection:* the proper component types and sizes must be selected. Component sizing requires power calculations to make sure the size of components is properly matched and that they have the capacity in terms of power, pressure, and flow rate to meet the performance requirements.

4. *Control algorithm design:* in computer controlled EH systems, there is always a controller hardware and real-time control software involved in control of the system.

5. *Modeling and simulation:* if necessary, the EH system hardware and control algorithm may be modeled and simulated off-line on a computer to predict the performance of the overall system.

6. *Hardware test:* finally, a prototype system must be built and the operation of the EH system, under the control of the designed control algorithm and hardware, must be tested. If the desired performance objectives are not met, the design process is iterated for refinements.

Measured variables in an EH system for control purposes may include the following:

1. load position and speed measured at the actuator (cylinder) and/or load end,

2. load pressure measured as pressure differential at the actuator ports or at the outlet ports (A & B) of the valve,

3. load force or torque,

4. spool displacement,

5. flow rate,

6. pump output pressure,

7. pump displacement,

8. pump speed.

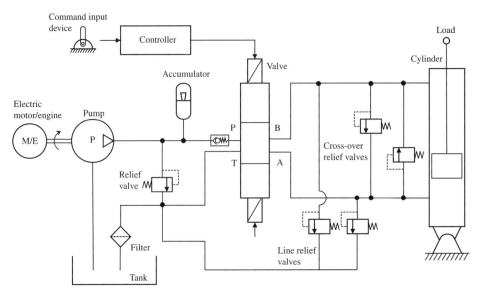

FIGURE 7.7: Safety valves typically added to a hydraulic motion system: pump side relief valves and load side relief valves (line-to-tank relief valves, cross-over relief valves), and check valves to limit the flow in a line to one direction only (i.e., flow is allowed from P to A or B ports, but not back to P port from A or B ports).

These measured quantities may be used in real-time as part of a closed loop control algorithm. Intermediate variables can be derived from the measured variables and regulated. For instance, if the flow rate and load pressure are measured, the output power can be calculated. Then it can be used to implement a closed loop control algorithm on the desired output power. Hence, the EH control system would operate so as to track a commanded output power level.

Figure 7.7 shows possible safety related additions to a typical hydraulic motion system. In general, every hydraulic circuit should include safety relief valves to protect the lines against excessive pressures at both (i) pump side and (ii) actuator (line) side. Typical pressure safety valves designed into the system are as follows:

- pump pressure relief valve which limits the maximum output pressure of the pump,
- line relief valve which opens the line to the tank if the line pressure exceeds a set limit, and/or cross-over relief valves which open one side of the actuator to the other side if the pressure differential between the two sides of the actuator exceeds a certain level.

ON/OFF valves are used in applications where the controlled flow is only needed to be turned ON or OFF. In this type of application, the actual speed is determined by the supply and load conditions. The spool position is controlled to one of two discrete positions: one position to completely block the flow (OFF position) and one position to fully open the orifice for flow (ON position). Proportional and servo valves are used on variable control applications where the valve spool position, and hence the opening of the orifice area, is controlled "proportionally," not just "fully ON" or "fully OFF," but in proportion to the command signal. Proportional and servo valves are used in applications where variable speed, position, or force control is needed. The orifice opening area versus the spool position function of the proportional and servo valves, especially around the *null-position*, is an important factor in closed loop control performance.

If there is only a finite number of desired speed or force control levels, this may be accomplished by a few ON/OFF valves in parallel at a lower cost and with simpler design. For instance, if an application requires only ON/OFF action, a single flow control valve between the hydraulic supply line and actuator is needed. If the application requires three different speed levels in addition to zero speed, for example 0%, 33.33, 66.66%, and 100% of maximum speed capacity, we can accomplish this with two ON/OFF valves connected in parallel between the hydraulic supply and actuator lines. Valve 1 would be sized to have a flow rate capacity of 1/3 of the maximum flow rate and valve 2 would be sized to have a flow rate capacity of 2/3 of the maximum flow rate needed to support the desired 100% actuator speed. Then, by turning ON/OFF both or either one of the valves, we can realize one of these three different speed ranges. Clearly, ON/OFF valves have limited flow metering capabilities. However, they do provide simpler and low cost motion control solutions in discrete speed or flow rate control applications.

Accumulators The *accumulator* serves as a pressurized fluid storage component. The accumulator is the hydraulic analogy of an electrical system battery or capacitor. It can help the control system in two ways in the transient response,

- maintain the hydraulic line pressure in case of a sudden drop on the hydraulic line pressure due to a sudden increase in demand or decrease in the the pump output,
- it can provide a damping effect and shock absorber function in case of large pressure spikes, that is as a result of sudden load changes.

On the other hand, an accumulator also reduces the open loop natural frequency of the EH system as a result of the reduced stiffness of fluid (smaller effective bulk modulus due to elasticity of the accumulator and increased fluid volume). As a result, the maximum bandwidth the control system can achieve is lower. There are three main types of accumulators: weight loaded, spring loaded, and hydro-pneumatic types. Hydro-pneumatic type accumulators use dry nitrogen as the compressed gas.

There are three major designs of hydro-pneumatic accumulators which are categorized in terms of the way hydraulic pressure and pressure storage components interact with each other: (i) piston type, (ii) diaphragm type, and (iii) bladder type (Figure 7.8). An accumulator is rated with the hydraulic fluid volume it can store (also called *working volume*), maximum and minimum operating pressures, precharge pressure, and maximum shock pressure it can tolerate. The compression ratio of an accumulator is the ratio between maximum and minimum operating pressure.

An accumulator can be sized either based on a required discharge volume and to limit the maximum pressure due to shocks. For a given discharge volume (V_{disch}), the required accumulator size (volume of the accumulator, V_{acc}) can be determined as follows,

$$V_{acc} = V_{disch} \cdot \frac{p_{min}}{p_{pre}} \left/ \left[1 - \left(\frac{p_{min}}{p_{max}} \right)^{1/n} \right] \right. \tag{7.1}$$

where $p_{pre}, p_{min}, p_{max}$ are precharge pressure of accumulator, minimum and maximum line pressure, n is an empirical number between 1.2 to 2.0.

The precharge pressure is about 100 psi less than the minimum line pressure requirement,

$$p_{pre} = p_{min} - 100 \, \text{psi} \tag{7.2}$$

After the accumulator is precharged, the line, which has a pressure presumably between minimum and maximum pressure, charges the accumulator. Fluid flows into the accumulator

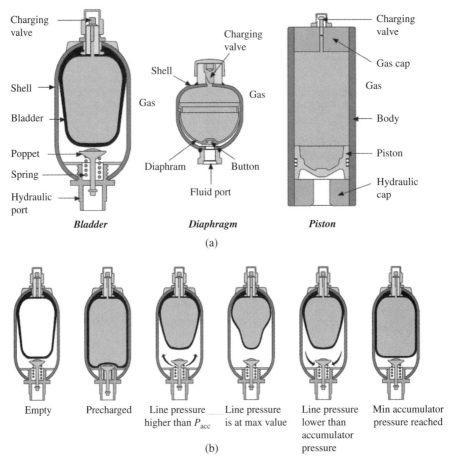

FIGURE 7.8: (a) Three different types of hydro-pneumatic accumulators: bladder type, diaphragm type, and piston type [13], and (b) various operating conditions of an accumulator (from left to right): 1) empty (uncharged), 2) precharged, 3) line pressure exceeds the precharge pressure and fluid flows into the accumulator, 4) line pressure is at maximum and accumulator is filled with fluid to its design capacity and the line relief-valve should open, 5) line pressure drops below the current accumulator gas pressure and fluid flows out of the accumulator, 6) line pressure is at its minimum and the accumulator has discharged all of its designed discharge volume. Reproduced with permission from Parker Hannefin.

until the pressure rises to the maximum pressure setting, at which point the relief valve opens. When the stored pressure in the accumulator is larger than the line pressure, that is due to sudden pressure drop in the line, the fluid flows from the accumulator to the line. This continues until the accumulator pressure reaches the minimum pressure setting, at which point the poppet valve would be closed. When the line pressure exceeds the accumulator pressure, fluid again flows into the accumulator and charges it. The pressure in the accumulator is limited to be between minimum and maximum pressure settings,

$$p_{\min} \leq p_{\text{acc}} \leq p_{\max} \tag{7.3}$$

Whenever $p_{\text{acc}} \leq p_{\text{line}}$, the accumulator is charged, and whenever $p_{\text{acc}} \geq p_{\text{line}}$, the accumulator is discharged. When $p_{\text{acc}} = p_{\min} \geq p_{\text{line}}$, the poppet valve closes (or in case of a piston type accumulator, the end of travel is reached). When $p_{acc} = p_{max} < p_{line}$, a system relief valve opens to protect the accumulator.

Typically, the accumulator volume is about three, four or more times the discharge volume. In general, the higher the dynamic response requirements, the larger the accumulator size is relative to the discharge volume. The full discharge volume of a typical accumulator can be discharged within a few hundred milliseconds. For instance, consider a piston type accumulator with a discharge volume of 20 l, and that it can discharge at 3600 l/min at 3000 psi pressure differential between the accumulator and the line pressure. Therefore, the accumulator would discharge at 60 l/s rate and the full volume would be discharged within about 1/3 s. Bladder and diaphragm type accumulators have a slightly faster response time than the piston type due to less friction between moving parts.

Figure 7.8b shows six selected conditions of a bladder type accumulator. In the first figure, the accumulator is empty both on the fluid and gas side. In the second figure, accumulator has been *precharged* with dry nitrogen gas to the *precharge pressure*, p_{pre}. The third figure shows the case when *line pressure* (p_{line}) exceeds the precharge pressure and fluid starts to fill-in the accumulator. The fourth figure shows the condition when the line pressure reaches its maximum value, $p_{line} = p_{max}$, and the line relief valve should open. The fifth figure shows the condition that the line pressure drops below the accumulator pressure, $p_{min} \leq (p_{line} < p_{acc}) \leq p_{max}$, and accumulator pressure adds fluid to the line. Finally, the sixth figure shows the condition when accumulator pressure reaches the *minimum pressure*, p_{min}, setting. For a correctly sized accumulator in a hydraulic circuit and under proper circuit operating conditions, the operation of the accumulator should cycle between the minimum and maximum pressure setting without opening the line relief valve (the third and sixth state shown in Figure 7.8b). If the minimum operating pressure is lowered without correspondingly lowering the precharge pressure, the accumulator will operate between second and third states in Figure 7.8b. This condition can lead to accumulator failure due to the contact between the bladder and poppet valve assembly. Therefore, the line pressure should not be allowed to drop below the minimum operating pressure of the accumulator. Otherwise, the accumulator operates between conditions two and six in the figure, which is not the intended, nor a desirable, operating condition.

In most circuits, the accumulator is placed on the pump side of the circuit, as close as possible to the valve in order to provide the best dynamic response. However, in some circuits the accumulator may be placed on the actuator side (cylinder or hydraulic motor) side of the circuit between the valve and the actuator. For example, let us consider an application where the external load changes so quickly (i.e., shock load, such as those experienced in rock crushing drives in mining applications) that neither the main valve nor the line pressure relief valve can act fast enough to limit the pressure on the circuit. In such a case, an accumulator is placed between the valve and the cylinder to absorb the suddenly rising pressure. In such applications, the minimum operating pressure of the accumulator should be higher than the normal operating pressure of the line so that it would be inactive during normal operating pressure ranges except when the shock load occurs.

Very small volume (i.e., 1 to 10 in^3) bladder type accumulators (also referred to as in-line hydraulic shock absorbers) are also used between the valve and cylinder to limit the maximum pressure, which is beyond the normal operating range of the line pressure. The purpose of such tiny accumulators is to absorb very small volumes of fluid which would otherwise cause very large pressure spikes.

Pressure sensors and the control valve should be placed to the cylinder ports as close as possible in forcel (or pressure) control applications. Pressure waves travel in oil at about 4.5 ft/ms. In other words, a 45 ft distance between actuation (valve) and sensing point (pressure sensor) would lead to a 10 ms pure time delay in the control system. This would severely limit the closed loop bandwidth capability of the system. This is an example of actuator-sensor non-colocation and the resulting transport delay (pure time delay) in a control system.

Example   Consider a hydraulic motion system where an accumulator is to be selected. Let the maximum and minimum line pressures be 3000 psi and 90% of that (2700 psi), respectively. Let $n = 1.4$ nominally for the nitrogen gas. Let the discharge volume to be 1 gallon. Calculate the necessary accumulator volume.

Let the precharge pressure to be 90% of the minimum pressure. Hence

$$p_{pre} = 0.90 \cdot p_{min} = 0.9 \cdot 2700 = 2430 \, \text{psi} \tag{7.4}$$

Using the empirical formula, the accumulator volume required,

$$V_{acc} = 1.0 \cdot \frac{2700}{2430} / (1 - (2700/3000)^{1/1.4}) \tag{7.5}$$

$$= 15.32 \, \text{gallons} \tag{7.6}$$

An accumulator size of 20 gallons is a good design choice for this case.

Let us further consider that we allow a larger pressure variation for the accumulator operation, with the same amount of discharge volume needed. Then it is expected that the required accumulator size will smaller since larger pressure variation is allowed. Let $p_{max} = 3000 \, \text{psi}$, $p_{min} = 2000 \, \text{psi}$, $p_{pre} = 0.9 \cdot p_{min} = 1800 \, \text{psi}$, $V_{disch} = 1.0 \, \text{gallon}$. The required accumulator volume, based on the empirical formula,

$$V_{acc} = 1.0 \cdot \frac{2000}{1800} / (1 - (2000/3000)^{1/1.4}) \tag{7.7}$$

$$= 4.41 \, \text{gallons} \tag{7.8}$$

An accumulator size of 5 gallons is a good design choice for this case.

Filtration   *Filtration* removes the solid particles from the hydraulic fluid. Filters may be located at the suction line, pressure line, return line or drain line (Figure 7.9). In addition, there could be a dedicated loop for continuous filtration of the fluid in the reservoir. This is called *kidney loop filtration*. If the pump is the most sensitive element in the circuit against dirt, the filter is placed at the inlet port of the pump between the reservoir and the pump. In this case it is important to maintain a pressure differential across the filter in order to avoid cavitation at the pump inlet port. If the valve is the most sensitive component (i.e., servo valve), then the filter is placed between the pump outlet port and the valve. Notice that such a filter must be rated to handle the maximum pressure and flow rate it is exposed to.

A filter has replaceable elements to remove particles above a rated size, such as 3-micron. As a safety measure, each filter usually has a parallel bypass valve (pressure switch). In case the filter becomes clogged due to excessive dirt accumulation before a scheduled filtering element replacement, the pressure build up in the clogged filter will open the bypass valve, and allow the flow to continue albeit it may not be filtered. Unfiltered fluid and the resulting dirt is perhaps the most common cause of component failure in high performance hydraulic servo systems where the servo components have precision manufacturing tolerances. The three most important specifications for a filter are:

1. minimum size of the particles it can filter (i.e., 10-micron, 5-micron),

2. maximum pressure it can withstand (i.e., in supply line or in return line),

3. maximum flow rate it can support.

Some Hydraulic Circuit Concepts   Figure 7.10a shows a simple hydraulic system where the hydraulic power is provided by the lever motion. In this case, the lever is

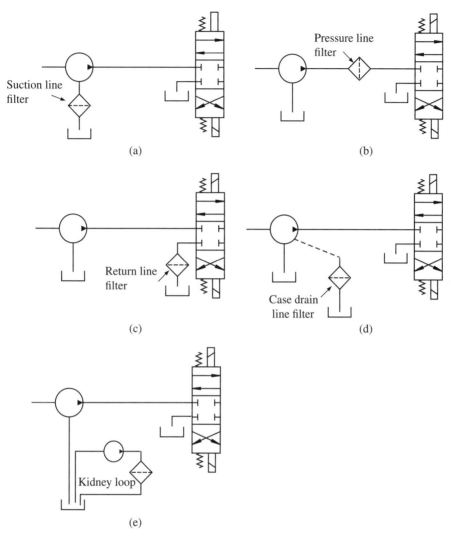

FIGURE 7.9: Filter locations in a hydraulic circuit: (a) filter in suction line, (b) filter in pressure line at the pump output, (c) filter in return line to the tank, (d) filter in case drain line, (e) kidney loop filtration. There is generally a by-pass check valve in parallel with the filter in order provide a flow path in case the filter is clogged up.

moved by an operator. As the lever moves up, due to the pressure differences, check valve 1 opens and check valve 2 closes. Hence, the lever sucks the fluid from the reservoir. In the down stroke, check valve 1 closes and check valve 2 opens. The pressure between the lever side and cylinder side are almost equal. Hence, the force is multiplied by the area ratio, while the speed of linear travel of the load and lever is divided by the same ratio. This is the result of conservation of power.

Figure 7.10b shows a simple hydraulic system where the source of hydraulic power is a fixed displacement gear type pump. Input mechanical power to the pump is provided by an electric motor. An operator controls the flow to the cylinder by a manually actuated valve. Notice that the valve is *open-center* type which means that the flow circulates from pump port to tank port when the valve is in neutral position.

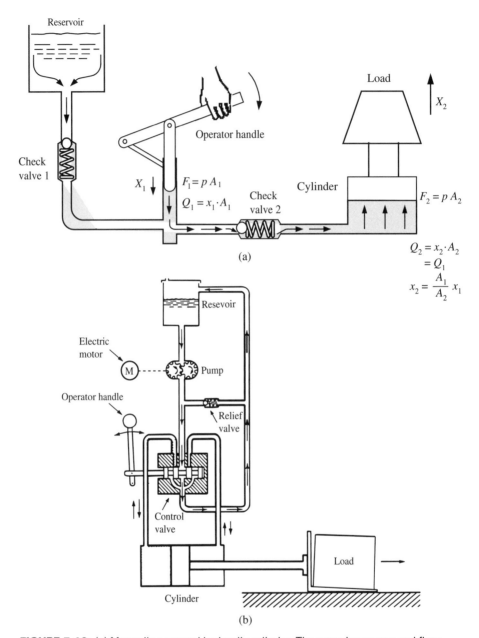

FIGURE 7.10: (a) Manually powered hydraulic cylinder. The pumping power and flow metering is controlled by the manual power provided to the lever arm. The flow rate is determined by the travel of the pumping action and rate of cycles per unit time. (b) Electric motor powered hydraulic system. The pump is a fixed displacement type gear pump type. Flow rate is controlled manually by the valve lever.

Figure 7.11 shows the components of a hydraulic circuit for the motion of a cylinder of an excavator. The power source is the internal combustion engine, the hydraulic pump is mechanically geared to the engine crankshaft for the mechanical power source. The directional valve is a manually controlled type.

Similarly, Figure 7.12 shows the components of an electrohydraulic circuit for the motion of a rotary hydraulic motor for a service truck. The hydraulic motor receives power

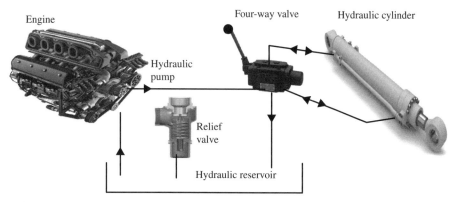

FIGURE 7.11: A manually operated hydraulic circuit for the cylinder motion of an excavator linkage. The main directional valve is actuated via a mechanical pedal by an operator.

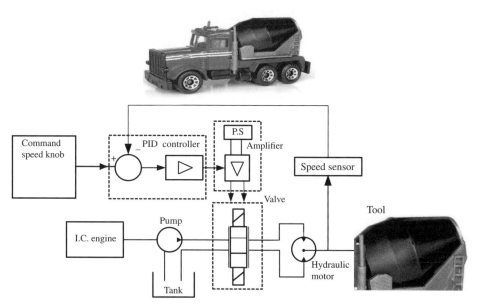

FIGURE 7.12: An electrohydraulic closed loop speed control system example for a service truck.

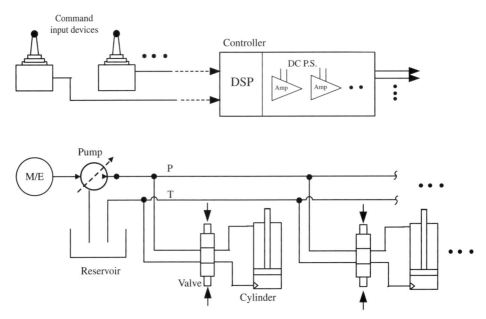

FIGURE 7.13: Multi degrees of freedom (multi function) hydraulic circuit. This example shows a parallel connection of two hydraulic actuator-valve circuits to supply lines. The valves are generally of closed center type in this configuration. The pump may be a variable or fixed displacement type (most likely variable displacement type).

from the pump, which receives its power from the engine, and its speed is controlled by a directional flow control valve and operator input switch. Notice that the hydraulic motor speed is closed loop controlled using a tachometer speed sensor as the feedback signal.

A multi function (multi degrees of freedom motion) hydraulic power based motion system is shown in Figure 7.13. In general one pump is used (large enough to support all functions), and a valve-actuator pair is used for each function. This figure shows a *parallel connection* of the each hydraulic valve-actuator pair to the pump and tank lines. In some applications, multiple circuits may have *series connections*, where a preceeding circuit has priority over the following circuits. Such implementation is necessary when it is not practical to provide a large enough pump to support the full load of each circuit and one of the circuits is more important than the others. One such implementation is used in the implement (bucket) hydraulic circuit for wheel type loaders. The valve-cylinder pair for lift and tilt motion are connected in series (Figure 7.14). The pump is not large enough to support both functions at their maximum flow capacity. The tilt circuit is connected ahead of the lift circuit. As a result, if the tilt valve is shifted to its maximum displacement, the tilt cylinder gets the maximum flow, and the remaining small amount of flow is passed to the lift circuit through the tilt valve. Such a hydraulic circuit is said to have tilt function priority over lift function.

All of the electrohydraulic circuit examples given above circulate the flow from reservoir to pump, to valve, then to the actuators and back to the reservoir. Such hydraulic systems are called *open circuit* systems. There are also the so called *closed circuit* hydraulic systems, such as those used in hydrostatic transmissions. A hydraulic circuit is called closed circuit if the main flow in the power line circulates in a closed line with small leakage and replenishment of fluid from the reservoir. Figure 7.15 shows the hydraulic circuit and its components for a typical closed circuit transmission. There are three major fluid flow

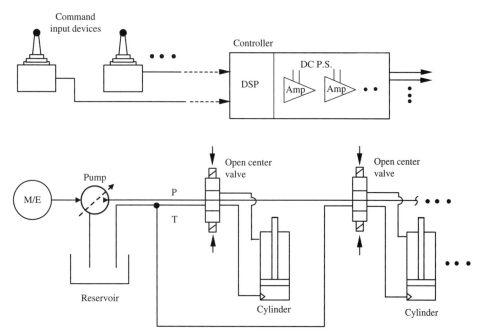

FIGURE 7.14: Multi degrees of freedom (multi function) hydraulic circuit. This example shows a series connection of two hydraulic actuator-valve circuits to supply lines. Valves are necessarily open-center type in order to supply the circuits in the serial connection when an earlier valve is in neutral position. The valves closer to the pump have priority over the ones that are connected later in the series circuit. The pump may be of variable or fixed displacement type.

circuits: the main flow loop between the pump and motor, the charge loop which supplies replenishment flow to the main loop as well as supply pressurized flow power used to actuate the pump displacement, and finally the the drain loop which sends the leakage flow through the components to the reservoir. The pump control types are discussed in more detail later. Suffice it to say that the pump displacement can be controlled to regulate various variables, such as pump output pressure, flow rate. The power necessary to actuate the main pump displacement control system is provided by the charge pump. The control

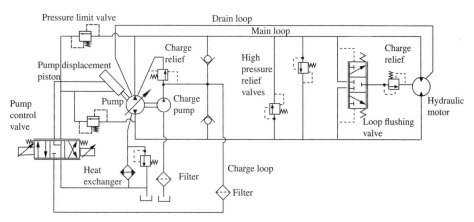

FIGURE 7.15: A typical closed circuit hydrostatic transmission circuit. Reproduced with permission from Danfoss Power Solutions.

of pump displacement is performed by the control valve which is actuated based on the control objective. In closed circuit hydraulic circuits, ideally there is no leakage nor need for replenishment fluid. In reality, it is desirable to have a certain amount of leakage. It is through the leakage and the replenishment circuits that fluid under high pressure is removed from the main loop, cooled by the heat exhanger, cleaned by filters and reinjected back into the main loop. Therefore, the leakage and charge pump relenishment is an essential part of all practical closed circuit hydraulic systems. In fact, the leakage rate is planned into the circuit design and regulated at a planned rate by a so called *loop flushing valve*. A typical system also includes various pressure relief valves between different sides of hydraulic lines.

The pump is a variable displacement type bidirectional pump. By changing the direction of fluid flow, the direction of speed of the hydraulic motor is changed. By controlling the pump, the flow rate and hence the speed of the hydraulic motor, is controlled. Similarly, if the pump is controlled to maintain a certain pressure, the torque output at the hydraulic motor is controlled.

A hydraulic system can be categorized in terms of the power source used to control the hydraulic valve. That is the actuation power used to move the spool. There are four major valve actuation methods (Figure 7.16):

1. mechanically direct actuated valve,
2. electrically direct actuated valve,

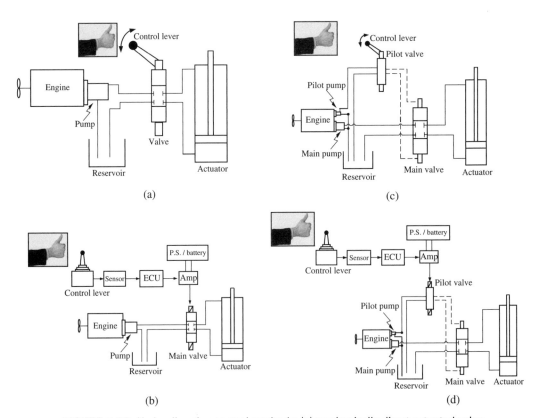

FIGURE 7.16: Hydraulic valve control methods: (a) mechanically direct actuated valve (single-stage valve), (b) electrically direct actuated valve (single-stage valve), (c) mechanical plus pilot hydraulics actuated valve (two-stage valve), (d) electrical plus pilot hydraulics actuated valve (two-stage valve).

3. mechanical plus pilot hydraulics actuated valve,

4. electrical plus pilot hydraulics actuated valve.

The focus in the classification is the source of actuation power for shifting the valve spool. In mechanically controlled valves, the operator pushes a lever which is connected to the main spool via a mechanical linkage. This is used in relatively small size mobile equipment applications. The same concept is extended to large size machine applications by using an intermediate pilot pressure circuit to actuate the main valve. The motion of the pilot valve is still controlled by direct mechanical motion of the lever. The pilot valve is essentially a proportional pressure reducing valve. It has a constant pilot pressure supply port and a tank port. The output pilot pressure is proportional to the lever displacement at a value between the tank port pressure and pilot supply port pressure. The main valve spool is shifted in proportion to the output pilot pressure.

Finally, if the spool displacement power comes from the electromagnetic force which is generated by an electric current in the solenoid winding, then it is an electrically controlled valve. Such hydraulic systems are called electrohydraulic (EH) systems. The EH valves may be single stage for small power applications or multi stage for large power applications. In multi stage EH valves, the electrical current in the solenoid moves the first stage valve spool (pilot spool), and then the motion of the pilot spool is amplified via a pilot output pressure line to move the main valve spool. In short, pilot-actuated multi stage valves may be controlled by the mechanical connection to the control lever (mechanically controlled multi stage hydraulic valve (Figure 7.16b)) or by an electrical actuator (electrically controlled (EH) multi stage valve, Figure 7.16d).

7.2 FUNDAMENTAL PHYSICAL PRINCIPLES

In this section we briefly review the fundamental principles of physics that govern hydraulic circuits. Namely, we discuss the *Pascal's law* and *Bernoulli's equation.*

Pascal's Law   Pascal observed that a confined fluid transmits pressure in all directions in equal magnitude. The implication of this observation is illustrated in Figure 7.17. The pressures on both sides on the hydraulic circuit are equal to each other.

$$p_1 = p_2 \tag{7.9}$$

$$\frac{F_1}{A_1} = \frac{F_2}{A_2} \tag{7.10}$$

$$F_2 = \frac{A_2}{A_1} \cdot F_1 \tag{7.11}$$

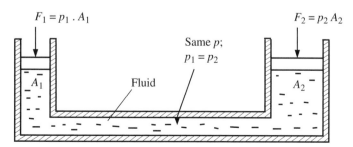

FIGURE 7.17: Fundamental principles of fluid flow in hydraulic circuits: Pascal's law.

By using a smaller weight on side one, we can move a much larger weight on side two. The circuit acts like a force multiplier. The conservation of energy principle also requires that

$$Energy\ In = Energy\ Out \tag{7.12}$$

$$F_1 \cdot \Delta x_1 = F_2 \cdot \Delta x_2 \tag{7.13}$$

$$F_1 \cdot \Delta x_1 = F_1 \cdot \frac{A_2}{A_1} \cdot \Delta x_2 \tag{7.14}$$

$$A_1 \cdot \Delta x_1 = A_2 \cdot \Delta x_2 \tag{7.15}$$

$$\Delta x_2 = \frac{A_1}{A_2} \cdot \Delta x_1 \tag{7.16}$$

which confirms the conservation of fluid volume. As an analogy, this hydraulic circuit acts like a gear reducer. That is, neglecting the loss of energy due to friction, the energy is conserved (input energy equal to output energy) and the force is increased while displacement is reduced.

Conservation of Mass: Continuity Equation   The principle of conservation of mass in hydraulic circuits is used between any two cross-section points (Figure 7.18). Assume that the mass of the fluid in the volume between the two points does not change. Then, the net incoming fluid mass per unit time is equal to the net outgoing fluid mass per unit time,

$$\dot{m}_{in} = \dot{m}_{out} \tag{7.17}$$

$$Q_1 \cdot \rho_1 = Q_2 \cdot \rho_2 \tag{7.18}$$

If the compressibility of the fluid is negligable, then $\rho_1 = \rho_2$, then the continuity equation for a non-compressible fluid flow between two cross-sections

$$Q_1 = Q_2 \tag{7.19}$$

Conservation of Energy: Bernoulli's Equation   Consider any two cross-section points in a hydraulic circuit (Figure 7.18). It can be shown that the conservation of energy principle leads to Bernoulli's equations. If there is no energy added or taken from

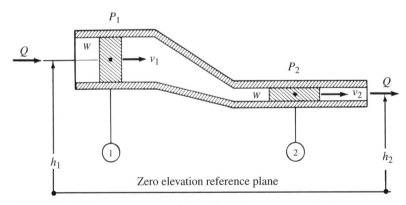

FIGURE 7.18: Fundamental principles of fluid flow in hydraulic circuits: conservation of mass and energy (Bernoulli's equation).

the flow and frictional losses are neglected, Bernoulli's equation can be expressed in its standard form as

$$h_1 + \frac{p_1}{\gamma_1} + \frac{v_1^2}{2g} = h_2 + \frac{p_2}{\gamma_2} + \frac{v_2^2}{2g} \tag{7.20}$$

where h_1, h_2 are the nominal height of the fluid from a reference point, p_1, p_2 are the pressures, v_1, v_2 are the velocities, γ_1, γ_2 are the weight densities at points 1 and 2, g is the gravitational acceleration ($9.81 \text{ m/s}^2 = 386 \text{ in/s}^2$). If the fluid compressibility is neglected, then $\gamma_1 = \gamma_2$.

If there is energy added between the two points (i.e., by a pump), E_p, or energy taken (i.e., by an motor), E_m, and if we take energy losses due to friction into account as E_l, Bernoulli's equation can be modified as

$$h_1 + \frac{p_1}{\gamma_1} + \frac{v_1^2}{2g} + E_p - E_m - E_l = h_2 + \frac{p_2}{\gamma_2} + \frac{v_2^2}{2g} \tag{7.21}$$

where the added or removed energy terms (E_p, E_m, E_l) are energy units per unit weight of the fluid. The energy unit is equal to force times the displacement. Energy per unit weight is obtained by dividing that unit by weight (*weight* = $m \cdot g$) which has force units. Hence, E_p, E_m, E_l have effective units of length. The energy stored in the fluid has three components: kinetic energy due to its speed, potential energy stored due to its pressure, and potential energy stored due to gravitational energy due to the elevation of the fluid. Quite often, the change in energy due to the change in height between two points is negligable and the h_1 and h_2 terms are dropped from the above equations in hydraulic circuit analysis.

Viscosity of Hydraulic Fluid   Viscosity is the resistance of a fluid to flow (Figure 7.24). It is a measure of how thick the oil is and how hard it is to move it. There are two definitions of viscosity: *kinematic* and *dynamics* (or absolute viscosity). Dynamic viscosity, μ, is defined as the ratio between the force necessary to move two mechanical surfaces relative to each other at a certain velocity when they are separated by a viscous fluid,

$$F = \mu \frac{A}{\delta} \dot{x} \tag{7.22}$$

where A is the surface area filled by the fluid between two pieces, δ is the fluid film thickness, $\dot{x}$ is the relative speed between two pieces (also it is the shear rate of fluid). Kinematic viscosity, v, is defined relative to the dynamic viscosity as

$$v = \frac{\mu}{\rho} \tag{7.23}$$

where ρ is the mass density of the fluid. Kinematic viscosity of a given fluid is measured as the time period it takes for a standard volume of fluid to flow through a standard orifice. The absolute viscosity is obtained by multiplying the kinematic viscosity with density of the fluid. There are many different standards for measuring the kinematic viscosity. The Saybolt Universal Seconds (SUS) measure is the most commonly used standard. The SUS viscosity measure of a fluid indicates the amount of time (in seconds) it takes for a standard volume of fluid to flow through a standard orifice, called a Saybolt Viscosimeter, at a standard temperature (i.e., $40 \,^\circ\text{C}$ or $100 \,^\circ\text{C}$). Other standards used for viscosity index are SAE (i.e., SAE-20, SAE-30W) and ISO standards. In general, the SUS viscosity of a hydraulic fluid should be within 45 to 4000 SUS.

The hydraulic fluid viscosity varies as function of temperature, especially during the startup phase of operation [14]. Therefore, the oil temperature must be kept within a certain range. Air-to-oil or water-to-oil heat exchangers are used to control the temperature of the hydraulic fluid. The *heat exchanger* transfers heat between two separate fluids (i.e., hydraulic fluid and water, or hydraulic fluid and air). If the heat is transferred from the hydraulic fluid to the other fluid (i.e., to cold water), it functions as a cooler. If the heat is transferred from the other fluid (i.e., from hot water) to the hydraulic fluid, it is used as a heater. In most applications, the hydraulic fluid needs to be cooled. However, in very cold environment temperatures, it is necessary to heat the hydraulic fluid instead of cooling it. A typical heat exchanger uses a temperature sensor in the hydraulic fluid reservoir in order to control the temperature in closed loop control mode. A regulator (analog or digital) controls a valve (which modulates the flow rate of the second fluid) based on the temperature sensor signal and the desired temperature value. In general, the viscosity of oil increases (oil gets thicker and resists flow more) as temperature decreases. As the oil viscosity increases, the flow rate of pumps reduces due to the fact that oil is less fluid and the suction action of the pump cannot pull-in as much oil.

Bulk Modulus   Bulk modulus, β, represents the compressibility of the fluid volume, and is defined as

$$\beta = -\frac{dp}{dV/V} \tag{7.24}$$

where the negative sign indicates that the volume gets smaller as the pressure increases. The bulk modulus has the same units as pressure. It is an indication of the stiffness of the fluid. The higher the bulk modulus of the fluid is, the less compressible it is. It is also a function of pressure and temperature. The nominal value of bulk modulus for hydraulic fluids is around $\beta = 250\,000$ psi. Entrapped air in fluid can drastically reduce the bulk modulus of a hydraulic line. Hence, the hydraulic circuit becomes softer and the system motion bandwidth gets slower. Therefore, it is important to take care in design and operation of hydraulic circuits that there is no entrapped gas in the hydraulic fluid lines. The overall system would lose its stiffness significantly.

Example   Consider a hydraulic fluid with bulk modulus of $\beta = 250\,000$ psi, a nominal volume of $V_0 = 100$ in^3. If the fluid is compressed from atmospheric pressure level to the 2500 psi level, find the change in the fluid volume.
Since,

$$\beta = -\frac{dp}{dV/V_0} \tag{7.25}$$

Then,

$$dV = -\frac{V_0 \cdot dp}{\beta} \tag{7.26}$$

$$= -\frac{100 \cdot 2500}{250\,000} \tag{7.27}$$

$$= -1\,\text{in}^3 \tag{7.28}$$

$$= 1\%V_0 \tag{7.29}$$

which shows that a typical hydraulic fluid will contract about one percent in volume under a 2500 psi increase in pressure. The negative sign indicates that the volumetric change is a decrease.

Entrapped gas (i.e., air) in hydraulic oil significantly reduces the effective bulk modulus of the hydraulic fluid. The more entrapped gas there is, the lower the bulk modulus will be. For instance, consider this example with entrapped air in the hydraulic fluid such that the bulk modulus is $1/2$ the previous case without entrapped gas. For the same load conditions, the change in the volume of the fluid would be twice (x2) as much as the previous case. If this was a hydraulic cylinder, the change in hydraulic volume due to load and finite bulk modulus of the fluid would result in a change in the cylinder piston position.

$$\Delta l \cdot A_{cyl} = \Delta V \tag{7.30}$$

$$\Delta l = \frac{\Delta V}{A_{cyl}} \tag{7.31}$$

For the second case, the piston position change is twice as large as the first case simply due to entrapped gas in the hydraulic fluid and its resulting effect in lowering the bulk modulus.

As the effective bulk modulus increases, the hydraulic stiffness increases, hence the frequency content of the pressure in the line increases, but the volumetric change (hence the cylinder position change) decreases for a given load condition. Similarly, as the effective bulk modulus decreases, the hydraulic stiffness decreases, hence the frequency content of the pressure in the line decreases, but the volumetric change increases for a given load condition.

If this volume of fluid is inside a hydraulic cylinder with with cross-sectional area of $A_{cyl} = 10\,in^2$, the nominal height of the piston would be $h = 10\,in$. The change in the volume due to pressure would result in a corresponding change in the height of the fluid since we can assume that the cylinder cross-sectional area does not change,

$$\Delta V = A \cdot \Delta h \tag{7.32}$$

$$1\,in^3 = 10\,in^2 \cdot \Delta h \tag{7.33}$$

$$\Delta h = 0.1\,in \tag{7.34}$$

7.2.1 Analogy Between Hydraulic and Electrical Components

An analogy between the components used in a hydraulic circuit and an electrical circuit is shown in Figure 7.20. Notice the fact that for the majority of electrical circuit components, there is a corresponding hydraulic component for the same functionality. The electrical circuits have voltage, current, resistors, capacitors, inductors, and diodes. The hydraulic circuits have pressure, flow, orifice restriction, accumulators, and check valves. However, it should be noted that there are three main differences between hydraulic and electrical circuits:

1. The flow-pressure (P, Q) relationship is nonlinear, whereas the current-voltage (i, V) relationship is linear,

$$Q = K(x_s) \sqrt{\Delta P} \tag{7.35}$$

$$i = (1/R)\, V \tag{7.36}$$

where x_s is the displacement of the valve spool, $K(x_s)$ represents the effective orifice area and discharge coefficient function as a function of x_s, ΔP is the pressure differential across the valve, and R is electrical resistance, V is voltage potential. Resistance (R) to current and hydraulic resistance (R_H) to fluid flow are similar. The R_H is a function of orifice area opening and its geometry. In the above flow–pressure relationship, the $K(x_s)$ is equivalent inverse resistance to flow and function of the orifice opening (x_s).

2. The medium (fluid or electrons) of power transmission is compressible in hydraulic systems, and its properties vary among different fluids.

3. Voltage is a relative quantity, there is no absolute zero voltage. However, there is absolute zero pressure, which is the vacuum condition.

The capacitor (C) and accumulator (C_H) analogy is as follows,

$$V = \frac{1}{C} \int i(t) \cdot dt \quad p = \frac{1}{C_H} \int Q(t) \cdot dt \tag{7.37}$$

where C_H is determined by the volume and stiffness of the accumulator, which has the units of Length^5/Force (i.e., m^5/Nt as a result of volume/pressure $= [\text{m}^3]/[\text{Nt/m}^2]$). The capacitance of a hydraulic circuit is defined as the ratio of volumetric change to pressure change.

$$\frac{dp(t)}{dt} = \frac{1}{C_H} \cdot Q(t) \tag{7.38}$$

$$C_H = \frac{Q(t)}{dp(t)/dt} \tag{7.39}$$

$$= \frac{(dV/dt)}{(dp/dt)} \tag{7.40}$$

$$= \frac{dV}{dp} \tag{7.41}$$

For a given volume of fluid, V, and bulk modulus, β, capacitance is

$$C_H = \frac{V}{\beta} \tag{7.42}$$

The inertia of moving fluid in a pipe acts like an inductor. Consider a pipe with cross-sectional area A, length l, mass density of fluid ρ, pressures at the two ends p_1, p_2, and flow rate Q through it. Assume the following,

1. the friction in the pipe is neglected,
2. the compressibility of the fluid in the pipe is neglected.

The motion of the fluid mass in the pipe can be described as,

$$(p_1 - p_2) \cdot A = m \cdot \ddot{x} \tag{7.43}$$

$$= (\rho \cdot l \cdot A) \cdot \frac{\dot{Q}}{A} \tag{7.44}$$

where $m = \rho \cdot V = \rho \cdot l \cdot A$, and $Q = \dot{x} \cdot A$. The pressure and flow relationship is

$$p_1 - p_2 = \left(\frac{\rho \cdot l}{A} \right) \cdot \dot{Q} \tag{7.45}$$

$$p_1 - p_2 = L \cdot \dot{Q} \tag{7.46}$$

where the hydraulic inductance is defined as $L = \rho \cdot l/A$. Notice the analogy between pressure, flow rate, and hydraulic inductance versus the voltage, current, and self inductance,

$$\Delta p(t) = L \cdot \frac{dQ(t)}{dt} \tag{7.47}$$

$$\Delta V(t) = L \cdot \frac{di(t)}{dt} \tag{7.48}$$

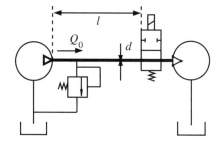

FIGURE 7.19: Hydraulic circuit where sudden closure of valve results in pressure spikes due to the inertia of the moving fluid.

Similarly, the *diode* and the *check valve* analogy is that they allow flow of electricity and fluid in one direction, and block it in the opposite direction,

$$i_o = i_i \quad V_i \geq V_o \tag{7.49}$$

$$= 0.0 \quad V_i < V_o \tag{7.50}$$

$$Q_o = Q_i \quad p_i \geq p_o \tag{7.51}$$

$$= 0.0 \quad p_i < p_o \tag{7.52}$$

where i_i, i_o are input and output current, V_i, V_o are input and output voltage, repectively. Similar notation applies for Q_o, Q_i, p_i, p_o for flow rate and pressure.

Example Consider the hydraulic circuit shown in Figure 7.19[1]. The dimensions of the pipe are shown in the figure, $d = 20$ mm, $l = 10$ m. Assume that there is a constant flow rate in steady-state, $Q_0 = 120$ l/min $= 2$ l/s $= 0.002$ m³/s. The mass density of fluid is $\rho = 1000$ kg/m³. Then, the valve closes suddenly, over a period of $\Delta t = 10$ ms, which results in a pressure spike. This phenomenon is also known as *water hammering*. Assume that the relief valve does not open.

The change in pressure due to the inertial deceleration of the fluid due to sudden closure of the valve is determined by

$$\Delta p = p_1 - p_2 = L \cdot \dot{Q} \tag{7.53}$$

$$= \frac{\rho \cdot l}{A} \frac{Q_0}{\Delta t} \tag{7.54}$$

$$= \frac{1000 \cdot 10}{\pi(0.02)^2/4} \frac{0.002}{0.01} \tag{7.55}$$

$$= \frac{0.2}{\pi} 10^8 \text{ Nt/m}^2 \tag{7.56}$$

$$= \frac{20}{\pi} \text{ MPa} = 6.36 \text{ MPa} \tag{7.57}$$

Notice that in this example we neglected the pressure change due to the compressibility of the fluid. A long pipe (inductor equivalent) and accumulator (capacitor equivalent) pair can be used as the LC filtering effect on the pressure in hydraulic circuits.

7.2.2 Energy Loss and Pressure Drop in Hydraulic Circuits

Anytime there is a pressure drop and flow in a hydraulic circuit, there is energy loss. The lost energy is converted to heat at the loss point. Recall that hydraulic power is equal to pressure difference times the flow rate between two points. Hence, the energy differential between

[1] Courtesy of Mr. Daniele Vecchiato.

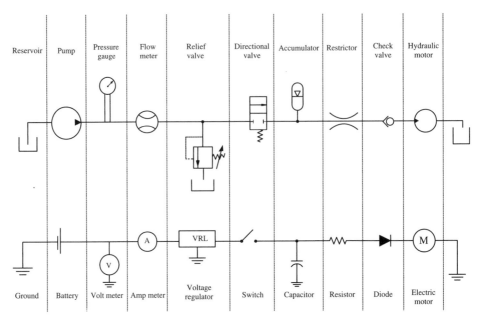

FIGURE 7.20: Analogy between hydraulic circuit components and electrical circuit components.

two points over a period of time is pressure differential times the total flow between the two points. In a typical hydraulic circuit, hydraulic fluid leaves the pump at a high pressure. There are pressure drops at the transmission pipes, valves, and actuators (Figure 7.21a). This means energy is lost during the transmission of fluid from pump to the load. In order to maximize the utilization of hydraulic power, and hence increase the efficiency of the hydraulic system, pressure drops between the source (pump) and load should be minimized.

$$Power_{12} = \Delta P_{12} \cdot Q_{12} \qquad (7.58)$$

$$Energy_{12} = \int_{t_1}^{t_2} \Delta P_{12}(t) \cdot Q_{12}(t) \cdot dt \qquad (7.59)$$

Pressure drop along hydraulic pipes is a function of the following parameters:

1. viscosity of the fluid, which is highly a function of temperature,
2. pipe diameter,
3. pipe length,
4. number of turns and bend in the pipe circuit,
5. surface roughness inside the pipes,
6. flow rate.

Manufacturers provide empirical data tables for pressure drop estimation as a function of the above parameters.

Another common "rule of thumb" in hydraulic circuit design is that the linear speed of the fluid should be kept below about 15 ft/s in order to minimize excessive pressure drop. Hence, the pipe diameter should be selected such that in order to support the maximum flow rate, the linear speed of the fluid should not exceed 15 ft/s.

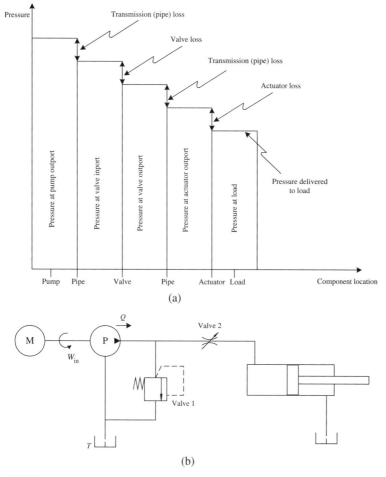

FIGURE 7.21: Hydraulic pressure and power loss in a hydraulic circuit: (a) pressure drop across different components in a hydraulic circuit, (b) an example circuit where pressure drop in each valve results in power loss in the form of heat.

Example   Consider the example hydraulic circuit shown in Figure 7.21b. Assume that the pump input speed (w_{in}) is constant, and hence the output flow is constant, $Q_p = 120\,l/min = 2\,l/s$. The pressure relief valve (valve 1) is set to a constant pressure value, $p_{relief} = 2\,MPa = 2 \times 10^6\,N/m^2$. Assume that the flow metering valve (valve 2) is sized to handle a maximum of $1/2$ of total pump flow, $Q_{v,max} = 0.5 \cdot Q_p$. Let us assume that the cylinder head-end cross-sectional area is $A_{he} = 0.01\,m^2$ and the load force is $F_l = 10\,000\,N$. Determine the total heat loss rate at the two valves under this operating condition in steady-state.

The heat loss rate is the pressure drop times the flow rate across each valve. Since the main flow valve (valve 2) can handle only half of the pump flow, the other half has to go through the pressure relief valve (valve 1).

$$Q_{v1} = 0.5 \cdot Q_p = 1.0\,l/s = 10^{-3}\,m^3/s \qquad (7.60)$$

$$Q_{v2} = Q_{v1} \qquad (7.61)$$

The pressure drop across valve 1 is the same as the pump outlet pressure since the output port of valve 1 is connected to the tank. The pressure drop across valve 2 is the input pressure (p_s) minus the load pressure, p_1.

$$\Delta p_{v1} = p_s - p_t = p_s = 2 \times 10^6 \, \text{N/m}^2 \tag{7.62}$$

$$\Delta p_{v2} = p_s - p_1 = 2 \times 10^6 - \frac{100\,000}{0.01} = 1 \times 10^6 \, \text{N/m}^2 \tag{7.63}$$

The total heat loss rate is

$$\begin{aligned} P_{\text{loss}} &= p_{v1} \cdot Q_{v1} + p_{v2} \cdot Q_{v2} &\tag{7.64} \\ &= 2 \times 10^6 \cdot 1.0 \times 10^{-3} \, \text{N/m}^2 \cdot \text{m}^3/\text{s} + 1 \times 10^6 \cdot 1.0 \times 10^{-3} \, \text{N/m}^2 \cdot \text{m}^3/\text{s} &\tag{7.65} \\ &= 2000 \, \text{W} + 1000 \, \text{W} &\tag{7.66} \\ &= 3000 \, \text{W} &\tag{7.67} \end{aligned}$$

which indicates that in this case the heat loss over the relief valve is twice as much as the heat loss over the main flow valve. Since the total power output of pump is

$$\begin{aligned} P_{\text{pump}} &= p_s \cdot Q_p = 2.0 \times 10^6 \cdot 2.0 \times 10^{-3} \, \text{N/m}^2 \cdot \text{m}^3/\text{s} &\tag{7.68} \\ &= 4000 \, \text{W} &\tag{7.69} \end{aligned}$$

The efficiency of the overall hydraulic circuit is 25%. Only 25% of the hydraulic power generated by the pump is delivered to the load. The rest is wasted at various pressure drop components. This example also illustrates one of the drawbacks of fixed displacement pumps. They have low efficiency. Since pump outputs flow at a constant rate as a function of input shaft speed regardless of the hydraulic flow demand, unused flow must be dumped (wasted) over the relief valve.

Figure 7.22 shows the list of ANSI/ISO standard symbols used for describing hydraulic components in circuits. Readers should get familiar with the symbols in order to comfortably interpret hydraulic circuits.

Example   Consider a hydraulic cylinder with 10 in bore diameter, 10 in full stroke length, and 5 in rod diameter. The cylinder is supplied by a hydraulic pump and flow is controlled by a valve. The cylinder is to make a full stroke extend motion in 7.5 s, wait for 7.5 s, make the retract motion again in 7.5 s, and wait for 7.5 s. This defines a complete repetative cycle (Figure 7.23). The minimum line pressure required is 1500 psi and maximum line pressure is 3000 psi. Select a pump size based on the average flow rate requirement, then select a bladder type accumulator to meet the above requirements.

The average flow rate that is needed from the pump is the total volume traveled during a full cycle divided by the cycle time,

$$Q_{\text{ave}} = \frac{(V_{\text{cyl}_{\text{up}}} + V_{\text{cyl}_{\text{down}}})}{t_{\text{cycle}}} \tag{7.70}$$

where $t_{\text{cycle}} = 7.5 + 7.5 + 7.5 + 7.5 \, \text{s} = 30 \, \text{s}$. The cylinder volume is

$$V_{\text{cyl}_{\text{up}}} = 10 \cdot (\pi/4)(10^2) = 785 \, \text{in}^3 \tag{7.71}$$

$$V_{\text{cyl}_{\text{down}}} = 10 \cdot (\pi/4)(10^2 - 5^2) = 589 \, \text{in}^3 \tag{7.72}$$

$$Q_{\text{ave}} = \frac{(785 + 589)}{30} = 45.8 \, \text{in}^3/\text{s} \tag{7.73}$$

The pump is selected to provide the average flow rate. Let us assume that the pump will provide the Q_{ave} flow rate. During the time period when the cylinder is extending and retracting, the pump will not be able to support the flow volume needed. The difference

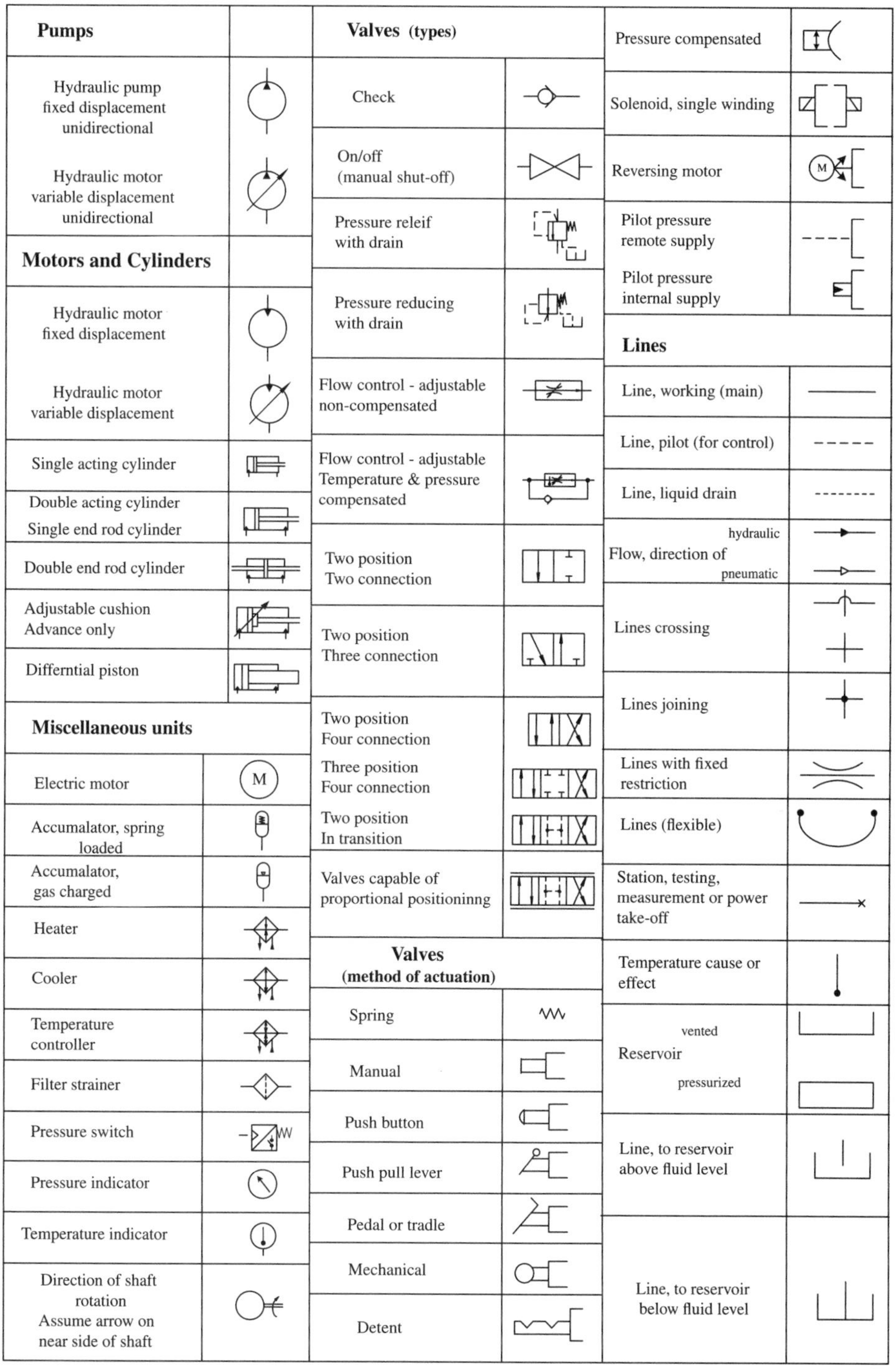

Pumps			Valves (types)			Pressure compensated	
Hydraulic pump fixed displacement unidirectional			Check			Solenoid, single winding	
Hydraulic motor variable displacement unidirectional			On/off (manual shut-off)			Reversing motor	
Motors and Cylinders			Pressure releif with drain			Pilot pressure remote supply	
						Pilot pressure internal supply	
Hydraulic motor fixed displacement			Pressure reducing with drain			**Lines**	
Hydraulic motor variable displacement			Flow control - adjustable non-compensated			Line, working (main)	
						Line, pilot (for control)	
Single acting cylinder			Flow control - adjustable Temperature & pressure compensated			Line, liquid drain	
Double acting cylinder Single end rod cylinder						Flow, direction of hydraulic pneumatic	
Double end rod cylinder			Two position Two connection				
Adjustable cushion Advance only			Two position Three connection			Lines crossing	
Differntial piston						Lines joining	
Miscellaneous units			Two position Four connection			Lines with fixed restriction	
Electric motor	M		Three position Four connection				
Accumalator, spring loaded			Two position In transition			Lines (flexible)	
Accumalator, gas charged			Valves capable of proportional positioninng			Station, testing, measurement or power take-off	
Heater			**Valves (method of actuation)**			Temperature cause or effect	
Cooler			Spring			Reservoir vented pressurized	
Temperature controller			Manual				
Filter strainer			Push button			Line, to reservoir above fluid level	
Pressure switch			Push pull lever				
Pressure indicator			Pedal or tradle			Line, to reservoir below fluid level	
Temperature indicator			Mechanical				
Direction of shaft rotation Assume arrow on near side of shaft			Detent				

FIGURE 7.22: ANSI/ISO standard symbols for hydraulic components. These component symbols are used in the schematic circuit design of hydraulic systems.

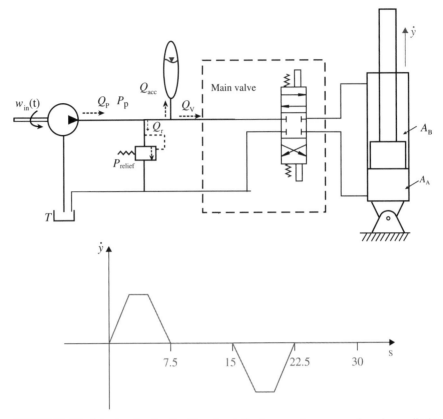

FIGURE 7.23: A hydraulic circuit with a fixed displacement pump, main valve, cylinder, and accumulator. The desired cyclic motion profile is shown below the figure. The pump is sized for average flow requirement, and the accumulator is sized to support the peak flow requirement.

has to be made up by the accumulator. The discharge volume needed from the accumulator during up and down motion is the difference between the volume of fluid needed and the volume of fluid the pump can provide,

$$V_{\text{disch}} = (V_{\text{cyl}_{\text{up}}} + V_{\text{cyl}_{\text{down}}}) - (Q_{\text{ave}} * (7.5 + 7.5)) \tag{7.74}$$

$$= 687 \, \text{in}^3 \tag{7.75}$$

The necessary accumulator size is calculated from the empirical formula, where we will assume a precharge pressure of 90% of minimum pressure, and $n = 1.4$. Then,

$$V_{\text{acc}} = 687 \cdot \frac{1}{0.9} / (1 - (1500/3000)^{1/1.4}) \tag{7.76}$$

$$= 1954 \, \text{in}^3 = 8.5 \, \text{gallon} \tag{7.77}$$

Selecting a 10 gallon accumulator would be a good choice with sufficient safety margin for the needed volume. During the part of the cycle when there is no motion in the cylinder, the pump flow is used to recharge the accumulator. It is important to check that the pump flow and charge time available are sufficient to recharge the accumulator for continuous cyclic operation. The pump can charge the following volume when the cylinder is in waiting mode,

$$V_{\text{charge}} = 45.8 \, \text{in}^3/\text{s} * 15 \, \text{s} \tag{7.78}$$

$$= 687 \, \text{in}^3 \tag{7.79}$$

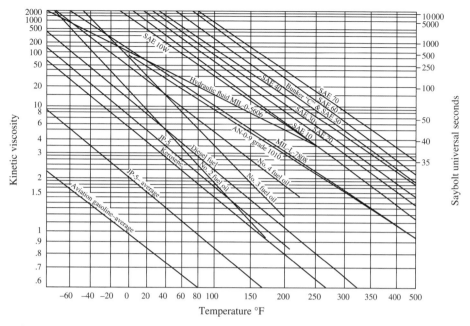

FIGURE 7.24: Viscosity of various hydraulic oils as a function of temperature. Notice the nonlinear scale in the *y*-axis. In all cases, as temperature increases, viscosity decreases. At a given temperature, different hydraulic oils have different viscosity. Higher viscosity increases the friction losses (hence reduces the efficiency of the hydraulic power system), but reduces leakage. Reproduced with permission from Parker Hannifin.

For a good design, $V_{\text{charge}} \geq V_{\text{disch}}$. Due to the symmetric ratio in the cycle times and cylinder head-end and rod-end diameters, as well as the pump size selection, $V_{\text{charge}} = V_{\text{disch}}$ in this example. It may be better to select a slightly bigger pump, such as $Q_{\text{ave}} = 50\,\text{in}^3/\text{s}$, which would provide an extra capacity for the pump to be able to charge the accumulator. In this case $V_{\text{charge}} > V_{\text{disch}}$, which is a desirable result in design.

7.3 HYDRAULIC PUMPS

The functional block diagram and operating principle of a pump are shown in Figure 7.25a–d. The pump is the device used to convert mechanical power to hydraulic power. A positive displacement pump concept is shown in Figure 7.25b–d. During the in-stoke $p_3 < p_1$, oil is sucked in from the "tank." During the out-stroke $p_3 \geq p_2$, oil is pushed out to the load. Notice that p_3, pump output pressure, is determined by the load pressure $p_3 \approx p_2 = p_{\text{load}} + p_{\text{spring}}$. If there is no load resistance, then the pump cannot build up pressure. In a hydraulic system including pump-valve-cylinder-tank, the pressure difference between the pump outlet port and tank is determined by the pressure drop on the valve and the load pressure created by the cylinder and load interaction. In this concept figure, the check valves control the direction of the flow. The line relief valve (Figure 7.25d) limits the maximum allowed line pressure as protection and returns the flow back to the tank if the line pressure tries to exceed a set limit. Hence, the relief valve assures that the line pressure stays less than or equal to maximum relief pressure set, $p_3 \leq p_{\text{max}}$. Notice that during the in-stroke, the volume is expanding. Similarly, during the out-stroke, the volume is contracting. This phenemenon provides the suction and pumping action for the pump.

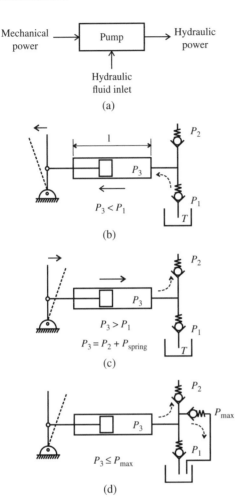

FIGURE 7.25: Concept of a positive displacement pump: (a) pump input–output block diagram, (b) in-stroke (suction), (c) out-stroke (pumping), (d) pressure limiting valve.

The volume spanned by the piston per stoke of the pump is called the "pump displacement per stroke." If that volume is fixed (not adjustable) for a given pump, the pump is called a *fixed displacement pump*. If that volume is adjustable, the pump is called a *variable displacement pump*. If the pump is of fixed displacement type, the flow volume pumped per stroke is fixed. Therefore, the only way to control the flow rate is to control the rate of the stroke (mechanical input speed). If the pump is of variable displacement type, the flow rate can be controlled by either input speed and/or by adjusting the displacement of the pump per stroke.

7.3.1 Types of Positive Displacement Pumps

Different types of positive displacement pumps are briefly discussed below (Figrue 7.26) and the differences among them are pointed out.

(i) Gear Pumps Gear pumps work by the principle of carrying fluid between the tooth spaces of two meshed gears (Figure 7.27). One of the two shafts is the drive shaft, called the drive gear, and the other, driven by the first, is called the driven or idler gear. The pumping chambers are formed by the spaces between the gears, enclosed by the pump housing and the side plates. The gear pump is always unbalanced by design since the high-pressure

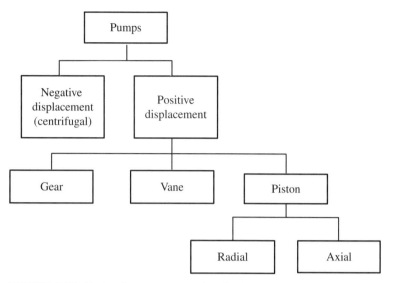

FIGURE 7.26: Hydraulic pump categories: Positive displacement pump types: gear, vane, piston (radial and axial).

side pushes the gear towards the low-pressure side. The relation between the driven and driving shaft causes imbalance that will have to be taken up by the bearings. Gear pumps are always of fixed displacement type. Despite these drawbacks gear pumps are popular because of their simplicity, low cost, and robustness. The unmeshing of gears around the inlet port creates a low pressure and results in the suction action. Similarly, the squeezing action of the gears on the fluid between the teeth and pump housing as well as when the meshing between teeth begins around the outlet port results in the increased pressure at the outlet port. The flow rate is determined by the input shaft speed of the pump and its fixed displacement. The pressure is determined by the load as long as it is not too high to the point of stalling the input shaft or breaking the pump housing. However, as the pressure increases, the leakage increases. Hence, the flow rate versus pressure curve drops down a

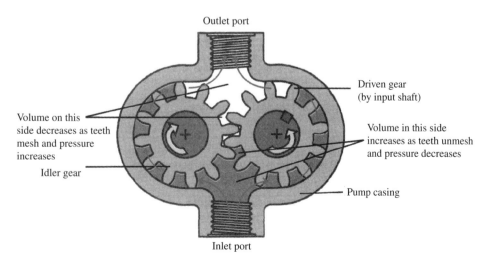

FIGURE 7.27: Gear pump cross-section views: two-dimensional cross-sectional view. Reproduced with permission from Eaton. Copyright 1999 Eaton. All Rights Reserved.

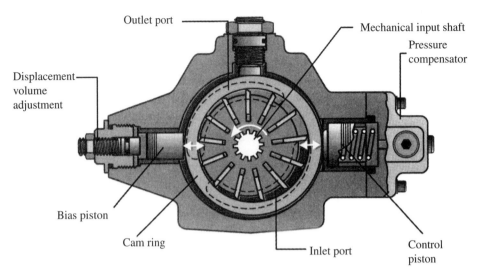

Outlet port

Mechanical input shaft

Pressure compensator

Displacement volume adjustment

Bias piston

Cam ring

Inlet port

Control piston

FIGURE 7.28: Vane pump: cross-section. Reproduced with permission from Eaton. Copyright 1999 Eaton. All Rights Reserved.

little bit as output pressure increases. Ideally, if there is no leakage, the flow rate would be independent of the pressure.

(ii) Vane Pumps Vane pumps are known to operate more quietly than gear pumps and used to have a cost benefit over piston pumps. More recent development has made piston pumps both quieter and cheaper so they now totally dominate the market for mobile hydraulics. A variable displacement vane pump is shown in (Figure 7.28). It works on the principle that oil is taken in on the side with an increasing volume and the vanes connected to the rotor then push the oil in to the part of the pump with decreasing volume, therefore increasing the pressure of the oil before it reaches the outlet port. The so-called cam ring is eccentrically mounted compared to the rotor of the pump, making the design unbalanced. Changing the position of the cam ring in relation to the rotor changes the displacement of the pump. The cam ring rests on a control piston on one side and the other side has a bias piston. By changing the pressure in the control piston (Figure 7.28) the displacement of the pump is changed. A control piston is used to adjust the pump displacement. The maximum displacement is set by the displacement volume adjustment that limits the travel of the bias piston.

(iii) Radial Piston Pumps A radial piston pump (Figure 7.29) is built on a rotating shaft with the cylinder block rotating on the outside. As the piston follows the outer housing of the pump on slippers, the offset from the central position creates the pumping motion. The pistons maintain contact with the outer ring at all times. The input shaft rotation axis is perpendicular to the plane of axes along which the pistons slide. The porting in the pivot pin allows intake at low pressure into the cylinder (from the inner diameter side) as the cylinder block passes by and pressurizes outflow as the cylinder block passes the outlet port (through the inner diameter side). The number of pistons, diameter of the pistons, and the length of their stroke determine pump displacement. The piston and cylinder sleeves are manufactured and custom fitted to high tolerances for each pump. Radial piston pumps are built so that more than one port discharges fluid at the same time and more than one port is open to intake at the same time. This design attempts to reduce cyclic flow rate and pressure pulsation in the output flow as a function of rotor angular position per revolution.

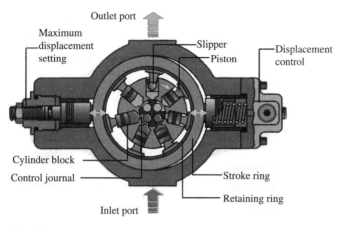

FIGURE 7.29: Radial piston pump: cross-section. Reproduced with permission from Eaton. Copyright 1999 Eaton. All Rights Reserved.

Typical radial piston pumps provide output pressure up to about 350 bar, and flow rates of 200 lpm.

(iv) Axial Piston Pumps Axial piston pumps are the most widely used pump types in mobile hydraulic applications. There are two major types:

- inline piston pumps (Figure 7.30),
- bent-axis axial piston pump (Figure 7.31).

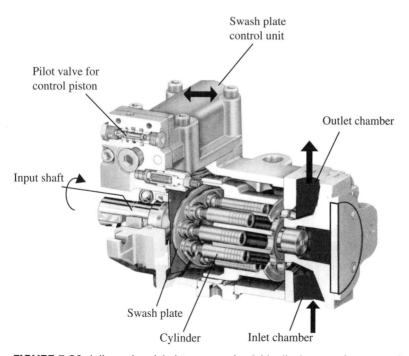

FIGURE 7.30: Inline-axis axial piston pump (variable displacement) cross-sectional view.

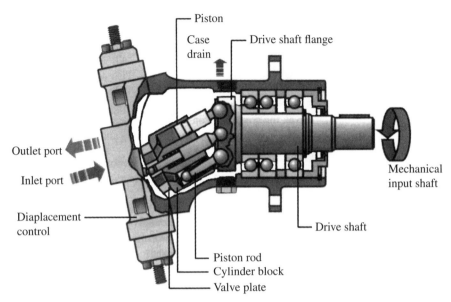

Piston
Case drain
Drive shaft flange
Outlet port
Inlet port
Mechanical input shaft
Diaplacement control
Drive shaft
Piston rod
Cylinder block
Valve plate

FIGURE 7.31: Bent-axis axial piston pump: variable displacement type. Reproduced with permission from Eaton. Copyright 1999 Eaton. All Rights Reserved.

The only difference between them is the orientation of the rotation axis of the pistons with respect to the input shaft. In the case of the inline piston pump, the mechanical input shaft and piston rotation axes are inline, whereas in the case of the bent-axis type, they are not inline. Varying the angle of the swash plate in the pump changes the displacement. The typical range of swash plate angle rotation is about 15°. The displacement of the pump is controlled by controlling the swash plate angle between its limits, that is between 0 and 15°. The oil is pushed into the pump through the intake side of the port plate, sometimes called the kidney plate because of the shape of the ports. The cylinders are filled when they pass the inside area and the oil is pushed out on the other side of the kidney plate. When the cylinders travel along, the swash plate pushes the trapped oil towards the kidney plate and therefore raises the pressure. When the swash plate is perpendicular to the rotation axis, the displacement is zero. Then, the pump is in its stand-by position and does not provide any flow. More accurately, at stand-by the pump provides flow only to compensate for leakage flow. It is this capability (adjustable displacement) that accounts for the energy saving potential of variable displacement pumps by providing adjustable flow rate based on demand.

Axial piston pumps can be fixed displacement or variable displacement, unidirectional or bidirectional, and may have over-center control (Figure 7.32). In a fixed displacement pump, the swash plate angle is constant. In a variable displacement pump, the swash plate angle can be varied by a control mechanism. A unidirectional pump is intended to be driven by an input shaft in either a clockwise or counter-clockwise direction. If the pump has the so called *over-center* control capability, which is the ability to change the swash plate angle about its neutral position in either direction, the pump output flow direction can be changed by the over-center control even though input shaft speed direction is same. A bidirectional pump can be driven by an input shaft in either direction and the flow direction changes with the input speed direction. An analogous operating principle applies for hydraulic motors. In the case of motor, the output is the shaft rotation and input is the fluid flow. A unidirectional motor is intended to receive flow in one direction and rotate in one direction only. Bidirectional motors can receive flow in either direction, hence can

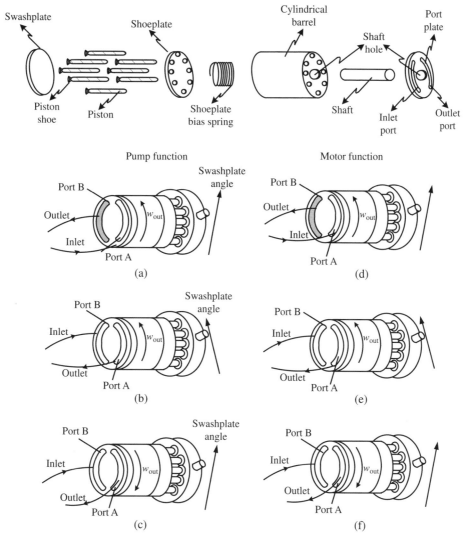

FIGURE 7.32: Axial piston pump and motor: components and operating principles. (a) unidirectional pump function, (a) and (b) over-center controlled pump functionality, (a) and (c) bidirectional pump functionality, (d) unidirectional motor function, (d) and (e) over-center contolled motor functionality, (d) and (f) bidirectional motor functionality.

provide output shaft rotation in either direction. Over-center control allows the motor output speed direction to be changed even though the input flow direction does not change.

7.3.2 Pump Performance

A variable displacement pump provides the most flexibility, but with additional control complexity. In this case, the input speed of the pump can be left to be determined by other conditions (i.e., operator may control engine speed for other considerations, such as vehicle travel speed), and the desired pump output is controlled by manipulating the swash-plate angle. The swash-plate actuation logic may be based on regulating the flow rate, output pressure, or other variables of interest (Figure 7.33). Figure 7.34 shows the typical steady-state performance characteristics of a pump in terms of its size and efficiency.

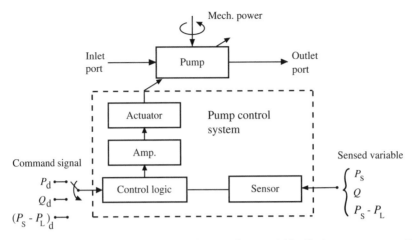

FIGURE 7.33: Control system block diagram for a variable displacement pump. The pump can be controlled to regulate the output pressure, flow, or other derived variable. The control element is the swash plate angle positioning mechanism of the pump.

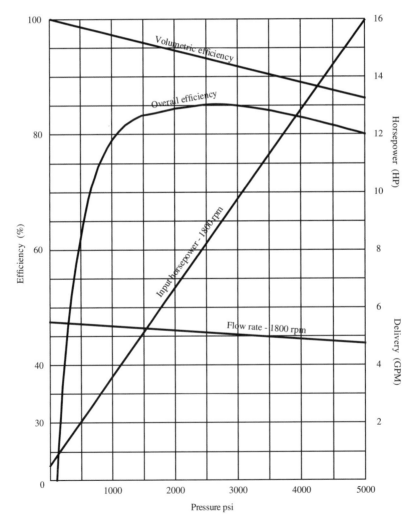

FIGURE 7.34: Steady-state performance characteristics of a pump: flow rate, power, and efficiency as a function of its operating pressure range at different input shaft speeds.

The primary performance measures of a pump are specified by the following parameters:

1. displacement in in^3/rev which defines the size of the pump, $D_p(\theta)$,
2. rated flow, Q_r, at rated speed and rated pressure,
3. rated pressure, p_r,
4. rated power (derived quantity from rated pressure and rated flow),
5. dynamic response bandwidth of the pump displacement control system if the pump is the variable displacement type.

The secondary peformance specifications include:

1. efficiency: volumetric, mechanical, and overall,
2. maximum speed,
3. weight,
4. noise level,
5. cost.

The volumetric efficiency (η_v) refers to the ratio of the output fluid flow to the volumetric displacement of the pump. It is a measure of the leakage in the pump. Mechanical efficiency (η_m) refers to the losses due to factors other than leakage, such as friction and mechanical losses. The overall efficiency (η_o) refers to the ratio of the output hydraulic power to the input mechanical power of the pump. Let the units of each term be as follows, $D_p \, m^3/rev$, $w_{shaft} \, rev/s$, $p \, N/m^2$. These efficiency measures are functions of the pressure and speed of the pump,

$$\eta_v = \frac{Volume_{out}}{Volume_{disp}} \tag{7.80}$$

$$= \frac{Q_{out}}{D_p \cdot w_{shaft}} \tag{7.81}$$

and

$$\eta_m = \frac{Power^*_{out}}{Power_{in}} \tag{7.82}$$

$$= \frac{p_{out} \cdot D_p \cdot w_{shaft}}{T \cdot w_{shaft} \cdot (2\pi)} \tag{7.83}$$

$$= \frac{p_{out} \cdot D_p}{2\pi T}; \quad for \; D_p \; in \; units \; of \; m^3/rev \tag{7.84}$$

and

$$\eta_o = \frac{Power_{out}}{Power_{in}} = \frac{p_{out} \cdot Q_{out}}{T \cdot w_{shaft}} \tag{7.85}$$

$$= \eta_v \cdot \eta_m \tag{7.86}$$

Notice that if the unit of D_p is in m^3/rad, then mechanical efficiency would be defined as

$$\eta_m = \frac{p_{out} \cdot D_p}{T}; \quad if \; D_p \; unit \; is \; m^3/rad \tag{7.87}$$

The difference in the definition of the mechanical and overall efficiency is the definition of output power. The mechanical efficiency defines the output power as the power if there were zero leakage, that is 100% volumetric efficiency. The overall efficiency takes into account both the leakage and other mechanical efficiency.

One of the main operational concerns is the noise level of the pump. The noise due to cavitation can be very serious. Low pressure levels at the pump input may result in allowing air bubbles to enter the pump inlet port. Under high pressure, the bubbles collapse which leads to high noise levels. Therefore, some pumps may require higher than normal levels of inlet port pressure, that is so called boost inlet pressure using another small pump (called charge pump) between the main pump and tank. Cavitation may also be caused by too high hydraulic fluid viscosity, as a result, the pump is not able to suck in enough fluid. This may happen especially in cold start-up conditions. Heaters may be included in the pump inlet to limit the fluid viscosity.

Example Consider a pump with fixed displacement $D_p = 100 \, \text{cm}^3/\text{rev}$ and following nominal operating conditions: input shaft speed $w_{shaft} = 1200 \, \text{rpm}$, torque at the input shaft $T = 250 \, \text{Nt} \cdot \text{m}$, the output pressure $p_{out} = 12 \, \text{MPa}$, and output flow rate $Q_{out} = 1750 \, \text{cm}^3/\text{s}$. Determine the volumetric, mechanical, and overall efficiencies of the pump at this operating condition.

$$\eta_v = \frac{Q_{out}}{D_p \cdot w_{shaft}} \qquad (7.88)$$

$$= \frac{1750 \, \text{cm}^3/\text{s}}{100 \, \text{cm}^3 \cdot 20 \, \text{rev}/\text{s}} = 87.5\% \qquad (7.89)$$

$$\eta_m = \frac{p_{out} \cdot D_p}{T} \qquad (7.90)$$

$$= \frac{12 \times 10^6 \, \text{N}/\text{m}^2 \cdot 100 \times 10^{-6} \, \text{m}^3/\text{rev}}{2 \cdot \pi \cdot 250} = 76.4\% \qquad (7.91)$$

$$\eta_o = \eta_v \cdot \eta_m \qquad (7.92)$$

$$= 0.875 \cdot 0.764 \cdot 100\% = 66.85\% \qquad (7.93)$$

Notice that if any of the efficiency calculations results in being larger than 100%, it is an indication that there is an error in the given information or measured variables since the efficiencies cannot be larger than 100%. In order to confirm the results, let us calculate the input mechanical power to the pump and output hydraulic power from the pump.

$$Power_{in} = T \cdot w_{shaft} \qquad (7.94)$$

$$= 250 \, \text{N} \cdot \text{m} \cdot 20 \, \text{rev}/\text{s} \cdot 2\pi \, \text{rad}/\text{rev} \qquad (7.95)$$

$$= 31.41 \, \text{N} \cdot \text{m}/\text{s} = 31.41 \, \text{kW} \qquad (7.96)$$

$$Power_{out} = p_{out} \cdot Q_{out} \qquad (7.97)$$

$$= 12 \cdot 10^6 \, \text{N}/\text{m}^2 \cdot (1750 \, \text{cm}^3/\text{s}) \cdot (1 \, \text{m}^3/10^6 \, \text{cm}^3) \qquad (7.98)$$

$$= 12 \cdot 1750 \, \text{N} \cdot \text{m}/\text{s} \qquad (7.99)$$

$$= 21 \cdot 10^3 \, \text{N} \cdot \text{m}/\text{s} = 21 \, \text{kW} \qquad (7.100)$$

$$\frac{Power_{out}}{Power_{in}} = \frac{21}{31.41} \qquad (7.101)$$

$$= 66.85\% \qquad (7.102)$$

$$= \eta_v \cdot \eta_m \qquad (7.103)$$

$$= \eta_o \qquad (7.104)$$

Example Consider the hydrostatic transmission shown in Figure 7.35. Assume that the pressure drop across the connecting hydraulic line is 200 psi. The pump variables and

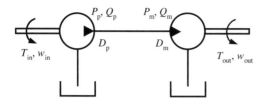

FIGURE 7.35: Hydraulic circuit with a hydraulic pump and hydraulic motor.

motor variables are shown in the table below, along with the volumetric and mechanical efficiencies.

Pump	Motor
$D_p = 10\,\text{in}^3/\text{rev}$	$D_m = 40\,\text{in}^3/\text{rev}$
$\eta_v = 0.9$	$\eta_v = 0.9$
$\eta_m = 0.85$	$\eta_m = 0.85$
$w_p = 1200\,\text{rpm}$	$w_m = ?$
$p_p = 2000\,\text{psi}$	$p_m = ?$
$Q_p = ?$	$Q_m = ?$
$T_{in} = T_p = ?$	$T_{out} = T_m = ?$
$PumpPower_{in} = ?$	$MotorPower_{out} = ?$

Let us first focus on the pump and its input power source. The net flow rate from the pump at the given input speed is

$$Q_p = \eta_v \cdot D_p \cdot w_p \tag{7.105}$$

$$= 0.9 \cdot 10 \cdot 1200\,\text{in}^3/\text{rev} \cdot \text{rev}/\text{min} \tag{7.106}$$

$$= 10\,800\,\text{in}^3/\text{min} \tag{7.107}$$

The input torque required to drive the pump is

$$\eta_0 = \eta_v \cdot \eta_m = \frac{PumpPower_{out}}{PumpPower_{in}} \tag{7.108}$$

$$= \frac{p_p \cdot Q_p}{T_p \cdot w_p} \tag{7.109}$$

$$T_p = \frac{1}{\eta_0} \cdot p_p \cdot \frac{Q_p}{w_p} = \frac{1}{\eta_0} \cdot p_p \cdot D_p \tag{7.110}$$

$$= \frac{1}{0.9 \cdot 0.85} \frac{2000\,\text{lb}/\text{in}^2 \cdot 10\,800\,\text{in}^3/\text{min} \cdot \frac{1\,\text{min}}{60\,\text{s}}}{20 \cdot 2\pi\,\text{rad}/\text{s}} \tag{7.111}$$

$$= 3744.8\,\text{lb} \cdot \text{in} \tag{7.112}$$

and the input power that must be supplied to the pump is

$$PumpPower_{in} = T_p \cdot w_p \tag{7.113}$$

$$= 3744.8 \cdot 20 \cdot 2\pi\,\text{rad}/\text{s} \cdot \frac{1\,\text{HP}}{6600\,\text{lb} \cdot \text{in}/\text{s}} \tag{7.114}$$

$$= 71.3\,\text{HP} \tag{7.115}$$

Now let us focus on the motor. The pump output pressure and flow is connected to the motor. We assumed that there is a 200 psi pressure drop in the hydraulic line and no leakage in the transmission line. Hence, the input pressure and flow rate to the motor are

$$p_m = p_p - 200\,\text{psi} = 1800\,\text{psi} \qquad (7.116)$$

$$Q_m = Q_p = 10\,800\,\text{in}^3/\text{min} \qquad (7.117)$$

The volumetric efficiency of the motor indicates that less than 100% of the input flow is converted to displacement,

$$\eta_v = \frac{Q_a}{Q_m} \qquad (7.118)$$

$$Q_a = \eta_v \cdot Q_m = 0.9 \cdot 10\,800 = 9720\,\text{in}^3/\text{min} \qquad (7.119)$$

and we can determine the output speed of the motor,

$$Q_a = \eta_v \cdot Q_m = D_m \cdot w_m \qquad (7.120)$$

$$w_m = \frac{Q_a}{D_m} = 243\,\text{rpm} \qquad (7.121)$$

The output power of the motor is determined from the ratio of its input power and output power, given the overall efficiency of the motor,

$$\eta_o = \eta_v \cdot \eta_m = \frac{T_m \cdot w_m}{p_m \cdot Q_m} \qquad (7.122)$$

$$T_m = \frac{\eta_o \cdot p_m \cdot Q_m}{w_m} \qquad (7.123)$$

$$= \frac{0.9 \cdot 0.85 \cdot 1800\,\text{lb/in}^2 \cdot 10\,800\,\text{in}^3/\text{min}}{243\,\text{rev/min} \cdot 2\pi\,\text{rad/rev}} \qquad (7.124)$$

$$= 61\,200/(2\pi)\,\text{lb} \cdot \text{in} = 9740\,\text{lb} \cdot \text{in} \qquad (7.125)$$

Hence, the mechanical power delivered at the output shaft of the motor is

$$MotorPower_{out} = T_m \cdot w_m = \frac{(61\,200/(2\pi)) \cdot 243 \cdot 2\pi}{6600} = 37.55\,\text{HP} \qquad (7.126)$$

7.3.3 Pump Control

The pump control element is the actuation mechanism that controls the angular position of the swash plate. The mechanism which controls the swash plate angle is called the *compensator* or the *pump controller*.

A pump control system has the following components (Figure 7.33),

1. One or a pair of small control cylinders (also called *displacement pistons*) used to move the swash plate.

2. A proportional valve which meters flow to the control cylinders, and is controlled by electrical or hydro-mechanical means for a certain control objective (i.e., constant pressure output from the pump, load sensing pump).

3. Sensor or sensors (either hydraulic pressure sensing lines or electrical sensors) used to measure the variables of interest, such as pressure, flow rate, position.

4. A controller (i.e., a digital electronic control module (ECM) or a hydro-mechanical control logic configured to control the proportional valve) used to implement the control logic (i.e., if pump is controlled to provide a constant output pressure).

The swash plate angle is limited to the range

$$\theta_{min} \leq \theta \leq \theta_{max} \tag{7.127}$$

where θ_{min} is the minimum and θ_{max} is the maximum angular displacement of the swash plate. If the pump has two output ports, it is called a *bidirectional pump*. Output flow is directed to one or the other port by the swash plate angle control. Such pump control is referred to as *over-center* pump control. If the swash plate displacement is on one side of the center, output flow is in one direction. If the swash plate displacement is on the other side of the center, then output flow is in the opposite direction. Hence the flow direction can be changed by pump control alone in a closed-circuit hydraulic systems such as hydrostatic transmission applications.

Figure 7.38b shows the logical relationship between the load-sensing valve and pressure limiting valve. The same control is shown in Figure 7.39 as a cross-sectional view of a variable displacement piston pump with a load sensing control valve and a pressure limit valve. If the order of the two valves were swapped (load sensing valve and pressure limiting valve in Figure 7.38), the pump would not be able to destroke when the cylinder hits a load that it can not move (stalled load condition). Therefore, it is important to note that the pressure limiting valve is placed between the load sensing valve and swash plate control piston. For further discussion, if we eliminate the load sensing valve and signal, and connect the pressure limiting valve to pump output and tank ports as well as the swash plate control piston port (three port connection), the pump control becomes a *pressure regulating* (also called *pressure-compensated* or *pressure controlled*) pump where the output pressure of the pump is set by the spring (Figure 7.36). In almost all hydro-mechanical pump controls, the preloads on the springs are adjustable by a screw, which allows the user to adjust the desired pressure limit and load-sensing pressure differential. Typical load pressure differential setting (the pump output pressure minus the load pressure feedback signal) is about 150 psi to 200 psi range in construction equipment applications.

In most pumps, the maximum and minimum swash plate angles can be mechanically adjusted by a set screw. In hydro-mechanically controlled pumps, the pressure feedback signals are provided by hydraulic lines with orifices. The command signal (i.e., desired output pressure) is implemented by an adjustable spring and screw combination. The actuator which moves the swash plate is called the control cylinder or control piston. In some pumps, the actuator provides the power to move the swash plate in both directions under the control of the proportional valve, whereas in others, the actuator provides power for one direction and the power for the other direction is provided by a preloaded spring. In most cases, at startup the preloaded spring will move the swash plate to maximum displacement. As the output pressure of the pump builds up, the compensator mechanism (proportional valve and control piston) provides control power to reduce the swash plate angle.

Pump control, that is the control of the pump displacement ($D_p(\theta)$) which is a function of the swash plate angle, may be based on different objectives such as (Figure 7.33)

1. pressure compensating and limiting,
2. flow compensated,
3. load sensing,
4. positive flow control (matched flow supply and demand),
5. torque limiting,
6. power limiting.

Among different pump control methods, pressure compensated valve control and load sensing valve control are the two most common methods.

(**1**) *Pressure compensated pump control:* control the pump displacement such that the actual pump output pressure (P_s) is regulated about a desired pump output pressure (P_{cmd}),

$$\theta_{cmd} = \theta_{offset} + K \cdot (P_{cmd} - P_s) \tag{7.128}$$

where θ_{cmd} is the commanded (desired) swash plate angle, θ_{offset} is the offset value of the swash plate angle, K is the proportional gain for the pressure regulator. For simplicity, we illustrate a proportional control logic here. Clearly, more advanced control logic can be implemented in the relationship between the desired variables, measured variables, and the output of the controller.

This type of pump control is called a *pressure compensated pump control*, and provides constant pressure output. The flow rate will be determined by whatever is needed to maintain the desired pressure, up to the maximum flow capacity (Figure 7.37a). The commanded pressure does not have to be constant and typically can be set between a minimum and a maximum value for a given pump. As the flow rate increases, the regulated pressure tends to reduce a little in hydro-mechanically controlled pumps. But this effect can be eliminated by software control algorithms in digitally controlled pumps. When the external load is so large that it cannot be moved, the pump under pressure compensated control will provide the desired regulated output pressure at almost zero flow rate. This is called the *dead head condition*.

Figure 7.36 shows the hydraulic diagram for a *pressure-compensated* pump control system. The valve has a spring through which the desired output pressure is set by preloading the spring. The pump output pressure is compared to that pressure setting by the valve. Based on the difference, either the pump or the tank ports are connected to the control piston line to up-stroke or down-stroke the swash plate of the pump. When $p_s = p_{desired}$ with a certain error margin, where the allowed error margin is defined by the overlapped part of the orifice opening of the valve, the output flow to the control piston is zero and the pump's swashplate maintains that angular position. If the pump output pressure is less than the desired pressure setting, the valve moves to the left and the control piston port is connected to the tank, hence as the flow moves out from control piston to tank, the swash plate stokes up, which should result in increased output flow and increased pump pressure. If the pump pressure is higher than the desired pressure setting, the valve moves to the right, connecting the control piston port to the pump pressure port. As the flow goes from pump port into the control piston, the swash plate destrokes, and the pump output flow decreases as does the pump output pressure. As a result, the pump output pressure is regulated to be equal to the desired pressure setting by the preloaded spring. Typical hydro-mechanical pump controls have a pressure control valve where the preloads on the springs are adjustable by a screw, which allows the user to adjust the desired pressure. If the desired output pressure is variable (i.e., based on operator command or control logic), then the preloaded spring is replaced by a remote pilot pressure which is variable and controlled by operator commands or control logic.

Pressure limiting control is established by feeding back the pressure in the outlet to a check valve. When a preset pressure limit value is reached, the check valve opens to let oil in to the control piston and therefore destroke the pump swash plate to lower the output pressure. Pressure limiting is installed, as a safety mechanism to make sure that the pump does not output a pressure higher than the maximum allowed pressure for the system.

(**2**) *Flow compensated pump control:* control the pump displacement such that the actual pump output flow rate (Q_p) is regulated about a desired pump output flow rate (Q_{cmd}),

$$\theta_{cmd} = \theta_{offset} + K \cdot (Q_{cmd} - Q_P) \tag{7.129}$$

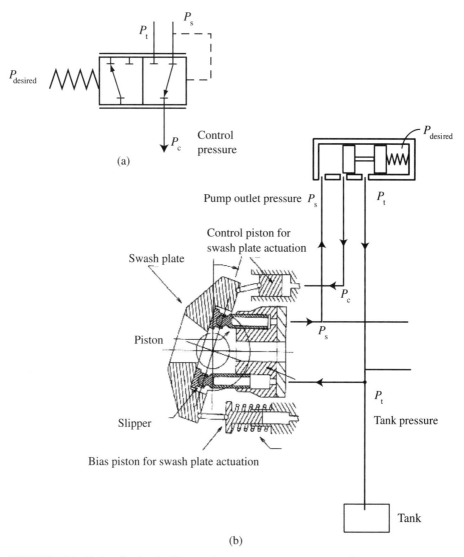

FIGURE 7.36: Hydraulic circuit diagram for a pressure compensated (pressure controlled) variable displacement piston type pump using a pressure regulating valve. (a) Symbolic representation of the pressure control valve. (b) Hydraulic circuit diagram of the pressure control valve and variable displacement piston pump. The desired pressure setting can either be adjusted by the screw-spring to a constant value, or through a remote pilot pressure control mechanism for variable pressure regulation applications.

The flow-pressure curves for a flow compensated pump control look as shown in (Figure 7.37b). In order to maintain a stable output flow, the mechanical input speed of the pump may also have to be above a minimum value.

(3) *Load compensated (load sensing) pump control:* control the pump displacement such that the difference between the actual pump output pressure and the load pressure is regulated about a desired differential pressure (Figure 7.38),

$$\theta_{cmd} = \theta_{offset} + K \cdot (\Delta P_{cmd} - (P_s - P_L)) \qquad (7.130)$$

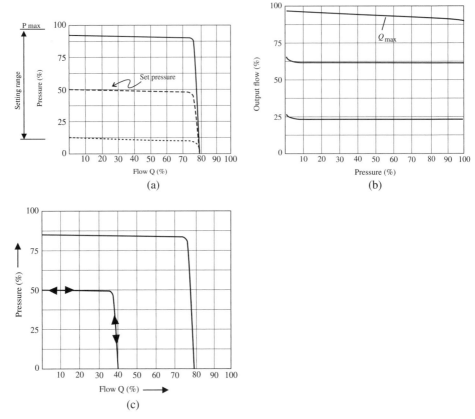

FIGURE 7.37: Performance characteristics of a pump under various control methods: (a) pressure regulated, (b) flow regulated, (c) flow regulated with pressure limit or pressure regulated with flow limit.

where ΔP_{cmd} is the desired pressure differential, P_s is the pump outlet pressure, P_L is the load pressure, which is the larger of the two pressures sensed from the two ports of the actuator (cylinder or hydraulic motor ports A and B). Typical values of ΔP_{cmd} are in the range of 10 bar to 30 bar. Notice that since this so-called load compensated (or load sensing) pump control method maintains a constant pressure differential across the valve, the speed of the cylinders is independent of the load (as long as maximum pump output pressure limit is not reached) and proportional to the valve spool displacement. This can be seen by examining the flow equations for a valve. The pump supplies the necessary flow (up to its saturation value at maximum displacement) in order to maintain the desired pressure differential between the pump outport pressure and the load pressure. Most load sensing pump control mechanisms also include a valve to limit the pump output pressure to a maximum value set by a spring loaded valve. This valve output flow is zero until the pump pressure exceeds the maximum pressure limit.

Load sensing control is used to reduce energy waste when the pump is not needed to operate at full system pressure and flow rate. It requires load pressure sensing and pump output pressure sensing. Based on these two signals, the swash plate angle is controlled by a valve and piston mechanism. Figure 7.38 shows a load sensing pump control system with completely hydro-mechanical components. This system uses a completely hydro-mechanical sensing, feedback, and actuation mechanism to perform a closed loop control function on the swash plate angle in order to regulate a constant pressure drop across the

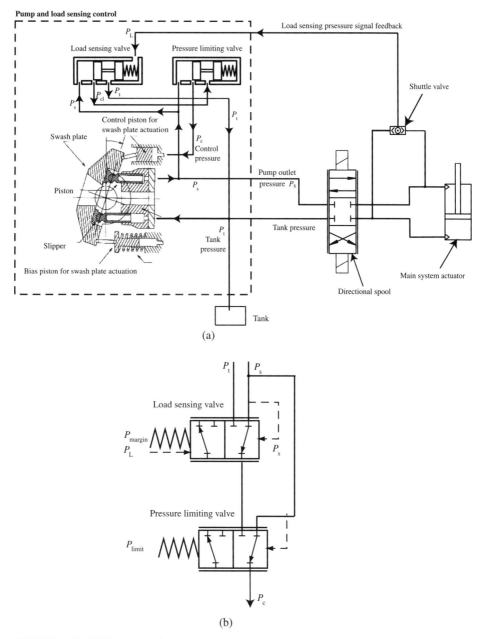

FIGURE 7.38: (a) Principle of pressure limiting and load sensing control using a hydro-mechanical feedback on an axial piston pump; control element is the swash plate angle, (b) symbolic representation and logical relationship between the load sensing valve and pressure limiting valve for the pump control.

valve. There are no electrical sensors nor digital controls involved. The load sensing valve controls the flow rate to the control piston in proportion to the difference between the pump outlet pressure (P_s) and the load pressure (P_L). As the pressure difference $P_s - P_L$ gets larger (i.e., P_L is lower), the flow from the load sensing valve to control piston increases. The movement of the control piston to the left reduces the swash plate angle, hence the pump output flow rate and the pump outlet pressure. This reduces the P_s. If the pressure difference $P_s - P_L$ gets smaller (i.e., P_L is larger), the flow rate from the load sensing valve gets smaller, hence the control piston moves to the right to increase the swash plate angle. This in turn increase the pump output (pressure and flow) to increase the P_s.

Maintaining a constant pressure drop across the valve for all load conditions up to the maximum pump outlet pressure limit makes the flow rate through the valve proportional to its spool displacement regardless of the load pressure as long as maximum pump outlet pressure is not reached. This gives good controllability since the flow rate through the valve, and hence the cylinder (actuator) speed, will be proportional to the displacement of the directional spool regardless of the load. Load sensing is usually mounted together with pressure limiting devices. The swash plate is controlled so that the pressure difference between the pump output pressure and load is maintained at a constant value. Hence, this type of control increases or decreases the pump output pressure depending on the load, up to a maximum limit set by the pressure limiting valve.

The combination of the load sensing valve and pressure limiting valve is referred as *the load sensing valve* (also called the load sensing compensator valve, Figures 7.38, 7.39). The load sensing compensator has four ports: pump output pressure input, load pressure input, tank port, and output port to swash plate control piston. The pressure limiting valve overrides the load sensing valve control only when the pump outlet pressure reaches a pre-set limit set by the spring in the pressure limiting valve. When $p_s > p_{limit}$ (i.e., the directional valve is in metering position and the load is very large), the flow from the pressure limiting valve to control piston will de-stroke the swash plate, and reduce the output pressure. Furthermore, if the cylinder has reached its travel limit and the directional valve is not in neutral position, pump output pressure will reach its limit but it will also de-stroke to almost zero flow. This is called the *high pressure cut-off condition*. Otherwise, the load sensing valve controls the swash plate angle and the pressure limit valve does not make a change in the swash plate angle control as long as the pump output pressure is less than the maximum limit. The pump output pressure is constantly fed to the bias piston and to the valves for the load sensing and pressure limiting. The bias piston puts the pump at maximum displacement until no flow is restricted by the actuator or the directional valve

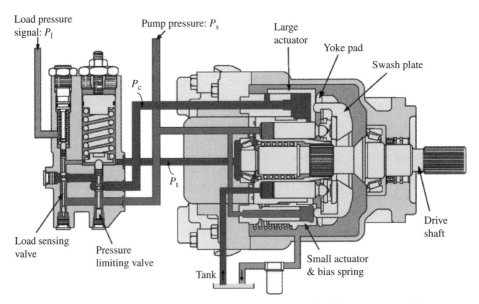

FIGURE 7.39: Cross-sectional component drawing of a variable displacement axial piston type pump and its load sensing control system components (load sensing valve and pressure limiting valve, Source: Sauer-Danfos.com).

of the system that will create a pressure at the discharge port of the pump. The pressure downstream the directional valve is fed back via a shuttle valve that always gives the highest pressure of the two sides of the actuator.

Although not shown in the figure, the load sensing shuttle valve is enabled by a check valve only when the directional control valve is not in the neutral closed center position. When in the neutral position, the cylinder is stationary and there is no connection between pump and cylinder ports. Therefore, the pump does not need to be concerned with supporting the load. As a result, the signal from the shuttle valve is blocked and the holding pressure of the load is not fed back to the pump control. In the neutral position of directional control valve, the pump will go to its *low pressure stand-by condition*. The load sensing signal line is connected to the tank line. As a result, pump output pressure is controlled to a value which is the tank pressure plus the pressure differential.

Figure 7.39 shows a cross-section drawing of a variable displacement axial piston type pump along with its two valves (load sensing valve and pressure limiting valve). The pump and load sensing control valve pair are generally packaged together. For the variable displacement pump to function, the load pressure feedback line must be connected to the proper location associated with the load.

Figure 7.40 shows a two-axis hydraulic circuit with a variable displacement axial piston pump which has load sensing hydro-mechanical controls. Notice that the load sensing

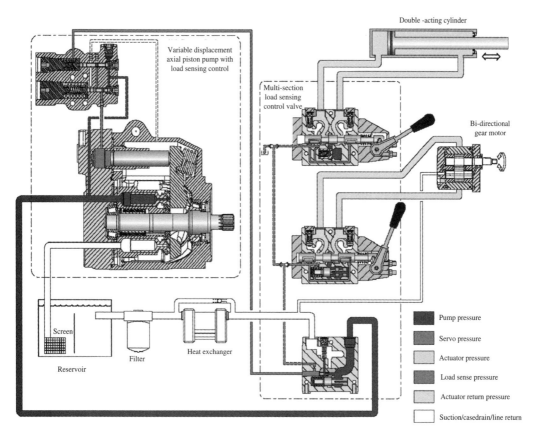

FIGURE 7.40: Two axis hydraulic motion control system: the pump is a variable displacement axis piston pump with a load sensing control (implemented hydro-mechanically), a linear cylinder, a bidirectional gear motor, and two manually controlled proportional valves. Pump and tank port connections to two proportional valves are not shown in the figure for simplicity. Reproduced with permission from Danfoss Power Solutions.

control valve on top of the pump has three ports: load pressure sensing port, pump output pressure sensing port, and output port to control the swash plate angle of the pump. The load pressure is the maximum of the two pressures sensed between the cylinder axis and rotary gear motor. Flow to each actuator is regulated by a manually operated proportional flow control valve. Two flow control valves are typically built on a single frame valve block with internal porting for P and T ports between them. Internal porting may connect P and T lines to each valve either in a parallel or series connection. P and T connections are now shown in the figure. Each valve has built-in ports to sense the maximum pressure at its output ports. Using two resolver valves, maximum pressure is fed back to the pump control valve. In this configuration, the smallest load pressure signal that can be fed back to the control valve is the tank pressure, which is shown as one of the inputs to the resolver valve next to the valve that controls the cylinder.

(4) *Positive flow control (PFC) of pump:* control the pump displacement such that it matches the flow demand of the line. Let Q_p be the flow demand by the line, and w_{eng} is the input shaft speed of the pump. The necessary pump displacement to provide the desired flow rate can be calculated from the pump performance characteristics,

$$Q_P = w_{eng} \cdot D_P(\theta) \tag{7.131}$$

$$\theta = D_P^{-1}(Q_P/w_{eng}) \tag{7.132}$$

It is desirable to implement the PFC algorithm without a flow rate sensor. Therefore, let us generate the desired pump displacement based on the predicted flow demand and using the pump map,

$$Q_{Pcmd} = Q_P(x_{s1}, x_{s2}, \ldots) \tag{7.133}$$

where $(x_{s1}, x_{s2}, \ldots)$ are the spool displacements of multiple valves supplied by the pump. The desired pump displacement is determined as (an offset value is added to account for initial nominal operating point),

$$\theta_{cmd} = \theta_{offset} + D_P^{-1}(Q_{Pcmd}/w_{eng}) \tag{7.134}$$

Notice that in order to implement the PFC method, we need the pump map function $D_P^{-1}(Q_P, w_{end})$ for the specific valve or valves supplied by the pump. The accuracy of this pump map is important since mis-match between the flow demand and flow supply can result in serious dynamic performance degradations. The price paid for the increased energy efficiency benefit of the PFC closed center EH systems is the increased complexity of the control algorithm.

Torque limiting and power limiting controls are implemented in pumps to prevent them from stalling when both high pressure and high flow rate occur in the system at the same time. In a mobile application the stalling will occur when the pump requires more power than the diesel engine can output and this will eventually bring the diesel engine to a dead stop. Torque limiting destrokes the pump to the point where the diesel engine is not stalled, but simply accepts the available power from the engine instead of overloading it.

Transient Response of the Pump If the delay associated with the dynamic response of the pump displacement controller is to be taken into account, instead of assuming that the pump displacement is equal to the commanded pump displacement, a first or second-order filter dynamics should be included in the pump model,

$$\theta = \frac{1}{(\tau_{p1}s + 1)(\tau_{p1}s + 1)} \cdot \theta_{cmd}(s) \tag{7.135}$$

where τ_{p1} and τ_{p2} are the time constants of the dynamic relationship between the pump displacement command and the actual pump displacement. This effective time constant of the pump dynamics is very critical in closed-center EH systems (closed center valve and variable displacement pump EH systems). The reason is the fact that the bandwidth of the main flow control valve is much faster than the bandwidth of the pump control. During any kind of valve closure, if the valve reaches the null position (hence almost zero flow demand) much faster than the pump can de-stroke, the pump flow will have no place to go and result in very large pressure spikes. This will most likely blow the pressure relief valves and result in low performance operation.

A mathematical model of a pump can be derived based on

1. the physical principles of fluid and inertial motion, or
2. based on the input–output (I/O) relationship using empirical data and modeled as a static gain (possibly nonlinear) plus dynamic filter effects.

Input variables of the pump are:

1. swash plate angle (or equivalent control element variable), and
2. input shaft speed.

And output variables of interest are:

1. outlet pressure,
2. outlet flow rate.

Some of the non-ideal characteristics of hydraulic pump (and motors) are:

1. Variation of displacement as a function of rotor position within one revolution. Since there is a finite number of fluid cavities (cylinder-piston pairs in a piston pump), the displacement has ripple as a function of the rotor position and number of piston–cylinder pairs (this is in principle the same as the commutation ripple in a brush-type DC motor).
2. Every hydraulic pump, valve, motor, and cylinder has leakage and it increases with the pressure.

7.4 HYDRAULIC ACTUATORS: HYDRAULIC CYLINDER AND ROTARY MOTOR

The translational cylinder and rotary hydraulic motor are the power delivering actuators in translational and rotary motion systems, respectively. The basic functionality of the actuator is to convert the hydraulic fluid power to mechanical power, which is the opposite of the pump function (Figure 7.41). Unidirectional pumps and motors are optimized to work in one direction in terms of reduced noise and increased efficiency. Bidirectional hydraulic pumps and motors have symmetric performance in either direction. In general, a pump can operate both in pumping or motoring mode. Similarly, a hydraulic motor can operate in motoring or pumping mode. However, there are exceptions. Some pump designs incorporate check

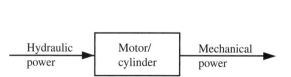

FIGURE 7.41: Hydraulic actuator functionality: convert hydraulic power to mechanical power. The hydraulic cylinder converts hydraulic power to translational motion power, the hydraulic motor converts it to rotational motion power.

valves in their design which makes it impossible for them to operate in motoring mode. Similarly, some motor designs are such that the hydraulic motor cannot operate in pumping mode even under *over-running (regenerative) load* conditions (Figure 7.32).

An *over-center* motor means that when the swash plate is moved in the opposite direction relative to its neutral position, the direction of the output shaft speed of the motor changes even though the input–output hydraulic ports stay the same. An *over-center* pump means that when the swash plate is moved in the opposite direction relative to its neutral position, the direction of the hydraulic fluid flow changes (input port becomes output port, and output port becomes input port), while the direction of the mechanical input shaft speed stays the same.

Given the displacement (D_m, fixed or variable) of a rotary hydraulic motor, the motor output speed (w) is determined by the flow rate (Q) input to it,

$$w = Q/D_m \tag{7.136}$$

Similarly, for linear cylinders, the same relationship holds by analogy,

$$V = Q/A_c \tag{7.137}$$

where w is the speed of motor, V is the speed of cylinder, D_m is the displacement of the motor (volume/rev), A_c is the cross-sectional area of the cylinder. If we neglect the power conversion inefficiencies of the actuator, the hydraulic power delivered must be equal to mechanical power at the output shaft, for the rotary motor

$$Q \cdot \Delta P_L = w \cdot T \tag{7.138}$$

and for the cylinder

$$Q \cdot \Delta P_L = V \cdot F \tag{7.139}$$

where ΔP_L is the load pressure differential acting on the actuator (cylinder or motor) between its two ports (A and B), T is the torque output, F is the force output. Hence, the developed torque/force, in order to support a load pressure ΔP_L, is

$$T = \Delta P_L \cdot D_m \tag{7.140}$$

$$F = \Delta P_L \cdot A_c \tag{7.141}$$

If the rotary pump and motor are the variable displacement type, and an input–output model is desired from the commanded displacement to the actual displacement, a first or second-order filter dynamics can be used between the D_m and D_{cmd},

$$D_m(s) = \frac{1}{(\tau_{m1} \cdot s + 1)(\tau_{m2} \cdot s + 1)} D_{cmd}(s) \tag{7.142}$$

There are applications which require extremely high pressures with a small flow rate which cannot be directly provided by a pump. In these cases, *pressure intensifiers* are used. The basic principle of the pressure intensifier is that it is a hydraulic power transmission unit, like a mechanical gear. Neglecting the friction and heat loss effects, the input power and output power are equal. The only function it performs is that it increases the pressure while reducing the flow rate. It is the analog of a mechanical gear reducer (increases the output torque, reduces the output speed). Figure 7.42 shows a pressure intensifier in a hydraulic circuit. The ideal power transmission between B and A pressure chambers means,

$$Power_B = Power_A \tag{7.143}$$

$$F_B \cdot V_{cyl} = F_A \cdot V_{cyl} \tag{7.144}$$

$$p_B \cdot A_B \cdot V_{cyl} = p_A \cdot A_A \cdot V_{cyl} \tag{7.145}$$

$$p_B \cdot A_B = p_A \cdot A_A \tag{7.146}$$

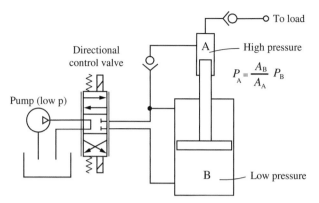

FIGURE 7.42: Pressure intensifier circuit.

Notice that the intensified pressure in the forward cycle of the cylinder is equal to the area ratios between the cylinder head-end and the intensifier ram (i.e., 10),

$$\frac{p_A}{p_B} = \frac{A_B}{A_A} \tag{7.147}$$

During the forward stroke, the pressure in chamber A is amplified by the area ratios defined by the above equation. During the retraction stroke, the intensifier and rod-end of the cylinder are filled with hydraulic fluid and are not considered a work cycle.

Example Consider a pump that supplies $Q_p = 60\,1/\mathrm{min}$ constant flow to a double acting cylinder. The cylinder bore diameter is $d_1 = 6\,\mathrm{cm}$ and rod diameter $d_2 = 3\,\mathrm{cm}$. Assume the rod is extended through both sides of the cylinder. The load connected to the cylinder rod is $F = 10\,000\,\mathrm{Nt}$. The flow is directed between the pump and the cylinder by a four way proportional control valve. Neglect the pressure drop and losses at the valve. Deterimine the pressure in the cylinder, velocity, and power delivered by the cylinder during the extension cycle. Assuming 80% overall pump efficiency, determine the power necessary to drive the pump.

We need to determine Δp_L, V, and P for both the extension and retraction cycle. Let us determine the areas of the cylinder.

$$A_c = \frac{\pi\left(d_1^2 - d_2^2\right)}{4} = \frac{\pi(6^2 - 3^2)}{4}\,\mathrm{cm}^2 = 19.63\,\mathrm{cm}^2 \tag{7.148}$$

$$= 19.63 \times 10^{-4}\,\mathrm{m}^2 \tag{7.149}$$

The pressure during the extension stroke that must be present as differential pressure between the two sides of the cylinder in order to support the load force,

$$F = \Delta p_L \cdot A_c \tag{7.150}$$

$$\Delta p_L = \frac{F}{A_c} = \frac{10\,000}{19.63 \times 10^{-4}} \tag{7.151}$$

$$= 5.09 \times 10^6\,\mathrm{N/m}^2 = 5.09\,\mathrm{MPa} \tag{7.152}$$

The linear velocity is determined by the conservation of flow,

$$Q = A_c \cdot V \tag{7.153}$$

$$V = \frac{Q}{A_c} = \frac{60\,1/\mathrm{min}\ 10^{-3}\,\mathrm{m}^3/1 \cdot 1\,\mathrm{min}/60\,\mathrm{s}}{19.63 \times 10^{-4}\,\mathrm{m}^2} \tag{7.154}$$

$$= 0.509\,\mathrm{m/s} \tag{7.155}$$

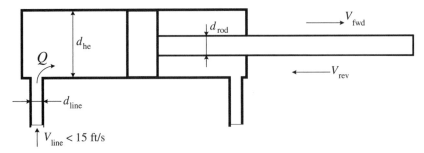

FIGURE 7.43: Maximum cylinder speed in forward and reverse direction, given the maximum linear speed of fluid flow (15 ft/s) and dimensions of the cylinder and connecting lines.

The power delivered to the load by the cylinder, is

$$Power_m = F \cdot V = \Delta p_L \cdot Q \tag{7.156}$$
$$= 10\,000\,\text{Nt} \cdot 0.509\,\text{m/s} = 5090\,\text{W} = 5.09\,\text{kW} \tag{7.157}$$

and the necessary pump power rating is

$$Power_p = \frac{1}{\eta_o} \cdot Power_m \tag{7.158}$$

$$= \frac{1}{0.8} \cdot 5.09\,\text{kW} = 6.36\,\text{kW} \tag{7.159}$$

Example Consider a double acting cylinder and its hydraulic lines connecting to its ports (Figure 7.43). Let us assume that the maximum linear speed of fluid in the transmission lines is to be limited to 15 ft/s in order to reduce flow turbulance and the resulting flow resistance. Calculate the forward and reverse speed of the cylinder that can be achieved for different values of the line diameter, rod diameter, and cylinder head-end diameter. Assume the rod diameter is half of the cylinder inner diameter. Consider the line diameters for $d_{line} = 0.5$ in, 0.75 in, 1.0 in, 1.5 in, and $d_{he} = 4.0$ in, 6.0 in, 8.0 in (see Table 7.1)
From flow continuity,

$$Q = A_{line} \cdot V_{line} = A_{he} \cdot V_{fwd} \quad \text{in forward motion} \tag{7.160}$$
$$= A_{re} \cdot V_{rev} \quad \text{in reverse motion} \tag{7.161}$$

where, we limit $V_{line} = 15$ ft/s. Then,

$$V_{fwd} = \frac{A_{line}}{A_{he}} \cdot V_{line} \tag{7.162}$$

$$V_{rev} = \frac{A_{line}}{A_{re}} \cdot V_{line} = \frac{A_{he}}{A_{re}} \cdot V_{fwd} \tag{7.163}$$

TABLE 7.1: Cylinder speed (forward and reverse direction in units of in/s) for different values of the line pipe diameter and cylinder diameter. It is assumed that the rod diameter is half of the cylinder inner diameter. The linear line speed of fluid is limited to 15 ft/s.

	d_{line}			
d_{he}	0.5	0.75	1.0	1.5
4.0	(3.75, 5.0)	(8.44, 11.25)	(15.0, 20.0)	(33.75, 45.0)
6.0	(1.67, 2.23)	(3.75, 5.0)	(6.67, 8.89)	(15.0, 20.0)
8.0	(0.94, 1.25)	(2.11, 2.81)	(3.75, 5.0)	(8.44, 11.25)

If we focus on the case where the cylinder inner diameter is twice the rod diameter,

$$V_{\text{fwd}} = \frac{\pi d_{\text{line}}{}^2/4}{\pi d_{\text{he}}{}^2/4} \cdot V_{\text{line}} \tag{7.164}$$

$$V_{\text{rev}} = \frac{d_{\text{he}}{}^2}{d_{\text{he}}{}^2 - d_{\text{re}}{}^2} \cdot V_{\text{fwd}} \tag{7.165}$$

$$= \frac{4}{3} \cdot V_{\text{fwd}} \tag{7.166}$$

7.5 HYDRAULIC VALVES

Valves are the main control components in hydraulic circuits. They are the *metering* component for the fluid flow. The metering of the fluid is done by moving a spool to adjust the orifice area. There are two main valve output variables of interest which can be controlled by the spool movement:

1. flow (rate and direction) and
2. pressure.

If the movement of the spool is determined in order to maintain a desired pressure, the valve is called a *pressure control valve*. If the movement of the spool is determined in order to maintain a desired flow rate, it is called a *flow control valve*. If the flow direction is changed between three or more ports, they are called *directional flow control valves*.

If the spool position of the valve is controlled only to two discrete positions, they are called *ON/OFF valves* (Figure 7.44). If the spool position of the valve can be controlled to be anywhere between fully open and fully closed position (i.e., in proportion to a signal), then they are called *proportional valves*. The ON/OFF type valves are used in applications which require discrete positions, that is open the door and close the door. Here, we are interested primarily in proportional and servo control valves.

A valve is sized based on three major considerations:

1. flow rating: the flow rate it can support at a certain pressure drop across the valve port when it is fully open (i.e., maximum flow at 1000 psi pressure drop for servo valves, at 150 psi for proportional valves),
2. pressure rating: the rated pressure drop across the valve and the maximum supply port pressure,
3. speed of response: the bandwidth of the valve from current signal to spool displacement.

The main differences between ON/OFF valves and proportional valves are as follows,

1. Solenoid design: in proportional valves, the gain of the current–force relationship is fairly constant in the travel range of the solenoid, whereas in ON/OFF valves, the main thing is to generate the maximum force and the linearity current–force gain relationship is not important.
2. Centering spring constant in proportional valves tends to be larger than than the comparable size ON/OFF valves.
3. While the valve body of a proportional and ON/OFF valve may be almost identical, the spool designs are different. The proportional valve spools are carefully designed with desired orifice profiles to proportionally meter the flow.

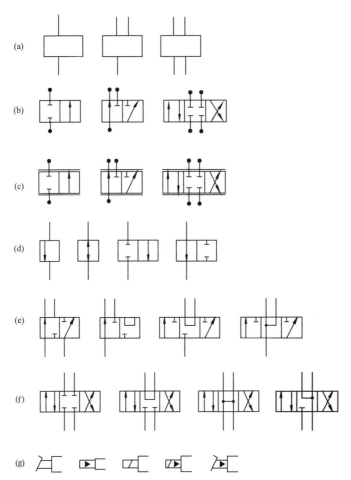

FIGURE 7.44: Valve categories: (a) number of ports (also called *ways*): two port (or two-way), three port, four port valves, (b) two, three, and four port valves with two, three, and four position configurations, (c) same valves with proportional flow metering capability (proportional valves), (d) normally closed, normally open conditions, that is the flow condition of the valve, when it is in default condition, (e) three port valve with various discrete positions, (f) various flow conditions when a four port valve is in the center position (closed center, tandem center, open center, float center), (g) actuation method of the valve: manual, pilot, solenoid or the combinations of them.

Most proportional valve geometries are designed so that the flow is approximately proportional to the spool displacement under a constant pressure drop across the valve.

Valves can be categorized according to the following:

1. number of ports, that is two, three, four ports,
2. number of discrete positions of the spool, that is one, two, or three discrete positions or ability to position the spool proportionally within its travel range,
3. flow condition of the valve at the center position, that is open center, closed center,
4. actuation method of valve, that is manual, electric actuator, pilot actuated, or a combination of them.
5. metering element type: sliding spool, rotary spool, poppet (or ball),

6. main control purpose of valve: pressure or flow rate,

7. mounting method of the valve into the circuit: stand alone body, multi function integrated valve block, subplate mounting, manifold block, stacked (sandwich) block. The mounting standards specify the mechanical dimensions of the ports, their locations as well as screw holes on the mounting plate. Mounting standards are specified by NFPA, ISO, and DIN. NFPA standards are referred to with letter D, and DIN standards are referred to by NG, ISO standards are referred to by CETOP. Equivalent standard specifications are DO3/NG6/CETOP 03, D05/NG10/CETOP 05, D07/NG16/CETOP 07, D08/NG25/CETOP 08, D010/NG32/CETOP 10.

7.5.1 Pressure Control Valves

Pressure control valves involve two or three port connections. The spool which controls the orifice area is actuated based on a sensed pressure. Relief valves, pressure reducing valves, counterbalance valves, sequence valves, brake valves are examples of pressure control valves (Figures 7.45–7.50). *Relief valves* limit the maximum pressure output of the valve by venting the flow to tank. In a mechanical relief valve, a spring sets the maximum pressure. In an EH controlled relief valve, the current level of the solenoid sets the maximum pressure, hence an EH controlled relief valve can be used to control a variable line pressure. There are two major types of pressure relief valves: direct and indirect acting (Figure 7.46). In a direct acting relief valve, the relief pressure is set by the spring. The line pressure directly acts on the area of the poppet spool. When the line pressure exceeds the set pressure, the poppet moves up and opens the line to tank.

$$p_{\text{line}} \cdot A_{\text{poppet}} \leq k_{\text{spring}} \cdot x_{\text{spring}}; \quad \textit{valve closed} \tag{7.167}$$

$$p_{\text{line}} \cdot A_{\text{poppet}} \approx k_{\text{spring}} \cdot x_{\text{spring}}; \quad \textit{valve open} \tag{7.168}$$

$$p_{\text{line}} \approx p_{\text{spring}} = (k_{\text{spring}} \cdot x_{\text{spring}})/A_{\text{poppet}}; \tag{7.169}$$

where $x_{\text{spring}} = x_{\text{preload}} + \Delta x_{\text{spring}}$

The indirect acting relief valve uses an intermediate orifice and spring between the poppet valve section and the line pressure. The function of the orifice in the indirect pressure relief valve is as follows: when the set pressure of the pilot stage is exceeded (spring 2 in Figure 7.46), the poppet spool opens and flow crosses the orifice. The role of the orifice is effectively to create a pressure drop between A and B locations, and hence allow the flow to

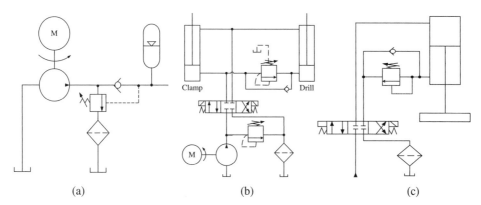

| (a) | (b) | (c) |

FIGURE 7.45: Pressure relief valve applications in a hydraulic circuit: (a) unloading valve, (b) sequence valve, (c) counter balance valve.

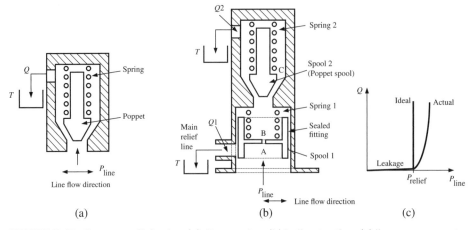

FIGURE 7.46: Pressure relief valve: (a) direct acting, (b) indirect acting, (c) line pressure, set relief pressure, and flow rate to tank relationship in steady state (ideal case and actual case shown).

overcome the preload of spring 1 to open the main relief line. When the poppet valve opens, the line pressure is relieved to the tank not through the poppet valve orifice, but through a separate orifice on the side. The poppet section has a small (or almost zero) leakage flow to tank. Compared to the direct acting relief valve, the indirect acting valve operation is less sensitive to flow rate, is more stable, but has a slightly slower response due to the orifice. It is not unusual that a direct acting relief valve can operate in a high frequency open and close motion due to pressure spikes in the line, hence resulting in valve failures in a matter of minutes in large power applications.

Poppet valves are commonly used as pressure regulators or pressure relief valves (Figure 7.47). The main components of a poppet valve are: poppet (spherical ball or conical shaped), seat, valve body, and poppet actuator which may include a spring. Poppet valves are easy to manufacture, have lower leakage and are insensitive to clogging by dirt particles compared to other types of spool valves. As a result, poppet valves are often used in load-holding applications to minimize leakage and hence movement of the load. Depending on the flow rate supported by the valve, it may be actuated directly by a solenoid for small flow rate valves or via a secondary pilot stage amplification of actuation force for large flow rate valves. Poppet valves are used as screw-in cartridge valves with manifold blocks. A poppet valve performance is rated in terms of its maximum operating pressure (p_{max}), rated flow through the valve at a given pressure drop across it (Q_r at Δp_v). In recent years, poppet valves have been developed as proportional flow control valves. However, since the valve actuation force directly acts against the dynamic flow forces, the dynamic stable control of the valve is a difficult task. The cross-sectional shape of the poppet and its seat make a significant difference in flow forces and proportional flow metering ability of a poppet valve.

Pressure reducing valves maintain an output pressure which is lower than the input pressure by venting the excess flow to tank or limiting the orifice area based on desired and actual output pressure sensing. Figure 7.48 shows the basic design, symbol, and steady-state input–output relation of the valve. When the output pressure is less than the output pressure set by the spring, the spool does not move. Output pressure is very close to the input pressure.

$$p_{out} \cdot A_{out} < F_{spring,0} \qquad (7.170)$$

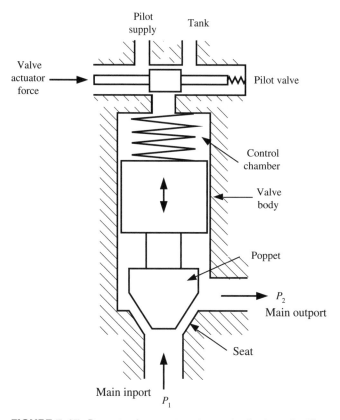

FIGURE 7.47: Poppet valve construction: valve body and orifice seat, poppet, control chamber with a mechanism to control the pressure or direct actuation of the poppet.

As the output pressure increases due to the increase in the input pressure or load, the spool force due to the pressure feedback will increase and start to move the spool against the spring until the spring force balances the output pressure feedback force,

$$p_{out} \cdot A_{out} \approx k_{spring} \cdot x_{spring} = F_{spring} \tag{7.171}$$

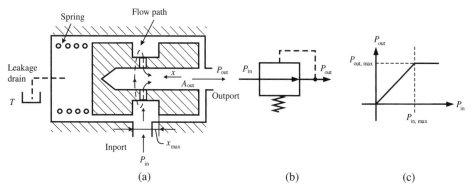

FIGURE 7.48: Pressure reducing valve: (a) valve design, (b) valve symbol, (c) steady-state input and output pressure relationship. Notice that only the output pressure is fed back to affect the movement of the spool and hence the metering orifice. Input pressure ideally does not affect the force on the spool as a result of the way the flow is routed around the spool.

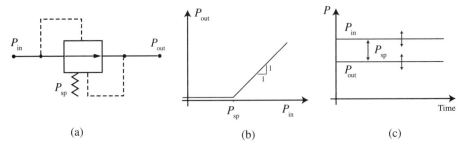

FIGURE 7.49: Pressure compensator valve: (a) symbol with pressure feedback and spring pressure, (b) input and output pressure relationship in steady state, (c) input, output, and spring pressure relationship over time.

As the spool moves, it restricts the flow orifice between the input and output port. Eventually, when the output pressure is high enough to further move the spool to close off the orifice between the input and output ports, the output pressure is limited to the maximum value set by the spring and orifice design of the valve. The valve then regulates within the vicinity of this point and maintains a constant output pressure which is set by the spring as long as the input pressure is larger than the set (desired output) pressure.

A modified version of the same valve concept (pressure reducing valve) is used as an adjustable pressure reducing valve in lever or solenoid operated *pilot valves*. Figure 7.50 shows a modification of the pressure reducing valve. The base of the spring is moved by a lever (or an electric actuator in the case of an EH type pressure reducing valve), hence changing the set spring pressure. Therefore, the output pressure will be proportional to the lever displacement as long as the input pressure is larger than the set pressure. A pair of such pressure reducing valves is used as the pilot stage in two-stage proportional valves. In mechanically actuated versions, a lever moves a pair of pressure reducing valves. In forward motion of the lever, one of the pilot valves is actuated, in reverse motion the other is actuated. Similarly, in an EH conrolled version, each pressure reducing valve is actuated by a solenoid based on lever command. The net result of such a valve is that the output

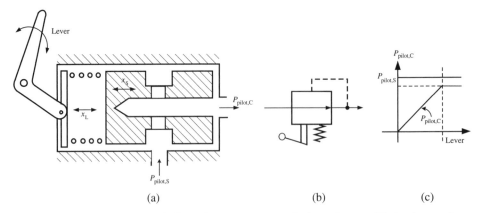

FIGURE 7.50: Lever controlled pilot valve: a pressure reducing valve modified to be used as a pilot valve. The output pressure is always smaller or equal to the input pressure. The output pressure is proportional to the lever displacement, $P_{pilot,c} \propto x_L$, while $P_{pilot,c} \leq P_{pilot,s}$. (a) Valve design concept, (b) symbol, (c) inport and outport pressures and control lever relationship. The lever movement sets the desired output pressure. The feedback from the output pressure to the spool moves to balance that and regulate to the set pressure.

pressure is proportional to the lever displacement with its maximum value being the input pilot pressure. This output pressure is then used to shift the main spool against a centering spring in order to generate a main spool displacement that is proportional to the lever displacement.

In pilot valve applications, the input pressure is the pilot supply pressure ($p_{pilot,s}$), and output pressure is the pilot control pressure ($p_{pilot,c}$) which is used to shift the spool of a larger flow control valve,

$$p_{pilot,c} = (k_{spring} x_{spring})/A_{out} \leq p_{pilot,s} \tag{7.172}$$
$$\approx K \cdot x_{spring} \leq p_{pilot,s} \tag{7.173}$$

A different version of the pilot valve has two inports: pilot pressure supply and tank ports. The output pilot control pressure is regulated by the lever motion to be between the tank pressure and pilot supply pressure.

Spring force is due to preloaded displacement ($x_{preload}$) plus the displacement during pressure regulation, Δx_v,

$$F_{spring} = k_{spring} \cdot (x_{preload} + \Delta x_v) \tag{7.174}$$

While

$$A_{out} \cdot p_{out} \leq k_{spring} \cdot x_{preload} \tag{7.175}$$

the valve orifice is fully open with negligible restriction and

$$p_{out} \approx p_{in} \tag{7.176}$$

When,

$$A_{out} \cdot p_{out} > k_{spring} \cdot x_{preload} \tag{7.177}$$

The force balance to define the position of the spool

$$A_{out} \cdot p_{out} = k_{spring} \cdot (x_{preload} + \Delta x_v) \tag{7.178}$$
$$p_{out} = p_{in} - K_{vp} \cdot \Delta x_v \tag{7.179}$$

where the last equation states that as the valve spool moves, Δx_v, due to increased output pressure, the orifice restriction increases (orifice starts to close, gets smaller), and the output pressure becomes less than the input pressure due to a pressure drop across the smaller orifice. The gain between the spool displacement and pressure drop, K_{vp}, is large and typically much larger than the spring constant, $K_{vp} \gg k_{spring}$. This means that the movement of a spool makes a large difference in the output pressure but not much change in the spring force. If we eliminate the Δx_v in the above equations, we can obtain a steady-state relationship between input and output pressures, which can be shown to be

$$A_{out} \cdot p_{out} = k_{spring} \cdot \left(x_{preload} + \frac{1}{K_{vp}} \cdot (p_{in} - p_{out}) \right) \tag{7.180}$$

$$\left(A_{out} + \frac{k_{spring}}{K_{vp}} \right) \cdot p_{out} = k_{spring} \cdot x_{preload} + \frac{k_{spring}}{K_{vp}} \cdot p_{in} \tag{7.181}$$

$$p_{out} \approx \frac{k_{spring}}{A_{out}} \cdot x_{preload} \tag{7.182}$$

$$p_{out} \approx p_{preload} \tag{7.183}$$

by approximating $k_{spring}/K_{vp} \approx 0.0$.

Note that $x_{spring} = x_{preload} + \Delta x_v$ and $x_{preload}$ is adjusted by the lever motion. As the operator moves the lever, $x_{preload}$ changes, hence $p_{preload}$ changes. This means that the output pressure,

$$p_{pilot,c} = p_{out} = p_{preload}(x_l) \tag{7.184}$$
$$\approx K \cdot x_l \tag{7.185}$$

changes as a function of lever displacement, where x_l is the lever displacement. This valve functions as a proportional pressure control valve using a pressure reducing valve. The output pressure limits are defined by the pilot supply pressure and tank pressure, and the actual value of the output pressure is proportional to the lever displacement,

$$p_{tank} \leq [p_{pilot,c} = p_{out} = p_{preload}(x_l)] \leq p_{pilot,s} \tag{7.186}$$

taking on a value between pilot supply pressure ($p_{pilot,s}$) and tank pressure (p_{tank}).

A pressure compensator valve that maintains a constant pressure drop between its input and output ports is shown in Figure 7.49. Force balance between feedback pressures and spring determines the steady-state position of the spool

$$p_{out} = p_{in} - p_{spring} = p_{in} - k_{spring} \cdot (x_{preload} + \Delta x_v) \tag{7.187}$$

and the spool position effect on the output pressure due to the change in orifice opening (notice the difference between this equation and the similar equation for pressure reducing valve),

$$p_{out} = p_{in} - K_{vp} \cdot \Delta x_v \tag{7.188}$$

If we eliminate the Δx_v from the above two equations to find a steady-state relationship between input and output pressure, it can be shown that

$$\left(1 - \frac{k_{spring}}{K_{vp}}\right) p_{out} = \left(1 - \frac{k_{spring}}{K_{vp}}\right) p_{in} - k_{spring} \cdot x_{preload} \tag{7.189}$$

$$p_{out} \approx p_{in} - k_{spring} \cdot x_{preload} \; ; \; K_{vp} \gg k_{spring} \tag{7.190}$$

Hence, the valve spool will move to regulate the output pressure to maintain the following relationship,

$$p_{in} - p_{out} = k_{spring} \cdot x_{preload} \tag{7.191}$$

If p_{in} increases, p_{out} has to increase in order to maintain the force balance. This is accomplished by the small movements of the valve spool.

An *unloading valve* is a pressure relief valve where the pilot pressure which activates the valve to open comes from a line past the valve. For instance, a fixed displacement pump may need to charge an accumulator which maintains a line pressure. When the accumulator pressure exceeds the pump output pressure, the pilot pressure opens the relief valve, the check valve closes, and the pump flow is sent to the tank without having to work against the load, hence the name *unloading* valve (Figure 7.45a).

A *sequence valve* is a pressure relief valve used in a circuit to make sure that a hydraulic line does not open until a certain pressure requirement at another location is met (Figure 7.45b). For instance, a pressure relief valve can be used to open or close a hydraulic line between two points based on the pressure feedback from another third location in the circuit. Until the pressure in the third location reaches a preset value, the line between location one and two is blocked. Such a valve may be used between a single directional control valve and two cylinders: clamp and drill cylinders. The objective is that the drill cylinder should not move until the clamp cylinder pressure reaches a certain value. This is accomplished by a pressure relief type sequence valve between the drill cylinder and

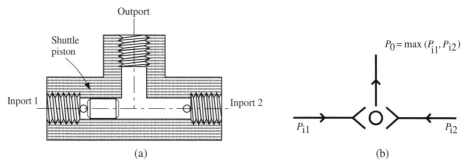

FIGURE 7.51: A shuttle valve is used as a maximum function. It selects the larger of the two pressures at its input ports and feeds it to its output port.

directional control valve output line. The control signal for the sequence valve comes from the pressure in the clamp cylinder. When the directional valve opens, the flow first goes to the clamp cylinder since the sequence valve makes the flow path to the drill cylinder have more resistance. When the clamp cylinder extends and pressure in the line increases, then the sequence valve opens and guides the flow to the drill cylinder. It should be noted that the sequence of motion between the two cylinders is based on pressure sensing, not based on position. Therefore, if the pressure increases in the clamp cylinder for some reason before the actual clamping operation is done, the drill cylinder will still actuate.

The *counterbalance valve* (also called load holding valve) is a pressure relief valve. It does not open until a desired pressure differential is met in order to avoid sudden movements (Figure 7.45c). They are used in load lowering applications such as lift trucks and hydraulic presses. The valve can be used to control the back pressure between the cylinder and tank port so that the speed of the cylinder does not move uncontrollably due to large load, but is controlled by the valve on the cylinder to tank line.

The so-called *shuttle* or *resolver* valves are commonly used as selector valves in hydraulic circuits (Figure 7.51). The valve has two input ports and one output port. The output port has the larger of the two input port pressures,

$$p_o = \max(p_{i1}, p_{i2}) \tag{7.192}$$

7.5.2 Example: Multi Function Hydraulic Circuit with Poppet Valves

Figure 7.52 (also see Figure 7.13) shows the hydraulic circuit schematics of the bucket control for the wheel loader model L120 by Volvo. The load sensing signal is used in the same way to control the pump displacement. That is the maximum load pressure signal is resolved by use of a series of shuttle valves, and the resulting pressure is used to control the displacement of the pump. The maximum load signal is determined by comparing the load signals from the bucket hydraulics and steering hydraulic system (the steering hydraulic system is not shown in the figure). That way, the pump is able to support the circuit with the most demanding load.

The tilt and lift valves are closed-center, proportional flow control valves (MV1, MV2). The lift valve also has a so-called *float position* which is used when the bucket is riding on the ground surface and desired to track the natural contour of the ground. At the float position, both sides of the cylinder are connected to the tank. This way, the cylinder follows the natural contour of the ground surface. The shuttle valve (SV1) is used to actuate the main valve spool to the float position when the command signal is larger than a certain

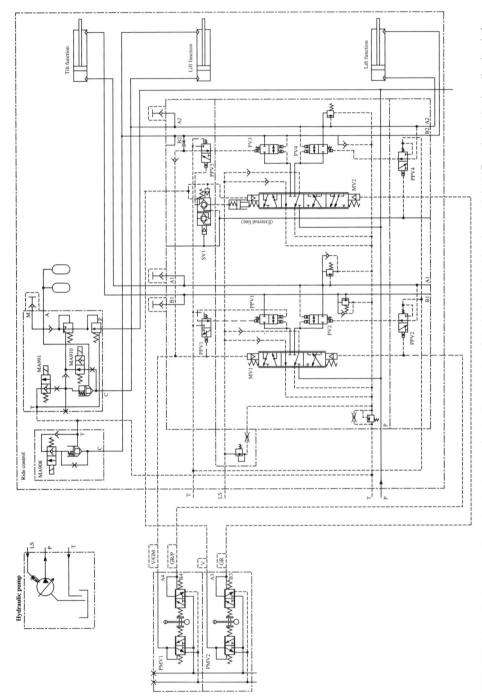

FIGURE 7.52: Hydraulic system circuit diagram for bucket motion of the Volvo wheel loader model 120E: only the hydraulic circuit for bucket motion is shown. The pump is a variable displacement axial piston type. The main flow control valves are connected to the hydraulic supply in parallel configuration and are of closed-center, non-pressure compensated type. Poppet valves are used to seal the load at each end of the cylinders.

set value. Also shown in the figure is the so called *ride control* or *boom suspension* circuit which reduces the oscillations transmitted to the machine frame between road surface and lift cylinder.

The flow between each service port (A and B) of the main flow control valve and the cylinder ports (head-end and rod-end) passes through a poppet valve. There are two poppet valves per valve–cylinder pair, one for each of the valve–cylinder port connection. Poppet valves provide a leakage free sealing on a line. Unlike a spool type valve, where some leakage is inevitable, poppet valves are excellent for leakage free sealing. In this configuration, the poppet valves are used as load holding valves, that is when the main flow valve is in neutral position, the poppet valve blocks the flow and seals the line. Thus, the load does not drift in position due to leakage problems. Hence, for lift and tilt circuits combined, there are four poppet valves (PV1, PV2, PV3, PV4). The spool displacement of the poppet valve is controlled by a pilot valve (PPV1, PPV2, PPV3, PPV4). When a pilot control pressure from the lever operated pilot (PMV1) valve acts on the main spool (MV1), it also acts on the pilot valve (PPV1) which then actuates the poppet valve.

7.5.3 Flow Control Valves

Flow rate through an orifice or restriction is a function of both the area of opening and the pressure differential across the orifice,

$$Q = K \cdot A(x_s) \cdot \sqrt{\Delta p} \tag{7.193}$$

where Q is the flow rate, Δp is the pressure drop across the valve, $A(x_s)$ is the valve orifice opening area as function of spool displacement x_s, and K is a proportionality constant (discharge coefficient). Non-compensated flow control valves set the orifice area only by moving a spool based on a command signal. Figure 7.53a shows a needle valve used as a flow control valve where the needle position is manually adjusted. The orifice area is approximately proportional to the needle position. If the input or output pressure change, the flow rate changes for a set needle position in accordance with the above orifice equation.

If it is desired that the flow rate should not change with pressure variations, the standard flow control valve can be modified with a pressure compensator spool and orifice. Such a valve is called *pressure compensated flow control valve*. There are two types of pressure compensated flow control valves: the restrictor type and by-pass type (Figure 7.53b,c).

There are two spool and two orifice areas in a pressure compensated flow control valve: one pair is the needle–orifice pair which sets the nominal orifice opening. Another pair modulates the second orifice opening based on input–output pressure feedback signals in order to maintain a constant pressure drop across the needle–orifice area. As a result, a constant flow rate is maintained at a constant setting of the needle even though input and output pressure may vary (since the second spool would compensate for it) as long as valve operating conditions do not reach saturation. This type is called a *restrictor type* pressure compensated flow control valve since flow is regulated against pressure variations by adding restriction in the flow line (Figure 7.53b). The valve regulates the pressure drop (tries to maintain it at a constant value) across the needle orifice.

Another type is the *by-pass* type where an orifice opening by-passes excess flow to the tank port as a function of pressure feedback signal. The output pressure is maintained at the load pressure plus the spring due to pressure, that is $p_{out} = p_1 + p_{spring}$ (Figure 7.53c).

Notice that the desired flow rate is set by the main orifice opening which is shown in the figures as being controlled by a manually moved needle-screw. This mechanism can also

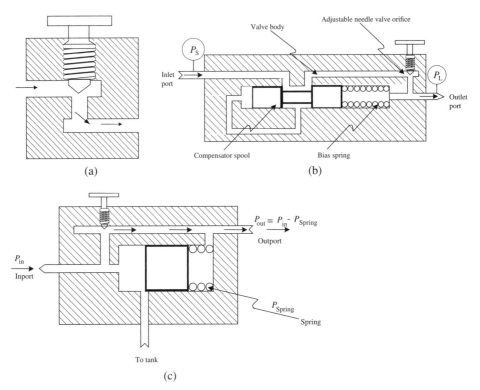

FIGURE 7.53: Flow control valves: (a) needle type flow control valve, (b) pressure compensated needle type flow control valve, (c) by-pass type pressure compensated flow control valve. The needle position may be manually or remotely controlled.

be proportionally controlled by an electric actuator, such as a proportional solenoid, which is the case in many pressure compensated EH flow control valves (see www.HydraForce.com).

In a *pressure compensated valve*, a compensator spool moves in proportion to the difference between two pressure feedback signals. It is also called *adjustable pressure compensated flow control valve*. In the so-called *pre-compensator valve configuration*, one pressure is fed back from the output of the compensator valve and the other pressure is fed back from the output of the flow control valve (Figure 7.54). In the so-called *post-compensator valve configuration*, one pressure feedback is from the input pressure to the compensator valve, and the other pressure feedback is from the load pressure (Figure 7.55).

The objective of the compensator valve is to maintain a constant pressure drop across the main flow control valve. Hence, when the main flow control valve spool is at a certain position, the flow rate across it would not be affected by the pressure variations at the pump supply or load side as long as the pressure variations do not reach the saturation levels. The function of compensator valve is to add restriction (pressure drop) in a hydraulic circuit. It is the analogy for variable resistance in series with another resistance in electrical circuits. A directional flow control valve in series with a pressure compensator valve is called a *pressure compensated directional flow control valve* or just a *pressure compensated directional valve*. Compensator valves are used in

1. circuits where load is variable and that the flow rate should not vary as a function of the load as long as the valve is not saturated (fully open or fully closed), and

2. multiple parallel hydraulic circuits where all circuits share the same pump supply pressure, but the load pressure in each cylinder may be different. If a compensator

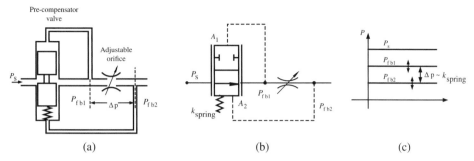

FIGURE 7.54: Pre-compensator valve configuration. A compensator valve is used to maintain a constant pressure drop across an orifice. This is accomplished by adding hydraulic resistance (pressure drop) in a circuit. A pre-compensator valve uses two pressure feedback signals for its control: the compensator output port pressure signal and the output of the flow control valve following the compensator valve (i.e., maximum of two pressures of the cylinder if the flow control valve is connected to a cylinder). The pressure drop regulated across the main flow control valve is proportional to the spring constant of the pre-compensator valve. (a) Component cross-section circuit diagram, (b) symbolic circuit diagram, (c) pressure relationships.

valve is not used, most of the flow will go the valve which has the lowest load, and other circuits will get very little flow. In order to provide equal flow on demand to each circuit regardless of the load pressure in other circuits, a compensator valve is used in each circuit to add additional restriction (hence pressure drop, just as a larger load would do) in the circuit so that all circuits effectively see the same load pressure.

The pressure relationship in Figure 7.54 is as follows,

$$p_s \geq p_{fb1} \geq p_{fb2} \tag{7.194}$$

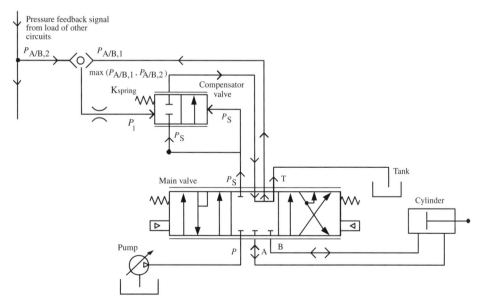

FIGURE 7.55: Post-compensator valve configuration. The two pressure feedback signal sources are the input pressure signal to the compensator valve and the load pressure. In multi circuit configurations, the load signal is the maximum of all load signals among circuits.

that is the outlet pressure of compensator valve is smaller or equal to the input pressure, and similarly, the output pressure of the flow control valve is smaller or equal to the pressure at its input. The p_{fb2} pressure is determined by the load conditions downstream. The task of the compensator pressure is to maintain a constant pressure differential between the input–output ports of the main flow control valve, labeled as the adjustable orifice in Figure 7.54,

$$\Delta p_{set} = p_{fb1} - p_{fp2} \tag{7.195}$$

by controlling the intermediate pressure p_{fb1} through the movement of the compensator valve spool,

$$k_{spring} \cdot (x_{preload} + \Delta x_{cs}) = p_{fb1} \cdot A_1 - p_{fb2} \cdot A_2 \tag{7.196}$$

$$p_{fb1} = p_s - K_{vp} \cdot \Delta x_{cs} \tag{7.197}$$

where Δx_{cs} is the spool displacement of the compensator valve. Again, if we eliminate the Δx_{cs} from the above equations, and note that $K_{vp} \gg k_{spring}$, and let $A_1 = A_2$ for simplicity, it can be shown that

$$\frac{k_{spring}}{K_{vp}A_1} \cdot (p_s - p_{fb1}) = (p_{fb1} - p_{fb2}) - \frac{k_{spring}}{A_1} \cdot x_{preload} \tag{7.198}$$

$$p_{fb1} \approx p_{fb2} + \frac{k_{spring}}{A_1} \cdot x_{preload} \tag{7.199}$$

where the left hand-side of the above equation is approximated as zero since $\frac{k_{spring}}{K_{vp}} \approx 0.0$. If load pressure p_{fb2} increases, so does the p_{fb1} by reducing the pressure drop relative to p_s, in order to maintain a constant pressure drop across the main flow control valve as long as $p_{fb1} \leq p_s$. When $p_{fb1} = p_s$, if p_{fb2} continues to increase, the compensator valve saturates and cannot maintain the pressure differential. If the load pressure (p_{fb2}) is so large that the pressure differential between p_s and p_{fb2} is less than the desired pressure differential, then the compensator valve modulation saturates and it tries to do its best by making p_{fb1} as close as possible to p_s (minimizes the restriction). Hence, when supply pressure is constant and the load pressure increases to a level so large that the desired pressure differential between two ports is not possible, the compensator valve fully opens, trying to minimize the pressure drop,

$$p_{fb1} \approx p_s \tag{7.200}$$

$$\Delta P_v \approx p_s - p_{fb2} \leq \Delta P_{set} \tag{7.201}$$

This is the condition where pump pressure supply has reached its maximum (saturation) level and load pressure is very high.

The post-compensator configuration (Figure 7.55) uses two pressure feedbacks to its spool: input port pressure feedback to one side and output (i.e., maximum load) pressure feedback to the other side.

The spool movement of the compensator valve is controlled by the following relationship (Figure 7.55),

$$k_{spring} \cdot (x_{preload} + \Delta x_{cs}) = p_c \cdot A_1 - p_1 \cdot A_2 \tag{7.202}$$

$$p_c = p_s - K_{vp} \cdot \Delta x_{cs} \tag{7.203}$$

where p_c is the pressure at the input port of the compensator valve, p_1 is the load pressure feedback signal. Again, if we eliminate the Δx_{cs} from the above equations, and note that $K_{vp} \gg k_{spring}$, and let $A_1 = A_2$, $(k_{spring}/K_{vp}) \approx 0.0$, it can be shown that

$$p_c = p_1 + \frac{k_{spring}}{A_1} \cdot x_{preload} \tag{7.204}$$

which is the same relationship as the previous case. However, there are functional differences between pre-pressure compensators and post-pressure compensators in multi function circuits. In post-pressure compensated systems, the pump is controlled to maintain a margin pressure (pressure difference) between the pump output pressure and load pressure,

$$p_s = p_1 + \Delta p_{margin} \tag{7.205}$$

The pressure differential across the main flow control valve is

$$p_s - p_c = p_1 + \Delta p_{margin} - \left(p_1 + \frac{k_{spring}}{A_1} \cdot x_{preload} \right) \tag{7.206}$$

$$p_s - p_c = \Delta p_{margin} - \frac{k_{spring}}{A_1} \cdot x_{preload} \tag{7.207}$$

which shows that the pressure drop across the main flow control valve is independent of the load pressure. It is important to emphasize that pressure compensator valves are primarily used to balance the flow distribution in multiple parallel hydraulic functions supported by a single pump.

In short, the compensator valve adds restriction (pressure drop) between the pump line and the output line (A or B) as a function of the difference between the pump pressure and the load pressure.

In multiple circuits, a set of shuttle valves are used to select the maximum load pressure among all circuits. The maximum load pressure is needed for two purposes: (i) to control the pump displacement so that the pump is able to support the circuit with the most load (worst case scenario), (ii) to control multiple compensator valves and try to equalize the flow distribution among circuits. There are two major types of pressure compensation in multi function circuits: the pre-compensator type and post-compensator type (Figures 7.56, 7.57).

Figure 7.56 shows a multi function hydraulic circuit supplied by a variable displacement pump. Each circuit has a pre-compensated valve along with the proportional directional flow control valve. Each compensator valve uses two pressure feedbacks to regulate the pressure differential across the main valve ahead of it. The pressure feedback signal for each compensator valve is local to that circuit: output pressure of the compensator and load pressure of the cylinder (maximum pressures of cylinder sides A and B are selected by a shuttle valve). Finally, the pump displacement is controlled by the maximum load pressure among all circuits so that the pump runs at a displacement that can support the largest load among all circuits.

Figure 7.57 shows a similar multi circuit. The only difference is the location of the compensator valves: in post-compensator configuration. In each circuit, the sources of the two pressure feedbacks to the compensator valve are not all local: one of the pressure feedbacks is the pump pressure line which is the input line pressure to the post-compensator and the second load pressure feedback is the maximum load pressure among all circuits. Hence, in the post-compensated configuration, the load pressure feedback (the second pressure feedback) to each compensator valve is determined by the rest of the circuits. It is always the maximum of all load pressures. The same maximum load pressure signal is used to control the pump displacement.

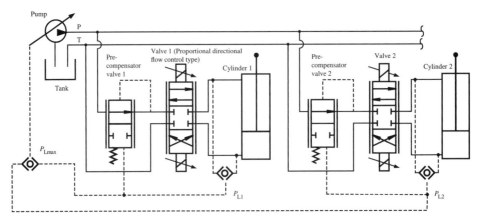

FIGURE 7.56: Pre-compensator valve configuration in a multi function hydraulic circuit. The pump is of the variable displacement type. Each function is connected to the supply line in parallel. Each compensator valve uses two pressure feedback signals for its control: the compensator output port pressure signal and the load pressure signal (i.e., maximum of two pressures of the cylinder it is controlling). Each compensator uses its own circuit's load pressure signal as feedback. In multi function circuits, the pump is controlled by the maximum of all load pressure signals among multiple circuits. When the pump saturates (maximum flow capacity is reached), the flow will go to the circuit with the lowest load pressure, and the circuits with higher load pressures will get lower or no flow.

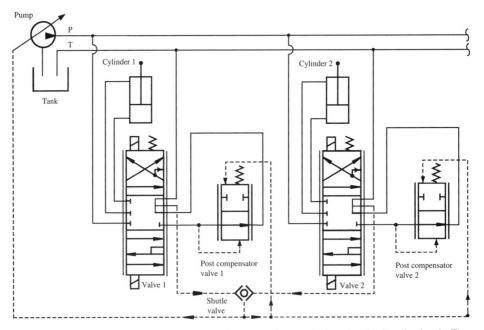

FIGURE 7.57: Post-compensator valve configuration in a multi function hydraulic circuit. The pump is of the variable displacement type. Each function is connected to the supply line in parallel. Each post-pressure compensator valve uses two pressure feedback signals for its control: the input port pressure to it and the maximum of all the load pressures among multiple circuits. Therefore, one of the two pressure feedback signals to a post-compensator valve may originate from a different circuit in multi circuit applications. The same pressure signal is used for the pump control. Post-pressure compensator valves are used to equalize the flow distribution among multiple circuits. When the pump is saturated, all circuits gets proportionally less flow (unlike the pre-compensated case). For instance, the hydraulic implement lift and tilt circuits of John Deere wheel loader model 644H (Figure 7.61) use the same type of post-compensators along with each proportional directional flow control valve.

Under non-saturated conditions, the performances of pre-compensated and post-compensated configurations of multi circuit hydraulic systems are very similar. When the pump capacity is saturated, the post-compensator configuration backs-off the flow-demand from each line proportionally and tries to maintain even flow delivery based on demand to each circuit. Whereas in pre-compensated multi circuits, when the pump saturates, the circuit which has the largest load pressure gets less flow and may eventually get zero flow. In three or more function circuits, the higher load pressure circuits get progressively less flow and only the lowest pressure circuit gets most of the flow. The orifice in the pressure feedback line from the load has increased the dampening effect (in addition to the added hydraulic resistance) so that high frequency pressure spikes are filtered from the feedback signal to the compensator valve.

The pressure compensated flow control valves work on the basic operating principle of maintaining constant flow despite pressure variations. Examining Equation 7.193 shows that for a given spool displacement, the flow rate will change as a function of pressure differential across the valve which may vary as a function of load and supply pressure variations. The basic principle is to change the valve opening $A(x_s)$ by shifting the spool as a function of pressure feedback. As a result, the effective spool displacement is not only a function of the current but also the pressure drop. Hence, if the pressure drop increases, feedback moves the spool to reduce the orifice area. Similarly, if the pressure drop reduces, feedback moves to increase the orifice area. The end result is to maintain an almost constant flow for a constant current signal under varying load conditions. The feedback mechanism to shift the spool as a function of pressure drop can be implemented by hydro-mechanical or electronic means. Normally, a pressure compensated flow control valve is implemented with two valves in series: one valve is the proportional directional flow control valve and the other valve is the pressure compensator valve. Examples of such implementations are shown in Figures 7.55–7.57. It should be noted that if two pressure sensors were available at the input and output ports of the proportional valve, the solenoid current could be controlled to perform the function of the second valve, while at the same time metering the flow. In other words, the control algorithm that decides on the solenoid current to shift the spool can take not only the desired command signal but also the two pressure signals into account. Hence, we can implement a pressure compensated flow control valve using a solenoid controlled proportional valve, two pressure sensors, and digital control algorithm.

The Need for Pressure Compensation in Multi Function Circuits

When a load-sensing pump is used used to supply two or more parallel hydraulic circuits, the flow distribution between the circuits is load dependent (Figure 7.58). More flow will go to the circuit with lower load resistance. Load check valves in each circuit are used to make sure that the load does not drive the flow back to the pump when the load pressure is larger than the supply pressure. Let us assume that both circuits have the same valve size and cylinders. The flow across each valve is given by

$$Q_{v1} = C_d \cdot A_{v1}(x_{v1}) \cdot \sqrt{p_s - p_{l1}} \tag{7.208}$$

$$Q_{v2} = C_d \cdot A_{v2}(x_{v2}) \cdot \sqrt{p_s - p_{l2}} \tag{7.209}$$

where

$$Q_s = Q_{v1} + Q_{v2} \; ; \; for \; p_s < p_{\text{relief}} \; (Q_{\text{relief}} = 0.0) \tag{7.210}$$

Since the valves are the same size, and assuming that both valve spools are displaced to the same value (same valve orifice openings), $C_d A_{v1}(x_{v1}) = C_d A_{v2}(x_{v2})$, the flow rate that

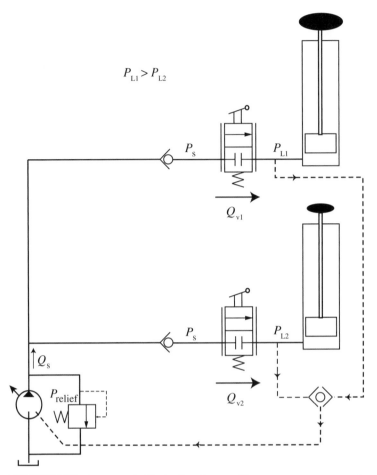

FIGURE 7.58: An example illustrating the need for pressure compensator valves in multi function parallel circuits.

goes into each cylinder depends on the load pressure. More flow will go to the circuit with the smaller load pressure, and the flow ratio is

$$\frac{Q_{v1}}{Q_{v2}} = \frac{\sqrt{p_s - p_{l1}}}{\sqrt{p_s - p_{l2}}} \tag{7.211}$$

For instance, let $p_s = 3000\,\text{psi}$, $p_{l1} = 2000\,\text{psi}$, $p_{l2} = 1000\,\text{psi}$, same size valves, and same valve spool displacement in each circuit. The flow rate distribution betwen the two circuits is

$$\frac{Q_{v1}}{Q_{v2}} = \frac{\sqrt{3000 - 2000}}{\sqrt{3000 - 1000}} = \frac{1}{\sqrt{2}} \tag{7.212}$$

If the load condition changes, the flow rate will change: let $p_s = 3000\,\text{psi}$, $p_{l1} = 1000\,\text{psi}$, $p_{l2} = 2000\,\text{psi}$, and same size valves and valve spool displacement in each circuit. The flow rate distribution betwen the two circuits under this load condition is

$$\frac{Q_{v1}}{Q_{v2}} = \frac{\sqrt{3000 - 1000}}{\sqrt{3000 - 2000}} = \frac{\sqrt{2}}{1} \tag{7.213}$$

Even though the commands to each valve do not change, the flow rate to each circuit, and hence the speed of each cylinder, will vary due to the variation in the load in each circuit.

If it is desired to modify circuit design *so that the flow rate distribution between circuits is independent of the load*, the pressure drop across each valve must be same even if the load pressures are different,

$$p_s - p_{l1} = p_s - p_{l2} \tag{7.214}$$

Under this condition, flow rate distribution to each circuit is independent of the load difference between the circuits and would be controlled by specific spool displacement of each valve (x_{v1}, x_{v2}). When both valves are same size and have the same displacement, flow rate distribution would be equal, 50% each of total flow, or $Q_{v1} = Q_{v2}$, $Q_s = Q_{v1} + Q_{v2}$. When load pressures are different $(p_{l1} \neq p_{l2})$, the equal pressure drop across the valve for each circuit can be accomplished by:

1. Adding restriction to the circuit which has the lower load so that the total hydraulic load seen by the valve is the same as the load on the other circuit. This means adding "positive hydraulic resistance" to the circuit with lower load.

2. Removing restriction from the circuit which has the higher load so that the total hydraulic load seen by the valve is same as the load on the other circuit. This means adding "negative hydraulic resistance" to the circuit with higher load.

The second option is not physically possible. Negative hydraulic resistance cannot be added to a hydraulic circuit. Therefore, in order to equalize the effective hydraulic resistance (making the total pressure drop equal in each circuit), we need to add hydraulic resistance into the circuit which has lower load.

The electrical analogy of this concept is shown in Figure 7.59a. Consider that we have a constant voltage supply (analogy of pump), and two parallel resistors (analogy of valve-cylinder of each circuit). The current (analogy of flow rate) is divided between the two circuits as follows,

$$i = i_1 + i_2 \tag{7.215}$$

$$i_1 = \frac{V}{R_{v1}} \tag{7.216}$$

$$i_2 = \frac{V}{R_{v2}} \tag{7.217}$$

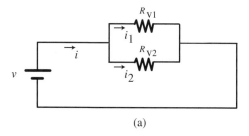

(a)

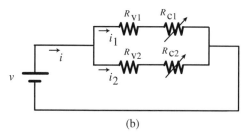

(b)

FIGURE 7.59: (a) Electrical analogy for a two function hydraulic circuit. (b) Same analogy with adjustable resistors in each path to control the current (flow rate) distribution between the two parallel paths.

In order to make current division to each path equal, $i_1 = i_2$, the resistance in each path must be the same, $R_{v1} = R_{v2}$. This can be accomplished by adding an adjustable resistor in each path (Figure 7.59b), where R_{c1}, R_{c2} are adjustable resistors. If $R_{v1} > R_{v2}$, we can make $R_{c1} = 0.0, R_{c2} = R_{v1} - R_{v2}$ to make the current in each path equal. Similarly, if $R_{v1} < R_{v2}$, we can make $R_{c1} = R_{v2} - R_{v1}, R_{c2} = 0.0$ to make the current in each path equal,

$$i = i_1 + i_2 \tag{7.218}$$

$$i_1 = \frac{V}{R_{v1} + R_{c1}} \tag{7.219}$$

$$i_2 = \frac{V}{R_{v2} + R_{c2}} \tag{7.220}$$

If we make, $R_{v1} + R_{c1} = R_{v2} + R_{c2}$, then $i_1 = i_2$, even if $R_{v1} \neq R_{v2}$.

The same concept can be accomplished in hydraulic circuits by adding an adjustable orifice ("hydraulic resistor") in series with each circuit. There are two methods for this:

1. pre-pressure compensated valves,
2. post-pressure compensated valves.

In multi function circuits as shown in Figure 7.60, both pre- and post-pressure compensated circuits provide the same behavior until the pump saturates or relief pressure is reached. Figure 7.60a shows a circuit with pre-pressure compensation. Figure 7.60b shows the post-pressure compensated circuit. We assume that both circuits have the same pump, same valves and cylinders and pressure compensator valves, except that the location of the pressure compensator valves and their feedback control signals are different. The pump is controlled in load-sensing mode and has the same margin pressure for both circuits.

Consider the pre-pressure compensated case, and the *pump is not saturated*. Let us assume that the preload setting in the pressure compensator valves are equal, $p_{sp1} = p_{sp2}$. The pre-pressure compensator valve spools move to control the restriction (adding hydraulic resistor from zero to infinity by opening and closing the compensator valve orifice by spool movement) so that the pressure difference across the main valve is constant. In other words, the pressure compensator valves work to maintain the following relationship as long as they are not saturated (fully closed or fully open),

$$p_{cp1} - p_{l1} = p_{sp1} \tag{7.221}$$

$$p_{cp2} - p_{l2} = p_{sp2} \tag{7.222}$$

Since flow rate across the main valves is determined by the following relationships,

$$Q_{v1} = C_d A_{v1}(x_{v1})\sqrt{p_{cp1} - p_{l1}} = C_d A_v(x_{v1})\sqrt{p_{sp1}} \tag{7.223}$$

$$Q_{v2} = C_d A_{v2}(x_{v2})\sqrt{p_{cp2} - p_{l2}} = C_d A_v(x_{v2})\sqrt{p_{sp2}} \tag{7.224}$$

which shows that the flow rates are *independent of the load* pressures (p_{l1}, p_{l2}). Instead, the flow rates Q_{v1} and Q_{v2} are functions of p_{sp1} and p_{sp2}, which are the pressure settings by the preloaded springs in the pressure compensator valves, respectively. Notice that $p_{c1} - p_{l1}$ is the pressure drop across main flow control valve number one, and $p_{c2} - p_{l2}$ is the pressure drop across main flow control valve number two, and they are equal to the preload pressure setting of the pre-pressure compensator valves, p_{sp1}, p_{sp2}, respectively. If they are equal by design, $p_{sp1} = p_{sp2}$, then the pressure drop across each main valve is regulated to that same value as long as saturation is not reached. Hence, the flow rate distribution between the circuit is load independent. The flow rate distribution is controlled by the spool position of the main flow control valves.

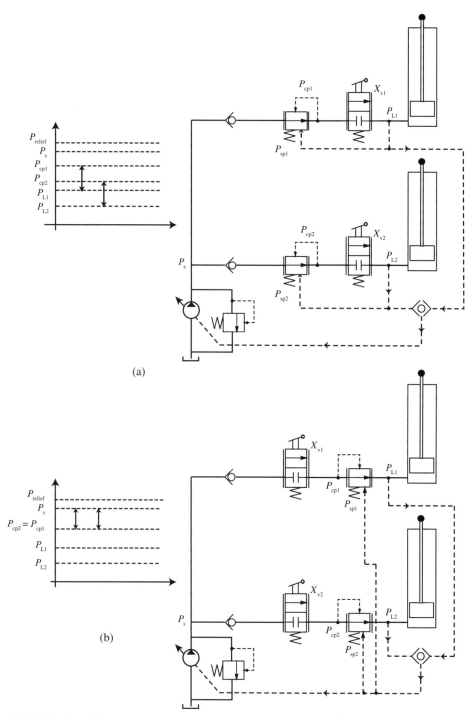

FIGURE 7.60: (a) Pre-pressure compensated multi function parallel hydraulic circuit, (b) post-pressure compensated multi function parallel hydraulic circuit.

Consider the case when the *pump saturates*, $p_s = p_{max}$, pump output pressure cannot increase anymore. Let us assume that the load pressures are such that $p_{l1} > p_{l2}$ and p_{l1} increases to a value that is very close to the p_s. Since p_s cannot be increased anymore, and p_{l1} is defined by the load, the compensator simply cannot increase the pressure differential between the two, since $(p_{c1} - p_{l1}) < (p_s - p_{l1}) < p_{sp1}$. In the meantime, the compensator in circuit two still works to maintain the pressure differential because the load is very low in that circuit and the $(p_{c2} - p_{l2}) = p_{sp2} < (p_s - p_{l2})$. So as a result, less flow will go to the circuit with higher load since the net pressure drop available across the main flow control valve of that circuit is lower. Eventually, if $p_s - p_{l1} = 0.0$, that is $p_{l1} = p_s$, the flow rate to circuit one will be zero, while flow to the other circuit will continue in proportion to the main valve opening. In summary, in pre-pressure compensated parallel multi circuit hydraulic sytems, when the pump saturates, flow distribution to the highest load circuit will eventually be reduced to zero, then the flow to the next highest load circuit will be reduced to zero, while other circuits will continue to get the flow.

In post-pressure compensated systems (Figure 7.60b), the flow rate distribution is the same as in the pre-pressure compensated case, when the pump is not saturated.

$$Q_{v1} = C_d A_{v1}(x_{v1}) \sqrt{p_s - (p_1^* + p_{sp1})} \qquad (7.225)$$

$$Q_{v2} = C_d A_{v2}(x_{v2}) \sqrt{p_s - (p_1^* + p_{sp2})} \qquad (7.226)$$

$$p_{sp1} = p_{cp1} - p_1^* \qquad (7.227)$$

$$p_{sp2} = p_{cp2} - p_1^* \qquad (7.228)$$

$$p_1^* = \max(p_{l1}, p_{l2}) \qquad (7.229)$$

Notice that both post-pressure compensator valves use the maximum load signal among multi circuits as its feedback signal (p_1^*), unlike the pre-pressure compensator valves where each circuit uses its own load signal as its feedback signal. Since $p_{sp1} = p_{sp2}$ by the preload settings of the compensator valves, flow rate distribution is independent of the load. When the pump saturates, the $p_s - p_{cp1} = p_s - p_{cp2}$ will still be equal, because from the above equations we have the same p_1^* and $p_{sp1} = p_{sp2}$. When the pump saturates, the pressure differential between the $p_s - p_{cp1} = p_s - p_{cp2}$ cannot be made as large as the case when the pump is not saturated. However, as the load pressure gets closer to the pump pressure, the pressure drop across each of the main flow control valves in the top and bottom circuits is still the same. As a result, the flow rate will be reduced in each circuit by the same ratio. When the pump saturates, as load pressure increases in one of the circuits, flow rate to each circuit is reduced by the same ratio because the same term

$$\sqrt{p_s - p_{cp1}} = \sqrt{p_s - p_{cp2}} \qquad (7.230)$$

(because $p_{sp1} = p_{sp2}$, hence $p_{cp1} = p_{cp2}$) defines the flow rate equations for both circuits. Hence, the flow rate does not go to zero for the circuit with higher load. This is the main functional difference between pre- and post-pressure compensated load sensing hydraulic systems.

7.5.4 Example: A Multi Function Hydraulic Circuit using Post-Pressure Compensated Proportional Valves

Figure 7.61 (also see Figure 7.13) shows the hydraulic circuit schematics of the bucket control for the wheel loader model 644H by John Deere. This circuit serves as a good example of post-pressure compensated proportional flow control valves. The figure shows only the hydraulic system relevant to control the bucket (lift, tilt, and auxiliary functions).

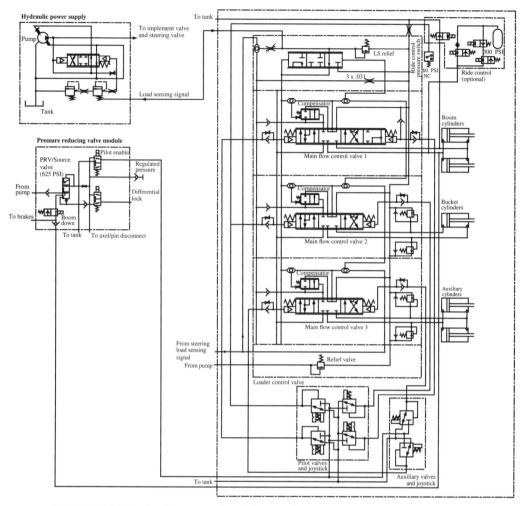

FIGURE 7.61: Hydraulic system circuit diagram for the bucket motion of the John Deere wheel loader model 644H: only the hydraulic circuit for bucket motion is shown. Hydraulic circuits for the steering, brake, and cooling fan are not shown. The pump is of the variable displacement axial piston type. The main flow control valves are connected to the hydraulic supply in parallel configuration. They are closed-center type and post-compensated for pressure drop.

The rest of the hydraulic circuit for the machine (steering, brakes, cooling fan drive) is not shown. The key features of this hydraulic circuit are

1. the pump is of variable displacement type and controlled by a load sensing hydraulic mechanism,

2. the valve–cylinder pairs for each function are of closed center type and are connected to the hydraulic supply lines (P and T) in parallel configuration,

3. each proportional directional main flow control valve has a post-compensator valve in order to maintain constant pressure drop across each main valve.

The hydraulic circuits (steering, implement hydraulics, and brake hydraulics) are supplied by a single variable displacement axial piston pump. The pump is controlled by a load sensing hydraulic circuit. Since the same pump supports multiple hydraulic circuits which may have different loads, it is necessary to sense all loads and use the highest load signal in control of the pump. Therefore, the pump displacement is determined by the maximum load pressure signal among the steering circuit (not shown in the figure), boom circuit, bucket, and auxiliary circuits. The selection of the maximum load pressure is made by a series of shuttle valves.

The valve–cylinder pair for each function is connected in parallel configuration to the pump and tank lines. Furthermore, each directional flow control valve (proportional type) is accompanied by a post-compensator valve. As discussed in Figure 7.57, the post-compensator valve works to maintain a constant pressure drop across the main valve so that the function speed is proportional to the lever command regardless of the load as long as the load is not so large as to saturate the pump. The load pressure feedback is the maximum load pressure among all circuit load pressures. This selection is made by a series of shuttle valves. Notice that due to the load sensing signal needed for the compensator, the main valve has three additional lines (in addition to the P,T, A, B main lines): two of them are for the input and output of the compensator valve connection and one of them is the load pressure sensing line for feedback control of the compensator. In addition, there are two pilot pressure ports, one for each side of the spool, which come from the pilot valves. The check valve and fixed orifice in parallel with it between the pilot control valve signal to the main valve spool on both sides have the effect of dampening the pressure oscillations on the pilot control line and result in smoother operation of the main valve.

The pilot pressure supply for the main valve control of the implement hydraulic is derived from the output of the main pump pressure using a pressure reducing valve. The output of the pressure reducing valve is a constant pilot supply pressure. The pilot valves allow the operator to control the pilot output pressure to the main valves. For each function (lift, tilt, auxiliary) there are two pilot valves. Each valve has two inports (pilot pressure supply port and tank pressure port) and one output port (output pilot pressure used to control the main valve). The output pilot pressure is approximately proportional to the mechanical movement of the lever controlled by the operator. As the operator moves the lever, that is to lift the bucket, the pilot valve connected to the lift lever sends pilot pressure (proportional to the lever displacement) to the main valve spool's control port. The main valve spool is spring centered and its displacement is proportional to the pilot pressure differential between the two sides (one side is at the tank pressure, the other side is modulated by the pilot valve output). Hence, if we assume a constant pressure drop across the main valve, the cylinder speed will be proportional to the main spool displacement which is in turn proportional to the lever displacement. In approximate terms, the cylinder speed is proportional to the lever displacement.

7.5.5 Directional, Proportional, and Servo Valves

Directional flow control valves are categorized based on the following design characteristics (ISO 6404 standard):

1. Number of external ports: two-port, three-port, four-port. Number of ports refer to to the plumbing connections to the valve which can be 2, 3, 4 or more. A four port valve connects the pump and tank (P, T) ports to two load ports (A, B) (Figure 7.44).

2. Number of discrete or continuously adjustable spool positions: ON/OFF two-position, ON-OFF three position, proportional valves.

3. Neutral spool position flow characteristics: open center where one or more ports are connected to the tank (i.e., P to T, A to T, B to T, A and B to T, P and A to T, P and B to T) or closed center where all ports are blocked. Closed center valves are generally

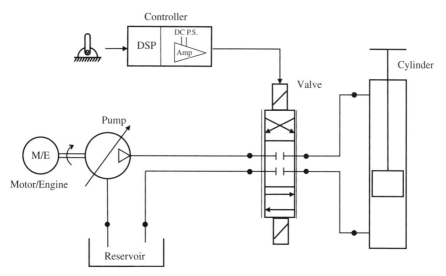

FIGURE 7.62: Closed center EH motion system for one axis motion: the valve is the closed center type, the pump is the variable displacement type.

used with variable displacement pumps, whereas open center valves are used with fixed displacement pumps (Figures 7.3, 7.62).

4. Number of actuation stages: single-stage, two-stage and three-stage valve spool actuation.

5. Actuation method: referring to the first stage actuation whether by mechanical lever (manual), electrical actuator (i.e., solenoid, torque motor, linear force motor), or by another modulated pilot pressure source.

The "servo" valve and "proportional" valve names both refer to controlling the spool displacement of the valve *proportional* to the command signal (i.e., solenoid current). A servo valve has a custom matched spool and sleeve, machined to higher tolerances than the spool for a proportional valve. A proportional valve has a spool and valve body. The fitting clearance between the spool–sleeve of servo valves is in the range of 2–5 microns, whereas the clearance between the spool and body of the proportional valve is in the range 8–15 microns or more. For all practical purposes, the only differences between these two types of valves are that

- Servo valves use feedback from the position of main spool to decide on the control signal (i.e., the feedback may be implemented by completely hydro-mechanical means or by electrical means), whereas proportional valves do not use main spool position feedback. However, it should be pointed out that more and more proportional valves have started to use spool position feedback in recent years. As a result, the differentiation between servo and proportional valves based on spool position feedback has started to disappear.

- Servo valves generally refer to higher bandwidth valves.

- Servo valves provide better gain control around the null position of the spool. As a result, servo valves can provide higher accuracy positioning control. The average deadband in servo valves is in 1 to 3% of maximum displacement, whereas it can be as high as 30 to 35% in proportional valves.

Otherwise, their construction and control are very similar.

The performance specifications of a proportional directional EH valve includes:

1. Flow rating at a certain standard pressure drop (i.e., 180 l/min at a standard pressure drop across the valve, when the valve is fully open. The standard pressure drop for the flow rating of the servo valves is 1000 psi and of proportional valves is 150 psi).
2. Maximum and minimum operating pressures, p_{max}, p_{min}.
3. Maximum tank line pressure, $p_{t,max}$.
4. Number of stages (single stage, double stage, triple stage).
5. Control signal type: manual, pilot pressure, or electrical signal controlled.
6. Pilot pressure range $(p_{pl,max}, p_{pl,min})$ and pilot flow (Q_{pl}) required if pilot pressure is used.
7. Open center or closed center type spool design.
8. Linearity of current–flow relationship at a constant pressure drop (Figure 7.77).
9. Symmetry of current–flow relationship between plus and minus side of current.
10. Nominal deadband.
11. Nominal hysterisis.
12. Maximum current from the amplifier stage if solenoid operated (and PWM frequency and dither frequency if used).
13. Bandwidth of the valve at a specified supply pressure (w_v Hz) or rise time for a step command change in flow at a certain percentage (i.e., 25%) of maximum flow rate.
14. Operating temperature range.

The specifications for the electronic driver (power supply and amplifier circuit) include:

1. Input power supply voltage (i.e., 24 VDC nominal, or ± 15 VDC) and current rating to the drive power supply section.
2. Control signal type and range (i.e., ± 10 VDC, ± 10 mA, or 4 to 20 mAmp, analog or PWM signal),
3. Recommended *dither* signal frequency and magnitude (i.e., less than 10% of rated signal at a higher frequency than the valve bandwidth) in order to reduce the effect of friction between the spool and valve body.
4. Feedback sensor signal type (if any for a driver which uses local valve spool position feedback to control the spool position).

For flow rates under 20 to 50 gallon/min (gpm), a single stage direct actuated valve is generally sufficient. For flow rate ranges between 50–500 gpm, a two stage valve is used. For flow rates over 500 gpm, typically a three stage valve is used. In a single stage valve, the electrical current sent to the solenoid (or linear torque motor), which creates electromagnetic force, directly forces the main spool and moves it (Figure 7.63). Multi stage valves (two or three stage) use pilot pressure (which is a lower pressure than the main supply line pressure) to amplify the electrical actuator signal to shift the main spool. This provides a very large amplification gain between the first stage electrical actuator signal and second stage pilot force. This gain cannot be matched by any direct drive electric motor technology currently. The large pilot stage gain generated shifting power also makes the valve less sensitive to main stage contamination problems since the large shifting pressure of the pilot stage is very likely to able to force through contamination problems. However, the contamination problem is more likely to occour at the pilot stage of the valve since the orifices at the pilot stage are much smaller than those in main stage.

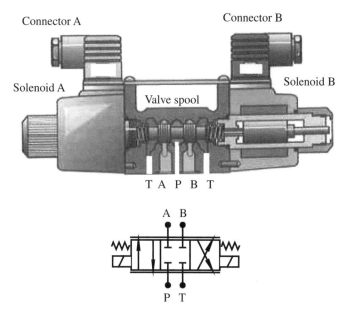

FIGURE 7.63: Single stage EH valve: two solenoids directly move the valve spool. Courtesy Moog Inc.

There are three main types of two-stage valve designs:

1. spool-spool (double spool) design,
2. double nozzle flapper design, and
3. jet pipe design.

In a two stage valve, the electrical current moves an intermediate spool, which then amplifies power using the pilot pressure line to move the second main spool (Figure 7.64). In multi stage valve cases, the valve is so large that the electrically generated force by the solenoid (or electric actuator) is not large enough to move the main spool.

The same concept applies to the three stage valves (Figure 7.65) where there are three spools in the valve, the first two acting as the amplifiers to move the final third stage main valve spool. The great majority of valves used in mobile equipment applications are two stage valves. From a control system perspective, the functionality of single stage and multi stage valves are the same: input current is translated proportionally into the main spool displacement (with some dynamic delay effects, of course) which is then proportional to the flow rate under constant pressure drop across the valve.

Notice that the second stage spool position (main spool) of a two-stage valve is integral of the first stage spool position (pilot spool) if there is no feedback from the main spool position to the pilot spool position. A two-stage spool valve is basically a single stage direct acting valve connected to a second valve spool. The second valve can be viewed as a small cylinder connected to a single-stage valve. Therefore, the current–main spool displacement relationship is not proportional, but an integral relationship.

$$x_{\text{main}}(s) = \frac{1}{s} K_{\text{mp}} \cdot x_{\text{pilot}}(s) \qquad (7.231)$$

$$= \frac{1}{s} K_{\text{mi}} \cdot i_{\text{sol}}(s) \qquad (7.232)$$

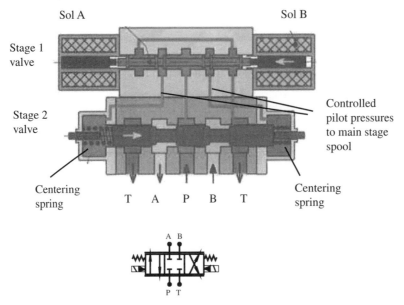

FIGURE 7.64: Two-stage valve: the first-stage spool is for the pilot stage, the second-stage spool is for the main stage.

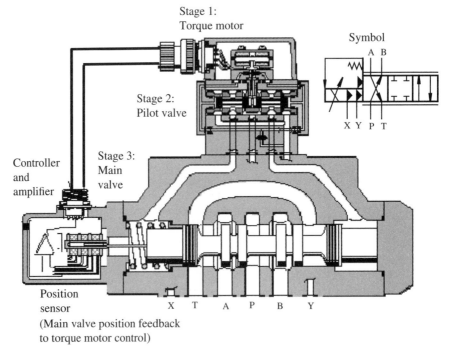

FIGURE 7.65: Three-stage valve: first stage is torque motor and double nozzle flapper, second stage spool is the pilot valve, and the third stage is the main valve. Second stage amplifies the actuation force of the first stage to move the third stage (main spool). The X and Y letters in the valve symbol indicate that the valve has two pilot stages. Hence, it is a three-stage valve, two pilot stages plus the main stage. The arrow on top of the solenoid signal around the spring indicates a valve position sensor feedback signal. The arrow over the solenoid indicates that it is a proportionally controlled solenoid, not just ON/OFF, although this representation is often omitted in symbol drawings. This particular valve can support flow rates upto 1500 l/min at 10 bar pressure drop across the valve (Moog series D663). Courtesy Moog Inc.

In order to make the current–main spool displacement a proportional relationship, like the case in a single stage valve, there has to be a "feedback mechanism" on the main spool position. The pilot spool position is determined by the solenoid force and the feedback from the main spool position. The feedback from the main spool to pilot spool may be in the form of a mechanical linkage, spring, pressure, or in the form of an electronic sensor signal to a controller. Let us consider a two stage valve which uses pressure feedback between the pilot and main stage (Figure 7.64). The pressure feedback from the main spool position acts like a balancing spring against the solenoid force

$$x_{\text{main}}(s) = \frac{1}{s} K_{\text{mp}} \cdot x_{\text{pilot}}(s) \tag{7.233}$$

$$x_{\text{pilot}}(s) = K_{\text{pi}} \cdot i_{\text{sol}}(s) - K_{\text{pf}} \cdot x_{\text{main}}(s) \tag{7.234}$$

$$x_{\text{main}}(s) = \frac{K_{\text{mp}} \cdot K_{\text{pi}}}{s + (K_{\text{pf}} \cdot K_{\text{mp}})} \cdot i_{\text{sol}}(s) \tag{7.235}$$

This feedback mechanism can be implemented either by hydro-mechanical means or by using spool position sensor and closed loop control algorithms.

Let us discuss how this feedback is physically accomplished by hydro-mechanical means (Figure 7.64). When solenoid B is activated, its corresponding force pushes directly against the pilot spool to open the pilot port A to pilot pressure supply. The pressure built up in the chamber on the right side of the main spool is also directly fed back to the end cap (A) of the pilot spool. When the pressure in the end cap equals or exceeds the force from the solenoid in B, the spool shifts back and closes the opening to pilot supply. The solenoid–feedback combination will now hold a pressure in port A that is proportional to the input current to the solenoid. The main spool maintains its spring-centered position until the pressure in the end gap is equal or exceeds the force in the centering spring. The main spool will shift to a metering position (if the deadband region is passed). This movement will cause port A to increase its volume and hence the pressure will drop. This pressure drop will cause the feedback force in the pilot spool to drop allowing the pilot spool to shift open to supply pressure to once again reach the pressure that corresponds to the solenoid input. When the current to the solenoid is cut, the pilot spool will shift back to open to tank and lower the pressure back to tank pressure. Accordingly, the main spool will shift back because of the force in the centering spring until the whole valve is back to neutral position. In summary, the pilot spool moves in proportion to the solenoid current, then the built-up pressure in pilot port moves the main spool. The same pilot pressure is fed back to the pilot spool end caps. In steady-state, the main spool is at a position that the spring force balances the built-up pilot pressure. The built-up pilot pressure is proportional to the solenoid current. The pilot spool returns to neutral position (closed) after the transients when a constant current is applied and the main spool reaches the proportional position. Without the pressure feedback to the pilot spool end caps, under constant current, the pilot spool would shift a constant amount, and the main spool position would keep increasing as an integral of the pilot spool, and hence that of the solenoid current. The key to making the current and main spool position proportional in steady-state is the pressure feedback between the main spool and pilot spool end caps.

There are also two-stage proportional valves where the main spool is shifted by two pressure reducing valves, one on each side (Figure 7.66). Such valves are widely used in construction equpment applications (Figures 7.52, 7.117). A pair of pressure reducing valves acts as a pilot valve for the main stage. The output pressure of the pilot valve is proportional to the lever displacement or solenoid current. A proportional main spool displacement is developed as a result of the balance between the pilot pressure output of the pressure reducing valve and the centering spring of the main spool. The input–output

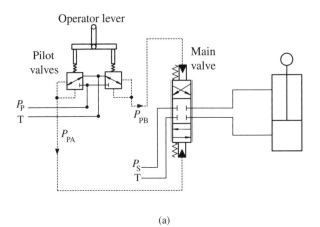

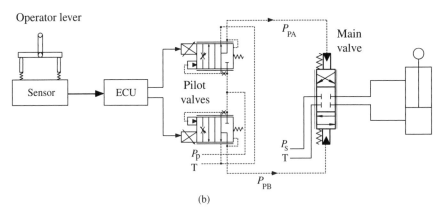

FIGURE 7.66: Two-stage proportional valve with two pressure reducing valves at the pilot stage. Pilot stage output is a pressure proportional to the input command (mechanical or electrical). The main spool displacement is proportional to the pilot output pressure (control pressure) as a result of the centering springs. (a) Mechanically actuated pilot valves (for instance, this type of valve is used on Komatsu wheel loader model WA450-5L for its bucket control system), (b) electrically actuated pilot valves.

relations in two stage proportional valves where the first stage is a pair of pressure reducing valve are as follows,

$$p_{pc} = \frac{K_{pi}}{(\tau_{pi}s + 1)} \cdot i_{sol} \tag{7.236}$$

$$x_{main} = \frac{1}{K_{spring}} A_{main} \cdot p_{pc} \tag{7.237}$$

$$= \frac{K_{pi} A_{main}}{K_{spring} (\tau_{pi}s + 1)} \cdot i_{sol}(s) \tag{7.238}$$

where p_{pc} is the pilot pressure output from the pressure reducing pilot valves to the main spool end caps, i_{sol} is the solenoid current, x_{main} is the main spool displacement, K_{spring} is the centering spring constant of the main spool, A_{main} is the cross-sectional area of the main spool at the end caps where the pilot control pressure acts, τ_{pi} is the time constant of the transient response between. Notice that each pilot control valve (pressure reducing pilot valve) may be controlled electrically by a solenoid or manually by a mechanical

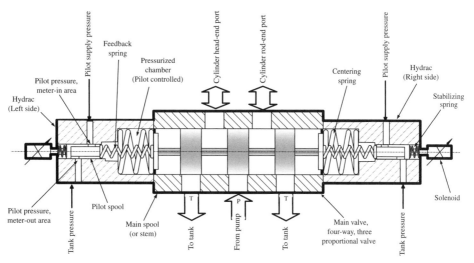

FIGURE 7.67: Two-stage proportional valve with two pressure reducing valves at the pilot stage in line with the main spool. Pilot stage output is a proportional pressure to the solenoid current. The main spool displacement is proportional to the pilot control pressure as a result of the centering spring. The feedback between the main spool and pilot spools (one on each side of main spool) has mechanical springs in addition to the hydraulic feedback.

lever (Figure 7.50). In the mechanically actuated pilot valve version, the pressure reducing pilot valve pair is mounted right under the operator control lever. The output pilot control pressure line is then routed to the main valve which is likely to be closer to the cylinder. In the electrically controlled version (EH version), the main valve and the pilot valve along with solenoid pair can all be located at the same location and control signals from the operator lever sent to the solenoids electrically. Therefore, the EH version has fewer transient delays associated with the pilot control pressure transmission.

A variation of the two-stage proportional valve with pressure reducing pilot valve pair is shown in Figure 7.67. The differences between the two versions (Figure 7.66 and Figure 7.67) are as follows:

1. The pilot valve pair (pressure reducing valve pair) is in line with the main spool and there is a mechanical spring feedback between the pilot stage spools (one on each side of main spool) in addition to the hydraulic pressure feedback.

2. The main spool is actuated by reducing the pilot control pressure on one side as oppose to increasing it in the previous version. In order to move the main spool to left, the solenoid on the left side is energized and pilot pressure is dropped to a lower value proportionally on that side. Hence, the main spool moves to the left until the centering spring on the left side balances the forces on both sides. When the current to the solenoid is reduced, the pilot control pressure increases and pushes back the main spool to neutral position. The operation for shifting the main spool to the right follows a similar relationship.

3. The centering springs follow the main spool to the pilot side but not into the main spool body side. As a result, when the main spool shifts to left, it works against only the centering spool on the left side. When it shifts to right side, it works against only the centering spool on the right side.

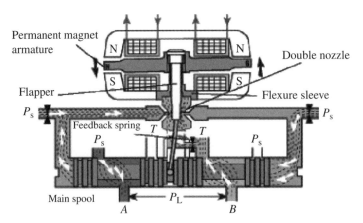

Permanent magnet
armature

N N

S S

Flapper

Double nozzle

Flexure sleeve

P_s

P_s

Feedback spring T T

P_s

P_s

Main spool

P_L

A B

FIGURE 7.68: Double nozzle flapper type two-stage servo valve – the most popular two-stage servo valve type. Courtesy Moog Inc.

The double nozzle flapper design is the most common two-stage servo valve used in high performance applications. This is the most accurate, but expensive, type of servo valve. Stage 1 is the torque motor plus the double nozzle flapper mechanism, stage 2 is the main valve spool. Figure 7.68 shows the cross-sectional picture and operating principle of a two-stage double nozzle flapper type valve. A torque motor is a limited rotation permanent magnet electric motor with a coil winding. The direction of the current in the winding determines the direction of the torque generated. The permanent magnet strength and air gap between the armature and the winding determines the current to torque gain. Hence, the current magnitude determines the torque magnitude. The armature and flexure tube are mounted around a low friction pivot point about which they rotate as the motor generates torque. When current is applied to the torque motor winding, the permanent magnet armature rotates in the direction based on the current direction, and by an amount proportional to the current magnitude. The flexure sleeve allows the armature–flapper assembly to tilt. As a result, the nozzle at the tip of the flapper is changed and hence a different pressure differential is created between the two sides. The pressure on the side of the nozzle where the flapper is closer gets larger. This pressure differential moves the main spool and along with it the feedback spring until the pressure is balanced on both sides of the flapper at the double nozzle point. At this time, the main spool is positioned in proportion to the input current. The input current is translated proportionally into the main spool displacement. Notice that due to the small nozzles at the double nozzle and flapper interface, the filtration of the fluid at the pilot stage is very important in order to avoid contamination related failures of the valve operation at the pilot stage.

The jet pipe valve design is very similar to the double nozzle flapper design, except that the pilot stage amplification mechanism is different (Figure 7.69). Stage 1 is the torque motor plus the jet pipe nozzle and two receiver nozzles. Stage 2 is the main spool. The pilot pressure is fed through the jet pipe nozzle which directs a very fine stream of fluid to two receivers. When the current in the torque motor is zero (null position), the jet pipe directs this very fine stream of "jet" flow equally to both receivers and there is a balance between the two control ports holding the main spool in null position. When the current is applied, the torque motor deflects the armature and the jet pipe proportionally. As a result, the jet pipe directs different amounts of fluid stream to two receivers, and hence creates a pressure differential in the control ports (two sides of the main spool). This pressure differential moves the main spool until the pressure differentials in the control ports are stabilized. The feedback from the main spool movement to the jet pipe is provided either mechanically by

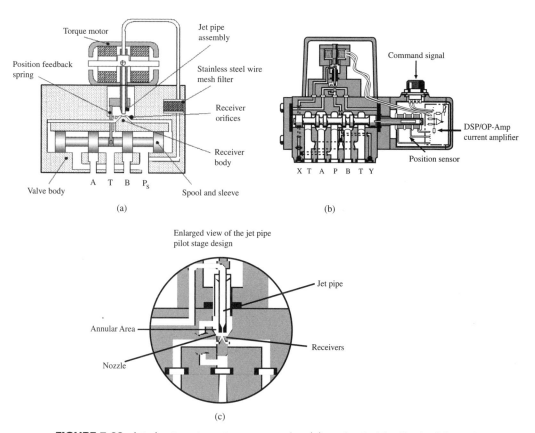

FIGURE 7.69: Jet pipe type two-stage servo valve: (a) mechanical feedback of the main stage spool position, (b) electrical sensor feedback of main stage spool position, (c) enlarged view of the jet pipe and receiver nozzles. In steady-state, the current of the torque motor is proportional to the main spool displacement. Courtesy Moog Inc.

a flexure spring between the nozzle and the main spool position or electrically by a spool position sensor (typically an LVDT) and a closed loop spool position control algorithm used to determine the torque motor current (Figure 7.69).

In general, the jet pipe design has lower bandwidth, requires lower manufacturing tolerances, and is able to tolerate more contamination compared with the double nozzle flapper servo valve. The jet pipe pilot stage can support a larger pilot force than the double-nozzle flapper valves, and hence tend to be used in large valves more often. In high flow rate applications (i.e., 1500 l/min), a double nozzle flapper or jet pipe valve is used as a pilot stage of a three-stage servo valve. In modern three-stage proportional or servo valves, the position loop on the main spool (third stage) is generally closed using an electrical position sensor. Notice that in both types of servo valves, the pressure for the hydraulic amplification stage can be provided either from a separate pilot pressure supply source or can be derived from the main supply line pressure using a pressure reducing valve or orifice. Filtration of the fluid, especially in the hydraulic amplifier stage (pilot stage), is very important since the nozzle and receivers have very small diameter dimensions and are prone to contamination. Flow orifices in the main stage spool are rather large and less prone to contamination. Therefore, filtration of the fluid is most important for the pilot stage.

Single-stage servo direct drive valves (DDV) currently support up to about 150 l/min (lpm) flow rate (i.e., Moog D633 through D636 series). Two-stage jet pipe

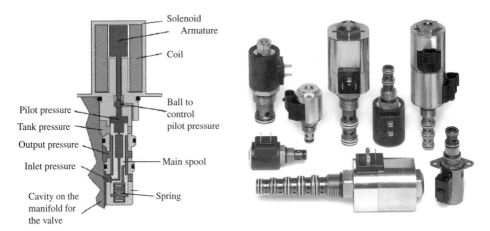

FIGURE 7.70: Examples of cartridge valves. The cross-section figure shows a cartridge valve with a sliding spool type metering element. Reproduced with permission from HydroForce (www.hydraforce.com).

servo valves support flow rates up to about 550 lpm at 70 bar pressure drop across the valve (i.e., Moog D661 through D664 series). A three-stage valve where the first two stage is either a double nozzle flapper or jet pipe type servo valve can support up to 1500 lpm flow rate (i.e., Moog D665 and D792 series).

Cartridge valves are designed to be assembled on a manifold. A manifold can be made of a single cartridge valve (single function manifold) or multiple valves (multi function manifold) which is generally the case. A manifold block may typically hold multiple cartridges and other types of valves (Figure 7.70, and Figure 7.71). Cavity sizes (diameter, depth, tread) on the manifolds are standardized so that cartridge valves from different manufacturers can be used interchangeably.

Cartridge valves can be categorized in terms of different criteria as follows:

1. Mechanical connection to the manifold:
   (a) Screw-in type which is installed by screwing valve threads into manifold cavity threads.
   (b) Slip-in type which is installed in the manifold by a bolted cover to the manifold. Screw-in type cartridge valves support flow rates up to about 150 lpm, and slip-in types support flow rates above 150 lpm. The slip-in type has the advantage over the screw-in type in that it does not squeeze the ports and hence achieves better repeatability in the assembly. There are seven standard slip-in cartridge valve sizes (specified by ISO 7368 and DIN 24342) where the nominal valve port diameter is 16, 25, 32, 40, 50, 63, and 100 mm, supporting flow rates in the range of 200–7000 lpm at about 5 bar pressure drop across the valve.

2. Metering component:
   (a) Spool type: the flow metering element can be a spool similar to a standard spool valve-body assembly. Spool type cartridge valves can be two-way, three-way, four-way or more.
   (b) Poppet type: the flow is controlled by a poppet and its seat. Poppet type cartridge valves are typically two-way valves. Cartridge valves use O-rings on the stationary component of the valve body in order to seal the valve ports from each other and minimize leakage. O-rings also help increase the damping effect on the valve, but add hysteresis to the valve input current–flow characteristics.

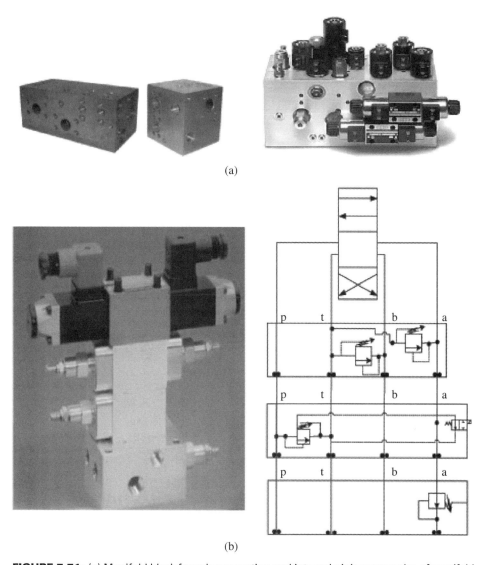

(a)

(b)

FIGURE 7.71: (a) Manifold block for valve mounting and internal piping: examples of manifold blocks, and a manifold block with valves mounted on it. (b) Sandwich style mounting blocks: typically used to integrate a directional or proportional valve function with additional functions such as relief valve, check valve, and pressure reducing valve functions. Reproduced with permission from Parker Hannifin.

3. Valve actuation methods:
  (a) Directly actuated by a solenoid with ON/OFF or proportional control.
  (b) Two-stage actuation with the pilot pressure in the final actuation stage being proportional to the first stage command signal. The command signal may be generated by manual controls or solenoid.
  (c) Three-stage actuation for very large flow rate applications.

In general, a manifold base plus a set of cartridge valves results in integrated hydraulic valve systems with low leakage and reliable operation. Cartridge valves can perform the directional flow and pressure control valve functions (i.e., proportional flow metering, pressure relief, check valve, unloading valve).

Cartridge valves for small flow rates can be directly actuated by a solenoid. Cartridge valves for large flow rates typically use a two-stage actuation mechanism which includes a first-stage electrical actuator and a second-stage pilot amplifier. For instance, let us consider a proportional spool type cartridge valve (Figure 7.70). The valve has three ports: inlet port (main input pressure, i.e., pump supply), tank port, and output port. The output port pressure is regulated to be somewhere between the inlet and tank port pressures. It is essentially a pressure reducing valve. The outlet pressure is regulated to a value between the two limit pressures (inlet and tank pressures) by the pilot stage presssure. The pilot stage supply pressure is obtained by a pressure reducing valve section from the main inlet pressure. The actual pilot pressure in the pilot control chamber is regulated by the poppet ball which is controlled by the solenoid. Therefore, the solenoid current regulates a pilot chamber pressure proportional to it which in turn positions the main stage spool proportional to the pilot pressure with the balancing force from the spring. The output pressure is proportional to the main spool position, hence to the solenoid current.

For very large cartridge valves, the valve actuator may take the form of a double nozzle flapper or jet pipe valve as the pilot stage. Such a valve can support very high flow rates (i.e., up to 9600 lpm by the Moog DSHR series). However, the addition of a pilot stage servo valve along with poppet position feedback makes the cost of the valve significantly higher than the cost of proportional spool-type valves. Screw-in cartridge valves provide check, relief, flow, pressure, and direction control functionalities for two, three, four or more port configurations.

The recent design of the *direct drive valve* concept uses electrical actuators with higher power density to directly shift the main spool of a two-stage valve. The valve then becomes a single stage with the flow capacity of a two-stage valve. The electrical actuator is either a rotary motor (hybrid permanent magnet stepper motor or bushless DC motor coupled with a rotary to linear motion conversion mechanism, i.e., helical cam, ball-screw), or a direct drive linear electric motor (i.e., linear force motor, linear brushless DC, linear stepper motor). The main benefit of the direct drive valve is that it eliminates the pilot pressure stage between the electrical actuator (solenoid in current designs) and the main spool motion. Due to power density and cost, current applications are still limited to valves with low flow rates. As the power density of electric actuators (permanent magnets and power amplifiers) gets larger and costs gets lower, the application of direct drive valve technology is likely to increase in the coming years.

7.5.6 Mounting of Valves in a Hydraulic Circuit

In hydraulic motion systems, a set of valve functions can be machined into a single *manifold block* (Figure 7.71). The manifold block, a set of cartridge valves, as well as other non-cartridge valves provide a very reliable and leakage free design for hydraulic valve groups. The manifold block provides the hydraulic connections from P and T ports to different valves internally. The hydraulic plumbing functions between P and T and service ports (A1, B1, A2, B2, etc.) are machined into the block. For instance, in a multi function circuit, parallel or series hydraulic connection of valve ports to P and T lines can be machined into the manifold. For each function, there is a valve (typically a screw-in cartridge valve) that controls (meters) the hydraulic flow between the ports. Technically, a single P port and T port is sufficient for the whole manifold block. There are as many A and B ports as there are the number of valve functions used. Furthermore, the manifold block is machined to accomodate open-center or closed-center functions by selectively plugging certain flow orifices on the manifold. Check valves, relief valves, and filter connections can also be built into the manifold design. The main advantages of the manifold block approach to multi valve hydraulic circuits instead of individual valves are

- the manifold block modularizes and simplifies hydraulic plumbing in installation and maintenance,
- it reduces leakage and can support higher pressure circuits,
- their compact size.

The manifold block port locations and sizes are standardized by ISO-4401 standards (i.e., ISO-4401-03, -05, -06, -07, -08, - 10 specify different number of ports, sizes, and locations for external connections). These are also refered as CETOP-03, ..., CETOP-10 standards. For pressures up to 3000 psi, aluminum manifolds, and for higher pressures (5000 psi) cast iron manifolds are recommended. In addition to manifold blocks, valves are also made with *stackable standard mounting plates* which makes connecting supply and tank lines between valves easier.

Other mounting methods for hydraulic plumbing are *sub-plates, inline bar manifolds, mounting plates, valve adaptors*, and *sandwich style mounting plates* . In particular, sandwich style mounting is typically used to integrate a directional or proportional valve and a number of relief, check, and pressure reducing functions in one stack. Sandwich type mounting plates also have DIN, ISO, and NFPA standard interface dimensions.

7.5.7 Performance Characteristics of Proportional and Servo Valves

The spool and orifice geometry around the null position is an important factor in servo valve performance. The spool may be machined so that at null position it overlaps, zero-laps, or underlaps the flow orifices. The zero-lapped spool is the ideal spool, but difficult to accomplish due to tight manufacturing tolerances. The overlapped spool results in a mechanical deadband between the current and flow relationship. The underlapped spool provides a large gain around the null position (Figure 7.72).

The deadband helps reduce the leakage in a valve, and hence allows the actuator to hold position better in open loop control in neutral position. In operator controlled open-loop speed control applications, where the speed of the actuator is desired to be proportional to the operator command signal which is proportional to the valve spool displacement, the deadband is quite often a desirable feature of the valve. Because small changes in the operator command due to human motion resolution or vibrations in the environment (hence the small changes of the valve spool position around neutral position) are desired to not create any motion, so that operator hand vibrations do not create unintended motion. On the other hand, in *closed loop* position or force or velocity control applications, the valve deadband acts as a nonlinearity and delay in the system response. This limits the closed loop control system bandwidth. The gain of the valve (flow rate divided by the spool displacement under a constant pressure drop condition across the valve) about null-position is *zero* if there is a finite deadband, that is 5 or 10% of total valve spool travel. Likewise, the gain of under-lapped spools is higher than those of zero-lapped or over-lapped spools. Hence, the closed loop system stability and dynamic performance of an electrohydraulic motion control system will be different for valves with different null-position characteristics. In a closed loop control system, having a different null-position gain for a valve is equivalent to having a different loop gain in the root locus analysis of the closed loop system. Different gains will result in different stability margins and closed loop pole locations.

Figure 7.73 shows various valve spool position-orifice area characteristics. The spool position versus the orifice area opening is directly determined by the way the valve is

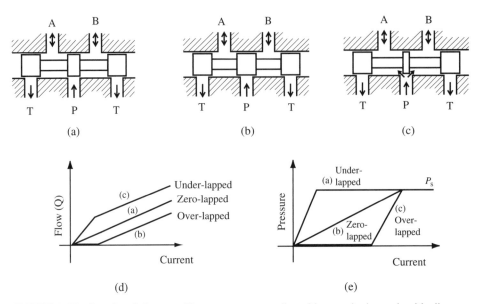

FIGURE 7.72: Spool and sleeve orifice geometry at null position: under-lapped, critically lapped, over-lapped. The resulting flow versus current and pressure versus current gains are also shown.

designed and machined. It is a mechanical geometric property of the valve. Let us assume that the spool actuation mechanism (i.e., solenoid or solenoid motion amplified by a pilot stage) generates a spool displacement proportional to the control (current) signal to the solenoid and that the pressure drop across the valve is constant. Then, the current signal to flow-rate relationship would have the same shape as the spool position versus the orifice area opening function. In closed loop precision position and force regulation applications,

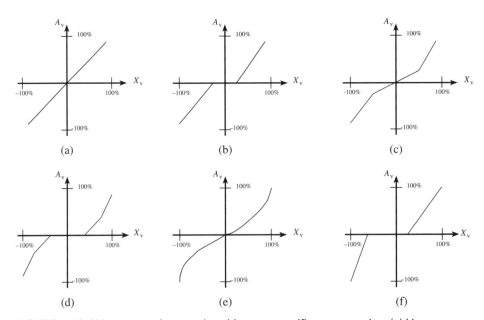

FIGURE 7.73: Valve geometries: spool position versus orifice area opening. (a) Linear (proportional valve), (b) valve with deadband and single gain, (c) valve with dual gain, (d) valve with deadband and dual gain, (e) nonlinear gain, (f) non-symmetric (asymmetric) gain.

the valve operates mostly in the vicinity of the null-position. However, the valve also operates in its full range of spool displacement when it is commanded to make a fast, large displacement or force changing motion. As a result, the valve spool position versus orifice opening area function of a valve for the full range of spool displacement is important. In ideal proportional and servo valves, this relationship is linear as shown in Figure 7.73a. Many open loop applications, especially the ones with an operator in the loop, use valves with deadband, as shown in Figure 7.73b. Other applications require valve spool geometry with dual gain, as in Figure 7.73c and d or nonlinear gain, as in Figure 7.73e. It is the application requirement that dictates the type of valve that should be used. Furthermore, the spool position versus orifice area opening curve does not have to be symmetric between positive and negative direction of the spool displacement, as in Figure 7.73f. For instance, injection molding applications commonly require valves with non-symmetric spool geometry. It is important to note that the nonlinear relationship in the valve spool geometry (i.e., nonlinear function of orifice opening as a function of spool displacement, Figure 7.73e) can be inverted in real-time valve control software so that the command signal to flow rate relationship can be made linear by effectively using a control gain that is inversely proportional to the valve spool position-flow rate gain. In other words, we can achieve the steady-state valve characteristics shown in Figure 7.73a by using a valve shown in Figure 7.73e plus a real-time control algorithm that adjusts the current signal to the valve in a way that is inversely proportional to the actual valve gain. The transient response would be a little different due to the control algorithm and spool motion transient delays. The adjustable gain in the control algorithm is calculated as the ratio of the desired gain (Figure 7.73a) divided by the actual valve gain (Figure 7.73e) at the current signal level. This is referred to as the *inverse valve gain compensation* in control.

The null-position is defined as the position of the spool where the pressure versus spool position curve (ΔP_L versus x_s or ΔP_L versus i) goes through zero value at the pressure axis. A null position test is conducted on a valve by fully blocking the ports between P-T and A-B. The null position of a valve is usually adjusted mechanically under a zero current condition. The mechanical adjustment of the null position may be provided by an adjustable screw on the valve which is used to move the spool around the neutral position by a small amount. This is part of the valve calibration procedure.

Null position performance is highly dependent upon the machining tolerances of the spool lands, sleeve and valve orifices, pressure, and temperature. Even high accuracy servo valves exhibit variation in their flow gain as a function of input current (under constant pressure drop) in the null-position vicinity. Recall that the flow is function of spool position and pressure,

$$Q = Q(x_{spool}, \Delta P_v) \qquad (7.239)$$

The valve characteristics around the null-position are represented by *flow gain, pressure gain, flow-pressure gain* (also called the *leakage coefficient*). The flow gain

$$K_q = \frac{\partial Q}{\partial x_{spool}} \qquad (7.240)$$

can vary between 50 to 200% of the nominal gain within the ±2.5% of maximum current value. The *pressure gain* of a valve is defined as the rate of change of output pressure as a function of solenoid current when output ports (A, B) are blocked (Figure 7.74),

$$K_p = \frac{\partial P_L}{\partial x_{spool}} \qquad (7.241)$$

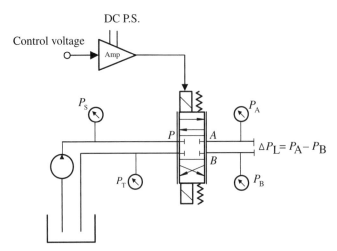

FIGURE 7.74: Test setup for measuring spool null position characteristics: pressure gain when valve output ports are blocked.

The *flow-pressure gain* is defined as

$$K_{pq} = \frac{\partial Q_L}{\partial P_{spool}} = \frac{K_q}{K_p} \tag{7.242}$$

The output pressure under this condition reaches the supply pressure very quickly. For a under-lapped spool, the output pressure reaches the supply pressure within 1 to 2% of maximum current. For zero-lapped spools, the same results are achieved within about 3 to 4% of maximum current. The over-lapped spool has similar behavior to that of the zero-lapped spool after it passes over the deadband region where the pressure gain is zero (Figures 7.72, 7.74). The pressure gain of a valve around its null-position is very important in servo position control applications since it strongly affects the stiffness of the positioning loop against external forces.

It is important to note that the null spool position is affected by the variations in supply pressure and temperature. The variation in null-position as a function of supply pressure and temperature is called *pressure null shift* and *temperature null shift*, respectively.

At the null-position, if the valve connects the pump port to the tank port and hence continues to circulate the fluid between pump and tank, even though the load does not require any, it is called a *open-center* valve (Figure 7.3). If the valve blocks the flow from pump to tank at null position, it is called a *closed-center* valve (Figure 7.62). An actual closed-center valve still exhibits some open-center characteristics due to leakage. Notice that an open-center valve can be used with a fixed displacement pump (Figure 7.3). The control task is simpler, but the system is not energy efficient since it continuously circulates flow in the system even if the load does not require it. A closed-center valve requires a variable displacement pump so that when the valve is at null-position, the pump is de-stroked to stop the flow or provide just enough flow to make up for the leakage (Figure 7.62). It is energy efficient since it provides pressurized fluid flow on demand and shuts it off when there is no demand. The control task is a little more complicated since both valve and pump must be controlled and coordinated to avoid large pressure spikes (i.e., closing the valve to null-position but keeping the pump running at high displacement will result in pressure spikes in the system and will most likely blow relief valves). Because of this, some degree of open-center characteristics is built into the valve, circulating a small percentage of its rated flow at null-position in order to simplifiy the coordinated control of valve and pump and increase system fault tolerance against the control timing errors between valve and pump.

The output flow, Q, from a valve is a function of orifice area (A_{valve}) and pressure differential across the port (ΔP_{valve}). The orifice area is a function of the spool displacement (x_{spool}) and the geometric design of the spool–orifice geometry. The spool displacement is proportional to the solenoid current (i_{sol}). Hence,

$$Q = Q(A_{valve}, \Delta P_{valve}) \tag{7.243}$$

where,

$$A_{valve} = A_{valve}(x_{spool}) \tag{7.244}$$

$$x_{spool} = K_{ix} \cdot i_{sol} \tag{7.245}$$

Let us consider a proportional valve, with a flow rating of Q_{nl} at maximum current, i_{max}, and no-load conditions (all of the supply pressure is dropped across the valve). Then the flow rate at any pressure drop (ΔP_{valve}) and solenoid current (i_{sol}) can be expressed as (Figure 7.75 and 7.103a type valve)

$$Q/Q_{nl} = (i_{sol}/i_{max})\sqrt{\Delta P_{valve}/P_s} \tag{7.246}$$

$$= (i_{sol}/i_{max})\sqrt{1 - (\Delta P_l/P_s)} \tag{7.247}$$

where the pressure drop across the valve is difference between the pump and tank pressure (net supply pressure) minus the load pressure (Figure 7.75),

$$\Delta P_{valve} = P_s - P_t - \Delta P_l \tag{7.248}$$

$$= P_s - P_t - |P_A - P_B| \tag{7.249}$$

$$= (P_s - P_A) + (P_B - P_t) \ or \tag{7.250}$$

$$= (P_s - P_B) + (P_A - P_t) \tag{7.251}$$

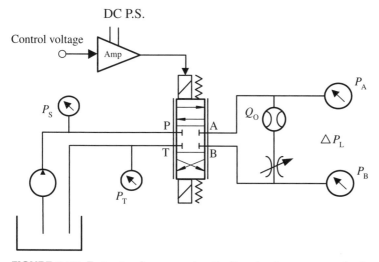

FIGURE 7.75: Test setup for measuring the flow, load pressure, and solenoid current relationship. In order to measure the flow-current relationship (approximate performance number to represent this relationship is flow gain, K_p), set the load pressure to a standard value. A standard way of measuring the flow-current relationship for servo and proportional valves is: set $P_s - P_T$ to a standard value (i.e., 1000 psi for servo valves, 150 psi for proportional valves), connect port A to port B directly (hydraulic short circuit, $\Delta P_L = 0$), hence pressure drop across the valve $\Delta P_v = P_s - P_T - \Delta P_l = P_s - P_T$. If $P_s - P_T$ cannot be set to a standard desired value, then connect port A and port B with an adjustable restriction ($\Delta P_l \neq 0$) so that the pressure drop across the valve is a standard value.

Let us assume that the supply pressure is constant. When the load pressure is constant, the flow rate is linearly proportional to the current in the solenoid. Similarly, for a constant solenoid current (equivalently a constant spool orifice opening), the flow rate is related to the load pressure with a square root relationship. As the load pressure increases, the available pressure drop across the valve decreases and hence the flow rate decreases. When the load pressure is the same as the tank pressure, all of the supply pressure is lost across the valve and maximum flow rate is achieved, which is called the *no-load flow*. However, the flow at the output of the valve has no pressure to exert force or torque to the actuator (cylinder or rotary hydraulic motor). Therefore, in a good design, part of the supply pressure is used to support the necessary flow rate across the valve, and part of it is used to provide a load pressure to generate actuator force/torque,

$$P_s - P_t = \Delta P_{valve} + \Delta P_l \tag{7.252}$$

In practice, a valve flow rating (Q_r) is given at a standard pressure drop across the valve ($\Delta P_r = 1000$ psi for servo valves and 150 psi for proportional valves) when the spool is fully shifted. The flow rate of the valve (Q) for a different pressure drop (ΔP_{valve}) across the valve when the spool is fully shifted can be obtained approximately by

$$Q = Q_r \cdot \sqrt{\Delta P_{valve}/\Delta P_r} \tag{7.253}$$

For valve measurements, P_s and P_t are set to constant values, and pressure drop is set to a number of discrete values (Figure 7.75). For each value of the load pressure (ΔP_L), current (i) is varied from zero to maximum value (i_{max}), and flow rate (Q) is measured. The results are plotted as valve flow rate, load pressure, and current relationship as shown in Figure 7.76 for a fixed P_s and P_t values. In Figure 7.76, the valve deadband is clearly observed.

The flow–current–pressure relationship given above neglects the leakage flow in the valve. Servo and proportional valves used in precise positioning applications generally

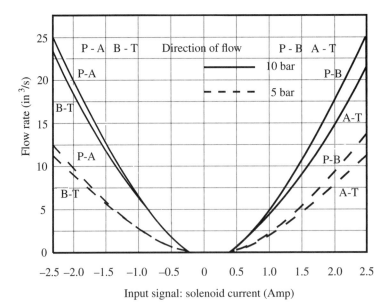

FIGURE 7.76: Valve flow rate versus solenoid current for different pressure drops across ports. A four-way proportional directional valve is considered for different pressure drop values. Servo valves have much smaller deadband and the flow rate-current relationship for a constant pressure drop is more linear compared to those of proportional valves.

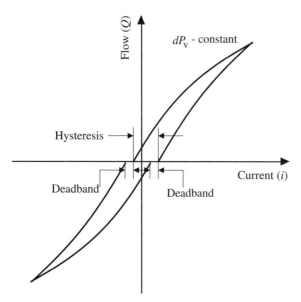

FIGURE 7.77: Non-ideal characteristics of a valve current–flow rate relationship under constant pressure drop across the valve: hysteresis, deadband, offset.

have critically lapped (almost zero-lapped) spools, hence have measurable leakage at zero spool displacement. When the current is zero and the spool is nominally at null-position, as the pressure drop across the valve is increased, the leakage flow increases. Taking this fact into account, the flow–current–differential pressure relationship can be expressed as (Figure 7.78a, $i = 0$ curve)

$$Q/Q_{nl} = (i/i_{max})\sqrt{1 - (\Delta P_L/P_s)} - K_{pq} \cdot (\Delta P_L/P_s) \qquad (7.254)$$

where the last term accounts for the leakage when the spool is around the null-position, K_{pq} is the leakage coefficient. For different values of constant current, flow versus the load pressure curves are shown in Figure 7.78a. When the load pressure equals the pump pressure, the flow rate is the leakage flow at all values of spool displacement. Let us focus on the flow–current–differential pressure relationship around the two extreme conditions:

1. no-load condition, $\Delta P_L = 0$, and
2. blocked-load condition.

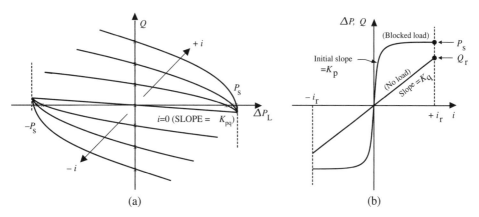

FIGURE 7.78: (a) Valve flow–current–differential pressure drop relationship, (b) no-load flow-current and blocked load–pressure–current relationship.

The blocked-load condition data plotted in Figure 7.78b is also conveyed in Figure 7.78a. The data for the port pressure versus current (Figure 7.78b) is same as the data on Figure 7.78a for flow rate equal to zero cases ($Q = 0$) for different values of i and ΔP_L. These are the data points obtained for ΔP_1 and i along the x-axis intersections. The data for the no-load flow versus current on Figure 7.78b is obtained from Figure 7.78a for zero values of load pressure, $\Delta P_L = 0$. These are the data points obtained for Q and i along the y-axis intersection.

The current versus the no-load flow, and the current versus the pressure drop across the valve when the load is blocked are shown in Figure 7.78b. Notice that the K_q is the no-load case current-to-flow gain (or flow gain), and K_p is the blocked load case current-to-pressure gain (or pressure gain). The flow-pressure gain (leakage coefficient) is $K_{pq} = K_q/K_p$.

Some of the non-ideal characteristics of a proportional valve are illustrated in Figure 7.77. They are

- deadband due to friction between the spool and sleeve, and leakage,
- hysteresis due to the magnetic hysteresis in the electromagnetic circuit of the solenoid (or torque motor), and
- zero-position current bias due to manufacturing tolerances in the spool and feedback spring.

In fact, the source of deadband and hysteresis in the valve cannot be exactly separated into friction and magnetic hysteresis. It is their combined effects that create the deadband and hysteresis in the valve input–output behavior.

Valve control can be based on regulating flow, pressure, or both. The control variable of an EH valve is the current. The transient response relation between the control signal (solenoid current command) and flow rate (or spool displacement) for a valve is characterized by small signal step response and frequency response (magnitude and phase between current command for an EH valve) and output signal (flow rate or spool displacement) as a function of frequency (Figure 7.79). The frequency response of a valve is also a function of the magnitude of the input signal as well as the supply pressure. The frequency response is typically measured under no-load conditions. Notice that the small signal step response of a servo valve is in the order of a few milliseconds and the bandwidth in the order of a few hundred Hertz. As the valve flow rating gets larger, the bandwidth of a given valve type gets smaller. The dynamic bandwidth between the electric actuator (solenoid, torque

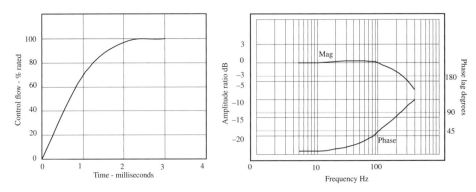

FIGURE 7.79: Dynamic characteristics of a servo valve. (a) Step response and (b) frequency response.

motor, or linear force motor) current and the main spool displacement is determined by two factors:

1. the bandwidth of the electric actuator,
2. the bandwidth of the pilot amplification stage (in direct drive valves, there is no pilot stage).

For a given valve, as the pilot stage supply pressure increases, the bandwidth of this stage also increases slightly.

The pilot stage may be supplied either by an external pilot pump or derived internally from main line pressure. When the pilot supply is derived internally from the main line, the flow goes through a pressure reducing valve in order to regulate a constant pilot pressure supply so that the pilot pressure does not fluctuate with the supply pressure (i.e., if the main pressure supply is a load sensing pump, as the load varies, the main supply pressure varies). The pilot supply would vary by a small amount as the supply pressure varies. However, if the modulating bandwidth of the pressure reducing valve is much higher than the main valve and the actuator, the transient effect of such variation in the pilot pressure is insignificant. For instance, when the pilot pressure is derived from a main line pump which is controlled using load sensing feedback, the lowest stand-by pressure setting of the pump should be above a certain minimum value to make sure the pilot supply is properly maintained.

Pressure Regulating Servo Valves   A servo valve can be locally controlled to regulate the load pressure. Figures 7.80 and 7.81 show two types of valves controlled to regulate the load pressure: one is a completely mechanical pressure feedback system (Figure 7.80), the other includes a sensor and an embedded digital controller (Figure 7.81). Under a constant solenoid current condition, the main spool is positioned in steady-state such that

$$\Delta P_{12} = K_{\mathrm{pi}} \cdot i_{\mathrm{sol}} \tag{7.255}$$

$$\Delta P_{\mathrm{AB}} \cdot A_{\mathrm{fb}} = \Delta P_{12} \cdot A_{12} \tag{7.256}$$

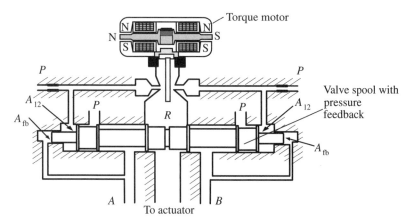

FIGURE 7.80: Pressure controlled valve with hydro-mechanical means of feedback control (Series 15 servo valve by Moog Inc). The pressure differential between the output ports A and B, which is the pressure differential applied to the load, is proportional to the current applied to the torque motor.

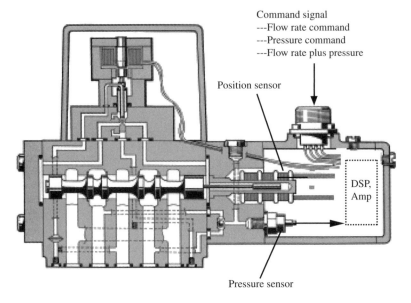

FIGURE 7.81: Pressure and flow controlled valve with electrical means of control actuation.

where ΔP_{12} is the pilot pressure differential acting on spool area A_{12}, ΔP_{AB} is the load pressure differential between the two ports A and B, and A_{fb} is the cross-sectional area of the spool where the load pressure differential acts. The objective for a pressure control valve is to provide an output pressure differential that is proportional to the solenoid current, as achieved by this valve,

$$\Delta P_{AB} = \frac{A_{12}}{A_{fb}} \cdot K_{pi} \cdot i_{sol} \tag{7.257}$$

In general, the dynamic characteristics of an EH system with a pressure control valve are highly dependent on the load dynamics.

The digitally controlled version of the valve can easily be modified in software to implement a flow control plus a programable pressure limit control logic (Figures 7.81, 7.82). Once the spool position and pressure sensors are available, the spool actuation of the

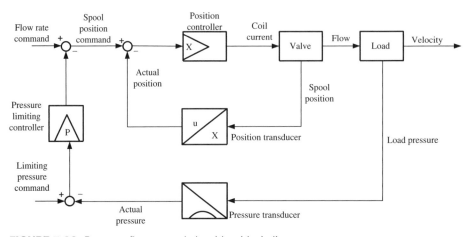

FIGURE 7.82: Pressure-flow control algorithm block diagram.

valve can be controlled to achieve either flow or pressure regulation by simply selecting a different software mode. The trend in valve technology is to implement the mechanical control mechanisms by mechatronic control components, that is instead of using hydro-mechanical control mechanisms, use electrical sensors, actuators, and embedded digital controllers.

7.6 SIZING OF HYDRAULIC MOTION SYSTEM COMPONENTS

Let us consider the component sizing problem for a one-axis hydraulic motion control system shown in Figure 7.83. The question of component sizing is to determine the size of the

1. pump(s),
2. valve(s) which includes main flow control valves, pilot valves, as well as safety valves,
3. hydraulic motor(s) and/or cylinder(s),
4. the size of the hydraulic tank volume,
5. filtration circuit and filters,
6. hydraulic pipe diameters and their routing (minimizing bends and length),
7. hydraulic oil cooler if needed.

Over-sizing of components results in lower accuracy, lower bandwidth, and higher cost. Under-sizing of components results in lower than required power capacity.

As a rule of thumb, a servo valve size should be selected such that the pressure drop at the valve at maximum flow should be about 1/3 of the supply pressure. In general, such

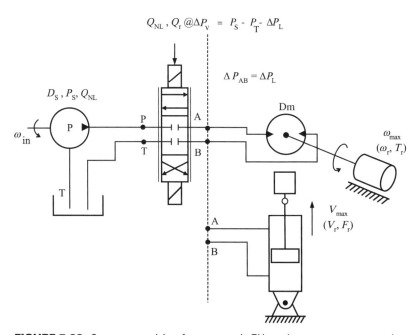

FIGURE 7.83: Component sizing for a one-axis EH motion system: pump, valve, cylinder (or rotary hydraulic motor) sizing for given load requirements.

a balanced distribution of supply pressure between valve and load is possible in servo valve applications. In proportional valve applications, especially in large construction equipment, the pressure drop across the valve can be much less than $1/3$ of the supply pressure. In such applications, precision motion control is not critical. It is desirable that more of the hydraulic power is delivered to the load and less is lost in the process of metering it. If the valve is sized too large and the pressure drop across it is not large enough, the flow will not be modulated well until the valve almost closes. As a result, resolution of the motion control will be low. If the valve is too small, it will not be able to support the desired flow rates or the pressure drop across it will be too large. The valve is the flow metering component. It is desirable to have a good resolution in metering the flow. In general, the higher the metering resolution required, the higher the pressure drop that must exist across the valve. Higher pressure drops lead to higher losses and lower efficiency. Therefore, metering resolution and efficiency are two conflicting variables in a hydraulic control system. A good design targets an acceptable metering resolution with minimal pressure drop.

There are two main variables which determine the component size requirements: *required force* and *speed*. Another way of stating that is *required pressure* and *flow rate*. Based on these two requirements, supply (pump) and actuator (cylinder and motors) are sized. The valve, which modulates and directs the flow, is sized to handle the flow rate and pressure required. The control question in hydraulic systems always involves the control of flow rate and direction, and/or the pressure.

In general, the load speed requirement dictates the flow rate, and the load force/torque requirement dictates the operating pressure. Consider a cylinder as the hydraulic actuator. A given load force/torque requirement can be met with different pressure and cylinder cross-sectional areas. For instance, in order to provide a certain force, many pressure and cylinder area combinations are possible, $F = p \cdot A$. Similarly, a given load speed requirement can be met with different combinations of flow rate and cylinder cross-sectional areas, $Q = V \cdot A$. The trend in industry is to use higher operating pressures which result in smaller components. However, high pressure circuits require more frequent maintenance and have a lower life cycle.

The application requirements typically specify the load conditions in terms of three variables: the no-load maximum speed (w_{nl} for a rotary actuator or V_{nl} for a translational actuator) and speed at rated load (w_r at T_r or V_r at F_r). For rotary and linear system the specifications are, respectively,

$$\{w_{nl}, (w_r, T_r)\} \quad or \quad \{V_{nl}, (V_r, F_r)\} \tag{7.258}$$

Recall that the relationships between the hydraulic variables and mechanical variables for *the actuator* (hydraulic rotary motor or hydraulic linear cylinder) are,

$$w = Q/D_m \quad or \quad V = Q/A_c \tag{7.259}$$

$$T = \Delta P_L \cdot D_m \quad or \quad F = \Delta P_L \cdot A_c \tag{7.260}$$

where ΔP_L is the pressure difference between the two sides of the actuator (two sides of the cylinder), D_m is the hydraulic motor displacement (volumetric displacement per revolution), A_c is the cross-sectional area of the cylinder (assuming it is a symmetric cylinder).

When the load force (or torque) is specified, the total force output from the actuator should include the force to be exerted (F_{ext}) on the external load (i.e., pressing force in a press or testing application), plus the force needed to accelerate the total moving mass (cylinder piston, rod, and load inertia, $m \cdot \ddot{x}$), and finally some force needed to overcome friction, (F_f),

$$F = F_L = m \cdot \ddot{x} + F_f + F_{ext} \tag{7.261}$$

Estimating the friction force, F_f, is rather difficult. One common way to take it into account is to use a safety factor, SF, as follows,

$$F = F_L = SF \cdot (m \cdot \ddot{x} + F_{ext}) \tag{7.262}$$

where SF is the safety factor (i.e., $SF = 1.2$). For rotary actuators, analogous relationships hold,

$$T = T_L = SF \cdot (J \cdot \ddot{\theta} + T_{ext}) \tag{7.263}$$

where J is the total mass moment of inertia of rotary load, $\ddot{\theta}$ is angular acceleration, T_{ext} is the torque to be delivered to the load. The flow–current–pressure differential relation for a proportional or servo valve which is positioned between the pump supply line and load line,

$$Q/Q_{nl} = (i/i_{max})\sqrt{1 - (\Delta P_L/P_S)} \tag{7.264}$$
$$= (i/i_{max})\sqrt{(\Delta P_v/P_S)} \tag{7.265}$$

where ΔP_L is the load pressure (i.e., in the case of a linear cylinder this is the pressure difference between the two sides of the cylinder, and in the case of a rotary hydraulic motor this is the inport and outport pressure differential). In other words, ΔP_L is the pressure differential between the A and B ports of the valve. $\Delta P_v = P_S - \Delta P_L - P_T$ is the pressure drop across the valve, P_S is supply pressure from pump, P_T tank return line pressure. Quite often, P_T is taken as zero pressure. Q_{nl} is the flow through the valve when it is fully open ($i = i_{max}$) and the pressure drop across the valve is P_S. When there is no-load, all of the supply pressure is dropped across the valve. In order to determine the flow rate (Q_r) of a valve at a particular pressure drop ($P_S - \Delta P_L$) when it is fully open, we set $i = i_{max}$ in the above equation,

$$Q_r/Q_{nl} = \sqrt{1 - (\Delta P_L/P_S)} \tag{7.266}$$
$$= \sqrt{(\Delta P_v/P_S)} \tag{7.267}$$

A proportional valve is rated in terms of its flow rate capacity in catalogs for a pressure drop of $\Delta P_v = 150\,\text{psi}$ across the valve. A servo valve is rated for $\Delta P_v = 1000\,\text{psi}$.

Component Sizing Algorithm 1

Step 1: Determine the performance specifications based on load requirements:

$$\{w_{nl}, (w_r, T_r)\} \quad \text{or} \quad \{V_{nl}, (V_r, F_r)\}$$

Step 2: Pick either displacement per unit stroke for the actuator (D_m or A_c) or the supply pressure, P_s. Let us assume we pick the displacement (D_m or A_c). This selection specifies the actuator size (hydraulic motor or cylinder).

Step 3: Calculate $Q_{nl}, Q_r, \Delta P_L$ using Equations 7.259 and 7.260 for the specifications.

$$Q_{nl} = D_m \cdot w_{nl} \qquad Q_{nl} = A_c \cdot V_{nl} \tag{7.268}$$
$$Q_r = D_m \cdot w_r \qquad Q_r = A_c \cdot V_r \tag{7.269}$$
$$\Delta P_L = T_r/D_m \qquad \Delta P_L = F_r/A_c \tag{7.270}$$

Step 4: Assume $i = i_{max}$, calculate the P_s from Equation 7.266. Add 10–20% to the P_s as a safety factor.

$$Q_r/Q_{nl} = \sqrt{1 - (\Delta P_L/P_S)} \tag{7.271}$$

Step 5: Given the calculated value of Q_{nl}, and given the operating input shaft speed of the pump w_{in} which is determined by the mechanical power source device driving the pump (i.e., electric motor in industrial applications, internal combustion engine in mobile applications), calculate the pump displacement requirement, D_p, using Equation 7.259. Select the proper pump given Q_{nl}, P_s, D_p.

$$D_p = Q_{nl}/w_{in} \qquad (7.272)$$

Step 6: Given the calculated Q_{nl}, P_s, Q_r, calculate the rated flow for the valve, Q_v, at its rated pressure ΔP_v for $i = i_{max}$ using Equation 7.266. Notice that rated flow for a valve is defined at a standard pressure drop across the valve (i.e., typically $\Delta P_v = 1000$ psi for servo valves, $\Delta P_v = 150$ psi for proportional valves) and at the fully shifted spool position ($i = i_{max}$). Select the proper valve size based on the calulated Q_v and assumed ΔP_v.

$$Q_v = Q_{nl}\sqrt{\Delta P_v/P_S} \qquad (7.273)$$

In summary, the load conditions determine the specifications:

- Given $\{w_{nl}, (w_r, T_r)\}$ (for linear actuator $\{V_{nl}, (V_r, F_r)\}$).
- As design choices, pick D_m hydraulic motor displacement (for linear actuator A_c cylinder cross-sectional area) and pick ΔP_v valve pressure rating.
- Then calculate the rest of the sizing parameters: $Q_{nl}, Q_r, \Delta P_L, P_s, D_p, Q_v$, which are the size requirements for pump, valve, and motor/cylinder using Equations 7.259, 7.260, 7.266.

Component Sizing Algorithm 2

Step 1: Determine the performance specifications based on load requirements:
$$\{w_{nl}, (w_r, T_r)\} \text{ or } \{V_{nl}, (V_r, F_r)\}$$

Step 2: Pick either displacement per unit stroke for the actuator (D_m or A_c) or the supply pressure, P_s. Let us assume we pick pump pressure, P_s.

Step 3: Pick the pressure differential to be delivered to the actuator to drive the load ΔP_L, calculate hydraulic motor displacement (D_m) or cylinder cross-sectional area (A_c).

$$D_m = T_r/\Delta P_L \qquad A_c = F_r/\Delta P_L \qquad (7.274)$$

Step 4: Calculate Q_{nl} from Equation 7.259 and Q_r at $\Delta P_L, P_s$ and $i = i_{max}$ from Equation 7.266.

$$Q_{nl} = D_m \cdot w_{nl} \qquad Q_{nl} = A_c \cdot V_{nl} \qquad (7.275)$$
$$Q_r = Q_{nl}\sqrt{1 - (\Delta P_L/P_s)} \qquad (7.276)$$

Step 5: Given the calculated value of Q_{nl}, and given the operating input shaft speed of the pump w_{in}, calculate the pump displacement requirement, D_p, using Equation 7.259. Select the proper pump given Q_{nl}, P_s, D_p.

$$D_p = Q_{nl}/w_{in} \qquad (7.277)$$

Step 6: Given the calculated Q_{nl}, P_s, Q_r, calculate the rated flow for the valve, Q_v, at its rated pressure ΔP_v. Select the proper valve size based on the calulated Q_v and assumed ΔP_v.

$$Q_v = Q_{nl}\sqrt{\Delta P_v/P_s} \qquad (7.278)$$

In summary, the load conditions determine the three specifications:

- Given $\{w_{nl}, (w_r, T_r)\}$ (for linear actuator $\{V_{nl}, (V_r, F_r)\}$).
- As design choices, pick P_s hydraulic pump rated pressure output and pick ΔP_L load pressure differential.
- Then calculate the rest of the sizing parameters: $D_m, D_p, Q_{nl}, Q_r, \Delta P_L, Q_v$, which are the size requirements for pump, valve and motor/cylinder.

Notice that the pump and motor displacement and pressure ratings are available in standard sizes, such as

$$D_p, D_m = 1.0,\ 2.5,\ 5.0,\ 7.5\,\text{in}^3/\text{rev} \tag{7.279}$$
$$P_s = 3000\text{–}5000\,\text{psi} \tag{7.280}$$

Similarly, the valves are available at standard rated flow capacity (measured at $\Delta P_v = 1000\,\text{psi}$ and at maximum current $i = i_{max}$ for servo valves) such as

$$Q_v = 1.0,\ 2.5,\ 5.0,\ 10\,\text{gpm}. \tag{7.281}$$

Based on the calculated sizes and safety margins, one of the closest sizes for each component should be selected with appropriate safety margin.

Example Consider a single axis EH motion control system as shown in Figure 7.85, which may be used for a hydraulic press. The system will be controlled in one of two possible modes:

- Mode 1: closed loop speed control where the commanded speed is obtained from a programmed command generator or from the displacement of the joystick. The cylinder speed is measured and a closed loop control algorithm is implemented in the electronic control unit (ECU) which controls the valve.
- Mode 2: closed loop force control where the force command is obtained from a programmed command generator or from the displacement of the joystick. The cylinder force is measured using a pair of pressure sensors on both ends of the cylinder.

In order for the operator to tell the ECU which mode to operate (speed or force) and from where to obtain the command signals, there should be two two-position discrete switches. Switch 1 OFF position means speed control mode, and ON position means force control mode. Switch 2 OFF position means the source of the command signal is programmed into the ECU, ON position means the joystick provides the command signal (Figure 7.84).

Select the proper components (components with appropriate technology such as variable displacement or fixed displacement pumps, ON/OFF or proportional or servo valves), and the proper size (calculate component size requirements) for pump, valve, and cylinder in order to meet the following specifications:

1. Maximum speed of the cylinder under no load conditions is $V_{nl} = 2.0\,\text{m/s}$.
2. Provide an effective output force of $F_r = 10\,000\,\text{Nt}$ at the cylinder rod while moving at the speed of $V_r = 1.5\,\text{m/s}$.
3. The desired regulation accuracy of the speed control loop is 0.01% of maximum speed.
4. The desired regulation accuracy of the force control loop is 0.1% of maximum force.

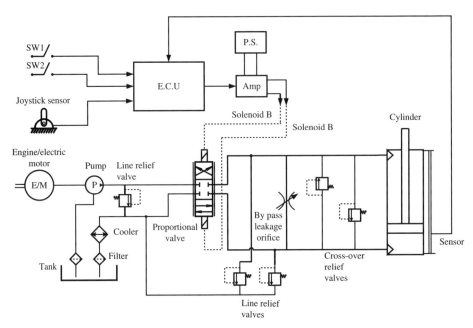

FIGURE 7.84: Control system components and circuit diagram for a single axis EH control system which operates in two modes using closed loop control: (i) speed mode, (ii) force mode.

In addition, select an operator input device (joystick for motion command) with the proper sensor, electronic control unit (ECU) with the necessary analog–digital signal interfaces, position and force sensors on the cylinder for closed loop control (Figure 7.85).

The range and resolution requirements for the measured variables determine the sensor and ECU interface requirements (DAC and ADC components). Let us assume that the maximum speed will be mapped to a maximum analog sensor feedback signal (i.e., 10 VDC), and the required accuracy is 0.01% of the maximum value which is 1 part in 10 000. In general, the measurement resolution should be at least 2 to 3 times better than the required regulation (control) accuracy. Therefore, we need a speed sensor with a data acquisition component that will have 1 part in 30 000 resolution. Let us assume that we have selected a magnetostrictive position sensor (also called *temposonic sensor*) from which speed information is digitally calculated. The output of the sensor can be scaled so that it gives 10 VDC when the speed is at 2.0 m/s. The sensor nonlinearity should be better than 0.01% of the maximum range. In addition, the analog to digital converter (ADC) must be at least 16-bit resolution in order to give a sampling resolution of better than 1/30 000 parts over the whole range (Figure 7.85).

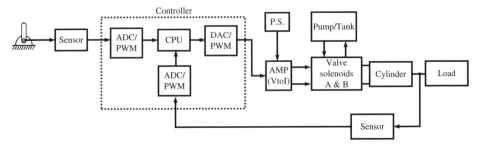

FIGURE 7.85: Components and block diagram of a single axis electrohydraulic motion control system. The controller is a programmable embedded computer.

The command signal (either in speed or force control mode) is obtained either from a real-time program or joystick. The real-time program can command motion to any accuracy desired, for example 1 part in 10 000 depending on the data type used in the control algorithm. If two-byte signed integers are used and scaled to cover the whole range, the commanded signal resolution can be 1 part in $(2^{15} - 1)$. If four-byte signed long integers are used in the servo control algorithm and numbers are scaled to cover the whole motion range, the commanded signal resolution can be 1 part in $(2^{31} - 1)$.

If the command signal is obtained from the joystick sensor, it is important to note that a human operator cannot command motion without shaking his/her hand better than 1% of the maximum displacement range of the joystick. Therefore, the resolution requirements of the joystick command are much smaller. Even if we provided a high resolution sensor and ADC, the human operator cannot actually change the command in finer resolutions than about 1 part in 100. Therefore, an 8-bit ADC converter with an analog speed sensor that can provide 10 VDC with an accuracy of 1% is sufficient. If an incremental encoder is used for the sensor, an encoder with 512 lines/rev also would have the sufficient range and resolution for the command signal.

The ECU should have resources in the following areas:

1. Speed of the CPU (microprocessor or DSP chip) in order to implement the control logic fast enough. Generally, this is not a problem with the current state of art.

2. Memory resources to store control code and sensory data for control purposes as well as for off-line analysis purposes (ROM or battery-backed RAM for program storage, RAM for real-time data storage and program execution).

3. I/O interface circuit: analog to digital converter (ADC), digital to analog converter (DAC), interface for discrete I/O (DIO), encoder interface, PWM signal input, and output interface. Notice that the ECU does not have to have all of these types of I/O interface, only the types required by the the selected sensor inputs and amplifier outputs.

4. Interrupt lines and interrupt handling software. Interrupts are key to the operation of most real-time systems, even though this application may be solved without use of interrupts.

In general, the higher the resolution (number of bits) of the ADC and DAC converter circuit, the better it is. If the programming is to be done for the servo loop by the designer, then it is preferable to have the ECU based on a microprocessor or DSP which has a C-compiler so that we can use a high level programming language to implement the servo control loops and use integer, long integer, or floating point data types. If we have to program the ECU in assembly language, then we have to manage the data and its size explicitly. In general, the speed, memory, and signal interface resources of the ECU are not limiting factors in the design.

Let us assume that the valve rating is specified for $\Delta P_v = 1000\,\text{psi}$ (6.8948 MPa $=$ 6.8948×10^6 Pa $= 6.8948 \times 10^6\,\text{Nt/m}^2$ pressure drop), and that the input shaft speed at the pump is $w_{in} = 1000\,\text{rpm}$. Following the component sizing algorithm, let us pick a cylinder diameter size, $d_c = 0.05\,\text{m}$, which determines the cross-sectional area, A_c. Using Equations 7.259–7.260, calculate the Q_{nl}, Q_r, and ΔP_L,

$$A_c = \pi d_c^2/4 = 0.0019635\,\text{m}^2 \approx 0.002\,\text{m}^2 \qquad (7.282)$$

$$Q_{nl} = V_{nl} \cdot A_c = 240\,\text{l/min} \qquad (7.283)$$

$$Q_r = V_r \cdot A_c = 180\,\text{l/min} \qquad (7.284)$$

$$\Delta P_L = F_r/A_c = 5 \times 10^6\,\text{Nt/m}^2 = 5\,\text{MPa} \qquad (7.285)$$

Assume $i = i_{max}$, and using Equation 7.266, calculate P_s and add 10% safety margin to it. Then calculate the necessary servo valve flow rating,

$$Q_r/Q_{nl} = \sqrt{1 - (\Delta P_L/P_s)} \tag{7.286}$$

$$\Delta P_L/P_s = 1 - (Q_r^2/Q_{nl}^2) = (Q_{nl}^2 - Q_r^2)/Q_{nl}^2 \tag{7.287}$$

$$P_s = \frac{Q_{nl}^2}{Q_{nl}^2 - Q_r^2} \cdot \Delta P_L \tag{7.288}$$

$$= 11.42\,\text{MPa} = 1658\,\text{psi} \tag{7.289}$$

$$Q_v = Q_{nl}\sqrt{\frac{\Delta P_v}{P_s}} \tag{7.290}$$

$$= 240 \cdot \sqrt{1000/1658}\,\text{l/min} \tag{7.291}$$

$$= 186\,\text{l/min} \tag{7.292}$$

where we assumed a pressure drop of 1000 psi across the valve for its flow rating. Since we know the maximum flow rate needed and the input speed to the pump, we can calculate the pump displacement, D_p, from Equation 7.259,

$$D_p = 0.140\,\text{l/rev} \tag{7.293}$$

In summary, the following component sizes are required: cylinder with bore diameter $d_c = 0.05$ m (bore cross-sectional area $A_c = 0.002\,\text{m}^2$), valve with flow rating of $Q_v = 200\,\text{l/min}$ or higher at 1000 psi pressure drop, and pump with displacement $D_p = 0.140\,\text{l/rev}$ and output pressure capacity of $P_s = 2000$ psi or higher value.

Finally, an amplifier with proper power supply is needed to drive the valve. An amplifier type which has a current feedback loop is preferred so that the current sent to the solenoid is maintained proportional to the signal from the ECU regardless of the variations in the power supply or solenoid resistance as a function of temperature. A commercial amplifier has tunable parameters to adjust the input voltage offset, gain, and maximum output current values (i.e., Parker BD98A or EW554 series amplifiers for servo valves plus a ±15 VDC power supply, such as Model PS15, which provides a DC bus voltage of ±15 VDC up to 1.5 Amp to the amplifier using a regular 85–132 VAC single phase line power). Similarly, if a variable displacement pump is used and the valve that controls the pump displacement is EH controlled, we would need an another amplifier-power supply matched to the size of the pump control valve. Notice that if the EH motion axis was to be controlled only through the joystick inputs, and no programming was necessary, the closed loop control algorithm can be implemented with analog OP-AMP circuits in the amplifier and the need for the ECU can be eliminated. The drive can have a PID circuit where the command signal from the joystick sensor and the feedback signal from the cylinder motion connects to the the input of the OP-AMP circuit (Figure 7.86). The error signal output of the OP-AMP is amplified as a current signal to the solenoids. However, the software programmability of the motion system is not possible in this case. One such drive is the Parker Series EZ595 used together with a Series BD98A amplifier, where the EZ595 implements the PID function and interfaces and BD98A implements the voltage to current amplification function.

List of components selected for this design:

1. *Pump and Reservoir:* Oil Gear model PVWH-60 axial piston pump, with a pressure rating of 2000 psi, and flow rate of 60 gpm when driven by an electric motor (i.e., AC induction motor) at 1800 rpm. The reservoir should handle 150 gallons (3 to 5 times the rated flow capacity per minute of the pump) of hydraulic fluid, fitted with filters

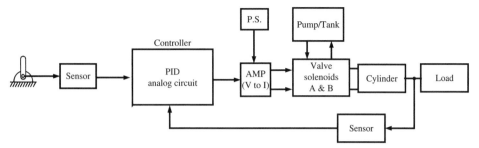

FIGURE 7.86: Components and block diagram of a single axis electrohydraulic motion control system. The controller is an analog op-amp which implements a closed loop PID control between the operator command, sensor signal, and valve-drive command.

in both return line and pump input line from the reservoir, as well as a cooler and pressure relief valve.

2. *Cylinder:* Parker industrial cylinder ("Series 2H") with 2.0 in bore, 40 in stroke, with threaded rod end for load connection, with base lugs mounting options at both ends,

3. *Valve:* MOOG servo valve model D791, rated flow upto 65 gpm at 1000 psi pressure drop across the valve when the valve is fully open, equivalent first-order filter time constant is about 3.0 ms.

4. *Valve Drive (Power Supply and Amplifier):* MOOG power supply and amplifier circuit matched to the servo valve: MOOG model "snap track" servo valve drives.

5. *Sensors:* Position sensor: linear incremental encoder. DYNAPAR series LR linear scale, range 1.0 m, 5.0 μ (0.0002 in) linear resolution, maximum speed 20 m/s, frequency response 1 MHz, quadrature output channels $A, B, C, \bar{A}, \bar{B}, \bar{C}$.

Pressure sensor: differential pressure sensor, Schaevitz series P2100, up to 5000 psi line pressure, and up to 3500 psi differential pressure, bandwidth in the 15 KHz range. Pressure sensors can be placed on the valve manifold.

6. *Controller:* PC based data aquisition card (National Instruments: NI-6111) with I/O interface capabilities for encoder interface, two channels of 16-bit DAC output, two channels of 12-bit ADC converter, 8 channels of TTL I/O lines. The control logic can be implemented on the PC and the data aquisition card is used for the I/O interface.

7. *Operator Input–Output Devices:* Joystick and Mode Selection Switches (Switch 1 and 2). One degree of freedom joystick with a potentiometer position sensor and analog voltage output in the 0 to 10 VDC range – Penny and Giles model JC 150 or ITT Industries model AJ3.

Example As an extension of the previous example, let us assume that we have selected two sensors for the hydraulic cylinder (Figure 7.87):

1. a position sensor (i.e., an absolute encoder or an incremental encoder plus two discrete proximity sensor to indicate mechanical travel limits and establish an absolute reference position: ENC1, PRX1, PRX2), and

2. a pair of pressure sensors which measure the pressure at both ends of the cylinder (P1, P2).

It is desired that the actuator is to operate in programmed mode and ignore the joystick input. The objective is that the actuator is to approach the load at a predefined speed until a certain amount of pressure differential is sensed between P1 and P2 signals.

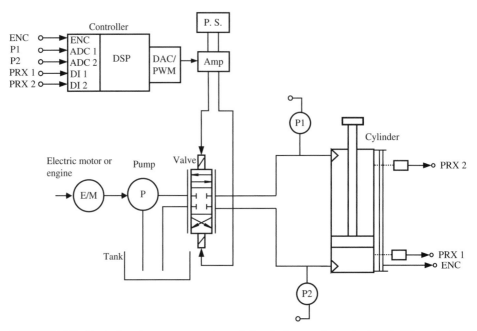

FIGURE 7.87: Control system components and circuit diagram for a single axis EH control system which operates in two modes using closed loop control: 1. speed mode, 2. force mode.

The position range of the actuator around which that is expected is known. When a certain pressure differential is sensed, the control system is supposed to automatically switch to force (pressure) regulation mode, maintain the desired pressure until a certain amount of time is passed or another discrete ON/OFF state sensor input is high. Let us assume that we will use the time period option to decide how long to apply the pressure. Then reverse the motion, and return to the original position under a programmed speed profile. Such EH motion control systems are used in mechanical testing of products, injection molding, nip roll positioning, and press applications.

 (i) Modify the design of the EH system used in the previous example.

 (ii) Draw a block diagram of the control logic and closed loop servo algorithms for both position and force servo modes. Suggest ways to make a "bumpless transfer" between the two modes.

There are two closed loop control modes: 1. position servo mode where the actual position is sensed using an encoder (ENC, PRX1, PRX2) and the desired position is programmed, 2. force servo mode where the force information is derived from the measured differential pressure (P1 and P2). The desired force profile (constant or time varying) is pre-programmed and its magnitude can be changed for each product. Therefore, we need two servo control algorithms (i.e., of PID type). Each loop has its commanded signal (desired position or desired force) and feedback signal (encoder signal or differential pressure sensor signal). At the power-up, the cylinder makes a home search motion sequence in order to find an absolute reference position (also called *home position*) so that every cycle is started from the same position. Using an absolute encoder, the controller can read the current position at power-up, and command the axis to move to the desired position for the home reference (using the position servo loop). If an incremental encoder is used, then an external

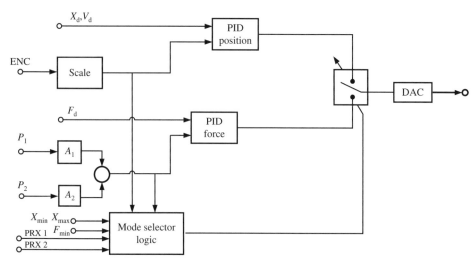

FIGURE 7.88: Real-time control algorithm for the EH control system using both position and force feedback and implement a "bumpless mode transfer."

ON/OFF sensor (i.e., proximity or photoelectric sensor) can be used to establish the absolute reference.

Once the starting position is accurately established after power-up (home motion sequence is completed), the system is to operate in position servo mode, followed by a force servo mode and then followed by position servo mode again to return to the starting position. Therefore, all that is needed is the logic to decide when to switch between the two modes (Figure 7.88).

Let us assume that we have the parameters for motion, and the I/O registers mapped to variables as follows,

Given: Programmable motion parameters

v_{d-fwd} desired forward motion speed

a_{d-fwd} desired forward motion acceleration

v_{d-rev} desired reverse motion speed

a_{d-rev} desired reverse motion acceleration

x_{min} minimum value of cylinder position where pressure is expected to rise

x_{max} maximum value of cylinder position where pressure is expected to rise

t_{press} time period for which to apply pressure.

I/O signal registers:

Enc accumulated encoder counts since the end of home cycle

P_1 pressure sensor signal 1

P_2 pressure sensor signal 2

Prx$_1$ ON/OFF sensor used to establish home reference

Prx$_2$ ON/OFF sensor used to establish end of force cycle

DAC digital to analog converter register.

Below is a pseudo-code for the control algorithm,

```
Home_Motion_Sequence() ;
while(true)
{
    1.  Wait until cycle start command is received, or
        trigger signal is ON or a pre-defined time
        period expires,
    2.  Enable position servo mode and move forward
        with desired motion profile,
    3.  Monitor pressure differential and switch mode
        if it reaches above a certain value when
        cylinder position is within the defined limits,
    4.  Operate in force servo mode until the pre-defined time
        period expired,
    5.  Then swicth to position servo mode and command
        reverse motion to home position.

}

void function PID_Position(x_d, v_d)
{
    static float K_p = 1.0,  K_d = 0.0, K_i=0.01 ;
    static float u_i = 0.0 ;
    static Scale = 1.0 ;

      u_i = u_i + K_i * (x_d - Scale * ENC)  ;
      DAC = K_p  * (x_d - Scale * ENC) + u_i ;

    return ;
}

void function PID_Force(F_d)
{
    static float K_pf = 1.0,  K_df = 0.0, K_if=0.01 ;
    static float u_if = 0.0 ;
    static float A1=  0.025, A2=0.0125 ;

      u_if = u_if + K_if * (F_d - (P1*A1-P2*A2))  ;
      DAC = K_pf * (F_d - (P1*A1-P2*A2)) + u_if ;

    return ;
}
```

7.7 HYDRAULIC MOTION AXIS NATURAL FREQUENCY AND BANDWIDTH LIMIT

As a general "rule of thumb," a control system should aim to limit the bandwidth of the closed loop system (w_{bw}) to less than 1/3 of the natural frequency (w_n) of the open loop system,

$$w_{bw} < \frac{1}{3} w_n \qquad (7.294)$$

Otherwise, too large closed loop control gains will result in closed loop instability or a very lightly damped oscillatory response, therefore, it is important to estimate the open loop natural frequency of a given EH motion axis. The most significant factors affecting the natural frequency of an EH axis are the compressibility of the fluid (its spring effect) and the inertial load on the axis. The fluid spring effect is modeled by its bulk modulus and fluid volume. The lowest natural frequency is approximated as

$$w_{\mathrm{n}} = \sqrt{K/M} \qquad (7.295)$$

where M is the total inertia moved by the axis. For varying load conditions, the worst case inertia should be considered. The spring coefficient, K, of the axis is approximated by

$$K = \beta \left(\frac{A_{\mathrm{he}}^2}{V_{\mathrm{he}}} + \frac{A_{\mathrm{re}}^2}{V_{\mathrm{re}}} \right) \qquad (7.296)$$

where, A and V represents cross-sectional area and volume, $_{\mathrm{he}}$ and $_{\mathrm{re}}$ subscripts refer to the head-end and rod-end of the cylinder, respectively. The fluid bulk modulus, β is defined as

$$\beta = \frac{-\Delta P}{\Delta V/V} \qquad (7.297)$$

The typical range of values for β is 2 to $3 \cdot 10^5$ psi.

It is important to note that entrapped air in hydraulic fluid reduces the effective bulk modulus very significantly, whereas dissolved air has very little effect. Furthermore, bulk modulus is a function of the temperature, and generally reduces with increasing temperature. For mineral oil type hydraulic fluid, bulk modulus is about 30×10^4 psi at $50\,°\mathrm{F}$, and reduces to about 15×10^4 psi at about $200\,°\mathrm{F}$ temperature [14].

The V_{he} and V_{re} are functions of cylinder position, hence the effective stiffness is a function of the cylinder position. The natural frequency of the axis varies as a function of the cylinder stroke. The minimum value of the natural frequency is around the middle position of the cylinder stroke. Unless advanced control algorithms are used, closed loop control system gains should be adjusted so that the closed loop system bandwidth stays below $1/3$ of this open loop natural frequency. In general the natural frequency of the valve is much larger than the cylinder and load hydraulic natural frequency. Therefore, the valve natural frequency is not the limiting factor in closed loop performance of most applications.

When the open loop system and load dynamics has low damping, the closed loop system bandwidth using position feedback is severly limited due to the low damping ratio, that is to much lower values than the $1/3\,w_{\mathrm{n}}$, where w_{n} is the open loop system natural frequency. Therefore, it is necessary to add damping into the closed loop system in order to achieve higher closed loop bandwidth. This can be achieved in two ways:

1. By-pass the leakage orifice at the valve between two sides of the actuator. This, however, has two drawbacks: 1. it wastes energy, 2. the static stiffness of the closed loop system against load disturbance is reduced.

2. Velocity and/or pressure feedback in the valve control.

The standard pressure compensated flow control valve increases the effective damping in the closed loop system by modifying the pressure-flow characteristics of the valve. The fact that the valve affects flow as a function of pressure, that is a constant times the pressure, it indirectly affects the velocity since flow rate and actuator velocity are closely related. Hence, the feedback in the form of *velocity* or *pressure* adds damping. However, the cost is the reduced static stiffness of the closed loop system. The pressure feedback alone makes the pressure-flow (PQ) characteristics of the valve linear, which has the effect of increasing the damping.

Example Let us consider an electrohydraulic (EH) motion axis with the following parameters for the cylinder and load.

$$A_{he} = 2.0 \, in^2 \tag{7.298}$$

$$A_{re} = 1.0 \, in^2 \tag{7.299}$$

$$L = 20 \, in \tag{7.300}$$

$$W = 1000 \, lb \tag{7.301}$$

Let us consider the natural frequency at the mid-stroke point,

$$V_{he} = A_{he} \cdot L/2 = 20 \, in^3 \tag{7.302}$$

and

$$V_{re} = A_{re} \cdot L/2 = 10 \, in^3 \tag{7.303}$$

Let us approximate the bulk modulus of hydraulic fluid to be 2.5×10^5 psi. The weight of the rod and load $W = 1000 \, lb$, hence the mass, $M = W/g = 1000 \, lb/386 \, in/s^2$. The open loop natural frequency of the EH axis when the cylinder is at mid-stroke point is

$$w_n = \sqrt{K/M} \tag{7.304}$$

$$M = \frac{1000}{386} \, lb/(in/s^2) = 2.59 \, lb/(in/s^2) \tag{7.305}$$

$$K = 2.5 \cdot 10^5 \cdot \left(\frac{2^2}{20} + \frac{1^2}{10} \right) \, lb/in^2 \, (in^2)^2/in^3 \tag{7.306}$$

$$= 7.5 \cdot 10^4 \, lb/in \tag{7.307}$$

Then,

$$w_n = \sqrt{K/M} \tag{7.308}$$

$$= \sqrt{7.5 \cdot 10^4/2.59 \, (lb/in)/(lb - s^2/in)} \tag{7.309}$$

$$= 170 \, rad/s \tag{7.310}$$

$$= 27 \, Hz \tag{7.311}$$

Therefore, a closed loop controller should not attempt to reach a bandwidth higher than

$$w_{bm} < (1/3 \cdot w_n) = 9.0 \, Hz \tag{7.312}$$

This particular EH motion axis cannot accurately follow cyclic small motion commands with a frequency higher than 9.0 Hz.

7.8 LINEAR DYNAMIC MODEL OF A ONE-AXIS HYDRAULIC MOTION SYSTEM

Linear dynamic models assume that the hydraulic system operates about a nominal operating condition, that is the valve is operating about the null position with small movements (Figure 7.64). Figure 7.80 shows a block diagram of a closed loop electrohydraulic (EH) control system. Figures 7.81, 7.89, and 7.90 shows a closed loop position control (position servo control) version of the system. Figure 7.82 shows a closed loop force control (force servo loop) version of the system. Figure 7.91 shows a one-axis EH motion system where the load pressure is regulated by the control element instead of the load position.

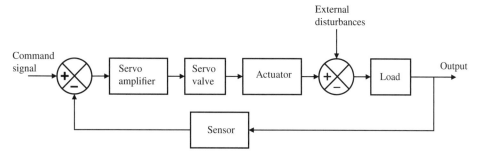

FIGURE 7.89: Block diagram of a closed loop controlled one-axis electrohydraulic motion system.

The following simplifying assumptions are made for the linear dynamics model of the hydraulic system.

1. The valve dynamics and its nonlinear flow rate–current–pressure relationship is approximated with a linearized flow rate–current relationship (K_q). Notice that this assumption is accurate only to the extent that the orifice area versus spool displacement geometry of the valve is linear and that the pressure drop across the valve is constant.

2. The pump pressure is assumed to provide a constant pressure drop across the valve, to justify the above approximation (($p_s - p_l$) is constant).

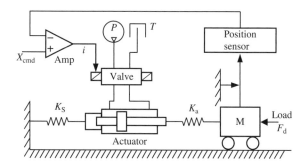

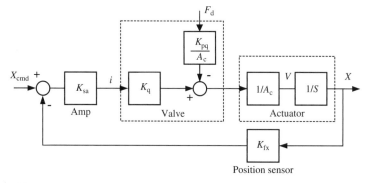

FIGURE 7.90: (a) Components, and (b) linear block diagram model of a postion servo controlled one axis electro-hydraulic motion system.

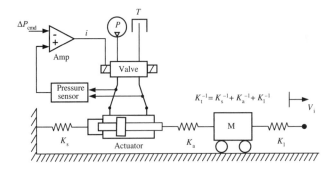

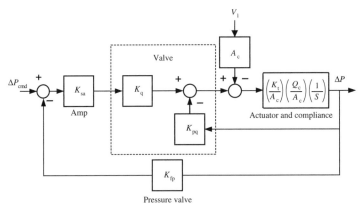

FIGURE 7.91: (a) Components, and (b) linear block diagram model of a force (load pressure) servo controlled one-axis electrohydraulic motion system.

3. Leakage flow rate in the valve is also modeled as a linear relationship between load pressure and leakage flow rate (K_{pq}).

4. Fluid compressibility is neglected.

5. The load inertia and cylinder inertia are neglected.

6. Compliance of the actuator to base mounting, and compliance of the actuator to load connection are neglected for closed position controlled hydraulic system model ($K_s \longrightarrow \infty$, $K_a \longrightarrow \infty$). In a closed loop force control system, the combined effect of all the compliances (K_s, K_a, K_l) are taken into account with the K_t term.

In the linear block diagram models, the symbols have the following physical meaning:

K_{sa} is the amplifier command voltage to current gain,

K_q is the valve solenoid current to flow rate gain,

K_p is the pressure gain,

K_{pq} is the leakage gain,

A_c is the cylinder cross-sectional area (assumed to be equal areas on both sides),

K_{fx} is the position sensor gain,

K_{fp} is the pressure sensor gain,

Q_c is the compliance flow.

7.8.1 Position Controlled Electrohydraulic Motion Axes

The controlled output, position (x), is determined by the commanded position (x_{cmd}) signal and load force (F_d). Using basic block diagram algebra, the transfer function from x_{cmd} and F_d to x can be calculated using the linear models for each component,

$$X(s) = \frac{1}{A_c \cdot s} \cdot \left(K_q \cdot i(s) - \left(\frac{K_{pq}}{A_c} \right) \cdot F_d(s) \right) \tag{7.313}$$

$$X(s) = \frac{1}{A_c \cdot s} \left(K_q \cdot K_{sa} \cdot (X_{cmd}(s) - K_{fx} \cdot X(s)) \right) \tag{7.314}$$

$$- \left(\frac{K_{pq}}{A_c} \right) \cdot F_d(s) \right) \tag{7.315}$$

$$(A_c \cdot s + K_{sa} \cdot K_q \cdot K_{fx}) \cdot X(s) = K_{sa} \cdot K_q \cdot X_{cmd}(s) - \left(\frac{K_{pq}}{A_c} \right) \cdot F_d(s) \tag{7.316}$$

$$X(s) = + \frac{(K_{sa} \cdot K_q / A_c)}{s + (K_{sa} \cdot K_q \cdot K_{fx}/A_c)} \cdot X_{cmd}(s) \tag{7.317}$$

$$- \frac{K_{pq}/A_c^2}{s + (K_{sa} \cdot K_q \cdot K_{fx}/A_c)} \cdot F_d(s) \tag{7.318}$$

Using the last expression for the transfer function and the final value theorem for Laplace transforms, we can determine the following key characteristics of the closed loop position control system.

- Steady-state position following error in response to a step position command and no disturbance force condition, $X_{cmd} = X_o/s$ and $F_d(s) = 0$

$$\lim_{t \to \infty} x(t) = \lim_{s \to 0} s \cdot X(s) \tag{7.319}$$

$$= \lim_{s \to 0} s \cdot \frac{(K_{sa} \cdot K_q / A_c)}{(s + (K_{sa} \cdot K_q \cdot K_{fx}/A_c)} \cdot X_o/s \tag{7.320}$$

$$= (1/K_{fx}) X_o \tag{7.321}$$

When a constant displacement is commanded and there is no disturbance force, the actual position will be proportional to the commanded signal by the feedback sensor gain. By scaling the commanded signal, the effective ratio between the desired and actual position can be made to be unity. In other words, the output position is exactly equal to the commanded position in steady-state when the commanded position is a constant value.

- What is the maximum steady-state position error in response to a ramp command signal

$$X_{cmd}(s) = V_0/s^2 \tag{7.322}$$

$$F_d(s) = 0 \tag{7.323}$$

The tracking error transfer function is

$$X_e(s) = X_{cmd}(s) - K_{fx} X(s) \tag{7.324}$$

$$= X_{cmd}(s) - \frac{K_{fx} \cdot K_{sa} \cdot K_q / A_c}{s + (K_{sa} \cdot K_q \cdot K_{fx}/A_c)} \cdot X_{cmd}(s) \tag{7.325}$$

$$= \frac{s}{(s + K_{fx} \cdot K_{sa} \cdot K_q / A_c)} \cdot V_0/s^2 \tag{7.326}$$

Using the final value theorem, it can be determined that the steady-state position following error while tracking a ramp position command (constant speed command) is

$$\lim_{t \to \infty} X_e(t) = \lim_{s \to 0} s \cdot X_e(s) \tag{7.327}$$

$$= \frac{1}{(K_{sa} \cdot K_q \cdot K_{fx}/A_c)} \cdot V_o \tag{7.328}$$

- Another important characteristic is the closed loop stiffness of the positioning system. What is the steady-state change in position if there is a constant disturbance force acting on the system, $F_d(s) = F_{do}/s$, $X_{cmd}(s) = 0.0$, $K_{cl} = F_{do}/X$? Using the final value theorem,

$$X(s) = -\frac{(K_{pq}/A_c^2)}{s + (K_{sa} \cdot K_q \cdot K_{fx}/A_c)} \cdot F_d(s) \tag{7.329}$$

$$\lim_{s \to 0} s \cdot X(s) = -s \cdot \frac{(K_{pq}/A_c^2)}{s + (K_{sa} \cdot K_q \cdot K_{fx}/A_c)} \cdot F_{d0}/s \tag{7.330}$$

$$x(\infty) = (K_{pq}/A_c^2)/K_{vx} \cdot F_{do} \tag{7.331}$$

$$\frac{F_{do}}{x(\infty)} = K_{cl} \tag{7.332}$$

The closed loop steady-state stiffness is

$$K_{cl} = \frac{K_{vx} \cdot A_c^2}{K_{pq}} \tag{7.333}$$

$$= \frac{\frac{K_{sa} \cdot K_q \cdot K_{fx}}{A_c} \cdot A_c^2}{K_{pq}} \tag{7.334}$$

$$= \frac{K_{sa} \cdot K_q \cdot K_{fx} \cdot A_c}{K_{pq}} \tag{7.335}$$

The leakage coefficient for servo valves is approximately

$$K_{pq} = 0.02 \cdot \frac{Q_r}{P_S} \tag{7.336}$$

Hence, the stiffness of the closed position control servo with such a valve is approximately,

$$K_{cl} = 50 \cdot K_{vx} \frac{A_c^2 \cdot P_S}{Q_r} \tag{7.337}$$

which clearly indicates the effect of the leakage coefficient on the closed loop stiffness.

It is important to recall that the natural frequency of the electrohydraulic system and its load imposes an upper limit on the closed loop system bandwidth. The position servo control system described above has approximately the closed loop bandwidth (closed loop transfer function pole) equal to K_{vx} and the value of it is limited by the natural frequency of the system due to stability considerations,

$$w_{bw} = K_{vx} \leq \frac{1}{3} \cdot w_n \tag{7.338}$$

The characteristics of the valve around its null position significantly affects the steady-state positioning accuracy of the closed loop system.

Example Consider a single axis electrohydraulic motion control system as shown in Figure 7.90 and its block diagram representation in Figure 7.89. We assume that the cylinder is rigidly connected to its base and load. Let us consider the small movements of the valve around its null position, that is a case that is maintaining a commanded position. Determine the cylinder positioning error for the following conditions:

1. the amplifier gain is $K_{sa} = 200\,\text{mA}/10\,\text{V} = 20\,\text{mA/V}$,
2. the valve gain around the null position operation, $K_q = 20\,(\text{in}^3/\text{s})/200\,\text{mA} = 0.1\,(\text{in}^3/\text{s})/\text{mA}$,
3. the cross-sectional area of the cylinder is $A_c = 2.0\,\text{in}$ on both sides (rod is extended to both sides),
4. the sensor gain is $K_{fx} = 10\,\text{V}/10\,\text{in} = 1\,\text{V/in}$,
5. the deadband of the valve is 10% of the maximum input current, $i_{db} = 0.1 \cdot i_{max} = 0.1 \cdot 200\,\text{mA} = 20\,\text{mA}$.

As long as the error magnitude times the gain of the amplifier is less than the deadband, there will not be any flow through the valve and no position correction will be made. Only after the error times the amplifier gain is larger than the deadband of the valve will there be flow through the valve and correction to the position error. Hence, the error in the following range is an inherent limitation of this system, since it does not result in any flow or motion in the system,

$$X_{e,max} \cdot K_{sa} = i_{db} \tag{7.339}$$

$$X_{e,max} = \frac{i_{db}}{K_{sa}} \tag{7.340}$$

$$= \frac{20\,\text{mA}}{20\,\text{mA/V}} \tag{7.341}$$

$$= 1\,\text{V} \tag{7.342}$$

which is 10% of the total displacement range.

Any positioning error less than that value, while maintaining a commanded position (by the regulating motion of the valve around the null-position), is inherently not corrected,

$$X_e \leq X_{e,max} \tag{7.343}$$

$$\leq \frac{i_{db}}{K_{sa}} \tag{7.344}$$

$$\leq 1\,\text{V} \tag{7.345}$$

which is equal to 1.0 in in physical units since $K_{fx} = 1\,\text{V/in}$.

Notice that if the amplifier gain is increased, say by 10, the positioning error due to the same valve deadband is reduced by the same factor.

$$X_e \leq \frac{i_{db}}{K_{sa}} = \frac{20\,\text{mA}}{200\,\text{mA/V}} = 0.1\,\text{V} \tag{7.346}$$

which is equal to 0.1 in in physical units since $K_{fx} = 1\,\text{V/in}$.

In order to ensure that positioning error due to the valve deadband is less than a desired value, the amplifier gain should be larger than a minimum value as follows,

$$K_{sa} \geq \frac{i_{db}}{X_{e,max}} \tag{7.347}$$

$$\geq \frac{20\,\text{mA}}{X_{e,max}\,\text{V}} \tag{7.348}$$

However, the amplifier gain cannot be made arbitrarily large due to the closed loop bandwidth limitations set by the natural frequency of the open loop system,

$$f_1(i_{db}, e_{max}) \le K_{sa} \le f_2(w_{wb}) \tag{7.349}$$

$$w_{bw} = K_{vx} = \frac{K_{sa} \cdot K_q \cdot K_{fx}}{A_c} \le \frac{1}{3} \cdot w_n \tag{7.350}$$

which indicates that the lower limit of the amplifier gain is set by the deadband and desired positioning accuracy, and the upper limit is set by the open loop bandwidth of the hydraulic axis.

7.8.2 Load Pressure Controlled Electrohydraulic Motion Axes

Let us also consider the same EH motion system, except that this time the closed loop control objective is to control the load pressure. The load pressure is measured as the differential pressure between the two output ports of the valve. We will assume that the pressure dynamics between the valve output ports and the actuator ports is negligable (Figure 7.91). The commanded signal represents the desired load pressure, ΔP_{cmd}. Using basic block diagram algebra, the transfer function from the commanded load pressure and the external load speed to the load pressure can be obtained as

$$\Delta P(s) = \frac{K_t}{A_c} \frac{Q_c}{A_c} \frac{1}{s} \cdot (K_q \cdot i(s) - K_{pq} \cdot \Delta P_L(s) - A_c \cdot V_1(s)) \tag{7.351}$$

Notice that the internal leakage term due $K_{pq} \cdot \Delta P(s)$ is relatively small for servo valves and can be neglected in the analysis. Assuming the leakage term is neglected, the closed loop transfer function can be expressed as

$$\Delta P_L(s) = + \frac{(K_t \cdot Q_c / A_c^2) \cdot K_{sa} K_q)}{(s + K_{fp} K_{sa} K_q K_t Q_c / A_c^2)} \cdot \Delta P_{cmd}(s) \tag{7.352}$$

$$- \frac{(K_t Q_c / A_c)}{(s + K_{fp} K_{sa} K_q K_t Q_c / A_c^2)} \cdot V_1(s) \tag{7.353}$$

In terms of command signal and output signal relationship, the transfer function behavior of the force servo is similar to the position servo. In the above linear models, the only dynamics included is the integrator behavior of the actuator. The transient response of the valve from current to flow is not modeled. Experimental studies indicate that the transient response of the proportional or servo valve from current to flow can be approximated by a first-order or a second-order filter depending on the accuracy of approximation needed. The transient response model of a valve is then,

$$\frac{Q_o(s)}{i(s)} = \frac{K_q}{(\tau_{v1} s + 1)(\tau_{v2} s + 1)} \tag{7.354}$$

where τ_{v1} and τ_{v2} are the two time constants for the second-order model. One of them is set to zero for a first order model.

Remark   The flow rate through an orifice, that is valve, is expressed in two different common equation forms,

$$Q(t) = C_d \cdot A(x_v) \cdot \sqrt{\Delta p(t)} \tag{7.355}$$

$$Q(t) = Q_r \cdot \frac{A(x_v)}{A_{max}} \cdot \sqrt{\frac{\Delta p(t)}{p_r}} \tag{7.356}$$

The first form of the equation is the classic orifice equation relating the pressure differential across the orifice, orifice area, and discharge coefficient to the flow rate. The second form relates the flow rate for a given valve based on the valve ratings as percentage of the rated flow. The second form of the flow rate equation is more useful in practice since for a given valve, its performance ratings are measured and given by the manufacturer (or can be determined experimentally):

$$Q_r, p_r, A_{max} \tag{7.357}$$

Then for a given orifice opening A and pressure differential across the valve Δp, we can find the flow rate across the valve.

7.9 NONLINEAR DYNAMIC MODEL OF ONE-AXIS HYDRAULIC MOTION SYSTEM

A nonlinear dynamic model for a single axis linear hydraulic motion system, where the actuator is a cylinder, is discussed below. The same identical equations apply for a rotary hydraulic motion system, where the actuator is a rotary hydraulic motor, with analogous parameter replacements. The fluid compressibility is also taken into account in the model (Figures 7.3 and 7.90). Let us consider the motion of the piston-rod-load assuming they are rigidly connected to each other. Using Newton's second law for the force–motion relationships of the cylinder and the load,

(i) Extension motion:

$$m \cdot \ddot{y} = P_A \cdot A_A - P_B \cdot A_B - F_{ext}; \quad 0 \le y \le l_{cyl} \tag{7.358}$$

and the pressure transients in the control volumes on both sides of the cylinder,

$$\dot{p}_A(t) = \frac{\beta}{(V_{hose,VA} + y(t) \cdot A_A)}(Q_{PA}(t) - \dot{y}(t) \cdot A_A) \tag{7.359}$$

$$\dot{p}_B(t) = \frac{\beta}{(V_{hose,VB} + (l_{cyl} - y(t)) \cdot A_B)}(-Q_{BT}(t) + \dot{y}(t) \cdot A_B) \tag{7.360}$$

$$\dot{p}_P(t) = \frac{\beta}{V_{hose,pv}}(Q_P(t) - Q_{PA}(t) - Q_{PT}(t) - Q_r(t)) \tag{7.361}$$

where

$$Q_P = w_{pump} \cdot D_p(\theta_{sw}) \tag{7.362}$$

$$Q_{PT} = Q_r \cdot \frac{A_{PT}(x_s)}{A_{PT,max}} \cdot \sqrt{(P_P - P_T)/P_r} \tag{7.363}$$

$$Q_{PA} = Q_r \cdot \frac{A_{PA}(x_s)}{A_{PA,max}} \cdot \sqrt{(P_P - P_A)/P_r} \tag{7.364}$$

$$Q_{BT} = Q_r \cdot \frac{A_{BT}(x_s)}{A_{BT,max}} \cdot \sqrt{(P_B - P_T)/P_r} \tag{7.365}$$

$$Q_r = Q_r(p_{relief}, p_p) \tag{7.366}$$

(ii) Retraction motion:

$$m \cdot \ddot{y} = P_A \cdot A_A - P_B \cdot A_B - F_{ext}; \quad 0 \le y \le l_{cyl} \tag{7.367}$$

and the pressure transients in the control volumes on both sides of the cylinder,

$$\dot{p}_A(t) = \frac{\beta}{(V_{\text{hose,VA}} + y(t) \cdot A_A)}(-Q_{AT}(t) - \dot{y}(t) \cdot A_A) \tag{7.368}$$

$$\dot{p}_B(t) = \frac{\beta}{(V_{\text{hose,VB}} + (l_{\text{cyl}} - y(t)) \cdot A_B)}(Q_{PB}(t) + \dot{y}(t) \cdot A_B) \tag{7.369}$$

$$\dot{p}_P(t) = \frac{\beta}{V_{\text{hose,pv}}}(Q_P(t) - Q_{PB}(t) - Q_{PT}(t) - Q_r(t)) \tag{7.370}$$

where

$$Q_P = w_{\text{pump}} \cdot D_p(\theta_{\text{sw}}) \tag{7.371}$$

$$Q_{PT} = Q_r \cdot \frac{A_{PT}(x_s)}{A_{PT,\max}} \cdot \sqrt{(P_P - P_T)/P_r} \tag{7.372}$$

$$Q_{PB} = Q_r \cdot \frac{A_{PB}(x_s)}{A_{PB,\max}} \cdot \sqrt{(P_P - P_A)/P_r} \tag{7.373}$$

$$Q_{AT} = Q_r \cdot \frac{A_{AT}(x_s)}{A_{AT,\max}} \cdot \sqrt{(P_B - P_T)/P_r} \tag{7.374}$$

$$Q_r = Q_r(p_{\text{relief}}, p_p) \tag{7.375}$$

where the design parameters and variables for the hydraulic system are,

m – inertia,

A_A, A_B – cylinder head-end and rod-end cross-sectional area,

β – bulk modulus of hydraulic fluid due to its finite stiffness,

$V_{\text{hose,pv}}$ – volume of the hose between pump and valve,

$V_{\text{hose,AV}}$ – volume of the hose between valve and cylinder side A,

$V_{\text{hose,BV}}$ – volume of the hose between valve and cylinder side B,

l_{cyl} – the travel range of the cylinder,

D_p – flow rating of the pump (Volume/Revolution),

Q_r – rate flow at rated pressure P_r when valve is fully open,

P_r – rate pressure,

$A_{PA}(x_s), A_{PB}(x_s), A_{AT}(x_s), A_{BT}(x_s), A_{PT}(x_s)$ – valve flow areas as a function of spool displacement, between pump, cylinder, and tank,

$A_{PA,\max}, A_{PB,\max}, A_{AT,\max}, A_{BT,\max}, A_{PT}$ – maximum value of valve flow areas when valve is fully open between pump, cylinder, and tank,

$Q_P, Q_{PT}, Q_{PA}, Q_{PB}, Q_{AT}, Q_{BT}$ – flow rate from pump, pump-to-tank, pump-to-cylinder and cylinder-to-tank,

x_s – valve spool displacement,

θ_{sw} – swash plate angle, θ_{sw0} constant value of swash plate angle,

y – cylinder displacement,

P_P, P_A, P_B, P_T – pressure at pump, head-end, rod-end and tank,

w_{pump} – speed of the pump,

p_{relief} – the line relief pressure setting on the relief valve.

Relief Valves Relief valves are used to limit the pressure in hydraulic lines, and hence protect the lines against excessive pressure. The maximum pressure limit of relief valve is adjustable. Let us assume that the maximum pressure is set to $p_{max,p}$ for the pump side, and to $p_{max,l}$ for the line side. If $P_P > p_{max,p}$, $P_P = p_{max,p}$; the pump side relief valve opens and limits the maximum output pressure of the pump. If either of the cylinder side lines' pressure exceeds the relief valve settings, $P_A, P_B > p_{max,l}$, $P_A, P_B = p_{max,l}$; the cylinder side relief valve opens.

If the relief valve is modeled to take all excess flow between the supply and flow that goes through the main valve as soon as the pump pressure reaches the relief pressure (ideal relief valve model, see Figure 7.41c),

$$Q_r(t) = Q_r(p_{relief}, p_P(t)) \tag{7.376}$$
$$= Q_P(t) - (Q_{PA}(t) + Q_{PT}(t)) \ ; \ for\ p_P = p_{relief}\ and\ up\ motion \tag{7.377}$$
$$= Q_P(t) - (Q_{PB}(t) + Q_{PT}(t)) \ ; \ for\ p_P = p_{relief}\ and\ down\ motion \tag{7.378}$$
$$= 0.0 \ \ \ \ \ \ \ \ \ \ \ \ \ \ \ ; \ for\ p_P(t) < p_{relief} \tag{7.379}$$

then, when $p = p_{relief}$, *the fluid compressibility effect on the pump-valve side is effectively cancelled, since this type of relief valve model enforces that there is no net fluid volume change in the hose volume between pump and valve.* In other words, the right hand side of Equation 7.297 (or Equation 7.305) would be zero when $p = p_{relief}$, which effectively cancels the fluid compressibility. Fluid compressibility is still taken into account when $p < p_{relief}$.

A more realistic relief valve model (non-ideal relief valve model) includes a small increase in the pump pressure and the ability to dump the flow as a function of the difference between its input pressure (pump pressure) and relief pressure with a high gain,

$$Q_r(t) = Q_r(p_{relief}, p_P(t)) \tag{7.380}$$
$$= K_{relief} \cdot (p_P(t) - p_{relief}) \ ; \ for\ p_P(t) \geq p_{relief} \tag{7.381}$$
$$= 0.0 \ \ \ \ \ \ \ \ \ \ \ \ \ ; \ for\ p_P(t) < p_{relief} \tag{7.382}$$

where K_{relief} is a high gain approximating the behavior of the relief valve (see Figure 7.41c, where this equation is an approximation to the "actual" relief valve behavior). With this model, flow through the relief valve is determined by the pressure relationships and it does not cancel the right hand side of the pressure dynamics Equation 7.297 or 7.305 for $p_P(t) \geq p_{relief}$. Therefore, this type of relief valve model does not cancel the fluid compressibility effect on the pump-valve side for $p_P(t) \geq p_{relief}$.

The non-compressible fluid condition can always be imposed in any section of the hydraulic circuit by setting the left hand side of the pressure dynamics equations for $p_P(t), p_A(t), p_B(t)$ equal to zero.

Further accuracy can be added to the relief valve model by including its transient dynamic response between pump pressure and and the flow rate through the relief valve, that is using a first (or second) order filter model,

$$if \ \ \ \ \ \ p_P(t) \geq p_{relief} \tag{7.383}$$
$$\tau_{relief} \frac{dQ_r(t)}{dt} = -Q_r(t) + K_{relief} \cdot (p_P(t) - p_{relief}) \tag{7.384}$$
$$else \tag{7.385}$$
$$Q_r(t) = 0.0 \tag{7.386}$$
$$end \tag{7.387}$$

where τ_{relief} is the time constant of the relief valve response when modeled as a first-order filter, $Q_r(t_0)$ initial condition would be needed to integrate this equation over time.

Further detail can be added to the relief valve dynamics by modeling it as a second-order filter, which means the inertial effects of the relief valve spool are taken into account.

$$if \quad p_P(t) \geq p_{relief} \tag{7.388}$$

$$\frac{d^2 Q_r(t)}{dt^2} = -2\xi w_n \frac{dQ_r(t)}{dt2} - w_n^2 \cdot Q_r(t) + K_{relief} \cdot (p_P(t) - p_{relief}) \tag{7.389}$$

$$else \tag{7.390}$$

$$Q_r(t) = 0.0 \tag{7.391}$$

$$end \tag{7.392}$$

where ξ and w_n represent the damping ratio and natural frequency of the second-order dynamic model for the relief valve.

Directional Check Valves   In many EH motion applications, it is not desirable to allow flow from the cylinder back to the pump which can happen if $P_A > P_P$ or $P_B > P_P$. This is accomplished by a one directional load check valve on each line. The load check valve closes to prevent back flow from cylinder to pump. During extension, if $P_A > P_P$, then $Q_{PA} = 0.0$; During retraction, if $P_B > P_P$, then $Q_{PB} = 0.0$.

Open-Center EH Systems   An "open-center" EH system has a fixed displacement pump and an open-center valve, where there is an orifice between pump and tank and the pump displacement is constant,

$$A_{PT}(x_s) \neq 0 \tag{7.393}$$

$$Q_P = w_{pump} \cdot D_p(\theta_{sw0}) \tag{7.394}$$

Closed-Center EH Systems   A "closed-center" EH system has a variable displacement pump and closed-center valve, where there is no orifice directly between pump and tank, and the pump displacement is variable,

$$A_{PT}(x_s) = 0; \tag{7.395}$$

$$Q_P = w_{pump} \cdot D_p(\theta_{sw}) \tag{7.396}$$

Dynamic Model of an Accumulator   The accumulator state at any given instant of time is defined by two variables: accumulator pressure and fluid volume in the accumulator, $p_{acc}(t)$, $V_{acc}(t)$. The operating parameters of the accumulator are precharge, minimum, and maximum pressures ($p_{pre}, p_{min}, p_{max}$) and the maximum discharge volume of the accumulator (the fluid volume difference in the accumulator between minimum and maximum pressures, that is V_{disch}). If we assume that the flow between the line and accumulator follows the same relationship as the flow through an orifice or valve, and that the pressure changes linearly as a function of the change in volume, the dynamic state of the accumulator can be described by the following equations,

$$If \quad V_{acc}(t) \leq 0.0 \quad and \quad p_{line}(t) \leq p_{acc}(t) \tag{7.397}$$

$$Q_{acc}(t) = 0.0 \tag{7.398}$$

$$else \tag{7.399}$$

$$Q_{acc}(t) = sign(p_{line}(t) - p_{acc}(t)) \cdot K_{acc} \cdot \sqrt{|p_{line}(t) - p_{acc}(t)|} \tag{7.400}$$

$$end \tag{7.401}$$

$$\frac{dp_{acc}(t)}{dt} = \left(\frac{p_{max} - p_{min}}{V_{disch}}\right) \cdot Q_{acc}(t) \tag{7.402}$$

$$\frac{dV_{acc}(t)}{dt} = Q_{acc}(t) \tag{7.403}$$

where the initial conditions of the accumulator pressure $p_{acc}(t_0)$ and fluid volume in the accumulator $V_{acc}(t_0)$ should be specified, K_{acc} is the flow rate gain across the valve between the accumulator and the line. Then, the dynamic behavior of the accumulator state (its pressure and fluid volume at any given time) can be predicted by solving the initial value problem.

As an analogy to electrical systems, the capacitance C_{acc} of the accumulator (as an energy storage device) is defined as

$$C_{acc} = \left(\frac{V_{disch}}{p_{max} - p_{min}} \right) \tag{7.404}$$

similar to electrical capacitance defined as $C = (\int i(t)dt)/V = Q/V$, where fluid volume in the accumulator is analogous to the electron charge accumulated in a capacitor, and pressure is analogous to voltage.

Notice that the *sign* function and absolute value of the pressure differential in the flow rate equation expresses that flow can flow into the accumulator from the line or from accumulator to the line. The pressure p_{line} is the line pressure connected to the accumulator. If the accumulator is placed on the pump-valve side, this can be taken as the pump pressure. Likewise, if the accumulator is connected to the cylinder-valve side, then the line pressure is the pressure of the corresponding cylinder line.

In summary, the accumulator design variables and operating variables are as follows: The accumulator is designed to operate within the range of $[p_{min}, p_{max}]$ pressure, with initial precharge pressure of p_{pre}, and accumulator discharge volume capacity V_{disch} and total accumulator volume $V_{acc,g}$. Given the accumulator valve flow gain K_{acc}, and initial conditions of the accumulator volume and pressure, $(V_{acc}(t_0), p_{acc}(t_0))$, and the line pressure condition $p_{line}(t)$, the state of the accumulator can be calculated using the above flow rate $(Q_{acc}(t))$ volume $(V_{acc}(t))$ and pressure $(p_{acc}(t))$ relations.

- p_{max} – maximum operating pressure, above which the line relief valve should open.
- p_{min} – minimum operating presssure, below which the bladder would start to contact the accumulator valve and is not desirable.
- p_{pre} – precharge pressure is the gas (i.e., nitrogen) pressure in the accumulator before having any fluid on the liquid side.
- $V_{acc,g}$ – accumulator volume at the gas side at precharge condition (this is also the total accumulator volume).
- V_{disch} – accumulator discharge volume, the maximum fluid the accumulator can discharge.
- K_{acc} – flow coefficient of the accumulator poppet valve.
- $Q_{acc}(t)$ – flow rate in or out of the accumulator to/from the line.
- $V_{acc}(t), V_{acc}(t_0)$ – liquid volume in the accumulator and its initial condition value.
- $p_{acc}(t), p_{acc}(t_0)$ – pressure in the accumulator and its initial condition value.
- $p_{line}(t)$ – line pressure, the pressure of the hydraulic circuit line the accumulator is connected to.

Summary of Hydraulic Equations

The algebraic sum of flow rate in and out of a cross-section (a node) is equal to zero. This is the analogy of Kirchoff's current law: the algebraic sum of currents at a node is zero.

$$\sum Q_{in} - \sum Q_{out} = 0 \tag{7.405}$$

$$\sum i_{in} - \sum i_{out} = 0 \tag{7.406}$$

The algebraic sum of pressure differentials in a closed path of a hydraulic circuit is zero, flow rate in and out of a cross-section (a node) is equal to zero. This is the analogy of Kirchoff's voltage law: the algebraic sum of voltage differentials in a closed circuit is zero

$$\sum \Delta P_i = 0 \tag{7.407}$$

$$\sum \Delta V_i = 0 \tag{7.408}$$

Inertance (inertia) of the hydraulic fluid is neglected in most cases since it is small compared to hydraulic pressure force. The only exception is in sudden valve opening/closing conditions where high frequency dynamics is desired to be captured, such as the water-hammering problem.

1. **Flow rate:** $Q(t)$, through an orifice area, A, and pressure differential across the orifice ports, $p_1 - p_2$, discharge coefficient of the orifice, C_d

$$Q(t) = C_d \cdot A(x_v) \cdot \sqrt{p_1(t) - p_2(t)} \tag{7.409}$$

$$= Q_r \cdot \frac{A(x_v)}{A_{max}} \cdot \sqrt{(p_1(t) - p_2(t))/p_r} \tag{7.410}$$

where for SAE 19 fluid, $C_d = 0.0582$, area unit is mm^2, pressure unit is kPa, and flow rate unit is l/min.

2. **Pressure change:** $(dp(t)/dt)$ of a compressible fluid in a finite volume ($V(t)$) due to net in–out flow rate difference ($Qin(t) - Q_{out}(t)$),

$$\frac{dp(t)}{dt} = \frac{\beta}{V(t)} \cdot (Q_{in}(t) - Q_{out}(t)) \tag{7.411}$$

where β is the bulk modulus. If the fluid is considered incompressible, then the net flow difference must be equal to zero in and out of the volume considered.

3. **Pump flow rate:** $Q_p(t)$ as function of displacement, $D_p(t)$, input shaft speed, $w_{shaft}(t)$, volumetric efficiency, η_v,

$$Q_p(t) = \eta_v \cdot D_p(t) \cdot w_{shaft}(t) \tag{7.412}$$

Torque input of the pump ($T_{shaft}(t)$) is

$$T_{shaft}(t) = D_p(t) \cdot p_p(t)/\eta_m \tag{7.413}$$

where $p_p(t)$ is the pressure developed at the pump output to overcome the load (assuming zero input port pressure), η_m is the mechanical efficiency of the pump.

Mechanical efficiency is defined as the ratio power output to power input (assuming 100% volumetric efficiency), whereas volumetric efficiency is the ratio of output volume of the fluid pumped to input volume spanned by the pump pistons

$$\eta_m = \frac{P_{out}}{P_{in}} = \frac{p_p(t) \cdot D_p(t) \cdot w_{shaft}(t)}{T_{shaft}(t) \cdot w_{shaft}(t)} = \frac{p_p(t) \cdot D_p(t)}{T_{shaft}(t)} \tag{7.414}$$

$$\eta_v = \frac{V_{out}}{V_{in}} = \frac{Q_{out} \cdot \Delta t}{D_p(t) \cdot w_{shaft}(t) \cdot \Delta t} = \frac{Q_{out}}{D_p(t) w_{shaft}(t)} \tag{7.415}$$

4. **Motor flow rate:** $Q_m(t)$, as function of displacement, $D_m(t)$, input shaft speed, $w_{out}(t)$, volumetric efficiency, η_v,

$$Q_m(t) = \frac{D_p(t) \cdot w_{out}(t)}{\eta_v} \tag{7.416}$$

Torque output of the motor ($T_m(t)$) is

$$T_m(t) = \eta_m \cdot D_m(t) \cdot p_m(t) \tag{7.417}$$

where $p_m(t)$ is the pressure differential developed at the motor input–output ports.

5. **Cylinder motion:** speed and flow rate, as well as the force–pressure relationship (neglecting inertia effects and fluid compressibility)

$$Q_c(t) = A_c \cdot \frac{dy(t)}{dt} \tag{7.418}$$

$$F_c(t) = A_{he} \cdot p_{he}(t) - A_{re} \cdot p_{re} \tag{7.419}$$

where $y(t)$ is the cylinder displacement, A_c is the cylinder cross-sectional area, Q_c flow rate into the cylinder, F_c force generated due to pressure difference on both sides of the cylinder.

6. **Leakage flow through a clearance:**

$$Q_{leak}(t) = f(\Delta p(t)) \approx K_{leak} \cdot \Delta p(t) \tag{7.420}$$

7. **Pressure drop:** across pipes and hydraulic hoses obtained from standard tables as constant pressure drop as a function of the hose parameters, fluid viscosity (μ) and flow rate (Q)

$$\Delta p = f(\mu, Q, l, d, \theta, RF) \tag{7.421}$$

where l, d, θ, RF are hose length, hose diameter, bending angle (if hose is not a straight line), and roughness factor.

Example: Modeling Hydraulic Circuits

Consider the hydraulic circuit shown in Figure 7.92. Part (a) of the figure shows the hydraulic circuit in symbolic form. Part (b) shows the shape and dimensions of the pipes and fittings which can be used in pressure loss estimation. Part (c) shows the various operating conditions considered and simulated for the circuit. In varying complexity, we will analyze this circuit and simulate its operation.

The hydraulic oil is SAE 10W, operating at a nominal temperature of $T = 25\,°C$. The pump, main valve (valve 1), and relief valve (valve 2) are connected to the circuit via straight pipes, a T-connector, and a right-angle (90°) elbow connector.

Given hydraulic fluid viscosity, μ (which is determined by the hydraulic fluid type and temperature), the geometric parameters (length and diameter, l, d), surface roughness (roughness factor, RF, of the pipes), and shape of the connectors (straight, T-connector, right-angle elbow connector), and nominal flow rate; we can estimate the approximate pressure loss in each section of the pipe,

$$p_{loss} = f(\mu, Q, l, d, RF, Shape) \tag{7.422}$$

where for each section of the connectors this formula can be applied, as follows (Figure 7.93).

Although the pressure losses are not constant, and are a function of the flow rate and temperature, and hence change as the flow rate and temperature change, it is a reasonable approximation to use constant values based on nominal flow rate values in the analysis. In a simulation for very detailed accurate prediction purposes, the dependence of pressure loss in the lines as a function of flow rate can be modeled and simulated. However, in the analysis below, we will use the constant pressure drop estimation based on nominal flow rate, for a given constant nominal operating temperature condition. In a given application, the hydraulic fluid type and nominal operating temperature are known as constant. The pressure

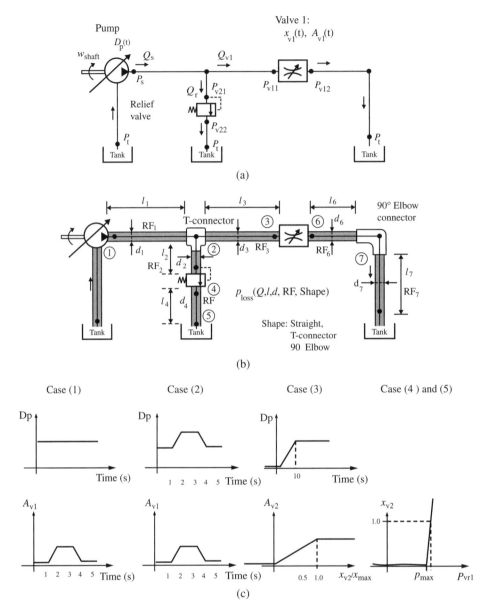

FIGURE 7.92: An example hydraulic circuit: (a) pump, main valve, relief valve, (b) including the pipe and fitting information for pressure drop estimation, (c) simulated conditions.

loss is estimated as a function of flow rate for different geometric (d, l, bend shapes) and surface roughness parameters (RF). Hence, the above general functional relationship is expressed as follows for a given fluid type and nominal operating temperature (i.e., SAE 10 W at 25 °C),

$$p_{loss} = f(Q, l, d, RF, Shape) \qquad (7.423)$$

and this data is given as numerical graphs or data tables by hydraulic fittings and pipe suppliers (Figure 7.93).

Pressure drop (psi)

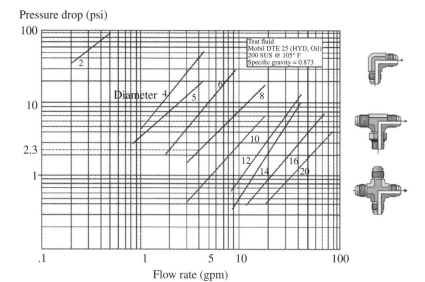

FIGURE 7.93: Pressure drop (loss) in pipes and fittings (turns, elbows, T-connections, cross-connections) as a function of fluid viscosity, pipe diameter and length, flow rate, fitting shape, and pipe inner surface roughness). There are similar graphs available from hydraulic pipe and fiting manufacturers based on measured data. Reproduced with permission from Parker Hannifin.

For the pressure drop from pump output to the input of the main valve,

$$p_{\mathrm{loss}-1-3} = p_{\mathrm{loss}}(Q, l, d, RF, straight\,pipe\,1) \tag{7.424}$$

$$+ p_{\mathrm{loss}}(Q, l, d, RF, T\,connector) \tag{7.425}$$

$$+ p_{\mathrm{loss}}(Q, l, d, RF, straight\,pipe\,3) \tag{7.426}$$

For the pressure drop from pump output to the input of the relief valve,

$$p_{\mathrm{loss}-1-2} = p_{\mathrm{loss}}(Q, l, d, RF, straight\,pipe\,1) \tag{7.427}$$

$$+ p_{\mathrm{loss}}(Q, l, d, RF, T\,connector) \tag{7.428}$$

$$+ p_{\mathrm{loss}}(Q, l, d, RF, straight\,pipe\,2) \tag{7.429}$$

For pressure drop from output of the relief valve to tank

$$p_{\mathrm{loss}-4-5} = p_{\mathrm{loss}}(Q, l, d, RF, straight\,pipe\,4) \tag{7.430}$$

For pressure drop from output of the main valve to tank

$$p_{\mathrm{loss}-6-7} = p_{\mathrm{loss}}(Q, l, d, RF, straight\,pipe\,6) \tag{7.431}$$

$$+ p_{\mathrm{loss}}(Q, l, d, RF, Right\,angle\,connector) \tag{7.432}$$

$$+ p_{\mathrm{loss}}(Q, l, d, RF, straight\,pipe\,7) \tag{7.433}$$

For each section, the pressure losses can be estimated as a constant value once the nominal flow rate and geometric properties of the piping are determined using the look-up tables and graphs provided by pipe-fitting suppliers.

Therefore, for the rest of the analysis, if we know the pressure at the pump output $(p_s(t))$ and the tank (return reservoir) pressure (p_t), we can determine the pressures at the relief valve input and output, main valve input and output. That way, we estimate the flow rate across the valves more accurately as a function of the pressure difference across them, compared to the case where the pressure loss (drop) along the piping is neglected.

$$p_2 = p_s - p_{\text{loss}-1-2} \tag{7.434}$$
$$p_3 = p_s - p_{\text{loss}-1-3} \tag{7.435}$$
$$p_4 = p_t + p_{\text{loss}-4-5} \tag{7.436}$$
$$p_6 = p_t + p_{\text{loss}-6-7} \tag{7.437}$$

In short, the pressure drop can be considered as:

1. zero (neglected), or
2. estimated as constant throughout the whole simulated condition as a function of a constant nominal viscosity, and flow rate, and geometric properties, or
3. can be estimated as a variable as a function of the variable flow rate and viscosity (which varies with temperature).

For this example, the length and diameters of the straight pipes and connectors are as follows,

$$l_1 = l_2 = l_3 = l_4 = l_5 = l_6 = 100\,\text{mm} \tag{7.438}$$
$$d_1 = d_2 = d_3 = d_4 = d_6 = d_7 = 10\,\text{mm} \tag{7.439}$$

The nominal flow rate from the pump is

$$Q_p(t) = D_p(t) \cdot w_{\text{shaft}}(t) \tag{7.440}$$

The input shaft speed to the pump is constant for all the cases considered for simulation,

$$w_{\text{shaft}}(t) = 2000\,\text{rpm} \tag{7.441}$$

The nominal pump flow rate is

$$Q_p(t) = D_p(t) \cdot w_{\text{shaft}}(t) \tag{7.442}$$
$$= 5\,\text{ml/rev} \cdot 2000\,\text{rev/min} \tag{7.443}$$
$$= 10\,\text{l/min} = 10/60 \cdot 10^{-3}\,\text{m}^3/\text{s} \tag{7.444}$$

Given this information, and the pressure drop data graphs provided by pipe-connector suppliers, specific to the used pipes and fittings, we can estimate the nominal (constant) pressure drops.

The displacement of the pump is different as a function of time for the different simulation cases considered (Figure 7.92c). The relief valve reacts to the line pressure. If the line pressure is below the maximum pressure setting, then the valve is fully closed. When the line pressure reaches the maximum pressure value, the relief valve opens to full orifice opening position within a small increment of the pressure above the maximum pressure setting. The line pressure causes the relief valve spool position to change. Due to the preset load in the spring, the spool does not move until the line pressure reaches the maximum pressure setting (determined by the preset load on the spring). When the line pressure reaches the maximum pressure and a little above, the relief valve spool moves quickly to fully open the valve. Let us assume the relief valve spool position to orifice area is a linear relationship plus saturation, as shown in the figure (Figure 7.92c, Case (3)). If we model the relief valve position as being instantaneously determined by the line pressure, we can model it as a static relationship between line pressure and relief valve spool position. Notice that the valve orifice opening is zero until the line pressure reaches the maximum pressure setting. Then it moves to fully open position within a small increase of pressure above the maximum pressure setting.

$$x_r(t) = K_{px} \cdot (p_s(t) - p_{max}) \; ; \; for \, p_s(t) > p_{max} \tag{7.445}$$
$$= 0.0 \; ; \; else \tag{7.446}$$
$$A_r(t) = K_{ar} \cdot x_r(t) \; ; \; for \, x_r(t) > 0 \tag{7.447}$$
$$= 0.0 \; ; \; else \tag{7.448}$$

For a certain value of pressure above maximum pressure, $p_s(t) - p_{max} = \Delta p_{ra}$, the relief valve will be fully open, that is $A_r(t) = A_{r,max}$. Given the performance specifications for the relief valve, we would know $A_{r,max}$ and Δp_{ra}. Then,

$$K_{ar} \cdot K_{px} = \frac{A_{r,max}}{\Delta p_{ra}} \tag{7.449}$$

If we wanted to model the inertial dynamics of the relief valve, then the relationship is a differential equation,

$$m_r \ddot{x}_r(t) + c_r \dot{x}_r(t) + k_r x_r(t) = A_{fb} \cdot (p_s(t) - p_{max}) \tag{7.450}$$
$$A_r(t) = K_{ar} \cdot x_r(t) \; ; \; for \, x_r(t) > 0 \tag{7.451}$$
$$= 0.0 \quad ; \; else \tag{7.452}$$

where $p_{max} = k_r \cdot x_{preload}/A_{fb}$, maximum pressure setting is defined in the circuit by the preload on the relief valve spring. In steady-state, the relief valve spool displacement is

$$x_r(\infty) = \frac{A_{fb}}{k_r} \cdot (p_s(t) - p_{max}) \tag{7.453}$$

$$A_r(\infty) = K_{ar} \cdot \frac{A_{fb}}{k_r} \cdot (p_s(t) - p_{max}); \; for \, x_r(t) > 0 \tag{7.454}$$

In addition, when the relief valve is fully open at the pressure $p(t) = p_{max} + \Delta p_{ra}$, the orifice area would saturate to $A_{r,max}$. Hence,

$$K_{ar} \cdot \frac{A_{fb}}{k_r} = \frac{A_{r,max}}{\Delta p_{ra}} \tag{7.455}$$

The dynamics of the main valve is not modeled as an inertia–force relationship in this example. Rather, we assume that the spool displacement of the main valve is specified as a function of time, and the dynamics of the actuation mechanism and the valve are neglected. Various conditions of the main valve orifice opening ($A_v(t)$) are specified below for the simulated conditions. The consideration of the inertial dynamics of the main valve is left as an exercise for further study of this example.

The governing equations for flow rate across each component (pump, relief valve, main valve) are as follows (pressure losses are not included since it is discussed above and their inclusion is very simple)

$$Q_p(t) = D_p(t) \cdot w_{shaft}(t) \tag{7.456}$$

$$Q_r(t) = Q_{rr} \cdot (A_r(t)/A_{r,max}) \cdot \sqrt{\frac{p_s(t) - p_t}{\Delta p_{rr}}} \tag{7.457}$$

$$Q_v(t) = Q_{vr} \cdot (A_v(t)/A_{v,max}) \cdot \sqrt{\frac{p_s(t) - p_t}{\Delta p_{vr}}} \tag{7.458}$$

where Q_{rr}, Q_{vr} are rated flow rates for relief valve and main valve, $A_{r,max}$ maximum orifice opening of the relief valve, $A_{v,max}$ maximum orifice opening of the main valve, Δp_{rr} rated pressure difference above the maximum line pressure for the relief valve to open fully, Δp_{vr} rated pressure differential across the valve used to measure the rated flow through the main valve. In practice, it is more common to use the rated flow equations instead of the one with orifice coefficients.

Let the parameters of the relief valve be

$$K_{ar} = 10.0 \tag{7.459}$$
$$K_{px} = 0.01 \tag{7.460}$$
$$A_{fb} = 1.0\,\text{mm}^2 \tag{7.461}$$
$$k_r = 100\,\text{kN/mm} \tag{7.462}$$
$$m_r = 0.1\,\text{kN/(mm/s}^2) \tag{7.463}$$
$$c_r = 100\,\text{kN/(mm/s)} \tag{7.464}$$
$$A_{r,max} = 100\,\text{mm}^2 \tag{7.465}$$
$$Q_{rr} = 200\,\text{ml/s} = 200 \cdot 10^{-6}\,\text{m}^3/\text{s} \tag{7.466}$$
$$p_{max} = 20\,000\,\text{kPa} \tag{7.467}$$
$$\Delta p_{ra} = 1000\,\text{kPa} \tag{7.468}$$
$$\Delta p_{rr} = p_{max} = 20\,000\,\text{kPa} \tag{7.469}$$

Notice that these parameters indicate that when $p_s(t) - p_{max} = \Delta p_{ra} = 1000\,\text{kPa}$, the relief valve will be fully open, $x_r = 10, A_r = 100 = A_{r,max}$.

Let the parameters of the main valve be:

$$A_{v,max} = 100\,\text{mm}^2 \tag{7.470}$$
$$Q_{vr} = 200\,\text{ml/s} = 200 \cdot 10^{-6}\,\text{m}^3/\text{s} \tag{7.471}$$
$$\Delta p_{vr} = 6895\,\text{kPa} = 1000\,\text{psi} \tag{7.472}$$

The tank pressure is

$$p_t = 100\,\text{kPa} \tag{7.473}$$

The bulk modulus and volume of the fluid in the space between pump and two valves are,

$$\beta = 1.5 \cdot 10^9\,\text{Pa} = 1.5 \cdot 10^6\,\text{kPa} \tag{7.474}$$
$$V_{hose} = \pi(d^2/4) \cdot (l_1 + l_2 + l_3) = 0.001\,\text{m}^3 \tag{7.475}$$

Initial conditions for the dynamic models are as follows. For relief valve spool position and velocity,

$$x_r(t_0) = 0.0 \tag{7.476}$$
$$\dot{x}_r(t) = 0.0 \tag{7.477}$$

and for line pressure

$$p_s(t_0) = p_t \tag{7.478}$$

In all of the simulated cases, we are interested in the following variables and their interaction,

$$1.\,D_p(t),\ \ 2.\,w_{shaft}(t),\ \ 3.\,Q_p(t),\ \ 4.\,Q_v(t),\ \ 5.\,Q_r(t),\ \ 6.\,A_v(t),\ \ 7.\,A_r(t),\ \ 8.\,p_s(t) \tag{7.479}$$

Simulated Case 1: Constant pump displacement, variable main valve opening.

$$D_p(t) = 4\,\text{ml/rev} = 4 \cdot 10^{-6}\,\text{m}^3/\text{rev} \tag{7.480}$$

$$w_{\text{shaft}}(t) = 2000\,\text{rev/min} \tag{7.481}$$

$$A_v(t) = 10.0; \qquad\qquad\qquad t < 1.0 \tag{7.482}$$

$$= 10.0 + 40.0\,(t - 1.0); \quad 1.0 <= t < 2.0 \tag{7.483}$$

$$= 50.0; \qquad\qquad\qquad 2.0 <= t < 3.0 \tag{7.484}$$

$$= 50.0 - 40.0\,(t - 1.0); \quad 3.0 <= t < 4.0 \tag{7.485}$$

$$= 10.0; \qquad\qquad\qquad 4.0 <= t < 5.0 \tag{7.486}$$

The fluid compressibility in the volume between pump, relief valve, and main valve is neglected. Then, we have the net flow rate in and out of this volume as equal to zero,

$$Q_p(t) - Q_r(t) - Q_v(t) = 0 \tag{7.487}$$

The relief valve is modeled as a static spool displacement–pressure relationship (inertial dynamics is not modeled). These equations can be solved over time for the given simulation conditions above. Notice that if the line pressure does not reach the maximum pressure setting of the relief valve (if $p_s(t) <= p_{\max}$), then the relief valve will be closed and $Q_r(t) = 0.0$, then $Q_p(t) = Q_v(t)$, which results in the following equation for the pump pressure,

$$p_s(t) = p_t + \left(\frac{Q_v(t)}{C_d A_v(x_v(t))} \right)^2 \tag{7.488}$$

$$= p_t + \left(\frac{Q_p(t)}{C_d A_v(x_v(t))} \right)^2 \tag{7.489}$$

$$= p_t + \left(\frac{D_p \cdot w_{\text{shaft}}}{C_d A_v(x_v(t))} \right)^2 \tag{7.490}$$

The same relationship can be derived using the rated flow equations,

$$p_s(t) = p_t + \left(\frac{Q_v(t)}{Q_{vr}} \frac{A_{v,\max}}{A_v(x_v(t))} \right)^2 \cdot \Delta p_{vr} \tag{7.491}$$

$$= p_t + \left(\frac{Q_p(t)}{Q_{vr}} \frac{A_{v,\max}}{A_v(x_v(t))} \right)^2 \cdot \Delta p_{vr} \tag{7.492}$$

$$= p_t + \left(\frac{D_p(t) \cdot w_{\text{shaft}}(t)}{Q_{vr}} \frac{A_{v,\max}}{A_v(x_v(t))} \right)^2 \cdot \Delta p_{vr} \tag{7.493}$$

The above solution is valid for $p_s(t) < p_{\max}$. If $p_s(t) \geq p_{\max}$, then we must solve for a $p_s(t)$ iteratively from $Q_v(t)$ and $Q_r(t)$ equations such that $Q_p(t) - Q_r(t) - Q_v(t) = 0$. For this case the solution is, for $p_s > p_{\max}$

$$p_s(t) = p_t + \left(\frac{Q_v(t)}{C_d A_v(x_v(t))} \right)^2 = p_t + \left(\frac{Q_p(t)}{C_d A_v(x_v(t))} \right)^2 \tag{7.494}$$

$$= p_t + \left(\frac{(D_p \cdot w_{\text{shaft}})^2}{C_d^2 (A_v^2(x_v(t)) + A_r(x_r(t))^2)} \right) \tag{7.495}$$

The same relationship can be derived using the rated flow equations,

$$p_s(t) = p_t + \left[\left(Q_{vr}(t) \frac{A_v}{A_{v \cdot max}} \right)^2 \cdot \frac{1}{\Delta p_{vr}} + \left(Q_{rr}(t) \frac{A_r}{A_{r,max}} \right)^2 \cdot \frac{1}{\Delta p_{rr}} \right] \cdot Q_p^2(t) \quad (7.496)$$

In this case, there is a constant supply of flow from the pump. Since the main valve is initially open a small amount, the line pressure increases quickly in order to force the incoming flow rate through the valve. As the main valve gradually opens more, the line pressure gradually drops since a lower pressure differential is needed to pass the same flow rate through a larger orifice area. If the line pressure had to increase beyond the maximum pressure setting of the relief valve in order to pass the supplied flow rate through the main valve, then the relief valve would have opened.

In Simulink®, we simply impose the conservation of flow condition equation, and let the iterative algorithm solve for $p_s(t)$ using algebraic constraint equation solver, $Q_p(t) - Q_r(t) - Q_v(t) = 0$, instead of evaluating these explicit solution equations for $p_s(t)$ (Figure 7.94).

Simulated Case 2: Variable pump displacement, variable main valve opening (Figure 7.95). The same conditions as in the previous case except that the pump displacement is specified as function of time as follows. The dynamics of the pump displacement actuation mechanims and the pump response is not modeled.

$$
\begin{align*}
D_p(t) &= 0.0; & t < 1.0 & \quad (7.497) \\
&= 4.0\,(t - 1.0); & 1.0 <= t < 2.0 & \quad (7.498) \\
&= 4.0; & 2.0 <= t < 3.0 & \quad (7.499) \\
&= 4.0 - 4.0\,(t - 1.0); & 3.0 <= t < 4.0 & \quad (7.500) \\
&= 0.0; & 4.0 <= t < 5.0 & \quad (7.501)
\end{align*}
$$

Same equations as in Case 1 apply, only the input variables are different for the simulation. Notice that since the displacement of the pump matches the main valve opening, the line pressure does not fluctuate as high as the first case.

Simulated Case 3: Variable pump displacement, closed main valve, relief valve activates – relief valve modeled as a static relation between pressure and valve opening (Figure 7.96). Since in this case the only path the flow can go through is the relief valve, the line pressure will quickly build up to maximum line pressure to open the relief valve. Same equations as in Case 1 apply. The pump displacement as a function of time is the same as in Case 2. The main valve orifice area is different: $A_v(t) = 0.0$ for the whole simulation. The relief valve orifice opening is determined instantaneously by the line pressure. In other words, the inertial dynamics of the relief valve is not modeled, rather it is the orifice area (or spool position) versus line pressure that is modeled as a static relationship, as defined above.

Simulated Case 4: Fluid compressibility in the piping between components is taken into account (Figure 7.97). Let us take into account the finite stiffness (bulk modulus), that is the compressible nature of the fluid in the volume between the pump and the two valves. There is not much interest in taking into account the compressibility of the fluid between the output port of the valves and the tank connection since the overall pressure in the return lines is rather small and the fluid compressibility does not make much difference in the pressure in that line.

Due to the compressibility of the fluid, the net flow rate change in a volume results in a change (time derivative) of the pressure scaled by the bulk modulus divided by the total fluid volume in the section considered. When the fluid compressibility

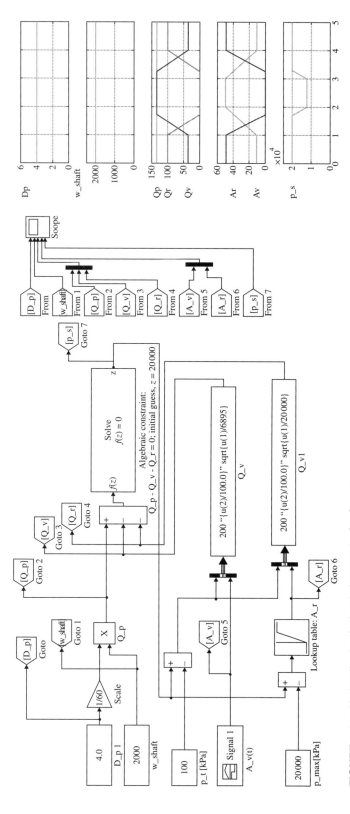

FIGURE 7.94: Hydraulic circuit simulation results for Case 1.

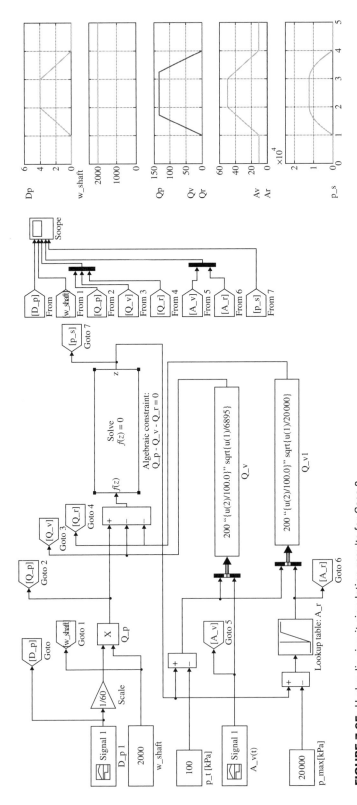

FIGURE 7.95: Hydraulic circuit simulation results for Case 2.

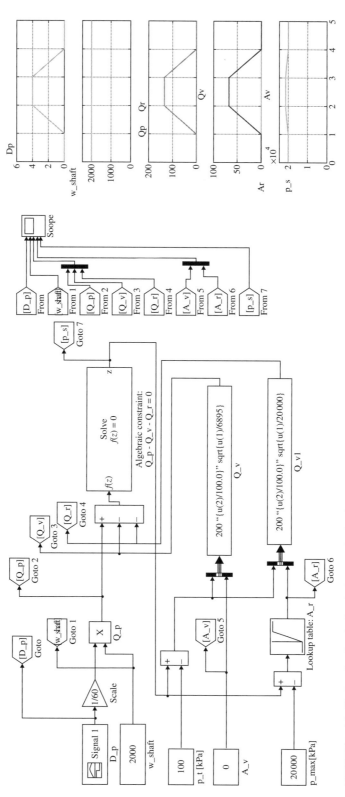

FIGURE 7.96: Hydraulic circuit simulation results for Case 3.

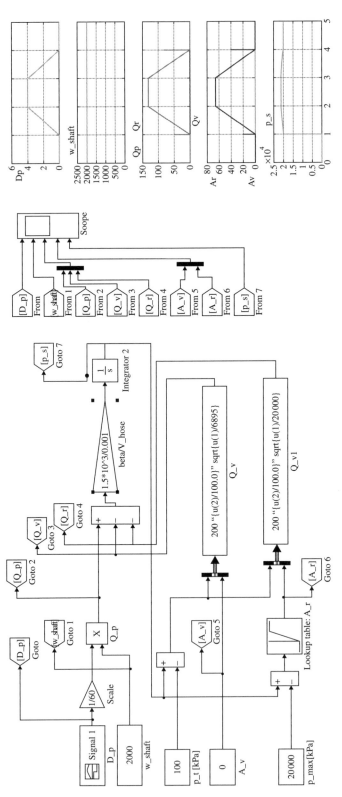

FIGURE 7.97: Hydraulic circuit simulation results for Case 4.

is not taken into account, the net flow rate into and out of a given section of volume (i.e., fluid volume in the piping between pump and relief valve and main valve) is equal to zero.

The pump output pressure will be determined by the dynamic interaction that involves the compressibility of the fluid. Once we calculate the pump output pressure from the differential equations, which takes the fluid compressibility into account, we can further subtract the pressure loss estimations from that in order to determine the dynamic pressures at the input ports of the two valves. In this case the main valve is modeled as a variable orifice where the orifice opening area is given as a function of time, that is the spool position is actuated by a solenoid. The dynamics of the actuator for the main valve spool motion is not included in the model. The relief valve is modeled as a variable orifice as a function of the line pressure. A more accurate model would be to model the inertia–force relationship for the relief valve spool position, and current–force–inertia relationship for the main valve.

$$\frac{dp_s(t)}{dt} = \frac{\beta}{V_{\text{hose}}} \cdot (Q_p(t) - Q_v(t) - Q_r(t)); \quad p_s(t_0) = 20\,000\,\text{kPa} \quad (7.502)$$

Simulated input variable conditions are the same as in Case 3, plus finite bulk modulus of the fluid in the volume between the pump and the two valves.

Simulated Case 5: Variable pump displacement, closed main valve, fluid compressibility is taken into account, the relief valve activates, where the relief valve is modeled as a dynamic inertia–force relationship. Simulated conditions are as Case 4, except that we modeled the relief valve dynamics more accurately. We take into account the dynamic delay in the relief valve spool position change as a result of the line pressure, inertia, spring, and damping in the valve (Figure 7.98).

Example: Dynamic Model of a Poppet Valve

Consider the poppet valve in a hydraulic circuit (Figure 7.99). Here, we will develop a dynamic model of this valve. We will consider the inertial effects of the poppet spool as well as the fluid compressibility effects inside the valve chamber. The basic operation of the valve is that input flow is in port 1 and output flow is in port 2 (Figure 7.99b). The poppet gets unseated when the flow force exceeds the force at the back of the poppet. Similar to sliding spool valves, poppet valves are also manufactured for multi port applications such as 2, 3, 4 port applications.

Let us express the force–acceleration relationship for the poppet, the spool (Figure 7.99b),

$$m_p \cdot \ddot{x}_p(t) + c_p \cdot \dot{x}_p(t) + k_p \cdot x_p(t) = -k_p \cdot x_{\text{preload}} + F_{\text{sol}}(t) + F_{\text{shear}}(t) + p_1(t) \cdot A_1 - p_3(t) \cdot A_3$$
$$(7.503)$$

Notice that the pressure $p_2(t)$ does not affect the spool motion in this model, because we consider that the pressure inside the valve body in region 2 is uniform. As a result, the net force applied to the spool by $p_2(t)$ is zero because it is acting both on the top and bottom surfaces of the poppet.

The relationship between the flow rate through the valve, valve position, pressure in the valve chamber (assuming no cavitation inside the valve), in-port pressure, and output (load) pressure is

$$\frac{dp_2(t)}{dt} = \frac{\beta}{V_2} \cdot (Q_2(t) - Q_{\text{leak}}(t) - Q_4(t)) \quad (7.504)$$

$$\frac{dp_3(t)}{dt} = \frac{\beta}{V_3} \cdot (Q_{\text{leak}}(t) - Q_{3t}(t)) \quad (7.505)$$

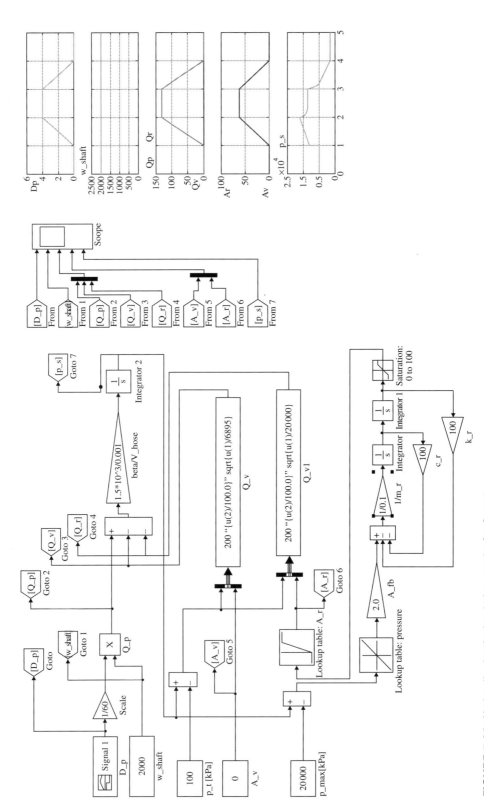

FIGURE 7.98: Hydraulic circuit simulation results for Case 5.

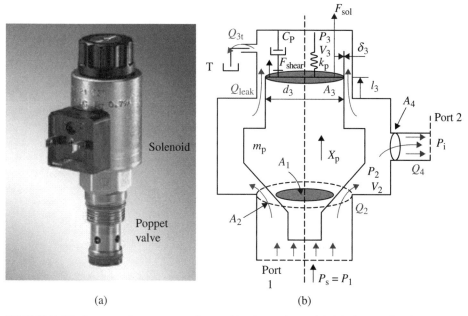

FIGURE 7.99: Poppet valve cross-section and its dynamic model: (a) picture of a 2/2 way poppet valve (by Rexroth), (b) cross-section of the 2/2 way poppet valve. Courtesy Bosch-Rexroth.

$$Q_2(t) = C_d \cdot A_2(x_p(t)) \cdot \sqrt{p_1(t) - p_2(t)} \qquad (7.506)$$

$$Q_4(t) = C_d \cdot A_4 \cdot \sqrt{p_2(t) - p_1(t)} \qquad (7.507)$$

$$Q_{3t} = C_d \cdot A_t \cdot \sqrt{p_3(t) - p_t} \qquad (7.508)$$

$$Q_{leak}(t) = Q_{leak}(l_3, d_3, \delta_3, p_2, p_3) \approx K_{leak} \cdot (p_2(t) - p_3(t)) \qquad (7.509)$$

where

$$F_{shear}(t) = F_{shear}(l_3, d_3, \delta_3, \dot{x}_p(t)) \; ; \; shear \, force \qquad (7.510)$$

which can be defined as a look-up table for the numerical simulation. The shear force is a function of the length l_3, diameter d_3, clearance δ_3, and the speed of the spool $\dot{x}_p(t)$.

The force generated by the electric actuator of the valve is modeled as,

$$F_{sol}(t) = K_{fi} \cdot i_{sol}(t) \qquad (7.511)$$

where $i_{sol}(t)$ is the current supplied to the solenoid (or to an electric actuator from the amplifier), K_{fi} is the current to force gain.

The parameters of the valve are;

$m_p = mass \, of \, the \, poppet$ (7.512)

$c_p = damping \, coeficient \, of \, the \, poppet \, motion$ (7.513)

$k_p = spring \, of \, the \, poppet. \, The \, spring \, is \, preloaded \, to \, hold \, the \, valve \, closed$ (7.514)

        $when \, the \, valve \, is \, not \, actively \, actuated \, by \, the \, electric \, actuator \, (F_{sol}(t))$ (7.515)

$A_1 = cross\text{-}sectional \, area \, of \, the \, valve \, where \, p_1(t) \, acts \, on$ (7.516)

$A_2 = orifice \, opening \, cross\text{-}sectional \, area \, of \, the \, valve$ (7.517)

$A_3 = area \, of \, the \, valve \, on \, the \, back \, side \, of \, the \, poppet$ (7.518)

A_4 = *output port cross-sectional area,* (7.519)

V_2 = *the volume between the poppet housing and the poppet* (7.520)

  (*approximately constant*) (7.521)

V_3 = *the volume between the poppet housing and the poppet in the section 3* (7.522)

β = *bulk modulus of the fluid* (7.523)

l_3 = *length of the overlap between the poppet and housing in area 3* (7.524)

d_3 = *diameter of the cross-section between the poppet and housing in area 3* (7.525)

δ_3 = *clearance between the poppet and housing along the back of* (7.526)

  the poppet (area 3) (7.527)

F_{shear} = *shear force function* (7.528)

Q_{leak} = *leakage flow rate function* (7.529)

K_{fi} = *current to force gain of the electric actuator (i.e., solenoid)* (7.530)

x_{preload} = *preload compression in spring* (7.531)

The initial conditions needed for a specific time domain simulation of the poppet valve behavior are

$x_p(t_0)$ = *initial position of the poppet (i.e., zero, closed valve condition)* (7.532)

$\dot{x}_p(t_0)$ = *initial speed of the poppet (i.e., zero, stationary valve condition)* (7.533)

$p_2(t_0)$ = *initial pressure inside the poppet chamber* (7.534)

$p_3(t_0)$ = *initial pressure inside the poppet chamber section 3.* (7.535)

  This pressure may be assumed to stay constant and equal (7.536)

  to tank pressure, by setting initial condition (7.537)

  to the tank pressure and defining the net flow rate (7.538)

  into the area as zero by setting $Q_{3t}(t) = Q_{\text{leak}}(t)$ (7.539)

  instead of using the orifice equation for $Q_{3t}(t)$ (7.540)

Given input pressure $p_1(t) = p_s(t)$ and load pressure $p_l(t)$ conditions, and the control signal $i_{\text{sol}}(t)$, we can determine the motion of the poppet ($x_p(t)$), flow rate through the valve ($Q_2(t), Q_4(t)$), pressure inside the valve $p_2(t)$, as well as leakage flow rate and shear forces $Q_{\text{leak}}(t)$, $F_{\text{shear}}(t)$ via numerical solution of the above equations.

In proportional control applications of poppet valves, the challenge is the high frequency content of the flow forces acting on the valve and the resulting difficulty in controlling the position of the poppet accurately. The control challenge is two fold:

- knowing (measuring and/or estimating) the flow force,
- having a fast and large enough valve actuator to be able to compensate for the flow force.

The benefits of poppet valves are that

- they seal the flow well in closed position (practically zero leakage, hence excellent position holding capability under load),
- reliability of operation against dirt in the hydraulic fluid (poppet does not easily get stuck due to dirt unlike the case in sliding spool valves), and
- low manufacturing cost of the poppet valves.

The MATLAB® code used to initialize the parameters and initial conditions for the poppet valve is shown below. The dynamic model of the poppet valve is implemented in Simulink® (Figure 7.100). Time domain simulation of the poppet valve model is shown in Figure 7.101. In this simulation we did not include the preload on the poppet valve spring in order to keep it sealed when the solenoid force is zero. As a result, when the solenoid force is zero (time period from 0 s to 1 s), the poppet valve opens due to the line pressure. In actual poppet valves, the spring would have initial preload to keep the poppet valve closed despite the line pressure. This model can be used to simulate other poppet valve sizes by simply specifying appropriate parameters. Load pressure is considered as constant in this simulation. In a real application, the load pressure, that is the output port pressure of the poppet valve, would be determined by the down-stream hydraulic circuit conditions.

```
% Poppet Valve parameters

    Kleak = 0.0 ;   % Leak coefficient
    At    = 0.0 ;   %
    Kfi   = 1.0 ;   % Solenoid current to force gain

    mp = 1.0  ;          % Poppet mass
    cp = 20.0  ;         % poppet damping
    kp = 1000.0 ;        % Poppet spring constant
    x_preload  = 0.0 ;   % Poppet spring preload displacement

    A1  = 0.01;    % Areas
    Ka2 = 0.01 ;   % with saturation at 0 and A2_max
                        (no negative orifice area)
    A3  = 0.01;
    A4  = 0.001 ;
    V2  = 0.10 ;   % Volumes inside the valve
    V3  = 1.0 ;

    beta = 250 * 10^3;  % Bulk modulus
    Cd = 100.0 ;        % Flow coeffient scaled sqrt(2/rho)

    ps = 3000 ;  % p1 = 3000 psi
    pt = 100  ;  % 100  psi
    pl = 100 ;   % 100 psi

    p20 = pt ;   % Initial conditions on pressures
    p30 = pt ;

    Xp0=0.0 ;    % Initial valve (spool) position and speed
    Xpdot0= 0.0 ;
```

An example of a poppet valve (by Hydraforce Inc), which is two-way, normally closed (when de-energized), proportional solenoid controlled, is shown in Figure 7.102. It shows the valve, its installation to a standard manifold cavity, its symbol, as well as its steady-state flow rate–solenoid current–differential pressure relationship (top right figure) and the dynamic response characteristics as the frequency response input–output relationship (gain and phase) between solenoid current and flow rate. Notice that for this particular poppet valve, flow-rate to current relationship under a specific pressure differential across ports 1 and 2 is not linear. When de-energized, flow from port 1 to port 2 is allowed, but flow from port 2 to port 1 is blocked by a check valve. This type of poppet valve is used in series with a proportional four-way valve output line for load-holding (sealing) function. It "seals" the

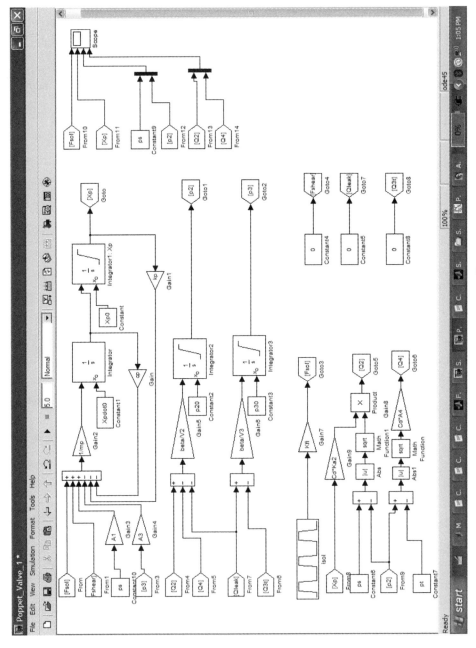

FIGURE 7.100: Poppet valve dynamic model implemented in Simulink®.

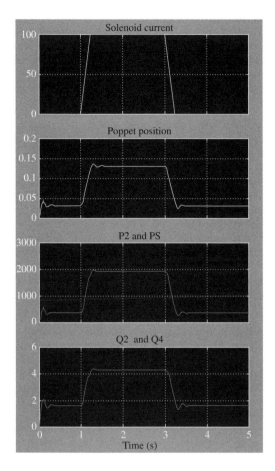

FIGURE 7.101: Simulation result of a poppet valve.

flow from load (port 2) back to pump or tank (port 1) when de-energized. Some of the key valve parameters are as follows

1. The rated operating pressure: 3000 psi
2. Rated flow rate: 5.8 gallons/min
3. Estimated transient response (bandwidth) from current to flow rate under rated pressure differential: 10 Hz.
4. Maximum internal leakage: 5 drops/min at 3000 psi
5. Temperature range of operation: −40–100 °C

Example: Closed Loop Control of an Electrohydraulic Servo Actuator

Consider a one degree of freedom electrohydraulic system (Figure 7.75 or Figures 7.103 and 7.104) where the main flow control valve is a servo valve (Figure 7.60) and its closed loop control (Figure 7.76). On the cylinder side, the line relief valve and line cross-over relief valves are not taken into account, and leakage is neglected in the analysis. We take the compressibility of the fluid into account both in the volume between the pump and the servo valve, and the servo valve and the cylinder. We assume 100% pump volumetric efficiency.

The pump is of fixed displacement type and is driven by an electric motor or engine at a constant speed. The excessive flow between the pump output flow and the flow through the servo valve is returned to the tank through the relief valve. The relief valve on the pump supply line is taken into account as an instantaneously acting valve; that is when the

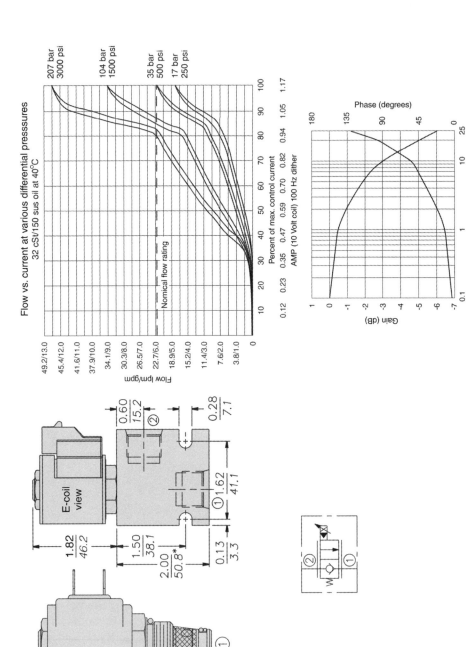

FIGURE 7.102: Poppet valve example: (a) Poppet valve outside dimensions and manifold cavity mounting, (b) ISO symbol for a two-way proportional valve, (c) Q-i (flow rate vs current) relationship for different pressure differentials across ports 1 and 2, (d) dynamic response characteristics in frequency response Bode plot for the flow rate/current ratio (output/input). Reproduced with permission from HydraForce (www.hydraforce.com).

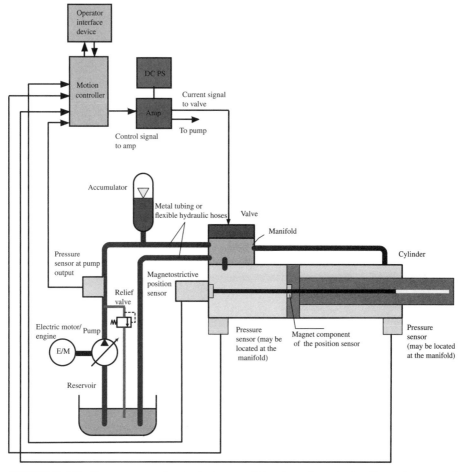

FIGURE 7.103: Components of a single degree of freedom electrohydraulic motion control system: magnetostrictive position sensor is integrated into the cylinder for closed loop position control, two-pressure sensors can be used for closed loop force control.

pump pressure is below maximum pressure setting, the relief valve is closed, but as soon as pump pressure is equal or exceeds the maximum pressure, the relief valve opens to dump the excess flow rate (flow rate difference between pump flow rate and the flow rate through servo valve) to the tank. One way to mathematically model the fluid compressibility and instantaneously acting relief valve is to limit the pressure integration output to the maximum pressure setting.

The dynamic model for pump, relief valve, and fluid compressibility effect in the volume between pump and servo valve is modeled as

$$Q_p(t) = D_p(t) \cdot w_{shaft}(t) \tag{7.541}$$

$$If \quad p_p(t) < p_{max} \tag{7.542}$$

$$Q_r(t) = 0.0 \tag{7.543}$$

$$Else \tag{7.544}$$

$$Q_r(t) = Q_p(t) - Q_v(t) \tag{7.545}$$

$$Endif \tag{7.546}$$

$$\frac{dp_p(t)}{dt} = \frac{\beta}{V_{hose,pv}}(Q_p(t) - Q_v(t)) \tag{7.547}$$

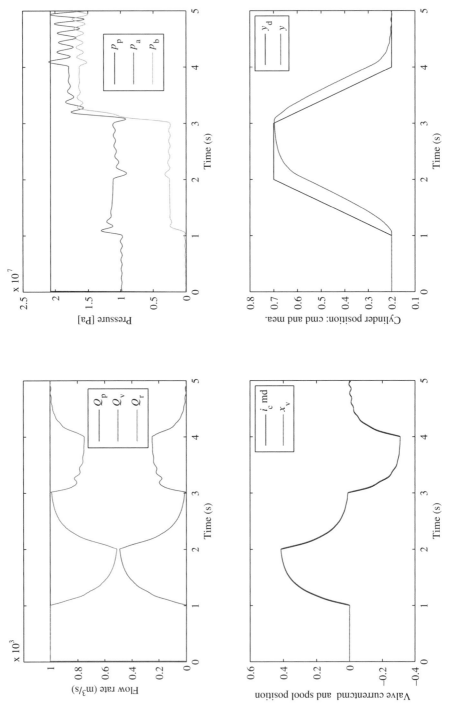

FIGURE 7.104: Simulation results of the closed loop control of an electrohydraulic (EH) servo actuator: flow rates, pressures, valve current command and spool position, cylinder position (desired and actual).

where $Q_p(t)$ is the flow rate from the pump, $Q_v(t)$ is the flow rate from pump to cylinder side of the valve (P to A or P to B, that is $Q_v(t) = Q_{PA}(t)$ for up motion or $Q_v(t) = Q_{PB}(t)$ for down motion, see below).

The two-stage servo valve (flapper nozzle type first stage, and spool type second stage (Figure 7.60)) is modeled as a current command and actual current dynamic relationship, torque generated from the current via a gain, and the spool displacement relationships of the second stage. Once the spool position is known, the flow rate through the valve can be determined as a function of the pressure differential across it and the spool position. The current amplifier and the electrical dynamics of the servo valve are modeled as the dynamic relationship between the commanded current and actual current. It is modeled as a first-order filter. The current to torque relationship is modeled as a constant gain. The torque to spool position is described as a second-order mass-force system, where the natural frequency of the model (w_n) varies in a range as a function of the magnitude of the spool displacement as a percentage of the maximum value,

$$\frac{x_v(t)}{x_{v,max}} \tag{7.548}$$

(or as the ratio of the commanded current to the maximum commanded current).

$$\frac{i_{cmd}(t)}{i_{cmd,max}} \tag{7.549}$$

In other words, the frequency response of the valve spool position to the current command (Figure 7.70) is different for different values of the magnitude of the current command. For instance, for 40% of maximum current command, corresponding to about 40% of maximum spool displacement in steady-state, the bandwidth of spool position is 75 Hz, whereas for 100% of the commanded current, the bandwidth is 25 Hz. The dynamic model of the valve and its current amplifier, relating the commanded current signal to the actual current, then to the valve actuation torque, and then to the valve spool position, and finally to the flow rate as function of spool position and pressure differential across the valve, are expressed below.

$$\frac{i_v(s)}{i_{cmd}(s)} = \frac{K_a}{\tau_a s + 1} \tag{7.550}$$

$$T_v(s) = K_t \cdot i_v(s) \tag{7.551}$$

$$\left(s^2 + \xi w_n s + w_n^2\right) \cdot x_v(s) = \frac{x_{vmax}}{T_{v,max}} \cdot w_n^2 \cdot T_v(s); \quad w_n(x_v/x_{vmax}) \tag{7.552}$$

$$For\ x_v \geq 0.0 \quad up\ motion \tag{7.553}$$

$$Q_{PA}(t) = Q_r \cdot \frac{|x_v(t)|}{x_{v,max}} \cdot \sqrt{\frac{(p_p(t) - p_A(t))}{\Delta p_r}} \tag{7.554}$$

$$Q_{BT}(t) = Q_r \cdot \frac{|x_v(t)|}{x_{v,max}} \cdot \sqrt{\frac{(p_B(t) - p_t(t))}{\Delta p_r}} \tag{7.555}$$

$$For\ x_v < 0.0 \quad down\ motion \tag{7.556}$$

$$Q_{PB}(t) = Q_r \cdot \frac{|x_v(t)|}{x_{v,max}} \cdot \sqrt{\frac{(p_p(t) - p_B(t))}{\Delta p_r}} \tag{7.557}$$

$$Q_{AT}(t) = Q_r \cdot \frac{|x_v(t)|}{x_{v,max}} \cdot \sqrt{\frac{(p_A(t) - p_t(t))}{\Delta p_r}} \tag{7.558}$$

where $x_{v,max}$ is the maximum spool displacement of the servo valve when maximum actuation torque is applied $T_{v,max} = K_t \cdot i_{v,max}$.

The cylinder and load dynamics, including the fluid compressibility effect taken into account,

$$(m_p + m_l) \cdot \ddot{y}(t) = -c_p \cdot \dot{y}(t) + p_A(t) \cdot A_A - p_B(t) \cdot A_B - F_{load}(t) \quad (7.559)$$

$$\textit{If up motion} \qquad x_v \geq 0.0 \quad (7.560)$$

$$\frac{d}{dt}(p_A(t)) = \frac{\beta}{V_{hose,VA} + y \cdot A_A}(Q_{PA}(t) - \dot{y}(t) \cdot A_A) \quad (7.561)$$

$$\frac{d}{dt}(p_B(t)) = \frac{\beta}{V_{hose,VB} + (l_{cyl} - y) \cdot A_B}(-Q_{BT}(t) + \dot{y}(t) \cdot A_B) \quad (7.562)$$

$$\textit{elseif down motion } x_v < 0.0 \quad (7.563)$$

$$\frac{d}{dt}(p_A(t)) = \frac{\beta}{V_{hose,VA} + y \cdot A_A}(-Q_{AT}(t) - \dot{y}(t) \cdot A_A) \quad (7.564)$$

$$\frac{d}{dt}(p_B(t)) = \frac{\beta}{V_{hose,VB} + (l_{cyl} - y) \cdot A_B}(Q_{PB}(t) + \dot{y}(t) \cdot A_B) \quad (7.565)$$

$$\textit{end} \quad (7.566)$$

The position sensor is modeled as a constant gain device,

$$y_{fb}(t) = K_{fb} \cdot y(t) \quad (7.567)$$

The closed loop control algorithm is a PID control,

$$i_{cmd}(t) = K_p(y_d(t) - y_{fb}(t)) + K_i \int_0^t (y_d(\tau) - y_{fb}(\tau))d\tau + K_d(\dot{y}_d(t) - \dot{y}_{fb}(t)) \quad (7.568)$$

Now, let us consider the simulated conditions: the size related parameters of the components, initial conditions of differential variables (i.e., pressures and spool position and velocity), and input variables as a function of time.

$$D_p(t) = 20 * 10^{-6} \, m^3/rev \quad \textit{pump displacement} \quad (7.569)$$

$$w_{shaft}(t) = 3000 \, rev/min \quad \textit{input shaft speed of the pump} \quad (7.570)$$

$$p_{max} = 3000 \, psi = 3000 \, psi \cdot 6.895 \cdot 10^3 \, (N/m^2)/psi \quad (7.571)$$

$$= 20\,685 \, kN/m^2 = 20\,685 \, kPa = 20.685 * 10^6 \, N/m^2 \quad (7.572)$$

$$\textit{relief valve max. presure setting} \quad (7.573)$$

$$Q_{rv} = 1.0 \cdot 10^{-3} \, m^3/s \quad at \quad \Delta p_r = 1000 \, psi = 6895 \, kPa \quad (7.574)$$

$$\textit{rated flow rate of servo valve} \quad (7.575)$$

$$p_r = 6.895 * 10^6 \, N/m^2 = 1000 \, psi \quad (7.576)$$

$$x_{v,max} = 10 \cdot 10^{-3} \, m \quad \textit{max. spool displacement of servo valve} \quad (7.577)$$

$$T_{v,max} = 1.0 \, N \quad \textit{max. actuation torque of servo valve} \quad (7.578)$$

$$\xi = 0.6; \quad \textit{damping ratio} \quad (7.579)$$

$$w_n = 75 \, Hz = 2\pi * 75 \, rad/s; \quad \textit{for} \quad \frac{x_v}{x_{v,max}} \leq 40\% \quad (7.580)$$

$$= 25 \, Hz = 2\pi * 25 \, rad/s; \quad \textit{for} \quad \frac{x_v}{x_{v,max}} = 100\% \quad (7.581)$$

$$\textit{natural frequency of valve spool motion} \quad (7.582)$$

$$\tau_a = 0.010 \, s \quad \textit{electrical time constant for current} \quad (7.583)$$

$$K_a = 1.0; \quad \textit{current amplifier gain} \quad (7.584)$$

$$K_t = 1.0; \quad \textit{current to torque gain} \quad (7.585)$$

$$\beta = 1.5 \cdot 10^9 \, \text{N/m}^2 \quad \textit{bulk modulus} \tag{7.586}$$

$$V_{\text{hose,pv}} = 1.0 \cdot 10^{-3} \, \text{m}^3 \quad \textit{volume between pump and valve} \tag{7.587}$$

$$V_{\text{hose,VA}} = 1.0 \cdot 10^{-3} \, \text{m}^3 \quad \textit{volume between valve and A side} \tag{7.588}$$

$$V_{\text{hose,VB}} = 1.0 \cdot 10^{-3} \, \text{m}^3 \quad \textit{volume between valve and B side} \tag{7.589}$$

$$A_{\text{A}} = 1.0 \cdot 10^{-3} \, \text{m}^2 \quad \textit{cylinder A side cross-section area} \tag{7.590}$$

$$A_{\text{B}} = 0.5 \cdot 10^{-3} \, \text{m}^2 \quad \textit{cylinder B side cross-section area} \tag{7.591}$$

$$l_{\text{cyl}} = 1.0 \, \text{m} \tag{7.592}$$

$$m_{\text{p}} = 10 \, \text{kg} \quad \textit{piston and rod mass} \tag{7.593}$$

$$m_{\text{l}} = 990 \, \text{kg} \, \textit{load mass} \tag{7.594}$$

$$c_{\text{p}} = 100.0 \quad \textit{cylinder damping} \tag{7.595}$$

$$F_{\text{load}}(t) = (m_{\text{p}} + m_{\text{l}}) \cdot g = 1000 \, \text{kg} \cdot 9.81 \, \text{m/s}^2 = 9810 \, \text{N} \tag{7.596}$$

$$\textit{PID gains:} \tag{7.597}$$

$$K_{\text{fb}} = 1.0 \quad \textit{position sensor gain} \tag{7.598}$$

$$K_{\text{p}} = 3.0 \quad \textit{PID : proportional gain} \tag{7.599}$$

$$K_{\text{i}} = 0.0 \quad \textit{PID : integral gain} \tag{7.600}$$

$$K_{\text{d}} = 0.0 \quad \textit{PID : derivative gain} \tag{7.601}$$

$$\textit{Initial conditions:} \tag{7.602}$$

$$p_{\text{p}}(t_0) = 101 \cdot 10^3 \, \text{N/m}^2 \quad \textit{pressures} \tag{7.603}$$

$$p_{\text{A}}(t_0) = F_{\text{load}}/A_{\text{a}} \, \text{N/m}^2 \tag{7.604}$$

$$p_{\text{B}}(t_0) = 101 \cdot 10^3 \, \text{N} \cdot \text{m}^2 \tag{7.605}$$

$$p_{\text{T}}(t_0) = 101 \cdot 10^3 \, \text{N} \cdot \text{m}^2 \tag{7.606}$$

$$i_{\text{v}}(t_0) = 0.0 \, \text{A} \quad \textit{current} \tag{7.607}$$

$$x_{\text{v}}(t_0) = 0.0 \, \text{m} \quad \textit{spool position} \tag{7.608}$$

$$\dot{x}_{\text{v}}(t_0) = 0.0 \, \text{m/s} \quad \textit{spool velocity} \tag{7.609}$$

$$y(t_0) = 0.2 \, \text{m} \quad \textit{cylinder–piston position} \tag{7.610}$$

$$\dot{y}(t_0) = 0.0 \, \text{m/s} \quad \textit{cylinder–piston velocity} \tag{7.611}$$

The desired cylinder–piston is a trapeziodal position command,

$$y_{\text{d}}(t) = 0.2 \quad ; \quad t < 1.0 \tag{7.612}$$

$$= 0.2 + 0.5 \cdot (t - 1.0); \quad 1.0 <= t < 2.0 \tag{7.613}$$

$$= 0.7 \quad ; \quad 2.0 <= t < 3.0 \tag{7.614}$$

$$= 0.7 - 0.5 \cdot (t - 3.0); \quad 3.0 <= t <= 4.0 \tag{7.615}$$

$$= 0.2 \quad ; \quad 4.0 <= t \tag{7.616}$$

The desired cylinder–piston velocity is

$$\dot{y}_{\text{d}}(t) = 0.0 \quad ; \quad t < 1.0 \tag{7.617}$$

$$= 0.5 \quad ; \quad 1.0 <= t < 2.0 \tag{7.618}$$

$$= 0.0 \quad ; \quad 2.0 <= t < 3.0 \tag{7.619}$$

$$= -0.5; \quad 3.0 <= t <= 4.0 \tag{7.620}$$

$$= 0.0 \quad ; \quad 4.0 < t \tag{7.621}$$

Remark on Component Size Matching in Design The above parameters represent the size of the components of the hydraulic system. It is important for a good design to have component sizes selected to match each other. That is the pump, valve, and cylinder sizing must be matched. A simple calculation can determine if the components are properly sized. For this purpose, one should compare the following size information:

- Pump flow capacity: $Q_p = D_p \cdot w_{shaft} = 1.0 \cdot 10^{-3} \, m^3/s$
- Servo valve rated flow at rated pressure drop across the valve:

$$Q_{rv} = 1.0 \cdot 10^{-3} \, m^3/s \; at \; p_r = 1000 \, psi \, (6.895 \, MPa) \tag{7.622}$$

- Speed and force requirements of the cylinder:

$$V_{max} <= Q_{rv}/A_a \tag{7.623}$$
$$F_{load} < (p_{max} - p_r) \cdot A_a + (m_p + m_l) \cdot \ddot{y}_{max} \tag{7.624}$$

At rated valve flow, the maximum cylinder speed is

$$V_{max} = Q_{rv}/A_a \tag{7.625}$$
$$= 1.0 \cdot 10^{-3}/0.001 \tag{7.626}$$
$$= 1.0 \, m/s \tag{7.627}$$

The difference between the

$$[(p_{max} - p_r) \cdot A_a - F_{load}] \tag{7.628}$$

is the hydraulic force available for accelerating the inertia and friction losses. Hence, at most the acceleration we can support is

$$\ddot{y}_{max} = [(p_{max} - p_r) \cdot A_a - F_{load}]/(m_p + m_l) \tag{7.629}$$

assuming no loss for friction. This "head-room" in hydraulic force available determines the maximum acceleration the EH system can support at rated conditions and affects the quality of transient response.

For instance if

$$Q_p = 1.0 \cdot 10^{-3} \, m^3/s \tag{7.630}$$
$$Q_{rv} = 10.0 \cdot 10^{-3} \, m^3/s \tag{7.631}$$

at rated pressure drop, this would mean that the valve is too large for this pump, or the pump is too small for the valve. Likewise, the relief valve should be set to limit the line pressure to a desired value, and should be sized to so that it has the rated flow capacity to be able to dump the pump flow capacity when the servo valve is in neutral position. Similarly, if the desired maximum cylinder speed is larger than the one afforded by rated flow rate, the pump–valve combination is too small to be able support the desired motion, that is if the required maximum cylinder speed is $V_{max} = 2 \, m/s$, and $Q_p = Q_{rv}$, matched pump–valve, but $Q_{rv}/A_a = 1 \, m/s$. In order to meet the cylinder speed requirement, we can either

1. Reduce the cylinder cross-sectional area by a factor of two, hence increase the cylinder speed by two for the same flow rate. But this in turn reduces the force output capacity of the cylinder by two.
2. Increase the pump and valve size such that their flow rate is increased by two.

Similarly, the maximum force the EH system can afford to exert even under a zero accelera-tion condition (no head-room for force) may be smaller than desired, that is $F_{load} = 15\,000$, and available force at rated conditions,

$$(p_{max} - p_r) \cdot A_a = (20.865 - 6.895) \cdot 10^6 \cdot 0.001 = 13\,590\,\text{N} \tag{7.632}$$

One way to determine the approximate proportional gain of the closed loop PID controller is to decide on the maximum position following error for which the valve should be fully open, neglecting transient effects. This relationship would give a good estimate of the proportional gain K_p. Then the rest of the PID gains K_i, K_d can be tuned, including the tuning of K_p around the estimated value until the desired closed loop performance is achieved. Let

$$(y_d(t) - y(t))_{max} = e_{max} \tag{7.633}$$

for which we would like the controller to open the valve fully, $x_v = x_{vmax}$. Neglecting the transient aspect of the amplifier and valve dynamics,

$$i_{cmd} \approx K_p \cdot e_{max} \tag{7.634}$$

$$x_v = x_{vmax} = x_{vmax} \cdot K_t \cdot K_a \cdot i_{cmd} \tag{7.635}$$

$$x_{vmax} = x_{vmax} \cdot K_t \cdot K_a \cdot e_{max} \cdot K_p \tag{7.636}$$

$$K_p = \frac{1}{K_t \cdot K_a \cdot e_{max}} \tag{7.637}$$

Note that this calculated value for K_p is only a guidance as a starting point, and a very useful one. The actual values of the PID gains (K_p, K_i, K_d) should be further tuned to achieve the desired dynamic performance on the actual hardware.

Remarks on Simulations In order to understand the effect of different circuit design parameters as well as to be able to confirm the accuracy of the simulation results by our qualitative physics based reasoning, it is instructive to examine the effects of the following different conditions in simulation;

1. In order to confirm that the overall dynamic model makes sense, simulate the system for a condition that the servo valve is always in neutral position (no flow through the servo valve), and initial conditions of the cylinder are stationary at a nominal position. The simulation results should confirm that the cylinder does not move, and pressures in A and B sides do not change, no flow through the servo valve. All of the pump flow should be dumped back to the tank through the relief valve.

2. The next level of check in the correctness of the model and simulation is to specifiy the servo valve spool position as a function of time, $x_v(t)$, instead of having it determined via a control current command and valve dynamics. Given a prescribed valve position as a function of time, we can approximately predict and expect that the cylinder velocity should have a similar profile to the motion of the valve spool, and the cylinder position should be similar to the integral of the valve spool displacement. This simulation excludes the effect of servo valve dynamics and the closed loop PID controller.

3. Much smaller and much larger hose volumes between pump and servo valve, and between servo valve and cylinder, can be simulated.

4. Different damping coefficient for the servo valve inertial dynamics: very small (i.e., $c_v = 0.1$) or very large (i.e., $c_v = 1.0$) can be simulated.

5. Different damping in the cylinder and load dynamics, c_p.

6. Gains of PID controller; especially K_d value since velocity can reach large values for very short periods of time due to high frequency content.

Remark on "Dither" Signal in Hydraulic Control   There are two physical sources of *hysteresis* in a hydraulic valve:

1. stiction friction,

2. electromagnetic hysteresis.

Electromagnetic materials have hysteresis. Let us assume that current is zero and magnetic field is zero in a solenoid. When current is applied from zero to a finite value, a magnetic field will develop approximately proportional to the current magnitude. When the current is reduced back to zero, the magnetic field will not be exactly zero, but some residual field will be left. This is the nature of electromagnetism and materials used in electromagnetic actuators such as solenoids.

Stiction friction presents a position control problem for valves. Due to stiction friction, when the control signal (current) is small, the spool movement may be prevented. In order to avoid this, it is very common in hydraulic circuits to add ("superimpose") a *dither* signal to the regular control signal in order to keep the valve spool moving about a nominal operating condition.

The dither signal is a periodic signal (i.e., sinusoidal) with a frequency typically in the range of 100 Hz to 300 Hz, and its magnitude is in the range of 2–10% of the maximum current signal,

$$i_{dither} = A_{dither} \cdot \sin(2\pi \, w_{dither} \, t) \tag{7.638}$$

where $A_{dither} = 0.02 \sim 0.10 \cdot i_{max}$, and $w_{dither} = 100 \sim 300$ Hz. The frequency and magnitude of the dither and its magnitude selection depends on the specific type of valve application and is mostly determined by experimentation. The key is that the dither frequency should be large enough not to affect the normal position response within the bandwidth of the valve, but small enough still to cause some motion to break the stiction friction. The dither signal can either be generated digitally in software or by an analog operational amplifier circuit and added to the normal current signal.

```
%%%%%%%%%%%%%%%%%%%%%%%%%%%%%%%%%%%%%%%%%%%%%%%%%%%%%%%%%%%%%%%%%%%%%%%%%%%%%%
% Filename: EH_Servo_Sim.m

global  D_p   w_shaft p_max
global  Q_rv  x_vmax T_vmax p_r c_v  w_n   tau_a  K_a  K_t  i_v  x_v
global  beta  V_hose_pv  V_hose_va  V_hose_vb  A_a A_b  l_cyl m_p  c_p m_l
             F_load

global  Q_p   Q_v    Q_r
global  p_p   p_a    p_b  p_t

global  y_d  ydot_d y  ydot  K_fb  K_p  K_i  K_d  u_i  i_cmd

% Parameters of the components of the  EH hydraulic circuit

   D_p = 20 * 10^-6  ;  % m^3/rev
   p_max = 20.685 *10^6 ;   % [N/m^2] = Pa
   x_vmax = 10 * 10^-3 ; %   m
```

```
c_v = 0.60 ; % damping ratio of the valve spool motion
tau_a =  0.01 ; %  sec
K_a =  1.0 ;    % current amp gain
K_t =  1.0 ;    % current to torque gain
Q_rv = 1.0 * 10^-3 ; % m^3/sec
p_r = 6.895 *10^6 ; %  Pa = N/m^2

beta = 1.5 *(10^9) ;  % Bulk modulus
V_hose_pv = 0.001 ; % [m^3]
V_hose_va = 0.001 ; % [m^3]
V_hose_vb = 0.001 ; % [m^3]

A_a = 0.001   ;
A_b = 0.0005 ;
l_cyl = 1.0  ; %  [m]
m_p =  10    ; % kg
m_l =  990 ; % kg
c_p =  100.0 ;

% PID controller parameters and initialization

K_fb =   1.0  ;
K_p  =  3.0  ;
K_i  =   0.0  ;
K_d  =   0.0  ;
u_i  =   0.0  ;

% Input conditions

   w_shaft = 3000/60 ; % rev/sec
   F_load = (m_p + m_l) * 9.81 ;  % kg m/sec^2 = N

% Initial conditions on head and rod-end volume of cylinder.

  i_v = 0.0  ;
  x_v = 0.0  ;
  xdot_v = 0.0 ;

  y = 0.2 ;           % Initial cylinder position
  ydot = 0.0 ;        % Initial cylinder velocity
  y_d = 0.2 ;
  ydot_d = 0.0 ;
                      % Initial pressures,  atmospheric pressure: 101 kPa
  p_p = 101 * 10^3;        % Pump pressure: Pa
  p_a = F_load/A_a ;
  p_b = 101 * 10^3 ;
  p_t = 101 *10^3  ;          % kPa Tank pressure
  Q_r = 0.0 ;
  Q_v = 0.0 ;

% Simulation over a time period: [t_0, t_f]

  t_0 =0.0;
  t_f =5.0 ;
```

```
      t_sample = 0.001 ;
      z=zeros(8,1);

      z(1) = p_p ;
      z(2) = i_v ;
      z(3) = x_v ;
      z(4) = xdot_v ;
      z(5) = p_a  ;
      z(6) = p_b  ;
      z(7) = y    ;
      z(8) = ydot ;

%   y_out = [ Q_p  Q_v  Q_r   p_p  p_a  p_b   i_cmd   x_v*10^3   y_d    y ];
      y_out = [ 0.0  0.0  0.0   p_p  p_a  p_b    0.0     0.0    0.2   0.2 ];

  for (t=t_0: t_sample:t_f)

      % Solve ODEs...

      t_span=[t,t+t_sample] ;
      i_cmd = EH_Servo_PID_Controller(t) ;
      [T,z1] = ode45('EH_Servo_Dynamics',t_span, z);
      [m,n]=size(z1);
      z(:)=[z1(m,:)] ;
      t
    y_out=[y_out ;
          Q_p   Q_v   Q_r   z(1)  z(5)  z(6)   i_cmd   z(3)*100   y_d   z(7) ];
  end

  [m,n]=size(y_out);
  t_inc = (t_f-t_0)/m ;
  tout=t_0:t_inc:t_f-t_inc;
  tout = tout' ;

figure(1) ;

subplot(2,2,1) ;
plot(tout, y_out(:,1), 'k',tout, y_out(:,2), 'b',tout, y_out(:,3), 'm');
xlabel('Time (sec)') ; ylabel('Flow rate (m^3/sec)') ;
legend('Q_p','Q_v', 'Q_r');

subplot(2,2,2) ;
plot(tout, y_out(:,4), 'k',tout, y_out(:,5), 'b',tout, y_out(:,6), 'g');
xlabel('Time (sec)') ; ylabel('Pressure [Pa]') ;
legend('p_p','p_a','p_b');

subplot(2,2,3) ;
plot(tout, y_out(:,7), 'k',tout, y_out(:,8), 'b');
xlabel('Time (sec)') ; ylabel('Valve currentcmd  and spool position') ;
legend('i_cmd','x_v');

subplot(2,2,4) ;
plot(tout,  y_out(:,9), 'k',tout,  y_out(:,10), 'b');
xlabel('Time (sec)') ; ylabel('Cylinder position: cmd and mea.') ;
legend('y_d','y');
```

```
%%%%%%%%%%%%%%%%%%%%%%%%%%%%%%%%%%%%%%%%%%%%%%%%%%%%%%%%%%%%%%%%%%%%%%%%%%%%%%
% Filename: EH_Servo_PID_Controller.m

function i_cmd = EH_Servo_PID_Controller(t)

global y_d ydot_d   y  ydot  K_fb  K_p  K_i  K_d   u_i

  %  Desired cylinder position command to PID controller

    if t<1.0
       y_d = 0.2 ;
       ydot_d = 0.0 ;
    elseif (t>= 1.0 && t<=2.0)
       y_d=  0.2 + 0.5 *(t-1.0) ;
       ydot_d = 0.5 ;
    elseif (t> 2.0 && t<=3.0)
       y_d  = 0.5 ;
       ydot_d = 0.0 ;
    elseif (t> 3.0 && t<=4)
       y_d = 0.7 - 0.5 * (t - 3.0) ;
       ydot_d = - 0.5 ;
    else
       y_d = 0.2 ;
       ydot_d = 0.0 ;
    end

  %  Cylinder position sensor reading; feedback signal.

    y_fb    = K_fb * y ;
    ydot_fb = K_fb * ydot ;

  % PID control algorithm

    u_i = u_i + (y_d - y_fb) ;
    i_cmd = K_p *(y_d - y_fb) + K_i * u_i + K_d * (ydot_d - ydot_fb)  ;

return ;

%%%%%%%%%%%%%%%%%%%%%%%%%%%%%%%%%%%%%%%%%%%%%%%%%%%%%%%%%%%%%%%%%%%%%%%%%%%%%%
%%%  Filename: EH_Servo_Dynamics.m

function zdot = EH_Servo_Dynamics(t,z)

global  D_p  w_shaft p_max
global  Q_rv  x_vmax  p_r c_v  w_n   tau_a  K_a  K_t  i_v  x_v
global  beta  V_hose_pv  V_hose_va  V_hose_vb  A_a A_b  l_cyl m_p  c_p m_l
             F_load

global  Q_p   Q_v    Q_r
global  p_p   p_a    p_b  p_t

global y_d  ydot_d  y  ydot  K_fb  K_p  K_i  K_d  u_i  i_cmd

    p_p    =  z(1) ;
    i_v    =  z(2) ;
```

```
   x_v     =   z(3) ;
   xdot_v  =   z(4) ;
   p_a     =   z(5) ;
   p_b     =   z(6)   ;
    y      =   z(7)  ;
   ydot    =   z(8)  ;


   zdot=zeros(8,1) ;

% ODEs....

%   Pump flow rate and  pressure, and relief-valve model (instantaneously
%   acting)

    Q_p = D_p * w_shaft ;

    if p_p < p_max
      zdot(1)= (beta/V_hose_pv)*(Q_p - Q_v) ;
      Q_r = 0.0  ;
    else
       zdot(1) = 0.0 ;
       Q_r = Q_p - Q_v ;
    end

% Amplifier and servo valve electrical dynamics

    zdot(2) = K_a * (1/tau_a)* (- z(2) + i_cmd) ;
    T_v = K_t * z(2) ;

% Servo valve stage two spool position and torque relationship

    if ((x_v/x_vmax) <= 0.4)
       w_n = 2*pi*75 ;  % rad/sec
    else
       w_n = 2*pi* (75 - ( ((x_v/x_vmax) - 0.4) * ((75-25)/(1.0-0.4)) ));
       % rad/sec
    end

    zdot(3) =  z(4) ;
    zdot(4) = (- 2 * c_v * w_n * z(4) - w_n^2 * z(3)) +  (x_vmax/T_vmax)
            * w_n^2 * T_v  ;


    if x_v >= 0.0
       Q_pa = Q_rv * abs(x_v/x_vmax) * ((p_p - p_a)/p_r)^0.5 ;
       Q_bt = Q_rv * abs(x_v/x_vmax) * ((p_b - p_t)/p_r)^0.5 ;
       Q_v = Q_pa  ;
       zdot(5) = (beta/(V_hose_va + y * A_a))*(Q_pa - ydot * A_a)  ;
       zdot(6) = (beta/(V_hose_vb + (l_cyl - y) * A_b))*(-Q_bt + ydot * A_b);
    else
       Q_at = Q_rv * abs(x_v/x_vmax) * ((p_a - p_t)/p_r)^0.5 ;
       Q_pb = Q_rv * abs(x_v/x_vmax) * ((p_p - p_b)/p_r)^0.5  ;
       Q_v = Q_pb  ;
       zdot(5) = (beta/(V_hose_va + y * A_a))*(-Q_at - ydot * A_a) ;
       zdot(6) = (beta/(V_hose_vb + (l_cyl - y) * A_b))*(Q_pb + ydot * A_b);
    end
```

```
% Piston and load inertia dynamics

    zdot(7) = z(8);
    zdot(8) = (1/(m_p+m_l))*(-c_p * z(8) + p_a * A_a - p_b * A_b - F_load) ;

    return;
%%%%%%%%%%%%%%%%%%%%%%%%%%%%%%%%%%%%%%%%%%%%%%%%%%%%%%%%%%%%%%%%%%%%%%%%%%%%%%%%%%%%%
```

Example: One Degree of Freedom Hydraulic Motion Control System with Flexible Base and Load Contact

Consider a single-acting hydraulic cylinder which is connected to a non-rigid base as well as a load (Figure 7.105). The base of the cylinder is not connected to a rigid ground, but to a flexible and moving base. This is the case in mobile work equipment where the hydraulic cylinder is connected to the machine frame and the machine frame rides on tires which are highly flexible. In addition, the rod-end is connected to the load through a non-rigid tool mechanism or the the load is not rigid, such as a pile of soil. The interaction between the cylinder and the load is modeled as an inertia, a spring, and a damper. The machine frame, tires, and the cylinder outer shell inertia are modeled as mass 1, m_1, and spring and damper constants k_1 and c_1. The piston and rod are modeled as moving inertia (mass) 2, m_2. The load contact and load inertia are modeled as mass 3, spring and damper constants, m_3, k_3, c_3, respectively. The same dynamic model principle is used in modeling hydraulically powered roll-thickness control systems in steel mills and other web-thickness control applications. In the roll-thickness

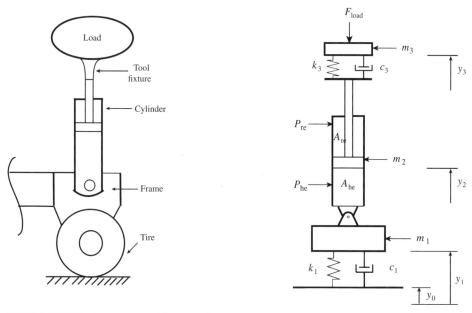

FIGURE 7.105: One degree of freedom hydraulic motion system with flexible base and load connection. A hydraulic cylinder is connected to a frame which rides on tires. The contact between the load and cylinder is a non-rigid tool fixture. The base frame and tire, as well as the tool fixture, are modeled as inertia with stiffness and damping. Cylinder inertia is included in the base inertia and modeled as mass 1, and piston and rod inertia is included as the mass 2, and load inertia is modeled as mass 3. This type of dynamic condition exists in many applications such as mobile equipment riding on tires, hydraulically controlled thickness control mechanisms in steel mill rolls. In steel mill roll thickness control systems, the base motion would be stationary ($y_0(t) = 0.0$).

control application case, the base motion would be zero, whereas the base motion is non zero in mobile equipment applications due to ground height variations as the equipment travels on the ground.

There are three inertias in the system: m_1, m_2, m_3, connected to each other via springs and dampers, as well as the hydraulic cylinder. The motion of the hydraulic system has three degrees of freedom, $y_1(t), y_2(t), y_3(t)$, which are the positions of each inertia relative to a reference ground. The ground height variation $y_0(t)$ is an input variable as a function of the actual ground conditions and travel speed of the equipment. In web-thickness roll control applications, it is zero, $y_0(t) = 0.0$. The motion of each mass is defined by a second-order differential equation. The coupling between the motion of the three inertias is defined by the springs and dampers, as well as the pressure in the cylinder. The cylinder pressure is modeled as a compressible fluid volume. The flow in and out of the cylinder is controlled by a flow control valve. The flow control valve is modeled to have a spool displacement versus orifice area relationship with a symmetric deadband. In other words, around the null-position of the spool, there is a deadband in equal amounts in both directions of the spool motion. While the spool position is in the deadband region, there is no orifice opening, hence no flow.

The resultant dynamic equations are the three coupled second-order and two first-order differential equations, where the flow and pressure conditions affect the motion of the inertia through pressure in the cylinder head-end and rod-end.

$$m_1 \ddot{y}_1(t) = -c_1(\dot{y}_1(t) - \dot{y}_0(t)) - k_1(y_1(t) - y_0(t)) - (A_{he} p_{he}(t) - A_{re} p_{re}(t)) \quad (7.639)$$

$$m_2 \ddot{y}_2(t) = -c_3(\dot{y}_2(t) - \dot{y}_3(t)) - k_3(y_2(t) - y_3(t)) + (A_{he} p_{he}(t) - A_{re} p_{re}(t)) \quad (7.640)$$

$$m_3 \ddot{y}_3(t) = -c_3(\dot{y}_3(t) - \dot{y}_2(t)) - k_3(y_3(t) - y_2(t)) - F_{load}(t) \quad (7.641)$$

where,

$$\frac{dp_{he}(t)}{dt} = \frac{\beta}{V_{he}(t)} \cdot (Q_{he}(t) - A_{he}(\dot{y}_2(t) - \dot{y}_1(t))) \quad (7.642)$$

$$\frac{dp_{re}(t)}{dt} = \frac{\beta}{V_{re}(t)} \cdot (-Q_{re}(t) + A_{re}(\dot{y}_2(t) - \dot{y}_1(t))) \quad (7.643)$$

For the spool position ranges in $x_s \geq x_{db}$, the valve spool has positive displacement past the deadband region and flow is from pump to head-end,

$$Q_{he}(t) = C_{d1} \cdot A_{he1}(x_s) \cdot \sqrt{|p_P(t) - p_{he}(t)|} \quad (7.644)$$

$$Q_{re}(t) = C_{d1} \cdot A_{re1}(x_s) \cdot \sqrt{|p_{re}(t) - p_t(t)|} \quad (7.645)$$

for the spool position ranges in $x_s \leq -x_{db}$, the valve spool has negative displacement past the deadband region and the flow is from pump to rod-end,

$$Q_{he}(t) = -C_{d2} \cdot A_{he2}(x_s) \cdot \sqrt{|p_{he}(t) - p_t(t)|} \quad (7.646)$$

$$Q_{re}(t) = -C_{d2} \cdot A_{re2}(x_s) \cdot \sqrt{|p_P(t) - p_{re}(t)|} \quad (7.647)$$

for the spool position ranges in $-x_{db} \leq x_s \leq x_{db}$, the valve spool is in the deadband region and there is no flow to either side of the cylinder,

$$Q_{he}(t) = 0.0 \quad (7.648)$$

$$Q_{re}(t) = 0.0 \quad (7.649)$$

and

$$V_{he}(t) = V_{he}(0) + A_{he} \cdot (y_2(t) - y_1(t)) \quad (7.650)$$

$$V_{re}(t) = V_{re}(0) + A_{re} \cdot (y_2(t) - y_1(t)) \quad (7.651)$$

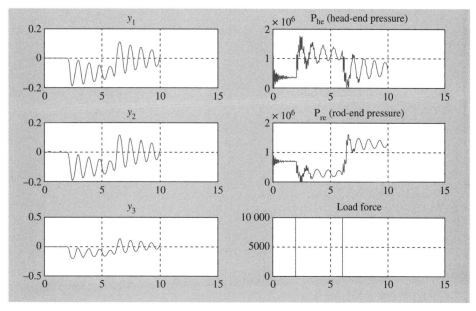

FIGURE 7.106: Simulation results: hydraulic cylinder connected to a flexible base and load. The position of each mass, pressure, and flow rate on both sides (head-end and rod-end) of the cylinder.

where $V_{he}(0)$, $V_{re}(0)$ are the initial conditions of the head-end and rod-end volume of the cylinder when $y_1(0) = y_2(0) = 0.0$

Simulation results for the following model parameters and initial conditions are shown in Figure 7.106. The model parameters used in the numerical simulations are

$$m_1 = 1000\,\text{kg} \tag{7.652}$$

$$m_2 = 100\,\text{kg} \tag{7.653}$$

$$m_3 = 1000\,\text{kg} \tag{7.654}$$

$$k_1 = 10^5\,\text{N/m} \tag{7.655}$$

$$k_3 = 10^6\,\text{N/m} \tag{7.656}$$

$$c_1 = 1000\,\text{N/(m/s)} \tag{7.657}$$

$$c_3 = 1000\,\text{N/(m/s)} \tag{7.658}$$

$$\beta = 7.0 \cdot 10^8\,\text{N/m}^2 \tag{7.659}$$

$$A_{he} = 0.01\,\text{m}^2 \tag{7.660}$$

$$A_{re} = 0.005\,\text{m}^2 \tag{7.661}$$

$$C_{d1}A_{he1}(x_s(t)) = 0.01 \cdot (x_s - x_{db}) \tag{7.662}$$

$$C_{d1}A_{re1}(x_s(t)) = 0.01 \cdot (x_s - x_{db}) \tag{7.663}$$

$$C_{d2}A_{he2}(x_s(t)) = 0.01 \cdot (x_s - x_{db}) \tag{7.664}$$

$$C_{d2}A_{re2}(x_s(t)) = 0.01 \cdot (x_s - x_{db}) \tag{7.665}$$

$$x_{db} = 10\%\ \textit{deadband in percentage of} \tag{7.666}$$

$$\textit{maximum spool travel} \tag{7.667}$$

The initial conditions are

$$y_1(0) = 0.0 \tag{7.668}$$
$$y_2(0) = 0.0 \tag{7.669}$$
$$y_3(0) = 0.0 \tag{7.670}$$
$$\dot{y}_1(0) = 0.0 \tag{7.671}$$
$$\dot{y}_2(0) = 0.0 \tag{7.672}$$
$$\dot{y}_3(0) = 0.0 \tag{7.673}$$
$$p_{he}(0) = (m_2 + m_3) \cdot 9.81 / A_{he} = (9.81) \cdot 110\,000\,\text{N/m}^2 \tag{7.674}$$
$$p_{re}(0) = 0.0 \tag{7.675}$$
$$V_{he}(0) = 0.01\,\text{m}^3 \tag{7.676}$$
$$V_{re}(0) = 0.01\,\text{m}^3 \tag{7.677}$$

The input conditions are

$$y_0(t) = 0.0 \tag{7.678}$$
$$F_{load}(t) = 0.0;\ \ 0 \le t \le 2.0\,\text{s} \tag{7.679}$$
$$= 10\,000\,\text{N};\ \ 2.0 < t \le 6.0\,\text{s} \tag{7.680}$$
$$= 0.0;\ \ 6.0 < t \le 10.0\,\text{s} \tag{7.681}$$
$$x_s(t) = 0.0 \tag{7.682}$$
$$p_P(t) = 20 \cdot 10^6\,\text{N/m}^2\ \textit{constant} \tag{7.683}$$

which simulates a case of response of all three inertias and pressures on both sides of the cylinder when the valve is in neutral position and a sudden change occurs on the load (Figure 7.106). Other conditions of valve control and the resulting behavior of the system can be easily simulated by specifying the valve spool motion as a function of time or as a control signal from a closed loop control logic. Here, we assumed that we used a valve with symmetric orifice geometry between all ports (pump to head-end, rod-end to tank, head-end to tank, pump to rod-end). In this particular simulation, the flow coefficients do not affect the simulation results because the valve is in neutral position and leakage through the valve at neutral position is neglected. However, the valve orifice geometry between ports has an important effect on the system transient response when the valve spool is in flow regulating mode of operation, that is when the valve spool is positioned outside the deadband region and allows flow between ports.

```
%%%%%%%%%%%%%%%%%%%%%%%%%%%%%%%%%%%%%%%%%%%%%%%%%%
%Filename: Sim381.m

% Sim381.m

t_0 =0.0;
t_f =10.0 ;
t_sample = 0.001 ;
z=zeros(8,1);
z(1) = 0.0 ;
z(2) = 0.0 ;
z(3) = 0.0 ;
z(5) = 0.0 ;
z(5) = 0.0 ;
z(6) = 0.0 ;
z(7) = 1100*9.81/0.01   ;
z(8) = 0.0 ;
```

```
  z_out = z' ;
  u_out = 0 ;

 for (t=t_0: t_sample:t_f)
     t_span=[t,t+t_sample] ;
     [t,z1] = ode45('cyl_flex',t_span, z);
     [m,n]=size(z1);
     z(:)=z1(m,:);
    if z(7)<0.0     % Pressure can not be negative.
         z(7)=0.0;
    end
    if z(8)<0.0
         z(8)=0.0;
    end

    if t<2.0      % Load force for plotting purposes.
       F_load= 0.0 ;
    elseif (t>= 2.0 & t<6.0)
       F_load = 10000 ;
    else
       F_load = 0.0 ;
    end
    u(1) = F_load ;

    z_out=[z_out ; z'] ;
    u_out=[u_out ; u];
 end
 t_out=t_0:t_sample:t_f+t_sample;
 t_out = t_out' ;
 subplot(321) ; plot(t_out,z_out(:,1));  title('y1'); grid on;
 subplot(323) ; plot(t_out,z_out(:,2)); title('y2'); grid on;
 subplot(325) ; plot(t_out,z_out(:,3)); title('y3');  grid on;
 subplot(322) ; plot(t_out,z_out(:,7)); title('Phe (Head end pressure)');
                grid on;
 subplot(324) ; plot(t_out,z_out(:,8)); title('Pre (Rod end pressure)');
                grid on;
 subplot(326) ; plot(t_out,u_out(:,1)); title('Load Force');  grid on;

%%%%%%%%%%%%%%%%%%%%%
%%% Filename: cyl_flex.m

function zdot=cyl_flex(t,z)

% Parameters
   m1 = 1000 ;
   m2 = 100 ;
   m3 = 1000 ;
   k1 = 1.0E5 ;
   k3 = 1.0E6 ;
   c1 = 1000 ;
   c3 = 1000 ;

   beta = 7*10^8 ;
   A_he = 0.01 ;
   A_re = 0.005 ;
```

```
   x_db = 10 ;

  C_he1 = 0.01 * abs(x_s - x_db) ;
  C_re1 = 0.01 * abs(x_s - x_db) ;
  C_he2 = 0.01 * abs(x_s - x_db) ;
  C_re2 = 0.01 * abs(x_s - x_db) ;

% Initial conditions on head and rod-end volume of cylinder.

  V_he0 = 0.01 ;
  V_re0 = 0.01 ;

% Input conditions

  x_s = 0.0 ;      % Valve spool position. Change this as function of time
                   %    to simulate a different condition.
  p_p = 20*10^6 ;  % Pump pressure
  p_t = 0.0 ;      % Tank pressure

  y0 = 0.0 ;  % Base motion
  doty0= 0.0 ;

  V_he = V_he0 + A_he*(z(2) - z(1));
  V_re = V_re0 + A_re*(z(2) - z(1));

  if x_s <= - x_db     %  Valve: flow-rate and pressure relation
    Q_he = -C_he2 * (abs(z(7)-p_t))^0.5 ;
    Q_re = -C_re2 * (abs(p_p- z(8)))^0.5 ;
  elseif (-x_db < x_s && x_s < x_db)
     Q_he = 0.0 ;
     Q_re = 0.0 ;
  elseif x_s >= x_db
    Q_he = C_he1 * (abs(p_p - z(7)))^0.5 ;
    Q_re = C_re1 * (abs(z(8)-p_t))^0.5 ;
  end

  if t<2.0      % Load force
     F_load= 0.0 ;
  elseif (t>= 2.0 & t<6.0)
     F_load = 10000 ;
  else
     F_load = 0.0 ;
  end

  zdot=zeros(8,1);
  zdot(1) = z(4) ;
  zdot(2) = z(5) ;
  zdot(3) = z(6) ;
  zdot(4) = (1/m1)*(-c1*(z(4)-doty0)-k1*(z(1)-y0)-(A_he*z(7)-A_re*z(8)));
  zdot(5) = (1/m2)*(-c3*(z(5)-z(6))-k3*(z(2)-z(3))+(A_he*z(7)-A_re*z(8)));
  zdot(6) = (1/m3)*(-c3*(z(6)-z(5)) -k3*(z(3)-z(2))-F_load)     ;
  zdot(7) = (beta/V_he)*(Q_he-A_he*(z(5)-z(4))) ;
  zdot(8) = (beta/V_re)*(-Q_re+A_re*(z(5)-z(4))) ;

%%%%%%%%%%%%%%%%%%%%%%%
```

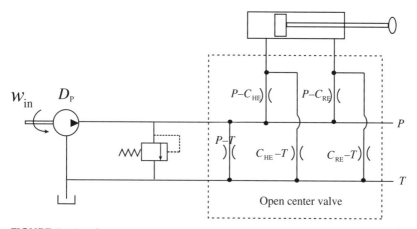

FIGURE 7.107: Open-center hydraulic system: fixed displacement pump, main flow control valve with pump-to-tank orifice at neutral position.

7.10 EXAMPLE: OPEN CENTER HYDRAULIC SYSTEM – FORCE AND SPEED MODULATION CURVES IN STEADY STATE

Consider the single degree of freedom hydraulic motion system shown in Figure 7.107[2]. An open center hydraulic system has a fixed displacement pump and a main flow metering valve which has an orifice opening between pump and tank ports (P–T) when its spool is in neutral position. Since the pump is of fixed displacement type, as long as the input shaft to the pump is rotating (i.e., typically the case in mobile equipment applications because the pump is driven by the engine), the pump generates flow according to

$$Q_s = \eta_v \cdot D_p \cdot w_{in} \tag{7.684}$$

where Q_s is the pump flow rate, D_p is the pump volumetric displacement, w_{in} input shaft speed of the pump, and η_v is the volumetric efficiency of the pump. The pump output pressure is determined by the load. When the main valve is in neutral position, this flow must go through the P–T port to the tank. Since this flow through the P–T port performs no useful work, it is a lost power. The lost power (P_{loss}) is the flow rate times the pressure drop across the valve,

$$Q_v = Q_s \tag{7.685}$$
$$P_{loss} = Q_v \cdot \Delta p_{PT} \tag{7.686}$$

where Q_v is the flow rate across the valve P–T orifice, Δp_{PT} is the pressure drop across the P–T orifice of the valve. In order to minimize the power loss, the pressure drop across the valve through the P–T port should be minimized. The design balance is obtained by a compromise between the desire to minimize the power loss, which requires as small a pressure drop as possible which in turn requires as large an orifice size as possible, while controlling accuracy and physical size (compactness requirement) limitations require as small a valve as possible. A typical compromise design choice in mobile equipment applications is that the valve is sized to support the rated flow with P–T pressure drop value in the range of 100–200 psi.

---

[2] This section is based on lectures by Dr. Richard Ingram.

The worst case power loss in open center systems occurs if the flow has to go though the relief valve. Relief valve pressure setting can be in the range of 3000–5000 psi. If we assume the relief valve pressure is set at $p_{\text{relif}} = 3000$ psi, and the main flow valve sized such that $\Delta p_{\text{PT}} = 100$ psi (to support full flow rate across the P–T orifice when the spool is in neutral position), hydraulic power loss (wasted power) can be 30 times higher if the flow has to go through the relief valve as opposed to P–T port of the main flow control valve.

$$P_{\text{loss}} = Q_s \cdot 100 \quad \text{flow through } P\text{–}T \tag{7.687}$$

$$P_{\text{loss}} = Q_s \cdot 3000 \quad \text{flow through relief valve} \tag{7.688}$$

Clearly, a hydraulic system with a fixed displacement pump, a relief valve, and a closed center valve would also be functional. However, since in a closed center valve, the P–T port does not exist ($A_{\text{PT}} = 0.0$), the flow would have to go through the relief valve when the main flow control valve is in neutral position. Such a system would be about 30 times less efficient than the same system with an open center main flow control valve.

Let us consider the orifice geometries as a function of the spool position. There are five orifice areas defined by the position of the spool: pump to tank (P–T), pump to cylinder head end (P–C_{HE}), pump to cylinder rod end (P–C_{RE}), cylinder head-end to tank (C_{HE}–T) and cylinder rod end to tank (C_{RE}–T) (Figures 7.107 and 7.108). Independently controlled valves (IMVs) control each one of these orifice areas by controlling a separate spool for each orifice. In a standard single spool valve, these orifice areas are directly a function of the spool. Figure 7.108 shows a possible set of orifice area curves which are directly a function of the geometry of the valve spool and valve body orifices. Let us consider the case that the spool moves to the right side in the following increments of its total displacement: 0, 25, 50, 100%. As that happens, the orifice openings between ports, pump, tank, cylinder head-end, and cylinder rod-end vary as follows,

$$A_{\text{PT}} \neq 0 \tag{7.689}$$

$$A_{\text{PC}_{\text{HE}}} \neq 0 \tag{7.690}$$

$$A_{\text{C}_{\text{RE}}\text{T}} \neq 0 \tag{7.691}$$

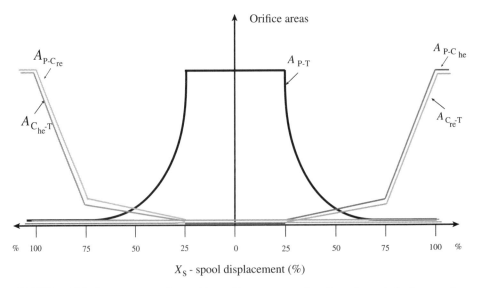

FIGURE 7.108: Open-center valve and its orifice areas as a function of spool displacement: P-T, P-C_{HE}, P-C_{RE}, C_{HE}-T, C_{RE}-T.

$$A_{PC_{RE}} = 0 \qquad\qquad (7.692)$$

$$A_{C_{HE}T} = 0 \qquad\qquad (7.693)$$

Notice that the valve has a large deadband. As the spool moves for up to about 25% of its total travel range, none of the orifices to either side of the cylinder open. In many cases, this deadband is built on purpose into the valve design in order to reduce leakage and unintended motion, at the expense of dynamic response. When the spool moves in the opposite direction, the orifice opening values change, establishing different hydraulic connection between ports P, T, C_{HE}, C_{RE}. Let us consider the case that spool moves to the left side in the following increments of its total displacement: $0, -25, -50, -100\%$. As that happens, the orifice openings between ports, pump, tank, cylinder head-end, and cylinder rod-end vary as follows,

$$A_{PT} \neq 0 \qquad\qquad (7.694)$$

$$A_{PC_{HE}} = 0 \qquad\qquad (7.695)$$

$$A_{C_{RE}T} = 0 \qquad\qquad (7.696)$$

$$A_{PC_{RE}} \neq 0 \qquad\qquad (7.697)$$

$$A_{C_{HE}T} \neq 0 \qquad\qquad (7.698)$$

resulting in cylinder motion in the opposite direction.

Flow rate across each orifice is defined by the following relationship,

$$Q_{v,ij} = C_D \cdot A_{ij}(x_s) \cdot \sqrt{\Delta p_{ij}} \qquad\qquad (7.699)$$

where i, j represent the ports P, T, C_{HE}, C_{RE} and their five possible connection possibilities in an open-center valve, $A_{ij}(x_s)$ represents the orifice area opening as a function of spool position between the ports, and Δp_{ij} represents the pressure drop across the port, C_D is the valve gain.

For a given physical valve, all five of the orifice areas as a function of spool displacement are known strictly based on the valve geometry: $A_{PT}(x_s), A_{PC_{HE}}(x_s), A_{PC_{RE}}(x_s), A_{C_{HE}T}(x_s), A_{C_{RE}T}(x_s)$. In addition, the valve gain C_D is known as a specification. If it is not available, C_D can be measured as a calibration parameter using the flow rate equation above: under known pressure drop conditiond, measured flow rate, and orifice opening, then calculate the C_D from the above equation. Let us assume that the pump parameters (volumetric displacement D_P and volumetric efficiency η_v) are specified and the input shaft speed of the pump (w_{pump}) is given.

Let us consider two different cases: case 1 is a blocked load, and case 2 is a moving load case with light and heavy load conditions (Figure 7.109).

Case 1: Blocked load case (force modulation curve). Since the load is blocked and the cylinder cannot move, the hydraulic system controls the force applied on the load.

Case 2: Moving load case for light load and heavy load. In this case, we will assume the force needed to accelerate/decelerate the load is negligable (inertial force is neglected) for the purpose of obtaining the steady-state force and speed modulation curves.

In both cases, consider that the spool is moved from left to right starting at center position. We can assume similar behavior in the opposite direction. Analysis of case 1 gives us the steady-state *force modulation curves* for the open-center hydraulic system. In other words, it defines in steady-state the force output at the cylinder rod (F) as a function of the spool position (x_v). As the valve moves from neutral position towards 25%, then 50% and 100% displacements, the orifice openings change according to the orifice functions shown in Figure 7.108. Yet, since the load is blocked, the cylinder will not move. All of the flow

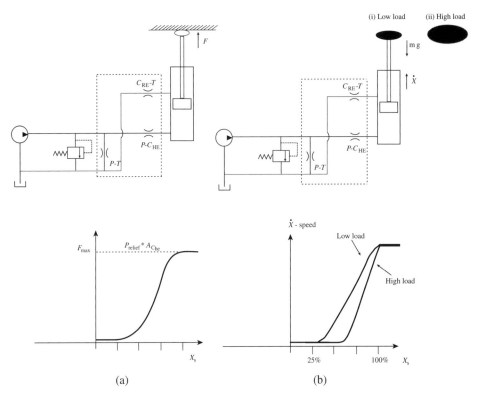

FIGURE 7.109: Open-center hydraulic system, operating conditions: (a) blocked load and steady-state force modulation curve, (b) moving load (low and high load cases) and steady-state speed modulation curves.

has to go through the P–T port, until the relief valve pressure is reached. As a result, from the flow equation, since we know the orifice opening and the flow rate that must go through the P–T orifice, we can determine the pressure drop developed across the P–T port. This same pressure is applied on the cylinder head-end as well. Hence, the applied force can be calculated as a function of spool displacement. The resulting force as a function of spool displacement is called the *force modulation curve.*

$$Q_s = \eta_v \cdot D_p \cdot w_{pump} \tag{7.700}$$

$$Q_{PT} = C_D \cdot A_{PT}(x_s) \cdot \sqrt{p_P - p_T} = Q_s \tag{7.701}$$

$$F_s(x_s) = A_{Cyl,HE} \cdot p_P(x_s) \tag{7.702}$$

From these equations, we can calculate the force modulation curve as a function of the spool displacement (Figure 7.109). For a given constant Q_s and spool position x_s, we know $A_{PT}(x_s)$. Then p_P can be determined as a function of x_s. Since $A_{PT}(x_s)$ is large when x_s is small, and gets smaller when x_s increases (Figure 7.108), the $p_P(x_s)$ gets larger as x_s gets larger (Figure 7.109a). Notice that $p_P \leq p_{relief}$, which defines the maximum force that can be applied,

$$(F_s)_{max} = A_{Cyl,HE} \cdot p_{relief} \tag{7.703}$$

Notice that the modulation curve is also a function of the engine speed, hence the modulation can be also affected by changing the engine speed. This is not possible in a load sensing (closed-center) hydraulic system.

Unlike in open-center hydraulic systems, in closed-center systems the operator cannot modulate the tool force.

For a moving load, Case 2, again we can trace the orifice openings as a function of the spool position. Let us neglect the inertia force needed to accelerate or decelerate the load. Then, the cylinder will start to move only after the developed pressure at the output of the pump is larger than the load pressure (load force/area). In the large load force case versus small load force case, the spool has to move more, restricting the P–T orifice area, making it smaller, hence increasing the pressure drop across P–T in order to support the flow rate. In other words, as the spool position x_s increases, A_{PT} decreases and P_P increases. At the same time, since the same pressure acts on the cylinder head-end side, it takes a larger spool displacement in order to initiate the motion against a larger load. Although not shown in the figure, we assume that there is a check valve between the pump and cylinder ports so that the flow can only move in one direction: from pump to cylinder when the pump pressure is larger than the cylinder port pressure. When the cylinder port pressure is larger than the pump pressure, the check valve blocks the flow from cylinder to pump direction. By analyzing the same relationship for different spool displacements, we can calculate data points to plot the *speed modulation curves* under *different loads*. The start of motion of the actuator at different spool displacements against different loads shows up as a *variable deadband* in the speed modulation curve. The effective deadband is a function of the load. Neglecting the force needed to accelerate the load and assuming the only load to overcome is the gravity load, the load will not move as long as the P–T orifice opening is such that the pump pressure developed, p_P, is smaller than $F_1/A_{Cyl,HE}$

$$Q_s = \eta_v \cdot D_p \cdot w_{pump} \tag{7.704}$$

$$Q_{PT} = Q_s = C_D \cdot A_{PT}(x_s) \cdot \sqrt{p_P - p_T} \tag{7.705}$$

$$F_s = A_{Cyl,HE} \cdot p_P < F_1 \tag{7.706}$$

As the P–T orifice continues to close, p_P will increase in order support the flow $Q_{PT} = Q_s$ through a smaller orifice until $A_{Cyl,HE} \cdot p_P$ is equal or larger than F_1, at which point the load starts to move. Since we neglected the inertial forces, the pressure developed at the pump output will be limited to $p_P = F_1/A_{Cyl,HE} + \Delta p_{PC}$, where Δp_{PC} is the pressure drop across the valve between pump and cylinder port, which is at most 100–200 psi range by design. As P–T closes and less orifice is available, more and more of the flow will go through the P–C_{HE} port, hence the speed of the cylinder will increase as spool displacement increases and P–T closes. For larger loads (F_L large), in order to develop larger p_P, the P–T port must get smaller compared to the lower loads. Hence, for larger loads, the first motion is achieved after larger spool displacement. In other words, the effective deadband (the start of motion as a function of spool displacement) is dependent on the load force, not just the geometric deadband of the spool. When the spool position is such that it fully closes the P–T port, all of the flow must go through the P–C_{HE} port, while $p_p < p_{relief}$, which means while the relief valve is closed. For large spool displacements, as long as $p_P < p_{relief}$ and $A_{PT} = 0.0$, the maximum speed of the cylinder for the low and high load condition would be the same.

Remarks Hydraulic systems which drive mechanisms such as those in construction equipment, have a linkage mechanism between the hydraulic cylinder and tool (e.g., bucket). Operator command signals are effectively translated to the speed of the cylinder. The linkage is simply a power conversion mechanism. If we assume it operates with 100% efficiency, the hydraulic power in the cylinder is converted to the tool's mechanical power. That is, power is converted from flow-rate times pressure to speed (angular or translational) times force (or torque if translational motion). The control "modulation" of the hydraulic cylinder refers to the gain between the lever displacement (operator command signal) and the change in the speed or force of the cylinder: speed modulation or force modulation. That gain is built into the hydraulic circuit including the control algorithm, control valves, pump, and cylinder parameters. Depending on the kinematic conversion relationship from the cylinder to the tool motion, that gain is further modified. Therefore, the linkage mechanism gain as a function of linkage angular position effects (multiplies) the control modulation capability from the operator commands to the motion/force delivered to the tool.

In load driven motion, such as lowering a bucket, if the downward motion speed is so high that the maximum pump flow cannot supply the other end of the cylinder, there will be cavitation in the other end of the cylinder. In order to maintain good controllability, cylinder cavitation should be avoided. Consider the lowering motion of the bucket lift cylinder in the load driven case. The maximum speed that the cylinder can move without causing cavitation in the rod-end of the cylinder is

$$V_{max} = \frac{Q_{max}}{A_{Cyl,RE}} \tag{7.707}$$

$$= \frac{\eta_v \cdot D_{p,max} \cdot w_{shaft}}{A_{Cyl,RE}} \tag{7.708}$$

Let us assume the load is W_{load}, and the head-end area of the cylinder is A_{HE}, and the valve cylinder-to-tank orifice area is fully open ($A_{v,CT}$) and that valve flow rate coefficient is known for the system, C_d, then the speed of the cylinder will be

$$p_{CT} = \frac{W_{load}}{A_{HE}} \tag{7.709}$$

$$Q = C_d \cdot A_{v,CT} \cdot \sqrt{p_{CT}} \tag{7.710}$$

$$V = \frac{Q}{A_{HE}} \tag{7.711}$$

If $V > V_{max}$, there will be voiding in the rod-end of the cylinder. If the speed of the cylinder exceeds that speed, there will be cavitation in the rod-end of the cylinder.

7.11 EXAMPLE: HYDROSTATIC TRANSMISSIONS

When mechanical power is transmitted from source (i.e., a diesel engine or electric motor) to a destination load (i.e., a track or wheel) via a pair of pump and motor, it is called a *hydrostatic transmission* or *hydrostatic drive*. Most common applications of hydrostatic transmissions are tracked vehicles, such as excavators, dozers, agricultural harvesters, military tanks (Figure 7.110). There are also wheeled vehicles where the wheels are powered via a hydrostatic transmission. In track-type vehicle applications, two pairs of pump-motor combinations drive the two tracks of the vehicle. Typically, there are three pumps connected to the diesel engine to support the two-track hydrostatic transmission system: two pumps to convert mechanical power to hydraulic power (one for each track), and one pump for the charge circuit to provide replenishing fluid into the circuit. Two hydraulic motors (one hydraulic motor for each track) convert the hydraulic power back to mechanical power (Figure 7.111). For simplicity, the charge circuit, flushing circuit, and line relief valves are not shown in Figure 7.111. Each hydraulic motor is connected to the track drive sprocket via a gear reducer. Engine speed can vary between its minimum (low idle) and maximum speed during normal operation as a function of operator commands and load conditions. Regardless of the variations of the engine speed within its operating range, our goal is to control position, and/or speed and/or force at the tracks. This clearly

Dozer (Track type tractor)

Excavator

Skid steer loader (Track type)

FIGURE 7.110: Hydrostatic transmission (drive) application in tracked vehicles.

requires a transmission system with adjustable gear ratio. The gear ratio of the hydrostatic transmission is determined by the ratio of the pump and motor displacement. In older hydrostatic transmission designs, the pump is variable displacement and the motor is fixed displacement. However, the current trend is towards designing hydrostatic transmissions with both pump and motor being of the variable displacement type.

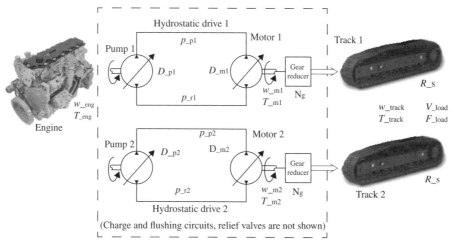

FIGURE 7.111: Hydrostatic transmission circuit and its components.

Let us consider a two track vehicle application and its two hydrostatic transmission designs and controls. For component sizing, we need the maximum load traction force and maximum translational speed requirements,

- maximum load traction force: $F_{l,max}$,
- maximum translational speed: V_{max}.

In other words, it is desired to have a tracked vehicle with a specified maximum traction force and maximum travel speed capacity.

Let us assume 100% efficiency in the components, and then take a safety factor in component sizing after the calculations. Furthermore, if we assume no pressure loss between pump and motor line, and that the charge circuit makes up for the flushing circuit flow perfectly, then the pump power is equal to the motor power. In other words, with the above assumptions, the hydraulic power of pump (converted by pump from the engine shaft's mechanical power into hydraulic power) is equal to the hydraulic power delivered to the motor.

We also assume that the compressibility of fluid in the hydraulic lines is negligable, which is a reasonably good assumption assuming that the load conditions are such that the frequency content of the load variations are low. However, if this assumption is not accurate enough under certain operating conditions (i.e., hitting a rigid wall or impact loading), then fluid compressibility may need to be taken into consideration in order to better predict the dynamic behavior of the hydrostatic transmission and accurate dynamic control.

Furthermore, we assume the flushing and charge circuit components are sized so that:

1. Flushing circuit components leak a percentage of rated flow of the circuit, so that fluid can be cooled and filtered, and then re-introduced to the circuit by the charge circuit,

$$Q_{flush} = 0.1 \cdot Q_{p,rated} \qquad (7.712)$$

2. The charge circuit is to make up the flushing circuit's leak at a desired pressure,

$$Q_{charge} = Q_{flush} \qquad (7.713)$$

$$p_{charge} = 2\,\text{MPa} \qquad (7.714)$$

Component sizing decisions are determined by the following two key observations:

- The pump determines how much power is converted from mechanical to hydraulic power, hence the hydraulic power in the circuit at any given time. Assume we have a constant engine speed, w_{eng}. The power converted by the pump and delivered to the hydrostatic transmission is determined by the pump displacement, which is controllable, and pressure in the hydrostatic line,

$$Power_p(t) = (p_p(t) - p_r(t)) \cdot D_p(t) \cdot w_{eng}(t) \qquad (7.715)$$

$$= T_{eng,p}(t) \cdot w_{eng}(t) \qquad (7.716)$$

where $p_p(t)$ is the pressure at the pump output line (high pressure side), $p_r(t)$ is the pressure at the pump input line (low pressure side), $D_p(t)$ is the pump displacement which has a range from negative to positive displacement through zero displacement in order to provide bidirectional flow, while the mechanical input shaft speed in unidirectional flow which is the engine speed, $T_{eng,p}(t)$ is the torque load on the engine due to the pump. As a result, such a pump has a swashplate angle control mechanism that requires two solenoids for an electronically controlled version. The pump power must meet the load power requirement. The pump power is determined by the engine speed, pump displacement (the controlled variable on the pump), and the line pressure which is determined by the load condition and motor displacement as shown below.

- The motor determines how this power is converted to mechanical power to meet the load demands. In other words, depending on the displacement of the motor, this power is converted to mechanical power in various combinations of torque and speed,

$$Power_m(t) = (p_m(t) - p_r(t)) \cdot D_m(t) \cdot w_m(t) \tag{7.717}$$

where $p_m(t)$ is the pressure on the high side of the motor port, $p_r(t)$ is the pressure on low side (return line), $D_m(t)$ is motor displacement, $w_m(t)$ motor output shaft angular speed.

For sizing calculations, we can use rated speed of the engine.

$$Power_p(t) = Power_m(t) = Power_{load}(t) \tag{7.718}$$

Pump displacement for each one of the two identical circuits is determined by the load power requirement,

$$Power_p(t) = (F_{l,max}/2) \cdot V_{max} \tag{7.719}$$
$$= p_{relief} \cdot D_p(t) \cdot w_{eng}(t) \tag{7.720}$$

Sizing should be done for worst case load conditions conditions,

$$D_p(t) \geq \frac{(F_{l,max}/2) \cdot V_{max}}{p_{relief} \cdot w_{eng}(t)} \tag{7.721}$$

Let us assume that pressure at the pump output and motor input is same, neglecting the pressure drop (pressure loss) in the hydraulic line between pump and motor.

$$p_p(t) = p_m(t) \tag{7.722}$$

In reality, the return line pressure is equal to the charge pressure, which is about 2 MPa in a typical hydrostatic drive,

$$p_{charge}(t) = p_r(t) = 2\,MPa \tag{7.723}$$

However, we will neglect the return line pressure

$$p_r(t) = 0 \tag{7.724}$$

since the pressure on the return line is much lower than the pressure on the powered line. For instance, $p_p(t)$ is around 40–50 MPa range.

In motor sizing and control, the main consideration is the load force. The load force is not under our control and is determined by operating conditions such as the ground traction conditions and the load the vehicle faces. The objective in motor control is to scale the available hydraulic power to mechanical power conversion so that the output torque is equal or greater than the load torque.

$$Power_m(t) = p_p(t) \cdot D_m(t) \cdot w_m(t) \tag{7.725}$$
$$= w_m(t) \cdot T_m(t) \tag{7.726}$$
$$= w_m(t) \cdot N_g \cdot R_t \cdot (F_l(t)/2) \tag{7.727}$$

where R_t is the radius of the track output sprocket connected to the motor driven final gear reducer, N_g is the gear ratio of the final gear reducer between the motor and track sprocket drive, F_l is the load force the two tracks must overcome. It is clear that

$$T_m(t) = p_p(t) \cdot D_m(t) = N_g \cdot R_t \cdot (F_l(t)/2) \tag{7.728}$$

From conservation of power between pump and motor (assuming perfect efficiency), the speed conversion ratio between engine speed and motor output shaft speed is,

$$Power_p(t) = Power_m(t) \tag{7.729}$$

$$p_p(t) \cdot D_p(t) \cdot w_{eng}(t) = p_p(t) \cdot D_m(t) \cdot w_m(t) \tag{7.730}$$

$$w_m(t) = \frac{D_p(t)}{D_m(t)} \cdot w_{eng}(t) \tag{7.731}$$

Similarly, the effective gear ratio between the load torque at each track and reflected load on the engine is

$$T_{eng,p}(t) \cdot w_{eng}(t) = T_m(t) \cdot w_m(t) \tag{7.732}$$

$$T_{eng,p}(t) = \frac{w_m(t)}{w_{eng}}(t) \cdot T_m(t) \tag{7.733}$$

$$= \frac{D_p(t)}{D_m(t)} \cdot T_m(t) \tag{7.734}$$

$$T_m(t) = \frac{D_m(t)}{D_p(t)} \cdot T_{eng,p}(t) \tag{7.735}$$

Notice that the ratio of pump displacement and motor displacement defines the effective transmission gear ratio between engine speed and motor speed. The gear ratio that can be realized with continuous variability is in the range of

$$\frac{D_{p,min}}{D_{m,max}} \leq N_{tr} = \frac{D_p}{D_m} \leq \frac{D_{p,max}}{D_{m,min}} \tag{7.736}$$

where $D_{p,min}, D_{p,max}$ are minimum and maximum pump displacements, and $D_{m,min}, D_{m,max}$ are minimum and maximum motor displacements. Maximum motor speed, when the engine is operating at a rated speed, can be determined by using the pump and motor displacements that are maximum. It should be checked whether it meets the specified maximum track speed requirement.

In short, pump displacement $D_p(t)$ must be controlled to meet the power demand (i.e., maximize the available hydraulic power at all times, or only to meet the current demand, or limit the load on the engine so that engine does not stall, or a combination of all of the above), whereas the motor displacement $D_m(t)$ is controlled to deliver the proper traction torque (or force) to meet the load torque demand or commanded speed.

For a given load force, each track needs to deliver half of it, and the torque each track motor must provide can be determined from,

$$T_m(t) = \frac{1}{N_g} \cdot R_t \cdot (F_l(t)/2) \tag{7.737}$$

$$= D_m(t) \cdot p_{line} \tag{7.738}$$

Since, we need to meet the maximum load demand with the maximum motor displacement condition, then the selected motor should have a maximum displacement that meets the following condition,

$$D_{m,max} \geq \frac{T_m(t)}{p_{relief}} \tag{7.739}$$

Once a particular motor is selected, then we would know the minimum and maximum displacements of the motor, $[D_{m,min}, D_{m,max}]$. Then for a given load force condition (i.e., worst case) F_l, we can calculate the line pressure developed when the motor is at its maximum displacement,

$$T_m(t) = \frac{1}{N_g} \cdot R_t \cdot (F_l(t)/2) \tag{7.740}$$

$$p_{line} = \frac{T_m(t)}{D_{m,max}(t)} \tag{7.741}$$

And the line pressure under the same load condition when the motor is at its minimum displacement is

$$T_m(t) = \frac{1}{N_g} \cdot R_t \cdot (F_l(t)/2) \tag{7.742}$$

$$p_{line} = \frac{T_m(t)}{D_{m,min}(t)} \tag{7.743}$$

In other words, when track conditions face a load condition, T_m, depending on the value of the motor displacement at any given time, line pressure will be at a value in the following range, never exceeding the relief pressure since relief valves would open at that condition,

$$\frac{T_m}{D_{m,max}} \leq p_{line} \leq min\left(\frac{T_m}{D_{m,min}}, p_{relief}\right) \tag{7.744}$$

Another way to view the dynamics of the system is in terms of the maximum traction that can be developed at the tracks and ground contact. If the ground conditions are such that a smaller value of traction force can be developed, then the tracks would start slipping when the the torque output from the motors exceeds the maximum torque that can be supported by the track-ground conditions. This may happen even before relief pressure is reached. For instance, in poor traction conditions (i.e., muddy and slippery ground conditions with poor friction, hence poor traction capacity), we quickly observe slipping of the tracks. Track slip condition should be minimized since it represents the wasted energy condition.

Component Sizing for Hydrostatic Transmissions Let us consider a numerical example for a hydrostatic transmission circuit component sizing. Given hydrostatic transmission specifications for an application, such as an excavator model,

- the hydrostatic drive should be able to deliver traction force of $F_{l,max} = 100\,000\,\text{N}$, and
- a linear speed of $V_{max} = 10\,\text{km/h}$.

Based on this specification, a designer must select a pair of hydraulic pump-motors, as well as the sprocket radius, final drive gear ratio, and charge pump.

The sprocket radius and final gear reducer ratio are typically chosen from a range of standard values depending on the power levels involved in the application. Let us assume that the final drive gear ratio and sprocket radius are

$$N_g = 50 \tag{7.745}$$

$$R_s = 0.5\,\text{m} \tag{7.746}$$

Let us assume that the flushing circuit is sized to approximately flush about 10% of the flow rate in the line. Then, the charge pump should be sized to provide this flow rate at the nominal charge pressure. Hence, the selection criteria for the charge pump is

$$Q_{flush} = 0.1 \cdot Q_{p,rated} \tag{7.747}$$

$$Q_{charge} = Q_{flush} \tag{7.748}$$

$$p_{charge} = 2\,\text{MPa} \tag{7.749}$$

These specifications (rated flow rate and rated pressure) then determine the size of the flushing valves, charge pump, and valves in the circuit.

One of the important engineering constraints is the maximum line pressure in the hydraulic lines, which sets the relief valve pressure settings,

$$p_{\text{relief}} = p_{1,\text{max}} \approx 50 \,\text{MPa} \tag{7.750}$$

The pump size and motor size selection is determined as follows: determine D_p, D_m.

$$T_m = \frac{R_f}{N_g} \cdot \frac{F_{1,\text{max}}}{2} = 500 \,\text{Nm} \tag{7.751}$$

$$w_m = \frac{N_g}{R_f} \cdot V_{\text{max}} = 277 \,\text{rad/s} = 2652 \,\text{rev/min} \tag{7.752}$$

Then, the motor size can be determined from

$$T_m = D_m \cdot p_{\text{line}} \tag{7.753}$$

$$w_m = \frac{Q_m}{D_m} \tag{7.754}$$

where we select $p_{\text{line}} = 50 \,\text{MPa} = 50 \times 10^6 \,\text{N/m}^2$, and solve for D_m and Q_m.

$$D_m = 2\pi \cdot 10 \cdot 10^{-6} \,\text{m}^3/\text{rev} = 6.28 \times 10^{-5} \,\text{m}^3/\text{rev} \tag{7.755}$$

$$Q_m = D_m \cdot w_m = 0.167 \,\text{m}^3/\text{min} \tag{7.756}$$

The pump size is calculated from (assuming 100% efficiency in the transmissions line and at the pump and motor)

$$T_p = D_p \cdot p_{\text{line}} \tag{7.757}$$

$$w_p = \frac{Q_p}{D_p} = w_{\text{eng}} \longrightarrow D_p = 6.65 \times 10^{-5} \,\text{m}^3/\text{rev} \tag{7.758}$$

where $Q_p = Q_m$ and $w_{\text{eng}} = 2500 \,\text{rpm}$ are known, then solve for D_p. Then, we can calculate the T_p to determine the load torque reflected on the engine by the pump to verify if it is acceptable for the engine.

Control System for Hydrostatic Transmissions   Components of the control system for a hydrostatic transmission are shown in Figure 7.112. Typical software development tools (a notebook PC, software, and communication adaptor) are shown in Figure 7.113. The control system components are grouped into three categories:

1. input and output devices for the operator (operator I/O devices in the cab),
2. input (sensors) and output (solenoids) for the hydrostatic transmission, generically referred to as the *machine I/O*,
3. electronic control module (ECM) to implement the digital logic and provide the interface circuit for the I/O.

I/O interface circuitry between the ECM and I/O (sensors and solenoids) will not be discussed here since they are standard interfaces integrated into the design of most ECMs. The control logic that relates the input signals (from operator and machine) to the output signals (solenoids and operator cab indicators) is developed using Simulink®, Stateflow, and C-functions incorporated to Simulink® and Stateflow as S-functions. The developed code can be simulated on a non-real-time computer using a dynamic model of the hydrostatic transmission. Then, the same code can be converted to a C-code by the

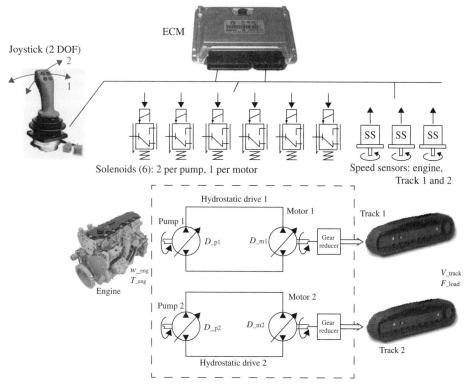

FIGURE 7.112: Control system: electronic control module, operator I/O, machine I/O.

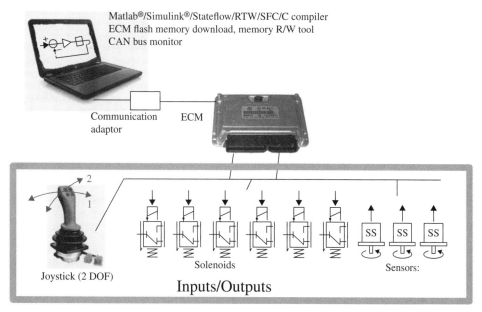

FIGURE 7.113: Control system with development tools: PC with code development, download and reads/write to ECM memory capability, electronic control module, operator I/O, machine I/O.

auto-code generation tools, and compiled and linked to generate the executable code for the target ECM. The executable code would then be downloaded to the flash memory of the ECM, and tested on the vehicle with the actual I/O hardware. The conditions of the control variables and control logic parameters can be continuously monitored and recorded for debugging purposes using direct memory access tools into the ECM memory map as well as by monitoring a common communication bus, that is CAN bus monitor software.

Basic operator I/O devices include two proportional command devices, such as a joystick connected to a position sensor for each degree of freedom (Figure 7.114). One joystick-position sensor combination is used to define the forward (positive) speed and reverse (negative) speed and neutral speed (zero speed) of the vehicle, a second joystick-position sensor combination is used to define the left and right turn steering command signal. Steering is achieved by controlling the speed of the two tracks to different levels. If the machine is commanded to turn left, then the left track is slowed down while the right track is speeded up proportional to the steering angle command as well as the speed of the machine.

The basic machine I/O includes six solenoids (two solenoids for bidirectional pump displacement swashpalte angle control, one solenoid for motor displacement swash-plate angle control per track), and two angular speed sensors (one per track).

An open loop speed control system for the hydrostatic drives is shown in Figure 7.114. The commanded speed of each track is determined from two joystick sensors: F/N/R speed command and the steering command,

$$V_{cmd1} = \theta_{1,cmd} \cdot (1 + \theta_{2,cmd}) \tag{7.759}$$
$$V_{cmd2} = \theta_{1,cmd} \cdot (1 - \theta_{2,cmd}) \tag{7.760}$$

Furthermore, for a realistic operator controlled system, there is always the following desired components in the control logic:

1. Deadband between the joystick position sensor and output signal in control logic so that small unintended movements of the joystick (due to vibrations and resolution of the operator control ability) do not cause motion in the hydrostatic drive.

2. Lowpass filter or rate limiter functions to limit the maximum accelerations and remove the sensor noise.

3. Gain (or look-up table for variable gain) to determine the commanded speed based on joystick sensors. In practice, this is generally a variable gain, which is a curve or multiple lines with different slopes or a look-up table.

4. Offset signal to solenoids, since there are always variations in the manufacturing tolerances of the solenoids and valves they actuate, the initial offset current command needed to just start the motion of the solenoid needs to be determined for each solenoid as part of a calibration process on the machine during manufacturing.

Remark 1 If the vehicle it to travel in a straight line, the steering command would be zero and both tracks would receive the same speed command (forward or reverse). The solenoids for both track's pumps and motors would also receive the same current commands, where the only difference might be if their offset currents are different, in order to obtain the same speed for the tracks. However, due to small variations in components (i.e., between pumps in two tracks), the actual speed of each track cannot be exactly equal in practical conditions. As a result, the vehicle travel path would not be a straight line. This can be corrected by the operator periodically by corrective steering commands. A better solution is to implement a closed loop control on the track speeds to make sure that both tracks run at the same speed, and hence assure a better straight path motion. The only

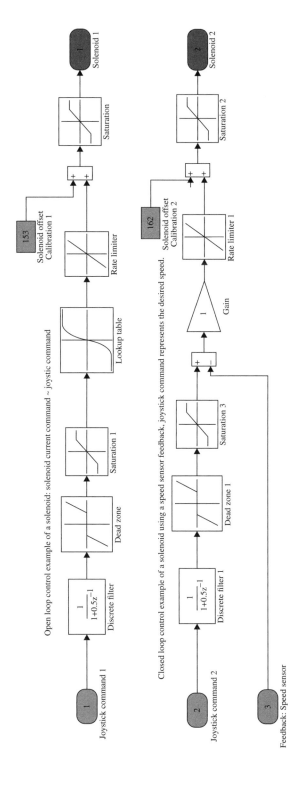

Open loop control example of a solenoid: solenoid current command ~ joystic command

Closed loop control example of a solenoid using a speed sensor feedback, joystick command represents the desired speed.

FIGURE 7.114: Control logic sample for open loop and closed loop control of hydrostatic transmission for dual track vehicle applications.

deviation from the straight path would be due to the control resolution of the track speeds and slip between the tracks and ground. A further level of path tracking accuracy can be added by closed loop path tracking using GPS signals.

Remark 2   It is assumed that the flushing circuit and charge circuit components are properly sized so that a certain percentage of the flow is flushed out of the circuit, then cooled and filtered, and reintroduced back to the circuit via the charge circuit,

$$Q_{\text{charge}} = Q_{\text{flush}} \tag{7.761}$$

Typically, the charge circuit pressure is used for the swash plate control circuit as the supply pressure. Hence, for predictable pump and motor displacement control, it is important to maintain the charge pressure as constant as possible.

Remark 3   In hydrostatic drives which have both their pump and motor as a variable displacement type, the control of pump and motor order depends on the application mode. When the vehicle just starts to move, we first up-stoke the pump to start generating hydraulic power with the motor at its maximum displacement to provide maximum torque, then slowly reduce motor displacement to get more speed (and less torque). For a given engine speed, pump displacement controls how much power the hydrostatic circuit draws from the engine. Motor displacement control is how that power is converted to speed and torque combination.

7.12   CURRENT TRENDS IN ELECTROHYDRAULICS

The current trend in future electrohydraulic technology is to increase

- the power/weight ratio to reduce physical size of components, hence to reduce cost,
- software programmable components.

In order to increase the power/weight ratio, the system pressure must be increased, which results in smaller size components to deliver the same power. However, increased system pressure reduces the resonant frequency of a hydraulic system due to oil compressibility, hence the control loop system bandwidth limit is lower. As the supply pressure gets higher, it is more important to minimize cavitation and air bubbles in the hydraulic lines. Otherwise, the system response will be significantly slower or even become unstable in the case of closed loop control. Furthermore, cavitation leads to damage to the hydraulic components and increases noise.

Another way of reducing the size of components and cost is to make more effective use of the components. Consider the hydraulic circuits (implement, steering, brake, cooling fan, pilot hydraulics) and the pumps used to support them in construction equipment applications. Traditional designs dedicate one or more pumps per circuit. Since all systems are not used in maximum flow demand at all times, all of the pumps are not used in their maximum capability. Furthermore, duplicate pumps are provided for safety backup reasons for critical systems (i.e., steering). The concept of sharing pumps among multiple circuits through a controllable power distribution valve has been emerging in recent years (Figure 7.115). Instead of dedicating a pump to each circuit in hardware, the total hydraulic power of all pumps is combined at a distribution valve, and under program control, the hydraulic power is distributed to the different subsystems based on demand. This approach has the promise of making better use of available component capability, reduced cost, and improved performance.

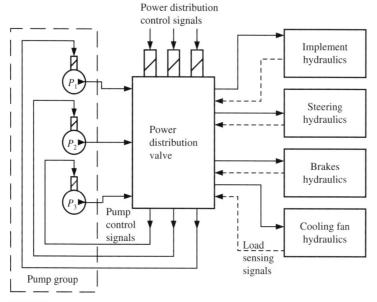

FIGURE 7.115: Programmable power allocation in multi pump multi circuit hydraulic systems. Instead of dedicating a pump to each circuit in hardware, the output of all pumps is brought into a controllable distribution valve. The valve directs the desired amount of flow to each sub-system based on demand.

The valve is the main critical component in a hydraulic system from a control system perspective. All of the valves we have discussed so far have a single spool for each stage. One spool geometry defines the orifice areas between the four ports of the valve: pump (P), tank (T), A and B side of the cylinder. The single variable, spool displacement x_{spool}, determines the orifice areas

$$A_{\mathrm{PA}}(x_{\mathrm{v}}), A_{\mathrm{PB}}(x_{\mathrm{v}}), A_{\mathrm{AT}}(x_{\mathrm{v}}), A_{\mathrm{BT}}(x_{\mathrm{v}}), A_{\mathrm{PT}}(x_{\mathrm{v}}) \qquad (7.762)$$

These geometric relationships between the spool displacement versus the orifice areas are designed and physically machined into each valve spool. Once the valve is machined, its orifice characteristics are fixed. In order to accomodate many different application specific requirements on orifice functions, many different spool geometry variations are often needed. This requires machining many variations of basically the same spool geometry for different applications. For instance, it has been estimated that one of the major construction equipment manufacturers alone machines over 1600 different valve spool geometries. It would be desirable to reduce the number of different spool geometries that must be physically machined. This idea had led to the development of the *independently metered valves* (IMV) concept. The idea is to define the orifice areas in software by actively and independently controlling each orifice area by a separate spool (Figure 7.116). The IMV valve has up to six independently operated spools and solenoids, one for each port connection orifice area,

$$A_{\mathrm{PA}}(x_1), A_{\mathrm{PB}}(x_2), A_{\mathrm{AT}}(x_3), A_{\mathrm{BT}}(x_4), A_{\mathrm{PT}}(x_5), A_{\mathrm{AB}}(x_6) \qquad (7.763)$$

where each spool position is proportional to the associated solenoid current. Therefore, the orifice area functions can be equally expressed as a function of solenoid currents.

$$A_{\mathrm{PA}}(i_1), A_{\mathrm{PB}}(i_2), A_{\mathrm{AT}}(i_3), A_{\mathrm{BT}}(i_4), A_{\mathrm{PT}}(i_5), A_{\mathrm{AB}}(i_6) \qquad (7.764)$$

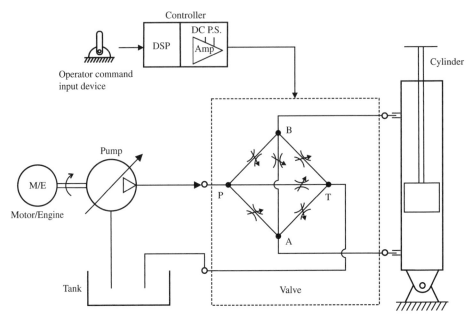

FIGURE 7.116: Independently metered valve (IMV) concept. Valve flow orifices are independently controlled and defined in real-time control software.

There are several patents on this concept [15]. The valve flow–orifice–pressure differential relationships are still the same, except that in the case of IMV each flow rate can be independently metered,

$$Q_{PA} = C_d \cdot A_{PA}(i_1) \sqrt{(P_P - P_A)} \qquad (7.765)$$

$$Q_{PB} = C_d \cdot A_{PB}(i_2) \sqrt{(P_P - P_B)} \qquad (7.766)$$

$$Q_{AT} = C_d \cdot A_{AT}(i_3) \sqrt{(P_A - P_T)} \qquad (7.767)$$

$$Q_{BT} = C_d \cdot A_{BT}(i_4) \sqrt{(P_B - P_T)} \qquad (7.768)$$

$$Q_{PT} = C_d \cdot A_{PT}(i_5) \sqrt{(P_P - P_T)} \qquad (7.769)$$

$$Q_{AB} = C_d \cdot A_{AB}(i_6) \sqrt{(P_A - P_B)} \qquad (7.770)$$

Since each solenoid pair is independently controlled, each orifice area can be controlled and their relationship to each other, like the case in single spool valve, can be defined in software. Therefore, the same IMV valve can be controlled to behave like a valve with a different geometry by changing the control software. In other words, the effective valve geometry is defined in software. For instance, if the IMV control is to emulate a standard closed center valve, then i_5, i_6 should be controlled such that

$$A_{PT}(i_5) = 0 \quad A_{AB}(i_6) = 0 \qquad (7.771)$$

If open-center valve emulation is desired, then i_6 should be controlled such that

$$A_{AB}(i_6) = 0 \qquad (7.772)$$

If regenerative power capability is desired in a closed-center valve emulation, then i_5, i_6 should be controlled such that

$$A_{PT}(i_5) = 0 \quad A_{AB}(i_6) \neq 0 \qquad (7.773)$$

The IMV concept offers the following advantages and disadvantages. The advantages are as follows.

- Valve geometry is defined in software. Therefore, only three or four mechanical valves would be needed to cover the low, medium, and high power applications. The application specific spool geometry would be defined in the software.
- Emulated valve geometry in software does not have be the equivalent of a single mechanical valve. It can function as the equivalent of different types of mechanical valve during different operating conditions to optimize the performance.
- Through the added flexibility in valve control, regenerative energy can be used to operate the EH system more energy efficiently.

The disadvantages are as follows.

- The control task is more complicated. While a standard valve has a single controlled solenoid, IMV has six controlled solenoids.
- Because of the increased number of electrical components, the number of possible failures is higher.

The IMV concept is one of the most significant new EH technologies in recent years. As embedded digital control of valves increases, the functionality of a valve component is defined in software. The mechanical design of it becomes simpler, since the functionality no longer needs to be machined into the valve. However, the software aspect of the valve becomes more complicated and without software the component would not be functional.

7.13 CASE STUDIES

7.13.1 Case Study: Multi Function Hydraulic Circuit of a Caterpillar Wheel Loader

Figure 7.117 (also see Figure 7.14) shows the complete hydraulic system schematics for a wheel loader model 950G by Caterpillar. The steering sub-system is discussed separately in detail below. The hydraulic system power is supplied by four major pumps: a variable displacement pump for the steering sub-system, and three fixed displacement pumps for implementing the hydraulics main pressure lines, for the brake charging and pilot pressure lines, and for the cooling fan motor. The steering system also has a secondary power source as a safety backup using a battery, electric motor, and pump (the fifth pump). This design uses:

1. Mostly fixed displament pumps (except the primary steering pump) and each pump is dedicated to one circuit (for example, the steering pump is dedicated to support the steering sub-system, the implement pump is dedicated to the implement sub-system).
2. lift, tilt, and auxilary function valve–cylinder functions are connected in series to the hydraulic supply lines (P and T) and are necessarily os open center type (see Figure 7.14). Furthermore, the function closer to the pump has priority over the functions following it.
3. The main flow control valves do not have pressure compensation (no compensator valves), hence the function speed will vary with load for a given command.

The valves of the implement hydraulics are of electrohydraulic (EH) type, meaning that the pilot control valves are controlled proportionally by an electric current sent to a solenoid for each pilot valve. The hydraulic system is also available with mechanically

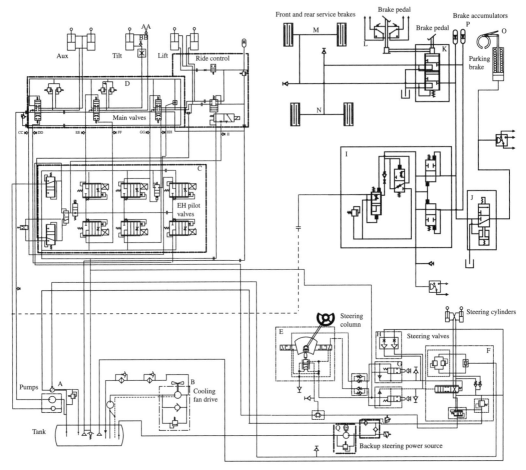

FIGURE 7.117: Hydraulic system circuit diagram of the Caterpillar wheel loader model 950G.

actuated pilot valves instead of EH pilot valves. Proportional flow control valves for mobile equipment are typically fabricated in stackable blocks, that is a valve stack may have up to ten valve sections for ten different functions (Figure 7.118).

In some cases, the pilot and main flow sections of the valves are designed as separate blocks. The advantage of this is that the main section of the valve accepts the pilot control pressure. The pilot control pressure may be controlled either by a mechanically actuated pilot valve or solenoid actuated EH valve. Hence, the main flow section of the valve can be used with either mechanically actuated pilot valves or electrically (solenoid) actuated pilot valves. It is instructive to note that the main flow control valve for multi function hydraulic circuits, such as the the lift and tilt circuit in Figure 7.52 (component V on the figure), is manufactured as a single valve block with all the individual valve functions incorporated into one block design that has the external port connections (pump, tank, service ports A and B for lift and tilt, and pilot pressure control ports, Figure 7.119).

Notice that the difference between mechanical lever controlled and electrical solenoid controlled pressure reducing pilot control valves is the source of actuation power to shift the spool of the pilot valve (see Figure 7.66). In a mechanical lever controlled system, as the operator moves the lever, the lever motion moves a spring in the pilot valve which then moves the pilot valve spool and changes the output pilot control pressure. In solenoid controlled pilot valves, the lever displacement is measured by a sensor, the electronic control

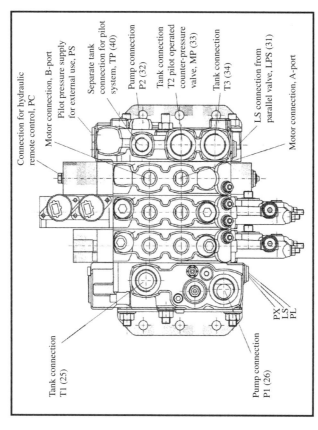

Connection for hydraulic
remote control, PC

Motor connection, B-port

Pilot pressure supply
for external use, PS

Separate tank
connection for pilot
system, TP (40)

Pump connection
P2 (32)

Tank connection
T2 pilot operated
counter-pressure
valve, MP (33)

Tank connection
T3 (34)

LS connection from
parallel valve, LPS (31)

Motor connection, A-port

Tank connection
T1 (25)

Pump connection
P1 (26)

PX
LS
PL

(b)

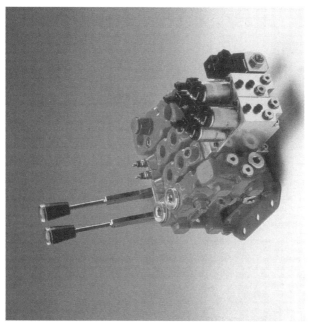

(a)

FIGURE 7.118: Stackable valve block for mobile applications. Reproduced with permission from Parker Hannifin.

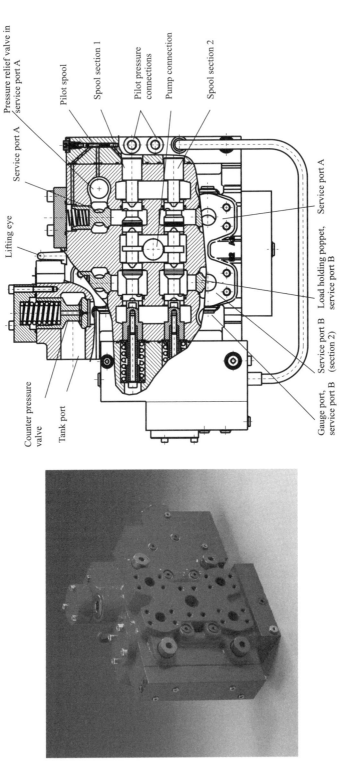

FIGURE 7.119: Valve block (Model M400LS by Parker Hannifin AB) used in various hydraulic circuits of mobile equipment applications. Notice that the external port connections are provided for pump (P), tank (T), service ports (A1, B1, A2, B2), pilot pressure ports. All other hydraulic connections (i.e., control of poppet valve, line relief valves, anti cavitation make-up valves) are internally machined into the valve block and do not require any plumbing. Reproduced with permission from Parker Hannifin.

module (ECM) samples the sensor signal and then sends a current to the pilot valve to move its spool. The rest of the power hydraulic circuit is the same.

There are two pilot valves per cylinder function, for example lift raise and lift lower pilot valves for the lift circuit. Each pilot valve is proportionally controlled by a solenoid in the EH system. Hence, there are two solenoid outputs per function. Each pilot valve has two inports: pilot pressure supply and tank ports. The outport carries the pilot control pressure to the main flow control valve. The outport pilot control pressure is proportionally controlled by the solenoid current between the tank pressure (minimum outport pilot pressure) and the pilot pressure supply (maximum outport pilot pressure). In short, these are proportional solenoid controlled pressure reducing valves. As the operator moves a lever (such as a lift lever), a sensor senses the displacement of the lever and the signal is sampled by the electronic control module (ECM). The ECM sends current approximately proportional to the lever displacement signal to the pilot valve solenoids, which in turn generate a proportional pilot control pressure to the main valves. Note that in wheel loader applications, the solenoid current sent to the solenoid by the ECM is not exactly proportional to the lever displacement, but includes a deadband around the neutral position of the lever and a modulation curve beyond the deadband region that is not necessarily perfectly linear. The pilot valve control is enabled/disabled by a solenoid controlled two position valve (E1). The operator select the desired function by a switch in the cab, and the ECM controls the solenoid (ON/OFF) based on that switch state (ON/OFF).

The displacement of the main flow control valve, and hence the flow rate, is proportional to the pilot control pressure since the force (due to pilot control pressure) acting on the sides of the valve spool is balanced by a pair of centering springs. It is important to note that the main flow control valves do not have any flow compensator valves. As a result, as the load changes, the cylinder speed changes for the same lever command. The cylinder speeds are load dependent. Furthermore, the lift, tilt, and auxilary function main flow control valves are hydraulically connected in series and they are of open-center type. The open-center hydraulic system (fixed displacement pump and open-center valves combination) are less energy efficient (waste more energy) than the closed-center hydraulic systems (variable displacement pump and closed-center valves combination). Since the tilt valve is closer to the pump in the serial connection, the tilt circuit has priority over the lift circuit. Tilt and auxilary function cylinders have anti cavitation make-up valves connected to the tank as well as line pressure relief valves. The lift cylinder has a ride control circuit that is engaged or disengaged by the operator using a selector switch.

7.14 PROBLEMS

1. Consider a pump with fixed displacement $D_p = 50\,\text{cm}^3/\text{rev}$ and the following nominal operating conditions: input shaft speed $w_{\text{shaft}} = 600\,\text{rpm}$, torque at the input shaft $T = 50\,\text{Nt} \cdot \text{m}$, the output pressure $p_{\text{out}} = 5\,\text{MPa}$, and output flow rate $Q_{\text{out}} = 450\,\text{cm}^3/\text{s}$. Determine the volumetric, mechanical, and overall efficiency of the pump at this operating condition. Consider another version of this problem statement where $p_{\text{out}} = 10\,\text{MPa}$, and all other data is the same. Is there any error in this pump data? If so, discuss what might be the source of error and suggest ways to correct the data.

2. Consider a bidirectional hydraulic motor with fixed displacement of $D_m = 100\,\text{cm}^3/\text{rev}$. The load torque it needs to provide at its output shaft is $100\,\text{Nt} \cdot \text{m}$ and to maintain a rotational speed of $600\,\text{rpm}$. Determine the necessary flow rate and differential pressure between the two ports (input and output) of the motor that must be supplied by a hydraulic circuit (i.e., a pump and valve preceeding the hydraulic motor). What is the power rating of the pump to supply this motor assuming an overall pump efficiency of 80% and neglecting the losses at the valve? Assume 100% efficiency for the hydraulic motor. See Figure 7.83.

3.   Consider the two-axis hydraulic motion system shown in Figure 7.13. The system is to operate under the control of an operator. The operator commands the desired speed of each cylinder by two joysticks. There is no need for sensors since this is an operator in the loop control system. Cylinder 1 needs to be able to provide a force of 5000 Nt at a rated speed of 0.5 m/s, and maximum no-load speed of 1.0 m/s. Cylinder 2 needs to be able to provide 2500 Nt force at the same rated speed, and has the same maximum no-load speed as cylinder 1.

(a) Select components and size them for a completely hydro-mechanical control of the system. There should be no digital or analog computers involved. Draw the block diagram of the control system, and indicate the function of each component in the circuit. Assume that we use a fixed displacement pump. As a designer make decisions regarding the anticipated realistic pressure that can be delivered at the cylinder ports and decide on a realistic cylinder bore size.

(b) Select components for an embedded computer controlled system. Discuss the differences between this design and the previous design in terms of the components and their differences.

4.   Consider hydraulic fluid in a hydraulic hose that has some entrapped air in it. Let the hose diameter be $D = 6$ in and wall thickness of the hose is $\delta x = 1$ in. Let the Young's modulus of the hose material be $E = 30 \times 10^6$ psi. Determine the effective bulk modulus for zero and 1% air entrapment, for operating pressures of 500 psi and 5000 psi.

5.   Let us consider a single axis EH motion control system with the following parameters. For the cylinder bore diameter, consider the following size information: $d_{bore} = 10$ cm, and the rod diameter $d_{rod} = 5$ cm, stoke is $L = 1.0$ m. Let us approximate the bulk modulus of hydraulic fluid to be 2.5×10^5 psi $= 1.723 \times 10^9$ Pa $= 1.723$ GPa. The mass carried by the cylinder (including the mass of the rod and piston and load) is $M = 200$ kg.

(a) Determine the lowest natural frequency of the system due to the compressibility of the fluid in the cylinder and the inertia.

(b) What is the maximum recommended closed loop control system bandwidth?

6.   Consider a single axis electrohydraulic motion control system as shown in Figure 7.90 and its block diagram representation in Figure 7.89. We assume that the cylinder is rigidly connected to its base and load. Let us consider the small movements of the valve around its null position, that is a case that is maintaining a commanded position.

    1. amplifier gain is $K_{sa} = 200$ mA$/10$ V $= 20$ mA/V,
    2. valve gain around the null position operation, $K_q = 20$ (in^3/s)$/200$ mA $= 0.1$ (in^3/s)/mA,
    3. cross-sectional area of the cylinder is $A_c = 2.0$ in^2 on both sides (rod is extended to both sides),
    4. the sensor gain is $K_{fx} = 10$ V$/10 = 1$ V/in,
    5. deadband of the valve is 2% of the maximum input current, $i_{db} = 0.02 \cdot i_{max} = 0.02 \cdot 200$ mA $= 4$ mA.

Assume that the total inertial load the cylinder moves is $W = 1000$ lb, $m = 1000$ lb$/386$ in/s$^2 = 2.59$ lb s^2/in, and the cylinder is currently at the mid position of its total travel range of $l = 10$ in. Note that the approximate bulk modulus of the hydraulic fluid is $\beta = 2.5 \times 10^5$ psi.

    In terms of the amplifier gain, what is necessary to reduce the positioning error? Is there an upper limit set on the value of the amplifier gain? If so, determine that limit. Discuss the limitations imposed on the amplifier gain by the small positioning error requirement and closed loop system stability. Determine the amplifier gain that provides the maximum practical closed loop bandwidth while minimizing the positioning error due to the deadband of the valve.

7.   Consider the single axis EH motion control system shown in Figure 7.85. The system needs to operate in two modes under the control of a programmed embedded computer:

    • Mode 1: closed loop speed control where the commanded speed is obtained from a programmed command generator. The cylinder speed is measured and a closed loop control algorithm implemented in the electronic control unit (ECU) which controls the valve.

- Mode 2: closed loop force control where the force command is obtained from a programmed command generator. The cylinder force is measured using a pair of pressure sensors on both ends of the cylinder.

A real-time control algorithm, programmed in the embedded computer, decides when to operate the axis motion in speed control or force control mode (Figure 7.84).

(a) Draw a block diagram of the components and their interconnection in the circuit.
(b) Select proper components in order to meet the following specifications:
1. Maximum speed of the cylinder under no load conditions is $V_{nl} = 1.0\,\mathrm{m/s}$.
2. Provide an effective output force of $F_r = 1000\,\mathrm{Nt}$ at the cylinder rod while moving at the speed of $V_r = 0.75\,\mathrm{m/s}$.
3. The desired regulation accuracy of the speed control loop is 0.1% of maximum speed.
4. The desired regulation accuracy of the force control loop is 1.0% of maximum force.

The components to be selected and sized include an electronic control unit (ECU) with necessary analog-digital signal interfaces, position and force sensor on the cylinder for closed loop control (Figure 7.85), pump, valve, and cylinder. Assume that the valve rating is specified for $\Delta P_v = 7\,\mathrm{MPa}$ ($1\,\mathrm{Pa} = 1\,\mathrm{Nt/m^2}$) pressure drop, and that the input shaft speed at the pump is $w_{in} = 1200\,\mathrm{rpm}$.

8. For the previous problem, write a pseudo-code for the real-time control software in order to implement the following logic for the operation of this hydraulic motion control system.

(a) First, make a list of all of the inputs and outputs from the controller point of view and give symbolic names to them.
(b) Then write the pseudo-code.

The objective is that the actuator is to approach the load at a predefined speed until a certain amount of pressure differential is sensed between P1 and P2 signals. The predefined speed is a trapezoidal speed, the cylinder is to start from stationary position and accelerate to a top speed, and run for a while at the top speed while monitoring the pressure sensors. The position range of the actuator around which that is expected is known. When a certain pressure differential is sensed, the control system is supposed to automatically switch to force (pressure) regulation mode; maintain the desired pressure until and move a defined distance at a lower speed. Then reverse the motion, and return to the original position under another programmed speed profile.

9. Consider the hydraulic circuit shown in Figure 7.120. a) Describe the operation of the circuit. b) What is the role of the emergency back up steering block and how does it work? c) What is the role of joystick and wheel steering and how do they work?

10. A Case Study: One Degree of Freedom Hydraulic Motion Control System with an Accumulator.
Consider the hydraulic system shown in Figure 7.121, which has a fixed displacement pump, a closed-center valve, a cylinder, a relief valve, and an accumulator. Notice the accumulator volume and flow rate are included in Equations 7.778 and 7.782 below, describing the pressure variation on the pump-valve line where the accumulator is located.
The dynamic model of the EH system is defined by the following equations in its general form, when the dynamics of cylinder-load inertia and the fluid compressibility are taken into account. Since the main valve is closed center type, $A_{PT}(x_s) = 0.0$, hence $Q_{PT}(t) = 0$ always. As a result, we have one less unknown and one less equation compared to the general treatment of this problem in Section 7.8. However, the same number of equations are displayed below to make the comparison easier.

$$(1) \quad m \cdot \ddot{y}(t) = -c \cdot \dot{y}(t) + p_A(t) \cdot A_A - p_B(t) \cdot A_B - F_{load}(t) \tag{7.774}$$

If up motion $\quad x_s \geq 0.0$ \hfill (7.775)

$$(2) \quad \frac{d}{dt}(p_A(t)) = \frac{\beta}{V_{hose,VA} + y \cdot A_A}(Q_{PA}(t) - \dot{y}(t) \cdot A_A) \tag{7.776}$$

$$(3) \quad \frac{d}{dt}(p_B(t)) = \frac{\beta}{V_{hose,VB} + (l_{cyl} - y) \cdot A_B}(-Q_{BT}(t) + \dot{y}(t) \cdot A_B) \tag{7.777}$$

$$(4) \quad \frac{d}{dt}(p_P(t)) = \frac{\beta}{V_{hose,pv} + V_{acc}}(Q_P(t) - (Q_{PA}(t) + Q_{PT}(t) + Q_r(t) + Q_{acc}(t))) \tag{7.778}$$

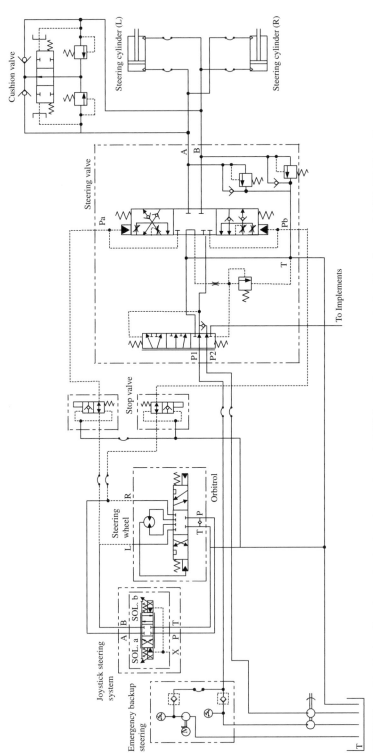

FIGURE 7.120: Hydraulic schematics for hand metering unit (HMU) steering system of wheel loader model WA450-5L by Komatsu.

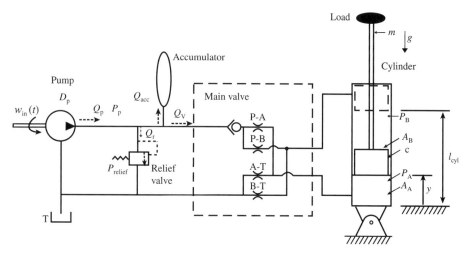

FIGURE 7.121: One degree of freedom hydraulic system with fixed displacement pump, closed center valve, a cylinder, and a relief valve.

$$elseif \ down \ motion \ x_s < 0.0 \tag{7.779}$$

$$(2) \ \frac{d}{dt}(p_A(t)) = \frac{\beta}{V_{hose,VA} + y \cdot A_A}(-Q_{AT}(t) - \dot{y}(t) \cdot A_A) \tag{7.780}$$

$$(3) \ \frac{d}{dt}(p_B(t)) = \frac{\beta}{V_{hose,VB} + (l_{cyl} - y) \cdot A_B}(Q_{PB}(t) + \dot{y}(t) \cdot A_B) \tag{7.781}$$

$$(4) \ \frac{d}{dt}(p_P(t)) = \frac{\beta}{V_{hose,pv} + V_{acc}}(Q_P(t) - (Q_{PB}(t) + Q_{PT}(t) + Q_r(t) + Q_{acc}(t))) \tag{7.782}$$

$$end \tag{7.783}$$

$$if \ ideal \ relief \ valve \tag{7.784}$$

$$(5) \qquad if \ Q_r(t) > 0.0 \ ; \ then \ p_P(t) = p_{relief} \tag{7.785}$$

$$(5) \qquad else \ if \ p_P(t) < p_{relief} \ ; \ then \ Q_r(t) = 0.0 \tag{7.786}$$

$$end \tag{7.787}$$

$$else \tag{7.788}$$

$$if \quad p_P(t) \geq p_{relief} \ and \ no \ transient \ dynamics \tag{7.789}$$

$$(5) \qquad Q_r(t) = K_{relief} \cdot (p_P(t) - p_{relief}) \tag{7.790}$$

$$elseif \ p_P(t) \geq p_{relief} \ and \ with \ transient \ dynamics \tag{7.791}$$

$$(5) \ \tau_{relief} \frac{Q_r(t)}{dt} = -Q_r(t) + K_{relief} \cdot (p_P(t) - p_{relief}) \tag{7.792}$$

$$else \tag{7.793}$$

$$(5) \qquad Q_r(t) = 0.0 \tag{7.794}$$

$$end \tag{7.795}$$

$$end \tag{7.796}$$

$$where \tag{7.797}$$

$$(6) \ Q_P(t) = D_p \cdot w_{pump}(t) \tag{7.798}$$

$$(7) \ Q_{PT}(t) = \sqrt{(p_P(t) - p_T(t))/\Delta p_r} = 0.0 \tag{7.799}$$

$$(8) \ Q_{PA}(t) = Q_r \cdot \frac{x_s(t)}{x_{s,max}} \cdot \sqrt{(p_P(t) - p_A(t))/\Delta p_r} \quad if \ x_s > 0 \tag{7.800}$$

$$(9) \ Q_{BT}(t) = Q_r \cdot \frac{x_s(t)}{x_{s,max}} \cdot \sqrt{(p_B(t) - p_T(t))/\Delta p_r} \quad if \ x_s > 0 \tag{7.801}$$

$$(8) \ Q_{PB}(t) = Q_r \cdot \frac{x_s(t)}{x_{s,max}} \cdot \sqrt{(p_P(t) - p_B(t))/\Delta p_r} \quad if \ x_s < 0 \tag{7.802}$$

$$(9) \ Q_{AT}(t) = Q_r \cdot \frac{x_s(t)}{x_{s,max}} \cdot \sqrt{(p_P(t) - p_T(t))/\Delta p_r} \quad if \ x_s < 0 \tag{7.803}$$

Notice that the ideal relief valve model cancels the fluid compressibility whenever $p_P(t) = p_{relief}$, by calculating the $Q_r(t)$ such that the right hand side of the pump pressure dynamics is equal to zero.

The accumulator acts both as a hydraulic power storage device which increases the line capacitance, hence absorb large pressure spikes, and as a hydraulic power source when line pressure drops below the pressure setting of the accumulator to support the pump. Especially when fast dynamic changes occur in demand for hydraulic power, the pump cannot react fast enough. But the accumulator can provide the transient power needed for a short period of time. The flow into or out of the accumulator is function of the line pressure, accumulator pressure and max-min pressure range settings of the accumulator.

Accumulator state (volume and pressure) is modeled as,

$$If \quad V_{acc}(t) \le 0.0 \ and \ p_P(t) \le p_{acc}(t) \tag{7.804}$$

$$(10) \quad Q_{acc}(t) = 0.0 \tag{7.805}$$

$$else \tag{7.806}$$

$$(10) \quad Q_{acc}(t) = sign(p_P(t) - p_{acc}(t)) \cdot K_{acc} \cdot \sqrt{|p_P(t) - p_{acc}(t)|} \tag{7.807}$$

$$end \tag{7.808}$$

$$(11) \quad \frac{dp_{acc}(t)}{dt} = \left(\frac{p_{max} - p_{min}}{V_{disch}} \right) \cdot Q_{acc}(t) \ ; \ p_{acc}(t_0) = p_{min} \tag{7.809}$$

$$(12) \quad \frac{dV_{acc}(t)}{dt} = Q_{acc}(t) \ ; \ V_{acc}(t_0) = 0.0 \tag{7.810}$$

where initial conditions of the accumulator pressure and fluid volume in the accumulator are specified ($p_{acc}(t_0)$ and $V_{acc}(t_0)$). On power-up, as accumulator is charged by the line pressure, the accumulator pressure increases. For a given accumulator, we know the discharge volume (V_{disch}), maximum, minimum, and precharge pressures ($p_{max} = p_{relief}, p_{min}, p_{pre}$) and initial conditions on pressure and fluid volume in the accumulator, that is typical values $p_{acc}(t_0) = p_{pre}$, $V_{acc}(t_0) = 0.0$.

For simulation purposes, we can consider the initial pressure of the accumulator as $p_{acc}(t_0) = p_{min}$ (or a value between p_{min} and p_{max}) and initial volume $V_{acc}(t_0) = 0.0$. Maximum discharge volume is reached at $p_{acc} = p_{max}$, and zero discharge volume is reached (no fluid volume left to discharge) at $p_{acc} = p_{min}$. The net result of adding an accumulator between the pump and the valve is to reduce the pressure variations by increasing the compliant fluid volume, while the accumulator acts both as energy storage (during $p_s > p_{acc}$) and energy source (during $p_s < p_{acc}$) device.

An engine or electric motor provides the mechanical power to the pump. The engine/motor speed is w_{pump} rev/min. The pump is of fixed displacement type, as indicated by its hydraulic symbol, and the volumetric displacement of the pump is D_p m^3/rev. The leakage flow from the pump is neglected. We assume that the pressure at the input port of the valve is the same as the pump output pressure, p_P, and the pressure at the output port of the valve is the same as the pressure at the cylinder. In other words, we neglect the pressure drop in the line between the pump and the main valve, and the main valve and the cylinder.

Let us assume the following values for the parameters of the system components:

$$D_p = 0.0001 \ m^3/rev \tag{7.811}$$

$$Q_r = 20.0 \times 10^{-4} \ m^3/s \tag{7.812}$$

$$\Delta p_r = 1000 \ psi \tag{7.813}$$

$$x_{db} = 10.0\% \tag{7.814}$$

$$x_{s,max} = 100.0\% \tag{7.815}$$

$$A_{PA}(x_s) = (20 \cdot 10^{-6})/(100 - x_{db})) \cdot (|x_s| - x_{db}) \ m^2 \ ; \ x_s \ge x_{db} \tag{7.816}$$

$$A_{PB}(x_s) = (10 \cdot 10^{-6})/(100 - x_{db})) \cdot (|x_s| - x_{db}) \ m^2 \ ; \ x_s \le -x_{db} \tag{7.817}$$

$$A_{AT}(x_s) = (40 \cdot 10^{-6})/(100 - x_{db})) \cdot (|x_s| - x_{db}) \, \text{m}^2 \, ; \, x_s \leq -x_{db} \tag{7.818}$$

$$A_{BT}(x_s) = (10 \cdot 10^{-6})/(100 - x_{db})) \cdot (|x_s| - x_{db}) \, \text{m}^2 \, ; \, x_s \geq x_{db} \tag{7.819}$$

$$A_{PT}(x_s) = 0.0 \ \ closed\text{-}center \ valve \tag{7.820}$$

$$A_A = 0.01 \, \text{m}^2 \tag{7.821}$$

$$A_B = 0.005 \, \text{m}^2 \tag{7.822}$$

$$m = 10\,000 \, \text{kg} \tag{7.823}$$

$$c = 0.0 \, \text{Nt/(m/s)} \tag{7.824}$$

$$l_{cyl} = 1.0 \tag{7.825}$$

$$\beta = 15.0 \cdot 10^8 \, \text{N/m}^2 \tag{7.826}$$

$$V_{hose,pv} = 0.0001 \, \text{m}^3 \tag{7.827}$$

$$V_{hose,VA} = 0.0001 \, \text{m}^3 \tag{7.828}$$

$$V_{hose,VB} = 0.0001 \, \text{m}^3 \tag{7.829}$$

$$K_{relief} = 1.0 \cdot 10^{-8} \tag{7.830}$$

$$\tau_{relief} = 0.025 \tag{7.831}$$

$$P_{relief} = 20.0 \cdot 10^6 \, \text{Nt/m}^2 \tag{7.832}$$

$$p_{max} = 20 \cdot 10^6 \, \text{N/m}^2 \tag{7.833}$$

$$p_{min} = 15 \cdot 10^6 \, \text{N/m}^2 \tag{7.834}$$

$$p_{pre} = 15 \cdot 10^6 \, \text{N/m}^2 \tag{7.835}$$

$$V_{disch} = 0.005 \, \text{m}^3 \tag{7.836}$$

$$K_{acc} = 1.0 \cdot 10^{-6} \tag{7.837}$$

$$C_{acc} = \frac{V_{disch}}{p_{max} - p_{min}} = 1.0 \cdot 10^{-9} \, \text{m}^3/(\text{N/m}^2) \tag{7.838}$$

The simulated input condition and initial conditions are

$$w_{pump}(t) = 25 \, \text{rev/s} \tag{7.839}$$

$$F_{load}(t) = 10\,000.0 \, \text{kg} \cdot 9.81 \, \text{m/s}^2 = 98.1 \, \text{kN} \ weight \tag{7.840}$$

$$x_s(t) = 0 \qquad ; \ 0 < t <= 1.0 \tag{7.841}$$

$$= (100/0.25) * (t - 1) \ ; \ 1.0 < t <= 1.25 \, \text{s} \tag{7.842}$$

$$= 100 \qquad ; \ 1.25 < t <= 3.0 \, \text{s} \tag{7.843}$$

$$= 100 - (100/0.25) \cdot (t - 3) \ ; \ 3.0 < t <= 3.25 \, \text{s} \tag{7.844}$$

$$= 0 \qquad ; \ \ 3.25 < t < 5.0 \tag{7.845}$$

$$p_T(t) = 0.0 \ ; \ tank \ pressure \tag{7.846}$$

$$y(0) = 0.10 \ ; \ initial \ cylinder \ position \tag{7.847}$$

$$\dot{y}(0) = 0.0 \ ; \ initial \ cylinder \ velocity \ is \ zero \tag{7.848}$$

$$p_P(0) = p_{relief} \tag{7.849}$$

$$p_A(0) = F_{load}(0)/A_A \ ; \ initial \ cylinder \ pressure \tag{7.850}$$

$$p_B(0) = p_T = 0.0 \tag{7.851}$$

$$p_{acc}(0) = p_{min} = 15 \cdot 10^6 \, \text{N/m}^2 \tag{7.852}$$

$$V_{acc}(0) = 0.0 \, \text{m}^3 \tag{7.853}$$

Note that the input shaft to the pump runs at a constant speed, the load on the cyclinder is simply a mass moving against gravity, the initial position of the cylinder is close to the bottom zero position. The valve opens and closes in a total of 3.0 s in a trapezoidal profile.

It is of interest to simulate the motion of the hydraulic system and plot the results:

1. $Q_s(t)$ – pump flow rate
2. $Q_{PA}(t)$ – flow rate from pump to A-side of the cylinder
3. $Q_{PB}(t)$ – flow rate from pump to B-side of the cylinder
4. $Q_{AT}(t)$ – flow rate from A-side of the cylinder to the tank
5. $Q_{BT}(t)$ – flow rate from B-side of the cylinder to the tank
6. $Q_r(t)$ – flow rate through the relief valve
7. $Q_{acc}(t)$ – flow rate into or out of the accumulator
8. $x_s(t)$ – main valve spool displacement
9. $F_{load}(t)$ – external load
10. $p_P(t)$ – pump pressure
11. $p_A(t)$ – pressure in the A-side of the cylinder
12. $p_B(t)$ – pressure in the B-side of the cylinder
13. $p_{acc}(t)$ – pressure in the accumulator
14. $V_{acc}(t)$ – fluid volume in the accumulator
15. $\dot{y}(t)$ – cylinder velocity
16. $y(t)$ – cylinder position, which is obtained from initial condition plus the integtration of cylinder velocity, $y(t) = y(t_0) + \int_{t_0}^{t} \dot{y}(\tau)d\tau$

Remark: The actual pressure developed at the outlet of the pump is determined by the load pressure, valve size, and relief pressure. Note that if the valve is undersized for the application, $A_v(x_s(t))$ is small, then p_P will be equal to p_{relief} most of the time to support the most flow rate it can. If, on the other hand, the valve is oversized, then $A_v(x_s(t))$ is too large, then p_P will not need to be as large as p_{relief} most of the time to support the flow, and the relief valve will not need to open, hence the system will be more efficient. However, the larger the $A_v(x_s(t))$ gain is, the worse the control accuracy of the flow rate, and hence the control accuracy of the EH motion control system.

It is instructive for the reader to change the valve size by changing the $A_v(x_s)$ and observe the effect of it on the p_P, that is what happens if we have sized the valve too small or too large for the application. As the valve size gets larger, the control accuracy gets poorer, while efficiency improves. The fundamental variable we control is the valve spool displacement, x_s, through which the valve orifice area $A_v(x_s)$ is determined. The smallest resolution that the valve is controllable is determined by the position control of the valve. Let us assume that it is a fixed quantity by the valve control accuracy, depending on the controller, amplifier, and valve position feedback accuracy (if any closed loop control is used on the valve position). For a given Δx_s, that is the smallest increment the valve position can be controlled, the change in flow rate, hence the cylinder speed is proportionally affected by the $A_v(x_s)$. The larger that gain is (the larger the valve size), the larger the smallest change in the cylinder velocity that we can affect, hence the worse the control resolution is (or control accuracy). Conversely, the smaller the valve size ($A_v(x_s)$), the smaller the flow rate change we can affect, hence the smaller cylinder velocity change we can affect, the better the control resolution (control accuracy). In contrast, energy efficiency requires a larger valve size, since energy efficiency requires us to minimize the power loss. It is desirable to support the largest flow rate with as little pressure drop as possible across the valve. In other words, as the valve size gets larger, the pressure drop needed to support a given desired flow rate gets smaller. Hence, the power loss

$$Q_{PA} \cdot (p_P - p_A) + Q_{BT} \cdot (p_B - p_T) \quad or \tag{7.854}$$

$$Q_{PB} \cdot (p_P - p_B) + Q_{AT} \cdot (p_A - p_T) \tag{7.855}$$

across the valve (due to pressure drop across the valve) is smaller. This is a fundamental design conflict between *control resolution (accuracy)* and *energy efficiency* in hydraulic systems. A good design must find a balance between these two conflicting requirements.

Case 1 – Simplified Model: We consider the following simplified version of the model

- neglect the cylinder and load inertial dynamics,
- neglect fluid compressibility throughout the hydraulic circuit,
- assume we have an ideal relief valve,
- assume we do not have any accumulator.

The only initial condition needed in this case is $y(t_0)$. The rest of the initial conditions ($p_P(t_0)$, $p_A(t_0)$, $p_B(t_0)$, $\dot{y}(t_0)$, $Q_r(t_0)$) do not need to be specified. Instead, they are calculated directly from the simultaneous solution of the algebraic equations. In this case the differential equations describing the variation in cylinder pressure and cylinder motion reduce to an algebraic equation where the cylinder velocity is directly determined by the flow rate into the cylinder. The solution can be obtained directly by substitution among the algebraic equations since there are as many equations as there are unknowns. However, a more useful method is to use a numerical method to solve these equations. The iterative solution of the algebraic equations requires initial guesses (not initial conditions), $p_P(t_0), p_A(t_0), p_B(t_0), \dot{y}(t_0), Q_r(t_0)$, which are used as the starting point for the numerical search algorithm for finding the roots. In order to simulate this case, simply set the left hand side of the differential Equations 7.776–7.778 to zero, turning them into algebraic equations, and solve the whole equation set (1)–(10) numerically.

Case 2 – Take into account cylinder and load inertial dynamics, and fluid compressibility on both pump and load side of the line: In this case we have the second-order differential equation as Equation (1), that defines the net force–acceleration relationship due to inertia and the differential Equations (7.776–7.782). Then we need $y(t_0)$ and $\dot{y}(t_0)$, $p_A(t_0)$, $p_B(t_0)$, $p_P(t_0)$ initial conditions to be specified. Numerical solution requires only the solution of a set of ordinary differential equations. All of the algebraic equations (5-9) can be solved directly (without any iterative algorithm) by using the available initial conditions on pressures.

We simulate this condition *without the accumulator* and *with the accumulator* in the circuit. The accumulator is added to the circuit model by including the V_{acc} and Q_{acc} terms in Equation (4), and the equations describing the dynamics of the accumulator state, $p_{acc}(t)$ and $V_{acc}(t)$ (Equations (10), (11), and (12)). We need to specify the initial pressure and volume of the accumulator ($p_{acc}(t_0)$, $V_{acc}(t_0)$), as well as the discharge volume capacity (V_{disch}) and operating pressure range (p_{min}, p_{max}).

CHAPTER 8

ELECTRIC ACTUATORS: MOTOR AND DRIVE TECHNOLOGY

8.1 INTRODUCTION

The term "actuator" in motion control systems refers to the component which delivers the motion (Figure 8.1). It is the component that delivers the mechanical power, which may be converted from an electric, hydraulic, or pneumatic power sources. In the electric power based actuator category, the motor and the drive are two power conversion components that work together. In a motion control system, when we refer to the performance characteristics of a motor, we always have to refer to it in conjunction with the type of "drive" that the motor is used with, for the type of drive determines the behavior of the motor. The term *drive* is generically used in industry to describe the *power amplification* and the *power supply* components together.

The discussion in this chapter is limited to motor-drive technologies that can be used in high performance motion control applications, that is involving closed loop position and velocity control with high accuracy and bandwidth. Low cost constant speed motor-drive components, which are used in mass quantities in applications such as fans and pumps, are not discussed. To this end, we will discuss the following motor-drive technologies:

1. DC motors (brush-type and brushless type) and drives.
2. AC induction motors and field oriented vector control drives.
3. Step motors and drives:
   (a) permanent magnet step motors,
   (b) hybrid step motors,
   along with full step, half step, and micro-stepping drives.

The operating principle of any electric motor involves one or more of the following three physical phenomenon:

1. opposite magnetic poles attract, and the same magnetic poles repel each other,
2. magnets attract iron and seek to move to a position to minimize the reluctance to *magnetic flux*,
3. current carrying conductors create electromagnets and act like a current-controlled magnet.

Every motor has the following components,

1. a rotor on a shaft (moving component),
2. a stator (stationary component),

Mechatronics with Experiments, Second Edition. Sabri Cetinkunt.
© 2015 John Wiley & Sons, Ltd. Published 2015 by John Wiley & Sons, Ltd.
Companion Website: www.wiley.com/go/cetinkunt/mechatronics

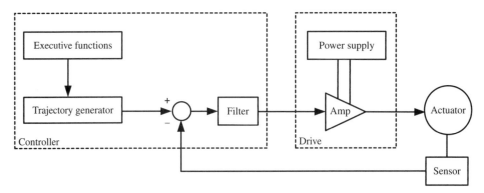

FIGURE 8.1: Functional blocks of a closed loop motion control system: actuator, sensor, amplifier, controller filters.

3. a housing (with end plates for rotary motors),

4. two bearings, one for each end, to support the rotor in the housing, including some washers to allow axial play between the shaft and the housing.

In addition, brush-type motors have commutator and brush assembly to direct current into the proper coil segment as a function of rotor position. Brushless motors have some type of rotor position sensor for electronic commutation of the current (i.e., Hall effect sensors or incremental encoders). Commutation means the distribution of current into appropriate coils as a function of rotor position.

Traditionally AC induction motors have been used in constant speed applications, whereas DC motors have been used in variable speed applications. With the advances in solid-state power electronics and digital signal processors (DSP), an AC motor can be controlled in such a way that it behaves like a DC motor. One way of accomplishing this is the "field oriented vector control" algorithm used in the drive for current commutation.

In the following discussion, a magnetic pole refers to a north (N) or south (S) pole, a magnetic pole pair refers to a N and a S pole. When we refer to a two pole motor, it means it has one N and one S pole. Likewise, a four pole motor has two N and two S poles.

An electric motor is a power conversion device. It converts electrical power to mechanical power. Input to the motor is in the form of *voltage* and *current*, and the output is mechanical *torque* and *speed*. The key physical phenomenon in this conversion process is different for different motors.

1. In the case of DC motors, there are two magnetic fields. In brush-type DC motors, one of the magnetic fields is due to the current through the armature winding on the rotor, and the other magnetic field is due to the permanent magnets in the stator (or due to field excitation of the stator winding if electromagnets are used instead of permanent magnets). In the case of brushless DC motors, the roles of rotor and stator are swapped.

2. In the case of AC motors, the first magnetic field is setup by the excitation current on the stator. This magnetic field in turn induces a voltage in the rotor conductors by Faraday's induction principle. The induced voltage at the rotor conductors results in current which in turn sets up its own magnetic field, which is the second magnetic field. The torque is produced by the interaction of the two magnetic fields. In the case of a DC motor and AC induction motor (with field oriented vector control), the two magnetic fields are always maintained at a 90 degree angle in order to maximize the torque generation capability per unit current. This is accomplished by commutating the stator current (mechanically or electronically) as a function of the rotor position.

3. Stepper motors (permanent magnet (PM) type) work on basically the same principle as brushless DC motors, except that the stator winding distribution is different. A given stator excitation state defines a stable rotor position as a result of the attraction between electromagnetic poles of the stator and permanent magnets of the rotor. The rotor moves to minimize the magnetic reluctance. At a stable rotor position of a step motor, two magnetic fields are parallel.

The torque generation, that is the electrical energy to mechanical energy conversion process, in any electric motor can be viewed as a result of the interaction of two *magnetic flux density* vectors: one generated by the stator ($\vec{B}_\mathrm{s}$) and one generated by the rotor ($\vec{B}_\mathrm{r}$). In different motor types, the way these vectors are generated is different. For instance, in a permanent magnet brushless motor the magnetic flux of rotor is generated by permanent magnets and the magnetic flux of stator is generated by current in the windings. In the case of an AC induction motor, the stator magnetic flux vector is generated by the current in the stator winding, and the rotor magnetic flux vector is generated by induced voltages on the rotor conductors by the stator field and resulting current in the rotor conductors. It can be shown that the torque production in an electric motor is proportional to the strength of the two magnetic flux vectors (stator's and rotor's) and the sine of the angle between the two vectors. The proportionality constant depends on the motor size and design parameters.

$$T_\mathrm{m} = K \cdot B_\mathrm{r} \cdot B_\mathrm{s} \cdot \sin(\theta_\mathrm{rs}) \tag{8.1}$$

where K is the proportionality constant, and θ_rs is the angle between the $\vec{B}_\mathrm{s}$ and $\vec{B}_\mathrm{r}$, and T_m is the torque.

Every motor requires some sort of *current commutation* by mechanical means as in the case of brush-type DC motors, or by electrical means as in the case of brushless DC motors. Current commutation means modifying the direction and magnitude of current in the windings as a function of rotor position. The goal of the commutation is to give the motor the ability to produce torque efficiently, that is to maintain $\theta_\mathrm{rs} = 90°$.

The design of an electric motor seeks to determine the following:

1. the three-dimensional shape of the effective magnetic reluctance of the motor by proper selection of materials and geometry of the motor,

2. the distribution of coil wires, coil wire diameter, and its material (i.e., copper or aluminum)

3. the permanent magnets (number of poles, geometric dimensions and PM material).

The engineering analysis is concerned with determining the resulting force/torque for a given motor design and coil currents. In addition, we also need to examine the flux density and flux lines in order to evaluate the overall quality of the design so that there is no excessive saturation in the flux path. These results are obtained from the solution of Maxwell's equations for electromagnetic fields. Modern engineering analysis software tools are based on the finite element method (FEM) to solve Maxwell's equations and used for motor design (examples include Maxwell 2D/3D by Ansoft Corporation, Flux2D/3D by Magsoft Corporation, and PC-BLDC by The Speed Laboratory).

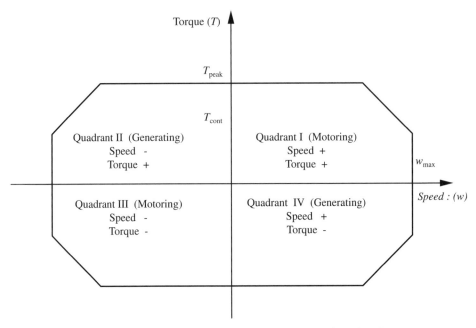

FIGURE 8.2: Four quadrant region torque-speed characteristics of an electric motor.

8.1.1 Steady-State Torque-Speed Range, Regeneration, and Power Dumping

Electric motors can act either as a *motor*, that is to convert electrical power to mechanical power, to drive loads, or as a *generator*, that is to convert mechanical power to electrical power, when driven externally by the load. Let us consider the steady-state torque versus speed plane (Figure 8.2). Motor-drive combinations that can operate in all four quadrants of the torque-speed plane are called *four-quadrant operation* devices and can act as motor and generator during different modes of an application. In the quadrants I and III of the torque-speed plane, the mechanical power output is positive.

$$P_m = T \cdot w > 0.0; \quad motoring\ mode \tag{8.2}$$

When the motor speed and torque are in the same direction, the device is in *motoring mode*. In quadrants II and IV of the torque-speed plane, the mechanical power output is negative. That means the motor takes mechanical energy from the load instead of delivering mechanical energy to the load. The device is in *generator mode* or *regenerative braking mode*.

$$P_m = T \cdot w < 0.0; \quad generating\ mode \tag{8.3}$$

This energy can either be dissipated in the motor-drive combination, stored in a battery or capacitor set, or returned to the supply line by the drive. This is precisely the energy recovered (and stored in batteries) while braking in a hybrid car, where the electric motor torque applied to provide the braking effect is in the opposite direction to the speed. By storing this energy in the batteries and using it later, instead of wasting it, the energy efficiency of the car is improved. Similarly, in industrial and factory applications, anytime this condition occurs for a motor operation (motor operating in II or IV quadrant, $T \cdot w < 0$), the energy can be stored in battery/capacitors, returned to the utility supply line or wasted as heat via resistors.

During acceleration, a motor adds mechanical energy to the load. It acts as a *motor*. During deceleration, the motor takes away energy from the load. It acts like a brake or *generator*. This means that energy is put into the load inertia during acceleration, and energy is taken from the load inertia (returned to the drive) during deceleration (Figure 8.2).

Some drives can convert the generated electric power and put it back to into the electric supply line while others dump the regenerative energy as heat through resisitors. The amount of regenerative energy depends on the load inertia, deceleration rate, time period, and load forces.

There are two different motion conditions where regenerative energy exists and satisfies the $T \cdot w < 0$ condition:

1. During deceleration of a load, where the applied torque is in the opposite direction to the speed of inertia.

2. In load driven applications, that is in tension controlled web handling applications, a motor may need to apply a torque to the web in the opposite direction to the motion of motor and web in order to maintain a desired tension. Another example for this case is where the gravitational force provides more than the needed force to move an inertia, and the actuator needs to apply force in the direction opposite to the motion in order to provide a desired speed.

Example   Consider the electric motor driven load shown in Figures 8.3 and 8.4. Assume that the load is a translational inertia and an electric motor is a perfect linear force generator. Consider an incremental motion that moves the inertia from position x_1 to position x_2 using a square force input. For simplicity, let us neglect all the losses. We will assume that the motor-drive combination converts electrical power ($P_e(t)$) to mechanical power ($P_m(t)$) with 100% efficiency in the motoring mode, and mechanical power to electrical power with 100% efficiency in the generator mode.

$$P_e(t) = P_m(t) \tag{8.4}$$

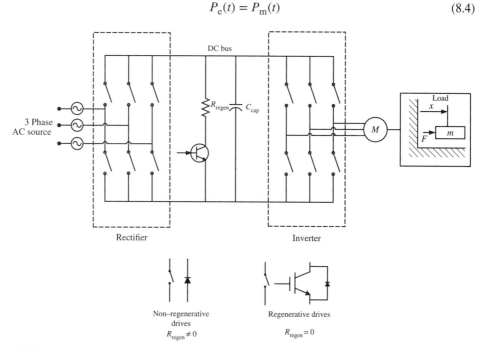

FIGURE 8.3: Regenerative energy in motion, its storage and dissipation.

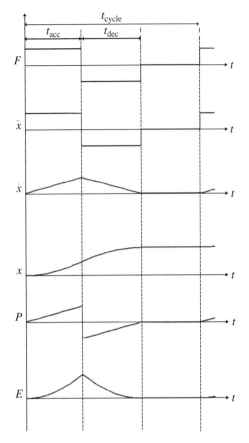

FIGURE 8.4: Typical output force to load and motion profile and regenerative energy opportunity for energy recovery during the deceleration phase of the motion.

The force–motion relationship from Newton's second law is

$$F(t) = m \cdot \ddot{x}(t) \tag{8.5}$$

The mechanical power delivered to the inertia is

$$P_m(t) = F(t) \cdot \dot{x}(t) \tag{8.6}$$

which is supplied by the electric motor and drive combination.

Notice that when the force and speed are in the same direction, the mechanical power delivered to the inertia is positive, and the motor-drive operates in motoring mode, that is to convert electrical energy to mechanical energy. Similarly, when the direction of force is opposite to the direction of speed, the mechanical power is negative, which means the inertia gives out energy intead of taking energy. This mechanical energy is converted to electrical energy by the motor since it acts like a generator under this condition.

$$P_m(t) = P_e(t) = F(t) \cdot \dot{x}(t) \tag{8.7}$$

$$P_m(t) = P_e(t) = F(t) \cdot \dot{x}(t) > 0 \quad \textit{motoring mode} \tag{8.8}$$

$$P_m(t) = P_e(t) = F(t) \cdot \dot{x}(t) < 0 \quad \textit{generating mode} \tag{8.9}$$

In motoring mode, the motor-drive provides energy to the load. In generator mode, the motor drive takes away energy from the load. This energy must be either stored, returned to line, or dissipated in resistors. One of the most common approaches in servo applications is to store a small portion of the energy in the DC bus capacitor and dissipate the rest as heat over external resistors added specifically for this purpose. In applications where this so-called *regenerative energy* is large, external resistors are added for the purpose of dumping it.

In load driven applications, such as tension controlled web handling or gravity driven loads, the motor continuously operates in regenerative power mode (generator mode). The tension, hence the torque, generated by the motor is always in the opposite direction of the speed. If all of the regenerative power is to be dissipated as heat using external resistors, the resistors should be sized based on the following continuous power dissipation cabability,

$$P_{cont} = RMS(F_{tension}(t) \cdot \dot{x}(t)) \tag{8.10}$$

In a given application, the amount of regenerative energy is a function of inertia, deceleration rate, and time period. In this example, the regenerative energy is is the time integral of the regenerative power as follows,

$$E_{reg}(t) = \int_0^{t_{dec}} P_m(t) \cdot dt \tag{8.11}$$

$$= \int_0^{t_{dec}} F(t) \cdot \dot{x}(t) \cdot dt \tag{8.12}$$

$$= \int_0^{x_{dec}} F(x) \cdot dx \tag{8.13}$$

$$= \int_0^{x_{dec}} m \cdot \ddot{x}(t) \cdot dx \tag{8.14}$$

$$= \int_0^{x_{dec}} m \cdot \dot{x}(t) \frac{d\dot{x}}{dx} \cdot dx \tag{8.15}$$

$$= \int_0^{x_{dec}} m \cdot \dot{x}(t) \cdot d\dot{x} \tag{8.16}$$

$$= \frac{1}{2} \cdot m \cdot \left(\dot{x}_1^2 - \dot{x}_2^2 \right) \tag{8.17}$$

If we consider the tension control application, then regenerative energy is always increasing which must be either stored or used or dissipated. Let us consider a tension control case where the web tension is constant $F(t) = F_0$ and the web speed is constant, $\dot{x}(t) = \dot{x}_0$. Then for any period of time Δt, the regenerative energy is

$$E_{reg}(\Delta t) = F_o \cdot \dot{x}_o \cdot \Delta t \tag{8.18}$$

This energy (E_{reg}) must be dissipated at the "regen" resistors and the motor winding due to its resistance (E_{ri}) and partially stored in the DC bus capacitors (E_{cap}).

$$E_{reg} = E_{cap} + E_{ri} \tag{8.19}$$

Let us assume that the regenerative resistors will be activated by an appropriate logic circuit whenever DC bus voltage reaches a voltage level V_{reg}, where $V_{nom} < V_{reg} < V_{max}$. V_{max} is the maximum voltage level above which the amplifier control circuit would disable the transistors and go into "fault" mode. Then, the amount of energy that can be stored in the capacitor is

$$E_{cap} = \frac{1}{2} \cdot C_{cap} \cdot \left(V_{reg}^2 - V_{nom}^2 \right) \tag{8.20}$$

$$C_{cap} = 2 \cdot \frac{E_{cap}}{\left(V_{reg}^2 - V_{nom}^2 \right)} \tag{8.21}$$

where C_{cap} is the capacitance of the capacitors. Clearly, the capacitor can store a finite amount of energy and its size grows as the required energy storage capacity increases.

Therefore, the remainder of the regenerative energy must either be returned to the supply line through a voltage regulating inverter or dissipated as heat at the regenerative resistor and the motor winding.

Let us neglect the energy dissipated to heat at the motor winding. Hence, the remaining energy must be dissipated at the "regen" resistors. The peak and continuous power rating of the regenerative resistors can be calculated from

$$P_{\text{peak}} = \frac{E_{\text{reg}} - E_{\text{cap}}}{t_{\text{dec}}} \tag{8.22}$$

$$P_{\text{peak}} = R_{\text{reg}} \cdot i^2 \tag{8.23}$$

$$= \frac{V_{\text{reg}}^2}{R_{\text{reg}}} \tag{8.24}$$

$$R_{\text{reg}} \leq \frac{V_{\text{reg}}^2}{P_{\text{peak}}} \tag{8.25}$$

where P_{peak} are peak, power dissipation capacity used to determine the regen resistor size R_{reg}, E_{cap} regenerative energy to be stored, V_{reg} is the nominal DC bus voltage over which the regenerative circuit is active (i.e., a value between the nominal DC bus voltage and maximum DC bus voltage, $V_{\text{nom}}, V_{\text{max}}$), to determine the regenerative storage capacitor size C_{cap}. In some applications, the regenerative power may be so small that the DC bus capacitor is large enough to store the energy without the need for dissipating it as heat over resistors. Notice that the capacitor (or battery) is sized based on maximum energy, whereas the resistor is sized based on maximum power.

8.1.2 Electric Fields and Magnetic Fields

There are two types of fields in electrical systems: *electric fields* and *magnetic* (also called *electromagnetic) fields*. Although we are primarily interested in electromagnetic fields for the study of electric motors, we will discuss both briefly for completeness. Electric fields ($\vec{E}$) are generated by static charges. Magnetic fields ($\vec{H}$, also called electromagnetic fields) are generated by moving charges (current).

An *electric field* is a distributed vector field in space whose strength at a location depends on the charge distribution in space. It is a function of the static location of charges and the amount of charges. By convention, electric fields start (emitted) from positive charges and ends (received) in negative charges (Figure 8.5a). Capacitors are commonly used to store charges and generate electric fields. The smallest known charge is that of an electron (negative charge) and a proton (positive charge) with units of *Coulomb*, C,

$$|e^-| = |p^+| = 1.60219 \times 10^{-19} \, \text{C} \tag{8.26}$$

The electric field $\vec{E}$ at a point in space (s) due to n many charges ($q_1, q_2, \ldots, q_n$) at various locations can be determined from (Figure 8.5b),

$$\vec{E}(s) = k_e \sum_{i=1}^{n} \frac{q_i}{r_i^2} \cdot \vec{e}_i \quad \text{N/C} \quad \text{or} \quad \text{V/m} \tag{8.27}$$

where $k_e = 8.9875 \times 10^9 \, \text{N} \cdot \text{m}^2/\text{C}^2$ is called the *Coulomb constant*, r_i is the distance between the location of charge (i) and the point in space considered (s), $\vec{e}_i$ is the unit vector between each charge location and the point s, q_i is the charge at location i. For negative charges, the unit vector is directed towards the charge, for positive charges it is directed

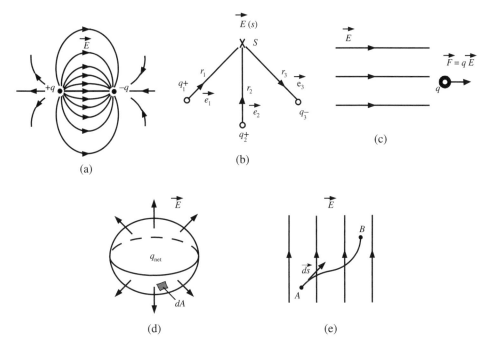

FIGURE 8.5: Electric fields due to stationary electric charges. (a) By convention, electric fields are assumed to eminate from positive charges and terminate at the negative charges. (b) Electric field at a point in space due to charged particles at other locations. (c) Force acting on a charged particle due to an electric field. (d) Integral of electric field over a closed surface is proportional to the net charge inside the volume spanned by the closed surface. (e) Voltage between two points in space (A and B) is equal to the line integral of electric field between the two points.

away from the charge. The Coulomb constant is closely related to another well known constant as follows,

$$k_e = \frac{1}{4 \cdot \pi \cdot \epsilon_o}$$

(8.28)

where $\epsilon_o = 8.8542 \times 10^{-12} \, \mathrm{C^2/N \cdot m^2}$ is the *permittivity of free space*.

Force ($\vec{F}$) exerted on a charge (q) which is in an electric field ($\vec{E}$) is (Figure 8.5c)

$$\vec{F} = q \cdot \vec{E}$$

(8.29)

and if the charge is free to move, the resulting motion is governed by

$$\vec{F} = m_q \cdot \vec{a}$$

(8.30)

where the generated force results in the acceleration ($\vec{a}$) of the charge mass (m_q) based on Newton's second law. This last equation is used to study the motion of charged particles in electric fields, that is the motion of electrons in cathode ray tube (CRT), or the motion of charged small droplets in ink-jet printing machines. The motion trajectory of a particle of a known mass can be controlled by controlling the force acting on it. The force can be controlled either by controlling its charge or the electric field in which it travels. Generally, the electric field is kept constant, and the charge on each particle is controlled before it enters the electric field. Another electric field quantity of interest is the *electric flux*, Φ_E,

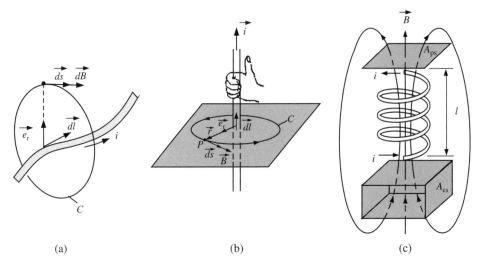

(a)   (b)   (c)

FIGURE 8.6: Magnetic fields due to current (moving charge): (a) magnetic field at any point P due to current over a general shape conductor, (b) magnetic field around an infinitely long straight current carrying conductor, (c) magnetic field inside a coil due to current.

which is the area integral of the electric field over a closed surface. The electric flux over a closed surface is proportional to the net electric charge inside the surface (Figure 8.5d),

$$\Phi_E = \oint_A \vec{E} \cdot d\vec{A} = \frac{q_{net}}{\epsilon_o} \tag{8.31}$$

where $d\vec{A}$ is a differential vector normal to the surface. The line integral of an electric field between any two points is the *electric potential* difference between the two points (voltage) (Figure 8.5e),

$$V_{AB} = -\int_A^B \vec{E} \cdot d\vec{s} \tag{8.32}$$

where the $d\vec{s}$ vector is a differential vector that is tangent to the path traveled from A to B.

The *magnetic fields* are generated by moving charges (Figure 8.6). There are two sources to generate and sustain a magnetic field:

1. current (moving charge) over a conductor,
2. permanent magnetic materials.

In the case of current carrying conductors, the magnetic field generated by the current (moving charges) is called an *electromagnetic field*. In the case of permanent magnet materials, the magnetic field is generated by orbital rotation of the electrons around the nucleus and the spin motion of electrons around their own axis of the permanent magnetic material. The net magnetic field of the material in macro scale is the result of the vector sum of the magnetic fields of its electrons.

In a non-magnetized material, the net effect of magnetic fields of electrons cancel out each other. Their alignment in a certain direction, by magnetizing the material, gives the material non-zero magnetization in a specific orientation. Either way, the magnetic field is a result of moving charges. The electric field vector starts at charges (i.e., positive charges) and ends in charges (negative charges). The magnetic field vector encircles the current that generates it (Figure 8.6). The vector relationship between the current and the magnetic field it generates follows the right hand rule. If the current is in the direction of the thumb, the magnetic field is in the direction of the fingers encircling the thumb. If the current changes direction, the magnetic field changes direction.

The *Biot-Savart law* states that the *magnetic field* (also called *magnetic flux density*), $\vec{B}$, generated by a current on a long wire at a point P with distance r from the wire is (Figure 8.6a and b)

$$d\vec{B} = \frac{\mu\, i\, \vec{dl} \times \vec{e}_r}{4\pi\, r^2} \tag{8.33}$$

where μ is the *permeability* of the medium around the conductor ($\mu = \mu_0$ for free space), $\vec{e}_r$ is unit vector of $\vec{r}$. We can think of permeability as the conductivity of the medium material (the opposite of resistance) for magnetic field. The permeability of free space is called μ_0

$$\mu_0 = 4\pi \cdot 10^{-7}\, \text{Tesla} \cdot \text{m/A} \tag{8.34}$$

Quite often, the permeability of a material (μ_m) is given relative to the permeability of free space,

$$\mu_m = \mu_r \cdot \mu_0 \tag{8.35}$$

where μ_r is the relative permeability of the material with respect to free space. If the Biot-Savart law is applied to a conductor over the length of l, we obtain the magnetic flux density $\vec{B}$ at point P at a distance r from the conductor due to current i (Figure 8.6a),

$$\vec{B} = \int_0^l \frac{\mu\, i\, \vec{dl} \times \vec{e}_r}{4\pi\, r^2} \tag{8.36}$$

The units of $\vec{B}$ in SI units is Tesla or T.

Ampere's Law states that the integral of a magnetic field over a closed path is equal to the current passing through the area covered by the closed path times the permeability of the medium covered by the closed path of integration (Figure 8.6a and b),

$$\oint_C \vec{B} \cdot \vec{ds} = \mu \cdot i \tag{8.37}$$

The vector relationship between the current, position vector of point P with respect to the wire and magnetic field follows the right hand rule. It describes how electromagnetic fields are created by a current in a given medium with a known magnetic permeability. The magnetic flux density is a continuous vector field. It surrounds the current that generates it based on the right hand rule. The magnetic flux density vectors are always closed, continuous vector fields.

By using either the Biot-Savart law or Ampere's law, the magnetic field due to current flow through a conductor of any shape in an electrical circuit can be determined. For instance, the magnetic field generated around an infinitely long straight conductor having current i in free space at a distance r from it can be calculated (Figure 8.6b)

$$|\vec{B}| = B = \frac{\mu_0 \cdot i}{2\pi \cdot r} \tag{8.38}$$

Similarly, the magnetic field inside a coil of solenoid is (Figure 8.6c)

$$|\vec{B}| = B = \frac{\mu_0 \cdot N \cdot i}{l} \tag{8.39}$$

where N is the number of turns of the solenoids, l is the length of the solenoid. It is assumed that the magnetic field distribution inside the solenoid is uniform and the medium inside the solenoid is free space. Notice that if the medium inside the coil is different than free space, such as steel with $\mu_m \gg \mu_0$, then the magnetic flux density developed inside the coil would be much higher.

Magnetic flux (Φ_B) is defined as the integral of magnetic flux density ($\vec{B}$) over a cross-sectional area perpendicular to the flux lines (Figure 8.6c),

$$\Phi_B = \int_{A_{ps}} \vec{B} \cdot d\vec{A}_{ps} \quad \text{Tesla} \cdot \text{m}^2 \quad \text{or} \quad \text{Weber} \tag{8.40}$$

where $d\vec{A}_{ps}$ is differential vector normal to the surface (A_{ps}). The area is the effective perpendicular area to the magnetic field vector. It is important not to confuse this relationship with Gauss's law.

Gauss's law states that the integral of a magnetic field over *a closed surface* that encloses a volume is zero (integration over the closed surface A_{cs} shown in Figure 8.6c),

$$\oint \vec{B} \cdot d\vec{A}_{cs} = 0 \tag{8.41}$$

where $d\vec{A}_{cs}$ is a differential area over a closed surface (A_{cs}), not a cross-sectional perpendicular area to flux lines. This integral is over a closed surface. In other words, net magnetic flux over a closed surface is zero. The physical interpretation of this result is that magnetic fields form closed flux lines. Unlike electric fields, magnetic fields do not start in one location and end in another location. Therefore, the net flow-in and flow-out lines over a closed surface are zero (Figure 8.6c).

Let us define the concept of *flux linkage*. Consider that a magnetic flux (Φ_B) is generated by a coil or permanent magnet or a similar external source. If it crosses one turn of conductor wire, the magnetic flux passing through that wire is called the *flux linkage* between the existing magnetic flux and the conductor,

$$\lambda = \Phi_B; \quad \text{for one turn coil,} \tag{8.42}$$

If the conductor coil had N turns instead of one, the amount of flux linkage between the external magnetic flux Φ_B and the N turn coil is

$$\lambda = N \cdot \Phi_B; \quad \text{for N turn coil,} \tag{8.43}$$

Magnetic field strength ($\vec{H}$) is related to the magnetic flux density $\vec{B}$ with the permeability of the medium,

$$\vec{B} = \mu_m \cdot \vec{H} \tag{8.44}$$

Magnetomotive force (MMF) is defined as,

$$MMF = H \cdot l \tag{8.45}$$

where l is the length of the magnetic field strength path.

Reluctance of a medium to the flow of magnetic flux is analogous to the electrical resistance of a medium to the flow of current.

The *reluctance* of a medium with cross-sectional area A and thickness l can be defined as

$$R_B = \frac{l}{\mu_m \cdot A} \tag{8.46}$$

where μ_m is the permeability of the medium, such as air, iron. *Permeance* of a magnetic medium is defined as the inverse of reluctance,

$$P_B = \frac{1}{R_B} \tag{8.47}$$

Reluctance is analogous to resistance, and permeance is analogous to conductivity.

The design task of shaping a magnetic circuit is the design task of defining the reluctance paths to the flow of magnetic field lines in the circuit. It is a function of material and geometry. Series and parallel reluctances are added following the same rules for electrical resistance. Iron and its various variations is the most commonly used material in shaping a magnetic circuit, that is in design of electric actuators. The material and geometry of the magnetic circuit determines the resistance paths to the flow of magnetic flux.

For instance, the magnetic flux in a coil (N turn coil with length l) can be determined as follows. The magnetic field due to current i has been given above,

$$B = \frac{\mu_0 \cdot N \cdot i}{l} \tag{8.48}$$

$$= \mu_0 \cdot H \tag{8.49}$$

$$H = \frac{N \cdot i}{l} \tag{8.50}$$

$$= \frac{MMF}{l} \tag{8.51}$$

$$MMF = N \cdot i \tag{8.52}$$

The magnetic flux is defined as the integral of the magnetic flux density over a surface perpendicular to the flux density vector (Figure 8.6c)

$$\Phi_B = B \cdot A = \frac{\mu_0 \cdot N \cdot i}{l} \cdot A \tag{8.53}$$

$$= \frac{N \cdot i}{[l/(\mu_0 \cdot A)]} \tag{8.54}$$

$$= \frac{MMF}{R_B} \tag{8.55}$$

For instance, a coil with N turns and current i, can be modeled in an electromagnetic circuit as an *MMF* source (similar to voltage source) and magnetic reluctance R_B (similar to electrical resistance) in series with the source (Figure 8.7),

$$MMF = N \cdot i \tag{8.56}$$

$$R_B = \frac{l}{\mu \cdot A} \tag{8.57}$$

Then, the magnetic flux (analogous to current) through the coil is

$$\Phi_B = \frac{MMF}{R_B} \tag{8.58}$$

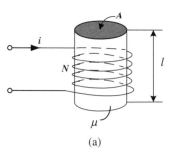

(a)

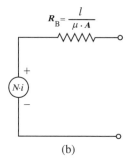

(b)

FIGURE 8.7: (a) A coil winding, and (b) its magnetic model. Coil is modeled as having an $MMF = N \cdot i$, magnetomotive source, and magnetic resistance (reluctance) $R_B = l/(\mu A)$ in series with the *MMF*.

TABLE 8.1: Analogy between electrical and electromagnetic circuits.

Electric circuits	Magnetic circuits
V	MMF or H
i	Φ_B or B
R	R_B or $\dfrac{1}{\mu_m}$
$i = \dfrac{V}{R}$	$\Phi_B = \dfrac{MMF}{R_B}$ or $B = \mu_m \cdot H$

By analogy to electrical circuits, there are three main principles used in analyzing magnetic circuits as follows (Table 8.1):

1. The sum of the MMF drop across a closed path is zero. This is similar to Kirchoff's law for voltages, which says the sum of voltages over a closed path is zero.

$$\sum_i MMF_i = 0; \quad \text{over a closed path,} \tag{8.59}$$

2. The sum of flux at any cross-section in a magnetic circuit is equal to zero (that is, the sum of incoming and outgoing flux through a cross-section). This is similar to Kirchoff's law for currents which says that at a node, the algebraic sum of currents is zero (sum of incoming and outgoing currents).

$$\sum_i \Phi_{Bi} = 0; \quad \text{at a cross-section,} \tag{8.60}$$

3. Flux and MMF are related by the reluctance of the path of the magnetic medium, similar to the voltage, current and resistance relationship,

$$MMF = R_B \cdot \Phi_B \tag{8.61}$$

Magnetic circuits typically have:

1. current carrying conductors in coil form which act as the source of the magnetic field,
2. permanent magnets (or a second current carrying coil),
3. iron based material to guide the magnetic flux, and
4. air.

The geometry and material of the medium uniquely determines the reluctance distribution in space. Current carrying coils and magnets determine the magnetic source. Interaction between the two (magnetic source and reluctance) determines the magnetic flux.

Motor Action: Force ($\vec{F}$) in a magnetic field ($\vec{B}$) and a moving charge (q) has vector relationship (Figure 8.8a),

$$\vec{F} = q\vec{v} \times \vec{B} \tag{8.62}$$

where $\vec{v}$ is the speed vector of the moving charge. This relationship can be extended for a current carrying conductor instead of a single charge. The force acting on a conductor of over length l due to the current i and magnetic field $\vec{B}$ interaction is (Figure 8.8b)

$$\vec{F} = l \cdot \vec{i} \times \vec{B} \tag{8.63}$$

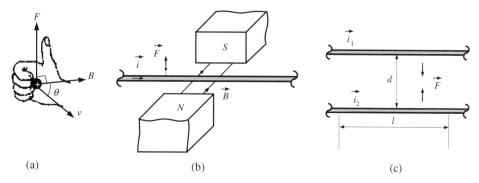

FIGURE 8.8: Magnetic forces: (a) magnetic force acting on a moving charge in a magnetic field, (b) magnetic force acting on a conductor with current in a magnetic field, and (c) magnetic force between two current carrying conductors.

This relationship is convenient to derive the Tesla or T unit of $\vec{B}$,

$$1 \text{ Tesla} = 1 \frac{\text{N}}{\text{C/s} \cdot \text{m}} = 1 \frac{\text{N}}{\text{A} \cdot \text{m}} \tag{8.64}$$

This is the basic physical principle for the electromechanical power conversion for *motor action*. Notice that the force is a vector function of the current and the magnetic flux density. $\vec{B}$ may be generated by a permanent magnet and/or electromagnet.

The force between two current carrying parallel conductors can be described as the interaction of the current in one conductor with the electromagnetic field generated by the current in the other conductor. Consider two conductors parallel to each other, carrying currents i_1 and i_2, separated from each other by a distance d and has length l. The force acting between them (Figure 8.8c)

$$|\vec{F}| = \mu_o \cdot \left(\frac{l}{2\pi \cdot d} \right) \cdot i_1 \cdot i_2 \tag{8.65}$$

where μ_o is the permeability of space between the two conductors. The force is attractive if the two currents are in the same direction, and repulsive if they are opposite.

Generator Action: Similarly, there is a dual phenomenon called the *generator action* which is a result of Faraday's law of induction. *Faraday's law of induction* states that an electromotive force (EMF) voltage is induced on a circuit due to changing magnetic flux and that the induced voltage opposes the change in the magnetic flux. We can think of this as the relationship between magnetic and electric fields: a changing magnetic field induces an electric field (induced voltage) where the induced electric field opposes the change in the magnetic field (Figure 8.9a).

$$V_{\text{induced}} = -\frac{d\Phi_B}{dt} \tag{8.66}$$

Note that the time rate of change in magnetic flux can be due to the change in the magnetic field source or due to the motion of a component inside a constant magnetic field strength which results in change of effective reluctance or both (Figure 8.9a,b,c,d). If we consider the induced voltage on a coil with N turns and the magnetic flux passing through each of the turns is Φ_B, then

$$V_{\text{induced}} == -N \cdot \frac{d\Phi_B}{dt} = -\frac{d\lambda}{dt} \tag{8.67}$$

where $\lambda = N \cdot \Phi_B$ is called the *flux linkage*, which is the amount of flux linking the N turns of coil.

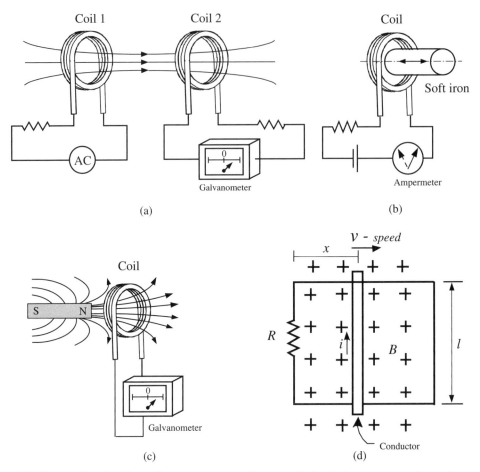

FIGURE 8.9: Faraday's law of induction: change in magnetic flux induces an opposing voltage in an electrical circuit affected by the field. The induced voltage is proportional to the time rate of change in the magnetic flux and in opposite direction to it. The source of the chance in magnetic flux can be (a) a changing current source which generates the electromagnetic field, (b) it can be due to the change in reluctance in a magnetic field (in other words, the inductance of the electrical circuit changes). In inductive circuits, the back EMF can be due to self-inductance itself and due to the change in the inductance), or (c) due to mechanical motion. The last two cases are observed as back EMF on solenoids and DC motors, respectively. The first case is observed in transformers. (d) Generator action as a result of Faraday's law of induction.

In electromagnetism, the current is the "cause" (source) and the magnetic field is the "effect" of it. Faraday's induction law states that the time rate of change in magnetic flux induces EMF voltage on a circuit affected by it. The source of electromagnetic induction (induced voltage) is the change in the magnetic flux. This change may be caused by the following sources (Figure 8.9):

1. A changing magnetic flux itself, that is changing magnetic flux created by a changing source current. In the case of a transformer, the AC current in the primary winding creates a changing magnetic flux. The flux is guided by the iron core of the transformer to the secondary winding. The changing magnetic flux induces a voltage in the secondary winding (Figure 8.9a, Figure 8.11).

2. A changing magnetic flux is created as a result of changing inductance. In a circuit, even though the current is constant, change in magnetic flux can be caused by the change in the geometry and permeability of the medium (change in reluctance). This results in induced back EMF. This phenomenon is at work in the case of solenoids and variable reluctance motors. In general this voltage has the form (Figure 8.9b),

$$V_{\text{bemf}} = -\frac{d\Phi_B}{dt} \tag{8.68}$$

$$= -\frac{d(L(x)i(t))}{dt} \tag{8.69}$$

$$= -L(x)\frac{di(t)}{dt} - \frac{dL(x)}{dx} \cdot i(t) \cdot \dot{x}(t) \tag{8.70}$$

where the first term is the back EMF due to self-inductance of the circuit ($L(x)$), and the second term is the back EMF induced due to the change in the inductance ($dL(x)/dx$).

3. A back EMF is induced in the conductors moving in a fixed magnetic field established by the field magnets (permanent magnets or electromagnets Figure 8.9d). As the conductor moves in the magnetic field, there will be a force acting on the charges in it just as there is a force acting on a charge moving in a magnetic field. Let us consider a constant magnetic field B, a conductor with length l moving in perpendicular direction to magnetic field vector, and its current position is x (Figure 8.9d). The induced back EMF because of this motion is

$$V_{\text{bemf}} = -\frac{d\Phi_B}{dt} = -\frac{d(B \cdot l \cdot x)}{dt} = -B \cdot l \cdot \dot{x} \tag{8.71}$$

This is referred to as the *generator action*. In the case of a brush-type DC motor, the conductor is the winding on the rotor which moves relative to stator magnets. In the case of brushless DC motors, the permanent magnets on the rotor move relative to the fixed stator windings. The resulting induced back EMF voltage effect is the same.

A coil of conductor has self-inductance, which opposes the change in the magnetic field around it. Through self-inductance, L, the coil generates an EMF voltage that is proportional to the rate of change of current in the opposite direction (Figure 8.10). If the circuit geometry and its material properties vary (i.e., in the case of a solenoid, the air gap varies), the inductance is not constant. The inductance is a function of the geometry (i.e., number of turns in a coil) and the permeability of the medium. If the core of an inductor winding has a moving iron piece and the permeability of the medium changes as the core moves, the inductance of the coil changes. Consider the self-inductance L of a solenoid coil with length l and a total of N turns, cross-sectional area of A. Let us assume the medium inside the coil is air. Let us assume that the magnetic flux inside the coil is uniform. Then,

$$B = \mu_0 \cdot (N/l) \cdot i \tag{8.72}$$

$$\Phi_B = B \cdot A = \mu_0 \cdot (N/l) \cdot i \cdot A \tag{8.73}$$

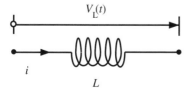

FIGURE 8.10: A coil of a conductor and its self-inductance. Self-inductance of a coil is a function of the number of turns in the coil, its geometry, and the material properties of the medium that the coil core and its surrounding.

$$V_{\text{bemf}} = -L\frac{di}{dt} = -N\frac{d\Phi_{\text{B}}}{dt} \tag{8.74}$$

$$L \cdot i = N \cdot \Phi_{\text{B}} \tag{8.75}$$

$$L = \frac{N \cdot \Phi_{\text{B}}}{i} \tag{8.76}$$

$$L = \frac{N \cdot \mu_o \cdot (N/l) \cdot i \cdot A}{i} \tag{8.77}$$

$$= \frac{\mu_o \cdot N^2 \cdot A}{l} \tag{8.78}$$

$$L = \frac{N^2}{R_{\text{B}}} \tag{8.79}$$

which shows that the inductance is a function of the coil geometry and permeability of the medium. If the coil core is iron, then μ_o would be replaced by μ_m for iron which is about 1000 times higher than the permeability of air. Hence, the inductance of the coil would be higher by the same ratio.

In electric actuator design, it is desirable to have a small magnetic reluctance, R_{B}, so that more flux ($\Phi_{\text{B}} = MMF/R_{\text{B}}$) is conducted per unit magnetomotive force (MMF). On the other hand, it is desirable to have small inductance (L) so that the electrical time constant of the motor is small. These are two conflicting design requirements. A particular design must find a good balance between them that is appropriate for the application.

Let us consider the transformer shown in Figure 8.11. A transformer has two windings, a primary and secondary winding, and a *laminated iron core* which magnetically couples them. The laminated design, as opposed to solid metal piece design, reduces eddy current losses. The iron core material of the laminations has a large magnetic permeability and a large flux saturation level, which helps conduct the generated magnetic flux through the circuit without saturation. In other words, it provides an efficient electromagnectic coupler between the two coils.

A transformer works based on Faraday's induction principle, that is, voltage is induced on a conductor due to a change in the magnetic field. In the case of transformers, the change in magnetic field is due to the alternating current (AC) nature of the source at the primary winding. An ideal transformer can be viewed as having pure inductance, although in reality there is some resistance and capacitance.

Faraday's law states that the voltage across the primary winding is proportional to the rate of change of the magnetic flux and opposes that change,

$$v_1(t) = V_1 \cdot \sin \omega t \tag{8.80}$$

$$v_1(t) = -N_1 \frac{d\Phi_{\text{B}}}{dt} \tag{8.81}$$

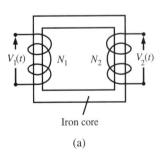

Iron core

(a)

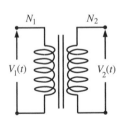

(b)

FIGURE 8.11: (a) An ideal transformer with primary and secondary coil windings, laminated soft iron core. (b) circuit diagram for a transformer.

where $v_1(t)$ is the source voltage applied to the primary winding, N_1 is the number of turns of the coil, and Φ_B is the magnetic flux. Assuming no loss in the magnetic flux, the induced voltage in the secondary winding is

$$v_2(t) = -N_2 \frac{d\Phi_B}{dt} \tag{8.82}$$

$$v_2(t) = \frac{N_2}{N_1} v_1(t) \tag{8.83}$$

Notice that when $N_2 > N_1$, the transformer increases the voltage (step-up transformer), and when $N_2 < N_1$, it reduces the voltage (step-down transformer, Figure 8.11). Notice that a transformer works on the AC voltage. The DC component of the voltage in the primary winding does not cause any change in the magnetic field, hence does not contribute to the voltage induced in the secondary winding. Therefore, a transformer is sometimes used to *isolate (or block) the DC component* of a source voltage in signal processing applications. When $N_1 = N_2$, there is no AC voltage level change between primary and secondary windings, and such a transformer is called the *isolation transformer*. The main purpose of an isolation transformer is to "isolate" the two devices on the primary side and secondary side from each other so that there are no physical ground loops between the two sides.

Example Consider the electromagnetic circuit shown in Figure 8.12. The core of the coil winding is made of a magnetically conductive material with a permeability coefficient of μ_c. The cross-sectional area, the length of the core material, and the total number of turns of the solenoid are A_c, l_c, N, respectively. Let the air gap distance be l_g. The cross-sectional area at the air gap is A_g. Determine the effective reluctance and inductance of the circuit.

The reluctance of the magnetically permeable core and air gap add in series like electrical resistance.

$$R = R_c + R_g \tag{8.84}$$

$$= \frac{l_c}{\mu_c \cdot A_c} + \frac{l_g}{\mu_0 \cdot A_g} \tag{8.85}$$

Notice that if $\mu_c \gg \mu_0$, then $R \approx R_g$.

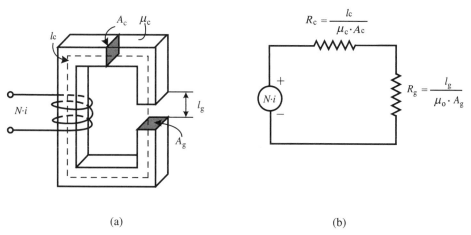

(a)   (b)

FIGURE 8.12: (a) An electromagnetic circuit example: coil wound over a core which has an air gap. (b) Magnetic circuit model.

The magnetomotive force (MMF) generated due to the coil and current is

$$MMF = N \cdot i \tag{8.86}$$

Flux circulating in the closed path along the core and through the air gap is

$$\Phi_B = \frac{MMF}{R} = \frac{N \cdot i}{R} \tag{8.87}$$

The flux linkage to the coil is

$$\lambda = N \cdot \Phi_B \tag{8.88}$$
$$= L \cdot i \tag{8.89}$$
$$= N \frac{N \cdot i}{R} \tag{8.90}$$
$$L = \frac{N^2}{R} \tag{8.91}$$

This shows that self-inductance of a magnetic circuit involving a coil is proportional to the square of the number of turns and inversely proportional to the reluctance of the circuit. In actuator applications, small reluctance is desirable in order to generate more magnetic flux, hence force or torque. However, smaller reluctance leads to large inductance, which results in a larger electrical time constant. In electric motor design applications, this conflicting design requirement must be balanced: for large force/torque, we want small reluctance and for small electrical time constant we want small inductance. However, inductance and reluctance are inversely related. As one increases, the other one decreases.

8.1.3 Permanent Magnetic Materials

Materials can be classified into three categories in terms of their magnetic properties:

1. paramagnetic (i.e., aluminum, magnesium, platinum, tungsten),
2. diamagnetic (i.e., copper, diamond, gold, lead, silver, silicon),
3. ferromagnetic (i.e., iron, cobalt, nickel, gadolinium) materials,
   - soft ferromagnetic materials,
   - hard ferromagnetic materials.

Of these, hard ferromagnetic materials are of interest to us as permanent magnetic materials. Soft ferromagnetic materials are used as lamination material for stator and rotor frames and transformer frames. The difference between these materials originates from their atomic structure. Magnetic field strength, $\vec{H}$, and magnetic flux density, $\vec{B}$, relationship in a given spatial location depends on the "permeability" of the surrounding material, μ_m,

$$\vec{B} = \mu_m \cdot \vec{H} \tag{8.92}$$

where, $\mu_m = \mu_r \cdot \mu_0$ is called the magnetic *permeability* of the material which is a measure of how well a material conducts magnetic flux, and μ_r is called the *relative permeability*, μ_0 is the permeability of free space. The relationship between the magnetic field strength ($\vec{H}$) and the magnetic flux density ($\vec{B}$) is almost linear for paramagnetic and diamagnetic materials. Although the same relationship can be used to describe the magnetic behavior of ferromagnetic materials, the relationship is not linear and exhibits large hysteresis (Figure 8.13). *Susceptibility*, χ, is defined as

$$\mu_r = 1 + \chi \tag{8.93}$$

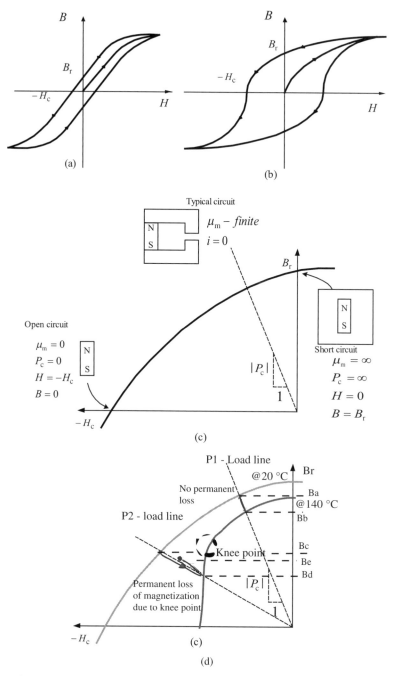

FIGURE 8.13: Magnetic properties of ferromagnetic materials are not linear. (a) In *soft magnetic materials,* the B_r and H_c are small, hence they have small hysteresis. They retain a very small portion of their magnetization when the external field is removed. They are temporarily magnetized, not permanently. (b) In *hard magnetic materials,* the B_r and H_c are large (with large hysteresis) which means the material retains a large magnetization after external field is removed. This is the case for permanent magnetic materials. (c) Permanent magnets (PM) in electric actuator applications operate primarily in the second quadrant of the B-H curve. Rare-earth PMs (i.e., samarium-cobalt mixtures, neodymium mixtures) have almost linear characteristics in the second quadrant. (d) Temperature effect on demagnetization curve of permanent magnets and the "knee point. *Source:* Adapted from www.hitachi-metals.co.jp/e."

The susceptibility

1. is positive but small (i.e., 10^{-4} to 10^{-5} range) for paramagnetic materials,

2. is negative but small (i.e., -10^{-5} to -10^{-10} range) for diamagnetic materials, and

3. is positive and several thousand times larger than 1 (i.e., in the range of 10^3 to 10^4) for ferromagnetic materials. Furthermore, the effective μ_m in the $\vec{B}$ and $\vec{H}$ relationship is not linear, but exhibits hysteresis nonlinearity. Ferromagnetic materials are categorized into two groups based on the size of the hysteresis:

    (a) materials that exhibit small hysteresis in their B-H curves are called *soft ferromagnetic* materials (Figure 8.13a),

    (b) materials that exhibit large hysteresis in their B-H curves are called *hard ferromagnetic* materials (Figure 8.13b).

In the case of *soft ferromagnetic materials*, the material goes though the full cycle of magnetization and demagnetization at the same frequency of the external magnetic field. For instance, the stator and rotor material in an electric motor are made of soft ferromagnetic material. As the stator current changes direction in a cyclic way, the B-H curve for the steel goes through the hysteresis loop. The energy in the hysteresis loop is lost as heat. *Hysteresis loss* is proportional to the maximum value of magnetic field intensity magnitude and its frequency. Therefore, in order to minimize the energy loss in motor and transformer cores, lamination material is chosen among the soft ferromagnetic materials. Soft ferromagentic materials have less hysteresis losses. But, as a result of the same property, they have a small residual magnetization when the external magnetic field is removed. Therefore, we can think of them temporarily magnetized materials.

Hard ferromagnetic materials have a large residual magnetization, which means they maintain a strong magnetic flux density even when the external field is removed, hence the name permanent magnets. But, as a result of the same property, they have large hysteresis losses if they operate in a full cycle of external magnetic field variation. The maximum value of the product of $B \cdot H$ on the hysteresis curve indicates the magnetic strength of the material. It is easy to confirm that the BH term has energy units as follows,

$$\text{Tesla} \cdot \text{A/m} = \frac{\text{Nt}}{\text{A m}}\text{A/m} = \frac{\text{Nt}}{\text{m}^2} = \frac{\text{Nt} \cdot \text{m}}{\text{m}^3} \tag{8.94}$$

$$= \frac{\text{Joule}}{\text{m}^3} \tag{8.95}$$

or in CGS units, BH has units of Gauss $\cdot$ Oerstead GOe $= \dfrac{1}{4\pi} \times 10^3 \text{Joule/m}^3$.

The residual magnetization (B_r) left after the external magnetic field is removed, which is a result of the magnetic hysteresis, is used as a way to permanently magnetize ferromagnetic materials. This permanent magnetization property is called *remanent magnetization* or *remanence* (B_r) of the material, which means the "remaining magnetization." As the magnitude of the residual magnetic flux density (B_r) increases, the capability of the material to act as a permanent magnet (PM) increases. Such materials are called "hard ferromagnetic materials" compared to the "soft ferromagnetic materials" which have small residual magnetization. The value of the external magnetic field strength (H_c) necessary to remove the residual magnetism (to totally demagnetize it) is called the *coercivity* of the material. It is a measure of how hard the material must be "coerced" by the external magnetic field to give up its magnetization. Notice that the area inside the hysteresis curve is the energy lost during each cycle of the magnetization between $\vec{H}$ and $\vec{B}$. This is called the *hysteresis loss*. This energy is converted to heat in the material. In electromagnetic actuator applications such as electric motors, a permanent magnet (PM) usually operates in the second quadrant of B-H curve. As long as the external field is below the H_c, the magnet state moves back and forth along the curve on the second quadrant. In this case, there is a very small hysteresis loss (Figure 8.13c).

The power of a permanent magnet is measured in terms of the magnetic flux and MMF it can support,

$$\Phi_B = B \cdot A_m \tag{8.96}$$

$$MMF = H \cdot l_m \tag{8.97}$$

where l_m is the length of the magnet in the direction of magnetization, and A_m is the cross-section of the magnet perpendicular to the magnetization direction. In order to increase the MMF for a given PM with a specific B-H characteristics, it must have a large thickness in the direction of magnetization (l_m). Similarly, in order to increase the flux, it must have a large surface area that is perpendicular to its magnetization (A_m).

If a permanent magnet is placed in an infinitely permeable medium, no MMF would be lost and the magnetic field intensity coming out of the magnet would be $B = B_r$, $H = 0.0$ (Figure 8.13c). If, on the other hand, the permanent magnet is placed in a medium with zero permeability, no magnetic flux can exit it. The operating point of the magnet would be $B = 0$ and $H = -H_c$.

In a real application, the effective permeability of the surrounding medium is finite. Hence, the permanent magnet operates somewhere along the curve between the two extreme points in the second quadrant of the B-H curve. The nominal location of the operating point is determined by the permeance of the surrounding medium. The absolute value of the slope of the line connecting the nominal operating point to the origin is called the *permeance coefficient*, P_c and the line is called the *load line*. The applied coil current shifts the net MMF (or H), and hence the operating point of the magnet along the H-axis. It is important that the applied coil current should not be large enough to force the magnet into the demagnetization zone. In permanent magnet motors, the electromagnetic circuit of the motor generally results in a load line that is $P_c = 4$ to 6 range. In magnetic circuits where the closed path of the flux is made of an air gap, highly permeable material, and a permanent magnet, it can be shown that

$$P_c \approx \frac{l_m}{l_g} \tag{8.98}$$

where l_m is the permanent magnet thickness in the direction of magnetization, and l_g is the effective air gap length.

The B-H curve of a permanent magnet is strongly a function of the operating temperature. Consider a permanent magnet in an electromagnetic circuit. We will consider two operating temperature conditions 20 °C and 140 °C. Figure 8.13d shows the B-H demagnetization curves in the second quadrant for a particular permanent magnet for these two temperatures. On the figure on the left of Figure 8.13d, B-H curve A is for 20 °C, and B-H curve B is for 140 °C. Notice the drop in the magnetic strength as the temperature increases. In particular we want to focus on the "knee point" effect. Let us assume that the current operating condition is defined by the load line P_1 and its intersection with the B-H curve. At 20 °C temperature, the nominal flux density is B_a. Now, assume that under this operating condition, the temperature has risen to 140 °C. Then, the nominal operating point is now the intersection between the load line P_1 and the B-H curve for the 140 °C temperature. Under this condition, the nominal magnetic flux density is B_b. Then, let us assume that the operating temperature returns to 20 °C. The nominal magnetic flux density will return to the B_a value. Under this operating condition, temperature variation did not result in permanent demagnetization. When the temperature increases, the magnetic field strength reduces. But when the temperature returns to its original lower value, the original

TABLE 8.2: Temperature dependence of the magnetic properties of various permanent magnetic materials.

Material	$\Delta B_r / \Delta T$	$\Delta H_c / \Delta T$
Neodymium	−0.10	−0.60
Samarium	−0.04	−0.30
Alnico	−0.02	+0.01
Ferrite	−0.18	+0.30

magnetic field strength (and magnetic flux density) is recovered. This is called reversible loss in magnetization. The loss on B value as a function of temperature is defined as temperature sensitivity constant for a permanent magnet material. For instance, Neodymium magnets lose about 10% of their magnetic strength when the temperature rises from room temperature 20 °C to 120 °C, whereas samarium cobalt loses only 4%.

There are also non-reversible losses due to temperature and operating condition variations, such as knee point, Curie temperature (Table 8.2). Next, let us consider the same temperature cycling between 20 °C and 140 °C and back to 20 °C while operating on the load line P_2 (i.e., the same magnet is used in a slightly different circuit, such as the air gap in this circuit is larger). When the temperature is 20 °C, the nominal magnetic flux density is B_c. When the temperature is 140 °C, the nominal flux density is B_d. It is important to note that this operating condition falls below the "knee point." Some of the magnetic strength will be permanently lost. When the temperature returns to 20 °C, the nominal magnetic flux density will be a value B_e between B_c and B_d. The original B_c value will not be recovered. This means that if the operating conditions (such as a combination of load line and temperature) bring the magnet to a point below the "knee point," there is some permanent loss of magnetic strength. In electromechanical actuators, the electromagnetic circuit should be designed such that PM never reaches the *knee point*, that is the point of permanently losing some of its magnetization. The lost magnetization can only be recovered by re-magnetizing the magnet.

In an electromagnetic circuit, a permanent magnet can be modeled as a flux source Φ_r and a reluctance R_m in parallel with it (Figure 8.14)

$$\Phi_r = B_r \cdot A_m \tag{8.99}$$

$$R_m = \frac{l_m}{\mu_r \cdot \mu_0 \cdot A_m} \tag{8.100}$$

where l_m is the length along the magnetization direction, and A_m is the cross-section area perpendicular to the magnetization direction, μ_r is the recoil permeability of the magnet material.

In short, remanence B_r, coercivity H_c, and $(BH)_{max}$ maximum energy are three nominal parameters which characterize the magnetic properties of a ferromagnetic material.

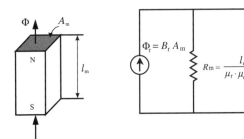

FIGURE 8.14: Magnetic model of a permanent magnet. The permanent magnet acts like a flux source and a parallel reluctance. The actual flux, Φ, which leaves the magnet is determined by the rest of the load circuit.

TABLE 8.3: Comparison of four major permanent magnetic material types.

Permanent magnet material	Max. magnetic energy (MGOe)	Curie temp (°C)	Max. operating temp (°C)	Cost $
Alnico	5	~1000	~500	Low
Ceramic	12	~450	~300	Moderate
SmCo	35	~800	~300	High
NdFeB	55	~350	~200	Medium

There are four major types of natural hard ferromagnetic materials that can be used as permanent magnets:

1. Alnico which is an aluminum (Al)-nickel (Ni)-cobalt (Co) ("AlNiCo") mixture.

2. Ceramic (hard ferrite) magnetic materials which consists of strontium, barium ferritite mixtures.

3. Samarium cobalt (samarium and cobalt mixtures, $SmCo_5, Sm_2Co_{17}$).

4. Neodymium (neodymium, iron, and boron are the main mixture components with small amounts of other compounds). The ideal mixture is $Nd_2Fe_{14}B_1$.

Alnico and ceramic ferrite permanent magnet materials are the lowest cost types and have lower magnetic strength compared to samarium and neodymium. The maximum magnetic energy of each type is shown in Table 8.3. Today, Alnico permanent magnets (PM) are used in automotive electronics, ceramic PM materials are used in consumer electronic, and samarium and neodymium are used in high performance actuators and sensors. The cost of the PM material increases as the magnetic energy level increases. Notice that the energy levels given in the table are the maximum currently achievable levels. Lower energy level versions are available at lower cost. For instance, the cost of NdFeB at 45 MGOe is twice that of NdFeB at 30 MGOe. The biggest advantage of samarium-cobalt PM material over neodymium PM material is the fact the samarium-cobalt PM material can operate at higher temperatures.

The manufacturing process for making permanent magnets from one of the above materials has the following steps (it is important to note that small variations in composition and the manufacturing process make a difference in the final magnetic and mechanical properties of the magnet):

1. Mix the proper amount of elements to form the magnet compound and melt it in furnace, and make ingots.

2. Process the ingots to turn them into fine powder and mix the powder.

3. Place the mixed powder in a die cavity, apply an initial electromagnetic field to orient the magnetic directions (pre-alignment of magnetic field), and press it down to about 50% of its powder state size. This is a powder metallurgy process. The product at this state is called "green."

4. Heat the "green" PM in a furnace (i.e., vacuum chamber at 1100–1200 °F for neodymium-iron-boron) which will result in further shrinkage in size. This process may be followed by a lower temperature heat treatment around 600 °C.

5. Saw and grind to the desired shape (rectangular, cylindrical) and size.

6. Coat the surface of the magnet piece if desired.

7. Magnetize each piece to magnetic saturation by an external electromagnetic field pulse (i.e., generally a few milliseconds duration of external magnetic field pulse with a high enough H value to make the magnet reach its saturation level). Pulse

duration can also be in the order of seconds depending on the magnet material being magnetized. Since the handling of magnetized permanent magnet pieces is difficult, it is desirable to magnetize the magnets as late as possible in the manufacturing process for a given magnetic product application.

8. Finally, the permanent magnet batch should be treated with stabilization and calibration processes. The calibration process makes sure that the magnetic strength of each piece is within a certain tolerance of the desired specifications (i.e., ±1%).

Typical shapes and magnetization directions of permanent magnets are shown in Figure 8.15. There is a critical temperature, called the *Curie temperature* for each ferromagnetic material above which the material losses its ferromagnetic properties and becomes a paramagnetic material. The Curie temperature for iron is 1043 °K, for cobalt 1394 °K,

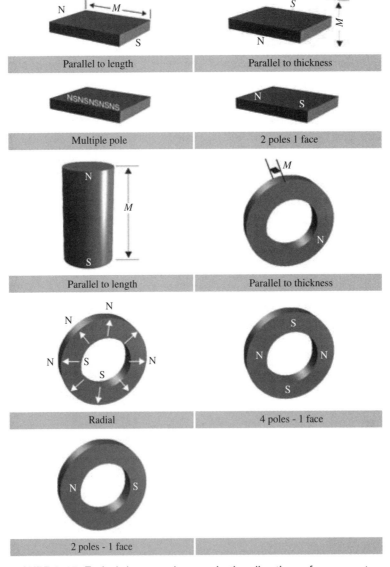

FIGURE 8.15: Typical shapes and magnetization directions of permanent magnets. The letter M with arrows indicates the magnetization direction of the piece.

and for nickel 631 °K. In practice, the allowed maximum operating temperature for a PM material is much lower than the Curie temperature.

The external magnetic field (also called the magnetizing force) should be strong enough to fully magnetize the material and saturate it so that the permanent magnet material attains its maximum possible magnetic capacity. This value is at least 3 to 5 times the coercivity (H_c) of the permanent magnet material.

Each PM material has a different susceptibility to environmental conditions. In addition to temperature, the chemical composition of the environment is most important. Alnico, ceramic, and samarium magnets are corrosion resistant, whereas neodymium magnets are very susceptible to corrosion. In the selection of a permanent magnet (PM) material, the environmental issues to consider are:

1. oxidation and humidity level,
2. acid content,
3. salt content,
4. alkaline content,
5. radiation level.

The following specifications are typically used by designers in selecting a proper permanent magnet for an application:

1. permanent magnet material,
2. remanance, coercivity, and maximum energy, B_r, H_c, $(BH)_{max}$,
3. maximum operating temperature without irreversible loss of magnetization,
4. mechanical dimensions (shape and size),
5. magnetization direction (radial, axial, etc.),
6. whether the permanent magnet is sintered or bonded,
7. surface coating (i.e., 2 μm to 20 μm range thickness, aluminum, nickel, or titanium nitride coating material).

The selected PM material largely determines the temperature coefficient for the loss of magnetic field as the temperature rises, and the thermal expansion coefficient of mechanical dimensions, which is important in a device assembly. Neodymium magnets expand in the magnetized direction and contract in the other direction with increasing temperature.

A permanent magnet piece is usually bonded to a steel backing material using adhesives in motor applications, that is in the case of electric motors on a steel rotor (3M Adhesives). The bonding material types include thermosetting epoxies, and structural adhesives. There are many adhesives specifically designed for motor applications (3M Adhesives). The strength of the bond between the PM and and the rotor is a function of the contact surface area. Typically, the surface area is roughened in order to provide a good bonding. After the application of adhesive, the PM is clamped on the rotor and cured at high temperature. The curing temperature for the bonding material (the temperature at which the bonded PM is baked for a certain time period) should be well below the demagnetization temperature of the PM, if the PM was magnetized before the bonding process.

8.2 ENERGY LOSSES IN ELECTRIC MOTORS

An electric actuator is a device that converts energy from electrical to mechanical form (Figure 8.16). The conversion process is not 100% efficient, hence there are losses.

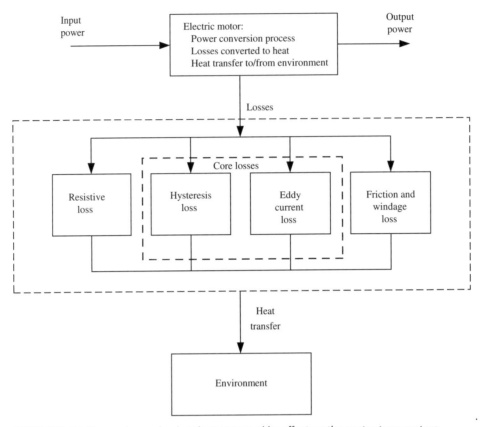

FIGURE 8.16: Energy losses in electric motors and its effect on the motor temperature.

These losses are in the form of heat. The energy losses can be categorized into three groups:

1. resistance losses (also called copper losses): $R \cdot i^2$,
2. core losses (hysteresis and eddy current losses),
3. friction and windage losses.

Resistance and core losses are electrical losses, whereas the friction and windage losses are mechanical losses.

The steady-state operating temperature of the actuator is determined by these losses, the heat transferred to the surrounding medium through conduction, convection, and radiation type heat transfer mechanisms. The maximum temperature the actuator can operate in (i.e., before losing its insulation material or magnetic properties) determines the maximum power capacity of the motor.

The energy balance (Q_{net}) is the difference between the generated heat due to the losses (Q_{in}) and the transferred heat from the actuator to its surrounding environment (Q_{out}) is the energy that must be absorbed by the actuator in the form of temperature rise,

$$Q_{net} = Q_{in} - Q_{out} \qquad (8.101)$$
$$= c_t \cdot m \cdot (T - T_0) \qquad (8.102)$$

where m is the mass of the actuator which absorbs heat, c_t is the specific heat of the material of actuator in units of Joule/(kg · °C), T_0 is the initial temperature and T is the steady-state temperature. The generated heat due to losses, Q_{in},

$$Q_{in} = Q_R + Q_C + Q_F \tag{8.103}$$

where Q_R is the resistive loss, Q_C is the core loss, and Q_F is the friction and windage loss. The transferred heat is estimated with approximation as follows,

$$Q_{out} = c_{out} \cdot (T - T_{amb}) \cdot \Delta t \tag{8.104}$$

where T_{amb} is the ambient temperature, and c_{out} is the effective heat transfer coefficient between the actuator and its surrounding with units (Joule/s)/°C = Watts/°C, and Δt is the time period.

The energy balance equation can also be expressed as power balance in differential equation form,

$$\frac{d}{dt}[Q_{net}] = \frac{d}{dt}[Q_{in} - Q_{out}] \tag{8.105}$$

$$P_{net} = P_{in} - P_{out} \tag{8.106}$$

$$c_t \cdot m \cdot \frac{dT}{dt} = P_{in} - c_{out} \cdot (T - T_{amb}) \tag{8.107}$$

$$c_t \cdot m \cdot \frac{dT}{dt} + c_{out}(T - T_{amb}) = P_{in} \tag{8.108}$$

This models the electric actuator thermal behavior like a first-order dynamic system. Notice that the actuator reaches a steady-state temperature, when $P_{net} = 0$, then $dT/dt = 0$. The difficulty in using such an analytical model is in the difficulty of accurately estimating c_T, c_{out} and P_{in}. Note that

$$P_{in} = \frac{d}{dt}Q_{in} \tag{8.109}$$

$$= \frac{d}{dt}Q_R + \frac{d}{dt}Q_C + \frac{d}{dt}Q_F \tag{8.110}$$

$$= P_R + P_C + P_F \tag{8.111}$$

For a given motor design and its expected operating conditions, if we can estimate the total losses $P_{in} = P_R + P_C + P_F$ and the heat transfer coefficient between the motor and environment c_{out}, then we can estimate the steady-state operating temperature of the motor. The steady-state value of the operating temperature is easy to calculate from the above differential equation

$$c_t \cdot m \cdot \frac{dT}{dt} + c_{out} \cdot (T - T_{amb}) = P_{in} \tag{8.112}$$

$$\frac{dT}{dt} = 0 \quad \text{in steady-state, then} \tag{8.113}$$

$$0 + c_{out} \cdot (T - T_{amb}) = P_{in} \tag{8.114}$$

$$T(\infty) = T_{amb} + \frac{1}{c_{out}} \cdot P_{in} \tag{8.115}$$

8.2.1 Resistance Losses

Electric actuators have coils which are windings of current carrying conductors. Coils act as current controlled electromagnets. The conductor material, such as copper or aluminum, has finite electrical resistance. In order to generate an electromagnetic effect, we need to pass a certain amount of current. As the electrical potential pushes the current through the conductor, there is energy loss due to resistance. This is similar to the energy loss in pushing a certain fluid flow rate through a restriction. The power (P_R) lost as heat due to resistance (R) of a conductor when current i is conducted,

$$P_R = R \cdot i^2 \tag{8.116}$$

Therefore, in order to minimize the heating of the motor, it is desirable to minimize the resistance and current. However, a large current is needed to generate a large force or torque. The energy (Q_r) that is converted to heat over a period of time t_{cycle} is

$$Q_R = \int_0^{t_{cycle}} P_R \cdot dt \tag{8.117}$$

$$= \int_0^{t_{cycle}} R \cdot i(t)^2 \cdot dt \tag{8.118}$$

Resistive loss, $P_R = R \cdot i^2$, is fairly well estimated. The temperature dependence of resistance can also be taken into account for better accuracy of thermal prediction,

$$R(T) = R(T_0)[1 + \alpha_{cu}(T - T_0)] \tag{8.119}$$

where $R(T), R(T_0)$ are resistances of coil at temperature T and T_0, respectively and $\alpha_{cu} = 0.00393$ for copper which is the property of the conductor material. Notice that for copper,

$$R(125\,°C) \approx 1.4 \cdot R(25\,°C) \tag{8.120}$$

The resistance of the coil increases by 40% when its temperature increases by $100\,°C$.

8.2.2 Core Losses

Core losses refers to the energy lost, in the form of heat, in electric actuators at their stator and rotor structure due to electromagnetic variations. There are two major physical phenomenon that contribute to core losses: *hysteresis losses* and *eddy current losses*.

Hysteresis loss is due to the hysteresis loop in the B-H characteristics of the core (rotor and stator) material which is typically a soft ferromagnetic material. As the magnetic field changes, the hysteresis loop is traversed and energy is lost. The hysteresis loss is proportional to the magnitude of the magnetic field change and its frequency. This loss can be minimized by selecting a core material that has small hysteresis loops, such as a high silicon content in the steel used for the core material.

Eddy currents and related losses can be summarized as follows. If a bulk piece of metal (e.g., iron, copper, aluminum) moves through a magnetic field or is stationary in a changing magnetic field, a circulating current is induced on the metal. This current is called *eddy current* and it results in heat loss due to the resistance of the metal. The induced current is a result of Faraday's law of induction. That is the changing magnetic field (whether due to change in $\vec{B}$ or motion of the metal conductor relative to $\vec{B}$ or both) induces a voltage on the conductor. This induced voltage generates the current. Power loss due to eddy currents is proportional to the square of magnetic flux density and its frequency of variations. Therefore, it is expected that the hysteresis losses will be more dominant at low frequencies, whereas eddy current losses will be more dominant at high frequencies.

Laminations of thin layers of metal which are insulated and stacked together, as oppose to bulk metal, are used to reduce the eddy currents in motor applications. Hence eddy current-related heat losses are reduced, increasing the efficiency of the motor. The stators of DC and AC motors are built using laminated iron instead of bulk iron in order to reduce the eddy current related loss of energy into heat. For different lamination materials, hysteresis and eddy current losses are given based on measured data as a function of field intensity and frequency,

$$P_C^* = f(B_{max}, w) \tag{8.121}$$

where B_{max} is the maximum flux density, and w is the frequency of change in magnetic flux. The nonlinear function $f(\cdot)$ is defined as numerical graphs for different lamination materials by manufacturers. Notice that the core loss data for different materials is given per unit mass of the material. Total core loss for a given mass of motor application,

$$P_C = m \cdot P_C^* \tag{8.122}$$

Hence, the designer can estimate the core losses for a given design using the manufacturer supplied data.

8.2.3 Friction and Windage Losses

These losses are significant at very high speeds due to the air resistance between the rotor and stator. The energy loss due to air resistance is called the windage loss. Energy loss due to bearing friction is called the friction loss. They are taken into account by increasing the resistive loss by a safety factor, since it is very difficult to accurately model friction and windage losses, that is

$$P_F = 0.1 \cdot P_R \tag{8.123}$$

8.3 SOLENOIDS

8.3.1 Operating Principles of Solenoids

A *solenoid* is a translational motion actuator with a rather limited motion range. Solenoids are used in fluid flow control valves and small range translational displacement actuators. A solenoid is made of (Figure 8.17),

1. a coil,
2. a frame which is a material with high permeability to guide the magnetic flux,
3. a plunger which also is made of high permeability material,
4. a stopper (and a centering spring in most cases), and
5. a bobin, which is a plastic or non-magnetic metal on which coil is wound. It is non-magnetic so that there is no short circuit for the flux between the coil and plunger.

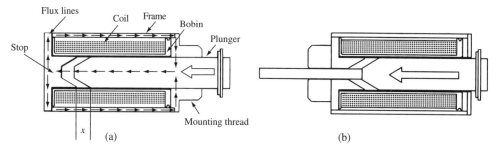

FIGURE 8.17: Solenoid components and its design: pull and push types.

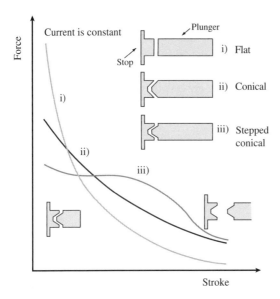

FIGURE 8.18: Solenoid force versus plunger displacement under constant coil current condition. As the plunger closes the air gap between itself and the stopper, the force capacity increases. This is called the holding force, which is generally higher than the average pull-push force.

The operating principle of solenoids is based on the tendency of the ferromagnetic plunger and coil generated magnetic flux to seek the minimum reluctance point. As a result, when the coil is energized, the plunger is pulled in towards the stopper. The higher the magnetic field strength ($n_{coil} \cdot i$, number of turns in the coil times current) and the better the magnetic permeability of the medium that guides the flux to the plunger, the higher the force generated. The plunger works on the pull principle. However, by mechanical design we can obtain pull or push motion from the solenoid (Figure 8.17). The mechanical connection between the plunger, made of a ferromagnetic material, and the tool must be via a nonmagnetic material. For instance, in the case of a push-type solenoid (Figure 8.17b), the push-pin is made of a nonmagnetic material. The quality of the magnetic circuit and its ability to guide magnetic flux lines depends on the permeability of the coil frame, the air gap between the plunger and coil (fixed gap), and the air gap between the plunger and stopper (variable gap). For a given current, the force generated by the solenoid varies as a function of the air gap between the plunger and the stopper. The smaller this gap is, the smaller the effective reluctance of the magnetic flux path, and hence the higher the force generated. The force as a function of plunger displacement under constant current varies as shown in Figure 8.18. Notice that the shape of the force-displacement curve for a constant current can be affected by the shape of the plunger and stopper head. For high performance applications, in order to reduce the eddy current losses in the solenoid, the iron core of the winding and the plunger may be constructed from laminated sheets of iron that are insulated.

Sometimes, solenoids are referred as *single acting* type and *double acting* types. A solenoid is a single acting type device: that is, when current is applied the plunger moves in one direction in order to minimize the reluctance, regardless of the direction of the current. Force is always generated in one direction. A double acting solenoid can move in both directions by generating force in both directions, both pull and push directions, by using two solenoids in one package. Therefore, a double acting solenoid is basically two solenoids with one plunger, two coils, and two stops.

A directional flow control valve may have one or two solenoids to position it in two or three discrete positions. For instance a double acting solenoid (two solenoids in one package) can have three positions: center position when both the solenoids are deenergized, left position when the left solenoid is energized, and right position when the right solenoid is energized. If the current in each solenoid is controlled proportionally, instead of fully

ON or fully OFF, then the displacement of the plunger can be controlled proportionally instead of two or three discrete positions. This is the method used in proportional valves. The solenoids are rated in terms of their coil voltage (i.e., 12 VDC, 24 VDC), maximum plunger displacement (i.e., $1/4$ in), and maximum force (i.e., 0.25 oz to 20 lb range). There are two major differences between ON/OFF type and proportional type solenoids:

1. the solenoid mechanical construction where the plunger, coil, and frame design provides a different flux path (Figure 8.18i-iii, the proportional solenoids are generally of type shown in Figure 8.18iiic)).
2. the current in the coil is controlled either in ON/OFF mode or in proportional mode.

The coil acts as the electromagnet in all electric actuators. Current (i), number of turns (n_{coil}), and the effective permeance of the magnetic medium (core material, air gap, etc.) determine the strength of the electromagnetic field generated by the coil. At the same time, there are mechanical size and thermal considerations. The rated current determines the minimum diameter requirement of the conductor wire. The wire diameter and the number of turns determines the mechanical size of the coil. In general, the insulation material increases the effective conductor diameter by about 10%. Different insulation materials have different temperature ratings (i.e., 105 °C for formvar, 200 °C for thermalex insulation material compounds commonly used industry). Once the coil wire diameter, number of turns, and mechanical size is known, the resistance of the coil is determined. Hence, the resistive heat dissipation is known. In order to make sure the temperature of the coil stays within the limits of its coil insulation rating, the thermal heat conducted from the coil should balance the resistive heat. The coil design requires the balancing of electrical capacity (current and number of turns), mechanical size, and thermal heat. In other words, the designer must find a balance between the mechanical dimension constraint, thermal heat dissipation capacity, and electrical power conversion capability for a coil design. In order to increase the MMF, we need a large number of turns, n_{coil}, which means diameter, d, of the coil wire should be small. However, as d gets smaller, resistance R increases, hence the maximum current we can pass through must be smaller in order to stay within the heat dissipation capacity of the solenoid,

$$MMF = n_{coil} \cdot i \tag{8.124}$$

$$R \cdot i^2 < P_{rated} \tag{8.125}$$

In order to increase *MMF*, we can decrease d so that n_{coil} is large. But this results in an increase in R which in turn requires lower i, which then reduces *MMF*.

The force generated by a solenoid is a function of the current in the coil (i), the number of turns in the coil (n_{coil}), the magnetic reluctance (R_B, which is a function of the plunger displacement, x, the design shape, and material permeability, μ), and temperature (T)

$$F_{sol} = F_{sol}(i, n_{coil}, R_B(x, \mu), T) \tag{8.126}$$

For a given solenoid, the n_{coil} is fixed, $R_B(x, \mu)$ varies with the displacement of the plunger and the air gap between the winding coil and the plunger. The main effect of temperature is the change in the resistance of the coil. This leads to a change in current for a given terminal voltage. If the control system regulates the current in the coil, the effect of temperature on force other than its effect on resistance is negligible. Therefore, for a given solenoid, the generated force is a function of operating variables as follows,

$$F_{sol} = F_{sol}(i, x) \tag{8.127}$$

The basic mode of control is the control of current in the coil in order to control the force. Unlike a rotational DC motor where the current and torque relationship is constant and independent of the rotary position of the shaft, the force–current relationship of a solenoid is nonlinear and a function of the rotor (plunger) position (Figure 8.18).

In general the force capacity of the solenoid is rated at 25 °C. The force capacity would typically reduce to 80% value at around 100 °C temperature. At a rated current, the force varies as function of plunger displacement (Figure 8.18). Notice that there is always a residual flux left in the core when the current is turned OFF (current is zero) due to the hysteresis nature of electromagnetism. Use of *annealed* steel for the core and plunger material minimizes that effect.

In the basic mode of operation, a solenoid is driven by DC voltage, V_t (i.e., 12 V, 24 V, 48 V) at its coil terminals, and steady-state current i is developed by the resistance ratio (neglecting the inductance of coil),

$$i = \frac{V_t}{R_{\text{coil}}} \tag{8.128}$$

The physical size of the solenoid determines the maximum amount of power it can convert from electrical to mechanical power. The continuous power rating of the solenoid should not be exceeded in order to avoid overheating,

$$P = R_{\text{coil}} \cdot i^2 = \frac{V_t^2}{R_{\text{coil}}} < P_{\text{rated}} \tag{8.129}$$

In some applications, it is desirable to provide a larger current (the *in-rush current*) and after the initial movement, the current is reduced to a lower value, called the *holding current*. One way to reduce the in-rush current to the holding current is to divide the coil winding into two resistor sections and provide an electrical contact between the two series resistor. When a large in-rush current is needed, short circuit via a switch (i.e., electronic transistor switch) the second resistor section to increase current by reducing effective resistance. When it is desired to reduce the current, turn OFF the switch to include the second part of the resistor in the circuit in series, hence reduce the current for the given terminal voltage. Some solenoids can be driven by 50 Hz or 60 Hz AC voltages (i.e., 24, 120 VAC). Force is related to the square of current. Therefore, the direction of generated force does not oscillate with the AC current direction change. However, the magnitude oscillates at twice the frequency of the supply current frequency, that is if the AC supply is 60 Hz, the force magnitude would oscillate at 120 Hz. In other words, the electromagnetic circuit of the solenoid acts like a "rectifier" between supply current and generated force.

8.3.2 DC Solenoid: Electromechanical Dynamic Model

Consider the solenoid shown in Figure 8.17. The electromagnetic energy conversion mechanism generates the force as a result of the interaction between the coil generated electromagnetic field and variable reluctance of the plunger–air gap assembly. Let us consider the magnetic flux path in the solenoid (Figure 8.17). Assume that the permeability of the plunger, frame, and stop are very high compared to permeability of air gap, $\mu_c \gg \mu_0$. We can assume the magnetic energy is only stored in the air gap, neglecting the magnetic energy stored elsewhere. This model is similar to the magnetic circuit shown in Figure 8.12 except that in the case of solenoids, the air gap is variable. Then, the following relations hold for magnetic field strength, flux density, and flux itself,

$$H_g \cdot x = n_{coil} \cdot i \tag{8.130}$$

$$B_g = \mu_0 \cdot H_g = \mu_0 \cdot \frac{n_{coil} \cdot i}{x} \tag{8.131}$$

$$\Phi_b = B_g \cdot A_g = \mu_0 \cdot \frac{n_{coil} \cdot i}{x} \cdot A_g \tag{8.132}$$

The flux linkage and the inductance is defined as,

$$\lambda(x, i) = \Phi_b \cdot n_{coil} = L(x) \cdot i(t) \tag{8.133}$$

$$= \mu_0 \cdot \frac{n_{coil}^2 \cdot i(t)}{x} \cdot A_g \tag{8.134}$$

Then, the inductance as a function of plunger displacement is

$$L(x) = \mu_0 \cdot \frac{n_{coil}^2}{x} \cdot A_g \tag{8.135}$$

The coil has n_{coil} turns, and the voltage is controlled across the terminals of the coil, $V(t)$. The plunger moves inside in the direction of x. The electromechanical dynamic model of the solenoid includes three equations: (i) the electromechanical relationship which describes the voltage, current in coil, and motion of the plunger,

$$V(t) = R \cdot i(t) + \frac{d}{dt}(\lambda(x, i)) \tag{8.136}$$

where $\lambda(x, i)$ is the flux linkage. For an inductor type coil circuit $\lambda(x, i)$ is

$$\lambda(x, i) = L(x) \cdot i(t) \tag{8.137}$$

Hence, the voltage–current–motion relationship can be expressed as,

$$V(t) = R \cdot i(t) + \frac{dL(x)}{dx} \cdot \frac{dx}{dt} \cdot i(t) + L(x) \cdot \frac{di(t)}{dt} \tag{8.138}$$

$$V(t) = R \cdot i(t) + L(x) \cdot \frac{di(t)}{dt} + \left(\frac{dL(x)}{dx}\right) \cdot i(t)) \cdot \dot{x}(t) \tag{8.139}$$

The force is calculated from the so-called *co-energy equation*,

$$W_{co}(\lambda, i) = \frac{1}{2}\lambda(x, i) \cdot i = \frac{1}{2}L(x) \cdot i^2 \tag{8.140}$$

$$F(x, i) = \frac{\partial W_{co}(x, i)}{\partial x} \tag{8.141}$$

$$= \frac{1}{2}\frac{\partial L(x)}{\partial x} \cdot i^2 \tag{8.142}$$

$$F(x, i) = -\frac{1}{2}\mu_0 \cdot \frac{n_{coil}^2}{x^2} \cdot A_g \cdot i^2 \tag{8.143}$$

Notice that the force direction does not depend on the current and it is proportional to the square of the current, and inversely proportional to the square of the air gap. The shape of the force as a function of displacement can be shaped with design of the plunger and stopper cross-sections (Figure 8.18). Finally, the force–inertia relationship defines the motion of the plunger and any load it may be driving,

$$F(t) = m_t \cdot \ddot{x}(t) + k_{spring} \cdot (x(t) - x_0) + F_{load}(t) \tag{8.144}$$

where m_t is the mass of the plunger plus the load, F_{load} is the load force, x_0 is the preload displacement of the spring.

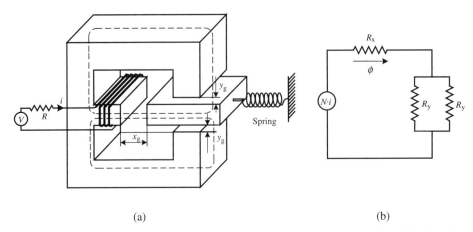

(a)   (b)

FIGURE 8.19: A solenoid example. The magnetic circuit is shown on the right. The force is a nonlinear function of the air gaps and current. The air gap x_g varies with motion, and the air gaps y_g are fixed.

Notice that the electrical time constant of the solenoid, $\tau = L/R$, can become large when L is large. In order to reduce the electrical time constant for fast response, one method is to increase the effective resistance by adding external resistance in series with the coil. As R increases, τ decreases. However, in order to provide the rated current at increased resistance, the supply voltage level must be proportionally increased since $i = V/R$. This in turn increases the resistive heat loss, $P_{Ri} = R \cdot i^2$.

Example   Consider the solenoid type shown in Figure 8.19. There are three air gap paths, one x_g and two y_g spaces. The winding has N turns, and a controlled current i is supplied to the winding. Let us determine the force generated as a function of current i and the variable air gap x_g. The air gaps y_g on both sides are constant. Assume that the permeability of the iron core is much larger than that of the air gaps and neglect the MMF loss along the path in the iron. Then, the total MMF is stored in the air gaps x_g and y_g. Let A_x and A_y denote the cross-sectional areas at the air gaps.

From conservation of the MMF, if we trace the flux path either on the top or bottom section, we obtain

$$H_x \cdot x_g + H_y \cdot y_g = N \cdot i \tag{8.145}$$

From symmetry,

$$\Phi_x = 2 \cdot \Phi_y \tag{8.146}$$
$$B_x \cdot A_x = 2 \cdot B_y \cdot A_y \tag{8.147}$$

Note that

$$B_x = \mu_0 \cdot H_x \tag{8.148}$$
$$B_y = \mu_0 \cdot H_y \tag{8.149}$$

Hence,

$$\Phi_x = 2 \cdot \Phi_y \tag{8.150}$$
$$\mu_0 \cdot H_x \cdot A_x = 2 \cdot \mu_0 \cdot H_y \cdot A_y \tag{8.151}$$

This results in the following magnetic field strength relation,

$$H_y = \frac{A_x}{2A_y} H_x \tag{8.152}$$

Then,

$$H_x = \frac{N \cdot i}{\left(x_g + \frac{A_x}{2A_y}y_g\right)} \tag{8.153}$$

$$H_y = \frac{A_x}{2A_y}\frac{N \cdot i}{\left(x_g + \frac{A_x}{2A_y}y_g\right)} \tag{8.154}$$

Another way to look at this is in terms of MMFs and reluctances. The flux due to MMF and effective reluctance in the circuit,

$$\Phi = \Phi_x = 2 \cdot \Phi_y \tag{8.155}$$

$$= \frac{MMF}{R_{eqv}} = \frac{N \cdot i}{R_x + R_{y,eqv}} \tag{8.156}$$

$$= \frac{\mu_0 N i A_x}{\left(x_g + \frac{A_x}{2A_y}y_g\right)} \tag{8.157}$$

where the effective reluctances

$$R_x = \frac{x_g}{\mu_0 \cdot A_x} \tag{8.158}$$

$$R_y = \frac{y_g}{\mu_0 \cdot A_y} \tag{8.159}$$

and $R_{y,eqv}$ from the parallel connection of two reluctances,

$$R_{y,eqv} = \left(\frac{1}{R_y} + \frac{1}{R_y}\right)^{-1} = R_y/2 \tag{8.160}$$

Flux linkage is a function of variable quantities i and x_g (y_g is constant)

$$\lambda(x_g, i) = \Phi_x \cdot N = L(x_g) \cdot i \tag{8.161}$$

$$= \mu_0 \frac{N^2 \cdot i}{\left(x_g + \frac{A_x}{2A_y}y_g\right)} A_x \tag{8.162}$$

$$L(x_g, i) = \mu_0 \frac{N^2}{\left(x_g + \frac{A_x}{2A_y}y_g\right)} A_x \tag{8.163}$$

The co-energy expression as a function of flux linkage and current is

$$W_{co} = \frac{1}{2}\lambda(x_g, i) \cdot i = \frac{1}{2}L(x_g) \cdot i^2 \tag{8.164}$$

$$F(x_g, i) = \frac{\partial W_{co}(x_g, i)}{\partial x_g} \tag{8.165}$$

$$= -\frac{1}{2}\frac{\mu_0 N^2 A_x}{\left(x_g + \frac{A_x}{2A_y}y_g\right)^2} \cdot i^2 \tag{8.166}$$

The generated force is a function of actuator geometry, A_x, A_y, x_g, y_g, permeability of gap, μ_0, coil turn, N, and current, i.

The complete electromechanical dynamic model of this actuator can be written as

$$V(t) = L(x_g) \frac{di(t)}{dt} + R_{coil} \, i(t) + \left(\frac{dL(x_g)}{dx_g} \right) \cdot i(t) \cdot \frac{dx_g(t)}{dt} \tag{8.167}$$

$$F(x_g, i) = -\frac{1}{2} \frac{\mu_o \, N^2 \, A_x}{\left(x_g + \frac{A_x}{2A_y} y_g \right)^2} \cdot i^2 \tag{8.168}$$

$$F(x_g, i) = m_p \frac{d^2 x_g(t)}{dt^2} + k_{spring} \cdot x_g(t) \tag{8.169}$$

where

$$L(x_g) = \mu_o \frac{N^2}{\left(x_g + \frac{A_x}{2A_y} y_g \right)} A_x \tag{8.170}$$

$$\frac{dL(x_g)}{dx_g} = -\mu_0 \cdot \frac{N^2}{\left(x_g + \frac{A_x}{2A_y} y_g \right)^2} \cdot A_x \tag{8.171}$$

In order to simulate this solenoid, we need the following:

- Parameters: $\mu_0, N, A_x, A_y, y_g, m_p, k_{spring}, R_{coil}$.
- Initial conditions: $i(t_0), x_g(t_0), \dot{x}_g(t_0)$.
- Variables of interest as simulation output: $F(t), x_g(t), V(t), i(t)$.

8.4 DC SERVO MOTORS AND DRIVES

DC servo motors can be divided into two general categories in terms of their commutation mechanism: (i) brush-type DC motors; (ii) brushless DC motors. The brush-type DC motor has a mechanical brush pair on the motor frame and makes contact with a commutator ring assembly on the rotor in order to *commutate* the current, that is to switch the current from one winding to another, as a function of rotor position so that the magnetic fields of the rotor and stator are always at a 90 degree angle relative to each other. In brush-type permanent magnet DC motors, the rotor has the coil winding, the stator has the permanent magnets.

The brushless DC motor is an inside-out version of the brush-type DC motor, that is, the rotor has the permanent magnets and the stator has the winding. In order to achieve the same functionality of the brush-type motor, the magnetic fields of the rotor and stator must be perpendicular to each other at all rotor positions. As the rotor rotates, the magnetic field rotates with it. In order to maintain a perpendicular relationship between the rotor and stator magnetic fields, the current in the stator must be controlled as a vector quantity (both magnitude and direction) relative to the rotor position. Control of current to maintain this vector relationships is called *commutation*. Commutation is done by solid-state power transistors based on a rotor position sensor. Notice that a rotor position sensor is necessary to operate a brushless DC motor, whereas a brush-type DC motor can be operated without any position or velocity sensor as a torque source. When a motor is controlled in conjunction with a position or velocity sensor, it is considered a "servo" motor.

The field magnetics in a brush-type DC motor can be established either by permanent magnets (hence the name *permanent magnet DC motor*) or electromagnets (hence the name *field-wound DC motor*). Field-wound DC motors are used in high power applications (i.e., 20 HP and above) where the use of permanent magnets are no longer cost effective. Permanent magnet (PM) DC motors are used in applications below 20 HP.

Coil winding (either on the stator in the case of brushless DC motors, AC induction motors, stepper motors or on the rotor in the case of brush-type DC motors) determines one of the magnetic fields essential to the operation of a motor. The coil design question is the question of how to distribute the coil around the perimeter of the stator or rotor. The design parameters are [16]

1. the number of electrical phases,

2. the number of coils in each phase,

3. the number of turns on each coil,

4. the wire diameter, and

5. the number of slots and how each coil is distributed over these slots.

There are two types of windings in terms of the spatial distribution of wire on the stator (Figure 8.20):

1. distributed winding, where each phase winding is distributed over multiple slots and one phase winding has overlaps with the other windings (i.e., AC induction motors, DC brushless motors),

2. concentrated winding where a particular winding is wound around a single pole (i.e., stepper motors).

Most common step motors have concentrated winding, whereas AC and DC motors have distributed winding. In concentrated winding, one coil is placed around a single tooth. By controlling the current direction in that particular coil, magnetic polarity (N or S) of that tooth is controlled. Hence, a desired N and S pole pattern can be generated by controlling each coil current direction and magnitude. In distributed winding, there are many variations on how to distribute the coils. The most common type is a three-phase winding. The coil can be distributed to generate two poles, four poles, eight poles, and so on the stator at any given current commutation condition. By controlling the current in each phase, both the magnitude and direction of the magnetic field pattern are controlled. It is common to view the coil distribution in slots in a linear diagram by considering the unrolled version of the motor stator and rotor.

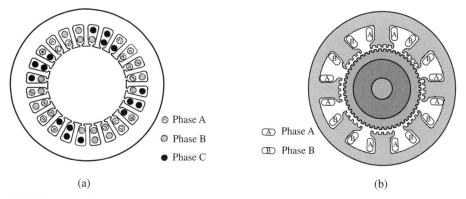

(a)   (b)

FIGURE 8.20: Winding types on the stator: (a) distributed winding (i.e., AC induction motors, brushless DC motors), (b) concentrated winding (i.e., stepper motors).

8.4.1 Operating Principles of DC Motors

There are three major classes of brush-type permanent magnet DC (PMDC) motors:

1. iron core armature,
2. printed-disk armature,
3. shell-armature DC motors.

Iron core armature PMDC motor is the standard DC motor where the stator has a permanent magnet and rotor has wound conductors (Figures 8.21a, 8.22, 8.23). The other two types are developed for applications which require a very large torque to inertia ratio, and hence the ability to accelerate and decelerate very fast (Figure 8.21). The use of the printed-disk and shell-armature type motors has significantly reduced in recent years since comparable or better torque to inertia ratios can be achieved by low inertia brushless DC motors with better reliability (Figures 8.21b, 8.24).

Brush-type DC motors have the permanent magnet as the stator (typically two pole or four pole configuration) and the windings on the iron core rotor (Figure 8.22). The rotor is supported by two ball bearings inside the housing. The ends of the housing are covered by the end plate and face mount plate. Small washers are placed between the bearings and the

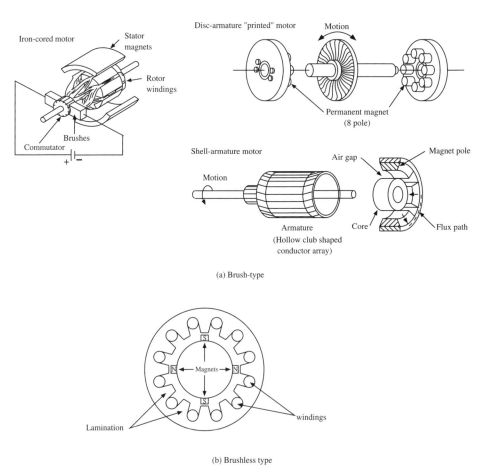

FIGURE 8.21: Permanant magnet DC motor types : (a) brush-types (iron-core, disc-armature, shell-armature) (b) brushless type.

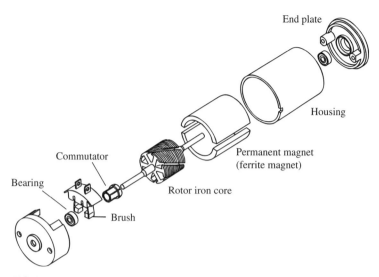

FIGURE 8.22: Components of a brush-type DC motor: rotor (typically made of laminated sheets of steel), stator, commutator, brush pair, housing, end and face plates, two bearings.

end plates in order to provide space for the rotor and housing expansion due to temperature variations. The rotor is typically made of laminated iron sheet metal and each lamination is insulated from each other. The laminated rotor has slots which house the windings. The core surface and windings are electrically insulated from each other by insulation material, coating on the core, or an insulating thin sheet of paper. The slots may be skewed on the perimeter in order to reduce the torque ripple at the expense of lower maximum torque.

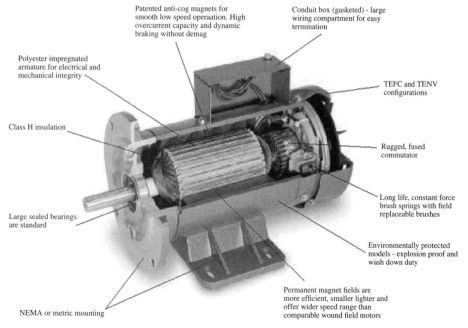

FIGURE 8.23: Brush-type iron core armature permanent magnet DC motor assembly
Source: Baldor.

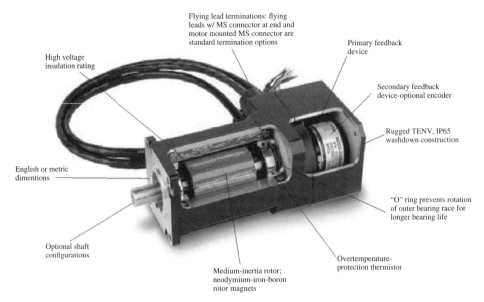

High voltage
insulation rating

Flying lead terminations: flying
leads w/ MS connector at end and
motor mounted MS connector are
standard termination options

Primary feedback
device

Secondary feedback
device-optional encoder

Rugged TENV, IP65
washdown construction

English or metric
dimentions

"O" ring prevents rotation
of outer bearing race for
longer bearing life

Optional shaft
configurations

Medium-inertia rotor;
neodymium-iron-boron
rotor magnets

Overtemperature-
protection thermistor

FIGURE 8.24: Brushless DC motor cross-sectional view showing the permanent magnet rotor, stator winding, and position sensor. The rotor has the permanent magnets glued on its periphery. In high speed and/or high temperature applications, a steel sleeve may be fitted over the magnets to hold them in place securely. The rotor may be manufactured from laminations fitted onto the solid shaft. The stator is made of laminations and houses the windings. Reproduced with permission from Parker Hannifin.

Almost identical mechanical components exists in the brushless DC motors with three exceptions:

1. there are no commutator or brushes since commutation is done electronically by the drive,

2. the rotor has the permanent magnets glued to the surface of the rotor and the stator has the winding,

3. the rotor has some form of position sensor (i.e., Hall effect sensors or encoder are the most common) which is used for current commutation.

In order to understand the operating principle of a permanent magnet DC (PMDC) motor, let us review the basics of electromagnetism (Figure 8.25). A current carrying conductor establishes a magnetic field around it. The electromagnetic field strength is proportional to the current magnitude, and the direction depends on the current direction based on the right hand rule. The magnetic field shape can be changed by changing the physical shape of the current carrying conductor, that is form loops of the conductor as in the case of a solenoid winding. When the current passes through the winding of a solenoid, the magnetic field inside the coil is concentrated in one direction, which in turn temporarily magnetizes and pulls the iron core of the solenoid. This is an example of electromechanical power conversion for linear motion.

Let us consider that a current carrying conductor is placed inside a magnetic field established by two permanent magnet poles (or the magnetic field can be established by a field-winding current in the case of field-wound DC motors). Depending on the direction of

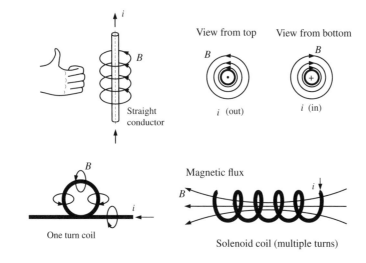

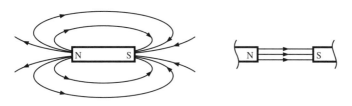

(a) Current carrying conductors and electromagnetic fields

(b) Magnetic field of permanent magnets

FIGURE 8.25: Basic principles of electromagnetism: (a) a current carrying conductor generates a magnetic field around it, (b) the magnetic field generated by permanent magnets.

the current flow, a force is generated on the conductor as a result of the interaction between the "stator magnetic field" and the "rotor magnetic field" (Figures 8.26a, 8.27).

$$\vec{F} = l\vec{i} \times \vec{B} \tag{8.172}$$

Next, let us consider that we place a loop of conductor into the magnetic field and feed DC current to it using a pair of brushes (Figure 8.27). Since the current directions in the two opposite sides of the conductor are in opposite directions, the exerted force on each leg of the conductor loop is in opposite directions. The force pair creates a torque on the conductor.

$$T_m = F \cdot d \tag{8.173}$$

Considering the fact that $\vec{B}$, l, d are constant, we can deduce that

$$T_m = K_t \cdot i \tag{8.174}$$

where $K_t(B, l, d)$, the torque constant, is a function of the magnetic field strength and size of the motor. For a practical motor, the conductor loop would contain multiple turns, not just one pair. As a result, the K_t constant is also a function of the number of conductor turns (n) or equivalently the surface area (A_c) over which flux density acts on the conductors, $K_t(B, l, d, n) = K_t(B, l, d, A_c)$. This is the main operating principle of a DC motor where

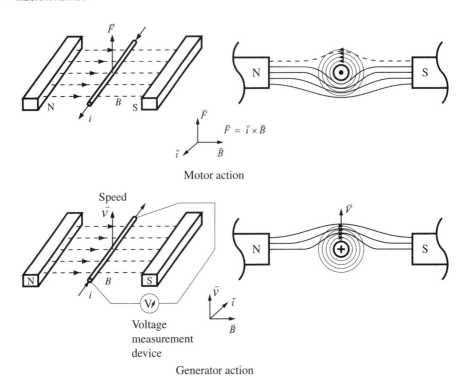

FIGURE 8.26: DC motor operating principles: a current carrying conductor in a magnetic field.

electrical power is converted to mechanical power. This is called the *motor action*. The current in the coil is controlled by controlling the terminal voltage and is affected by the resistive, inductive and back EMF voltages. The electrical circuit relationship is

$$V_t(t) = R \cdot i(t) + \frac{d\lambda(t)}{dt} \tag{8.175}$$

$$= R \cdot i(t) + L\frac{di(t)}{dt} + k_e \cdot \dot{\theta}(t) \tag{8.176}$$

where the $d\lambda(t)/dt$ is the induced voltage as a result of Faraday's law of induction. It has two components: the first is due to the self-inductance of coils, and the second is due to the generator action of the motor.

Notice that when the rotor turns 90 degrees, the moment arm between the forces is zero and no torque is generated even though each leg of the conductor has the same force. In order to provide a constant torque independent of the rotor position, for a given magnetic field strength and current, multiple rotor conductors are evenly distributed over the rotor armature. In order to switch the current direction for a continuous torque direction, a pair of brushes and commutators are used. Without the current switching commutation, the motor would only oscillate as the torque direction would oscillate between clockwise and counter clockwise direction for every 180 degree rotation. Consider the line connecting the two brushes and the coils above and below that. At any given position, the current in one half of the coil is in the opposite direction to the current in the other half of the coils.

Also shown in Figure 8.26 is the *generator action* of the same device. This is the result of the Faraday's induction principle which states that when a conductor is moved in a magnetic field, a voltage is induced across it in proportion to the speed of motion and

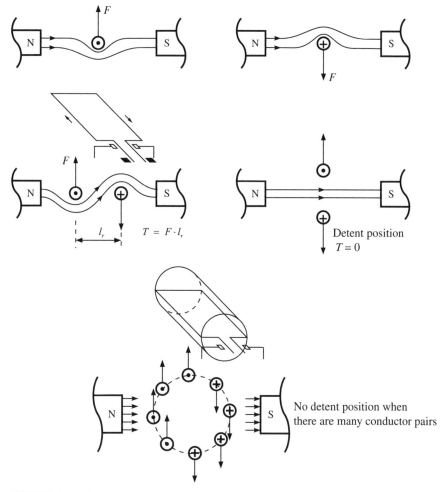

FIGURE 8.27: DC motor operating principles: obtaining a continuous torque by winding a coil around a rotor.

the strength of the magnetic field. This is represented by the $k_e \dot{\theta}(t)$ term in Equation 8.176. Therefore, both the motor and generator actions are at work at the same time during the operation of a DC motor.

Figure 8.28 shows the brush and commutator arrangement and torque as a function of rotor position for a different number of commutator segments. Ideally, the larger the number of commutators, the smaller the torque ripple is. However, there is a practical limit on how small the brush-commutator assembly can be sectioned. If we neglect the torque ripple due to commutation resolution, torque is proportional to the armature current for a given permanent magnet field and independent of the rotor angular position. In a PMDC motor, the magnetic field strength is fixed, and current is controlled by a drive (the term *drive* is used to describe the amplifier and power supply components together as one component).

A brushless permanent magnet DC (BPMDC) motor is basically an "inside-out" version of the brush-type PMDC motor (Figure 8.29). The rotor has the permanent magnets, and the stator has the conductor windings, usually in three electrically independent phases. The stator winding of brushless servo motors is similar to the stator winding of traditional

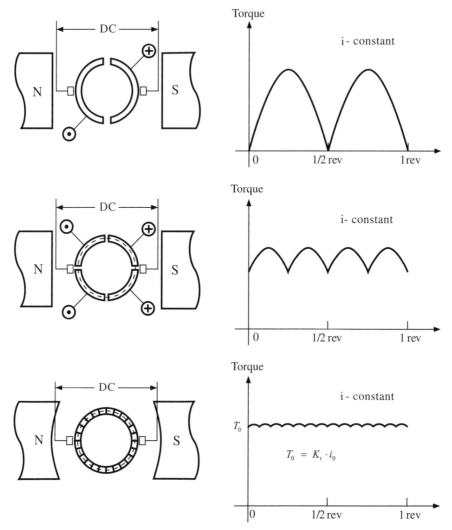

FIGURE 8.28: Commutation and torque variation as a function of the angular position of the rotor. Torque ripple magnitude and frequency is a function of the number of commutation segments. At any given time, coils on one side of the line between the brushes (and the coils on the other side) have a current in the opposite direction. The contribution of each coil to torque production under a constant current depends on its angular position relative to the magnetic field of the permanent magnets at that instant.

induction motors and lends itself to the same well established winding processes used in manufacturing induction motors.

The operating goal is the same: maintain the field (stator) and armature (rotor) magnetic fields perpendicular to each other at all times. If this can be accomplished, the electromechanical power conversion relationship and torque generation in a BPMDC motor would be identical to that of a brush-type PMDC motor. Of course, the difference is in the commutation (Figure 8.29). In a brush-type motor, the magnetic flux generated by permanent magnets (or electromagnets) of the stator is fixed in space. The magnetic field generated by armature is also maintained fixed in space by the mechanical brush-commutator assembly and perpendicular to that of the stator. In the case of the brushless

(a) Conventional DC motor

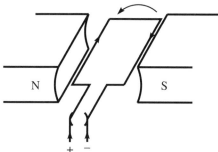

(b) Inside-out DC motor

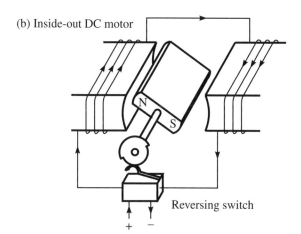

Reversing switch

(c) Brushless DC motor

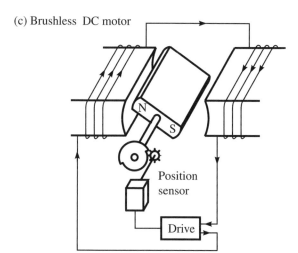

Position
sensor

Drive

FIGURE 8.29: DC motor types:
brush-type and brushless DC
motors.

PMDC motor, we have the same objective. However, the field magnetics is established by
the rotor and it rotates in space with the rotor. Therefore, the stator winding current has to
be controlled as a function of the rotor position so as to keep the stator generated magnetic
field always perpendicular relative to the magnetic field of the rotor, although both rotate in
space with the rotor speed. In other words, not only the magnitude of current in the stator
winding, but also the vector direction of it must be controlled. This switching of the currents
into the stator windings is achieved by controlling solid-state power transistors as a function

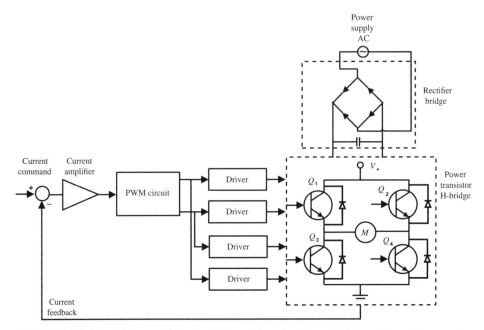

FIGURE 8.30: Block diagram of the brush-type DC motor drive: PWM amplifier with current feedback control.

of the rotor position. Hence, the brushless motor requires a rotor position sensor for its power stage.

8.4.2 Drives for DC Brush-type and Brushless Motors

Drive is considered as the power amplification stage of an electric motor. It is the drive that defines the performance of an electric motor. The most common type of power stage amplifier used for DC brush-type motors is an H-bridge amplifier (Figure 8.30). The H-bridge uses four power transistors. When controlled in pairs (Q1 & Q4 and Q2 & Q3), it changes the direction of the current, and hence the direction of generated torque. Notice that the pair Q1 & Q3 or the pair Q2 & Q4 should never be turned ON at the same time since it would form a short-circuit path between supply and ground. The diodes across each transistor serve the purpose of suppressing voltage spikes and provide a free-wheeling path for the current to follow. Large voltage spikes occur across the transistor in the reverse direction due to the inductance of the coils. If a current flow path is not provided, the transistors may be damaged. The diodes provide the alternative current path for inductive loads and let current pass through the coil. When a diode is ON, the current flows from the negative end of the power supply to the positive end due to the fact the inductive voltage raises the negative side of the potential temporarily (transient) higher than the positive side. The use of diodes in all power amplifiers for different motor types serves the same purpose.

Transistors take a longer time to change from the ON to OFF state than they do to change from OFF to ON state. In other words, transistors turn OFF slower than they turn ON. When transistor states in one leg of the H-bridge are changed (i.e., Q1 is ON, and Q2 is OFF, then we command Q1 to be OFF and Q2 to be ON, and visa versa), Q1 will stay partially turned ON when Q2 is fully turned ON. This difference is in the tens of microseconds range. As a result, this would result in a short circuit between supply and ground voltage. In order to eliminate this problem, when H-bridges are controlled (i.e., by a microcontroller), either in software or hardware, a *dead time* (or delay) is introduced between the ON to OFF command to one transistor and OFF to ON command to the other transistor (this one is delayed in microsecond range) in order to avoid a short circuit. Most microcontrollers have PWM output peripheral where this delay is programmable by writing to a setup register associated with the PWM output channel.

If the top two transistors are both turned ON and the bottom two are turned OFF (or the reverse; top two transistors are OFF, bottom two transistors are ON), the terminals of the motor winding are effectively shorted without connection to the power supply, $V_t(t) = 0.0$. This results in the so-called *dynamic braking*.

$$V_t(t) = R \cdot i(t) + \frac{d\lambda(t)}{dt} \tag{8.177}$$

$$0 = R \cdot i(t) + L \cdot \frac{di(t)}{dt} + k_e \cdot \dot{\theta}(t) \tag{8.178}$$

$$-k_e \cdot \dot{\theta}(t) = R \cdot i(t) + L\frac{di(t)}{dt} \tag{8.179}$$

Due to the back EMF voltage, there will be a current developed in the motor winding (hence torque) in the opposite direction to the rotation speed. As a result, it will brake (slow down) the motion. The generated current is approximately proportional to the speed, hence the dynamic braking torque is large at high speeds and small at low speeds. One of the transistors and the diode on the other side provides the conductive path. Let us assume the top two transistors were turned ON, and the bottom two were OFF. Then for CW motion of the motor, Q1 transistor and D2 diode provide the conductive path for current flow from left to right across the motor. For CCW direction, Q2 diode and D1 transistor provide the conductive path for current flow from right to left (opposite). Similar behavior occurs if the bottom two transistors were ON and the top two transistors were OFF.

By controlling the current magnitude through the power transistors, the magnitude of the torque is controlled. In very small size motors (fractional horsepower), linearly operated transistor amplifiers are used. The pulse width modulation (PWM) circuit operates the transistors in all ON or all OFF modes in order to increase the efficiency. The linear amplifiers provide lower noise but are less efficient than the PWM amplifiers. The most important difference between the *linear mode* and *PWM mode* of operating the power transistors is the efficiency. Power loss at the transistor is approximately the voltage drop across the transistor times the current it conducts,

$$P_{loss} = V_{CE} \cdot i_{CE} \tag{8.180}$$

In the linear mode of operation, that is 50% turned ON, the voltage V_{CE} across the transistor will be 50% of the supply voltage and the current will be the equal to the current amplification gain of the transistor (between the base current and output current). Both values are finite values, and there is significant power loss across the power transistor in linear operating mode. In PWM mode, the transistors are always in one of two states: fully ON or fully OFF. When the transistor is fully ON, voltage drop across the transistor is almost zero, $V_{CE} \approx 0.0$, hence $P_{loss} = 0$. When the transistor is fully OFF, the i_{CE} is almost zero, hence hence $P_{loss} = 0$. As a result, the PWM mode of operation of a transistor results in much less power loss than the linear mode of operation. The only drawback of the PWM mode of operation compared to linear mode of operation is that in PWM mode there will be more noise in the current due to the switching frequency. Typical switching frequencies for the PWM mode of operation for power transistors is in the range of 2–20 kHz. The PWM switching frequency should be significantly larger than the desired current control

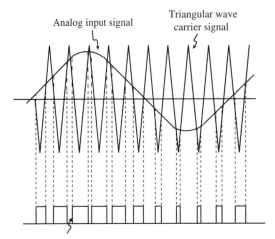

Analog input signal

Triangular wave carrier signal

PWM output signal (PWM equivalent version of the analog input signal)

FIGURE 8.31: PWM circuit function: the carrier signal is a high frequency triangular signal. The input signal is an analog signal value. The output pulse has a fixed frequency which is the carrier frequency. The ON/OFF pulse width is varied as a function of the value of the input signal relative to the carrier signal.

loop bandwidth (i.e., at least 10 to 25 times larger) in order to minimize the effect of PWM frequency on the overall performance of the control system.

The PWM circuit converts an analog input signal (i.e., amplified error signal from the current loop which is an analog signal) to a fixed frequency but variable pulse width signal. By modulating the ON-OFF time of the pulse width at a high switching frequency, hence the name "pulse width modulation," a desired average voltage can be controlled. The PWM circuit (Figure 8.31) uses a triangular carrier signal with a high frequency (also called the switching frequency). When the analog input signal is larger than the carrier signal, the pulse output is ON, when it is smaller, the pulse output is OFF.

PWM is another way of transmitting an analog signal which takes on a value between a minimum value and a maximum value, that is a value between 0–10 VDC or −10–10 VDC. Instead of transmitting the signal by an analog voltage level, PWM transmits the information as a percentage of the ON cycle of a fixed frequency signal. When the signal value is to have the minimum value, the duty cycle (ON cycle) of the signal is set to zero percent. When the signal value is to have the maximum value, the duty cycle (ON cycle) is set to 100 percent. In other words, the analog value of the signal is conveyed as the percentage of duty cycle of the signal.

Figure 8.32 shows the block diagram of a current controlled drive for a three-phase brushless DC motor. The switch set is based on the familiar H-bridge, but uses three bridge legs instead of two legs as is the case for an H-bridge drive for brush-type DC motors. Since each leg of the H-bridge has two power transistors, the brushless motor drive has six power transistors. The stator windings are connected between the three bridge legs as shown in Figure 8.32. The so-called Y-connection shown is the most common type of phase winding connection, while Δ-connection is used in rare cases. At any given time, three of the transistors are ON and three are OFF. Furthermore, two of the windings are connected between the DC bus voltage potential and have current passing through them in a positive or negative direction, whereas the third winding terminals are both connected to the same voltage potential (either V_{DC} or 0 V) and act as the balance circuit. The combination of the ON/OFF transistors determines the current pattern on the stator, hence the flux field

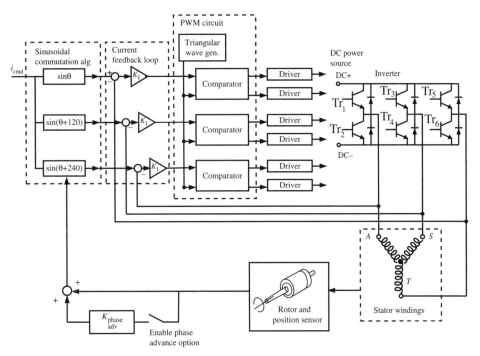

FIGURE 8.32: Brushless servo motor drive block diagram.

vector generated by the stator. In order to generate the maximum torque per unit current, the objective is to keep the stator's magnetic field perpendicular to that of the rotor. By controlling the phase currents in the stator phase windings, we control the stator's magnetic field (magnitude and direction, a vector quantity). Therefore, the torque direction and magnitude can be controlled by controlling the stator's magnetic field relative to that of the rotor.

There are two types of brushless drives based on the commutation algorithm:

1. sinusiodal commutation, and
2. trapezodial commutation.

If the winding distribution and the effective magnetic circuit of the motor is such that the back EMF function is a sinusoidal function of rotor angle, such a motor should be controlled using a drive that uses the sinusoidal commutation algorithm. Similarly, if the motor back EMF is of trapezoidal type, the drive should be the type which uses the trapezoidal current commutation. The variation of the back EMF voltage as a function of rotor angle is the same a function for torque gain as a function of the rotor position under constant current conditions.

The sinusoidal commutation drive provides the best rotational uniformity at any speed or torque. The primary difference between the two types of drives is a more complex control algorithm. For best performance, the commutation method of the drive is matched to the back EMF type of the motor. The back EMF of the motor is determined primarily by its winding distribution, lamination profile, and magnets. The feedback sensor and power electronics components remain the same. In recent years, as the cost of high performance digital signal processors has fallen, the use of the sinusoidal commutation brushless drive has exceeded that of all others.

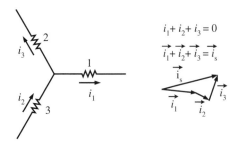

$$i_1 + i_2 + i_3 = 0$$

$$\vec{i}_1 + \vec{i}_2 + \vec{i}_3 = \vec{i}_s$$

FIGURE 8.33: Brushless motor current vector and phase current components (three-phase Y wound).

The current in each of the three phases is controlled by 120 degrees phase shift relative to each other. The rotor position is tracked by a position sensor. The vector sum of phase currents are commutated relative to the rotor position such that two magnetic fields are always perpendicular to each other. The algebraic sum of the currents in three phases is zero, but not the vector sum (Figure 8.33). A current feedback loop is used in the PWM circuit to regulate the current in each phase with sufficient dynamic bandwidth. Notice that the magnitude of the magnetic field generated by the permanent magnets is constant and not actively controlled and it rotates with the rotor. Whereas the stator field (a vector quantity which has magnitude and direction) is controlled actively by the drive current control loop.

In digital implementation, the current command, the commutation, and current control algorithms can reside either in the drive or in a higher level controller. In the latter case, the drive performs only the PWM modulation and is called "the power block." In most cases, current commutation and regulation algorithms are implemented in the drive.

The stator being static by definition, the angle of each stator phase is fixed on the diagram (Figure 8.33). The magnitude of each phase current (i_a, i_b, i_c) for a given total current vector $(\vec{i}_s)$ is the projection of the current vector onto the phase (Figure 8.33).

$$i_n = \vec{i}_s \cdot \vec{u}_n \tag{8.181}$$

$\vec{u}_n$ is the unit vector for the phase, $\vec{u}_a$ for phase a, $\vec{u}_b$ for phase b, $\vec{u}_c$ for phase c.

The derivation below shows that the sinusoidal commutation algorithm for a three-phase brushless motor (which is designed to have a sinusoidal back EMF) produces the same torque–current relationship as that of a brush-type DC motor. Let us assume that the brushless motor has three phase winding and each phase has a sinusoidal back EMF as a function of rotor position. As a result, the current-to-torque gain for each individual phase has the same sinusoidal function. For each phase, they are displaced from each other by a 120 degrees angle as a result of the physical distribution of the windings around the periphery of the stator.

Consider that rotor is at angular position, θ, and each phase of the stator has current values i_a, i_b, i_c. The torque generated by each winding is T_a, T_b, and T_c.

$$T_a = i_a \cdot K_T^* \cdot \sin(\theta) \tag{8.182}$$
$$T_b = i_b \cdot K_T^* \cdot \sin(\theta + 120°) \tag{8.183}$$
$$T_c = i_c \cdot K_T^* \cdot \sin(\theta + 240°) \tag{8.184}$$

Let us control the current (with commutation and current feedback control algorithms) such that each phase is 120 degrees apart from each other and sinusoidally modulated as a function of the rotor position.

$$i_a = i \cdot \sin(\theta) \tag{8.185}$$
$$i_b = i \cdot \sin(\theta + 120°) \tag{8.186}$$
$$i_c = i \cdot \sin(\theta + 240°) \tag{8.187}$$

The total torque developed as a result of the contribution from each phase is

$$T_m = T_a + T_b + T_c \tag{8.188}$$
$$= K_T^* \cdot i \cdot (\sin^2\theta + \sin^2(\theta + 120°) + \sin^2(\theta + 240°)) \tag{8.189}$$

Note the trigonometric relation,

$$(\sin(\theta + 120°))^2 = (\cos\theta \sin 120° + \sin\theta \cos 120°)^2 \tag{8.190}$$

$$= \left(\frac{\sqrt{3}}{2}\cos\theta - \frac{1}{2}\sin\theta \right)^2 \tag{8.191}$$

$$= \frac{3}{4}\cos^2\theta + \frac{1}{4}\sin^2\theta - \frac{\sqrt{3}}{2}\sin\theta \cos\theta \tag{8.192}$$

$$(\sin(\theta + 240°))^2 = \frac{3}{4}\cos^2\theta + \frac{1}{4}\sin^2\theta + \frac{\sqrt{3}}{2}\sin\theta \cos\theta \tag{8.193}$$

$$T_m = K_T^* \cdot i \cdot \frac{3}{2}(\sin^2\theta + \cos^2\theta) \tag{8.194}$$

Hence, the torque is a linear function of the current, independent of the rotor angular position, and the linearity constant torque gain (K_T) is a function of the magnetic field strength.

$$T_m = K_T \cdot i \tag{8.195}$$

where,

$$K_T = K_T^* \cdot \frac{3}{2} \tag{8.196}$$

Therefore, a sinusoidally commutated brushless DC motor has the same linear relationship between current and torque as does the brush-type DC motor. Most implementations make use of the fact that the algebraic sum of three phase currents is zero. Therefore, only two of the phase current commands and current feedback measurements are implemented. Third phase information for both the command and feedback signal is obtained from the algebraic relationship (Figures 8.32, 8.33).

Actual back EMF of a real motor is never perfectly sinusoidal nor trapezodial. The ultimate goal in commutation is to maintain a current-torque gain that is independent of the rotor position, that is a constant torque gain. In order to achieve that, the current commutation algorithm must be matched to the back EMF function of a particular motor. Clearly, if a trapezoidal current commutation algroithm is used with a motor which has a sinusoidal back EMF function, the current-to-torque gain will not be constant. The resulting motor is likely to have large torque ripple.

Let us consider the effect of commutation angle error on the current–torque relationship of the motor. Let θ_e be the measurement error in rotor angle. That is θ_a is the actual motor angle, and θ_m is the measured angle that is used for commutation algorithm, $\theta_e = \theta_a - \theta_m$. It can be shown that the current–torque relationship under non-zero commutation angle error conditions is

$$T_m = K_T \cdot i \cdot \cos(\theta_e) \tag{8.197}$$

Notice that

1. when $\theta_e = 0$, we have the ideal condition,
2. when $\theta_e = 90°$, then there is no torque generated by the current,
3. when $-90 \le \theta_e \le 90$, effective torque constant is less than ideal,

4. when $90 \leq \theta_e \leq 270$, effective torque constant is less than ideal and negative. If the normal position feedback and command polarity is used in the closed position control loop, the motor will run away. In other words, the closed loop position feedback control of the motor would be unstable.

At the most inner loop, voltage fed into each phase is controlled by the PWM amplifier circuit in such a way that the current in each phase follows the commanded current. The dynamic response lag between the voltage modulation and current response in the stator is small but finite. This lag becomes an important factor at high speeds, since

$$\phi_{lag} = \tan^{-1}(\tau_e \cdot \omega_m) \tag{8.198}$$

where ϕ_{lag} is the phase lag, τ_e is the electrical time constant of the current loop, and ω_m is the rotor speed. As the speed of the motor increases, the phase lag of the current control loop can become significant. As a result, the effective angle between the field and armature magnetic fields will not be 90 degrees. Therefore, the motor will be producing torque at a lower efficiency. Keeping this in mind, if the time constant of the current loop ($\tau_e \simeq L/R$) is known approximately or estimated in real-time, the commanded current can be calculated to make not a 90 degree phase with the rotor magnetic field vector but 90 degree plus the anticipated phase lag. In other words, anticipating the phase lag, we can feed the command signal with a phase lead to cancel out the phase lag due to the current regulation loop. This is called *phase advancing* in the brushless drive commutation algorithm. This can be accomplished in real-time by modifying the rotor position sensor signal as shown in Figure 8.32.

Finally, the brushless commutation algorithm requires the absolute position measurement of the rotor within one revolution. This is needed to initialize the commutation algorithm on power-up. On power-up, incremental position sensors (i.e., incremental optical encoders) do not provide this information, whereas resolvers and absolute encoders do. Over 70% of the brushless motors are used with incremental type encoders. Therefore, on power-up a *phase finding* algorithm is needed to establish the absolute position information when an incremental position sensor is used as the position feedback device.

Example: Drive Sizing   Motor and drive (amplifier and power supply) sizes must be matched in a well designed system. Let us assume that we have a motor size selected. It is characterized by its torque capacity (peak and RMS: T_{max}, T_{rms}), current to torque gain (which is the same as back EMF gain: $K_T = K_E$), winding resistance (R), and maximum operating speed (w_{max}). Let us determine the required drive size.

The drive size determination means the determination of DC bus voltage ($V_{DC,max}$) and current (i_{max}, i_{rms}) that the drive must supply.

Given motor specifications: $T_{max}, T_{rms}, K_T = K_E, R, w_{max}$

Determine matched drive size: $i_{max}, i_{rms}, V_{max}$

As a result, the maximum and RMS current requirements are calculated from,

$$i_{max} = T_{max}/K_T \tag{8.199}$$

$$i_{rms} = T_{rms}/K_T \tag{8.200}$$

The maximum DC bus voltage required at worst conditions, that is when providing maximum torque and running at maximum speed and neglecting the transient inductance effects,

$$V_{DC} = L \cdot \frac{di(t)}{dt} + R \cdot i(t) + K_E \cdot w(t) \tag{8.201}$$

$$V_{DC,max} \approx R \cdot i_{max} + K_E \cdot w_{max} \tag{8.202}$$

where the $L \cdot di/dt$ term in the electrical model of the motor is neglected. Then, a drive should be selected that can provide the required DC bus voltage and current requirements with some safety margin. Notice that at a given torque capacity, the so-called *head room of supply voltage* available limits the maximum speed capacity the drive can support,

$$V_{\text{head-room}} = V_{\text{DC,max}} - R \cdot i_{\text{max}} \tag{8.203}$$

$$= K_{\text{E}} \cdot w_{\text{max}} \tag{8.204}$$

$$w_{\text{max}} = \frac{(V_{\text{DC,max}} - R \cdot i_{\text{max}})}{K_{\text{E}}} \tag{8.205}$$

Top speed is limited by the back EMF of the motor. At any given torque level T_{r}, the current required to generate that torque is $i_{\text{r}} = T_{\text{r}}/K_{\text{T}}$. This means $R \cdot i_{\text{r}}$ portion of the available bus voltage is used up to generate the current needed for torque. The remaining voltage $V_{\text{DC,max}} - R \cdot i_{\text{r}}$ is available to balance the back EMF voltage. Hence, the maximum speed at a given torque output is limited by the available "head-room voltage."

Example: Brush-Type DC Motor
Consider a brush-type DC motor with stator coil resistance of $0.25\,\Omega$ at nominal operating temperature. The DC power supply is 24 VDC. The back EMF constant of the motor is 15 V/krpm. The voltage to the motor is turned ON and OFF by an electromechanical relay. Consider two cases: (i) motor shaft is locked and not allowed to rotate, (ii) motor speed is nominally at 1200 rpm. Calculate the steady-state current developed in the motor when the relay is turned on for both cases.

When the motor is locked and not allowed to rotate, there is not any back EMF voltage as a result of the generator action of the motor. This is the stall condition. For steady-state condition analysis, we neglect the transient effect of inductance, the electrical equations for the motor give

$$V(t) = R \cdot i(t) + L \cdot \frac{di(t)}{dt} + K_{\text{E}} \cdot w \tag{8.206}$$

$$\approx R \cdot i(t) \tag{8.207}$$

$$i = \frac{24\,\text{V}}{0.25\,\Omega} \tag{8.208}$$

$$= 96\,\text{A} \tag{8.209}$$

When the motor speed is non-zero, the terminal voltage minus the back EMF is available to develop current. Then,

$$V(t) = R \cdot i(t) + L \cdot \frac{di(t)}{dt} + K_{\text{E}} \cdot w \tag{8.210}$$

$$\approx R \cdot i(t) + K_{\text{E}} \cdot w \tag{8.211}$$

$$24\,\text{V} = 0.25 \cdot i + (15/1000) \cdot 1200 \tag{8.212}$$

$$i = \frac{24 - 18}{0.25} \tag{8.213}$$

$$= 24\,\text{A} \tag{8.214}$$

Notice that under no load conditions ($i = 0$), the maximum speed is limited by the back EMF and it is

$$V(t) = R \cdot i(t) + L \cdot \frac{di(t)}{dt} + K_{\text{E}} \cdot w \tag{8.215}$$

$$24 \approx R \cdot 0 + (15/1000) \cdot w_{\text{max}} \tag{8.216}$$

$$w_{\text{max}} = \frac{24 * 1000}{15} \tag{8.217}$$

$$= 1600\,\text{rpm} \tag{8.218}$$

In SI units, $K_T = K_E$, where as in CGS units K_T N m/A $= 9.5493 \times 10^{-3} K_E$ V/krpm. Then we can find the maximum torque developed at stall and at 1200 rpm speed.

$$T_{\text{stall}} = K_T \cdot i \qquad (8.219)$$

$$= 9.5493 \times 10^{-3} \times 15 \times 96 \, \text{Nm} \qquad (8.220)$$

$$= 13.75 \, \text{Nm} \quad \text{at stall} \qquad (8.221)$$

$$T_r = 9.5493 \times 10^{-3} \times 15 \times 24 \, \text{Nm} \qquad (8.222)$$

$$= 3.43 \, \text{Nm} \, \text{at} \, 1200 \, \text{rpm} \qquad (8.223)$$

Example: PWM Control H-Bridge IC Drive for Brush-Type DC Motors–TPIC0107B   The integrated circuit (IC) package TPIC0107B by Texas Instruments implements an H-bridge and switching control logic for brush-type DC motors (Figure 8.34). The drive supply voltage (V_{cc}) must be in the 27 VDC to 36 VDC range, and support up to 3 A continuous bridge output current. The two terminals of the DC motor

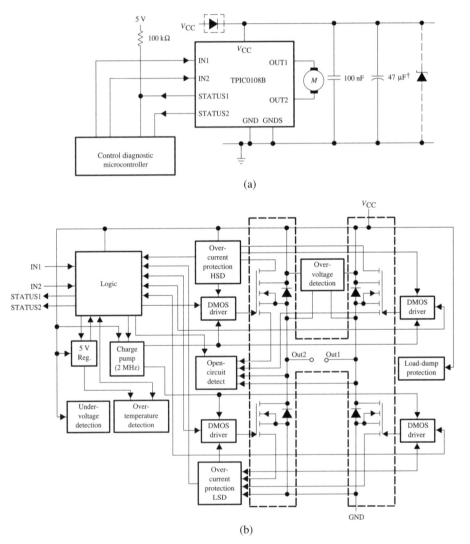

FIGURE 8.34: PWM controlled H-bridge integrated circuit chip for brush-type DC motors: TPIC01017B by Texas Instruments. *Source:* Texas Instruments.

are connected between the OUT1 and OUT2 ports. DC supply voltage and ground are connected to the V_{cc} and GND terminals.

Logic voltage is internally derived from the V_{cc}. The operation of the TPIC0107B is controlled by two input pins: DIR (IN1) and PWM (IN2). The PWM pin should be connected to the PWM output port of a microcontroller, whereas the DIR pin can be connected to any digital output. The PWM switching frequency is 2 kHz. The state of OUT1 and OUT2 (H-bridge output) follows the signal in the PWM pin. The actual PWM signal needs to be formed by the PWM port of the microcontroller. For instance, the current control loop must be implemented in the microcontroller. The IC is capable of sensing over-voltage, under-voltage, short-circuit, over and under current, over temperature conditions and shuts down bridge output and sets the status output pins to indicate the error code (STATUS1 and STATUS2 pins).

Example: PM DC Motor Consider a permanent magnet DC motor. The armature resistance is measured to be $R_a = 0.5\,\Omega$. When $V_t = 120$ V is applied to the motor, it reaches 1200 rpm steady-state speed and draws 40 A current. Determine the back EMF voltage, IR power losses, power delivered to the armature, and torque generated at that speed.

The basic relationship for the motor,

$$V_t(t) = L \cdot \frac{di(t)}{dt} + R \cdot i(t) + V_{\text{bemf}} \tag{8.224}$$

$$120\,\text{V} = 0 + 0.5 \cdot 40\,\Omega\,A + V_{\text{bemf}} \tag{8.225}$$

$$V_{\text{bemf}} = 100\,\text{V} \tag{8.226}$$

where the transient effects of inductance is neglected in steady-state. The IR power loss is

$$P_{\text{IR}} = R_a \cdot i^2 = 0.5 \cdot 40^2\,\Omega\,A^2 = 800\,\text{W} \tag{8.227}$$

and torque generated,

$$K_E = \frac{V_{\text{bemf}}}{\dot{\theta}} = \frac{100\,\text{V}}{1.2\,\text{krpm}} \tag{8.228}$$

$$K_T = 9.5493 \cdot 10^{-3} \cdot K_E\,\text{Nm/A} = 0.7958\,\text{Nm/A} \tag{8.229}$$

$$T = K_T \cdot i = 31.83\,\text{Nm} \tag{8.230}$$

and the electrical power converted to mechanical power,

$$P_m = T \cdot w \tag{8.231}$$

$$= 4000\,\text{W} \tag{8.232}$$

$$= P_e \tag{8.233}$$

$$P_e = V_{\text{bemf}} \cdot i_a \tag{8.234}$$

$$= 100\,\text{V} \cdot 40\,\text{A} = 4000\,\text{W} \tag{8.235}$$

Since the motor speed is constant, the motor torque generated by the torque must be used by a total load torque of equal magnitude but in the opposite direction.

8.5 AC INDUCTION MOTORS AND DRIVES

AC induction motors have been used in constant speed applications in huge quantities. They have been the work horse of the industrial world. In recent years, they have also been used as closed loop position servo motors with a sophisticated current commutation algorithm in the drive being the key to the increased capabilities. Three-phase AC motors are more common than single phase AC motors in applications that require high efficiency and large power. The most common type of multi phase AC induction motors are as follows.

1. The squirrel-cage type AC induction motor is the most common type of AC motor. The stator has phase windings, and the rotor is a squirrel-cage type conductor (copper or aluminum conductor bars) placed in the rotor frame. The conductor bars are short circuited with end rings. The rotor has no external electrical connection.

2. Wound-rotor AC induction motors differ from the squirrel-cage type in the construction of the rotor. The rotor has wound conductors. External electrical connections to the rotor winding are provided via slip rings.

3. Synchronous motors are used for constant speed applications. The motor design is such that the motor torque-speed curve in steady-state provides a constant speed for a wide range of load torque. If the load torque exceeds a maximum value, motor speed decreases abruptly. Compared to the squirrel-cage motors, the synchronous motor speed shows much less variation as load torque varies.

Stator and rotor cores are made from thin steel disk laminations. The purpose of the laminated core is to reduce the eddy current losses.

Conductors of the stator winding are covered with insulation material to protect them against high temperatures. Motor temperature rises primarily as a result of resistance losses, called IR or copper losses. There are four major insulation material classes: class A for temperatures up to 105 °C, class B for temperatures up to 130 °C, class F for temperatures up to 155 °C, and class H for temperatures up to 180 °C.

8.5.1 AC Induction Motor Operating Principles

The main components of an AC induction motor are the stator winding (single phase or three phase) and a rotor with conductors (Figure 8.35). The rotor of a squirrel-cage type AC induction motor is made of steel laminations with holes in its periphery. The holes are filled with conductors (copper or aluminum) and short-circuited to each other by conducting end rings (Figure 8.36). The air gap between the stator and rotor ranges from less than one millimeter (i.e., 0.25 mm) for motors up to 10 KW power to a few millimeters (i.e., 3 mm)

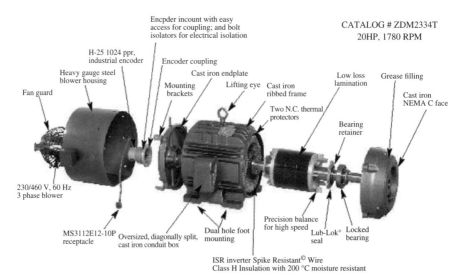

FIGURE 8.35: AC induction motor components and assembly view. *Source:* Baldor.

Conductor bars

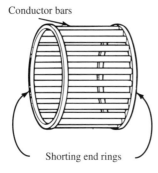

Shorting end rings

FIGURE 8.36: Rotor of a squirrel-cage type AC induction motor. The conductor bars are shorted together at both ends by two end rings.

for motors up to 100 KW (Figure 8.35). The air gap is even larger for motors which are designed for applications that have very large peak torque requirements.

The number of phases of the motor is determined by the number of independent windings connected to a separate AC line phase. The number of motor poles refers to the number of electromagnetic poles generated by the winding. Typical numbers of poles are $P = 2, 4$, or 6 (Figure 8.37). The coil wire for each phase is carefully distributed by design over the periphery of the stator to shape the magnetic flux distribution.

The torque in any AC or DC electric motor is produced by the interaction of two magnetic fields, with one or both of these fields produced by electric currents. In an AC induction motor, the current in the stator generates a magnetic field which induces a current in the rotor conductors. This induction is a result of relative motion between the stator magnetic field (rotating electrically due to AC current) and the rotor conductors (which are initially stationary). This is a result of Faraday's induction law. The stator AC current sets up a rotating flux field. The changing magnetic field induces EMF voltage, hence current, in the rotor conductors. The induced current in the rotor in turn generates its own magnetic field. The interaction of the two magnetic fields (the magnetic field of the rotor trying to keep up with the magnetic field of the stator) generates the torque on the rotor. When the rotor speed is identical to the electrical rotation speed of the stator field, there is no induced voltage on the rotor, and hence the generated torque is zero. This is the main operating principle of an AC induction motor.

In order to draw a visual picture of how torque is generated, let us consider a two-phase AC induction motor (Figure 8.38). The principle for other numbers of phases is similar. Let us consider the case where only phase 1 is energized and the current on the

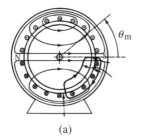

(a)

(b)

(c)

FIGURE 8.37: Stator windings of an AC induction motor: (a) two-pole ($P = 2$) configuration, (b) four-pole ($P = 4$) configuration, (c) six-pole ($P = 6$) configuration.

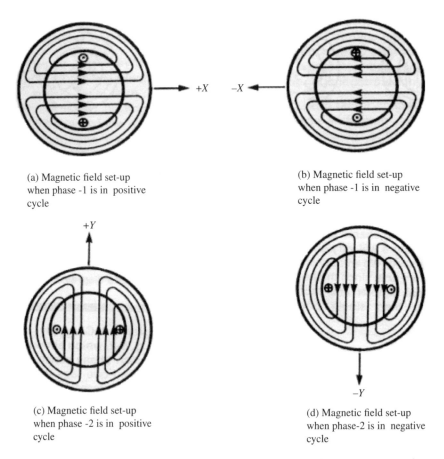

(a) Magnetic field set-up
when phase -1 is in positive
cycle

(b) Magnetic field set-up
when phase -1 is in negative
cycle

(c) Magnetic field set-up
when phase -2 is in positive
cycle

(d) Magnetic field set-up
when phase-2 is in negative
cycle

FIGURE 8.38: AC induction motor operating principle: a two-phase motor example.

phase is a sinusoidal function of time. The induced magnetic field changes magnitude
and direction as a function of current in the phase. It is basically a pulsating magnetic
field in the X direction (Figure 8.38a,b). Next, let us consider the same for phase 2 which
is spatially displaced by 90 degrees from the first phase. The same event occurs except
the direction of the magnetic field is in the Y direction (Figure 8.38c,d). Finally, if we
consider the case where both phases are energized but by 90 electrical degrees apart and
same frequency, the magnetic field would be the vector addition of two fields and it rotates
in space at the same frequency as the excitation frequency (Figure 8.39). Therefore, we
can think of the magnetic field as having a certain shape (that is distribution in space as a
function of rotor angle) as a result of the winding distribution and current, and it rotates in
space as a result of the alternating current in time. In other words, the flux distribution is a
rotating wave.

   In a three-phase motor, the windings would be displaced by $\pm$ 120 degrees spatially
and electrically excited with the same frequency source except $\pm$ 120 electrical degrees
apart. This rotating magnetic field, generated by the stator winding voltage, induces voltage
in the rotor conductors. As a result of Faraday's induction law, the induced voltage is
proportional to the time rate of change in the magnetic flux lines that cut the rotor. In other
words, if the rotor was rotating mechanically at the same speed as the electrical rotating
speed of the stator field, there would not be any torque generated.

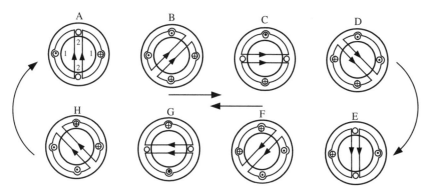

FIGURE 8.39: Progression of the magnetic field in a two-phase stator at eight different instants.

In a P-pole motor, the electrical angle (θ_e) and mechanical angle (θ_m) are related by (Figure 8.37),

$$\theta_m = \theta_e/(P/2) \tag{8.236}$$

Therefore, the frequency of the electrical excitation w_e, is related to the synchronous speed, w_{syn} (Figure 8.37)

$$w_{syn} = w_e/(P/2) \tag{8.237}$$

The difference between the electrical field rotation speed, w_{syn}, and the rotor speed, w_{rm}, is called the *slip speed* or *slip frequency*, w_s

$$w_s = w_{syn} - w_{rm} \tag{8.238}$$

$$= s \cdot w_{syn} \tag{8.239}$$

where s, slip, is defined as

$$s = \frac{w_{syn} - w_{rm}}{w_{syn}} \tag{8.240}$$

If we consider the case that the rotor is locked ($w_{rm} = 0$), then the slip is $s = 1$ and $w_s = w_{syn}$.

The AC motor operating principle is similar to a transformer. Stator windings serve as the transformer primary winding. The stator structure serves as the transformer "iron." The rotor serves as the transformer secondary winding. The only difference is that the secondary winding is the rotor conductor's and it is mechanically rotating. The rotor sees an effective magnetic flux frequency of $w_s = w_{syn} - w_{rm}$, slip frequency, due to the relative motion between the electrically rotating stator flux and mechanically rotating rotor. The induced voltage in the rotor is analogous to the induced voltage in the secondary winding of a transformer. The voltage in the primary winding generates a magnetic flux as follows. Let $P = 2$, hence $w_{syn} = w_e$. The stator AC voltage,

$$v_s(t) = V_s \sin(w_e t) \tag{8.241}$$

The resulting flux is

$$\Phi = -\frac{V_s}{N_1 \cdot w_e} \cos(w_e t) \tag{8.242}$$

where N_1 is the number of turns in the primary coil. The rotor sees a magnetic flux frequency of $w_s = w_{syn} - w_{rm}$. The induced output voltage as a result of the Faraday's law of induction is

$$v_r(t) = N_2 \frac{d\Phi}{dt} \tag{8.243}$$

$$= -\frac{N_2}{N_1} \frac{V_s}{w_e} \frac{d}{dt}[\cos(w_e - w_{rm})t] \tag{8.244}$$

$$= \frac{N_2}{N_1} \frac{(w_e - w_{rm})}{w_e} V_s \sin(w_s t) \tag{8.245}$$

The operating principle of an AC induction motor in terms of cause and effect relationship is illustrated below. The stator current is a result of the applied stator voltage ($v_s(t)$), and induced rotor current is a result of the induced rotor voltage ($v_r(t)$).

$$V_s(w_{syn}) \Longrightarrow i_s(w_{syn}) \Longrightarrow B_s(w_{syn}) \tag{8.246}$$

$$w_s \Longrightarrow V_r(t) \, induced \Longrightarrow i_r(t) \, induced \Longrightarrow B_r(w_{syn}) \tag{8.247}$$

$$B_s(w_{syn}) \, \& \, B_r(w_{syn}) \Longrightarrow T_m \, (torque) \tag{8.248}$$

Notice that B_s and B_r rotate with the synchronous speed, w_{syn}. The rotor mechanically rotates close to the synchronous speed with the difference being the slip speed, $w_s = w_{syn} - w_{rm}$. When the mechanical speed of the rotor is smaller than the synchronous speed, torque is generated by the motor (motoring action). Whereas if the mechanical speed of the rotor is larger than the synchronous speed, torque is consumed by the motor (it is in generating mode). In the vicinity of the rotor mechanical speed close to synchronous speed, the torque is proportional to the slip speed. When the slip is zero, the generated torque is zero (Figure 8.41).

The steady-state torque-speed characteristics of an AC induction motor can be summarized as follows:

1. In the vicinity of a small slip, the torque is proportional to the slip frequency for a given stator excitation. If the rotor speed is smaller than the synchronous speed (slip is positive), then the torque is positive. The motor is in *motoring mode*. If the rotor speed is larger than the synchronous speed (slip is negative), then the torque is negative. The motor is in *generator mode*.

2. At a certain value of slip, the torque reaches its maximum value. For slip frequencies larger than that, the inductance of the rotor becomes significant and the current is limited at the higher slip. As a result, torque drops after a certain magnitude of slip frequency.

3. The shape of the steady-state torque-speed curve can be modified for different applications by modifying the rotor conductor shape and stator winding distribution.

The torque generation in an AC induction motor can be viewed as the result of the interaction between the magnetic flux distributions of the stator and the rotor. From basic reasoning regarding the interaction of magnets and magnetic fields (Figure 8.40), it can be concluded that

$$T_m = K_m \cdot B_s \cdot B_r \cdot \sin(\theta_{rs}) \tag{8.249}$$

which states that the torque is proportional to the size and design parameters of the motor (K_m), the flux density in rotor and stator, and the angle between the two flux vectors. The so-called *vector control* algorithm discussed later in this chapter attempts to maintain a

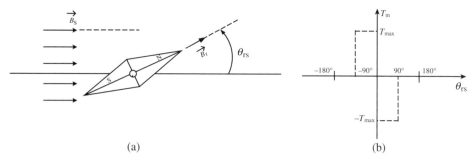

(a)   (b)

FIGURE 8.40: Torque production as a result of the interaction between two magnetic fields. (a) A permanent magnet in an external magnetic field, (b) torque as a function of the angle between the two flux vectors. In the case of AC induction motors, the external field flux density (B_s) is setup by the stator current and the equivalent field of the magnet (B_r) is setup by the rotor conductors as a result of the induced voltage.

$90°$ angle between the stator and rotor flux density vectors in order to maximize torque production per unit current.

Open loop steady-state torque-speed characteristics of an AC induction motor are shown in Figure 8.41 for the case of constant AC voltage magnitude and frequency supplied to the stator windings. Note that the shape of that curve can be changed by using different stator winding and slotting configurations. In fact, different motor designs are used to meet different steady-state torque-speed curve shapes needed by different applications. Most AC induction motors have maximum slip in the range of 5–20% of the synchronous speed. Maximum slip is defined as the slip speed at which maximum torque is reached. Clearly, the smaller the maximum slip is, the smaller the speed variations of the motor under varying loads. Likewise, if large speed variation is desired as the load varies, a motor with large maximum slip should be used.

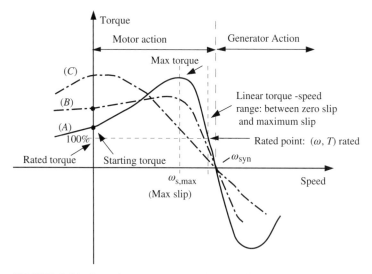

FIGURE 8.41: Open loop torque speed curve of an AC induction motor. Under a given constant AC voltage magnitude and frequency, the steady-state torque-speed relationship is nonlinear. This curve can be given different shapes by different electromagnetic designs. Curves (A), (B), and (C) show different AC motor torque-speed characteristics that can be achieved by varying designs.

It should be quickly pointed out that these performance characteristics are only for the motor when controlled from a line voltage supply directly without any active commutation by a drive. The current-torque characteristics of an AC motor can be made to behave like a DC motor using the so called "field oriented vector control" algorithm which is typically implemented in the drive. The performance of a motor should always be evaluated together with the drive it is used in. Depending on the drive type used, the performance characteristics of the motor-drive combination can be quite different.

Example Consider an AC induction motor driven by a line supply frequency of $w_e = 60\,\text{Hz}$. Assume that it is a two-pole motor, $P = 2$. The motor design is such that at the maximum load, the slip is 20% of the synchronous speed, $s = 20\%$. Determine the speed of the motor for the following conditions: (a) no-load speed, (b) speed at maximum load, (c) speed at a load that is 50% of the maximum load. Determine the speed variation (sensitivity) due to the variation of load as percentage of its maximum value. Let maximum load torque be $100\,\text{lb}\cdot\text{in}$.

Referring to Figure 8.41, let us assume that the curve connecting the no-load speed and maximum-load speed points is a linear line. In steady-state, the actual motor speed is determined by the intersection of the torque-speed curve with the load torque. As long as the load torque is less than the maximum load torque, the motor speed will be somewhere between the no-load speed and the speed at maximum load. As the load varies up to the maximum load torque, the steady-state speed of the motor variation follows the linear torque-speed line.

The no-load speed of the motor is

$$w_{rm} = \frac{w_e}{P/2} = \frac{60\,\text{Hz}}{2/2} = 60\,\text{rev/s} = 3600\,\text{rev/min} \tag{8.250}$$

At the maximum load, the motor specifications indicate that it has 20% slip,

$$s = \frac{w_{syn} - w_{rm}}{w_{syn}} = 0.2 \tag{8.251}$$

$$w_{rm} = w_{syn} - 0.2 \cdot w_{syn} = 0.8 \cdot w_{syn} = 2880\,\text{rev/min} \tag{8.252}$$

When the load is 50% of maximum load, the slip will be 50% of the maximum slip. Therefore, the steady-state rotor speed is

$$s = \frac{w_{syn} - w_{rm}}{w_{syn}} = 0.1 \tag{8.253}$$

$$w_{rm} = w_{syn} - 0.1 \cdot w_{syn} = 0.9 \cdot w_{syn} = 3240\,\text{rev/min} \tag{8.254}$$

The speed varies from synchronous speed at no-load to 20% slip at maximum load, hence

$$\frac{\Delta V}{\Delta T_1} = \frac{w_{syn} - ((1-s) \cdot w_{syn})}{100} \tag{8.255}$$

$$= \frac{s \cdot w_{syn}}{100} \tag{8.256}$$

$$= 7.2\,\text{rpm/(lb}\cdot\text{in)} \tag{8.257}$$

8.5.2 Drives for AC Induction Motors

The drive controls the electrical variables, which are the voltage and current, in the stator windings of an AC induction motor in order to obtain the desired behavior in mechanical variables, which are torque, speed, and position. In particular, frequency and magnitude of the voltage control are of interest. Major drive types are discussed below (Figure 8.43) where each drive type operates based on varying one or more of the electrical variables, that is voltage and current and their frequency and magnitude.

The steady-state torque-speed curve of an AC induction motor whose stator winding phases are fed directly from an AC line is shown in Figure 8.41. The motor synchronous speed (w_{syn}) is determined by the line voltage frequency (w_e). The actual speed of the rotor (w_{rm}) would be a little below that since there is a slip (w_s) between the synchronous speed and actual rotor speed in steady-state. The slip speed depends on the load torque.

$$w_{syn} = \frac{w_e}{(P/2)} \tag{8.258}$$

$$w_{rm} = w_{syn} - w_s \tag{8.259}$$

The maximum torque characteristic is a function of the motor design and the line voltage magnitude. This basic relationship indicates that an AC motor driven directly from a supply line has a speed that is largely determined by the frequency of the supply voltage. The slip frequency is a function of the load and the type of motor design (Figure 8.40). The exact mechanical speed of the rotor is determined by the load around the synchronous speed.

Scalar Control Drives   If the drive varies the magnitude of voltage applied to the motor, while keeping the frequency constant, the torque-speed characteristics of the motor are as shown in Figure 8.43a. Notice that as the voltage magnitude decreases relative to the rated voltage (V_r), the torque gets smaller. It can be shown that the maximum torque is proportional to the square of the applied voltage magnitude. If the load is a constant torque load, by varying the amplitude of the voltage (variable voltage method), we can obtain some degree of variable speed control in the vicinity of synchronous speed of the motor.

The next method of control is to vary the frequency of the applied voltage while keeping the magnitude of the voltage constant. The steady-state torque-speed performance of an AC motor with such a drive is shown in Figure 8.43b. Notice that the synchronous speed of the motor is proportional to the applied frequency of the voltage, that is if the applied frequency is 50% of the base frequency, then the synchronous speed is also 50% of the original synchronous speed. However, the effective impedance of the motor is smaller at lower frequencies. This leads to large currents and results in magnetic saturation in the motor. Therefore, in order to improve the efficiency of the motor, it is better to maintain a constant ratio of voltage magnitude and frequency.

Variable frequency (VF) drives are capable of adjusting the AC voltage frequency, w_e, as well as the magnitude of the AC voltage of each phase, V_o. For a three-phase AC induction motor, phase voltages may be

$$V_a = V_o \sin(w_e t) \tag{8.260}$$

$$V_b = V_o \sin(w_e t + 2\pi/3) \tag{8.261}$$

$$V_a = V_o \sin(w_e t + 4\pi/3) \tag{8.262}$$

where the VF drive can control both w_e and V_o from zero to a maximum value. Hence, the steady-state torque-speed curve of the motor can be made as shown in Figure 8.43c. The power electronics of the VF drive are identical to that of a brushless motor drive, that is a three-phase inverter (Figure 8.42). The only essential difference is in the real-time control algorithm that operates the PWM circuit. The PWM circuit is controlled in such a way that the frequency and the magnitude of each phase voltage is changed depending on where on the torque-speed curve we want the motor to operate. Since both frequency and magnitude of voltage is controlled, such drives are also called variable frequency and variable voltage (VFVV) drives. Such a drive control method is also referred to as the *Volts/Hertz (V/Hz)*

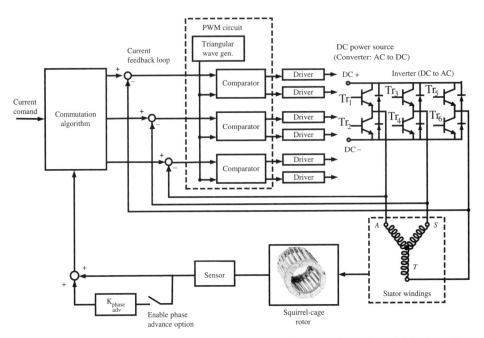

FIGURE 8.42: Current commutation and regulation in AC motor phases by a field oriented vector control (FOVC) drive.

method. It is classified as one of the *scalar control* methods as oppose to the *vector control* methods discussed next.

The power stage of a Volts/Hertz drive for a three-phase AC induction motor is a three-phase inverter. The most common power switching devices are power MOSFETs and IGBTs. Power MOSFETs are voltage controlled transistors with low power loss, but they are sensitive to temperature. IGBTs (insulated gate bipolar transistors) are essentially bipolar transistors where the base is controlled by MOSFETs. IGBTs have higher switching frequency, but are less efficient than MOSFETs.

For a three-phase AC induction motor control (Figure 8.42), a three-phase drive ("inverter") is needed. In each leg of the bridge, power transistors operate in opposite state to each other, that is Tr1 = ON and Tr2 = OFF or Tr1 = OFF and Tr2 = ON. If both are ON at the same time, the supply voltage potential would be effectively shorted and most likely damage the power transistors due to excessive current. Furthermore, it should be kept in mind that power transistors have a longer turn-OFF time than turn-ON time. Hence, when the state in the power transistors change, there should be some "dead-time" (delay or wait time) of the transistor to be commanded to go from an OFF to ON state so that the other one which is going from ON to OFF state has sufficient time to do so, in order to avoid the short circuit. PWM peripherals of modern microcontrollers and digital signal processors automatically handle this issue and dead-time insertion is programmable.

Vector Control Drive: Field Oriented Vector Control Algorithm

The roles of the drive are the commutation and the amplification of current in the stator windings (Figure 8.42)[1]. The only difference is that a DC brushless motor has a permanent magnet rotor, whereas an AC induction motor has a squirrel-cage rotor with no magnets. Therefore the magnetic field of the rotor of an AC motor is not physically locked to the rotor, unlike the

---

[1] This section can be skipped without loss of continuity. See [17] for more details.

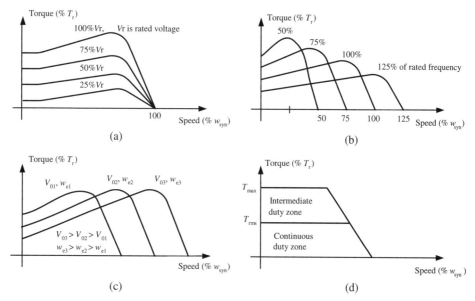

FIGURE 8.43: AC induction motor torque-speed performance in steady-state under various control methods for varying voltage, frequency, and current in the motor stator phase windings: (a) variable voltage amplitude, fixed frequency, (b) variable frequency, fixed voltage, (c) variable voltage and variable freqency, while keeping the voltage to frequency ratio constant for different frequency ranges (Volt/Hertz method), (d) field oriented vector control.

DC brushless motor. In order to commutate the current in the windings so that two magnetic fields are perpendicular for maximum torque generation per current unit, measuring rotor angle is not sufficient to know the relative angle between the magnetic field of the stator and the magnetic field of the rotor. The "field oriented vector control algorithm" is the name used for AC motor current commutation where the angle between the magnetic fields (the magnetic field of the stator and the induced magnetic field of the rotor) is estimated based on the dynamic model of the motor.

Assuming that this angle between the two magnetic fields is known, an AC motor can be commutated to provide essentially the same torque-speed characteristics of a DC brushless motor (Figure 8.43d). The only difference may be in the transient response. The field oriented vector control algorithm are a current commutation algorithm for AC induction motors. This current commutation algorithm attempts to make an AC induction motor behave like a DC motor, that is to have a linear relationship between the torque and commutated current. The hardware components of a drive for AC motors which can implement the vector control commutation algorithm are identical to that of a drive for DC brushless motors. Both drives attempt the same thing: to maintain a perpendicular relationship between the field magnetic flux vector and the controlled current vector.

The AC induction motor differs from the DC brushless motors in two ways. First, the controlled current is on the stator which induces current in the conductors of the rotor. That induced current generates its own magnetic field in the rotor. The induced magnetic field is not locked to the rotor. There is a slip between the rotor and the induced field. Second, there are two components of the controlled current that are of interest: the component that is parallel to the rotor field and the component that is perpendicular to it. It can be shown mathematically [17] that the parallel component (magnetization current) determines the torque gain of the motor, whereas the perpendicular component determines the current multiplier for torque generation. Let us express the torque–current relationship

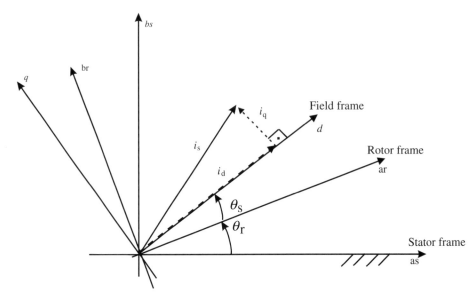

FIGURE 8.44: Coordinate frames used to describe the dynamics of an AC induction motor.

for a DC motor and the same desired torque–current relationship for a vector controlled AC motor,

$$T_m^{DC} = K_T \cdot i \tag{8.263}$$

$$T_m^{AC} = K_T(i_{ds}) \cdot i_{qs} \tag{8.264}$$

Notice that the same linear relationship of the DC motor between torque and current can be obtained. Let the "dq"-coordinate frame be fixed to the field vector of the rotor. Hence, it rotates relative to the rotor with slip frequency (Figure 8.44). The torque gain K_T is a function of the direct (parallel) component of the current in "dq" coordinate frame, and the torque producing current is the "q" (perpendicular) component of the current.

8.6 STEP MOTORS

The step motor, also called a stepper motor, electromechanical construction is such that it moves in discrete mechanical steps. A change in phase current from one state to another creates a single step change in the rotor position. If the phase current state is not changed, the rotor position stays in that stable position. In contrast, a brush-type DC motor keeps accelerating for a fixed supply voltage condition until the back EMF voltage due a top speed balances the supply voltage.

The basic position control of a stepper motor does not require a position sensor. It can be position controlled in open loop (Figure 8.45). Whereas a DC motor must use a position sensor in order to be controlled in position mode.

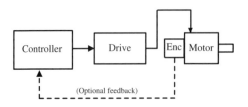

FIGURE 8.45: Stepper motor control system components. Position sensor feedback is optional.

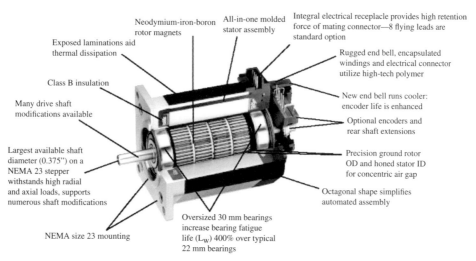

Neodymium-iron-boron rotor magnets

All-in-one molded stator assembly

Integral electrical receplacle provides high retention force of mating connector—8 flying leads are standard option

Exposed laminations aid thermal dissipation

Rugged end bell, encapsulated windings and electrical connector utilize high-tech polymer

Class B insulation

New end bell runs cooler: encoder life is enhanced

Many drive shaft modifications available

Optional encoders and rear shaft extensions

Largest available shaft diameter (0.375") on a NEMA 23 stepper withstands high radial and axial loads, supports numerous shaft modifications

Precision ground rotor OD and honed stator ID for concentric air gap

Octagonal shape simplifies automated assembly

NEMA size 23 mounting

Oversized 30 mm bearings increase bearing fatigue life (L_w) 400% over typical 22 mm bearings

FIGURE 8.46: Cross-sectional view of a four stack hybrid type stepper motor. *Source:* Reproduced with permission from Parker Hannifin.

The most significant advantages of step motors are their low cost, simplicity of design, and ruggedness. The disadvantages are that step motors have mechanical resonance and step loss problems, although most of these drawbacks have been largely eliminated by the "microstepping drive" technology. Figure 8.46 shows a picture of the most common type of stepper motor (hybrid-stepper motor). Figure 8.47 shows the construction of the rotor, where laminations with teeth are assembled over an axially magnetized permanent magnet pole pair. Furthermore, the north pole group of laminations is mounted on the rotor with one half pitch of a tooth angular phase from the south pole group of the laminations.

A stepper motor has a rotor (Figure 8.47) and stator (Figure 8.46). The rotor and stator are made of laminations of soft iron material. Each stator pole has a concentrated

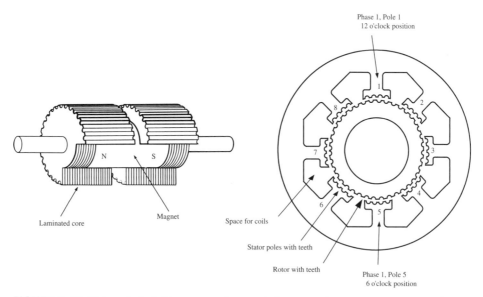

Phase 1, Pole 1
12 o'clock position

Laminated core

Magnet

Space for coils

Stator poles with teeth

Rotor with teeth

Phase 1, Pole 5
6 o'clock position

FIGURE 8.47: Rotor of a hybrid permanent magnet stepper motor: permanent magnet is polarized axially (north pole on one side and south pole on the other side along the shaft) and laminated iron core with teeth. *Source:* Reproduced with permission from Parker Hannifin.

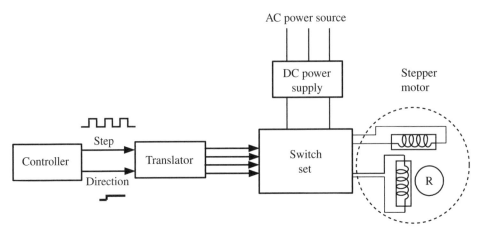

FIGURE 8.48: Stepper motor control system components: step motor, DC power supply, power switch set, translator, controller.

coil. The stator and rotor have teeth. For a given state of a stator current, the rotor moves to align its teeth with those of the stator by the natural tendency to minimize reluctance to magnetic flux. The air gap between the stator and rotor represents the resistance to the magnetic flux. The smaller the air gap is, the less the resistance to magentic flux, hence the higher the torque capacity of the motor. A typical air gap in stepper motors is in the range of 30–125 μm. The components of a stepper motor control system are shown in Figure 8.48. A given switch state of a stator represents a magnetic flux field. There is a corresponding stable rotor position for each stator phase current condition. The "switch set" block represents the power transistors. The "translator" block represents the logic block which determines the order and time the power transistors should be switched based on a planned motion which is generated by the controller. It is the responsibility of the translator block or the controller block to limit the maximum switching frequency so that the rotor is not left behind to the point of missing a step. Similarly, the translator block should minimize operating at a switching frequency range which might excite the natural resonance frequency of the stepper motor. Today, stepper motor sizes are standardized and the most common ones are NEMA 17, NEMA 23, NEMA 34, and NEMA 42.

A given phase current condition generates a certain stable rotor position, not continuous torque. Normally the drive or the controller does not need position feedback. But position feedback is commonly used as a way to detect possible step loss and compensate for it when necessary.

The translator, switch set, and DC power supply blocks are collectively called the "drive." Drive controls the current in each phase. Velocity or position feedback is used to control the speed or position of the motor.

8.6.1 Basic Stepper Motor Operating Principles

Let us consider the operating principles of a basic stepper motor. In our basic stepper motor, the rotor has one north and one south pole permanent magnet. The stator has four-pole, two-phase winding with four switches (Figure 8.49). At any given time either switch S1 or S2, and S3 or S4 can be ON to affect the polarity of electromagnets. For each switch state, there is a corresponding stable rotor position. In this concept figure, a stepper motor with unipolar winding is shown where each coil is center tapped to the ground.

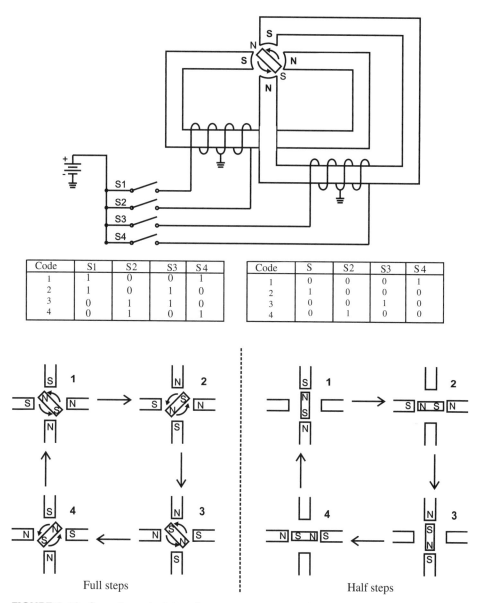

Code	S1	S2	S3	S4
1	1	0	0	1
2	1	0	1	0
3	0	1	1	0
4	0	1	0	1

Code	S	S2	S3	S4
1	0	0	0	1
2	1	0	0	0
3	0	0	1	0
4	0	1	0	0

Full steps

Half steps

FIGURE 8.49: Operating principles of a stepper motor: a unipolar step motor winding with a unipolar drive model is shown.

Let us consider the switching sequence shown on the left four illustrations at the bottom of Figure 8.49. In this case, at any given time all of the stator phases are energized. At all times, each rotor pole is attracted by two winding poles. Following the four switching sequences, the rotor would take the shown stable positions. The current pattern for these four discrete switch states are shown in Figure 8.50a. This type of phase current switching, where both phases are energized, is referred as the "full-step" mode of operation. As ON/OFF states of switches are changed in the order shown in the figure, the magnetic field generated by the stator rotates in space. Hence, the permanent magnet (PM) rotor follows it.

Now, let us consider the four sequences of switch states shown on the right hand side of Figure 8.49. In this case, only one of the stator phases is energized while the other phase

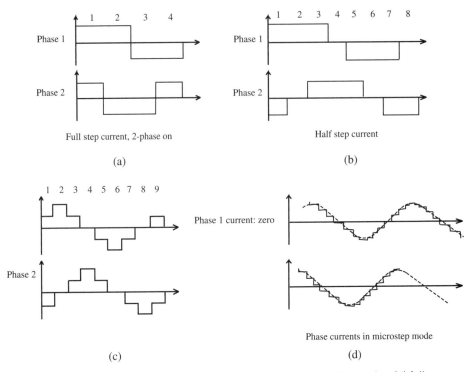

FIGURE 8.50: Phase currents in a stepper motor in different operating modes: (a) full step mode, (b) half step mode, (c) modified (increased current during alternate half steps) half step mode, and (d) microstepping mode.

is OFF (i.e., both S1 and S2 OFF, or both S3 and S4 OFF). The corresponding stable rotor positions are shown in the figure. However, notice that since the magnetic force pulling the rotor is provided by only one phase, the holding torque of the motor at these switch states is less than (approximately 1/2) that of the holding torque at the full-step mode. This mode of switching power transistors at the drive is referred to as the "half step" mode. We can, therefore, energize the motor in alternate modes (one full step and one half step) in every other step in order to increase its positioning resolution (from 4 stable positions to 8 stable positions). The current pattern in the phase windings in full and half step modes is shown in Figure 8.50. In order to make sure the torque capacity is similar at all steps, the current can be increased during the half step mode compared to the current during full step mode (Figure 8.50c). The step size is defined as the smallest rotor position change that can be achieved by switching the current from one state to another state in the stator winding using full step and half step modes. For this simple concept stepper motor, step size is

$$\theta_{step} = \frac{360}{8} = 45° \tag{8.265}$$

The standard hybrid permanent magnet (PM) step motor operates in the same way as our simple model, but has a greater number of teeth on the rotor and stator, giving a smaller basic step size. The rotor is in two sections axially. The two sections of the rotor, both sections with teeth, are separated by a permanent magnet (Figure 8.47). North and south poles are magnetized axially, so the N-pole is on one side and the S-pole is on the other side along the shaft. Hence, one side is magnetized as the north pole, the other side is magnetized as the south pole. Furthermore, there is a one-half tooth pitch angular displacement between the two sections (north and south sections) on the rotor. Let us consider an example hybrid step motor. The stator has 8 poles each with 5 teeth on each, making a total of 40 teeth (Figure 8.47), plus the space between each pole: there are 48 spaces. Let us assume that the rotor has 50 teeth. Using full step and half step modes, we can advance the teeth alignment between rotor and stator by the number of rotor teeth times the number of phases times 2. The step angle of an hybrid PM stepper motor is determined by the number of electrical phases (N_{ph}) and the number of rotor teeth (N_r),

$$\theta_{step} = \frac{360°}{2 \cdot N_r \cdot N_{ph}} \tag{8.266}$$

The number of steps of a PM stepper motor per revolution is the number of electrical phases times the number of rotor teeth times two,

$$N_{step} = 2 \cdot N_r \cdot N_{ph} \tag{8.267}$$

The switching of power transistors from one state to another from ON-to-OFF and OFF-to-ON state instantaneously results in an instantaneous change in the magnetic field. The motor behaves like a mass-spring system. We can take the concept of half step mode further to ratio (smoothly change) the current in phases instead of making transitions from full ON to full OFF states. This will result in smoother motion and electronically controlled finer step sizes. This is the main operating principle of the so-called "micro stepping drives" (Figure 8.50d). As a result of the smoother current switching between phases, the torque acting on the rotor shaft between steps is smoother, and the step motion of the rotor is less oscillatory. In addition, microstepping reduces the resonance and step loss problems associated with step motors which are operated full step and half step current control drives.

The stator windings typically form two phases. If the step motor is to be operated by a *unipolar drive*, each winding must be center tapped to ground and positive voltage is connected to both ends (Figure 8.51a). The opposite connection can also be made: the center tap is connected to the DC supply voltage, and the other two ends are connected to the ground. Per winding, only one of the connections at a time is switched ON in order to control the direction of the current and hence the generated electromagnetic pole type (north or south). Only half of a particular winding is used at a switched ON state (Figure 8.51a). If the step motor is to be operated by a *bipolar drive*, then the current direction can be controlled by the drive and all of each winding is used at a switched ON state (Figure 8.51b,c). Some step motors are wound with two separate windings per pole, and hence can be driven by a unipolar or bipolar drive by appropriately terminating the winding ends. For unipolar

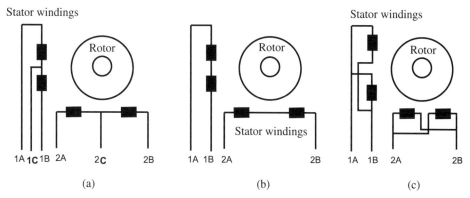

FIGURE 8.51: Stator winding connections of a two-phase, four winding step motor:
(a) unipolar drive configuration with center tapped connection, (b) bipolar drive configuration with series connection, (c) bipolar drive configuration with parallel connection.

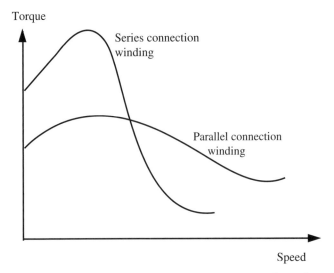

FIGURE 8.52: Steady-state torque and speed curves for series and parallel connected windings.

operation, two windings are connected in series and center tapped to ground. For bipolar operation, two windings can be either parallel or serial connected and the current direction is controlled by the drive (Figure 8.51b,c). The implications of connecting the windings in series or in parallel in terms of the steady-state torque-speed characteristics of the motor are shown in Figure 8.52.

Step loss and resonance are two fundamental problems inherent in stepper motors. If the phase currents are switched so fast that the rotor cannot keep up with it, then step loss occurs. The only way to correct the problem is to use a position sensor to detect the step loss and command additional steps to compensate for it. Therefore, the maximum switching frequency is limited. Higher switching frequencies can be used only if the rotor is accelerated to high rotational speeds under controlled acceleration profiles (Figure 8.53). As the switching frequency increases, there is less time for the phase current to develop due to the electrical time constant of phase winding ($\tau = L/R$). Therefore, at higher switching frequency the torque capacity of the motor drops compared to the torque capacity at lower speeds.

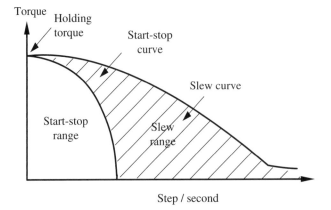

FIGURE 8.53: Torque capacity and step rate of stepper motors.

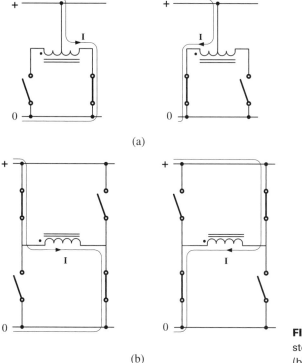

(a)

(b)

FIGURE 8.54: Drive types for step motor: (a) unipolar drive, (b) bipolar drive.

8.6.2 Step Motor Drives

Drive controls the direction and magnitude of current in each phase, and hence controls the direction and magnitude of torque. There are two types of drives which must be matched to the winding type of a step motor (Figure 8.54),

1. unipolar drive,

2. bipolar drive.

Unipolar drives require the motor windings to be center tapped (Figures 8.54a, 8.51a). By turning ON the current at one end of the winding versus the other end, the direction of current in the winding is changed, hence the flux direction. When a phase winding is energized, only 50% of it is used by a unipolar drive. Unipolar drives require two power transistor switches per phase. In a unipolar center-tapped motor, each phase has three leads: two sides and one center tap lead (Figure 8.51a).

Bipolar drives use an H-bridge (four power transistor switches) per phase. A bipolar drive for a two-phase step motor has two H-bridges ($2 \times 4 = 8$ transistors). The direction of the current is changed by controlling which two pairs of the H-bridge switches are turned ON. Although it requires twice as many power switches, it makes 100% use of the conductors of the energized coil. A stepper motor with bipolar winding has two leads per phase (Figure 8.51b,c). Some motors are wound such that they can be configured to be driven by either a unipolar or bipolar drive. In this case, each winding has two identical windings with four leads. Depending on how the leads are connected, the motor can be made to operate with a unipolar or a bipolar type drive.

The microstepping drive was developed to solve resonance problems. In microstepping, the current is not simply switched ON/OFF between phases, but gradually changed.

As a result, the current vector does not change in sudden jumps, but instead it is smoothly changed. Resonance is greatly reduced, and resolution is sharply increased, because there are many more equilibrium points as the current is ratioed between multiple phases. Microstepping is usually performed with two bidirectional phases. These two phases have similar torque equations as a brushless DC motor. Let us assume that the phase current-torque gain is a sinusoidal function of the rotor position, θ_r. The two phases are 90 degrees apart. Further, let us assume that the current to each phase is controlled with a sinusoidal function in order to implement the microstepping mode of current control. The phase current and torque relationships are,

$$T_a = K \cdot i_a \cdot \sin(\theta_r) = K\, i \cos(\theta_c) \cdot \sin(\theta_r) \tag{8.268}$$

where

$$i_a(t) = i \cdot \cos(\theta_c(t)) \tag{8.269}$$
$$T_b = K \cdot i_b \cdot \cos(\theta_r) = K\, i\, (-\sin(\theta_c)) \cdot \cos(\theta_r) \tag{8.270}$$

where

$$i_b(t) = i \cdot (-\sin(\theta_c(t))) \tag{8.271}$$
$$T_{total} = T_a + T_b = K\, i\, \sin(\theta_c - \theta_r) \tag{8.272}$$

where θ_r is the real position and θ_c is the commanded position, i_a, i_b are currents in phase a and b, K is a proportionality constant. In equilibrium, these two positions are identical, $\theta_c = \theta_r$, so the total torque equation becomes:

$$T_{total} = K\, i\, \sin(\theta_c - \theta_r) = 0 \tag{8.273}$$

and when $\theta_c - \theta_c = 90°$, the torque is maximized for a given current.

The step motor torque is developed as the rotor is forced to move away from the equilibrium position for a given state of winding current. The equilibrium position is the one that minimizes the magnetic reluctance (maximizes the inductance). When the rotor is at the exact position that minimizes the reluctance, the torque is zero. The holding torque is developed as the rotor position deviates from the ideal position due to load or current commutation in the windings.

Example: Unipolar Integrated Circuit (IC) Drive for Step Motors
An integrated circuit (IC) drive which can be used to drive a unipolar stepper motor is shown in Figure 8.55. The SLA7051M (by Philips Semiconductors) integrates two blocks of Figure 8.48, the translator and the power switch set. The translator section is made of a low-power CMOS logic circuit and handles the logic for sequencing, direction, full and half step operation. The translator decides on the firing sequence of windings for full step (AB, BC, CD, DA in forward or reverse direction) or for half step (A, AB, B, BC, C, CD, D, DA in forward or reverse direction) as a function of the input signals at the terminals STEP (also named Clock), FULL/HALF, CW/CCW. At each low to high transition of the STEP input signal, the translator checks the state of FULL/HALF pin to determine if *full step mode* or *half step mode* is commanded (high for full step, low for half step mode), and the CW/CCW input to determine the *direction* command. Then it decides which one or two of the windings to be fired according to the sequence defined above (AB, BC, CD, DA or A, AB, B, BC, C, CD, D, DA).

The PWM current controlled power stage uses FET output and can handle up to 2 A and 46 V per phase. The maximum current is controlled by the reference voltage and

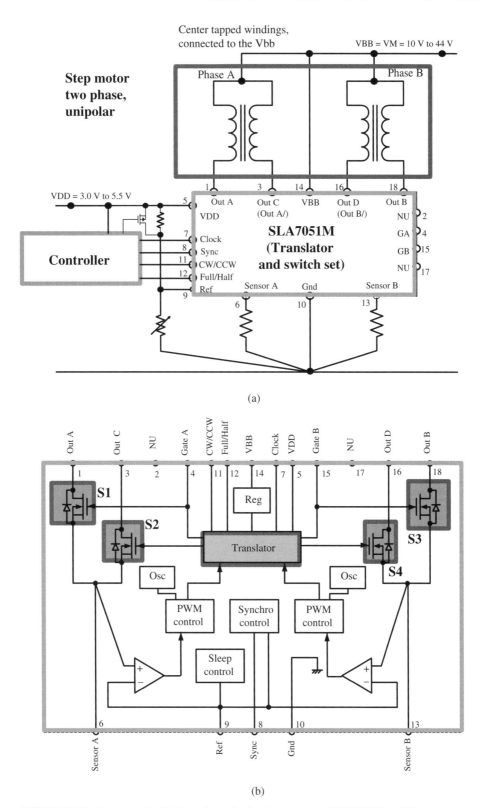

FIGURE 8.55: An example IC driver for unipolar step motors: SLA7051M by Philips Semiconductors. The unipolar step motor driver IC has the translator and output power stage integrated into one small IC package. *Source:* Adapted from Philips.

current sense resistors for each phase (pins REF, SENSA, SENSB). The output stages control the phase currents via terminals OUTA, OUTA\, OUTB, OUTB\ of the IC chip. Notice that this IC drive is capable of controlling only FULL and HALF step modes and does not support the microstepping mode. For driving step motors which will require power dissipation more than 20 W, an external heat sink must be added to the FET power transistor stage.

Example: Bipolar Integrated Circuit (IC) Drive for Step Motors The LM18293 (by National Semiconductor) integrates two H-bridges on the chip with 1 A continuous (2 A peak) current capacity per channel. Therefore, it can be used to drive a two-phase bipolar step motor (Figure 8.56). It can also be used to drive two separate DC brush motors bidirectionally. The maximum DC supply voltage is 36 V. A separate enable pin (ENABLE1 and ENABLE2) controls one of the two H-bridges. If the ENABLE pins

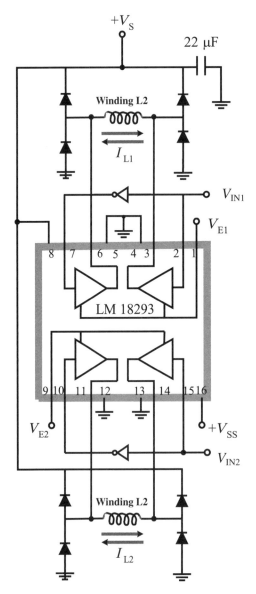

FIGURE 8.56: Integrated circuit drive chip (LM18293 by National Semiconductor) for bipolar step motors or brush-type DC motors. *Source:* Texas Instruments.

are floating (not connected), it is assumed that the H-bridges are enabled. Input channels (TTL level signals) INPUT1 and INPUT2 control the direction and magnitude of current in one of the H-bridges. Once an H-bridge is enabled, the current is controlled by the pair of INPUT (1/2 or 3/4) pins for that H-bridge. By modulating the pulse width of the INPUT lines, the average current can be controlled by an external microcontroller.

8.7 LINEAR MOTORS

In principle, all of the rotary electric motors can be designed and manufactured as linear electric motors by un-rolling the cylindrical shape of the stator and rotor into linear form. Such linear motors have rectangular cross-sections and are called *flat linear motors*. In addition, the un-rolled linear motor can be rolled back around the axis of linear motion to obtain a *tubular* linear motor construction. Such linear motors have circular cross-sections (Figure 8.57).

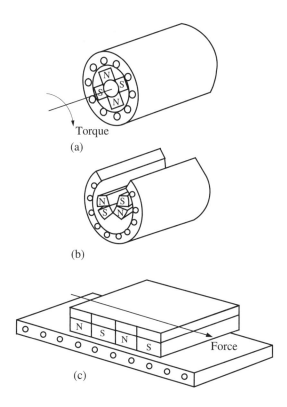

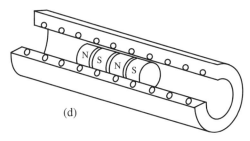

FIGURE 8.57: Principle of linear motor construction by un-rolling a rotary motor: (a) a rotary permanent magnet type electric motor (i.e., brushless DC), (b) un-rolling concept, (c) flat type linear brushless DC motor, and (d) tubular type linear motor by rolling again the flat type linear motor concept around the axis of its force.

The electromagnetic force generation between the two components follows the same physical principles as in the case of their rotary counterparts. Figure 8.57 shows the basic principle of a linear motor by un-rolling the cylindrical shape to a flat shape. The number of actively controlled stator phases the same. The commutation of the current in each phase is based on the cyclic linear distance (pitch) of the permanent magnet dimensions instead of the rotary angle. In a two pole rotary brushless DC motor, the commutation cycle is 360 degrees, whereas the commutation cycle in a linear brushless motor is the distance between two consecutive pole pairs (i.e., the distance between two north or two south pole magnets, total length of a north and a south pole magnet). In general, the same amplifiers used in the rotary version are used for the linear version of the motor. The feedback sensor (if used for commutation) is a linear displacement sensor instead of a rotary displacement sensor, that is Hall effect sensors are used to detect the relative position of stator with respect to the rotor within a cyclic distance of pitch. Similarly, linear encoders may be used in place of rotary encoders for position sensing.

Linear brushless DC motors have three basic types: iron core, ironless (air core), and slotless. In a brushless DC tubular linear motor, the permanent magnets are rolled into a cylindrical shape around the axis perpendicular to the rotary motor axis. The stator has three phases and is wound around the rotor. The controller and amplification stages of a linear motor are identical to those used in the rotary version. The commutation in the amplifier is based on the cyclic pitch distance of the magnet pairs in the rotor, instead of the angular position in the case of rotary motors.

The dynamic model of a linear DC motor is identical to that of rotary DC motors. The electrical dynamics of the motor is

$$V_t(t) = R \cdot i(t) + L \cdot \frac{di(t)}{dt} + K_e \cdot \dot{x}(t) \tag{8.274}$$

where $V_t(t)$ is the terminal voltage, R is the winding resistance, L is the self-inductance, k_e is the back EMF gain of the motor. The net electrical power converted to mechanical power is

$$P_e(t) = V_{bemf}(t) \cdot i(t) \tag{8.275}$$
$$= K_E \cdot \dot{x}(t) \cdot i(t) \tag{8.276}$$
$$= K_T \cdot \dot{x}(t) \cdot i(t) \tag{8.277}$$
$$= F(t) \cdot \dot{x}(t) \tag{8.278}$$
$$= P_m(t) \tag{8.279}$$

assuming 100% efficiency in converting the electrical power to mechanical power,

$$P_m(t) = F(t) \cdot \dot{x}(t) \tag{8.280}$$
$$= P_e(t) \tag{8.281}$$
$$= V_{bemf}(t) \cdot i(t) \tag{8.282}$$

where the force–current relationship is

$$F(t) = K_T \cdot i(t) \tag{8.283}$$

which also shows the force/torque gain is equal to the back EMF gain. The gain K_t is a function of the magnetic flux (flux density times the cross-sectional area linking the magnetic field to the conductors) and the number of turns of the coil. In other words, it is a function of the magnetic flux at the operating point of the permanent magnets and its size, plus the number of coil turns that links the flux. For a practical motor, the solution of flux is best obtained by finite element based software tools. Therefore, the force gain is

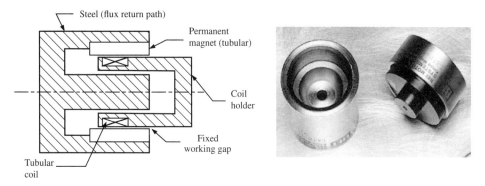

FIGURE 8.58: Voice coil actuator operating principle and components.

determined by the solution of FEA based software instead of analytical solutions which are only possible for simple and idealized motor geometries.

A *voice coil actuator* (also called a moving coil actuator) is made of a tubular permanent magnet (PM) and a coil winding (Figure 8.58). The interaction between the current carrying coil assembly and PM generates the linear force. In principle, this is identical to the brush-type DC motors. At any given instant in time, both motor action (force generation) and generator action (back EMF voltage) are in effect,

$$F = k \cdot l \cdot N \cdot B \cdot i = K_F \cdot i \qquad (8.284)$$
$$V_{bemf} = k \cdot l \cdot N \cdot B \cdot \dot{x} = K_E \cdot \dot{x} \qquad (8.285)$$

where F is linear force, V_{bemf} is the back EMF voltage, l is the length of the winding and N is the number of turns of the coil, B is the magnetic field strength across the air gap between the rotor and stator, i is the current in the coil, and $\dot{x}$ is the linear speed of the rotor. As long as the geometric overlap between the rotor and stator (coil and PM) is the same, the force is essentially independent of the displacement of the rotor and only a function of the current. In order to provide such a force–current–displacement relationship, the axial length of the PM and the coil must be different (one longer than the other). The current in the coil winding is non-commutated. There are no commutator-brush components. Only the direction and magnitude of the current is controlled, that is using an H-bridge amplifier which is the same type used for a brush-type DC motor. Moving coil actuators are precision motion versions of the solenoid-plunger or audio-speaker designs. Voice coil actuators typically have a small travel range (i.e., microns to a few inches range) and very high bandwidth.

8.8 DC MOTOR: ELECTROMECHANICAL DYNAMIC MODEL

The most commonly used model for a direct current (DC) electric motor is shown in Figure 8.59. This dynamic model includes electrical, electrical to mechanical power conversion, and mechanical dynamic relations.

The electrical relation between terminal voltage, current, and rotor speed is

$$V_t(t) = L_a \frac{di(t)}{dt} + R_a i(t) + k_e \dot{\theta}(t)$$

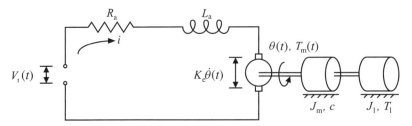

FIGURE 8.59: DC motor dynamic model.

where $k_e\dot{\theta}(t)$ term is the voltage generated by the back electromotive-force (back EMF) as a result of the generator action, L_a, R_a are inductance and resistance of the winding, respectively. The electrical to mechanical power conversion (current to torque) is given by

$$T_m(t) = K_T\, i(t)$$

where, K_T is the torque gain, and T_m is the torque generated by the motor. Finally the mechanical relation between torque, inertia, and an other load is given by

$$T_m(t) = (J_m + J_l)\,\ddot{\theta} + c\dot{\theta}(t) + T_l(t)$$

where J_m is rotor inertia, J_l is load inertia, c is damping constant, T_l is load torque. From these three basic relationships, we can derive various transfer functions, such as the transfer function between terminal voltage to motor speed or motor position, or the transfer function from armature current to motor speed. Physically, an amplifier manipulates the motor terminal voltage in order to control the motor motion. This voltage control may be based on current feedback, voltage feedback or both.

Let us derive the transfer function from motor terminal voltage to motor speed. Taking the Laplace transform of the above equations for zero initial conditions,

$$V_t(s) = (L_a s + R_a)i(s) + k_e\dot{\theta}(s) \tag{8.286}$$

$$\rightarrow i(s) = \frac{1}{L_a s + R_a}[V_t(s) - k_e\dot{\theta}(s)] \tag{8.287}$$

$$T_m(s) = K_T i(s)$$

$$(J_T s + c)\dot{\theta}(s)y = T_m(s) - T_l(s)$$

where $J_T = J_m + J_l$.

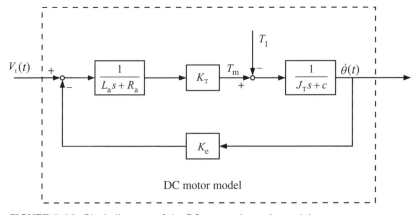

FIGURE 8.60: Block diagram of the DC motor dynamic model.

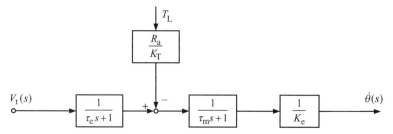

FIGURE 8.61: Block diagram of the relationship between motor terminal voltage and load torque to motor speed for a typical DC servo motor.

The transfer function describing the effect of terminal voltage and load torque on the motor speed can be found as (Figure 8.61)

$$\dot{\theta}(s) = \frac{K_T}{(J_t s + c)(L_a s + R_a) + K_T k_e} V_t(s) - \frac{(L_a s + R_a)}{(J_t s + c)(L_a s + R_a) + K_T k_e} T_l(s)$$

The transfer function from motor terminal voltage to motor speed is given by

$$\frac{\dot{\theta}(s)}{V_t(s)} = \frac{K_T}{(J_t s + c)(L_a s + R_a) + K_T K_e} \tag{8.288}$$

$$= \frac{K_T}{J_T L_a s^2 (L_a c + J_T R_a) s + (c R_a + K_T K_e)} \tag{8.289}$$

$$= \frac{K_T}{J_T L_a s^2 + \left(\frac{L_a c + J_T R_a}{J_T L_a}\right) s + \left(\frac{c R_a + K_T K_e}{J_T L_a}\right)} \tag{8.290}$$

The poles of the transfer function are given by

$$s^2 + \left(\frac{L_a c + J_T R_a}{J_T L_a}\right) s + \left(\frac{c R_a + K_T K_e}{J_T L_a}\right) = 0$$

Normally, this equation has two complex conjugate roots.

Special Case: DC Servo Motors DC servo motors have very low inductance (L small), and very low damping (c small). Using these facts for the DC servo motors case the transfer function can be approximated as

$$\frac{\dot{\theta}(s)}{V_t(s)} \simeq \frac{\frac{K_T}{J_T L_a}}{s^2 + \left(\frac{R_a}{L_a}\right) s + \left(\frac{K_T K_e}{J_T L_a}\right)}$$

where the poles are given by

$$p_{1,2} = -\frac{R_a}{2L_a} \pm \frac{\sqrt{\left(\frac{R_a}{L_a}\right)^2 - 4\left(\frac{K_T K_e}{J_T L_a}\right)}}{2} \tag{8.291}$$

$$= -\frac{R_a}{2L_a} \pm \frac{\sqrt{4K_T k_e J_T \left(\frac{R_a^2 J_T}{4K_T K_e} - L_a\right)}}{2L_a J_T} \tag{8.292}$$

$$\simeq -\frac{R_a J_T}{2L_a J_T} \pm \frac{R_a J_T \left(1 - \frac{2L_a K_T K_e}{R_a^2 J_T}\right)}{2L_a J_T} \tag{8.293}$$

where we used the approximation

$$\sqrt{1-x} \simeq 1 - \frac{x}{2}; \quad for \quad x \ll 1$$

The poles are

$$p_1 = -\frac{K_T K_e}{J_T R_a} \tag{8.294}$$

$$p_2 = -\frac{-2R_a J_T + \left(\frac{2L_a K_T K_e}{R_a}\right)}{2L_a J_T} \tag{8.295}$$

$$= -\frac{R_a}{L_a} + \frac{K_T K_e}{J_T R_a} \tag{8.296}$$

Since $\frac{R_a}{L_a} \gg \frac{K_T K_e}{J_T R_a}$ the second pole is approximately

$$p_2 \simeq -\frac{R_a}{L_a} \tag{8.297}$$

The motor terminal voltage to motor speed transfer function can be approximated as

$$\frac{\dot{\theta}(s)}{V_t(s)} = \frac{\left(\frac{K_T}{J_T L_a}\right)}{(s - p_1)(s - p_2)} \tag{8.298}$$

$$= \frac{\left(\frac{1}{K_e}\right)}{(\tau_m s + 1)(\tau_e s + 1)} \tag{8.299}$$

where

$$\tau_m = -\frac{1}{p_1} = \frac{J_T R_a}{K_T k_e} \quad \text{mechanical time constant} \tag{8.300}$$

$$\tau_e = -\frac{1}{p_2} = \frac{L_a}{R_a} \quad \text{electrical time constant} \tag{8.301}$$

and the mechanical time constant τ_m is much larger than the electrical time constant τ_e for most motors. The motor speed is also influenced by disturbance or load torque. The transfer function including the load torque can be derived as follows

$$\dot{\theta}(s) = \frac{\frac{K_T}{J_T L_a}}{s^2 + \left(\frac{R_a}{L_a}\right)s + \left(\frac{K_T K_e}{J_T L_a}\right)} V_t(s) - \frac{\left(\frac{L_a s + R_a}{J_T L_a}\right)}{s^2 + \left(\frac{R_a}{L_a}\right)s + \left(\frac{K_T K_e}{J_T L_a}\right)} T_1(s) \tag{8.302}$$

$$= \frac{\frac{1}{K_e}}{(\tau_m s + 1)(\tau_e s + 1)} V_t(s) - \frac{(\tau_e s + 1)R_a \frac{1}{K_T} \frac{K_T}{J_T L_a}}{s^2 + \left(\frac{R_a}{L_a}\right)s + \left(\frac{K_T K_e}{J_T L_a}\right)} T_1(s) \tag{8.303}$$

$$= \frac{\frac{1}{K_e}}{(\tau_m s + 1)(\tau_e s + 1)} V_t(s) - \frac{\frac{1}{K_e} \frac{R_a}{K_T}(\tau_e s + 1)}{(\tau_m s + 1)(\tau_e s + 1)} T_1(s) \tag{8.304}$$

$$\dot{\theta}(s) = \frac{\frac{1}{K_e}}{(\tau_m s + 1)(\tau_e s + 1)} V_t(s) - \frac{\frac{1}{K_e} \frac{R_a}{K_T}}{(\tau_m s + 1)} T_1(s) \tag{8.305}$$

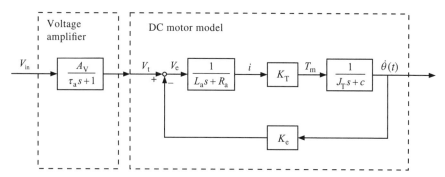

FIGURE 8.62: Block diagram of transfer functions for a DC servo motor driven by a voltage amplifier.

The transfer function from the terminal voltage and load torque to the motor speed in terms of time constants and dc gains are shown in block diagram form in Figure 8.61.

8.8.1 Voltage Amplifier Driven DC Motor

If a voltage amplifier is used to drive the motor, the transfer function between voltage amplifier input and motor speed is given by (Figure 8.62)

$$\frac{\dot{\theta}(s)}{V_{in}(s)} = \frac{\dot{\theta}(s)}{V_t(s)} \frac{V_t(s)}{V_{in}(s)} = G_{amp}(s) G_{motor}(s)$$

where the voltage amplifier is represented by a first-order filter model,

$$\frac{V_t(s)}{V_{in}(s)} = G_{amp}(s) = \frac{A_v}{(\tau_a s + 1)} \tag{8.306}$$

8.8.2 Current Amplifier Driven DC Motor

In the majority of cases, the DC motor terminal voltage is controlled by an amplifier which regulates armature current (Figure 8.63). Directly regulating current is essentially directly regulating the torque generated by the motor, since generated torque is proportional to the current. The transfer function of the current amplifier plus DC motor combination from the

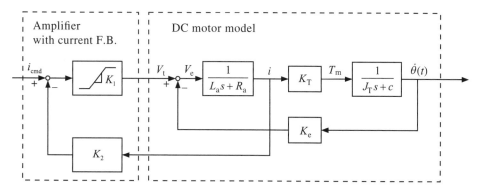

FIGURE 8.63: Block diagram of DC motor and amplifier using current feedback.

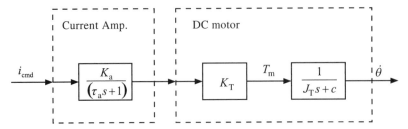

FIGURE 8.64: Current amplifier plus DC motor transfer function from commanded current to motor speed.

amplifier input (commanded current, $i_{cmd}(t)$) to the motor speed can be derived as follows (neglecting the effect of amplifier terminal voltage saturation, Figure 8.64)

$$\frac{\dot{\theta}(s)}{i_{cmd}(s)} = \frac{K_1 \frac{K_T}{(L_a s + R_a)(J_T s + c) + K_e K_T}}{1 + K_1 \frac{K_T}{(L_a s + R_a)(J_T s + c) + K_e K_T} K_2 \left(\frac{J_T s + c}{K_T}\right)} \tag{8.307}$$

$$= \frac{\frac{K_1 K_T}{L_a J_T}}{s^2 + \left(\frac{L_a c + R_a J_T s + K_1 K_2 J_T}{L_a J_T}\right) s + \left(\frac{K_e K_T + K_1 c + K_1 K_2 c}{L_a J_T}\right)} \tag{8.308}$$

$$= \frac{K}{s^2 + bs + c} \tag{8.309}$$

$$= \frac{K_a}{(\tau_a s + 1)} \frac{K_T}{(J_T s + c)} \tag{8.310}$$

If we consider the transfer function between armature current and motor speed;

$$T_m(t) = K_T i(t) \tag{8.311}$$

$$T_m(t) = J_T \ddot{\theta} + c\dot{\theta} - T_l(t) \tag{8.312}$$

$$\dot{\theta}(s) = \frac{K_T}{(J_T s + c)} \cdot i(s) - \frac{1}{(J_T s + c)} \cdot T_l(s) \tag{8.313}$$

8.8.3 Steady-State Torque-Speed Characteristics of DC Motor Under Constant Terminal Voltage

Consider the electrical and electrical to mechanical power conversion relations for a DC motor.

$$V_t(t) = L_a \frac{di}{dt} + R_a i + K_e \dot{\theta}(t) \tag{8.314}$$

$$T_m(t) = K_T i(t) \tag{8.315}$$

In steady-state the effect of L_a will be zero. If we set $L_a = 0$ for steady-state analysis, the torque-speed terminal voltage relationship is given by,

$$T_m(t) = \frac{K_T}{R_a} \cdot V_t(t) - \frac{K_T K_e}{R_a} \cdot \dot{\theta}(t) \tag{8.316}$$

This is a linear relationship of the type

$$y = -a\,x + b \tag{8.317}$$

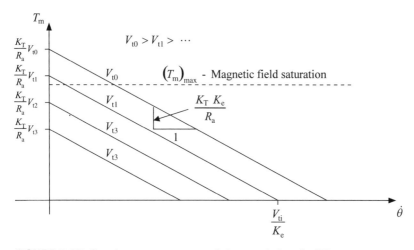

FIGURE 8.65: Steady-state torque-speed characteristics of a DC motor.

Let us consider this equation for various constant terminal voltage values, V_{ti} (Figure 8.65). The steady-state torque-speed curve has a negative slope of $\frac{K_T K_e}{R_a}$ for a given constant terminal voltage. A given motor maximum torque at stall will saturate at the magnetic field saturation point and the torque will not increase even if terminal voltage is increased beyond a certain voltage value. For a given constant terminal voltage, as the speed ($\Delta\dot\theta$) increases, the torque capacity of the motor decreases by $\left(\frac{K_T K_e}{R_a}\right)\Delta\dot\theta$.

8.8.4 Steady-State Torque-Speed Characteristic of a DC Motor Under Constant Commanded Current Condition

When we consider a DC motor driven by a current amplifier, we need to consider the following additional relation,

$$V_t(t) = K_1(i_{cmd}(t) - K_2\, i(t)); \; V_t(t) \le V_{t\,max}$$

when the amplifier saturates, $V_t(t) > V_{t\,max}$

$$V_t(t) = V_{r,t\,max}$$

Substituting the above current feedback control to the motor dynamics, and letting $L_a \approx 0.0$,

$$V_t(t) = L_a\frac{di}{dt} + R_a\, i + K_e\,\dot\theta(t) \tag{8.318}$$

$$K_1(i_{cmd}(t) - K_2 i(t)) = L_a\frac{di}{dt} + R_a\, i + K_e\,\dot\theta(t) \tag{8.319}$$

Solve for $i(t)$ in terms of $i_{cmd}(t)$ and $\dot\theta(t)$, and substitute in the torque-current equation to obtain the steady-state torque versus speed relationship under constant current condition,

$$T_m(t) = K_T\, i(t) \tag{8.320}$$

$$T_m = \frac{K_T K_1}{R_a + K_1 K_2}\, i_{cmd} - \frac{K_T K_e}{R_a + K_1 K_2}\,\dot\theta \tag{8.321}$$

When the amplifier is commanded a constant current, (i_{cmd}), the speed torque characteristic is linear with negative slope. However, the slope is much flatter than the case of the DC

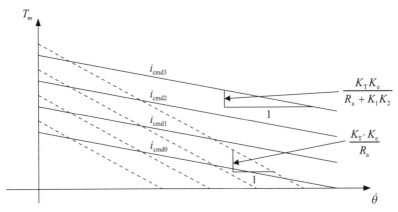

FIGURE 8.66: Speed-torque characteristics of a DC motor in steady state with current controlled amplifier.

motor with a constant terminal voltage (Figure 8.66). In other words, the steady-state torque produced by a DC motor under constant current conditions (i.e., controlled by a current feedback amplifier) decreases with increasing speed at a much slower rate that the case when the motor is controlled by a voltage amplifier under constant terminal voltage condition.

Example Consider a DC motor with the following parameters, $K_T \text{N m/A} = K_E \text{V/}$ $(\text{rad/s}) = 6.7 \times 10^{-2}$, $R_a = 0.5\,\Omega$, $L_a = 2\,\text{mH}$, $J_m = 4.8 \times 10^{-5}\,\text{kg m}^2$, $J_1 = J_m$. Determine the mechanical and electrical time constant of the motor (consider the case that the motor and the load inertia are rigidly connected). If the current loop control gains are $K_1 = 10.0$ and $K_2 = 1.0$, determine the speed-torque slopes under constant voltage and constant current control conditions.

The mechanical and electrical time constants of the motor with the load connected

$$\tau_m = \frac{J_T R_a}{K_T K_E} = \frac{9.6 \times 10^{-5} \cdot 0.5}{6.7^2 \times 10^{-4}} \tag{8.322}$$

$$= 0.010\,\text{s} = 10\,\text{ms} \tag{8.323}$$

$$\tau_e = \frac{L_a}{R_a} = \frac{2 \times 10^{-3}\,\text{H}}{0.5\,\Omega} = 4 \times 10^{-3}\,\text{s} = 4\,\text{ms} \tag{8.324}$$

The slope of the torque-speed curve in steady-state under constant voltage and constant current conditions is,

$$slope_v = \frac{K_T K_E}{R_a} \tag{8.325}$$

$$= \frac{6.7^2 \times 10^{-4}}{0.5} \tag{8.326}$$

$$= 89.7 \times 10^{-4}\,\text{N} \cdot \text{m/(rad/s)} \tag{8.327}$$

$$= 0.0857\,\text{N} \cdot \text{m/rpm} \tag{8.328}$$

$$slope_i = \frac{K_T K_E}{R_a + K_1 K_2} \tag{8.329}$$

$$= \frac{6.7^2 \times 10^{-4}}{0.5 + 10} \tag{8.330}$$

$$= 4.27 \times 10^{-4}\,\text{N} \cdot \text{m/(rad/s)} \tag{8.331}$$

$$= 0.00407 \, \text{N} \cdot \text{m/rpm} \tag{8.332}$$

The ratio of the slope change in the torque-speed curve is

$$\frac{slope_i}{slope_v} = \frac{4.27}{89.7} = \frac{1}{21} \tag{8.333}$$

For a given unit speed change, the torque variation under the current controlled amplifier operation is 21 times less than that under the voltage controlled amplifier.

8.9 PROBLEMS

1. Consider the solenoid shown in Figure 8.17. Assume that the permeability of the iron core is much larger that the permeability of the air gap (x). Hence, let us neglect the reluctance of the path in the iron core.

(a) Draw the equivalent electromagnetic circuit diagram.
(b) Derive the relationship for the inductance as a function of the air gap distance x which is variable. Let the air gap cross-sectional area $A_g = 100 \, \text{cm}^2$, number of turns of the coil $N = 250$. Plot the inductance as a function of the air gap x for $0.0 \, \text{mm} \geq x \geq 10.0 \, \text{mm}$.

2. Consider the linear motion mechanism shown in Figure 3.3. Draw a block diagram of the closed loop position control system components for this motion control system. Discuss the advantages and disadvantages of using a (i) DC motor, (ii) stepper motor, (iii) vector controlled AC induction motor as the actuator to provide the mechanical power to the ball-screw mechanism. Assume that the necessary power level is under 1.0 kW.

3. Consider a DC motor (or equivalent) motion control system. Assume that the DC bus voltage available is $V_s = 90 \, \text{V DC}$. The back EMF constant of the motor is $K_e = 20 \, \text{V/krpm}$. The nominal terminal resistance of the motor stator windings is $R = 10 \, \Omega$.

(a) Determine the maximum no-load speed and the maximum torque capacity at stall (zero speed) of the motor in steady-state. Discuss the factors that detemine the peak torque and RMS torque capacity of the motor and predict peak and RMS torque capacity of the motor.
(b) Assume that the DC motor is controlled by the velocity mode amplifier and speed sensor feedback (i.e., tachometer). The amplifier sends a voltage command to the motor proportional to the speed error. Input speed is commanded with a 0–10 V DC voltage source. Determine the sensor gain and amplifier gain so that the closed loop system has good closed loop response and that a 0–10 V DC command signal results in proportional output speed in the speed range of the motor.

4. Consider a hybrid PM stepper motor.

(a) What is the difference between the half step and full step mode. What is the main advantage and disadvantage of the half stepping mode?
(b) Discuss how microstepping is accomplished and its advantages/disadvantages.
(c) Discuss the major performance differences between step motors and brushless DC motors.

5. Consider a stepper motor with two phases (Figure 8.67). Phase 1 is wound with two identical coils. Phase 2 is also wound with two identical coils. Therefore, there are eight wires that come out from the stator winding, four for each phase.

(a) Connect the phase terminal wires in such a way that the motor can be used as a unipolar wound motor and draw a schematics for the amplifier circuit of the motor windings.
(b) Connect the phase terminal wires in such a way that the motor can be used as a bipolar wound motor with *series* connected windings (two identical windings for each phase are connected in series). Draw the amplifier circuit and its connections to motor windings.

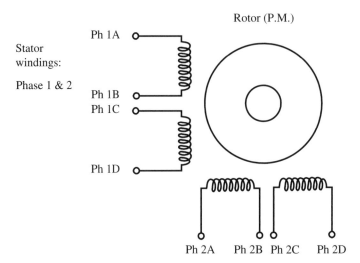

FIGURE 8.67: Two-phase stepper motor. Each phase is wound with two identical windings. Each phase has four terminal wires. The phase windings can be terminated to configure the motor for unipolar or bipolar operation.

(c) Connect the phase terminal wires in such a way that the motor can be used as a bipolar wound motor with *parallel* connected windings (two identical windings for each phase are connected in parallel). Draw the amplifier circuit and its connections to motor windings.

(d) What is the main performance difference between bipolar series and bipolar parallel configuration of the motor?

6. Consider the drive circuit for a three phase brushless DC motor shown in Figure 8.32. Assume that the rotor has one south and one north pole (like the one shown in Figure 8.29c). Let us focus on the ON/OFF state of six power transistors $(Tr_1, \ldots, Tr_6)$ in order to generate torque in the forward and reverse direction. Determine the sequence of the ON/OFF conditions of the transistors for forward and reverse torque generation. Assume a nominal rotor position, hence the magnetic field of the rotor, as the beginning of the commutation cycle. Hint: make a table with columns being the transistors 1 through 6 states, the position of the rotor at the beginning of each transistor state switching sequence, and the magnetic field vector generated by the stator at a particular switching pattern. The rows should be six different transistor states for forward torque generation in sequence and similar information for the reverse torque generation. In an actual motor, the power transistors would be controlled by a PWM signal to generate a sinusoidal or trapezoidal current as a function of rotor position as opposed to the ON/OFF switching discussed in this problem. Nonetheless, the ON/OFF switching of transistors is still useful in understanding the current commutation on DC brushless motors.

7. The torque versus rotor position under constant current conditions for a DC brush-type motor (Figure 8.28). Assume that for each coil connected to the commutators, the torque is a sinusoidal function of rotor position under constant current, as shown in the same figure. Plot the torque versus rotor position for one revolution, under constant current, for the following cases: a) two coils only (and two commutators), b) four coils (and four commutators), c) eight coils, and d) 16 coils. What is the benefit of having a large number of commutator segments? Assume that all coils are symmetrically distributed, that is in the four coil case, the second coil is 180° electrically (1/4 revolution mechanical degrees) out of phase with the first coil, in the eight coil case each coil is 1/8 of a revolution out of phase from the previous one). Use of MATLAB®/Simulink® for the plot and calculations is recommended.

8. Consider a DC motor, current amplifier, a closed loop PD type controller, and a position feedback sensor (Figures 8.59, 8.64, 2.42). Consider that the motor dynamics is described by its

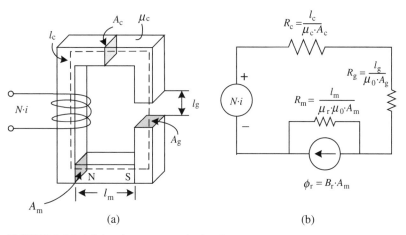

FIGURE 8.68: (a) An electromagnetic circuit example involving a permanent magnet, a coil wound over a core which has an air gap. (b) Magnetic circuit model.

rotary inertia (J_m kg m^2) and current to torque gain (K_t Nm/A). Let the current amplifier be modeled by a static gain of voltage command to current gain of K_a A/V. The PD controller has gains K_p and K_d which define the relationship between the position and velocity error and control signal (voltage command). Let the position feedback sensor also be represented by its gain and neglect any dynamic delays, K_s Counts/Rad. a) Let $J_m = 10^{-4}$ kg m^2, $K_t = 0.10$ N m/A, $K_a = 2.0$ A/V, $K_f = 2000/(2\pi)$ Counts/Rad, $K_p = 0.02$ V/Counts, $K_d = 10^{-4}$ V/(Counts/s). Find the loop transfer function (the transfer function from the error signal to the output of the sensor (sensor signal), then determine the frequency at which the magnitude of the loop transfer function is equal to unity. At that frequency, determine the phase angle of the loop transfer function. b) Similarly, determine the frequency at which the phase angle of the loop transfer function is equal to 180 degrees (if such a finite frequency exists) and find the magnitude of the loop transfer function at that frequency. In a real DC motor control system, such a frequency does exist even if the above analysis may indicate that such a frequency is not finite. Discuss why in real hardware such a frequency is finite (Hint: neglected filtering effects and time delays in the above analysis compared to the real system. A pure time delay can be modeled in the frequency domain as e^{-jwt_d}, where t_d is the time delay. It adds phase lag to the loop transfer function).

9.   Consider the electromagnetic circuit shown in Figure 8.68 where there are two magnetic field sources: the permanent magnet and the coil. The coil has N turns and current is i. Let the permeability constant of the core be μ_c. Let the air gap, permanent magnet, core cross-section areas be the same, $A_m = A_c = A_g$, for simplicity.

(a) Determine the inductance of the circuit and,

(b) the P_c, magnitude of the slope of load line, assuming $\mu_c = \infty$.

(c) Determine the operating point of the permanent magnet under a given non-zero current condition at the coil. Neglect the reluctance of the core relative to the reluctance of the magnet and air gap and let the magnet permeability be the same as that of air ($\mu_r = 1.0$). Also, let $A_m = A_g$, the cross-section areas of magnet and air gap be the same. Let the coil current be i and the number of turns of the coil be N.

CHAPTER *9*

PROGRAMMABLE LOGIC CONTROLLERS

9.1 INTRODUCTION

The programmable logic controller (PLC) is the de-facto standard computer platform used in industrial control, factory automation, automated machine and process control applications. PLCs were developed as a result of a need in the automotive industry. In the early 1960s, the General Motors (GM) Corporation stated that the factory automation systems based on hard-wired relay logic panels were not flexible enough for the changing needs of the industry. When a new car model required a difference sequence of control logic, re-wiring the control panel was taking too much time and the response time of the company was slow for new car models. With hard-wired relay logic panels, if there is a need to change the automation logic and functions of a line, the logic wiring between the input and output signals in the panel had to be physically changed. This is a time consuming and costly process. It was requested in a design specification that a general purpose wiring of all I/O devices be brought to a panel, but the logical relationship between the I/O be defined in software instead of being hard-wired. In other words, it was desired that the logic be *soft-wired* instead of *hard-wired*. That was the beginning of PLCs. PLCs play a very important role in the automated factories of the industrial world. PLCs replaced the hard-wired relay logic panels. The current trend is large scale networking via the internet between PLCs at the factory and enterprise level computers which may be physically distributed all over the world.

The physical shape of PLCs made by many different companies all have the same form: it is rack mounted with standard size slots to plug in I/O interface units (Figure 9.1). A typical PLC rack starts with a power supply and a CPU module plugged into a backplane of an interface bus. A rack may have a different number of slots,such as 4 slots, 7 slots, 10 slots, 15 slots and so on. A PLC may support multiple I/O racks (main rack that has the CPU and expansion I/O racks). Typical software development tools for a PLC include a notebook PC, a serial or ethernet communication interface and cable, and a software development environment for that particular PLC (Figure 9.2). Depending on the I/O capabilities, CPU speed and program functions, PLCs are categorized into three to four major sizes (Figure 9.1). Any of the I/O interface modules can be plugged into the slots. All of the interface and unit power lines are provided by the snap-on connection to the rack. Typical I/O units supported by virtually every PLC platform include digital input and output modules, analog input and output modules, a high speed counter and timer module, serial communication module, communication network interface modules (i.e., DeviceNet, CAN, ProfiBus), a servo motor control module, and a stepper motor control module. For a given application, the necessary I/O units are selected and inserted into the slots. Furthermore, if

Mechatronics with Experiments, Second Edition. Sabri Cetinkunt.
© 2015 John Wiley & Sons, Ltd. Published 2015 by John Wiley & Sons, Ltd.
Companion Website: www.wiley.com/go/cetinkunt/mechatronics

Main rack (CPU rack)

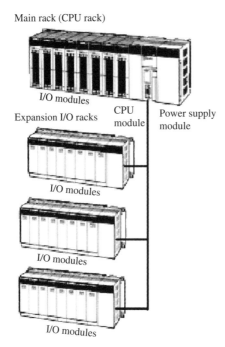

I/O modules

Expansion I/O racks CPU module Power supply module

I/O modules

I/O modules

I/O modules

FIGURE 9.1: The PLC hardware configuration and components: (a) Different PLC catagories in terms of their number of I/O capabilities and CPU speed. (b) A PLC CPU rack and three I/O expansion racks.

the application needs changing, additional I/O modules can be added by simply inserting them into the available slots on the main rack or the expansion rack. Each I/O module occupies a finite number of memory in the PLC's memory space. For instance, a 16-point discrete input module occupies a 2-byte space in the PLC's memory space. A four-channel 12-bit analog to digital (ADC) converter occupies $4 \times 12 = 48$ bit $= 6$ bytes of memory

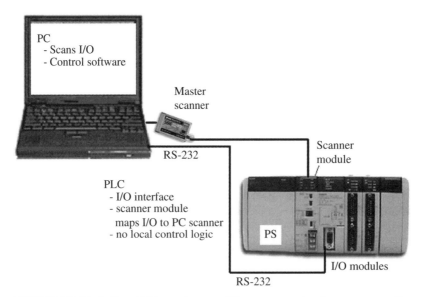

PC
- Scans I/O
- Control software

Master scanner

Scanner module

RS-232

PLC
- I/O interface
- scanner module
  maps I/O to PC scanner
- no local control logic

PS

I/O modules

RS-232

FIGURE 9.2: Typically, a PC is used as an offline program development, as well as an online debugging and monitoring tool (i.e., using the RS-232 serial interface). The PLC is used as the "brains" of the control system by implementing the control logic, and provides the means for I/O interface. In another mode (using master and slave scanner modules), the PC is involved in the real-time control logic implementation. The PLC merely acts as the I/O interface device, and maps the I/O to the communication bus between the PLC and PC using a scanner (slave) card.

space. In other words, once the location of each module in the PLC rack is known, the I/O of all modules are memory mapped to the PLC's memory space using the configuration functions of the PLC program development environment. Then, those memory locations (input/output data) are used in the PLC's application program logic.

Although there have been claims for the past 30 years that PLCs would be a thing of the past and that personal computer (PC) based control would take over the industrial control world, PLCs continue to be strong in the market. Let us compare PLC based control with the PC based control.

1. PLCs have modular design. If a different type of I/O signal needs to be processed, all we need to do is to add a different I/O interface module and modify the software. Furthermore, the I/O modules include the terminals necessary for field wiring. In PC based systems, for different I/Os, we need to insert a different PC card and provide a separate terminal block for wiring which connects to the PC card through a ribbon cable. This tends to be a messy and non-standard wiring process.

2. PLCs have a rugged design suitable for harsh industrial environments against high temperature variations, dust, and vibrations.

3. Programming of PLCs is mostly done using *ladder logic diagrams* which are understood by millions of technicians in the field. This proves to be one of the biggest advantages of PLCs. Even though ladder logic diagram (LLD) programming does not have the programming environment capabilities provided for PCs, it is well established, proven to work well, and a large base of technical personnel can work with it.

The real trend observed in industry is not competition between PLCs and PCs in industrial control, but rather the complementary use of them. PCs are used in conjunction with PLCs at two different levels,

1. PCs are used as networking and user interface devices (Figure 9.2).

2. PCs implement the control logic, replacing the role of the CPU on the PLC while the PLC provides the I/O interface (Figure 9.2). In this configuration, PLC has the I/O modules and a scanner module which updates the I/O between the PLC rack and the PC. The PC implements the control logic which may be developed using any of the programming tools under the PC platform, that is using C, Basic, or PLC specific graphical program development tools. The key issue is to guarantee hard real-time performance in the PC using a real-time operating system. As real-time operating systems become more robust and low cost over time, this model of PC and PLC combination may be widely accepted in industry.

9.2 HARDWARE COMPONENTS OF PLCs

9.2.1 PLC CPU and I/O Capabilities

Perhaps the biggest reason for the success of PLCs in the industrial control market is the fact that the hardware of almost all PLCs have very similar designs. The hardware design is based on a backplane which carries the power and communication bus. A snap-on input/output (I/O) card into any one of the slots makes the necessary electrical contacts for power as well as interface (Figures 9.1, 9.2). Each PLC needs:

1. rack(s) with slots: a backplane for the communication bus,

2. a power supply module,

3. and a CPU module (or a scanner card if PLC is used in conjunction with a master controller, i.e., PC),

4. as many I/O modules as needed by specific application.

The slots form the electrical interface between the I/O modules and the bus of the PLC in the backplane. The bus consists of four major group of lines: power lines, address lines, data lines, and control lines. The end-user does not need to be concerned with the details of the bus since the interface between the CPU and all of the I/O modules supported for a given PLC is already worked out and cannot be modified by the user. The real-time kernel on the PLC and the user program are stored in the memory. The memory can be ROM (read-only memory), EPROM (electrically programmable ROM), EEPROM (erasable electrically programmable ROM), or battery backed RAM (random access memory) type.

Each I/O point on each unit must have a unique address on the PLC bus. The I/O address is generally determined based on the

- rack number,
- slot number, and
- channel number.

Typically, there can be up to three to five racks supported by one PLC. In each rack, there are 4 to 15 slots. In each slot, there can be a single I/O module. For instance, the address of a 16-point discrete input module on the main rack, slot number 3, would be determined by the address code Rack-Slot-IO Channel: 1-3-n, where n is 1 to 16 representing the 16 I/O channels on the module. Similarly, the I/O modules that have analog signal interfaces map their I/O values into the memory of the PLCs. Notice that a PLC with 3 racks, 12 slots in each rack, and supporting 16 discrete I/O in each slot, can support a total of $3 \times 12 \times 16 = 576$ discrete I/Os. Similarly, the same PLC can also support the following combination of discete and analog I/Os:

1. 8 channels of 16-bit A/D converter ($8 \times 16 = 128$ bits),
2. 8 channels of 16-bit D/A converter ($8 \times 16 = 128$ bits),
3. 320 discrete I/O channels (320 bits).

It was noted above that one of the main advantages of PLCs is the fact that a great variety of I/O interface modules are available in standard form. A hardware interface of a simple discrete I/O module has the same difficulty as the interface of a special purpose I/O module. They all snap-on to one of the slots on the PLC rack. This standard hardware interface proves to be a very important asset. The I/O data associated with the I/O modules are memory mapped to the CPU's address space. Below is a list of I/O interface modules available for most PLCs.

1. Discete input modules for DC and AC type signals.
2. Discrete output modules for DC and AC type signals. Each discrete I/O module typically contains 8, 16, or 32 I/O points.
3. Analog input modules (ADC with various voltage OR current range and resolution, i.e., 0–5 VDC, 0–10 VDC, −10 to 10 VDC, 4–20 mA, 0–10 mA ranges, 10-bit, 12-bit, 16-bit resolutions).
4. Analog output modules (DAC with various voltages and current range and resolution).
5. Timer and counter modules (hardware timers, pulse and event counters). A hardware timer module can be programmed to generate an input to the PLC as well as to generate an output to a device when a time period is passed. The beginning (trigger)

of the timer may be program control, free running, or externally triggered by an input. Counter modules are used to count pulses. For instance, ON/OFF state transitions from a proximity sensor which counts the number of teeth on a gear can be used by the counter module to count the number of teeth on a gear for quality control purposes. The pulses from an optical encoder can be input to the counter module to measure displacement.

6. High speed counter modules are used for counting high frequency pulses and detect very short periods of trigger signals (i.e., a high resolution encoder signal input). For instance, this module can be used to measure position in high speed and high resolution encoders.

7. Programmable cam switch module is used to emulate the function of a mechanical cam switch set. A mechanical cam switch set turns ON/OFF a number of outputs as a function of a master cam shaft position. In the mechanical system, the state of outputs is determined by the shape of each cam switch shape that is machined into the cam. In a programmable cam switch module, these functions are programmable.

8. Thermocouple sensor interface modules (any many other special sensor interface modules).

9. PID controller modules (i.e., closed loop temperature controller, closed loop pressure regulator, closed loop liquid level regulator).

10. Motion control modules (servo motor, stepper motor, electrohydraulic valve control modules) used for the closed servo motion control. The actuators may be electric motor and drive or hydraulic valve and amplifiers. The motion control module sends out either the desired number of position pulses as the position command to the drive or sends a voltage command proportional to the desired speed or torque depending upon the mode of the amplifier.

11. Most PLCs support a standard module called a ASCII/BASIC module. This module provides a RS-232 serial interface as well as a separate processor that supports BASIC programming language. The BASIC program is stored in the module on a battery backed RAM. A PLC with an ASCII/BASIC module basically is a dual processor controller. The ladder logic running in the main CPU and the BASIC program running in the ASCII/BASIC module communicate with each other over a predefined memory for data exchange. Complicated mathematical calculations, that may be difficult to code in ladder logic, can be implemented in the ASCII/BASIC module.

12. Master and slave scanner modules when the PLC is used as an I/O interface station and the control logic is implemented by a master controller (another PLC or PC).

13. Network communication modules (DeviceNet, CAN, ProfiBus, Ethernet, RS-232-Cm etc.). PLCs are increasingly part of a larger networked control system. Many so-called fieldbus communication protocols are available.

14. Other special function modules such as fuzzy logic modules. New special function modules are being added to PLCs.

Network communication protocols suitable for real-time control systems, such as DeviceNet, have been changing the hardware configuration of PLC controlled systems in recent years (Figure 9.3). More and more I/O devices (individual sensors, motor starters, closed loop controllers) are being made available with a network interface (i.e., a proximity sensor or a motor starter with DeviceNet interface). Therefore, the I/O devices do not need to be wired into the modules on the PLC's I/O rack. Instead, each I/O device connects to a common communication bus using a T-type connector. This reduces the amount of wiring

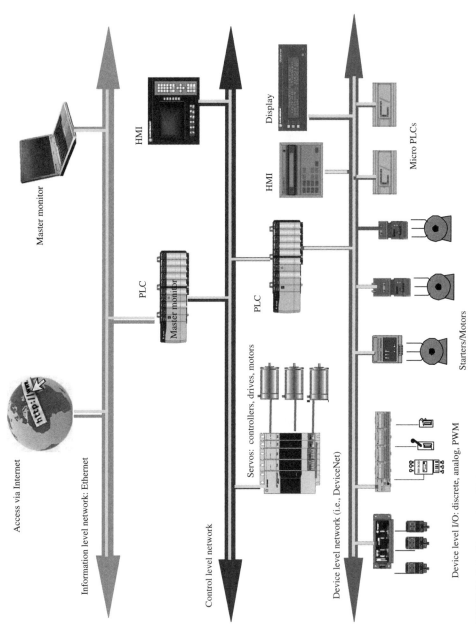

Master monitor

Access via Internet

Information level network: Ethernet

PLC

HMI

Control level network

Servos: controllers, drives, motors

PLC

HMI

Display

Micro PLCs

Starters/Motors

Device level network (i.e., DeviceNet)

Device level I/O: discrete, analog, PWM

FIGURE 9.3: Networked PLC system: three layers of communication network. At the lowest level, simple I/O devices connected to a network (i.e. DeviceNet) using T-connection cables. Adding a new I/O to the system simply requires the device and cables to connect to the common network cable. Second level is control network (i.e. Ethernet/IP). Finally the third level is enterprise wide information network (i.e., Ethernet). Network based systems reduce wiring costs, makes I/O expansion easier, and distribute intelligence.

needed between the I/O device and the PLC. The network communication I/O module (i.e., DeviceNet module) handles the data read/write operation between the bus and the memory of the PLC.

The application programming does not change either way, other than the fact that there is a real-time network communication driver running in the background, working with the communication module, which is transparent to the application program developer, and handles the data between the communication bus and PLC memory. The PLC application program accesses the memory for I/O data as if the I/O devices were directly wired to the I/O modules.

Networked PLC control reduces the wiring costs, distributes the intelligence to local devices, and makes the system I/O expansion easier (Figure 9.3). When a new I/O device is added, the electrical wires of the I/O device do not need to be run all the way to the physical location of the PLC rack, but rather, the communication wires of the I/O device simply need to be connected to the long communication bus cable using a T-type connector.

9.2.2 Opto-isolated Discrete Input and Output Modules

The most common I/O types used in PLC applications are discrete (two state: ON/OFF) type inputs and outputs. The discrete input can be a conducting or non-conducting state (ON/OFF) of a DC or AC circuit component. Similarly, the output can be turning ON or OFF of a DC or AC circuit component. The voltage levels of the input and output circuits are in the order of 12 V to 120 V either DC or AC. In order to electrically isolate the PLC hardware from the high voltage levels of the I/O devices, the interface between the PLC bus and the I/O devices are provided through optically coupled switching devices, namely LEDs, phototransistors, and phototriacs.

Figure 9.4 shows the four types of opto-isolated I/O modules that are used to interface two-state input/output (DC or AC circuit) devices to a PLC. Notice that in all cases, the signal coupling between the PLC side and I/O side is through the light (or optical coupling). When the conducting/non-conducting state of an AC input circuit is interfaced, a rectifier circuit is built into the opto I/O module to convert it to DC current and drive an LED. The resistor in series limits the amount of current flow through the LED. The light emitted by the LED turns ON a phototransistor on the PLC side. If the input circuit is already a DC type, then the rectifier would not be necessary. Similarly, the DC and AC outputs are turned ON/OFF by the PLC by driving an LED (photo-transistor) on the PLC side. For DC outputs, the LED drives a phototransistor. For AC outputs, the LED drives a phototriac. Although not shown in the figures, it is important to point out that the solid-state opto-I/O interface modules also incorporate other functions useful in practical control systems, such as a switch debouncing circuit or inductive voltage surge protections.

PLC I/O modules provide the same type of I/O interface in groups of 8, 16, or 32 points, that is 8-point DC input module, 16-point AC output module. Figure 9.5 shows the typical field wiring of the I/O modules. The DC I/O modules can be in current sinking or current sourcing mode. In current sinking mode, the DC I/O module "sinks" the current received from the I/O device to the ground. The I/O device is connected between the DC(+) and the I/O module terminal. In current sourcing mode, the DC I/O module "sources" the current to the I/O device. The I/O device is connected between the I/O module and the ground (or DC−) terminal.

9.2.3 Relays, Contactors, Starters

Relays, contactors, and starters are all electromechanical switches. Through an electromagnetic actuation principle, mechanical motion is obtained. The mechanical motion is used to open/close a switch which is used to control the continuity of an electric circuit.

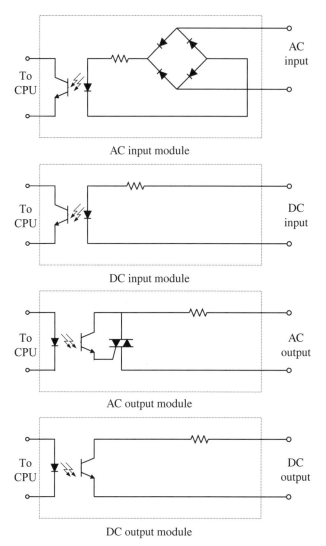

AC input module

DC input module

AC output module

DC output module

FIGURE 9.4: Opto-isolated input and output interface modules for two-state (ON/OFF, conducting/non-conducting) devices: DC input, AC input, DC output, AC output types.

Relays electromechanical switches. It is the most commonly used electrical switch in control circuits. It has two states, ON and OFF (Figure 9.6). A relay is used to connect an electrical line with contacts. It is the electrical equivalent of a manually operated switch in electrical circuits. It has two main parts:

1. contacts for connecting the main electrical circuit (to turn ON and OFF the main line),

2. coil to operate the plunger mechanism that moves the contacts.

The basic operating principle follows that of a solenoid winding. A conductor winding forms the coil. The control voltage is applied to the coil and hence develops a current proportional to the control voltage and resistance of the winding. The coil then generates an electromagnetic field. The plunger is pulled as a result of the electromagnetic field. When the coil is deenergized, the electromagnetic force goes to zero and the plunger moves back under the effect of a spring. In a relay, the plunger motion is connected to a mechanism so

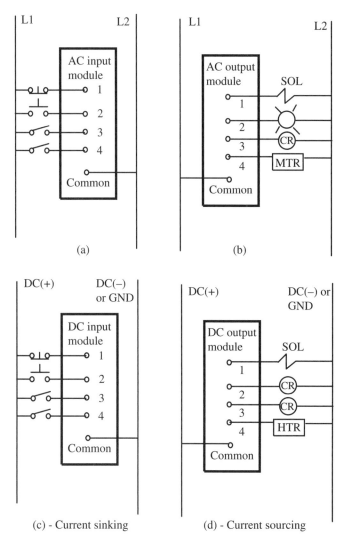

FIGURE 9.5: Wiring of PLC I/O modules to the field devices: (a) AC input module, (b) AC output module, (c) DC input module (shown wired as current sinking), (d) DC output module (shown wired as current sourcing).

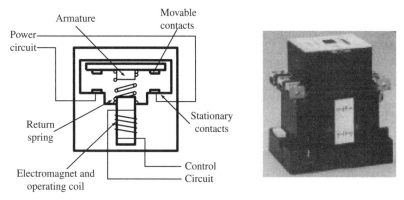

FIGURE 9.6: Operating principles of relay, contactors, starters.

that it makes or breaks the contact connections. The plunger may be connected to activate multiple output contacts in the main line.

The performance ratings of a relay are as follows:

1. Contact ratings: maximum voltage and current the contacts can carry, (V_{max}, i_{max}, i.e., 24 V to 600 V, 50 A) in the main circuit, the number of normally open (NO) and normally closed (NC) contacts operated by the single coil relay (i.e., 6 NO, 6 NC).

2. Coil ratings: nominal voltage (i.e., 6 V to 120 V range) to operate the control circuit coil.

A basic relay design requires the coil to be energized to keep its contacts engaged. Variations in the designs include the so-called *latching relay*. In this design, the relay uses two coils: one for latching and one for unlatching. The relay is energized to start the contacts and move the plunger. Then, a mechanical latch mechanism locks it in place and keeps the contacts connected. Then, the coil does not need to be kept energized in order to keep the contacts connected. In order to disconnect the contacts and release the mechanical latching mechanism, the other coil is energized.

Contactors operate on the same principle and have a similar design to the relays. The main difference being in their mechanical components so that the main line voltage and current capacity that is conducted by the contacts can be much larger than the contact ratings of a relay (Figure 9.6).

Starters are similarly designed to the relay principle except that they may have overcurrent protection as well as "smooth start" related current shaping and limiting mechanisms built in to the design. The overload protection is built in based on the current–temperature relationship in the contact material. When the current is too high for an extended period of time, a bimetallic material breaks the contact, just as is the case in a *circuit breaker*. Starters are used in motor control applications.

9.2.4 Counters and Timers

When a control logic requires that an action be taken after a certain delay, we use counters or timers. If the delay is based on counting something, a counter is used. If the delay is based on time, a timer is used. It should be emphasized that the counter and timer functions can be implemented either in hardware as PLC I/O modules (counter and timer modules) or in software by using general purpose I/O. Dedicated counter and timer hardware offers higher frequency counting and higher resolution timing functions than can be accomplished in software implementation. Hardware counters and timers may also be used as part of a hard-wired control circuit without any PLC software involvement.

Both counter and timer hardware modules have three major circuits: *the power circuit, control circuit, and output circuit*. The power circuit is needed to power the counter and timer modules. The control circuit is the signal used to trigger the module. For a counter, it is the signal transition from one state to another state (OFF to ON or ON to OFF) that is to be counted. For a timer, it is the signal that triggers the start of the process of timing a period. The logic about when to keep timing may vary. Once it is triggered, a timer can run until the present time value, or the timing operation continues only while the control signal is ON and is suspended when it is OFF. Both counter and timer have *preset values*. When the counter counts up to the preset value, the output circuit is turned ON. When the timer measures that the preset value of time has passed since the control signal triggered it, the output circuit is turned ON. The output circuit of a timer and counter is similar to the output circuit of a relay. An electrical contact is made (ON) or broken (OFF) as output action.

9.3 PROGRAMMING OF PLCs

Every PLC includes a program development software tool which allows the communication between a PC and the PLC, development of the PLC application software, debugging, downloads, and tests it (Figure 9.2). A notebook PC is used only as development tool in this case, not as part of the real-time control. The development tools for different PLC manufacturers currently are not interchangable. The application development engineer must use the development tools supplied by the specific PLC vendor.

Assuming that we have the program development tools (i.e., a notebook PC and a PLC specific ladder logic program development environment), the software that runs on the PLCs to control a specific industrial application is the issue discussed here.

There are three main types of programming languages available for PLCs:

1. *Ladder logic diagrams* (LLD) which emulate the same structure of the hard-wired relay logic diagrams. These have the widest use since most field technicians are familiar with hardware relay logic diagrams and can understand LLD programs.

2. *Boolean language* is a statement list and is similar to BASIC programming language.

3. *Flowchart language* uses graphical blocks. It is more intuitive than the other two languages. Although the use of flowchart languages has started to increase in recent years, LLD is still the dominant language.

The flowchart type languages may eventually gain more widespread acceptance. Today, every PLC has its own LLD language and they are not compatible across different PLCs. A flowchart language designed based on standards can generate different run-time programs for different PLCs simply by using a PLC specific compiler. The LLD programming environment for all PLCs looks very similar from the application program development point of view. In the rest of this chapter, only the LLD programs will be discussed, not the specific development environment for a particular PLC.

There is a fundamental difference between the way a PLC executes its program and the way a PC does. The program flow in a PC is controlled by the *flow control statements* such as *do-while, for* loops, *if-else* blocks, function calls, jump or go-to statements. In high level programming languages, it is possible that the program can be limited to a local loop. Whereas a PLC program runs in the so-called *scan mode* (Figure 9.7). The whole program

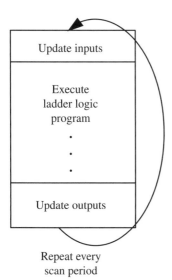

Repeat every
scan period

FIGURE 9.7: Ladder logic program execution model in a PLC. The PLC executes its program in scan mode. All of the code is checked for execution every scan period.

logic, from the beginning to the end, is scanned every scan period of the PLC. Depending on the computational power of a PLC, typical scan times are in the order of a few milliseconds or less per thousand lines of ladder logic code. The benchmark speed of scan time for per thousand ladder logic code is generally limited to basic logic functions such as AND, OR, NOT, flip-flop. Special functions, and trigonometric functions, PID control algorithms implemented in the ladder logic, take a longer computation time. As a result, the scan time will be slower for ladder logic programs that include many special functions. Therefore, the scan time estimates given for a PLC by its manufacturer should be interpreted with this in mind. The actual scan time of a ladder program for an application can be exactly measured by the software development tools of the PLC. The software development tools installed and run on a PC for offline program development and debugging purposes also include utilities that measure the scan time while the program is executing on the PLC. In the ladder logic programming, sections of the program can be conditional like the *if-else* blocks, and subroutines may be called or skipped based on the coded conditions. All of the PLC ladder logic program is scanned for execution every scan time. It is useful to imagine the PLC ladder logic as a circular instruction sequence where the CPU goes through the circle once every scan time period (Figure 9.7).

From a programming point of view, once the racks and slots of a PLC are populated with I/O interface modules, all of the I/O is memory mapped. There is a one-to-one memory map between the I/O points on the PLC hardware and the CPU memory addresses (Figure 9.8). The logic is implemented between the I/O (memory locations) using the logic functions provided by the ladder logic program. The ladder logic programming focuses on

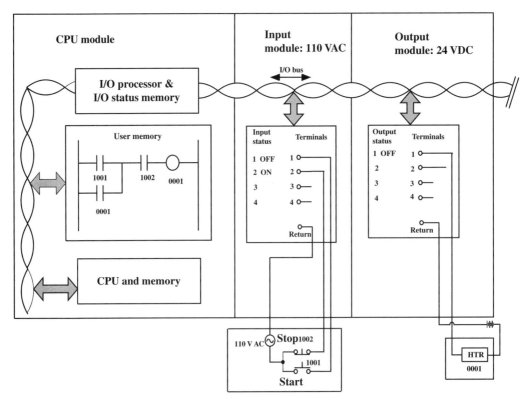

FIGURE 9.8: The PLC I/O interface, communication bus, CPU, and memory relationship.

the industrial control and logic, not high level object oriented data structures. Therefore, typical data structures supported include:

1. *bits* for discrete I/O,
2. *bytes* and words for analog I/O and charater data.

There are standard key words reserved for discrete input, output, timer, and counter funtions in the PLC. For instance, the DIN 19239 standard specifies the following key words for PLC ladder logic programming concerning I/O and memory addresses:

1. I: for discrete input lines, for instance I0 through I1023.
2. O: for discrete output lines, for instance O0 through O1023.
3. T: for timer functions, for instance T0 through T15.
4. C: for counter functions, for instance C0 through C31.
5. F: for "Flag" to store and recall a bit of data, that is flip-flop operations.

Once the actual I/O terminal points on the I/O modules in the PLC slots are decided, then there is a memory map established between the I/O variables in the memory and the actual I/O terminal points of the I/O modules, using the configuration software tools of the PLC program development enviroment. Furthermore, most PLC program development tools support *symbolic names*. In other words, if the discrete input lines 0 and 1 are connected to two input switches called "START" and "STOP," they can be symbolically defined to map to I0 and I1. Then, in the program all references to these switches can be made using the symbolic variable names "START" and "STOP" which is more descriptive and makes the program easier to understand.

The logic operators and statements supported by most PLC ladder logic programs include

1. logic functions: AND, OR, NOT,
2. shift functions: left shift, right shift,
3. math functions : add, substract, multiply, divide, sin, cos,
4. software implemented timer and counter functions,
5. software implemented flip-flops,
6. conditional blocks (similar to if-else),
7. loops (do-while, while, for loops). Since loops can tie up a logic and a PLC must scan all of the logic, loops in PLC ladder logic are treated differently then the loops in C-like high level programming languages,
8. functions (subroutines),
9. interrupt service routines to be executed on interrupt inputs.

The AND and OR logic functions between two state variables are implemented in ladder logic diagrams by series and parallel connections, respectively. The NOT function is implemented by a cross over the contact signal. Other functions are generally supported by a rectangular box with appropriate input and output lines.

Some of the syntax rules common to all ladder language programs are as follows:

1. A ladder logic program is a sequence of rungs. Each rung must begin with an input or local data.
2. Output channels can appear at the end of a rung only once.

3. The discrete input channel state is represented by two vertical lines like a contact.

4. The discrete output channel state is represented as a circular symbol.

5. Timers, counters, and other function blocks are represented by a rectangular shape with appropriate input source and output lines in the rung.

6. Logic AND is implemented by putting two contacts in series, and logic OR is implemented by putting them in parallel (Figure 9.8).

One of the standard logics in a ladder logic diagram (LLD) is a seal-in circuit. The idea is to turn ON an output channel (relay output) and keep it ON based on a momentary input (START switch). Then keep it ON until another momentary input (STOP Switch) is pressed. This is shown in Figure 9.8.

The following software concepts are commonly used in PLC programs: *shift register, sequencer, and drums*. Shift register and sequencer are both registers. The individual bit locations in the register are used to indicate the location of a part in a multi station assembly line or the number of the operation currently performed from a sequence. A bit is associated with the current product or operation. As the bit is shifted in the register, its location on the assembly sequence is advanced. Similarly, drum is the software analog of a mechanical cam. There is a master shaft and multiple cams on it. Each cam actuates (turns ON and OFF) an output line as a function of the input cam shaft position. This functionality is duplicated in software with the electronic drum concept. The benefit of the software drum compared to the mechanical drum is that it is programmable.

9.3.1 Hard-wired Seal-in Circuit

In original hard-wired relay circuits, all of the control logic is physically wired into the electrical connections of input, output, and intermediate control devices. When programmable logic controllers replaced the hardwired relay logic, the logic between the inputs/outputs is defined in software (programmed) in the PLC. Very little of the control logic is hard-wired. Still, some of the safety related functions are hard-wired in case the software fails. Most emergency shut-down and cycle start functions are hard-wired. One of the most frequently hard-wired logics is the *seal-in circuit* used between START and STOP push buttons. The same diagram is used as the one shown in Figure 9.9. The only difference is that the logic

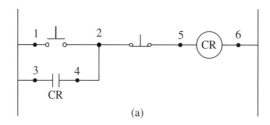

(a)

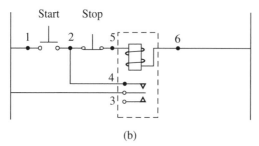

(b)

FIGURE 9.9: Hardwired seal-in circuit in a ladder logic diagram. (a) Ladder logic diagram of connections, (b) component connections between START button, STOP button, and the RELAY.

is hard-wired instead of programmed in the PLC. A hard-wired seal-in circuit includes the following components:

1. a START button which is a momentary contact switch,
2. a STOP button,
3. a relay or starter coil from the output device (i.e., coil of a starter used to start a motor)

When the START button makes a momentary contact, it energizes the coil of the relay (or starter). Then the contacts from that coil maintain the flow of current since it is wired in parallel with the START button even when the START button no longer makes contact since it is a momentary button. The power flow is cutoff (stopped) any time the STOP button is pressed. This is called the *three-wire* control. If for some reason the power is lost to the circuit, the START button must be momentarily pressed again in order to start the motor again. This is good for safety, but requires human intervention if the power flow in the circuit is interrupted. In other words, the re-start of the cycle is not automatic.

9.4 PLC CONTROL SYSTEM APPLICATIONS

9.4.1 Closed Loop Temperature Control System

Figure 9.10 shows a closed loop temperature control system in an oven using a PLC. The PLC based control system uses a temperature sensor (i.e., a thermocouple interfaced to the PLC using a thermocouple sensor interface module), a proportional valve which regulates the fuel into the furnace proportional to the displacement of the valve spool, a fuel supply line (i.e., gas line from utility supply), and a display to inform the operator of the process variables in real-time (i.e., the desired temperature setting, actual temperature, fuel rate). The PLC controls the valve with an analog output module. If the current capacity of the analog drive is not large enough to drive the valve, then there would be a current amplifier between the analog output module and the valve. The closed loop control algorithm is

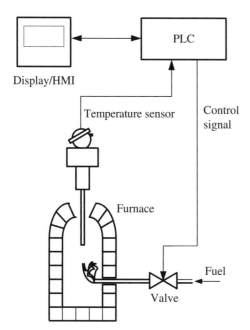

FIGURE 9.10: Example temperature control system with a PLC using a PID module.

implemented in the ladder logic diagram using a PID algorithm function block where the parameters of the algorithm are to be tuned by the application engineer. The PID control algorithm determines the command signal to the valve (how much it should open) based on the desired temperature (which may be programmed by the operator using the HMI device interface) and the actual temperature measurement from the temperature sensor. The PID algorithm can be implemented either in the ladder logic diagram in software or the PID implementation can be handled by a dedicated PID I/O module. A PID I/O module would be placed in the rack of the PLC, and has one analog input connected to the temperature sensor, and one analog output connected to the valve, and the desired temperature is a data that is defined by the ladder logic using the HMI and passed to the PID module via the backplane communication bus.

9.4.2 Conveyor Speed Control System

Conveyor speed control is a very common automation problem. Figure 9.11 shows a motion control application using incremental encoder input and pulse output modules in the PLC rack. The PLC controls the speed (and position if desired) of two conveyors. The PLC program measures the speed of two conveyors using incremental encoders and commands the drives which control the power to each motor. Typical applications include:

1. Independent speed control of two conveyors. The desired speed of each conveyor can be set by the operator through a user interface device, and the PLC controls both conveyor speeds independent of each other.

2. Master-slave speed control: the speed of one of the conveyors may be set as the master speed and the other conveyor commanded speed is determined in proportion

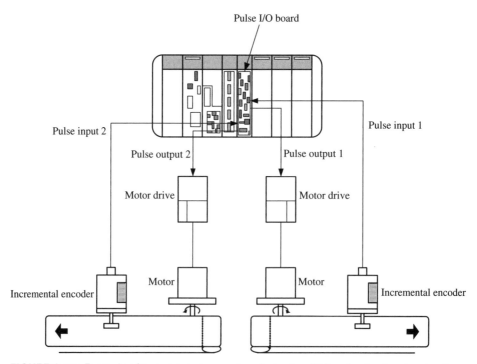

FIGURE 9.11: Example of two conveyor speed control using PLC and pulse input and pulse output modules.

to the first one. The second conveyor is controlled in slave mode. The speed of the first conveyor may be programmed or determined based on a sensor (i.e., speed of another station in the line).

3. Master-slave speed control with position phase adjustment: in addition to making one conveyor follow the speed of the other conveyor (master-slave relationship), the position of one slave conveyor relative to the master conveyor can be adjusted based on position sensor information. This is very common in packaging applications where material is transferred from one conveyor to another. In steady-state, two conveyors run in a certain gear ratio (i.e., 1:1 master-slave relationship). Due to the variation in material location on each conveyor (product on one and container on the other), in order to place the product on the container properly, the position phasing must be adjusted to correct for the placement variations.

9.4.3 Closed Loop Servo Position Control System

Servo positioning using a lead-screw (or ball-screw) drive is a very common high precision motion control application (Figure 9.12). The ball-screw positioning system is used in all CNC machine tools, XYZ positioning tables, gantry-type robots, printed circuit board

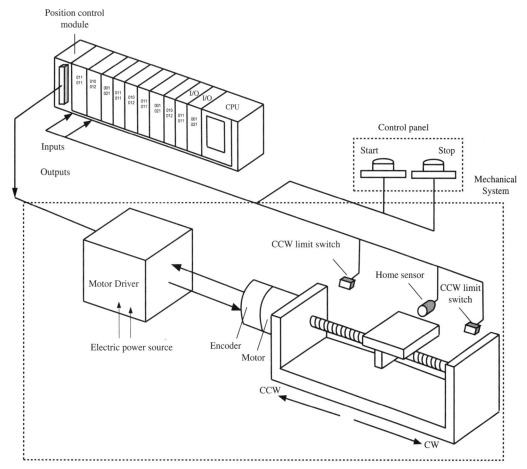

FIGURE 9.12: Example servo positioning control with a PLC and servo control module.

(PCB) assembly machines, and others. This example shows the components of a PLC based system to control the position of a linear stage using an electric motor (i.e., stepper motor, DC brushless motor, DC brush motor) and a rotary incremental encoder. Other real-world components are also shown, such as the limit sensors to indicate the mechanical limits of motion, a proximity sensor to establish a reference position (home position) after power-up, since an incremental encoder can not measure absolute positions, and operator command buttons (START and STOP). The components of such a closed loop translational position control system include:

1. Mechanical motion mechanism: lead-screw or ball-screw and moving table.
2. DC servo motor with integrally built incremental encoder on the same mechanical assembly.
3. DC motor drive: amplifier and DC power supply package.
4. PLC with a servo control module, where the encoder signal is feedback to the module and the output signal from the module is the command signal to the amplifier.
5. "Home" reference sensor (ON/OFF type) and CW and CCW limit switches (ON/OFF type) on the translational stage to define the motion limits and starting reference as an absolute position.
6. START and STOP bush buttons for operator interface.

The sequence of operations, from start up to automated operation is then coded in the ladder logic diagram. High performance motion profiles are defined in the servo control module.

9.5 PLC APPLICATION EXAMPLE: CONVEYOR AND FURNACE CONTROL

Consider the conveyor and heat furnace shown in Figure 9.13. The conveyor moves parts into a heating chamber. The part is kept inside the temperature controlled furnace for a certain amount of time (programmable) and at a certain temperature range (also programmable). Then the conveyor moves to bring in the next part. In the process, the PLC controls the temperature of the furnace and the speed of the conveyor. In addition, it controls the opening and closing of the heater doors as well as position of the part inside the furnace using a presence switch.

The system inputs are:

1. START switch (ON/OFF).
2. STOP switch (ON/OFF).
3. Part presence sensor (ON/OFF).
4. Door open switch (ON/OFF).
5. Temperature range sensor (analog).

The outputs are:

1. Analog output to heater element proportional amplifier.
2. ON/OFF output to motor starter.
3. ON/OFF output to door actuator relay control.

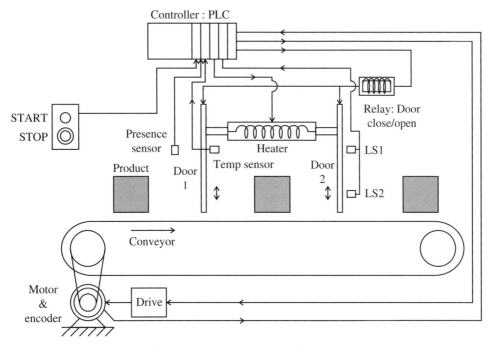

FIGURE 9.13: PLC control (logic and closed loop control) of a heater and a conveyor motor.

The desired control logic is expressed in pseudo-code as follows:

1. The operation is started when the START button is pressed. When the STOP button is pressed the cycle is stopped.

2. PLC opens the doors of the furnace and waits until the doors open switch is ON.

3. Then the motor is started. The motor is stopped when the part presence sensor is ON.

4. The furnace doors are closed.

5. The PID loop which controls the furnace temperature is enabled. The PID controls the heater output based on the preprogrammed desired temperature range and temperature sensor feedback. The PID control output has a deadband. In other words, when the temperature is within a desired range, the heater is turned OFF. These aspects of the PID control loop are specified as part of the PID block setup.

6. After the furnace temperature is within the desired range, a timer is started and the temperature control continues for a specified amount of time (i.e., 2 minutes).

7. When the heating time expires, the heater is turned OFF (PID loop is disabled).

8. Repeat starting with Step 2.

Figure 9.14 shows the ladder logic diagram program that implements the above described control logic. Notice that for each I/O, there is a memory location assigned (see the numbers under the input and output devices as illustration on left and right side of the diagram). Then the basic cycle logic is implemented in the ladder diagram using the input–output memory references, timer, and PID functions. Notice that in order to implement the desired logic, internal memory (local data) is used in addition to the memory mapped to the I/O channels. The desired temperature range and heating time data should be adjustable by an appropriate operator interface device, such as a PC communicating with the PLC over a serial communication bus.

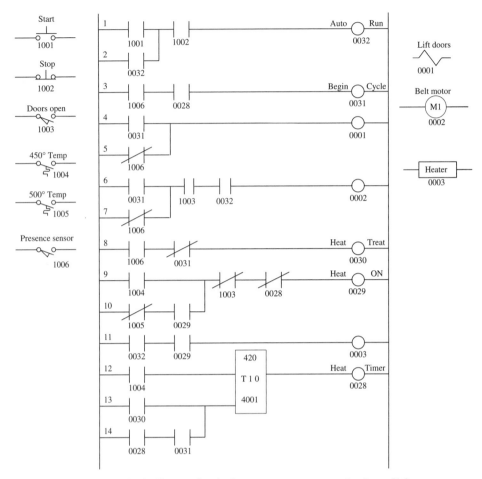

FIGURE 9.14: Ladder logic diagram for the heater–conveyor control using a PLC.

9.6 PROBLEMS

1. Consider the temperature control system shown in Figure 9.10. Specify the necessary PLC hardware (PLC plus the required I/O modules). Design a ladder logic diagram so that the PLC controls the temperature using the temperature sensor as a feedback signal and the proportional fuel valve as the controlled actuator. The logic should include input switches to start and stop the control system as well as display status information to the operator.

2. Consider the speed control system shown in Figure 9.11. The objective is to run the conveyor 1 at a programmable speed where that value is entered by the operator via an operator interface device. The conveyor 2 is suppose to run as if it is mechanically geared to conveyor 1 and the gear ratio is also programmable by the operator. Specify the necessary PLC hardware and design the LLD program.

3. Consider the position control system shown in Figure 9.12. Our goal is that when the operator pushes the "Start" button, the automatic cycle of the positioning table starts. When the "Stop" button is pushed, the table is supposed to stop to a predefined home position. When the start button is pushed, the motor is to move in a direction to seek the "home" sensor. If it reaches one of the limit switches before reaching the home sensor, then it must reverse direction, go past the home sensor, stop, and come back to turn ON the home sensor from a predefined direction. Then stop and establish a "home" reference. After that, the motor is suppose to make incremental forward and reverse motions, that is five in a row, 1.0 in displacement motions separated by 100 ms delay between the completion and start of each motion. Specify the necessary PLC hardware and design the program logic.

4. Consider the circuit shown in Figure 9.9. What happens if the seal-in circuit at points 3 and 4 was not connected? If the seal-in circuit is not connected, and we change the "Start" switch to a latching switch (not momentary), how is the operation of the circuit different than the seal-in circuit operation (consider that if we started the circuit and stopped it with the "Stop" switch, how would you re-start the system)?

CHAPTER 10

PROGRAMMABLE MOTION CONTROL SYSTEMS

10.1 INTRODUCTION

Programmable motion control systems (PMCS) are used in all mechanical systems that involve computer controlled motion. A robotic manipulator, an assembly machine, a CNC machine, XYZ table, and construction equipment tool control systems are all examples of applications of PMCS. As the name indicates, PMCS are motion mechanisms where the motion is controlled by a digital computer, and hence are programmable. PMCS are good examples of mechatronic systems in that they involve a mechanical motion system, various actuators and sensors, and computer control. Figure 10.1 shows the typical components of an electric motor based programmable motion system, that is a motor with position sensor, amplifier and power supply (drive), and controller. The figure shows different types of rotary motors and drives (brushless and brush-type DC, AC induction) as well as linear motors.

In the past, coordinated motion control in automated machines was achieved by mechanically connecting various machine components with linkages, line shafts, and gears. Once the master line shaft is driven by a constant speed motor or engine, the rest of the motion axes derive their motion from it based on the mechanical linkage relation. This was what is called "hard automation." The availability of low cost microprocessors and digital signal processors (DSP) as well as their high reliability has made it possible to control motion and coordinate various axes under computer control. The coordination between axes is not fixed by mechanical linkages, but coded in software. Hence, the coordination logic can be changed on the fly in software. The same machine can be used to perform with different coordination relations to produce different parts by simply changing software. In mechanically coordinated machines this may require a change of gears, linkages which may require very long setup times. Some of the complicated coordination functions may not even be feasible to achieve by mechanical coordination while they can be easily coded in software. The "programmable" aspect of motion control comes from the fact that the control logic is programmed. Therefore, it is a "soft automation" or "flexible automation." The single most significant advantage of "soft automation" over "hard automation" is the significantly reduced setup times for product changeovers. Figure 10.2 shows an example of a printing machine with both old mechanical automation and programmable automation versions. In the mechanical version, each station is coordinated with respect to the master shaft through mechanical gears. When a different product is required, the gear reducer ratios are mechanically changed. Therefore, for different products, different gear reducers have to be kept in the inventory. In addition, physically changing the gear reducers is labor intensive and time consuming. In the electronically coordinated version of the machine, the

Mechatronics with Experiments, Second Edition. Sabri Cetinkunt.
© 2015 John Wiley & Sons, Ltd. Published 2015 by John Wiley & Sons, Ltd.
Companion Website: www.wiley.com/go/cetinkunt/mechatronics

FIGURE 10.1: Components of an electric servo control system: motors (rotary and translational) with built-in position sensors, drives (DC power supply and amplifier), and controllers. *Source:* Baldor.

gear ratios between stations are defined in the application software. There are no physical gears between axes. The same functionality is achieved by software control of the motion of each axis. So, when a different product requires different gear ratios between axes, the only change needed is to select a different software parameter set from the database. The change-over time for different products is almost insignificant.

In a factory automation application, multiple PLCs and programmable motion controllers may be used to control the whole process. PLCs handle the general purpose I/O and act as the higher level supervisory controller relative to the lower level local controllers (Figure 9.4). Programmable motion controllers handle the high performance coordinated motion control aspects of individual stations and communicate to the PLCs. In small automation applications, a stand alone programmable motion controller may have enough I/O capability such that the machine control can be handled by it without the use of a PLC (Figure 10.3).

Every programmable motion control system has the following components (Figure 10.4):

1. controller,
2. actuators (motor, amplifier, power supply),
3. sensors (encoders, tachometers, tension sensors),
4. motion transmission mechanisms (gears, lead-screws),
5. operator interface devices.

Most of the technologies for the components of a PMCS have already been covered in this text in various chapters. Here we will focus on the system level control software features specific to automation applications of PMCS.

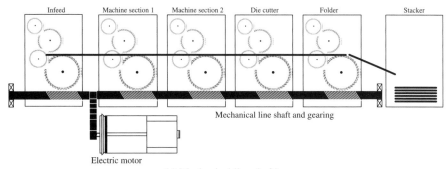

(a) Mechanical line shafting

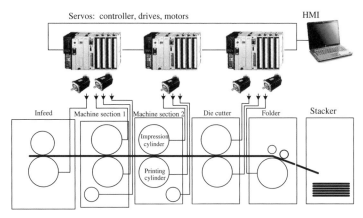

(b) Electronic line shafting

FIGURE 10.2: (a) An automated printing machine with mechanical coordination. A long master shaft is driven by a large electric motor. All other stations are geared to the master shaft mechanically. (b) An automated printing machine with electronic motion coordination. Each station is independently driven by a small actuator (i.e., electric servo motor with position sensors).

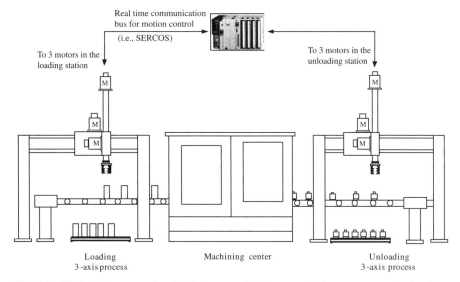

FIGURE 10.3: Motion coordination between a loading, machining center, and unloading station. The machining center typically has its on CNC controller. Custom loading and unloading stations need to be designed and controlled in coordination with the machine tool operation.

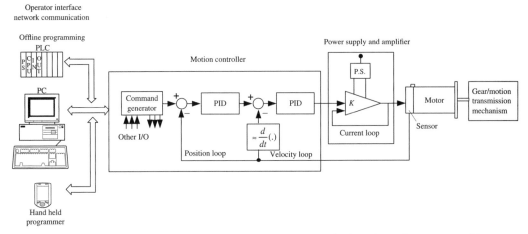

FIGURE 10.4: Components of a single-axis servo controlled motion axis: motor, position sensor, power supply and amplifier (drive), motion controller for servo loop, machine level controllers (PLCs), and operator interface devices.

Let us focus on the control of the motion of one axis only. The motor, amplifier, and power supply form the "muscle" of the axis which we call the "actuator." The controller handles the following primary tasks:

1. The closed loop motion control (position loop, velocity loop, and even current loop in some cases): The position and velocity loop control are typically implemented as a single closed loop control algorithm. The most dominant servo position and velocity control algorithm is a form of PID control algorithm with various feedforward compensation terms, and limits on the contribution of each term. Figure 10.4 shows a servo position and velocity control algorithm used in industry which is clearly a variation of the standard PID control algorithm.

2. It handles the other input and output signals associated with the machine control functions (Figures 10.4, 10.5, and 10.6).

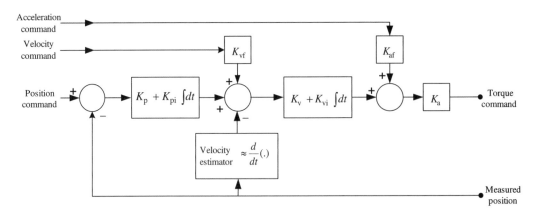

FIGURE 10.5: An example PID control algorithm implemented by a commercially available motion controller. Notice that practical PID control algorithms for servo motion control include feedforward terms in addition to the textbook standard PID gains. Other modifications may include addition of deadband, limits on integral control, activation and deactivation logic for integral action, integrator anti-windup, friction and backlash compensation logic.

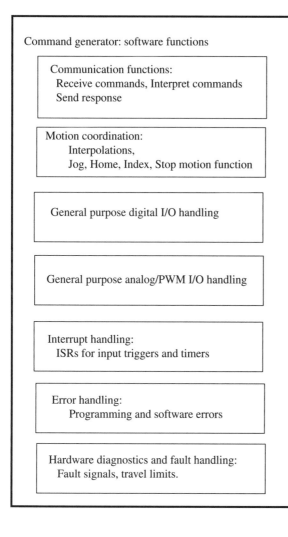

Command generator: software functions

Communication functions:
Receive commands, Interpret commands
Send response

Motion coordination:
Interpolations,
Jog, Home, Index, Stop motion function

General purpose digital I/O handling

General purpose analog/PWM I/O handling

Interrupt handling:
ISRs for input triggers and timers

Error handling:
Programming and software errors

Hardware diagnostics and fault handling:
Fault signals, travel limits.

FIGURE 10.6: Command generator block details: typical groups of functions implemented on a general purpose motion controller, in addition to the closed loop servo control algorithm.

3. The controller generates the commanded motion to the closed loop (servo loop) based on the logic of the application software (Figure 10.4).

4. The axis controller also communicates with the operator interface devices and communication bus to coordinate its operations with the rest of the control system (Figure 10.4).

In general, the motion of one axis needs to be coordinated to other external machine events or to the motion of other axes. In order to coordinate motion between multiple axes, the controllers need to communicate with each other. Figure 10.7 shows different hardware options to establish the communication between multi axis controllers. In the first case, there is a stand alone controller for each axis and the coordination data needed between axes are simple. In this case, the coordination signals between axes can be handled by digital I/O lines where one axis tells the other one that it is time to move (i.e., "Go" signal, Figure 10.7a). If there is more detailed data needed for the motion coordination, such as more than just the "Go" signal for example the actual speed and position of another axis), each stand alone axis controller must be on a common communication bus to exchange the necessary information in real-time. Sercos is one of the serial communication standards adapted specifically for high performance motion coordination applications (Figure 10.7b).

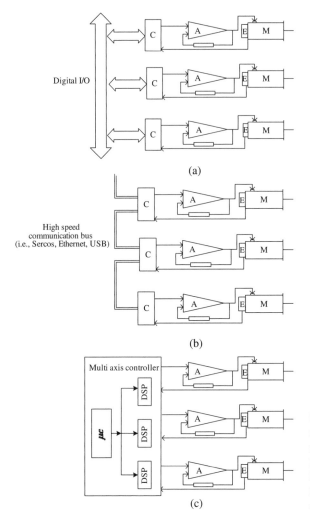

(a)

(b)

(c)

FIGURE 10.7: Coordination methods between multi axis motion axes: (a) using digital I/O lines as dedicated triggers, (b) high speed data communication bus, (c) multi axis motion controller.

Finally, if there is a multi axis controller that is interfaced to all of the tightly coordinated axes, the controller has access to the motion information of all axes (Figure 10.7c).

10.2 DESIGN METHODOLOGY FOR PMC SYSTEMS

The following is a practical methodology which outlines the typical steps involved in designing a PMC system.

Step 1:   The first step is to define the purpose and functionalities of the machine, that is what it does and how it does it. The operation sequence can be roughly defined without fine details. The required motion should be decomposed into various coordinated moves. Depending on the complexity of each motion axis, the proper actuation system for each axis should be selected. For instance, simple two position moves can be implemented more cheaply with solenoid actuated two position air cylinders than a servo motor driven actuation. However, the speed of air cylinders is much slower than the speed of servo or stepper motor actuated mechanisms.

Step 2: Describe the operation of the machine in more detail.
1. Required modes of operation (setup mode, manual mode, auto mode, power-up sequence things to do).
2. Required operator interface, that is what commands the operator will give, what status information the operator will need, what data the operator can change and monitor.

Step 3: Decide on the required hardware components and design integration of the electromechanical system (mechanical assembly and electrical wiring of components).
1. Actuators: servo motors, DC motors, hydraulic or pneumatic cylinders.
2. Sensors: encoder feedback, proximity switches, photo-electric switches.
3. Controllers needed: PLCs, servo controller, sensor controller.
4. Operator interface devices.
5. Motion transmission mechanisms: gears, lead-screws, timing belts.

Step 4: Develop application software
1. Top-down tree structure design of software.
2. Psuedo-code.
3. Code in specific programming language.

Step 5: Set up each programmable motion axis
1. Basic hardware check.
2. Power-up test.
3. Establish serial communication.
4. Servo tuning ($K_p, K_v, K_i, K_{vf}, K_{af}, \ldots$, values).
5. Default parameter set-ups (default acceleration, deceleration, jog speed, maximum motion planning parameters).
6. Simple moves: Jog, Home, Single incremental motion ("Index") parameters.

Step 6: Debug, test, verify performance.

Step 7: Documentation: document clearly so that later on someone else can debug and/or modify the machine.

10.3 MOTION CONTROLLER HARDWARE AND SOFTWARE

The heart of a PMC system is the controller. The same controller can be used with different actuators such as electric servo, hydraulic servo, or pneumatic servo power. The degree of intelligence and sophistication that can be designed into the system is largely dependent upon the capabilities of the controller. The custom application software in the controller uses the I/O hardware to define the logic between them and control the operation of the machine. The I/O hardware can be grouped into the following categories (Figure 10.3):

1. Axis level I/O
   (a) Servo control I/O:
      • Command signal to amplifier in the form of ± 10 VDC range analog DC voltage or PWM signal.
      • Feedback signal from the position sensor (encoder, resolver).
   (b) Axis I/O:
      • Travel limits: positive and negative direction limits.
      • Home sensor input.
      • Amplifier fault input.

- Amplifier enable output.
- High speed position capture input.
- High speed position trigger output (cam function outputs).

2. Machine level I/O:
   (a) Digital inputs (ON/OFF sensors: proximity, photo-electric, limit switches, operator buttons).
   (b) Digital outputs (ON/OFF output: relays, solenoids).
   (c) Analog inputs (+I OVDC, tension, temperature, force, position etc.).
   (d) Analog outputs (+10 VDC to other device amplifiers).

3. Communication:
   (a) Serial RS 232/422 communication with HMI.
   (b) Network communication between higher level intelligent devices.

The software functionalities control algorithm involves the following in a general purpose motion controller (Figure 10.6),

1. Communication functions: "receive commands", interpret them, "send response" and report "status" functions over a communication port such as USB, RS-232, CAN, Sercos, TCP/IP.

2. Coordinated motion functions: interpolation algorithms (linear interpolation, circular interpolation, cubic spline interpolation), point-to-motion profile generation based on trapezoidal and modified sine function velocity profiles, Jog command implementation, HOME command implementation, STOP command implementation, INDEX command implementation.

3. Servo loop control task: $U_{DAC} = PID$ (*desired and actual motion*).

4. General purpose digital input status monitor, control output decisions on digital output lines, including the limit switch states,

5. General purpose analog and PWM input and output ports: reading and writing (based on a customized application specific logic) of general purpose analog/PWM I/O, which are not used for the immediate servo loop.

6. Interrupt handling: based on a timer (periodic) or based on input signal triggers, setup interrupt service routines (ISR) which implement logic ("what to do") when that interrupt occurs.

7. Error handling is the functions used to handle programming and software related errors.

8. Diagnostics functions: maximum following error, fault status on the input signals and handling of faults and errors (i.e., fault signal from the amplifier, travel limits sensors, torque limits).

10.4 BASIC SINGLE-AXIS MOTIONS

The following motion types are so often encountered in automated machines that they are given specific industry standard names and are consider basic motion types: *home, jog, stop, index* motions. A typical motion is either a triangular or trapezoidal velocity profile as a function of time. Furthermore, it is also common to modify the trapezoidal velocity profiles with smoothed sine functions in order to reduce the jerk effect of the commanded motion. It is customary to define the motion profile in terms of velocity as a function of time.

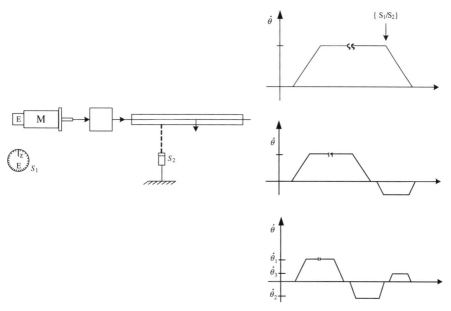

FIGURE 10.8: HOMEing routine (establishing a position reference) using motor and tool mounted sensors.

HOME: standard name used in the programmable motion control industry to describe the motion sequence used to establish a reference position for the motion axis. The objective of the homing motion is to bring the axis to a known reference position, called the "home" position. All other positions are referenced to that position for the axis. For instance, the machine tool axes must be at a pre-defined position before the beginning of a cycle. Figure 10.8 shows an example of homing motion. If there is an absolute position sensor for the motion axis, the homing can be performed from the current known position to the desired position in one move where the distance is known. Quite often, a motion axis uses an incremental position sensor. The position of the axis is not known at power-up. The axis must search for a reference position using a combination of sensors that indicate a known position. Typically the encoder index channel and/or external ON/OFF sensor are used to establish the reference position. Depending on the accuracy required in starting at the known reference position, different homing motion sequences can be performed. Figure 10.8 shows three different types of common homing motion sequences. Others can be designed to meet the needs of a particular application. The first home sequence simply moves at a constant speed until a sensor signal changes state (i.e., turns ON) and then stops the motion at a defined deceleration rate. The second home sequence does the same thing as the first home sequence, plus it moves backward by a predefined distance after stopping. The third home sequence adds another predefined motion distance in the forward direction. Notice the small wait periods between the stop and incremental moves. The purpose there is to allow the axis motion to settle to a complete stop. Parameters used to define a HOME motion are *home search speed, acceleration and decceleration rates, incremental move distances, and time periods to complete them.*

JOG: standard name used to move at a certain speed while the operator holds an input switch on. Using the JOG function, the machine axes can be moved around to any position under the operator control. This function is needed in machine setup and maintenance. JOG motion refers to a constant speed motion of an axis until a condition is met to stop it. The

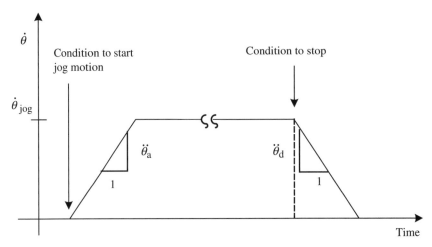

FIGURE 10.9: Motion executed on STOP command.

parameters used to define the jog motion are *jog speed, acceleration and decceleration rates* (Figure 10.9).

STOP: the rate at which the motion must be decelerated to a stop (zero speed) is defined by some PMC configuration parameters. When the STOP command (or equivalent) is executed, the motion is decelerated to zero speed from its current speed. The STOP motion may be initiated by an operator command or a sensor trigger or under program control. The only parameter needed to define the STOP motion is the *rate of decceleration* (Figure 10.9).

INDEX: a common name used to describe a move which causes a predefined positional change in the axes. The described move in position can be either in absolute or incremental units. Depending on the automation requirement, the typical automated cycle modes consist of various forms of INDEXes synchronized to other axes' motion and machine I/O status (Figure 10.10). Parameters used to describe the INDEX motion are *the distance, accelera-tion and decceleration rates, top speed or total index time period*. In some motion control applications an index motion may have multiple speeds. Such moves are called *compound indexes*. Compound indexes can be considered a combination of various trapezoidal and triangular motion profiles. It is also very common to define an index motion with only distance and total time period for the motion. In this case, the time period is used in three equal portions: one third for the acceleration period, one third for the constant speed run period, and one third for the decceleration period.

Example Consider a DC motor connected to a ball-screw (Figure 3.3). Let the ball-screw lead be 0.2 in/rev (or equivalently it has pitch of 5 rev/in). Assume that there are two limit switches, one at each end of the ball-screw, to indicate the mechanical limits of travel. In addition, there is a presence sensor at the mid point of the ball-screw's travel range to indicate a home position. Define a home motion sequence, a jog motion, a stop motion, and an incremental motion of 1.0 in distance movements.

A possible home motion may be defined as follows: the speed at which the home sensor trigger will be sought is 0.5 in/s. The acceleration rate is defined by the time allowed for the motor to reach the home speed, which is $t_{\text{acc}} = 0.5\,\text{s} = 500\,\text{ms}$. Then when the home sensor tigger turns ON, the motion will stop at the deceleration rate defined by the stop motion. Then the axis will wait for one second, move back 0.2 in in 1.0 s, wait for 1.0 s and

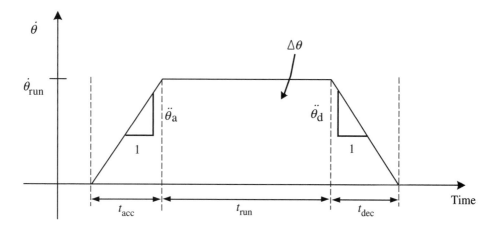

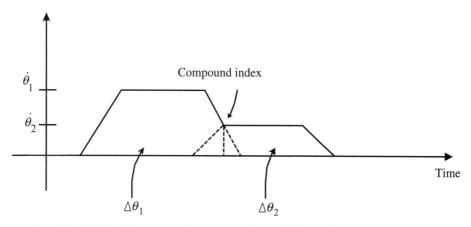

FIGURE 10.10: INDEX is a name given to a motion profile for a given distance (moving to an absolute position or moving to an incremental distance).

move with a much lower speed of 0.05 in/s until the home sensor turns ON again. Then, it stops.

The jog motion can also be defined using two motion parameters: the acceleration rate (or the time period t_{acc} to reach jog speed) and the jog speed value.

The stop motion definition requires only the decceleration rate and the condition on which the stop motion is initiated.

Finally, for index motion of 1.0 in increments in either direction, we can define the acceleration, run, and decceleration times. Notice that

$$v = a \cdot t_{acc} \tag{10.1}$$

$$\Delta x_{acc} = \frac{1}{2} a \cdot t_{acc}^2; \ during \ constant \ acceleration \ motion \ phase \tag{10.2}$$

$$\Delta x_{run} = v \cdot t_{run}; \ during \ constant \ speed \ motion \ phase \tag{10.3}$$

$$\Delta x_{dec} = \frac{1}{2} d \cdot t_{dec}^2; \ during \ constant \ deceleration \ motion \ phase \tag{10.4}$$

$$x = \Delta x_{acc} + \Delta x_{run} + \Delta x_{dec} \tag{10.5}$$

where a, d are acceleration and deceleration values, v is the top speed, $t_{acc}, t_{dec}, t_{run}$ are the time periods for acceleration, deceleration, and constant speed run phases of the index motion (Figure 3.13). Let us assume that the index time period is equally divided between acceleration, constant speed run, and deceleration time periods,

$$t_{acc} = t_{run} = t_{dec} = \frac{1}{3} \cdot t_{index} \tag{10.6}$$

Let $t_{index} = 3.0 \, s$. The motor rotation to make the 1.0 in move is $1.0 \, in \cdot 5 \, rev/in = 5 \, rev$. It can be shown that the total traveled distance is

$$x = \frac{1}{2} \cdot v \cdot t_{acc} + v \cdot t_{run} + \frac{1}{2} \cdot v \cdot t_{dec} \tag{10.7}$$

$$= \left(\frac{1}{2} \cdot \frac{1}{3} + \frac{1}{3} + \frac{1}{2} \cdot \frac{1}{3} \right) \cdot v \cdot t_{index} \tag{10.8}$$

$$= \frac{2}{3} \cdot v \cdot t_{index} \tag{10.9}$$

$$v = \frac{3}{2} \frac{x}{t_{index}} \tag{10.10}$$

$$= \frac{3}{2} \frac{1.0 \, in}{3.0 \, s} \tag{10.11}$$

$$= 0.5 \, in/s \tag{10.12}$$

The pseudo code to define these motion sequences may look as follows. The exact syntax of the motion control software will depend on the particular motion controller.

```
% Assume we have two functions "MoveAt(...)" and "MoveFor
   (....) " to generate the desired motion.
% If these motion command (trajectory) generator functions
   are defined in C/C++, they can be overloaded to accept
% different argument list.
%
%
%  Calculate or set parameters

     Home_Speed_1 = 2.5    % [rev/sec]
     Home_Speed_2 = 0.25   % [rev/sec]
     Home_Sensor =  1      % I/O channel number for the home
                             sensor
     Jog_Speed =    1.0    % [rev/sec]
     Jog_Stop  = 2         % I/O channel to indicate to stop
                             motion.
     Stop_Rate =    10.0   % [rev]/[sec^2]
     Index_Value = 5.0     % [rev]
     t_acc   = 1.0         % [sec]
     t_run   = 1.0         % [sec]
     t_dec   = 1.0         % [sec]


% Home motion

     MoveAt (Home_Speed_1)
     Wait until Home_Sensor = ON
     MoveAt (Zero_Speed, Stop_Rate)
     Wait for 1.0 sec
     MoveFor(Index_Value)
     Wait for 1.0 sec
     MoveAt (Home_Speed_2)
```

```
      Wait until Home_Sensor = ON
      MoveAt (Zero_Speed, Stop_Rate)

% Jog Motion

      MoveAt (Jog_Speed, t_acc )
      Wait until Jog_Stop=ON
      MoveAt (0, Stop_Rate)


% Stop motion

      MoveAt (0, Stop_Rate)

% Index motion

      MoveFor (Index_Value, t_acc, t_run, t_dec )

%
```

10.5 COORDINATED MOTION CONTROL METHODS

In PMCS that involve multiple actuators (also called *multi degrees of freedom* or *multi axis motion control system*), the motion coordination (synchronization) between different axes can be categorized as follows (Figure 10.11):

1. point-to-point control applications (insertion machines, assembly machines, pick-and-place machines),
2. speed ratio (electronic gearing) applications (coil winding, packaging, printing, paper cutting, web handling machines),
3. contouring applications (CNC machine tools, robots, laser cutting machines, knitting machines),
4. sensor based motion planning and autonomous motion control.

There are two very common motion coordination methods in the web handling automation industry and they use some of the above coordination methods. They are

1. motion coordination using a "registration" signal from an external sensor (also called *registration application*),
2. tension control applications (paper, plastic, wire handling machines).

Both of these applications use electronic gearing as the basis of motion coordination and add a motion modification to it based on the external sensor (registration sensor or tension sensor).

10.5.1 Point-to-point Synchronized Motion

Point-to-point position synchronization refers to the positioning of one axis or more with respect to another axis at a selected number of points during a cycle. In this type of application, there are a finite number of important points during each cycle where the relative positions are critical, that is in insertion applications, the pin and the housing must be properly positioned right before the insertion takes place. When an axis (master) is in a certain position, the other axis (slave) must be in a certain position relative to the master axis. The relative position of the slave with respect to the master is important at

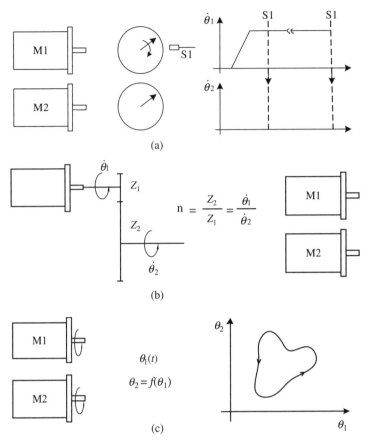

FIGURE 10.11: Motion coordination between two axes: (a) trigger at a certain position, (b) electronic gearing, (c) camming.

the synchronization reference points. During the motion between these points, the relative positions between axes is not critical. For instance, in pin-insertion applications, as long as the pin is in the correct position relative to the housing before the insertion starts, the motion profile of how each axis reaches that point is not important (Figure 10.11a).

Consider a two axes motion control system. Assume that axis one is the master and runs at a constant speed. During every cycle of the axis 1 motion, axis 2 is supposed to make predefined motion when axis 1 is at a specific position in a given cycle relative to the beginning of the cycle. If the move of axis 2 is defined as a function of time, then the only information needed to coordinate the motion is:

- Axis one: one cycle distance, zero reference at each cycle, current position within the current cycle in order to generate the trigger signal for the axis 2.
- Axis two: when triggered by axis 1 position, make a predefined move. In turn, at the completion of axis 2 motion (or at a certain position of axis 2 during its motion), it may trigger axis 1 to start a new cycle.

If the desired motion of axis 2 needs to be done within a certain distance of axis 1 motion, the motion of axis 2 is defined as a function of axis 1, and that requires "electronic gearing" or "electronic contouring."

10.5.2 Electronic Gearing Coordinated Motion

In mechanically coordinated designs, one axis motion can be tied to the other axis through a mechanical gear. The motion of axis one axis is directly proportional to the motion of the other axis in position, velocity, and acceleration (neglecting transmission imperfections due to backlash and tolerances). In mechanical designs, the relationship is fixed by the gear ratio mechanically.

The drawback of mechanical gearing is that if a different gear ratio is needed for a different product, the gear ratio between the shafts needs to be mechanically changed by installing different gears for every different gear ratio. This will increase the setup time. It may also be economically in feasible for applications involving many different gear ratios to keep many different gears in stock. The functionality needed here is to provide a motion to the second shaft which is proportional to the motion of the first shaft. The proportionality constant may vary for different products.

If both shafts have their independent actuation source, the commanded motion to the slave axis can be derived from the commanded or the actual motion of the master shaft. In the process of calculating the commanded motion for the slave shaft, any desired gear ratio can be used (Figure 10.11b). The gear ratio can be constant or variable. Furthermore, in certain positions in the cycle additional forward or backward moves can be implemented to change the phasing of the slave shaft with respect to the master shaft. This type of motion profile generation is called *electronic gearing* or *software gearing* since it is done electronically by software implementation as opposed to mechanical gears. The position tracking accuracy should be at least very close to those accuracy that would be obtained if the shafts were connected to each other through mechanical gears.

The gear ratio is set in software and easily changeable without any time consuming mechanical disassembly and re-assembly. Electronic gearing is perhaps the most common type of programmable motion control application encountered in industry today. Let us consider two axes: each are driven by a separate motor. Axis 1 is the master. The axis 2 desired motion can be generated from the motion of axis 1 using a fixed (or changing) ratio. As long as the servo loop delivers a good position tracking accuracy, the net result is a two-axis motion system making geared motion. The commanded motion to axis 2 can be generated from either the actual or the commanded motion of axis 1. The acceleration and deceleration rate of the slave axis can be defined as a function of the master axis position change instead of being defined as a function of time. In all electronic gearing applications, the slave axis motion profile can be defined as a function of one of two independent variables:

1. time, or
2. master axis position.

The accuracy of the electronic gear tracking is as good as the tracking accuracy of each slave axis whatever method of control is used. Therefore, for the electronic gearing to be successful, the servo loops should be properly tuned. Otherwise, the tracking errors may deteriorate the performance to the point that it does not resemble the performance that would be achieved had the shafts been connected to each other by mechanical gears.

Fixed Ratio Gearing   This is the simplest form of electronic gearing. It exactly emulates the relationship between two shafts connected to each other via a fixed gear ratio. During the cycle, the slave axis (or axes) are commanded to move at a certain gear ratio relative to the master axis motion. It is also very straightforward to use two or more fixed gear ratios between two axes where the gear ratio is changed from one fixed value to another at certain positions during a cycle. For instance, rotating knife applications where the web

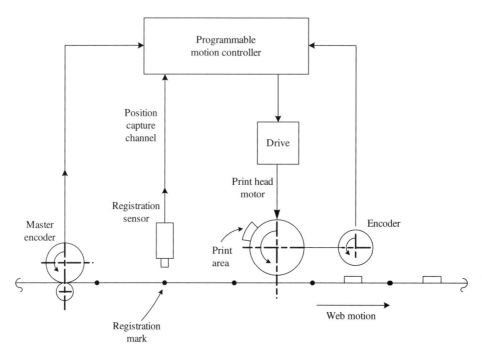

FIGURE 10.12: Web handling with registration application. A print head (or a cut head) follows the master encoder in order to match the speed of the web. It makes phase adjustments to its position based on a registration sensor. The position of the print head must be captured very accurately when the registration sensor triggers ON. The registration sensor must have good repeatability and fast response in order to have a good positioning accuracy for the print head.

needs to be cut at a certain length (cut length is programmable). During the position range when the knife is in contact with the web, the linear motion speed of the knife must match the linear motion of the web (gear ratio z_1), and during the position range when the knife is not in contact with the paper, the knife must speed up or slow down to a different gear ratio (z_2) to let the desired amount of web length move (Figures 10.12, 10.13b).

Gearing with Registration Printing, cutting, and sealing applications in web handling processes require motion coordination which involves gearing plus registration. This very common form of motion coordination involves the following (Figures 10.12, 10.13):

1. electronic gearing of a slave axis to a master axis with a certain gear ratio, and
2. for each cycle, after a sensor input condition (registration mark or registration signal) is true, the slave axis must make an incremental move during a certain distance of the master axis. This is called phase adjustment based on registration mark sensing. A variation of this approach is to calculate a new gear ratio and change the gear ratio in order to get the same amount of additional move.

There are many variations of the registration mark applications. The incremental motion distance may be either a predefined length or calculated based on the difference between the actual position captured when the registration sensor triggers and a desired position. The keys to achieving high accuracy in registration applications are to capture certain motion related conditions very quickly (i.e., less than 25 μs) and respond very quickly (i.e., a few millisecond).

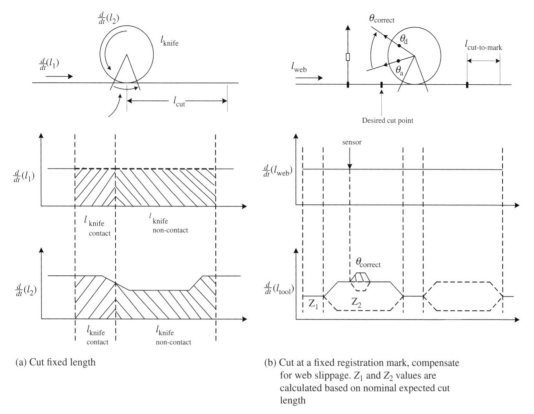

(a) Cut fixed length

(b) Cut at a fixed registration mark, compensate for web slippage. Z_1 and Z_2 values are calculated based on nominal expected cut length

FIGURE 10.13: Web handling with registration mark application: (a) Cut a web every given length (l_{cut}) and there is no registration mark. (b) Cut web always at a certain distance from a registration mark.

For instance, in paper cutting machines, the flying knife can be geared to the master axis moving the paper. The paper must be cut at a certain offset distance from a registration mark. However, the registration mark may not be perfectly printed on the machine at exact distances from each other. The paper may slip in the master axis drive such that the master encoder reading of paper movement cannot always be perfect. To overcome such unavoidable problems, a registration mark is always sensed on the paper at a fixed distance from the cutting point. Regardless of the printing or master ecnoder accuracy loss due to slippage, there has to be a fixed phase relationship between the cutting knife and the instant of the registration mark. Therefore, the solution is:

1. run the knife at nominal gear ratio relative to the master axis (i.e., using the master axis encoder position as the master position) feeding the paper,

2. at each registration mark sensor trigger, capture the knife axis position, compare the actual position with the desired position, and correct the phase error, if any, with superimposed incremental moves on top of the current gearing motion.

The position capture on the registration sensor input must be done very fast. This generally requires the use of axis level high speed I/O lines instead of the general purpose machine I/O. Let us consider a paper cutting machine example. The paper is cut at a certain distance from a registration mark. Based on the nominal expected value of the cut length, a nominal gear ratio is set between the paper line speed (master axis) and the rotation knife axis (slave axis). When the controller recieves the registration mark signal, it captures the knife position. Then it compares that captured position with the desired position that the knife should be at in order to cut the paper at the right location. Then, the knife is commanded to make an incremental move to make the phase correction. Let us consider that the paper cutting machine runs at 1200 ft/min line speed. When the registration mark is sensed, the controller will have to capture the knife axis position within a certain finite time. At 1200 ft/min rate, paper moves 20 ft/s or 240 in/s or 240/1000 in/ms. If the paper is to be cut with ±5/1000 in accuracy, the position must be accurately captured before that amount of paper passes. The 5/1000 in paper movement takes about 40 μs. Therefore, the position capture on the registration mark must be done within 40 μs. This type of response requires dedicated axis level high speed I/O. Notice that the correction of the position error does not (and cannot) need to be made within 4 μs, rather within the period remaining in the cycle which is typically in the order of tens of milliseconds.

10.5.3 CAM Profile and Contouring Coordinated Motion

Consider a shaft driven by an independent actuation source (i.e., a motor-drive) and a mechanical cam on the shaft. The cam can be on the face of the shaft or around the shaft. A tool is connected to the cam profile through a cam follower. As the shaft rotates, the tool moves up and down as function of the cam profile machined on the cam groove. For every point on one revolution of the shaft, there is corresponding point on the cam, hence there is a corresponding position of the tool (Figure 3.9). This relationship is mechanically determined by the cam design. For every revolution of the shaft, the cam relationship between the shaft and the tool is repeated. More than one cam can be connected to a shaft to drive more tools as slaved to the shaft motion. The slave motion synchronization for each tool is determined by the cam profile machined into each cam. This type of synchronization is more complicated than a constant gear ratio relationship.

The functionality in cam synchronization is that the slave axis position is directly a function of the master axis position. The relationship repeats every revolution of the master axis output shaft.

The same functionality can be more flexibly achieved by *electronic cam* or *software cam* motion coordination. Both master and slave must have their independent actuation source (motor-amplifier). As the master axis moves, the desired (commanded) motion for the slave axis can be derived based on the position of the master axis. The desired relationship in this case happens to be a cam profile. The cam profile is defined only in software as a mathematical equation or a look-up table. As the master axis moves, the position of the master axis is periodically sampled. Usually, the sampling rate is the same as the servo loop update rate. Then the corresponding position where the slave axis should be is derived from the mathematical cam profile equation or from the look-up table. Furthermore, in software implementation, the slave axis does not have to be driven by the cam relationship throughout the whole revolution or cycle. The cam following mode can be entered and exited at various points during a cycle based on some application dependent condition.

Contouring coordination is mostly used in machine tools and plasma cutting types of machines. The contour coordination is a more complicated form of cam coordination. The basic idea is that two or more axes must move in space in order to make a tool trace a certain path. In two-dimensional *x-y* motion, this corresponds to drawing an arbitrary curve in plane, in three-dimensional *x-y-z* motion, it corresponds to drawing an arbitrary curve in 3D space. At any given time all of the axes have a position that they must be at in order to trace the desired path. The way the commanded motion is generated for each axis can vary: one alternative is that the feedrate along the path (tangential linear speed) can be the indepedent variable. The feedrate can be planned based on application specific requirements. Then as the tool travels along the path, the desired motion of each axis is calculated and commanded. In this case, the feedrate acts as the master (although it is not an axis), and all of the motion axes are slaved to it. Another alternative is to select one of the axis as master and define its motion, then derive the desired motion of other axes so that the tool traces the desired path. CNC machine tools are an example of the coordinated motion control application where contour coordination is most commonly implemented.

Consider a two-axis motion control application where the two axes are required to trace a contour, that is an XY stage holding a tool, and the tool is required to trace a path in *x-y* plane. Gearing coordination would generate a path that is a straight line. Contouring is a more general form of coordination and requires the ability to trace any path in the motion space.

The path can be defined as function of:

1. Time, each axis motion is separately defined as a function of time to generate the path.

2. One axis is set as master and its motion for the path is defined, the other axis' motion is set as a function of the master axis position.

3. Path length or speed parameter – each axis motion is defined in terms of a path parameter.

CNC Programming Computer numeric control (CNC) programming is a programming language used to define the motion of machine tools. The most common CNC language in use today is the G-code. The G-code standard is defined by the Electronics Industry Association (EIA) standard 274-D. Although this standard exists, there are minor variations in the implementation of G-code from one manufacturer to another. The meaning of some codes also varies from one machine tool type to another. For example, G70 code means programming in units of inches in machining centers, whereas the same G70 code means edge finding in electric discharge machines (EDM).

10.5.4 Sensor Based Real-time Coordinated Motion

As the sophistication of computer controlled machines increases, the type of motion demands placed on them becomes more and more complicated. Particularly in robotic manipulator applications, the desired motion may not be known in advance. The machine is required to sense its environment (i.e., using vision systems), and decide on a motion strategy and generate the motion command profiles for each individual axis. The motion synchronization used may be different for different phases of its motion. Furthermore, this decision as to what kind of strategy to use is determined online by the control software. Interpretation of sensory data, and generating intelligent motion planning strategies not planned in advance, is one of the current challenges of programmable motion control in robotics devices.

10.6 COORDINATED MOTION APPLICATIONS

10.6.1 Web Handling with Registration Mark

The following applications have very similar motion coordination requirements (Figures 10.12, 10.13):

1. rotating cutting knife to cut a web (i.e., paper, plastic) to a fixed length,

2. rotating printing head to print over a web,

3. rotating sealing head to seal a web or bag.

There are three key issues in general web handling applications with registration marks:

1. match linear speeds between the web and tool during contact phase,
2. adjust tool speed to meet the cycle web length requirements during non-contact phase,
3. adjust tool position (phase) with respect to the registration mark.

The common key motion coordination requirement in these applications is that during a certain length of each cycle of web movement, the tool (cutting knife, print head, sealing head, etc.) must be moving at the same linear speed as the web. Usually, when the tool is in contact with the web (i.e., the knife is in the process of cutting the paper, the print head is in contact with the web), the linear speed of the tool and the web must be the same. Some printing heads print through the whole circumference of the print cylinder. In that case, the linear speeds must always be matched. The whole cycle is a contact phase. Only position phase corrections can be done in order to print at the correct position relative to the registration mark.

When the tool is not in contact with the web, the tool speed can be adjusted in order to let the proper amount of web distance pass. If the design was not a programmable motion control system and it was mechanically geared, a rotating knife machine could only cut one paper length only which is the length of the knife circumference. The ability to program the motion of the knife so that while it is in contact with the web it matches the linear speed for a correct straight cut, and speed up or slow down while it is not in contact with the web, provides us with the capability to cut many different lengths of web with one rotary knife. The same discussion applies to print heads and sealing head applications where the cycle length for which a new print or seal is to be made can be programmed.

If the web does not have a registration mark, the motion coordination problem is easier. There are only two phases to the coordination:

1. Contact phase – tool and web are in contact.
2. Non-contact phase – tool and web are not in contact.

The web axis motion is always the master since it is difficult to make the web do whatever we want it to do. Rather, we need to follow and track the web with the tool axis. In contact phase, the tool axis gear ratio needs to be adjusted so that the linear speed between the web and tool are equal. If the gear ratio is programmed in linear motion units, it will be 1 to 1. During the non-contact phase, the gear ratio will be different than 1 to 1 linear ratio. The gear ratio will be such that the remanning length of web will pass during the time that the tool moves over the non-contact phase. If the web cycle length happens to be equal to the circumference of the rotary tool, then the gear ratio will be 1 : 1 during non-contact phase in linear distance units as well (Figure 10.13a).

More often, a cut must be made relative to a registration mark (Figures 10.12, 10.13b). A print must be made repeatedly but at a certain offset distance from a registration mark every cycle. Similar needs exists for sealing type applications. In order words, the tool will cut/print/seal the web every cycle where the web length per cycle is programmable, but it must be done relative to a registration mark. If the registration mark and length requirements are in conflict, the registration mark requirement will supercede. In other words, let us assume that we would like to cut a paper from a registration mark at 1.00 distance and 12.00 total length. When we receive the registration mark, we record that only 11.5 inches of paper has passed the knife. The conflict is this: if we cut the paper at 1.0 inch distance, the paper length will be 12.5 inches. If we cut it at 12.0 inch length (0.5 inch after the registration mark), the offset distance from the registration mark will be 0.5 inches instead of 1.0 inches. There is no solution to this problem since with one measure, we can not satisfy two different requirements. Quite often, the offset distance from the registration mark takes priority over the cycle nominal length since the web can slip and the master axis reading may not be accurate. The registration mark provides us with a positive sensing mechanism for the actual web length passed.

From an electronic gearing motion coordination point of view, the registration mark requirement adds one more motion coordination problem. That is, the tool must be in correct positional phase relative to the mark in order to cut it at the right offset location. Assuming that there is no slip or variation in registration mark location, once the tool and paper are in phase, they will stay in phase for a fixed cycle length. However, to ensure accuracy in every cycle against web slippage or variations in web registration mark location, the tool-web position phase is checked and corrected if necessary.

```
Motion Coordination Algorithm: Rotating Tool (Cutting Knife,
Print Head, Seal Head) with Registration Mark Application:

   Main Program
   {
      while (TRUE)
      {
         Initialize the rotating knife with registration mark
         application algorithm
         Verify HOME motion sequence is done
         Check Task 1
         Check Task 2
      .....
            Check Print Phase and Registration Mark
            {
               When Enterered Contact Phase, execute
               Contact_Phase Function once
               When Entered Noncontact Phase, execute
               Non_Contact_Phase Function once
               When Registration mark is received, excecute
               Registration_Mark_Phase_Adjustment
            }
      }

   }


   Algorithm Initialize:
         web_cycle_length =    ... % web length per cycle
         tool_circumference = .... % pi * diameter
         contact_percentage = .....   % 0 to 100
         gear_ratio_contact_num = 1
         gear_ratio_contact_den  = 1
         gear_ratio_noncontact_num = web_cycle_length *(100-
         contact_percent)/100
         gear_ratio_noncontact_den  = tool_circumference *(100-
         contact_percent)/100

         define desired position phase for tool on registration
         mark
         setup fast position capture (actual or commanded) of
         tool axis on registration mark

   Return

   Contact_Phase:
         change gear ratio of tool (slave) to web (master) to
         ratios defined by
         z = gear_ratio_contact_num / gear_ratio_contact_den
      Return
```

```
Non_Contact_Phase:
      change gear ratio of tool (slave) to web (master) to
      ratios defined by
      z = gear_ratio_noncontact_num / gear_ratio_
      noncontact_den
Return


Registration_Mark_Phase_Adjustment:
      Get the position of the tool captured on registration
      mark with high speed capture
      Compare the actual position to the position desired
      for proper phase
      Command additional move (Position desired - Position
      captured) to complete fast
Return
```

10.6.2 Web Tension Control Using Electronic Gearing

Web handling is a generic name used to describe manufacturing processes where a continuous web must be moved and processed. The web material is generally one of the following (Figures 10.14, 10.15):

1. paper in printing machines, in paper cutting machines,

2. plastic in packaging and in labeling machines,

3. wire in winding and wire drawing processes,

4. sheets of steel in steel mills.

The web handling motion control problem has two components:

1. the web must be moved at a certain process speed, which is usually referred to as "line speed,"

2. while being moved, web tension must be maintained at a desired level.

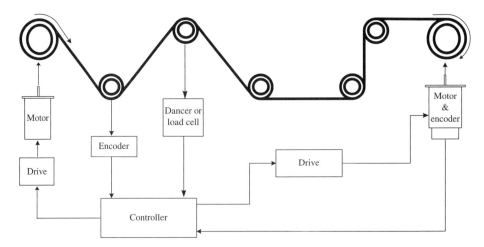

FIGURE 10.14: Web motion control with tension regulation. Either one of the two motors can be used to regulate tension. The encoder is used to determine the speed of the web. The dancer/load sensor is used to measure the web tension. The goal is to move the web at a desired speed while maintaining a desired tension during the motion.

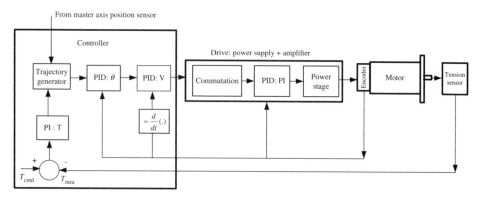

FIGURE 10.15: A servo control loop to control tension in the web. There are two different common implementations: (i) No master axis speed is provided. The commanded speed of this axis is modified (increased or decreased) based on the tension control loop: i.e., $\dot{\theta}_{cmd} = \dot{\theta}_{cmd0} + \Delta\dot{\theta}_{cmd0}(T_d - T_a)$. (ii) a master axis speed is provided and the gear ratio between the master axis and this (slave) axis is modified based on the tension control loop: i.e., $\dot{\theta}_{cmd} = z \cdot \dot{\theta}_{master}, z = z_0 + \Delta z(T_d - T_a)$.

In general, the production rate dictates the line speed at which the web must be moving nominally. While the web is moving, the tension of the web must also be measured and corrections must be made to the feed in order to maintain the desired tension. In order to both control the speed and tension on a web, there needs to be two actuation sources: one to set the nominal line speed, the other to track the first nominally but to maintain the desired tension.

Consider that the web speed is set by the nip roller (or similar mechanism) which is specified by the required process speed. The unwind/rewind roll speed needs to be adjusted in order to maintain the desired tension while the web is being moved. In unwind applications, if the tension is less than the desired tension, the unwind roll speed needs to be decreased. The opposite is true in the rewind applications. That is, if the tension is less than the desired tension, the rewind roll speed needs to be increased. This is referred to as the polarity of the tension control loop.

$$\Delta V = (sign) \cdot k \cdot (T_d - T_a) \tag{10.13}$$

where the sign is $(+1)$ for unwind applications, and (-1) for rewind applications. The polarity is defined as the sign (positive or negative) ratios of the change in the tension to the required change in the speed of the drive used to control the tension. The polarity of unwind and rewind tension control are opposite (10.14).

The differences between using a strain-gauge based tension sensor versus a dancer arm based tension sensor are as follows:

1. Strain-gauge based tension sensor results in larger loop gain since small changes in web in-feed and out-feed results in large tension changes compared to the dancer arm based tension sensor. On one hand, if the web is too stiff, the loop gain as a result of small differences in in-feed and out-feed web may be so large that large oscillations and closed loop instability may occur. With sufficiently flexible web, on the other hand, this method would provide the fastest closed loop response. The web stiffness and tension sensor sensitivity combination must be carefully judged to determine whether they will result in good tight closed loop performance with large loop gain or they will result in too large loop gain that will create instability problems.

2. Dancer arm based tension sensor results in lower loop gain since the same change in web in-feed and out-feed will result in smaller tension variations. This effectively reduces the loop gain of the closed loop system and instability problems are less likely to occur. The potential problem with this method is the inertia and spring effect of this type of tension sensor. The dancer mechanism has some inertia. It also usually has a preloaded spring. As the dancer moves under the web tension, it may tend to oscillate. The oscillations are more significant during high acceleration and deceleration. Therefore, the oscillations of the dancer arm itself causes fluctuations in actual tension. It may also excite natural resonance of the web and result in large oscillations. The only solution would be to move the web in such a way that the dancer arm is not displaced too fast, which means the variation in web in-feed and out-feed rate must be very slowly varying.

We will consider two different control ideas which differ only in their control algorithms. In both approaches, we use the same nip roller drive and tension transducer. Let us assume that the nip roller is commanded to run at a certain speed with predefined acceleration and deceleration rates. The only difference is in the way we control the unwind/rewind roll in order to regulate the tension. All of the following ideas are identically applicable to both unwind and rewind application with the only exception being polarity difference. A tension control algorithm for unwind tension control problem can be applied to a rewind tension control problem by changing the sign of the output of the tension control loop output.

- Approach 1: Tension Control with Adjusted Velocity Command – web tension control by controlling the roll drive with inner velocity servo loop and outer PI tension servo loop.
- Approach 2: Tension Control with Electronic Gearing – web tension control by controlling the unwind/rewind roll drive with inner position loop commanded by electronic gearing, and outer tension loop which modifies the electronic gear ratio.

Approach 1: Tension Control with Adjusted Velocity Command   The traditional approach is to control the rewind/unwind roll motor using two closed loops: (i) inner loop is a velocity PI type-loop (compare desired velocity and measured velocity and command the amplifier based on velocity error) which presumably makes the motor track the commanded velocity within its bandwidth capability; (ii) outer tension PI-loop which generates the commanded velocity based on the tension loop error.

This approach has two limitations:

1. closed loop stability problem due to large loop gain,
2. slow acceleration/deceleration rate limitation.

Let us explain these limitations by focusing on the tension control loop control system. Figure 10.15 shows the closed loop of the tension control from a different perspective. The key observation to be made in this block diagram is that the loop gain from web tension error to roll speed change is large. In other words, small changes in in-feed and out-feed rates will result in large tension errors, and that in turn will result in large commands and possibly transient oscillation on the roll speed. Both of the above mentioned limitations are the result of large loop gain. Large loop gain not only causes stability problems, but also the acceleration/deceleration rates must be slow enough to avoid large oscillations. Large acceleration/deceleration rates will require large changes in commanded speed, which means large changes in tension. This means that the system will not be able to accurately regulate tension (keep tension error small) under large accelerations/deceleration cases of motion.

Approach 2: Tension Control with Electronic Gearing   Electronic gearing refers to a position servo controlled motion where the desired motion command is generated based on a gear ratio multiplied by a master (reference) motion source. In this case, the master is the nip roller speed. In software, we monitor the motion of nip roller at the servo loop update rate, and command the desired motion to the unwind roll proportional to the master motion speed. The proportional value is the gear ratio defined in the software.

The key difference in this approach is that the variation in tension is not allowed to get too large as seen by the feedback loop. The gear ratio in software is updated based on the tension error. The tension error is passed through a PI control algorithm, and the result is the updated gear ratio. The feedback loop sees much smaller tension error levels. Hence, it can have larger gain and larger acceleration/deceleration rates without causing closed loop stability problems. Notice that the gear ratio initial starting value must be accurate for this approach to be more accurate than the first approach. This can be obtained by either directly sensing the roll diameter at the beginning of a cycle or always starting with a known diameter known in advance or entered by the operator as setup information.

```
Control Algorithm  for Tension Control Using Electronic
Gearing

Initialize once:
    {Define/Input/Read}   tension loop polarity (unwind/
                          rewind) sign =  +1  or -1
    {Define/Input/Read}   desired tension:
                          Td  = ....
    {Define/Input/Read}   nip roll diameter
                          d1  = ....
    {Define/Input/Read}   current unwind/rewind roll diameter
                          d2  = .....
    Calculate  initial gear ratio:      z  = (d_1 / d_2)

     Initialize the parameters of the tension control
     algorithm (PI in this example):
            z_ I  =  0.0
            Kp    =  0.01
            Ki    =  0.001

End_Initialize

Update Periodically: (i.e. every 10 msec)

        Read tension sensor:            Ta
        Tension error:                  e_T = Td - Ta
        Integral portion of control     dz_i =  dz_i +
                                        sign *  Ki * e_T
        Proportional portion of control dz = sign * Kp *
                                        e_T
        P + I portion        dz =  dz + dz_i
        Calculate the new gear ratio      z  = z  +  dz
        Update the gear ratio 'z' in the electronically geared
        motion between master and slave axes.

EndUpdate
```

10.6.3 Smart Conveyors

Conveyors are one of the most common mechanical systems used in industrial automation. They can be considered "the work-horse" of mass production. A typical conveyor runs at

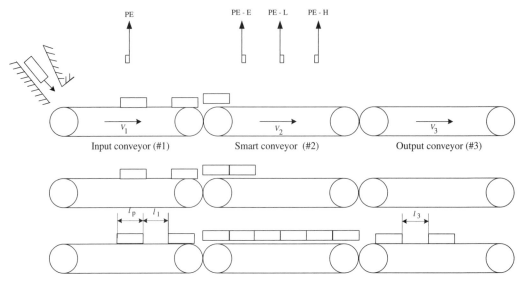

FIGURE 10.16: Smart conveyor operation.

a constant speed and continuously moves material from one end to the other. The "smart" conveyors differ from the typical conveyors in that in addition to running at a certain speed, they make position or speed adjustments in order to space parts uniformly even though the in-feed rate into the conveyor is not uniform.

A smart conveyor system typically involves three conveyors (Figure 10.16): 1. input conveyor, 2. smart conveyor, 3. output conveyor. In some cases, the smart conveyor and the output conveyor are the same conveyor.

The generic automation problem is as follows:

- The parts come onto the input conveyor at random intervals. There is a fixed average feed rate, i.e. is 300 parts per minute, but the spacing between them is not uniform. The objective is to adjust the spacing between the parts to a desired distance.

There are three basic questions about the application which affect the smart conveyor design:

1. Can the parts touch each other?
2. Does the speed of conveyors need to match the other adjacent conveyor speed during the part transfer between them?
3. Is there a limit on the acceleration and deceleration rate due to part considerations?

Depending on how the first two questions answered, the design can lead to one of two different approaches: (i) constant gear ratio spacing conveyors; (ii) position phase adjusting conveyors. The third question affects the number of smart conveyors to be used. If there is an acceleration/deceleration limit, multiple smart conveyors may need to be used in order to give more time for the acceleration/deceleration moves. Soft or liquid products have acceleration/deceleration limits so that their shape is not changed or spilled due to excessive inertial forces.

Constant Gear Ratio Spacing Conveyors with a Queue (Smart)
Conveyor

This design is appropriate when the parts can touch each other and the speeds of two adjacent conveyors do not have to match during the transfer. In this case, there will be three conveyors: 1. input conveyor, 2. queue conveyor (the smart conveyor), 3. output conveyor. Parts come to the input conveyor at a fixed average rate (N_1), but with non-uniform spacing. The queue (smart) conveyor runs slower than the input conveyor to provide a constant supply of continuous parts (Figure 10.16). The output conveyor runs faster to adjust the proper spacing. If the queue conveyor is always full and parts are touching, the spacing will be proportional to the speed ratio. Let us assume the part transfer rate is N_1. In continuous operation, the part rate in the conveyors must be same. Length of each part is L_p, and the average spacing is L_1. The speed of the queue conveyor will be determined by the requirement that the average part rate must be same in both conveyors:

$$N_1 = N_2 \tag{10.14}$$

$$\frac{V_1}{(L_p + L_1)} = \frac{V_2}{L_p} \tag{10.15}$$

$$V_2 = L_p * N_1 \tag{10.16}$$

$$= \frac{L_p}{L_p + L_1} \cdot V_1 \tag{10.17}$$

On the output conveyor, if the desired spacing between the parts is L_3, then the spacing conveyor speed needs to satisfy the same average part rate requirement,

$$\frac{V_2}{L_p} = \frac{V_3}{(L_p + L_3)} \tag{10.18}$$

$$V_3 = \frac{L_p + L_3}{L_p} \cdot V_2 \tag{10.19}$$

On the output conveyor, if the difference between V_2 and V_3 is so large that the acceleration rate is too high for the part integrity (i.e., soft parts or fluid content), then more than one spacing output conveyor would be used in order to reduce the maximum acceleration levels experienced by the part. From the queue conveyor point of view, the cycle time per part is

$$t_{cycle} = \frac{L_p}{V_2} \tag{10.20}$$

Let us assume that L_{pt} is the portion of the part length L_p during which the part is in contact with both conveyors and changes speed from V_2 to V_3. The acceleration rate is then

$$t_{transfer} = \frac{L_{pt}}{L_p} \cdot t_{cycle} \tag{10.21}$$

$$A = \frac{(V_3 - V_2)}{t_{transfer}} \tag{10.22}$$

The number of spacing output conveyors needed is determined by rounding the following result to the next highest integer value

$$N_{spacing} = Int\left(\frac{A}{A_{max}}\right) \tag{10.23}$$

where A_{max} is the maximum allowed acceleration/deceleration. If the maximum allowed acceleration or deceleration is a limiting factor, the conveyor design parameters can be calculated to satisfy this constraint. Given A_{max}, calculate either $L_{3,max}$ in t_{cycle} or the

t_{cycle} necessary for L_3. If we calculate the $L_{3,max}$, it will determine the number of spacing conveyors necessary following the queue conveyor. The result is same as

$$N_{spacing} = Int\left(\frac{A}{A_{max}}\right) \tag{10.24}$$

$$= Int\left(\frac{L_3}{L_{3max}}\right) \tag{10.25}$$

If we choose to use only one spacing conveyor and meet the acceleration/deceleration limits, then the t_{cycle} requirement is calculated for the desired L_3 spacing, which sets the maximum line speed achievable with one spacing conveyor while meeting the acceleration/deceleration limits.

In summary, the queue conveyor design is affected by the following parameters:

$$\{A_{max}, L_3, t_{transfer}, N_{spacing}\}.$$

If three of these parameters are specified, the fourth one can be calculated. The maximum allowed acceleration and deceleration rate and the desired spacing is given. Then, if we choose to use $N_{spacing}$ conveyor (i.e., $N_{spacing} = 1$ or 2), that determines the minimum $t_{transfer}$ (hence t_{cycle}) time period in order not to exceed the acceleration limit. That sets the maximum through-put rate for a conveyor with specifications: $1/t_{cycle}$

$$\{A_{max}, L_3, N_{spacing}\}$$

Let us consider the case that the input conveyor part feedrate may fluctuate. If it is desired that the conveyor system should be able to adjust to that condition, we need to add additional sensors. Assume we know L_p, and the desired spacing L_3. First, we need to measure the moving average part input rate, N_1. From that, we can calculate the nominal speed of the queue conveyor and the output conveyor. In addition, we can add three ON/OFF sensors on the queue conveyor to indicate the following conditions: queue high, queue low, queue empty. If the queue is low, the conveyor speed V_2 can be reduced by a percentage, if the queue is high it can be increased by a percentage. If the queue is empty, it should stop and go into a homing routine to fill the queue before continuing the normal cycle. As we modify the motion of the queue conveyor, the output conveyor will track it with the appropriate ratio since the output conveyor speed is gear ratioed to the queue conveyor as its master. A possible homing sequence before the automatic cycle begins is shown in Figure 10.16.

```
Algorithm: Constant Gear Ratio Spacing Conveyors with a
Queue Conveyor

  I/O Required:

    Sensors (Inputs):
        One presence sensor (Photo electric eye or proximity
        sensor) and control logic to count the part rate input

        Three presence sensors to detect queue conveyor part
        level: high, low, empty

    Actuators (Outputs):
        Queue conveyor speed control,

        Output conveyor speed control by electronically gearing
        (slaved) to the speed of the queue conveyor


Control Algorithm Logic:
```

```
Initialize:

   Given the process parameters: L_p - part length,
                                 L_3-desired part spacing,
   Given algorithmic parameter: V_percent  (i.e. 0.9 = 90 %
                                 speed reduction if queue is
                                 full)

   Output conveyor speed is electronically geared to the queue
   conveyor speed by:    V_3 = ((L_p+L_3)/L_p ) * V_2
   Where the gear ratio between queue conveyor (master) and
   the output conveyor (slave) is:     (L_p+L_3)/L_p

Repeat Every Cycle:
   Measure,  N_1, the part rate,
   Calculate queue conveyor speed:  V_20 = L_p * N_1
   Check queue level and modify V_2 accordingly,
   Update speed command to the queue conveyor (V_2)
   If V_2 = 0,  call homing routine to fill-up the queue
   conveyor.
Repeat Loop End

Return


Check queue level and modify queue speed algorithm:
      If (Queue is low (PRX-1 and PRX-2  are OFF) and
      PRX-3 is ON)
         V_2 =  V_20 / V_precent
      Else if  (Queue is High (PRX-1 and PRX-2 are ON )
      and PRX-3 is ON )
         V_2 =  V_20 * V_precent
      Else if (Queue is Empty: PRX-3 is OFF)
         V_2 =  0.0
      Else
         V_2 = V_20
      End if
Return




Homing Routine:
'   Queue conveyor is empty.
    Start input conveyor if it is stopped
    Disengage electronic gearing of  output conveyor
    While (PRX-3 is OFF)
      Wait until PRX-1 triggers, then move for L_p distance.
    Repeat
    Engage gearing of output conveyor
'   On return, queue conveyor is full and all parts are
    touching each other.
Return
```

Position Phase Adjusting Conveyor: Smart Conveyor Slaved to both the Input and the Output Conveyor

In most general form, the smart (also called shuttle) conveyor must coordinate its motion both with input and output conveyors at different phases of the motion. In some applications, coordination only to the input or only to the output conveyor may be sufficient. The smart conveyor acts as a very fast slave

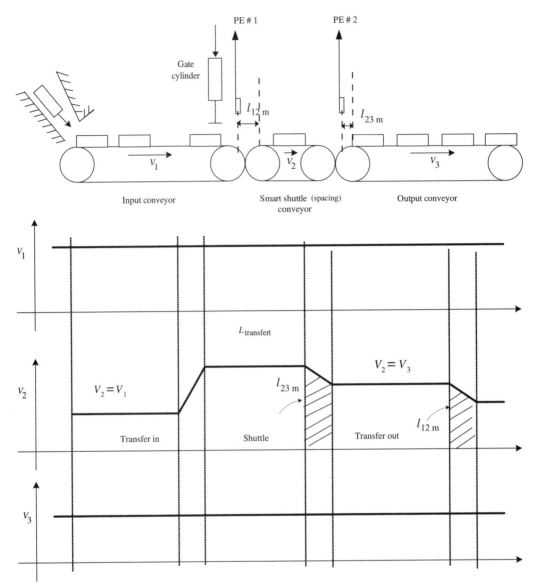

FIGURE 10.17: Another smart conveyor concept using a so-called "smart shuttle conveyor."

between two masters (Figure 10.17). The spacing conveyor is the slave. There are two masters, input and output conveyors, for the spacing conveyor to match linear speeds with during two phases out of three phases of each cycle.

Each cycle is divided into three phases:

1. Transfer phase from input conveyor to spacing conveyor (spacing conveyor matches linear speed to the input conveyor by considering the input conveyor as the master).

2. Non-transfer phase of spacing conveyor with no part contact to adjacent conveyors (spacing conveyor does not have to match linear speeds to any conveyor).

3. Transfer phase to output conveyor (spacing conveyor matches linear speed to output conveyor by considering the output conveyor as the master).

During each transfer (between input and smart conveyor, and smart conveyor and output conveyor), there are two phases of motion:

1. contact phase, and
2. non-contact phase.

During the contact phase, a part is on both conveyors and the speed of both conveyors must be matched during that phase. During the non-contact phase, a part is on only the smart conveyor and it can move independently of the other conveyors. It is this period during which the smart conveyor makes the position corrections. In order for this approach to work, the time period during which input and smart conveyors are coordinated should not have an overlap with the time period that the smart and output conveyors are coordinated. If the smart conveyor transfers one part at a time, this condition would be automatically met.

A possible design of a smart shuttle conveyor is shown in Figure 10.17.

1. A photo-electric sensor #1 (PE #1) is used to detect the beginning of the contact transfer phase of the part from the input conveyor to smart conveyor. This sensor must detect the incoming part before the physical contact between the part and smart conveyor. Over a distance from the detection point to the part contact to smart conveyor (l_{12m}), the smart conveyor must match the speed of the input conveyor. The PE#1 sensor must be located such that l_{12m} distance is long enough to satisfy the maximum acceleration/deceleration limits. If the l_{12m} distance is too short, the maximum acceleration/deceleration limit condition cannot be met.

2. After that, the smart conveyor must maintain the speed matching until it is time to prepare the transfer to the output conveyor, which is indicated by another sensor PE#2.

3. When the PE#2 signal indicates that the part is in the right position to start changing the speed to match the output conveyor's speed. Then the smart conveyor speeds up or slows down to bring the part to the output conveyor with proper spacing. The proximity sensor (PE#2) detects the trigger position when it is time to feed a new part to the output conveyor. Again, with a defined distance, the smart conveyor brings the part in contact with the output conveyor. Then, the smart conveyor matches speed with the output conveyor within the l_{23m} distance, and maintains that speed until the transfer is complete.

4. If before the transfer to the output conveyor is completed another part is about to enter the smart conveyor from the input conveyor, then the gate cylinder is actuated to stop the incoming part since the smart conveyor cannot match the conflicting motion requirements of two different conveyors at the same time. When the new transfer cycle begins, the gate cylinder is deactivated to start a new cycle.

10.7 PROBLEMS

1. Consider an xy table where x axis is stacked on top of the y axis. Each axis is driven by a brushless DC motor. Each ball-screw has 2.0 rev/in pitch. The motors have incremental encoders with 1024 lines/rev and the controller has ×4 encoder decoder circuits. a) Define the external sensors necessary so that each axis can be stopped before hitting the travel limits as well as a physical home reference position to be established after power-up. b) Assume that the homing search speed is 1.0 in/s. Write the pseudo-code of a homing motion. c) Calculate the parameters of an incremental move of 0.1 in increments made in 300 ms. Command this motion in forward direction five times and in reverse direction five times and allow 200 ms wait period between each incremental motion.

2. Consider two independedly driven rotary motion axes: axis 1 and axis 2. Our objective is to make axis 2 follow axis 1 as if they are connected with a mechanical gear ratio. Let the gear ratio be z. a) Draw the command motion generation block diagram for axis 2 for the case where the master motion it follows is the commanded motion of axis 1. b) Do the same, however this time the master motion it follows is the actual position of the axis 1. c) Further consider the case that axis 1 is the drive for an unwind roll and axis 2 is the drive for a rewind roll for a web application. Axis 2 needs to modify the gear ratio as a function of the sensed tension. Draw the block diagram and write the pseudo-code to accomplish this.

3. Consider the registration application shown in Figure 10.12. The rotating axis moves at a constant gear ratio relative to the speed of the web which is measured by an encoder. The objective is to make sure the rotating tool axis (print head or knife head) has to be at a certain position within one revolution when the registration mark passes under the registration sensor. When the registration mark sensor turns ON, we capture the position of the rotary axis and compare it with the desired position. Then, the rotating axis makes a corrective index motion on top of the gear ratio motion. Assume that the web is moving at 2000 ft/min. and a cycle is repeated every 1.0 ft. It is desired that the accuracy of the positioning and operation done by the rotating tool axis relative to the web must be with in $\pm 1/1000$ in. a) Determine the minimum psoition sensor resolution on the rotating axis. b) What happens to the accuracy if the registration mark sensor has 1.0 ms variation in its accuracy of responding to the mark? c) What is the maximum allowed variation in the registration mark sensor's response time? d) If the position capture is done under software polling and the application software may have 5 ms variation when it may sample the position of the rotating axis relative to the instant the registration mark sensor comes on, what would be the positioning error? e) What is the viable position capturing method in such applications?

4. Consider an xy table. Our objective is to design a programmable motion control system such that the xy table makes various contour moves. Assume the table x and y axes have already established a home position. a) Design a pseudo-code so that xy table draws a line with a given slope and length. b) Design a pseudo-code so the xy table draws a circle with a given center coordinates and diameter.

5. Consider the tension control system shown in Figures 1.7, 10.14 and 10.15. The example on page 7 shows a web control method that changes the speed command to one of the motorized rolls in proportion to the tension error. Assume that Figure 10.13 is a more complete and updated version of Figure 1.7 where the analog op-amp controller circuit is replaced with a digital controller. In addition, the wind-off roll is also a motorized drive, and there is a line position sensor (encoder). Update the control system using the electronic gearing method to control the wind-up roll. Assume the wind-off roll is running at a constant speed. The control algorithm should use the line speed encoder in Figure 10.13 as the master position to follow with a gear ratio, where the gear ratio is modified based on tension error through a PI type controller (Figure 10.14). Show the benefits of the electronic gearing based control method over the traditional direct speed control based on tension sensor method.

CHAPTER 11

LABORATORY EXPERIMENTS

11.1 EXPERIMENT 1: BASIC ELECTRICAL CIRCUIT COMPONENTS AND KIRCHOFF'S VOLTAGE AND CURRENT LAWS

Objectives

1. Confirm by measurements the input–output behavior of basic electrical and electronic circuit components: resistance measurement, capacitance measurement, diode test.

2. Confirm Kirchoff's voltage law and current law in a simple DC circuit with resistors.

3. Build a voltage divider circuit and verify the predicted results from Kirchoff's laws by measurement.

4. Build a current divider circuit and verify the predicted results from Kirchoff's laws by measurement.

5. Build an RC circuit and an RL circuit, measure their input output voltage relationships, and compare the differences between their responses. Measure the R, C, and L components in the RC and RL circuits.

Components

Item	Quantity	Part No.	Supplier
Resistors Assortment Kit	1	81832	Jameco Electronics (www.jameco.com)
Capacitors Assortment Kit	1	130232	Jameco Electronics (www.jameco.com)
Inductor Assortment Kit	1	388042	Jameco Electronics (www.jameco.com)
Breadboard	1	20722	Jameco Electronics (www.jameco.com)
Set of connection wires	1 set	20079	Jameco Electronics (www.jameco.com)

Theory

The passive components make up the main building blocks of electrical circuits: resistor, capacitor, and inductor. The input–output relationship of these "ideal" components are as follows.

An "ideal" resistor has the following current–voltage relationship, also called *Ohm's Law*,

$$V_{12}(t) = R \cdot i(t) \tag{11.1}$$

Mechatronics with Experiments, Second Edition. Sabri Cetinkunt.
© 2015 John Wiley & Sons, Ltd. Published 2015 by John Wiley & Sons, Ltd.
Companion Website: www.wiley.com/go/cetinkunt/mechatronics

where V_{12} is the voltage potential across the resistor, $i(t)$ is the current across the resistor, and R is the resistance of the component.

Resistors are marked with a color coding standard. Using this color code, one can determine the value of a resistor, including its possible tolerance (variation in value due to manufacturing tolerances). Using a DMM (digital multimeter), the exact value of a resistor can be easily measured.

An "ideal" capacitor has the following current–voltage relationship,

$$V_{12}(t) = V_{12}(t_0) + \frac{1}{C} \int_{t_0}^{t} i(\tau)d\tau \tag{11.2}$$

which says that the voltage across a capacitor is the initial voltage of the capacitor plus the integral of the current flowing through the capacitor scaled by the capacitance value C. $V_{12}(t)$ is the voltage across the capacitor at a given time t. Any capacitor will eventually saturate when it stores the maximum charge it can store. In order to limit the current coming into a capacitor from a DC power supply, a capacitor is never directly connected to a supply, but through a resistor. A mechanical analogy for a "capacitor" is a "water tank": initial water level (water height) in the tank is $V_{12}(t_0)$, water flow rate into the tank is $i(t)$, cross sectional area of the water tank is C, and water level at anytime is $V_{12}(t)$.

An "ideal" inductor has the following current–voltage relationship,

$$V_{12}(t) = L \cdot \frac{di(t)}{dt} \tag{11.3}$$

which says that the voltage across the inductor is proportional to the time rate of change of current. Another way of interpreting it is that the current is integral of voltage applied across it. If a constant voltage source is applied, the current would increase as integral of it, scaled by the inductance.

Kirchoff's voltage law states that the sum of voltages in a closed path of an electrical circuit is zero (conservation of voltage potential) at any given instant,

$$V_{14} = V_{12} + V_{23} + V_{34} \tag{11.4}$$

Kirchoff's current law states that the algebraic sum of currents at any point in an electrical circuit is zero, that is, the sum of incoming currents is equal to the sum of outgoing currents (conservation of electrons),

$$\sum i_1 + i_2 + i_3 = 0 \tag{11.5}$$

Consider the voltage divider circuit shown in Figure 11.1a and the current divider circuit shown in Figure 11.1b. Notice that a series of resistors and a voltage supply form a voltage divider. The voltage across each resistor is proportional to that resistor value relative to others. Whereas, a set of parallel resistors make up a current divider circuit. Larger current

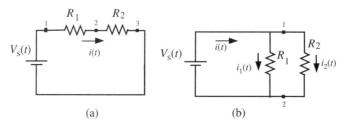

(a)   (b)

FIGURE 11.1: (a) Voltage divider circuit, (b) current divider circuit.

goes through the smaller resistor since there is smaller resistance to the flow of electrons. For the voltage divider circuit

$$V_s(t) = V_{12}(t) + V_{23}(t) \tag{11.6}$$

$$= R_1 \cdot i(t) + R_2 \cdot i(t) \tag{11.7}$$

$$i(t) = \frac{1}{R_1 + R_2} V_s(t) \tag{11.8}$$

$$V_{12}(t) = \frac{R_1}{R_1 + R_2} V_s(t) \tag{11.9}$$

$$V_{23}(t) = \frac{R_2}{R_1 + R_2} V_s(t) \tag{11.10}$$

The last two equations show how the voltage is divided between the series resistors.
     For the current divider circuit,

$$V_s(t) = V_{12}(t) \tag{11.11}$$

$$= R_1 \cdot i_1(t) \tag{11.12}$$

$$= R_2 \cdot i_2(t) \tag{11.13}$$

$$i(t) = i_1(t) + i_2(t) \tag{11.14}$$

$$= \frac{V_s(t)}{R_1} + \frac{V_s(t)}{R_2} \tag{11.15}$$

$$= \left(\frac{1}{R_1} + \frac{1}{R_2} \right) \cdot V_s(t) \tag{11.16}$$

$$V_s(t) = \frac{R_1 \cdot R_2}{R_1 + R_2} \cdot i(t) \tag{11.17}$$

$$= R_1 \cdot i_1(t) \tag{11.18}$$

$$= R_2 \cdot i_2(t) \tag{11.19}$$

$$i_1(t) = \frac{R_2}{R_1 + R_2} \cdot i(t) \tag{11.20}$$

$$i_2(t) = \frac{R_1}{R_1 + R_2} \cdot i(t) \tag{11.21}$$

Notice that the last two equations show how the current in the main line is divided over the two parallel branches of resistors. Current measurement requires the measurement instrument to be placed in series between the two points through which the current flow is being measured. This requires disconnecting the circuit at a point, and inserting the current measurement instrument (i.e., DMM in current measurement mode, ammeter) into the circuit. Such circuit modifications are not always possible nor convenient. Another way to measure current indirectly is to measure the voltage across a *known resistor*, then use Ohm's Law to calculate the current. If there is a resistor of known value at the circuit branch that we want to measure current, then we can measure the voltage across it, and divide the measured voltage by the resistor value to calculate the current that passes through that resistor. This method is easier and accurate enough for us to use in the experiments.

Procedure

1. Using the digital multi-meter (DMM), measure the resistance and capacitance of a few resistors and capacitors. Confirm your measurement with the specifications of the component.

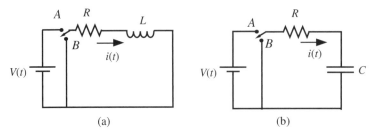

FIGURE 11.2: (a) RL circuit (b) RC circuit.

2. Using DMM, measure and determine the direction of current conduction of a diode.

3. Using DMM, measure the continuity across a mechanical switch by turning ON/OFF the switch (this is called the continuity test between two points in an electrical circuit).

4. Build a voltage divider circuit (Figure 11.1a). Confirm the Kirchoff's voltage and current law on the circuit.

5. Build a current divider circuit. Confirm the Kirchoff's voltage and current law on the circuit using various closed paths for voltage law and nodes for current law (Figure 11.1b).

6. Build the RL and RC circuits discussed in the Section 5.5 of the textbook (Figures 11.2 and 11.4), and duplicate the predicted results obtained from simulations. Use the R, L, C values you have on your experiment for the R, L, C values in your simulation model. To do that, measure R with DMM, and then the step response of the circuit voltages as discussed below, and estimate the C and L values from the step response measurements.

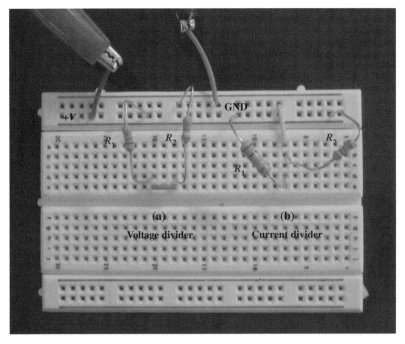

FIGURE 11.3: Picture of the circuits for: voltage divider and current divider RL circuit and RC circuit.

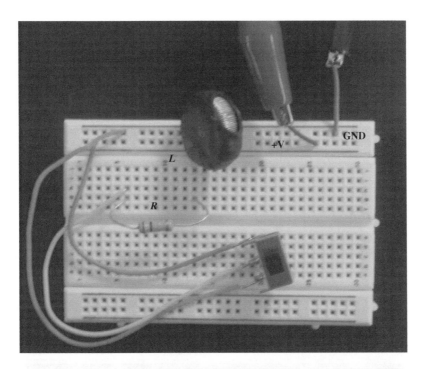

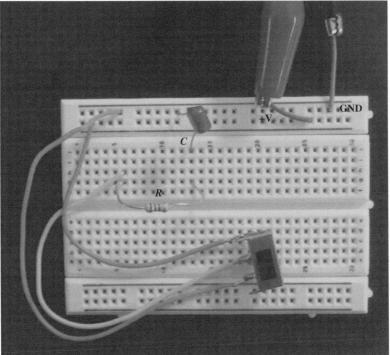

FIGURE 11.4: Picture of the circuits for: RL circuit and RC circuit.

7. In the RC circuit, measure the step response: input voltage is $V_s(t)$ which can be connected to a function generator to generate a square (or rectangular) wave signal instead of using battery and mechanical switch and output is the voltage across the capacitor. Measure the time constant of the response, τ: the time duration during which output voltage reaches 63% of its steady-state value. Note $\tau = R \cdot C$ for this RC circuit input–output transfer function.

8. Do the same for the RL circuit for step response. This time measure the voltage across L to measure the time constant. Note $\tau = L/R$ s for the RL circuit input–output transfer function.

9. Experiment with different values of the R, L, C in the RC and RL circuit and measure results in terms of input voltage, output voltage, and current. Verify these measurements with numerical simulations in MATLAB®/Simulink®.

Note that the mechanical switch can be removed from the circuit. Then, V_s supply voltage and ground can be connected to the function generator which is configured to generate a square (or rectangular) wave signal. This approach may be easier to capture the step response of the input–output relationships in the RC and RL circuit.

11.2   EXPERIMENT 2: TRANSISTOR OPERATION: ON/OFF MODE AND LINEAR MODE OF OPERATION

Objectives

1. Understand operating principles of a NPN type BJP transistor.

2. Design and build a circuit involving a NPN type BJP transistor in common emitter configuration.

3. Experimentally determine the ON/OFF mode and linear mode of operation, and measure the relevant voltages.

Components

Item	Quantity	Part No.	Supplier
Resistors as calculated, Assortment Kit	1	81832	Jameco Electronics (www.jameco.com)
BJP transistor, NPN type	1	38359PS	Jameco Electronics (www.jameco.com)
Breadboard	1	20722	Jameco Electronics (www.jameco.com)
Set of connection wires	1 set	20079	Jameco Electronics (www.jameco.com)

Theory

The transistor is like an "electron valve." It is the electrical analogy of the hydraulic valve. A hydraulic valve regulates the flow of fluid: it can be fully closed, fully open, or partially open. Rate of fluid flow is a function of the hydraulic pressure across the valve and the amount of opening of the valve. A transistor is an electrical valve. It regulates the flow of electrons. It can be fully OFF, fully ON, or partially ON. If a transistor is operated only in either fully ON or fully OFF mode, we call it ON/OFF mode operation. In the OFF mode, transistor state is called the cutoff state. In the fully ON mode, it is called the saturation state. In between them, it operates proportional to the base current and is said to be in linear mode.

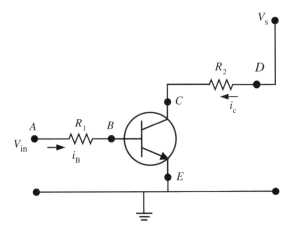

FIGURE 11.5: Circuit of a voltage amplifier using a transistor in a common emitter configuration.

The relationships between electrical variables (voltages and currents) and electrical parameters (resistance, transistor parameters) can be shown as follows (Figure 11.5). Kirchoff's voltage law between base (B) and emitter (E) indicates

$$V_{in} = V_{AB} + V_{BE} \tag{11.22}$$
$$= R_1 \cdot i_B + V_{BE} \tag{11.23}$$

where $V_{BE} = V_{FB} = 0.6\,V$ to $0.8\,V$ range. V_{FB} is the forward bias voltage between base and emitter and it can be up to 0.6 V to 0.8 V range typically. This value is a property of a transistor and is given in the datasheet of the transistor. If V_{in} is so small that that $V_{BE} < V_{FB}$, then no current flows at the base $i_B = 0$ and the transistor is in the cutoff (OFF state) mode. Similarly, applying Kirchoff's voltage law on the output side of the transistor, and noting that the transistor acts as a current amplifier between its fully OFF (cutoff) and fully ON states (saturation). The current gain (β) is a property of the transistor and provided in its datasheet. There can be variations in that gain up to 100% due to manufacturing variations. Robust circuit designs should not rely on the exact value of the current gain of the transistor.

$$i_C = \beta \cdot i_B \tag{11.24}$$
$$V_s = V_{DC} + V_{CE} \tag{11.25}$$
$$= R_2 \cdot i_C + V_{CE} \tag{11.26}$$
$$= R_2 \cdot (\beta \cdot i_B) + V_{CE} \tag{11.27}$$
$$V_s = R_2 \cdot \beta \left(\frac{V_{in} - V_{FB}}{R_1} \right) + V_{CE} \; ; \; for \; V_{in} \geq V_{FB} \tag{11.28}$$
$$V_{CE} = V_s - \left(\frac{R_2 \cdot \beta}{R_1} \right) \cdot (V_{in} - V_{FB}) \tag{11.29}$$
$$V_{DC} = V_s - V_{CE} \tag{11.30}$$
$$= \left(\frac{R_2 \cdot \beta}{R_1} \right) \cdot (V_{in} - V_{FB}) \tag{11.31}$$

Notice that V_{CE} can be between maximum value of V_s (when $i_C = 0$ for $i_B = 0$ when $V_{in} < V_{FB}$), transistor is in cutoff (OFF) mode), and minimum value of about $V_{CE,min} = 0.2\,V$ (when the transistor is saturated, fully ON) which is a property of the particular transistor.

We can measure the two input voltage values of interest: first, the minimum input voltage necessary to start making the transistor conduct, $V_{in,min}$. For input voltages below that the transistor will not conduct (OFF). Second, the value of the input voltage for which the transistor saturates (fully ON) and collector current and V_{CE} output voltage does not change, $V_{CE} = V_{CE,min}$ if input voltage is increased beyond that value ($V_{in,sat}$).

The $V_{in,min}$ is the voltage value which is necessary to provide the forward bias voltage plus just a bit more in order to make the transistor start conducting. So, this value is expected to be close to the forward-bias voltage of the transistor. This can be determined experimentally by slowly increasing V_{in} and monitoring the change in V_{CE}.

The $V_{in,sat}$ value is the value which provides a base current (i_B), and after being amplified results in $i_C = \beta i_B$ such that $V_{CE} = V_{CE,min}$. Voltages more than that will not result in any change in the output since the V_{CE} can be no less that $V_{CE,min} \approx 0.2$ V and the rest of the available supply voltage is used to generate the current i_C. From the voltage relations of the output and input circuits, we can calculate the input voltage saturation value and experimentally measure to confirm it approximately,

$$i_C = \frac{V_s - V_{CE}}{R_2} \tag{11.32}$$

$$= \frac{V_s - 0.2}{R_2} \tag{11.33}$$

$$i_B = \frac{1}{\beta} \cdot i_C \tag{11.34}$$

$$= \frac{1}{\beta} \cdot \frac{V_s - 0.2}{R_2} \tag{11.35}$$

$$V_{in,sat} = R_1 \cdot i_B + V_{BE} \tag{11.36}$$

$$= R_1 \cdot \frac{1}{\beta} \cdot \frac{V_s - V_{CE,min}}{R_2} + V_{FB} \tag{11.37}$$

$$V_{in,sat} \approx R_1 \cdot \frac{1}{\beta} \cdot \frac{V_s}{R_2} + V_{FB} \tag{11.38}$$

where $V_{BE} \approx V_{FB} = 0.6$ V to 0.8 V range, $V_{CE,min} \approx 0.2$ V. Exact values of these voltages may vary a little from transistor to transistor. However, we can accurately measure them experimentally by carefully recording when the transistor makes the transition from cutoff state to conducting state in the linear region and from linear region to fully ON saturation region.

Procedure

1. Design and build the circuit shown in Figures 11.5 and 11.6.

2. Choose $R_1 = 10\,k\Omega$, $R_2 = 1.0\,k\Omega$, $V_s = 9$ V. Other values for R_1 and R_2 can be chosen. And other available DC voltages can be used for V_s.

3. Provide a means of adjusting the voltage input at the base of the amplifier at point A. This can be done by using either a DC power supply or a battery (i.e., 9 V plus an adjustable series resistor (R_s) that would act as a voltage divider when used in series with the R_1 resistor) or through the DAC output channel of the PIC microcontroller.

4. Measure the voltage at the following points, V_{in}, V_{AB}, V_{BE}, V_{CE}, V_{DC}, and make a table of them, each voltage representing a column on Table 11.1.

5. Set V_{in} to different values in increments, that is, 0.0 V, 0.1 V, ..., 2.0 V, and measure the other four voltages and record them using your digital multimeter (DMM) or oscilloscope.

6. Plot the input voltage (V_{in}) and output voltage relationships (V_{CE}, V_{DC}) and conclude in what range of input voltage the transistor is fully OFF, fully ON, and in proportional (linear) amplifier mode.

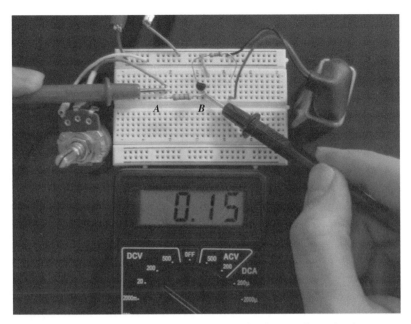

FIGURE 11.6: Picture of the experiment setup for the transistor circuit.

7. Discuss how the input voltage and output voltage relationship would change if the resistor values R_1 and R_2 were to change, that is $R_1 = R_2 = 10\,\text{k}\Omega$.

8. (Optional) Obtain a way to generate PWM input voltage at the V_{in} location, that is PIC microcontroller PWM output, or an op-amp circuit. Select the magnitude of V_{in} to be large enough to fully turn on the transistor. Select a PWM frequency that is large enough compared to a typical electromechanical actuator bandwidth (for actuator control applications, i.e., actuator bandwidth 100 Hz), yet small enough for the transistor bandwidth (i.e., 1 MHz). Then the PWM frequency maybe chosen

TABLE 11.1: Measurements for Experiment 2.

V_{in}	V_{AB}	V_{BE}	V_{CE}	V_{DC}
0.0				
0.1				
0.2				
0.3				
0.4				
0.5				
0.6				
0.7				
0.8				
0.9				
1.0				
1.1				
1.2				
1.3				
1.4				
1.5				
1.6				
1.7				
1.8				
1.9				
2.0				

1 kHz or 10 kHz. Then set the duty cycle (pulse width) of the signal to the following percentages: 0, 25, 50, 75, 100%. Record the $V_{in}(t)$ and V_{CE} on the dual-channel oscilloscope. Discuss your results. Discuss the following condition: if the output circuit of the transistor was connected to an electromechanical actuator's amplification circuit, what would be the voltage in the output circuit? Assume the actuator and amplification circuit combined have a bandwidth of 100 Hz. Simulate this condition using Simulink® or MATLAB®. What is the main advantage and disadvantage of PWM base signal control versus linear base signal control?

11.3   EXPERIMENT 3: PASSIVE FIRST-ORDER RC FILTERS. LOW PASS FILTER AND HIGH PASS FILTER

Objectives

1. Understanding the theory of filter circuits and their applications.
2. Circuit design of a passive low pass filter. Build and test the complete circuit.
3. Circuit design of a passive high pass filter. Build and test the complete circuit.
4. Getting familiar with standard measurement tools and signal generators. Measure the input and output voltage signals of the filter circuit and confirm the expectations with measurements.

Components

Item	Quantity	Part No.	Supplier
Resistor as calculated	1	81832	Jameco Electronics (www.jameco.com)
Capacitor as calculated	1	130232	Jameco Electronics (www.jameco.com)
Breadboard	1	20722	Jameco Electronics (www.jameco.com)
Set of connection wires	1 set	20079	Jameco Electronics (www.jameco.com)

Theory

Filters are used to "filter" the frequency content of the input signal and present the "filtered" or "cleaned-up" version of the input signal as its output signal. Low pass filters pass the low frequency content and remove (more accurately "attenuate" or reduce) the high frequency content of the input signal. High pass filters do the opposite: remove (filter-out, more accurately "attenuate" or reduce) the low frequency content and pass the high frequency content. Band pass filters pass the frequency content in a specific frequency range, and remove the frequency content below and above that range. Notch filters do the opposite: pass all frequency content except a selected range which is removed.

In this experiment, we will build and test a passive low pass filter and a passive high pass filter. The filters will be built using passive components: a resistor and a capacitor. Figure 11.7 shows the circuit diagram for a low pass and a high pass passive filter.

The input–output voltage relationship for the low pass filter can be derived as follows. The voltage across the capacitor is the output voltage and is related to the current and capacitance value as (assuming zero initial voltage at $t = 0$, $V_o(0) = 0.0$),

$$V_o(t) = \frac{1}{C} \int_0^t i(\tau)d\tau \tag{11.39}$$

$$V_o(s) = \frac{1}{Cs}i(s) \tag{11.40}$$

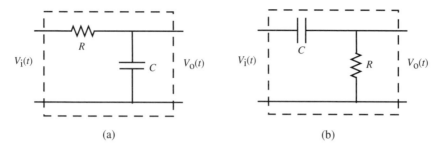

FIGURE 11.7: Circuit of a first-order passive filters: (a) passive low pass first-order filter, (b) passive high pass first-order filter.

The current in the circuit is

$$i(t) = \frac{V_i(t) - V_o(t)}{R} \tag{11.41}$$

$$i(s) = \frac{V_i(s) - V_o(s)}{R} \tag{11.42}$$

$$R \cdot i(s) = V_i(s) - \frac{1}{Cs} i(s) \tag{11.43}$$

$$\left(R + \frac{1}{Cs}\right) \cdot i(s) = V_i(s) \tag{11.44}$$

$$\frac{i(s)}{V_i(s)} = \frac{Cs}{RCs + 1} \tag{11.45}$$

Hence the transfer function between the output voltage and input voltage is

$$\frac{V_o(s)}{V_i(s)} = \frac{1}{RCs + 1} \tag{11.46}$$

$$\frac{V_o(jw)}{V_i(jw)} = \frac{1}{1 + jRCw} \tag{11.47}$$

Notice that when $w = \frac{1}{RC}$ rad/s, the magnitude ratio of the output voltage to input voltage is

$$\left| \frac{V_o(jw)}{V_i(jw)} \right|_{w=w_c=\frac{1}{RC}} = \left| \frac{1}{1 + jRCw_c} \right| \tag{11.48}$$

$$= \left| \frac{1}{1 + jRC\frac{1}{RC}} \right| \tag{11.49}$$

$$= \frac{1}{(1^2 + 1^2)^{1/2}} \tag{11.50}$$

$$= \frac{1}{\sqrt{2}} \tag{11.51}$$

$$= 0.707 \tag{11.52}$$

where the value $w_c = \frac{1}{RC}$ rad/s or $f_c = \frac{1}{2\pi RC}$ Hz is called the cutoff frequency of the filter, that is the frequency at which the output signal magnitude is 0.707 times the input signal magnitude in steady-state. In other words, the output signal is attenuated by 3 dB in comparison to the input signal.

$$20 \cdot \log_{10}(0.707) = -3.0 \, dB \tag{11.53}$$

In a MATLAB$^®$ command window, this can be calculated as shown,

```
>> 20*log10(0.707)

ans =

   -3.0116
```

Similarly, the input–output voltage relationship can be derived for the high pass filter. Notice that the location of resistor and capacitor on the circuit is swapped compared to the low pass filter. Following the similar derivation process, it is straightforward to derive the input–output voltage relationship.

$$V_o(t) = R \cdot i(t) \tag{11.54}$$

$$V_o(s) = R \cdot i(s) \tag{11.55}$$

The voltage across the capacitor, assuming zero initial voltage across the capacitor, $V_c(0) = 0$

$$V_c(t) = V_i(t) - V_o(t) \tag{11.56}$$

$$V_i(t) - V_o(t) = V_c(0) + \frac{1}{C}\int_0^t i(\tau)d\tau \tag{11.57}$$

$$V_i(t) - V_o(t) = 0 + \frac{1}{C}\int_0^t i(\tau)d\tau \tag{11.58}$$

$$V_i(s) - V_o(s) = \frac{1}{Cs}i(s) \tag{11.59}$$

By substituting the relationship for $i(s) = \frac{1}{R}V_o(s)$, it can be shown that

$$\frac{V_o(s)}{V_i(s)} = \frac{RCs}{1 + RCs} \tag{11.60}$$

$$\frac{V_o(jw)}{V_i(jw)} = \frac{jRCw}{1 + jRCw} \tag{11.61}$$

which represents the high pass filter transfer function. Again, notice that at $w = w_c$ where $w_c = \frac{1}{RC}$ rad/s or $f_c = \frac{1}{2\pi RC}$ Hz, the magnitude ratio is 0.707. Except that in the high pass filter, the filter passes the frequency content above this frequency and attenuates the frequency content below that frequency. The low pass filter does the opposite.

Procedure

1. Design and build a passive low pass filter and a passive high pass filter as shown in Figures 11.7, 11.8, and 11.9 with a cutoff frequency in the range of $f_c \approx 1.0$ kHz. Select proper R and C values.

2. Setup the function generator to produce a sinusoidal wave with an amplitude of 6 V (peak-to-peak). Notice that we do not need to provide any power to the circuit, only the input signal, since it is a passive circuit. When we build active filters with op-amps, we need to provide power to the op-amp in addition to providing the input signal.

3. Connect an oscilloscope to the input and output of the low pass filter. Sweep the input signal frequency range of 10 Hz to 100 kHz and measure the output voltage at each selected frequency. In particular, record 1) the magnitude ratio of output magnitude to input magnitude, and 2) phase shift between input and output signal at selected

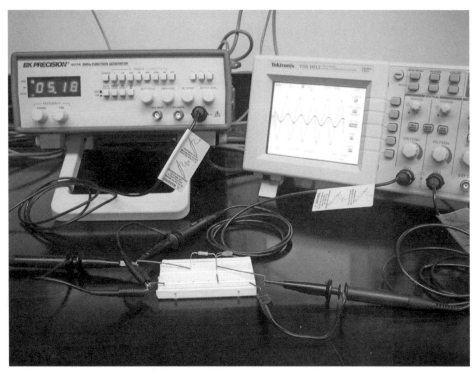

FIGURE 11.8: Picture of the experiment setup for a first-order passive filter.

frequencies. In order words, record a table of three columns: column 1 = frequency, column 2 = magnitude ratio, column 3 = phase shift.

4. Obtain oscilloscope screen images (or simply draw a sketch at selected frequencies) to show the input voltage and output voltage relationship of each filter in the time domain.

5. Compare your experimental measurements of frequency response with the analytical predictions. For analytical predictions, plot the transfer function magnitude and phase angle as a function of frequency, that is using MATLAB®.

6. Do the same for the high pass filter.

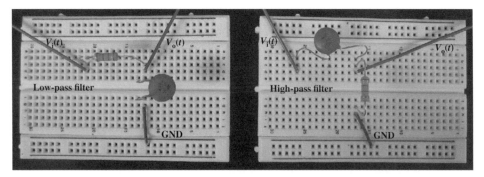

FIGURE 11.9: Picture of the assembled circuit on bread-board for a first-order passive low pass and high pass filter.

7. Compare the input–output voltage relationship for the low pass filter and high pass filter. Describe the differences.

8. Discuss how you can build a band pass filter using a combination of low pass and high pass filter.

9. Discuss how you can build a band reject (Notch) filter using a combination of low pass and high pass filter.

10. (Optional) Complete the same experiment for band pass filter and Notch filter.

11.4 EXPERIMENT 4: ACTIVE FIRST-ORDER LOW PASS FILTER WITH OP-AMPS

Objectives

1. Understanding the theory of active filter circuits and their applications.

2. Circuit design of an active op-amp low pass filter. Build and test the complete circuit.

3. Measure the input and output voltage signals of the filter circuit and confirm the expectations with measurements.

Components

Item	Quantity	Part No.	Supplier
LM358 Op-Amp IC	1	23966	Jameco Electronics (www.jameco.com)
Resistors as calculated	2	81832	Jameco Electronics (www.jameco.com)
Capacitor as calculated	1	130232	Jameco Electronics (www.jameco.com)
Breadboard	1	20722	Jameco Electronics (www.jameco.com)
Set of connection wires	1 set	20079	Jameco Electronics (www.jameco.com)

Theory

For our purpose, a filter is designed as an electronic circuit that produces a prescribed frequency response characteristic, of which the most common objective is to pass a certain frequency range while rejecting others. Filters may be classified into two major groups: passive filters and active filters. Passive filters consist of combinations of resistors, capacitors, and inductors. Passive RLC filters are capable of achieving relatively good filter characteristics in applications in the audio frequency range. But at the lower end of the audio frequency range, a problem occurs due to the high internal loss of inductors at low frequencies. Active filters consist of combinations of resistance, capacitance, and one or more active devices, such as op-amps, employing feedback. Since inductances are not required, the difficulties associated with them at low frequencies are eliminated.

The low pass filter built in this experiment is different from the one discussed in the Chapter on Electronics in that the DC gain of the low pass filter discussed in Figure 5.32 is equal to $+1$ if $R_f = R_i$, whereas the DC gain of the low pass filter discussed in this experiment is equal to -1 if $R_1 = R_2$. Otherwise, they both are first-order (single-pole) low pass active filters.

From the point of view of the amplitude of the frequency response, most filters can be classified as low pass, high pass, band pass and band rejection (notch) filters. In this experiment we will concentrate on a low pass filter design. Figure 11.10 shows an

Amplitude ratio (Output signal/input signal)
response

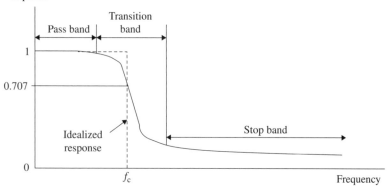

FIGURE 11.10: Representative amplitude response of a realistic low pass filter.

amplitude response of a realistic low pass filter. The quantity f_c represents the the cutoff frequency. In an ideal case, the amplitude response for $f < f_c$ is unity, so frequencies in this range are passed by the filter. For $f > f_c$, the amplitude response is zero, so frequencies in this range are completely eliminated by the filter. In reality, however, there is a *transition band* in between the *pass band* and the *stop band*, where the amplitude response decreases continuously.

Figure 11.11 shows a first-order active low pass filter circuit. The transfer function in the frequency domain between input and output voltages can be derived by following the op-amp idealized assumptions and Kirchoff's current and voltage laws. The voltage at the positive terminal is grounded, hence the voltage potential at the negative terminal is also grounded since $v^+ = v^-$,

$$v^+ = 0 \qquad (11.62)$$
$$v^- = 0 \qquad (11.63)$$

Then we can calculate the current over the resistor R_1.

$$i_1 = \frac{V_i}{R_1} \qquad (11.64)$$

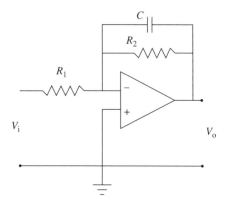

FIGURE 11.11: Circuit of a first-order active low pass filter.

Also note that no current would flow into the op-amp, then the same current must pass through the R_2 and C combination (noting Kirchoff's current law),

$$i_1(t) = \frac{V_i(t)}{R_1} = i_{R2}(t) + i_c(t) \tag{11.65}$$

$$= \frac{V_o(t)}{R_2} + i_c(t) \tag{11.66}$$

$$V_o(t) = \frac{1}{C} \int_0^t i_c(\tau)d\tau \tag{11.67}$$

where we use the fact that the current flow over the R_2 and C is determined by the output voltage and the ground voltage at the negative input (inverting) terminal. We will add the negative sign to the input–output voltage relation at the end of the derivation. If we take the Laplace transform of the above equations, we can easily find the transfer function between input voltage and output voltage,

$$i_1(s) = \frac{V_i(s)}{R_1} = \frac{V_o(s)}{R_2} + Cs \cdot V_o(s) \tag{11.68}$$

$$\frac{V_o(s)}{V_i(s)} = -\frac{R_2}{R_1} \cdot \frac{1}{1 + R_2 Cs} \tag{11.69}$$

where we added the negative sign to indicate the sign relationship between input voltage and output voltage is opposite. In order to obtain the frequency response relationship between input voltage and output voltage, let $s = jw$,

$$H(j\omega) = \frac{V_o(jw)}{V_i(jw)} = -\frac{R_2}{R_1} \cdot \frac{1}{1 + j\omega R_2 C} \tag{11.70}$$

For the case of $R_1 = R_2 = R$ we have

$$H(j\omega) = -\frac{1}{1 + j\omega RC} \tag{11.71}$$

The cutoff frequency f_c is defined as the point where

$$|H(j\omega)| = |\frac{V_o(jw)}{V_i(jw)}| = \frac{1}{\sqrt{2}} \tag{11.72}$$

$$20 \cdot \log_{10} |H(j\omega)| = 20 \cdot \log_{10} \frac{1}{\sqrt{2}} = -3 \, \text{dB} \tag{11.73}$$

Thus, we can write

$$1 = j\frac{\omega}{\omega_c} RC \tag{11.74}$$

from that it follows

$$\omega_c = 2\pi f_c = \frac{1}{RC} \, \text{rad/s} \tag{11.75}$$

and finally

$$f_c = \frac{1}{2\pi RC} \, \text{Hz} \tag{11.76}$$

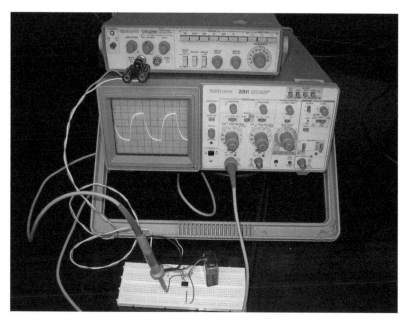

FIGURE 11.12: Picture of the experiment for a first-order active low pass filter: circuit on breadboard and the oscilloscope.

Procedure

1. Design an active low pass filter as shown in Figures 11.11, 11.12, 11.13 with a cutoff frequency of $f_c \approx 1.9\,\text{kHz}$ and a DC gain of $|G_{DC}| = 1$. Use $R_1 = 820\,\text{k}\Omega$ and calculate R_2 and C to meet the design specifications.

2. Build up the circuit and power the op-amp with a 9-V battery or a DC power supply.

3. Set up the function generator to produce a sinusoidal wave with an amplitude of 6 V (peak-to-peak). Adjust the offset voltage so that input excitation voltage range is within the DC power supply voltage range.

FIGURE 11.13: Picture of the circuit of a first-order active low pass filter on the solderless breadboard.

4. Connect an oscilloscope to the input and output of the low pass filter. Sweep the frequency range of 10 Hz to 100 kHz and measure the output voltage at each selected frequency, that is 10 Hz, 100 Hz, 1, 10, 50, 100 kHz. Create a table showing the ratio $|\frac{V_{out}}{V_{in}}| = |H(jw)|$ for the selected frequencies. Do the same for phase shift (phase angle) between $V_{out}(jw)$ and $V_{in}(jw)$.

5. Use logarithmic scale to plot the amplitude response and phase angle of the filter (Bode plot). Show $|H(j\omega)|$ over *frequency*. Mark in the $20\log_{10}|H(jw)| = -3$ dB (equivalently, $|H(jw)| = 0.707$) cutoff frequency. Compare the experimental results with analytical results obtained by plotting Equation 11.71 using MATLAB®.

6. What is the attenuation of the filter per decade in the transition band (that is the slope in log-log scale, $20\log_{10}|H(jw)|$ versus $\log_{10}w$)?

7. Set up the function generator to produce a square wave with a frequencies of 100 Hz, 2.0 kHz, 10 kHz. Measure the output signal and make a sketch in your solution sheet. Explain the result in terms of frequency content.

8. (Optional) If your digital storage oscilloscope is capable of taking FFT (Fast Fourier Transforms), take the FFT of both the input signal and output signal. Discuss the results. From the FFT of input and output signals, obtain the experimental transfer function of the low pass filter circuit in the frequency domain. Discuss the frequency response (the transfer function in the frequency domain, that this the ratio of FFT of output signal to the FFT of the input signal) compared to the analytical transfer function evaluated as a function of frequency.

9. (Optional) Discuss how you could take the FFT of the input and output signals using the PIC microcontroller. (Hint: Use ADC channels 0 and 1 for input and output signals. What is the limitation of the PIC microcontroller: available RAM memory for data?)

10. (Optional) Repeat the same experiment for an active high pass filter.

11.5 EXPERIMENT 5: SCHMITT TRIGGER WITH VARIABLE HYSTERESIS USING AN OP-AMP CIRCUIT

Objectives

1. Understanding the theory of the Schmitt trigger circuit (relay control with hysteresis) and its applications.

2. Circuit design of a variable hysteresis Schmitt trigger circuit using op-amp.

3. Measuring the hysteresis band of the Schmitt trigger circuit.

4. Application of this circuit as a relay control (with hysteresis) for closed loop control (i.e., as analog controller).

Components

Item	Quantity	Part No.	Supplier
LM358 Op-Amp IC	2	23966	Jameco Electronics (www.jameco.com)
Potentiometer (2 kΩ)	2	41865	Jameco Electronics (www.jameco.com)
Resistor 4.7 kΩ	1	107633	Jameco Electronics (www.jameco.com)
Resistor 1 kΩ	5	29663	Jameco Electronics (www.jameco.com)
Battery 9 V	2	198791	Jameco Electronics (www.jameco.com)
Breadboard	1	20722	Jameco Electronics (www.jameco.com)
Set of connection wires	1 set	20079	Jameco Electronics (www.jameco.com)

Theory

Comparator op-amp circuits employing positive feedback are widely known as Schmitt trigger circuits. The addition of positive feedback results in an effect called *hysteresis*. Hysteresis is a phenomenon in which the transition point of the output signal as function of the input signal is different when switching from the low state to the high state as compared with switching from the high state to the low state. In other words, the transition point is direction sensitive. Op-amp comparator circuits with hysteresis are used to reduce the possibility of undesirable state changes due to spurious noise pickup. Schmitt trigger circuits are also used in the digital I/O interfaces to remove the intended ON/OFF switching of the digital I/O, such as the so-called *switch debouncing*.

The hysteresis effect is often a desirable feature in many ON/OFF control systems. Such control systems are called *relay (ON/OFF) control with hysteresis*.

Consider an inverting Schmitt trigger op-amp (Figure 11.14). The voltage at the (+) terminal is the same as the voltage across R_2,

$$v^+ = V_{R2} = \frac{R_2}{R_1 + R_2} \cdot V_o = \frac{R_2}{R_1 + R_2} \cdot V_{sat} \tag{11.77}$$

Let $V_T = V_{R2}$. The output of the amplifier is essentially always saturated depending on the signals on its input terminals (+) and (−). The hysteresis band between input and output, V_T is

$$V_T = \frac{R_2}{R_1 + R_2} \cdot V_{sat} \tag{11.78}$$

Let us trace the operation on the op-amp along the input–output relationship curve that has the hysteresis loop. Note that the input–output relationship of the op-amp is

$$V_0 = V_{sat} \quad if \quad v^+ > v^- \tag{11.79}$$
$$V_0 = -V_{sat} \quad if \quad v^+ < v^- \tag{11.80}$$

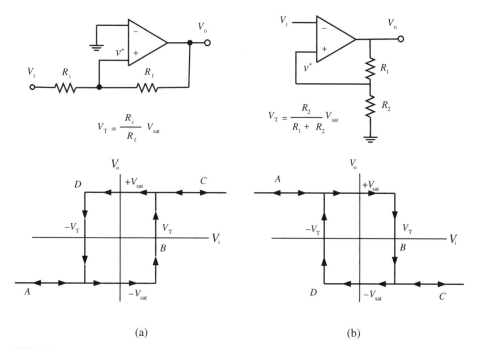

(a)   (b)

FIGURE 11.14: Schmitt trigger: (a) non-inverting, and (b) inverting configuration, ON/OFF output with hysteresis function.

Assume that initially we start on the curve from the left hand side,

$$V_o = V_{sat}, \quad v^+ = V_{R2}, \quad V_i < 0 \tag{11.81}$$

For the op-amp output to change state, $V_o = -V_{sat}$ (move to the C part of the curve), the inverting input (v^-) must slightly exceed the noninverting input (v^+), ($V_i = v^-$) > ($v^+ = V_{R2}$). In other words, we move along B on the input–output curve to C, neglecting the transient response details of the change in the output state of the op-amp, when $V_i = v^-$ becomes larger than the value of $v^+ = V_T$. At this point and for larger values of V_i, the output of the op-amp switches to the negative saturation value, $V_o = -V_{sat}$. In order for the op-amp output to return to the previous state, now the inverting input must be slightly more negative than the noninverting input which is at the value of $-V_T$, ($v_i = v^-$) < ($v^+ = -V_T$). Then, the state of the op-amp output will switch back to ON (D line along the hysteresis loop).

The noninverting Schmitt trigger circuit works with the same principle except the output polarity is different (Figure 11.14a), where,

$$V_T = \frac{R_i}{R_f} \cdot V_{sat} \tag{11.82}$$

This circuit can be used as an analog controller in a closed loop control system. In order to apply and test this circuit in such an application, we would need to provide

1. a subtraction function: *Reference Signal − Feedback Signal*, using a difference amplifier (Figure 5.31a)

$$e(t) = r(t) - y(t) \tag{11.83}$$
$$= V_{ref}(t) - V_{sensor}(t) \tag{11.84}$$

2. two potentiometers to simulate the "Reference Signal" and "Feedback Signal from Sensor," to generate (simulate)

$$V_{ref}(t) \tag{11.85}$$
$$V_{sensor}(t) \tag{11.86}$$

Then, moving the potentiometers by hand to simulate real-application conditions (i.e., fixed reference signal, varying sensor signal), we can monitor the output of the Schmitt trigger circuit working as a closed loop controller (Figure 11.15).

Procedure

1. Design the noninverting Schmitt trigger circuit in Figure 11.15 with hysteresis voltage V_T varying from 0.25 to 0.6 times V_{sat}.

2. Assemble the circuit on the breadboard as shown in the circuit diagram. Take care not to connect the 9-V batteries until the entire circuit has been assembled (Figures 11.16, 11.17).

3. Set up the function generator to produce a triangular wave with an amplitude of 6 V.

4. Connect the oscilloscope to the input and output of the noninverting Schmitt trigger circuit. Measure the input voltage and output voltage.

5. Sketch the input and output voltage signals. Verify the hysteresis effect between the input and output voltages.

6. Vary R_i value on the circuit and confirm its effect of the hysteresis band.

7. The students may also build and test an inverting version of this circuit as shown in Figure 11.14b.

8. Assemble a difference amplifier, and two potentiometers. Connect them together to the Schmitt trigger circuit, to effectively implement the closed loop analog controller.

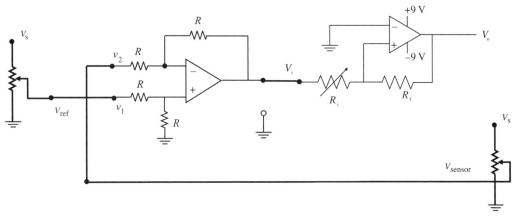

FIGURE 11.15: Noninverting Schmitt trigger circuit.

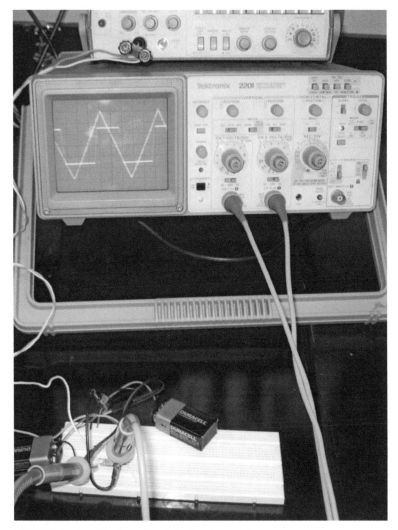

FIGURE 11.16: Noninverting Schmitt trigger lab setup.

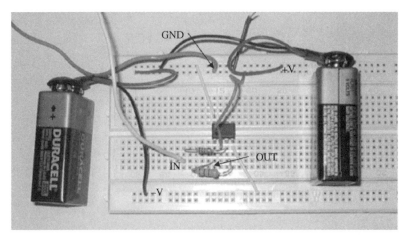

FIGURE 11.17: Picture of a noninverting Schmitt trigger circuit built on a breadboard (without the difference amplifier).

11.6 EXPERIMENT 6: ANALOG PID CONTROL USING OP-AMPS

Objectives

1. Understanding the theory of summing, inverting, differential, derivative, and integrator op-amps.

2. Build a complete analog PID control circuit.

3. Test the input–output signal relation of a PID circuit (i.e., P-only, D only, I only, PD, PI, PID versions of the circuit).

Components

Item	Quantity	Part No.	Supplier
LM358 Op-Amp IC	3	23966	Jameco Electronics (www.jameco.com)
Resistor 1 kΩ	8	29663	Jameco Electronics (www.jameco.com)
Resistor 4.7 kΩ	4	107633	Jameco Electronics (www.jameco.com)
Resistor 100 kΩ	4	29997	Jameco Electronics (www.jameco.com)
Resistor 470 Ω	1	107537	Jameco Electronics (www.jameco.com)
Capacitor 0.22 μF	2	25540	Jameco Electronics (www.jameco.com)
Battery 9 V	2	198791	Jameco Electronics (www.jameco.com)
Breadboard	1	20722	Jameco Electronics (www.jameco.com)
Set of connection wires	1 set	20079	Jameco Electronics (www.jameco.com)

Theory

The "pure" derivative has a large gain at high frequency and will amplify the noise in the closed loop, and hence lead to stability problems. In order to reduce the gain of the pure derivative at high frequency, a practical derivative op-amp circuit is modified so that it has a first-order pole in addition to the derivative, hence reducing the high frequency gain of the transfer function thereby reducing the problem of noise amplification. This is done by

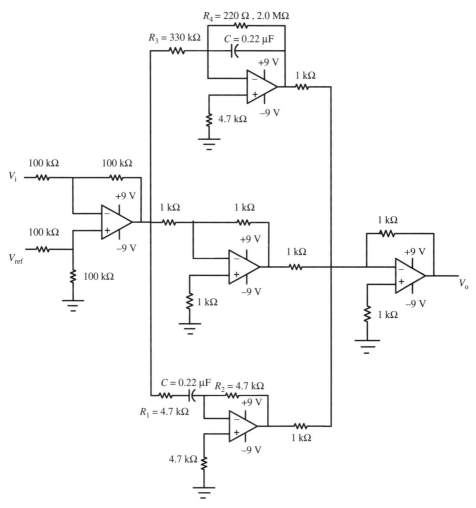

FIGURE 11.18: Analog PID control circuit.

adding a resistor R_1 in series with the capacitor C (Figures 11.18, 11.19, 11.20, and 11.21). Let us derive the transfer function for this practical derivative circuit. Notice that $v^+ = GND$ and $v^+ = v^-$ at the input terminals of the op-amp. Since there cannot be current drawn into the op-amp, then

$$i_1(t) = i_2(t) \tag{11.87}$$

where $i_1(t)$ is the current on the input side of the op-amp through R_1 and C, and $i_2(t)$ is the current on the feedback loop of the op-amp through R_2. It is easy to show that

$$V_i(t) = R_1 \cdot i_1(t) + \frac{1}{C}\int_0^t i_1(\tau)d\tau \tag{11.88}$$

$$V_i(s) = R_1 \cdot i_1(s) + \frac{1}{C}\frac{1}{s}i_1(s) \tag{11.89}$$

$$i_1(s) = \frac{Cs}{R_1 Cs + 1}V_i(s) \tag{11.90}$$

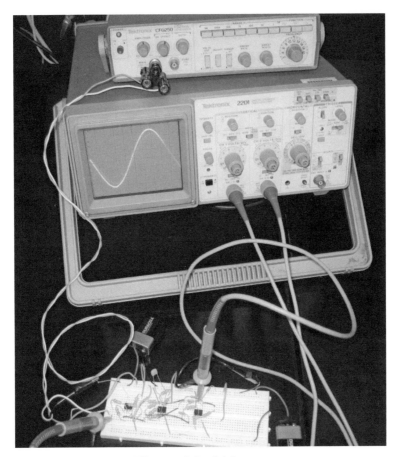

FIGURE 11.19: Analog PID control circuit lab setup.

Similarly,

$$i_2(t) = \frac{0 - V_o(t)}{R_2} \tag{11.91}$$

$$i_2(s) = \frac{-V_o(s)}{R_2} \tag{11.92}$$

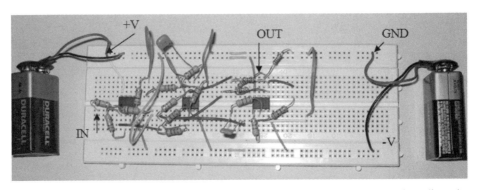

FIGURE 11.20: Picture of the analog PID control circuit based on op-amps on a breadboard.

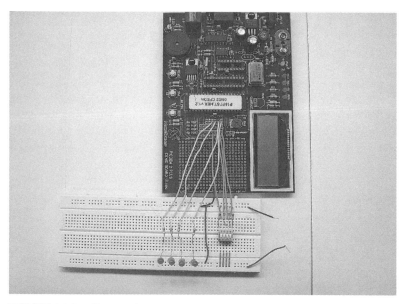

FIGURE 11.21: Picture of the complete circuit for LED control via the digital output of a PIC microcontroller.

Hence,

$$i_1(s) = i_2(s) \tag{11.93}$$

$$\frac{Cs}{R_1 Cs + 1} V_i(s) = \frac{-V_o(s)}{R_2} \tag{11.94}$$

$$\frac{V_o(s)}{V_i(s)} = -\frac{R_2 Cs}{R_1 Cs + 1} \tag{11.95}$$

which shows that the new transfer function is a modified version of the differentiator. It not only has a zero at the origin $s = 0$, but also has a pole at $s = -1/(R_1 C)$. This added pole eliminates the high noise amplification problem of the "pure" differentiator.

In practice, an integral op-amp circuit typically includes a parallel resistor with the capacitor (R_4 in Figure 11.18). The reason for this is that a "pure" integrator adds a -90 degree phase (phase lag) to the loop at all frequencies. If we could reduce that phase lag at least at lower frequencies, it would improve the stability of the closed loop system. Hence, a pure integrator may be modified with a resistor R_4 in parallel with the capacitor in the feedback path to accomplish this. Let us show the new transfer function by derivation. Let $i_1(t)$ be the current across R_1, $i_2(t)$ be the current across R_2, and $i_3(t)$ be the current across C. From the same op-amp relations, it follows that

$$i_1(t) = i_2(t) + i_3(t) \tag{11.96}$$

$$i_1(s) = i_2(s) + i_3(s) \tag{11.97}$$

It is straightforward to show the relationships between currents, voltages, and resistors/capacitor,

$$\frac{V_i(t)}{R_3} = \frac{0 - V_o(t)}{R_4} + C\frac{d(0 - V_o(t))}{dt} \tag{11.98}$$

$$\frac{V_i(s)}{R_3} = \frac{-V_o(s)}{R_4} + Cs(-V_o(s)) \tag{11.99}$$

$$\frac{V_o(s)}{V_i(s)} = -\frac{R_4}{R_3} \frac{1}{(R_4 C s + 1)} \qquad (11.100)$$

$$\frac{V_o(s)}{V_i(s)} = -\frac{1}{RCs + 1}; \quad for \quad R_3 = R_4 = R \qquad (11.101)$$

which shows that the addition of R_4 in parallel with the C capacitor on the feedback path changes the "pure integrtor" (which has a pole at $s = 0.0$) into a first-order filter with a pole at $s = -1/(R_4 C)$, where R and C parameters should be chosen such that the pole at $s = -\frac{1}{R_4 C} \approx 0.0$ is close to the origin in order to approximate the integrator function.

The textbook PID controller transfer function is

$$u(s) = \left(K_p + \frac{K_i}{s} + K_d s \right) \cdot e(s) \qquad (11.102)$$

$$u(t) = K_p \cdot e(t) + K_i \cdot \int e(\tau) d\tau + K_d \cdot \frac{de(t)}{dt} \qquad (11.103)$$

where $e(s)$ is the Laplace transform of the error signal, $s(s)$ is the PID controller output.

Procedure

1. Assemble the circuit on the breadboard as shown in Figure 11.18. Take care not to connect the 9-V batteries until the entire circuit has been assembled.

2. Derive the complete transfer function of the PID controller. Calculate the proportional, derivative, and integrator gains: K_p, K_d, K_i, and the additional pole location of the modified derivative term, and pole location and gain of the modified integral term. Experiment with both pure-derivative and pure integral, and modified derivative and modified integral, versions of the circuit. That is, remove R_4 from the integrator circuit to implement a pure integrator, and remove R_1 from the derivative circuit to implement a pure derivative function.

   For the measurements to be taken below, remove R_1 from the derivative circuit and remove R_4 from the integrator circuit. There are ten (10) results to capture.

3. Set up the function generator to produce two different wave forms with an amplitude of 6 V: square wave and sine wave. The frequency of the wave functions can be easily adjusted on the function generator, for example to 1.0 Hz. Connect the function generator signal to V_i (feedback signal). Connect the V_{ref} signal to the ground voltage level.

4. Connect the oscilloscope channel 1 to the input signal from the function generator, and oscilloscope channel 2 to the output of the differential op-amp, and sketch (or take a picture of the oscilloscope screen) the input and output waveforms.

5. Connect the oscilloscope channel 1 to the input signal of the proportional op-amp (output of the differential op-amp), and oscilloscope channel 2 to the output of the proportional op-amp, and sketch (or take a digital picture of the oscilloscope screen) the input and output waveforms.

6. Connect the oscilloscope channel 1 to the input signal of the derivative op-amp (output of the differential op-amp), and oscilloscope channel 2 to the output of the derivative op-amp, and sketch (or take a digital picture of the oscilloscope screen) the input and output waveforms.

7. Connect the oscilloscope channel 1 to the input signal of the integrating op-amp (output of the differential op-amp), and oscilloscope channel 2 to the output of the

integrator op-amp, and sketch (or take a digital picture of the oscilloscope screen) the input and output waveforms.

8. Connect the oscilloscope channel 1 to the input signal of the op-amp (output of the differential op-amp), and oscilloscope channel 2 to the output of the summing op-amp (which is the output of the PID controller circuit), sketch (or take a digital picture of the oscilloscope screen) the input and output waveforms. Confirm that the output voltage of the summing op-amp is the sum of the voltages from the P-I-D sections of the circuits (voltage outputs from P, I, and D circuits) multiplied with -1.

9. (Optional) Repeat the above 10-result test with the modified derivative (R_1 is in the circuit) and the modified integral term (R_4 is in the circuit). Note that the values of the added R_1 and R_4's affect the performance of the PID controller and must be selected properly for the intended performance.

10. (Optional) Repeat the above 10-result test with the modified derivative (R_1 is in the circuit) and the modified integral term (R_4 is in the circuit). Except that in this case, select a poor value for R_4 on purpose so that the circuit is not a good approximation to the integral function circuit.

11. (Optional) Closed loop control test: build a mathematical model of the PID circuit (using MATLAB® or Simulink®) and simulate for the same circuit parameters and input signal conditions, and compare your simulation results with the experiments. Comment on the causes of the differences between simulation and experimental results.

12. (Optional) Build a second-order system, that is double integrator (mass-force analogy), circuit to "simulate" the plant dynamics. Simply connect two integrators in series. Then connect the output of the PID controller circuit to the input of the double-integrator (plant), and the output of the double integrator into the negative port of the PID's comparator circuit at the input (Figures 2.35 and 2.36 in the textbook). Using a potentiometer, generate the desired (commanded) output voltage. Simulate the whole system in MATLAB® or Simulink®. Test the same simulated conditions on the actual circuit and compare the closed loop system response results (simulation results and hardware test results). Digital implementation of the PID controller in real-time is given by Equations 2.42–2.44 in the textbook. If the sampling period T is fast relative to the bandwidth of the system, the digital implementation should give almost identical results to the analog implementation. The difference between digital and analog implementation can be tested in simulation only using MATLAB®/Simulink® without any hardware involvement, in that we should be able to confirm a selection of a sampling period that is fast enough that the difference between analog and digital implementation of the PID controller is insignificant. In the simulation, we can also include the quantization errors due to finite word-length (finite resolution) of the ADC and DAC converters. Equations 2.42 and 2.44 do not include the effect of the quantization errors introduced by DAC and ADC.

11.7 EXPERIMENT 7: LED CONTROL USING THE PIC MICROCONTROLLER

Objectives

1. Learn to use the integrated development environment (IDE) software and PIC microcontroller development board.

2. Learn basic hardware interface between a microcontroller, digital input devices (DIP switches in this case), and digital output devices (LEDs in this case).

3. Learn the basic hardware and software features of PIC 18F452 or PIC 18F4431, and software control a digital I/O channel.

4. To read the status of digital inputs: status of the DIP switches under software control.

5. To drive LEDs under software control by using digital output pins of an I/O port of the PIC18F452 (or PIC 18F4431) microcontroller, and to change the status of LEDs under software control based on digital inputs to the PIC microcontroller.

Microcontroller discussion is based on PIC 18F452. However, PIC 18F4331 has a peripherial interface for quadrature encoder, and hence makes Experiment number 12 much easier to handle. There are minor differences between the two models of this microcontroller family. The reader should make the necessary adjustments depending on which model is being used.

Components

Item	Quantity	Part No.	Supplier
LED	4	119634	www.jameco.com
100 Ω and 100 kΩ resistors	12	107465	www.jameco.com
DIP Switch	1	38818	www.jameco.com
PIC demo board/connector	1 set	DV164006 or DM163022	www.microchipdirect.com

Theory

The PIC 18F452 microcontroller has five input–output ports, labeled PORTA, PORTB, PORTC, PORTD, and PORTE. These are used to interact with the world outside, consisting of various sensors and actuators. The interface circuit schematic is shown in Figures 11.22 and 11.23. Of these, Ports A and E are 6-bit ports, while B, C and D are 8-bit ports. These ports can be configured as either input or output through the use of the TRIS command.

The command syntax for output is:

```
TRISx = 0;
PORTx = <value>;
```

where x refers to any of the ports A,B,C,D or E.
The syntax for input is:

```
TRISx = 1;
value = PORTx
```

Once a particular port is configured as an output, any (binary) data that is sent to the port is placed on the port pins on the microcontroller chip. When used as an input, any binary signal value applied to the input ports is written into that port register in the memory.

The default state of digital inputs can be defined by either connecting the input port pin to a supply voltage (i.e., 5 VDC) or to ground via a resistor. If the connection is made to the supply voltage (i.e., default ON state), it is called *pulled-up* and the resistor is called the *pull-up resistor* (Figure 11.23). If the connection is made to the ground (i.e., default OFF state), it is called *pulled-down* and the resistor is called the *pull-down resistor* (Figure 11.22). The appropriate value for the pull-up and pull-down resistors are determined by the supply voltage and maximum current draw specifications of the microcontroller. In this experiment, it is recommended that both types of digital input connections be tested and the switch state is properly interpreted in the application software.

Application Software Description

The program code for all the experiments is written in C programming language and compiled with MPLAB C18 C-compiler. The code can be typed in the MPLAB Integrated Development Environment (IDE) using the built-in editor or in any standard ASCII editor such as Notepad. The code is then compiled using the C-18 compiler in the MPLAB IDE.

The program code must contain at least two sections: a setup section and logic section. Depending on the state of input switches selected by the user, the program execution is transferred to a particular section, and that section is executed. Points to be noted are:

1. Include proper header files provided with the C-18 compiler,
   if PIC 18F452 microcontroller is used

   ```
   #include <p18f452.h>
   #include <delays.h>
   ```

   or if PIC 18F4431 microcontroller is used

   ```
   #include <p18f4331.h>
   #include <delays.h>
   ```

2. Configure the correct ports as input and output using the TRISx command that matches your hardware interface choices for input and output. Clear any existing values on the ports by setting them to zero.

3. If you are using the "Watch" option while debugging in MPLAB to keep track of register values, note that the values are in hexadecimal (0-9, A-F).

4. As an example, if we wanted to send out a high signal on pin 0 and pin 5 of Port D, the code is:

   ```
   TRISD = 0; /* Set Port D as output */

   PORTD = 0; /* Clear existing Port D value */

   PORTD = 33; /* Decimal equivalent of 00100001 */

   /* Value for Port D in the Watch window in MPLAB is 21
   (hex equivalent of 33) */
   ```

Procedure

1. Assemble the circuit on the breadboard as shown in Figures 11.22 or 11.23.

2. Connect the four LEDs to each of four pins of PORTB of the PIC microcontroller on the demo board. The resistors in series with the LEDs are used to limit the maximum current through the LED, hence the current load (draw) from the PIC microcontroller per LED. Resistor values in the range of $100\,\Omega$ to $1\,k\Omega$ are typically used.

3. Connect four switches on the 8-pin DIP switch to pins on the PORTC of the PIC demo board. Notice the resistors (i.e., $10\,k\Omega$ or $100\,k\Omega$) connecting the input ports to the ground. As a result of this connection, when the input switch is open (OFF), the input pin is at ground (OFF) level. When the input switch is closed (ON), the input pin is at high level (5 V) in ON state. This is called the *pull-down resistor* configuration, because the resistor "pulls-down" the pin to the ground. If we had swapped the location of the switches and resistors, the ON/OFF relationship would be reversed. In that case, when a switch is OFF (open), the corresponding input pin status would be high (ON). When the switch is ON (closed), the corresponding

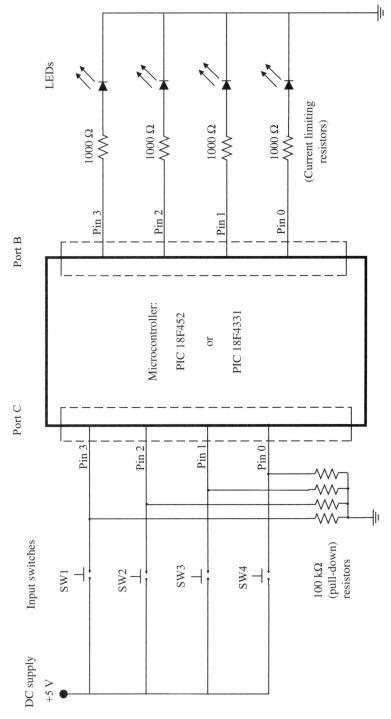

FIGURE 11.22: Circuit diagram for digital I/O: read status of input switches and turn ON/OFF LED outputs. Switch inputs are connected to the microcontroller with pull-down resistors (i.e., SW1 = OFF, Pin 3 = OFF (GND)).

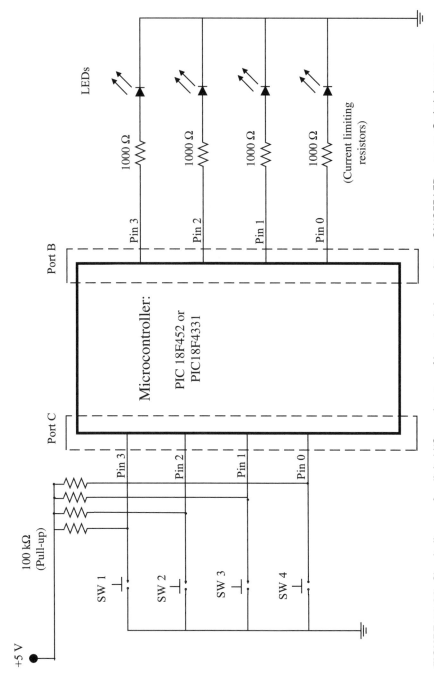

FIGURE 11.23: Circuit diagram for digital I/O: read status of input switches and turn ON/OFF LED outputs. Switch inputs are connected to the microcontroller with pull-up resistors (i.e., SW1 = OFF, Pin 3 = ON (5 V)).

input pin status would be low (OFF). This is called the *pull-up resistor* configuration, because the resistor "pulls-up" the pin to the supply voltage.

4. Open MPLAB environment, and load your project or open a new project. Your code for the project must contain two separate sections as described earlier: a setup section and logic section.

5. Polling method of programming: the logic implemented between the input DIP switches and output LEDs is left to the student. One simple logic implementation may be to display the same state of DIP switches on the LEDs.

6. Implement an interrupt driven program using a timer: use one of the timers modules (i.e., TIMER0, TIMER1, TIMER2) to generate an interrupt at a programmed frequency. Write the interrupt service routine (ISR) for the timer interrupt which will read the input switches, and write the status of the inputs to the output LEDs. Setup the priority of the interrupt. Direct the timer interrupt to the written ISR function.

7. Implement an interrupt driven program based on an external event (instead of a timer): a change of state in any of the input switches is defined as the interrupt. Perform the same function.

11.8 EXPERIMENT 8: FORCE AND STRAIN MEASUREMENT USING A STRAIN GAUGE AND PIC-ADC INTERFACE

Objectives

1. Build a complete circuit to interface a strain-gauge sensor to the A/D converter of the PIC microcontroller. This includes building a Wheatstone bridge and an operational amplifier to amplify the voltage output of the Wheatstone bridge, and interfacing it to one of the ADC channels of the PIC controller.

2. Develop application software to sample the strain-gauge output voltage using the ADC and estimate/measure the force and strain due to load applied (Figure 11.24).

FIGURE 11.24: Picture of the complete circuit for the strain gauge sensor experiment.

Components

Item	Quantity	Part No.	Supplier
Aluminum beam (2 mm x 15 mm x 1000 mm) (Cut into shorter lengths)	1	1663T12	McMaster
Strain gauge $G = 2$, $R = 120\ \Omega$	1	SG-6/120LY11	www.omega.com
(and bonding adhesive)	1	SG401	www.omega.com
LM358 Op-Amp IC	2	120862	(www.jameco.com)
Potentiometer (200 kΩ max)	1	241349	(www.jameco.com)
Resistor 120 Ω	2	30082	(www.jameco.com)
Resistor 100 kΩ	1	107764	(www.jameco.com)
Resistor 1000 Ω	1	30081	(www.jameco.com)
Breadboard	1	20722	(www.jameco.com)
Set of connection wires	1 set	20079	(www.jameco.com)
PIC Demo Board/connectors	1	DM163022	(www.microchip.com)

Theory

Force and Strain Relationship   The strain gauge setup consists of an aluminium beam that is fixed at one end to a frame, and is free at the other. A picture of the experiment setup is shown in Figure 11.24. A schematic of the mechanical setup is shown in Figure 11.25. The aluminum cantilever beam is loaded by deflecting the free end with the aid of a vertically mounted screw.

Application of a force, F, on the tip of a cantilever results in a bending moment

$$M = F \cdot l_s \tag{11.104}$$

where l_s is the distance between the force application point and the center of the measurement point (the strain gauge location). This moment is balanced by "stress," σ, that causes

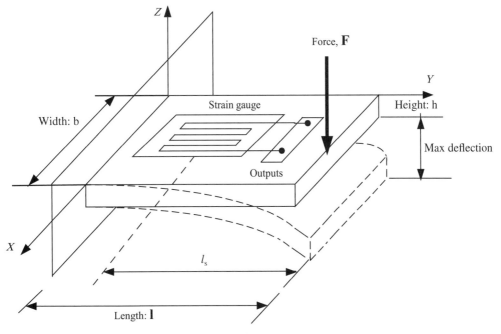

FIGURE 11.25: Strain gauge sensor experiment.

lengthening and shortening of fibers of material. The amount of extension is called "strain," ϵ, and is defined as

$$\epsilon = \frac{\Delta l}{l} \tag{11.105}$$

As long as the deformation of the material stays within the linear elastic region, the relation between stress and strain is provided by the constitutive law of the material:

$$\sigma = E \cdot \epsilon \tag{11.106}$$

where E is the Young's modulus of the material.

There is a linear relationship between the applied force and the induced strain, as long as the deformation is limited to the elastic region of the deformation of the material. For a cantilever beam with rectangular cross-section we have the following relations,

$$\sigma_{\text{max}} = \frac{M \cdot (h/2)}{I} \tag{11.107}$$

$$= \frac{F \cdot l_s \cdot (h/2)}{(1/12) \cdot b \cdot h^3} \tag{11.108}$$

$$= \frac{6 \cdot F \cdot l_s \cdot h}{b \cdot h^3} \tag{11.109}$$

$$= \frac{6 \cdot F \cdot l_s}{b \cdot h^2} \tag{11.110}$$

$$\epsilon_{\text{max}} = \frac{1}{E} \sigma_{\text{max}} \tag{11.111}$$

$$= \frac{6 l_s}{E b h^2} \cdot F \tag{11.112}$$

where b, h are the width and thickness of the cantilever beam used in this experiment, l_s is the distance from the force application point to the strain gauge (measurement) point. I is the area moment of inertia of the cross-section of the beam around the neutral axis of bending, $I = \frac{1}{12} \cdot b \cdot h^3$.

Equation 11.112 shows that measuring the strain results in an indirect measure of the applied force. If we can measure strain, then we can estimate/calculate force if we have the geometric and material property information of the beam (l_s, b, h, E).

$$F = \left(\frac{E b h^2}{6 l_s} \right) \epsilon_{\text{max}} \tag{11.113}$$

Strain Gauge The strain measurement may be performed by strain gauge sensors. These devices are applied on the test part in such a way that they are subjected to the same strain (deformation) as the test part. In other words, the strain-gauge is bonded (glued) on the test surface with a high strength adhesive. The resistance of the small wires that constitute the strain gauge increases when they lengthen, and decreases when they shorten. It follows that strain variations results in small resistance variations.

The resistance of a conductor of length L and cross-section A is given by:

$$R = \rho \frac{L}{A} \tag{11.114}$$

where parameter ρ is the *resistivity* of the material. Taking the logarithm of Equation 11.114 and differentiating, we obtain:

$$\ln R = \ln \rho + \ln L - \ln A \tag{11.115}$$

$$\frac{dR}{R} = \frac{d\rho}{\rho} + \frac{dL}{L} - \frac{dA}{A} \tag{11.116}$$

$$\frac{\Delta R}{R} = \frac{\Delta \rho}{\rho} + \frac{\Delta L}{L} - \frac{\Delta A}{A} \tag{11.117}$$

Assuming that the section of the conductor wire is circular, $A = \pi \frac{D^2}{4}$, it follows

$$\frac{\Delta A}{A} = \frac{2 \Delta D}{D} = -v 2 \frac{\Delta L}{L} \tag{11.118}$$

where it can be shown that

$$\frac{dA}{A} = \frac{d}{dA}(\ln A) \tag{11.119}$$

$$= \frac{d}{dD}(\ln(\pi D^2/4)) \tag{11.120}$$

$$= \frac{(\pi/4)2D \cdot dD}{(\pi/4)D^2} \tag{11.121}$$

$$= \frac{2\,\mathrm{dD}}{D} \tag{11.122}$$

Hence,

$$\frac{\Delta A}{A} = \frac{2\Delta D}{D} \tag{11.123}$$

In Equation 11.118 the coefficient of Poisson, v, of the material has been used in order to relate longitudinal and transverse deformations of the wire.

Substituting Equation 11.118 into 11.117 the following is obtained:

$$\frac{\Delta R}{R} = (1 + 2v)\frac{\Delta L}{L} + \frac{\Delta \rho}{\rho} \tag{11.124}$$

It is common to define the *gauge factor G* as:

$$G = \frac{\Delta R/R}{\Delta L/L} = (1 + 2v) + \frac{\Delta \rho/\rho}{\Delta L/L} \tag{11.125}$$

$$= \frac{\Delta R/R}{\epsilon} \tag{11.126}$$

Then,

$$\frac{\Delta R}{R} = G \cdot \epsilon \tag{11.127}$$

Equation 11.125 shows that the sensor's gauge factor G depends on a geometric term $1 + 2v$ and on a microstructural term $\frac{\Delta \rho/\rho}{\Delta L/L}$ that relates the variation of resistivity to deformation. This term characterizes the *piezoresistive* behavior of the material.

For some materials (like constantan), the piezoresistive component is much smaller than the geometric one. The strain gauges obtained from these materials are called "metallic." They have a relatively small gauge factor ($G = 2$). They are stable under temperature variations and linear in the operating range.

Other strain gauges, made from different materials (i.e., semiconductors) exhibit gauge factors in the range of $G = 70$–200. Despite the increased sensitivity, their use is more difficult since they are nonlinear and may need temperature compensation.

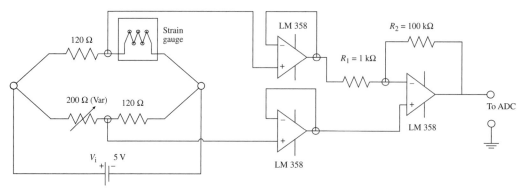

FIGURE 11.26: Circuit diagram to amplify the voltage output of the strain gauge sensor.

For a given strain gauge, the gauge factor G is known. Then, if we can measure the change in resistance ($\Delta R/R$), we can calculate the strain ϵ.

Signal Conditioning   The signal conditioner has to convert the variations of resistance of the sensor into a proportional electrical voltage signal that is compatible with the PIC microcontroller.

The resistance variation is converted to a voltage differential by a Wheatstone bridge. For the applied circuit, see Figure 11.26, the output voltage is

$$V_o = \frac{V_i}{4} \frac{\Delta R}{R_o} \tag{11.128}$$

Notice that the location of the variable resistor for the sensor in the Wheatstone bridge in the figure shown for the experiment is at location for R_2, whereas it was at location for R_1 in Chapter 6. It can be shown that the same output voltage and resistance change relationship holds except the polarity of the output voltage between these cases. It is instructive for the student to show that this is true. In Chapter 6, in Equation 6.20, let $R_1 = R_3 = R_4 = R_0$ and $R_2 = R_0 + \Delta R$. And substitute these in Equation 6.20 and make the same approximations in deriving Equation 6.23. Simply let $R_1 = R_3 = R_4 = R_0$ and $R_2 = R_0 + \Delta R$ in Equation 6.20.

Since V_0 is typically in the order of millivolts, an amplification stage has to be provided. An operational amplifier can be used for this purpose. Resistors $R_1(1 \text{ k}\Omega)$ and $R_2(100 \text{ k}\Omega)$ (see Figure 11.26) determine the gain

$$K_a = -\frac{R_2}{R_1} \tag{11.129}$$

The low impedance R_1 of the amplification stage will strongly affect the output of the bridge if it is directly connected to it. In order to avoid this effect, a voltage follower stage is provided (referred to also as voltage buffer stage or impedance isolation buffer, using two op-amps LM 358).

Application Software Description

The measurement principle here is that given the strain gauge properties, that is G, Wheatstone bridge parameters, V_i, R_0, and in the signal amplification stage, R_1, R_2, we can calculate the strain ϵ from the output voltage measurement V_{ADC}. Then, given the beam material and geometric parameters, E, b, h, l_s, we can calculate the force, F.

The software development task includes setting up the registers to configure the pin connected to the strain gauge sensor as one of the ADC channels to be used. This can be done directly by writing to the specific registers in C or assembly language or by calling the appropriate C library functions which hide the details of register configuration. The main purpose is to be able to read (and display) the voltage level at the ADC pin of the microcontroller. In this lab, no output operation at the I/O ports are performed. It is only an analog input experiment. Variations on the software can be made to handle the ADC conversion process either using

1. the polling method (i.e., by repeatedly checking if the ADC conversion is done), or

2. the interrupt method (i.e., a timer generate periodic signal triggers ADC conversion to start, and the ADC registers can be setup to generate interrupt every time an ADC conversion is completed). If the interrupt method is used, then an interrupt service routine (ISR) must be written and interrupt conditions must be configured by setting appropriate register bits.

Relevant header files to be included with C-18 compiler are as follows,

```
#include <p18f4331.h>
#include <adc.h>
#include <timers.h>
#include <delays.h>
```

Procedure

1. Assemble the circuit on the breadboard as shown in the circuit diagram. Take care not to connect the 5-V supply until the entire circuit has been assembled.

2. Glue the strain gauge onto the fixture using the strain gauge adhesive provided.

3. Connect the strain gauge outputs to one arm of the Wheatstone bridge setup and the operational amplifier. Notice that one arm of the Wheatstone bridge is made as an adjustable resistor (potentiometer). This will help us adjust the offset voltage output, hence it will be used in the calibration of the sensor. We will use the adjustable potentiometer to make sure the sensor output voltage is zero when there is zero deformation. Note that, in order to build an accurate Wheatstone bridge, the resistor values should be the same. As a result, for higher accuracy applications, resistors with higher tolerance (smaller variation in their value due to manufacturing variations) should be used.

4. Open MPLAB and load your project into the workspace. The source code must contain commands to configure the analog to digital converter on the PIC, and read input voltage at the ADC pin to which the sensor signal is connected to, and display the results.

5. Download and run the program on the PIC microcontroller. Then, deflect the aluminum beam on the strain gauge fixture by lowering the screw at the tip. Take care not to deflect it by more than a few millimeters (about an eighth of an inch at most).

6. Verify that the ADC now shows a value. Now raise the screw by a few turns. This will reduce the strain on the beam; read the value of the ACD again. The value in the ADC should now have reduced to reflect lower strain value.

7. Repeat the procedure for various deflections of the beam, all the while taking care not to deflect the beam by more than 2 or 3 mm. Record the deflection versus the strain measurements. Calculate the force applied at the tip of the beam. Note that for the calculation of force, you need the beam material and geometric parameters (E, b, h, l_s), strain gauge gauge factor (G), Wheatstone bridge parameters V_i, R_0, and signal amplifier gain $K_a = -R_2/R_1$. Plot the results:

(a) measured ADC voltage versus the deflection distance of the beam,

(b) calculated force versus the deflection distance of the beam.

In order to be sure of your force estimation accuracy, the deformation can be induced by a known set of weights at the tip of the beam instead of the screw mechanism. Then, the student can compare the estimated force to the actually applied weight ($weight = force = mass \cdot gravity$, $W = F = m \cdot g$). It is equally important to be able identify the sources of error in the measurement. Discuss how this principle can be used to design a digital weight measurement device (scale) used in homes.

8. *(Optional)* In order to improve the signal amplification quality against noise and temperature variations; design, build and test an *instrumentation amplifier* instead of the amplifier used above.

9. *(Optional)* Develop two different software versions to handle ADC conversion: 1. ADC conversion under program control, 2. ADC conversion under interrupt control. In version 1, the program logic must:

(a) Setup the ADC conversion channel.

(b) Start the ADC conversion.

(c) Wait for ADC conversion to complete.

(d) Read the ADC conversion result and display it.

(e) Repeat steps (b–d) for every ADC conversion under program control.

In version 2, the program sets up the ADC conversion process to be triggered by a timer, and every time the ADC conversion is completed, it generates an interrupt to the CPU, then the CPU simply handles the ADC ISR (interrupt service routine) which should just read and display the ADC conversion result.

(a) Setup the ADC conversion channel for timer driven periodic ADC conversion, and for ADC to generate an interrupt whenever an ADC conversion is completed.

(b) Write the ISR routine for "ADC conversion complete" interrupt handling: read the ADC register and display result. Setup this ISR as the ADC conversion complete interrupt handling routine.

(c) Enable interrupts and start normal program execution. In the background, ADC conversion will be periodically triggered by the timer, every time ADC conversion is completed, the ADC will interrupt the CPU. Then, the CPU will execute the defined ISR which reads and displays the result. Then returns to its previous task.

We can experiment with the functionality of ADC operation by simply providing a voltage to the ADC pins via a potentiometer which is connected to a 0–10 VDC or ± 10 VDC (or smaller range) DC power supply (i.e., batteries). In the process of making this experiment work, there are two aspects to debugging the potential problem:

1. First is the ADC operation of the microcontroller: does the ADC converter interface and software work – does it read the analog voltage present at the pins correctly. We can test that by proving a known voltage with a simple potentiometer and DC supply. We can measure the voltage with a DMM or oscilloscope, and confirm if the ADC reads it correctly.

2. The second aspect of the experiment is the strain gauge sensor, its signal processing via the Wheatstone bridge and amplification of that signal. We can test that part of the experiment simply by imposing deformation on the strain gauge mounted beam, then measuring the output voltage at the output of the amplifier as well as at the output of the Wheatstone bridge.

11.9 EXPERIMENT 9: SOLENOID CONTROL USING A TRANSISTOR AND PIC MICROCONTROLLER

Objectives

To control a DC solenoid using a power transistor and PIC 18F452 (or PIC 18F4331) microcontroller.

Components

Item	Quantity	Part No.	Supplier
On/Off (pull) type DC solenoid	1	142463	www.jameco.com
On/Off (push-pull) type DC solenoid	1	145314	www.jameco.com
Transistor: IRF510 (MOSFET)	1	06F8238	www.newark.com
Diode	1	76970	www.jameco.com
PIC Demo Board/Connectors	1 set	DM163022 or DV164006	www.microchipdirect.com

Theory

A solenoid is a linear displacement actuator. It has a coil, a plunger, and a core to guide the electromagnetic field between the coil (stator) and the plunger (rotor). When current is applied to the coil, force is generated in the direction to minimize the magnetic reluctance. The direction of the current does not affect the direction of the force. The force generated is proportional to the square of the current and inversely proportional to the square of the air gap between the plunger and the stopper. By design, some solenoids are designed to be operated in ON/OFF mode and some are designed to be operated in proportional mode. In the ON/OFF mode of operation, the plunger is intended to take one of two positions (fully OPEN or fully CLOSED) based on the current in its coil. In the proportional mode of operation, the solenoid can take intermediate positions as a function of the coil current. In this experiment we use a solenoid designed to be operated in ON/OFF mode. With careful real-time control software, it can still be operated in the proportional mode. To control the solenoid in ON/OFF mode, the output signal from the microcontroller to the transistor, which switches the load current to the solenoid, can be a digital output signal. If we wanted to control it in proportional mode, then a PWM output pin of the PIC should be used to drive the transistor. We can also use a digital output channel and drive it with a PWM type signal under software control, since the PWM frequency to drive the solenoid does not have to be very high and such low frequency PWM signals can be generated under software control at a standard digital output pin. For instance, if we wanted to generate a PWM signal at 50 Hz with 5% duty cycle control resolution, it requires us to update the digital output pin at the following rate: $50 \cdot \frac{100}{5} = 1000$ Hz.

Hardware

The circuit is designed to switch the voltage across the solenoid during the ON/OFF periods of the cycle. In every cycle, when the voltage is high (12 V) the coil is energized by the current flowing through it. A magnetic field, and the resulting actuation force, is produced due to the tendency of the ferromagnetic plunger and coil generated magnetic flux to seek the minimum reluctance point. This magnetic field pulls the plunger in towards the stopper. When the voltage is low (0 V) for the remainder of the cycle (OFF period) the base current of the transistor falls below a minimum value and it stops transmitting current to the solenoid. As a result the magnetic field collapses and the plunger is released. A protection diode is connected in parallel to the solenoid.

Application Software Description

The program code essentially configures the PIC microcontroller to give out a step signal that varies between high and low based under software control. The header files of interest to be included in the code are as follows,

```
#include <p18f4331.h>
#include <pwm.h>
#include <timers.h>
#include <delays.h>
```

Procedure

1. Assemble the circuit as shown in the circuit diagram (Figures 11.27 and 11.28). Take care not to connect the power supply until the entire circuit has been assembled.

2. Configure the pin 2 of Port C of as a digital output, and turn ON and OFF the output pin under software control and verify the solenoid motion and force in response to your software.

3. Configure the output pin as a PWM output (and rewire the transistor base signal to one of the PWM pins. Note that RC2 pin can be configured either as a digital I/O or PWM output). Then decide on the PWM frequency and duty cycle.

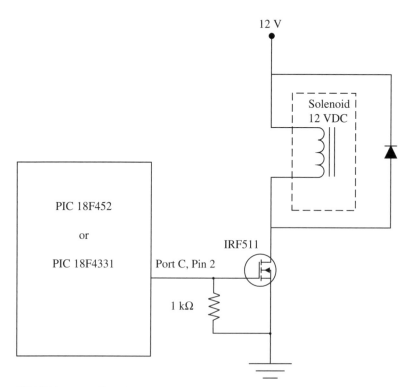

FIGURE 11.27: Circuit diagram for the DC solenoid control experiment.

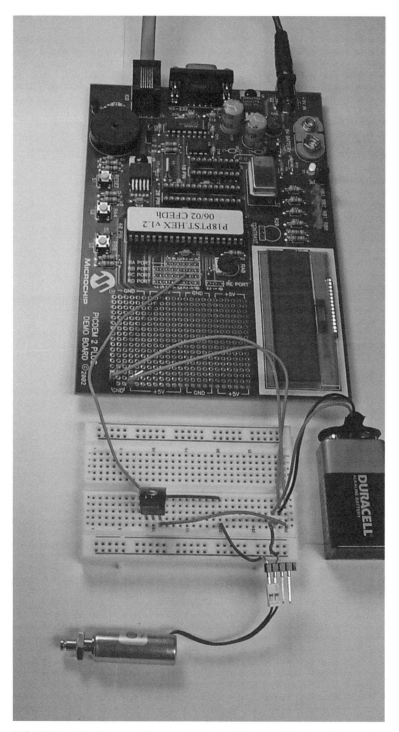

FIGURE 11.28: Picture of the complete circuit for the DC solenoid control experiment.

4. Experiment with different PWM duty cycles and PWM frequencies. Try to feel the net force generated as a function of PWM duty cycle while holding the solenoid at a fixed mid position.

5. (Optional) Obtain a proportional single stage valve with two proportional solenoids, spool position sensor, and centering spring. Connect two PWM output pins from the PIC microcontroller to drive two transistors to control two solenoids. Under software control, control either one of the solenoids proportionally under software control. Verify the spool position change as function of the solenoid (number 1 or number 2) and the magnitude of the command signal sent to the solenoids by the PIC microcontroller.

11.10   EXPERIMENT 10: STEPPER MOTOR MOTION CONTROL USING A PIC MICROCONTROLLER

Objective

Design a complete system for motion control of a stepper motor using the PIC microcontroller and a step motor controller integrated circuit (IC). Write software to control

1. magnitude of speed,
2. direction of speed,
3. run the motor in full-step mode and half-step mode.

Components

Item	Quantity	Part No.	Supplier
EDE1200 IC (translator chip)	1	141532	www.jameco.com
Stepper Motor	1	151861	www.jameco.com
Potentiometer (200 Ω)	1	181972	www.jameco.com
DIP Switch	1	38818	www.jameco.com
PIC Demo Board/connectors	1 set	DM163022 or DV164006	www.microchipdirect.com
ULN2003A - Transistor array	1	60K7049	www.newark.com
IN4744 - Zener diode	1	36185	www.jameco.com
Resonator: AWCR 4.00 MD	1	13J2002	www.newark.com

Theory

Stepper Motor   A stepper motor rotates one *step* per change in the energized state of its stator windings. The stepper motor used in the experiment is a unipolar, 2-phase (center tapped), 7.5 degrees/step, 5 VDC stepper motor.

EDE1200 Unipolar Stepper Motor IC and ULN2003A Driver   EDE1200 is an integrated circuit (IC) logic chip (also called the translator) for a unipolar step motor. The actual current control in the motor phases is done by another high current driver, which is ULN2003A chip in this case. EDE1200 accepts 5 V logic level signals and generates output commands which define how the current in each phase should be controlled. EDE1200 provides four TTL level output signals. These TTL-level output signals control the sequence of the amplifier drive circuit, which generally consists of standard power transistors or a transistor array integrated circuit (IC). ULN2003A is the stepper drive circuit (power transistor "switch set"). Hence, the ULN2003A high voltage and high current transistor array is used to connect between the EDE1200 integrated circuit chip (translator) and the stepper motor. It is capable of controlling full-step and half-step modes and direction as well. It does not have an in-built oscillator, so a 4 MHz external oscillator is connected to it.

The following are the input command signals to the translator EDE1200 IC that must be controlled by the microcontroller,

1. Step: Pin 9 = 1, should be maintained high for a step motion.
2. Direction: Pin 7, 1 = clockwise, 0 = counter-clockwise.
3. Full/Half Stepping: Pin 8, 1 = for full stepping, 0 = for half stepping.

When the EDE1200 is running at an external clock speed of 4 MHz, the "Step" pin signal can range up to 5 kHz resulting in a motor speed over 6250 RPM with a 7.5 degree per step motor. Very high acceleration rates should be avoided. That is, we should not instantly apply high speed requests to a stopped motor. In other words, commanded acceleration should not exceed a maximum value. Otherwise, the motor may not be able follow the very high acceleration rate and this results in "step loss," meaning that the motor does not make the commanded number of step motions. The only way to detect the step loss is to measure the rotor position with a position sensor (i.e., incremental encoder). Once we determine that the motor did not make the commanded number of steps, the additional missing number of steps can be re-commanded in a closed loop control algorithm.

Application Software Description

The stepper motor is driven by the stepper motor driver chip, which in turn is given a step input by the PIC microcontroller. A schematic of the arrangement is shown in Figure 11.29 and a picture of the assembled circuit is shown in Figure 11.30.

The code generates a pulse signal of the required frequency using one of the port output pins of the PIC. This is done by simply raising that particular pin to a *high* status, waiting a certain delay time, and then lowering it to *low*. The basic code layout is given below. Figure 11.29 shows that Port B, pins 0, 1, 2, 3 are used to interface the PIC microcontroller to the EDE 1200 IC.

1. Clear selected port values, and configure ports to act as outputs for Step, Direction, and Full/Half Step Mode commands.
2. Select "Direction" and "Full/Half Step" mode by writing 1 or 0 to the corresponding port pins (which should be connected to Pins 7 and 8 on the EDE1200 IC).
3. Raise the "Step" pin (which should be connected to Pin 9 on the EDE 1200) to *high*. Wait for a specified time (depending on the required stepping frequency).
4. Drop pin to *low*. Wait for a specified time (depending on the required stepping frequency).
5. Experiment with different numbers of steps. Verify that the step motor runs as expected and moves the expected number of steps.
6. Experiment with different directions.
7. Experiment with Full Step and Half Step mode.
8. Experiment with different rates of step signals (speed control).

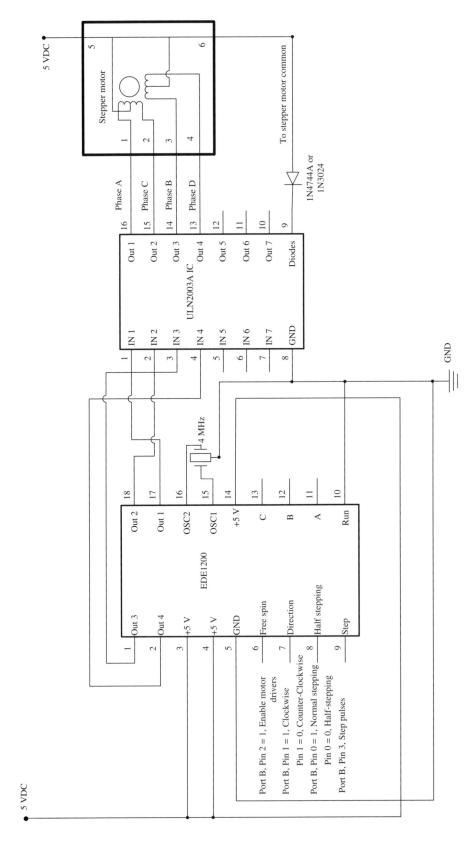

FIGURE 11.29: Circuit diagram for the stepper motor control experiment.

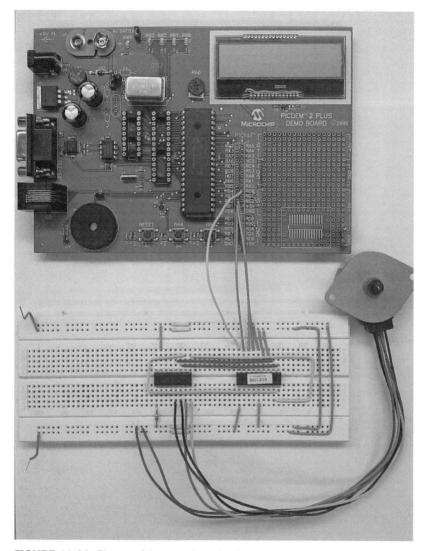

FIGURE 11.30: Picture of the complete circuit for the stepper motor control experiment.

Procedure

1. Assemble the circuit as shown in the circuit diagram.
2. Compile and run the code to rotate the stepper with a pre-specified speed and number of steps.
3. Change the stepping direction.
4. Change the status of the stepping-mode input signal to switch between Full and Half Step modes. Experiment and compare the difference in the motion between Full and Half Step modes. Note the difference for your lab report.
5. While holding a position in Full Step mode and in Half Step mode, test the holding torque magnitude with your hand by trying to move the shaft. Verify the difference in the holding torque magnitude between Full Step mode and Half Mode.

11.11 EXPERIMENT 11: DC MOTOR SPEED CONTROL USING PWM

Objectives

To control the speed of a DC motor using a PWM signal from a PIC 18F452 microcontroller in conjunction with an H-bridge amplifier circuit.

Components

Item	Quantity	Part No.	Supplier
DC Motor	1	154915	
Optoisolator	1	114083	www.jameco.com
Potentiometer (200 Ω)	1	181972	
IRF510 (MOSFET)	2	06F8238	www.newark.com
IRF9520 (MOSFET)	2	07B1521	www.newark.com
1N4003 Diode	4	76970	www.jameco.com
PIC Demo Board/connectors	1 set	DM163022 or DV164006	www.microchipdirect.com

Theory

Pulse Width Modulation   A PWM signal has two variables that must be decided by the control software:

1. PWM signal carrier (or base) *frequency*,
2. PWM signal *duty cycle*.

The PWM frequency is generally a constant frequency in the kHz range. The duty cycle is the value decided in real-time to indicate the equivalent analog value of the signal.

This is accomplished by generating an ON-OFF signal of a high frequency, and then varying the percentage of time that the signal is in the *ON* state versus the *OFF* state. This is called varying the *duty cycle*. The average value of the signal is equivalent to an analog signal, provided that the PWM switching frequency is much higher than the bandwidth frequency of the electromechanical system. For example, assume that a voltage signal between 2.5–3.0 V is required to drive a motor, while the supply voltage is a fixed 5 V. The PWM signal will consist of a voltage that varies at a high frequency (i.e., 1 kHz) between 0–5 V. If, in every time period (1 ms), the voltage level is kept high (5 V) for 50% of the period, and low (0 V) for the remaining 50%, then the average voltage seen at the output will be 2.5 V. In this case, the duty cycle is 50%. Similarly, if a high voltage is maintained for 60% of the period, and low for the remaining 40%, then the average voltage is 0.6×5 V = 3 V.

Care must be taken to ensure that PWM frequencies are kept reasonably high compared to the bandwidth of the control system, since a low frequency PWM signal may actually be seen as varying, and not continuous, voltages, especially for loads with a smaller electrical time constant. Typical PWM frequencies are of the order of a few kHz.

The PWM signal format is used in two different contexts:

1. Power transistor circuits, where the transistors are operated in all ON and all OFF alternating states, which is called the PWM mode. In this mode of operation, as opposed to the linear mode of operation where the transistor is partially turned ON/OFF, the efficiency of the power transistors is significantly improved. In high

power transistor applications, the PWM mode of operating the transistors is invariably more efficient than the linear (proportional) mode of operating them. However, due to the high switching frequency of the PWM circuit, there could be a small high frequency noise (in same cases, also audible) induced on the power output lines.

2. Low power sensor or signal processor signals: when a signal is transmitted as an analog voltage level, it is susceptible to noise. If it is transmitted as a PWM signal, small changes in the voltage due to noise do not affect the information coded in the signal magnitude unless the noise is so large that it changes the ON/OFF state of the signal, and the resulting duty cycle. Hence, the PWM mode of transmitting an analog signal is more immune to noise than the analog voltage mode of transmission.

H-Bridge Circuit The DC motor is driven using the H-bridge amplifier circuit comprising of four MOSFET power transistors (two of them are p-channel (IRF9520) and the other two are n-channel (IRF511/IRF510) MOSFET transistors), four diodes (1N4003) and the motor itself. Each MOSFET pair on each side of H-bridge consist of a p-channel on the top and a n-channel on the bottom as shown in the figure. The motor connects in between the two legs. The p-channel source terminals are connected to 9 V and the drain terminals are connected to motor leads, whereas the source terminals of n-channel are connected to ground and the drain terminals are connected to motor leads.

When PWM channel 1 is turned ON, the transistor pair on the top-left and bottom-right will be conducting, hence the current will flow from left to right direction. When the PWM channel 2 is turned ON, the opposite pair will be conducting (bottom-left and top-right), hence the current will flow through the motor winding in the opposite direction (from right to left).

When all four transistors are OFF, the motor is in uncontrolled "coast" state. When only the bottom two transistors are turned ON, then both leads of the motor are grounded and the motor is in "dynamic braking" state.

It is important that the two transistors on any one side of the H-bridge should not be turned ON at the same time since it would create a short-circuit between the DC bus supply voltage and ground, resulting in a very large current and possibly destroying the transistors. Keeping in mind that power transistors have longer turn-OFF time than turn-ON time, there should be some "dead-time" between turn-OFF and turn-ON time of the two transistors on any one side. This can be accomplished either in software or by H-bridge integrated circuits (IC) which have dead-time insertion and short-circuit protection capabilities. PWM output peripherals in microcontrollers have dead-time insertion capability as a programmable feature.

Application Software Description

The program code configures the PIC to output two PWM signals of constant frequency and variable duty cycle. These are controlled by the:

- CCP1RL and CCP1CON registers for PWM channel 1 on pin RC2, and
- CCP2RL and CCP2CON registers for PWM channel 2 on pin RC1.

PWM channel 1 (RC2) controls the current in one direction, while PWM channel 2 (RC1) controls the current in the opposite direction in the H-bridge. By choosing which one of the PWM outputs to turn on, we control the direction of current, and hence the direction of the torque generated by the motor. By controlling the magnitude of the current in the PWM pin, we control the magnitude of the current, and hence the magnitude of the torque.

Commercial H-bridge amplifiers accept two different types of PWM command signals and there is a "H-bridge driver/controller" section of the circuit which translates the input command signals to corresponding H-bridge drive signals to operate the transistor pairs:

1. PWM command signal (0–100% *duty cycle*, corresponding to 0–100% of maximum current command) and direction signal. The *direction* signal (ON for clockwise or positive, OFF for counterclockwise or negative) decides on the direction of the current, the PWM signal's duty cycle determines the magnitude of the commanded current. Using these two signals, the H-bridge driver circuit generates two command signals that drive one of the pair (two transistors) at a predefined PWM frequency.

2. PWM command signal only in (0–100% duty cycle), where 50% duty cycle is considered as zero current command. A PWM duty cycle less than 50% is interpreted as counterclockwise direction, and a PWM duty cycle above 50% is interpreted as clockwise direction. Then, the H-bridge controller activates the two PWM channels to control one of the two pair of power transistors.

The type of PWM command signal an H-bridge expects is a matter of interpreting the command signals in a clear way so that the direction and magnitude of the signal is defined. The direction information determines which one of the two pairs of transistors will be activated. The magnitude information determines what percentage of the time (duty cycle) the active transistor pair is kept ON and OFF.

If the H-bridge expects an analog voltage signal for current command, it again needs a way to determine the magnitude and direction. For instance, 0–10 VDC analog input signal defines the magnitude of desired current, and a direction input signal determines the sign (direction) of the desired current. Likewise, an analog voltage command range of -10 V to $+10$ V alone can be used to define both magnitude and direction of the commanded signal as well.

Regardless of the type of command signal used, the H-bridge controller needs to be able to determine the desired current direction and magnitude. Once that is known, then the corresponding transistor pair is activated with a proper duty cycle of a fixed frequency PWM signal.

Procedure

1. Assemble the circuit as shown in the circuit diagram (Figures 11.31 and 11.32).
2. Calculate the values of the PWM frequency and duty cycle, and implement the code for these values.
3. Download and run the program. The motor speed should change based on the duty cycle.
4. Vary the duty cycle value in the code gradually to force the motor to correspondingly increase or decrease its speed. Notice that if we simply control the PWM signal based on open loop logic, we are controlling the H-bridge amplifier in voltage mode. The average voltage applied across the motor terminals will be simply proportional to the PWM output signals from the microcontroller. We refer to this type of amplifier as a "voltage amplifier."
5. Change the direction and magnitude of the speed under software control.

Closed Loop Current Control If we can measure the current passing through the motor windings, and read it back to the microcontroller, then we can use it to implement a closed loop current control algorithm to decide on the PWM output. For instance,

$$PWM = K \cdot (i_{\text{cmd}}(t) - i_{\text{mea}}(t)) \qquad (11.130)$$

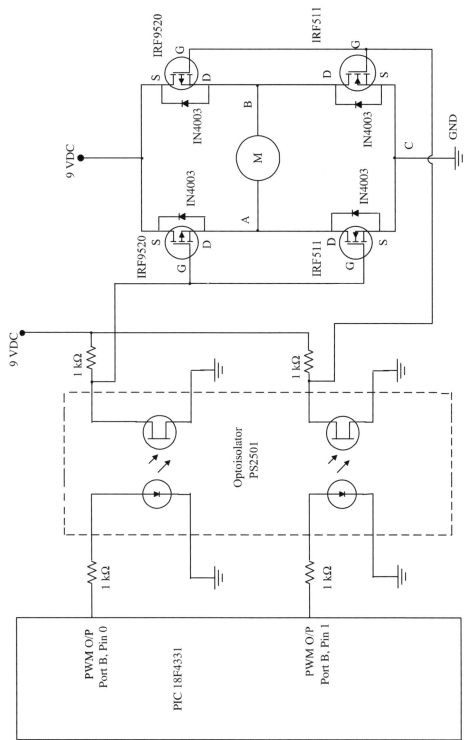

FIGURE 11.31: Circuit diagram for DC motor speed control experiment.

FIGURE 11.32: Picture of the complete circuit for the DC motor control experiment.

where $i_{cmd}(t)$ is the desired current, $i_{mea}(t)$ is the actual (measured) current, K is the current loop gain, PWM is the output signal value (duty cycle) of the PWM channel.

In order to implement such a closed loop current control algorithm, we need to measure the current through the motor and interface it to the PIC microcontroller's ADC input channel (one of the channels), and sample it at a high sampling rate (i.e., at least ten times faster than the desired current loop bandwidth).

The motor current is measured by inserting a precision resistor at the bottom leg and ground connection of the H-bridge (C and GND on the H-bridge circuit). The key here is that the resistor value should be accurate and stable (not changing much due to temperature) so that current times the resistor gives us a voltage which we can sample through the ADC converter. Since the motor amplifier is PWM controlled, the measured voltage may be passed through a low pass filter before being interfaced to the ADC converter. The low pass filter bandwidth should be low compared to the PWM frequency, but high enough not to filter out the current loop dynamics. For instance, if PWM frequency is 10 kHz, and current loop bandwidth desired is 100 Hz, we can pick a low pass filter with cross-over frequency of 500 Hz or 1000 Hz. The motor current and voltage across the precision resistor is related as

$$V_{mea}(t) = R_s \cdot i_{mea}(t) \tag{11.131}$$

where R_s is the precision resistor, $i_{mea}(t)$ is the motor current, and $V_{mea}(t)$ is the voltage proportional to the current. R_s should be as small a value as possible in order to not add additional resistance to the circuit for the purpose of current measurement. Since the current measurement resistor acts as a series resistor with the motor resistance, it should not be significantly increasing the total resistance. Relative to the motor winding resistance, R_m, R_s should be in the range of $1/1000 \cdot R_m$. For instance, in an application where $R_m = 10\,\Omega$, $i_m = i_{max} = 10\,A$, $V_{ss} = 100\,VDC$ DC bus voltage supply, the maximum power of the motor is $1000\,W = 1\,kW$. If we use a current measurement precision resistor $R_s = 0.1\,\Omega$, at maximum current, there would be $R_s \cdot i^2 = 10\,W$ power dissipation (wasted power) in order to accomplish the current measurement.

Furthermore, this voltage may be passed through a low pass filter before it is connected to the ADC,

$$V_{ADC}(jw) = LPF(jw) \cdot V_{mea}(jw) \qquad (11.132)$$

where $LPF(jw)$ is the frequency response of the low pass filter.

11.12 EXPERIMENT 12: CLOSED LOOP DC MOTOR POSITION CONTROL

Objectives

Closed loop position control of a DC motor using a position feedback sensor and PWM output signal in conjunction with an H-bridge amplifier circuit, and PIC microcontroller.

Components

Item	Quantity	Part No.	Supplier
DC Motor	1	154915	(www.jameco.com)
Optoisolator	1	114083	(www.jameco.com)
Potentiometer (200 Ω)	1	181972	(www.jameco.com)
IRF511 (MOSFET)	2	39C4310	www.newark.com
IRF9520 (MOSFET)	2	07B1521	www.newark.com
1N4003 Diode	4	76970	www.jameco.com
Opto-interrupter	2	273560	www.jameco.com
Disk with holes	1	Can be constructed in lab	A tick paper or plastic disk
PIC Demo Board/connectors	1 set	DM163022 or DV164006	www.microchipdirect.com

Theory

In this lab, we implement a closed loop position control of a DC motor using a PIC 18F452 (or PIC 18F4431) microcontroller. In order to perform closed loop position control, we need a position sensor. For the purpose of understanding the incremental encoder working principles and keeping the cost low, we will construct a very simple "home-made" incremental encoder. For that we need two things (Figure 11.33):

1. A disk that has evenly spaced holes that will block and pass the light alternately. For our purposes, we will have 4 holes on the disk. The higher the number of holes, the better the position sensor resolution, but the more difficult it is for us to mechanically assemble the experiment. So, we will experiment with a relatively low resolution home-made incremental encoder. This disk can be made from a thick paper or plastic or glass.

2. Two opto-interruptors. Opto-interruptors are placed over the disk. As the disk rotates, the light path is interrupted by the solid sections and passed by the hole sections of the disk. If we use only one opto-interruptor, we can detect the change of position, but we cannot detect the direction of position change, that is we cannot detect the direction of speed. So, the second opto-interruptor is used for that purpose. The second opto-interruptor is mechanically placed at a 90° mechanical phase angle relative to the first opto-interruptor position over the hole–solid sections of the disk. Note that 360° is considered a complete cycle of the one hole and one solid section of the disk. If the disk is rotating in a clockwise direction (forward), the digital output signal from

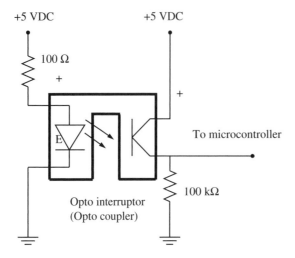

2 Opto-interrupters

"Home-made" incremental encoder

FIGURE 11.33: A "home-made" incremental encoder to sense the position change of the motor shaft. Two opto-interrupters and a disk are used to make a simple incremental encoder.

opto-interrupter #1 would lead the signal from the opto-interrupter #2 by 90° phase angle. If the disk is rotating in a counterclockwise direction (reverse), the opposite would happen.

We can also implement quadrature (times 4, ×4) resolution improvement in the position measurement accuracy by noting that fact that the two opto-interrupters are 90° out of phase. At any given time, by evaluating the state of two opto-interrupters within a cycle, we can determine where the position is within 1/4 of that cycle.

Below we provide two simple ways to keep track of encoder interface and position measurement without using a dedicated quadrate decoder interface perhipheral (without using QDEC) on the microcontroller. Modern microcontrollers for motion control have a quadrature decoder interface peripheral for an incremental encoder interface. Once the encoder A and B channels are connected to the QDEC pins, and software setup is done, one has to only read a register in the QDEC peripheral at any time to get the current position. This is the preferred method of incremental encoder interface since the QDEC peripheral handles the encoder interface and position measurement details, relieving the CPU for other tasks. For the CPU, the position reading from the encoder is simply accessing a register location.

The following is an interrupt-driven software driver to keep track of the encoder position (×1 decode, not ×4) without using QDEC peripheral. This method would work only at relatively low speeds of motor, that is the speed of the motor is so low that the time period between each pulse of the encoder is long enough for the microcontroller to have sufficient time to handle each pulse as an interrupt, execute the ISR (Interrupt Service Routine) routine to keep track of position count, and also perform other logic functions.

1. Hardware interface: Encoder channel A is connected to a digital input pin, Encoder channel B is connected to another digital input pin of the microcontroller.
2. Software driver:
   (a) Software setup: Digital input pin connected to Encoder channel A is setup to generate interrupt on rising-edge. Initially, the position count is set to zero. An ISR is setup for the interrupt handling.
   (b) Interrupt Service Routine (ISR): when Channel A has rising-edge, an interrupt is generated and ISR is called automatically.

```
ISR()
{
   If  Channel_B  == OFF
       Position = Position +  1 ;
   else
       Position = Position -  1 ;
   endif
}
```

A possible implementation for encoder position tracking with ×4 decoding under software control is as follows:

1. Hardware interface: Encoder channel A is connected to a digital input pin, Encoder channel B is connected to another digital input pin.
2. Software driver:
   (a) Software setup: Both digital input pins connected to Encoder channel A and B are setup to generate interrupts on rising-edge and falling-edge: ISR1 and ISR2. Initially, the position count is set to zero. Two ISRs are setup for the interrupt handling.
   (b) Interrupt Service Routine 1 (ISR1): when Channel A has rising-edge or falling-edge, an interrupt is generated and ISR1 is called automatically.

```
ISR1()
{
   If  (Channel_A== ON  && Channel_B  == OFF )
       Position = Position +  1 ;
   else if (Channel_A == OFF  && Channel_B == ON)
       Position = Position +  1 ;
   else if (Channel_A == ON  && Channel_B == ON)
       Position = Position -  1 ;
   else if (Channel_A == OFF  && Channel_B == OFF)
       Position = Position -  1 ;
   endif
}
```

   (c) Interrupt Service Routine 2 (ISR2): when Channel B has rising-edge or falling-edge, an interrupt is generated and ISR2 is called automatically.

```
ISR2()
{
   If  (Channel_A== ON  && Channel_B  == ON )
       Position = Position +  1 ;
```

```
        else if (Channel_A == OFF  && Channel_B == OFF)
            Position = Position +  1 ;
        else if (Channel_A == ON  && Channel_B == OFF)
            Position = Position -  1 ;
        else if (Channel_A == OFF  && Channel_B == ON)
            Position = Position -  1 ;
        endif
}
```

Application Software Description

In most microcontrollers for motion control applications, there is a dedicated chip to interface to the encoder and count the position change pulses as well as calculate the speed. However, in the PIC 18F452 microcontroller we have used so far, such a peripheral is not available, whereas PIC 18F4431 does have an incremental encoder interface peripheral. For PIC 18F452, we can implement the incremental encoder interface using discrete input lines. We can either connect the two opto-interrupter signals to two digital inputs and sample them fast enough so that we do not lose any pulse, or we can connect them to hardware interrupt lines. In the latter case, everytime a pulse transition occurs, an interrupt is generated. At the

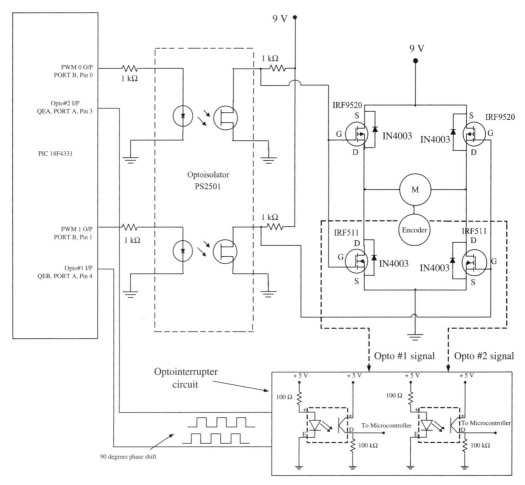

FIGURE 11.34: Circuit diagram for DC motor closed loop position control experiment.

interrupt service routine, we would determine the *direction* of the motion and increment or decrement the position count.

In microcontrollers with a dedicated incremental encoder interface (also called a *quadrature decoder circuit*), the microcontroller does not have to service the interface for each pulse. Rather, at any given time it would read the counter register since the interface circuit handles the counting process. PIC microcontroller 18F4431 has one incremental encoder interface (Figure 11.34). As a result, implementing this experiment with PIC 18F4331 is much easier than with PIC 18F452. The interface is capable of decoding the incremental position information with ×4 multiplication, as a result, it is commonly referred to as the "quadrature decoder" or QDEC interface. The pin numbers RA2, RA3, RA4 are programmable to interface to the INDEX, channel A, and channel B of the encoder, respectively. In our experiment, we do not use the INDEX signal. PWM output channels can be assigned to pin RB0 for PWM channel 1, and pin RB1 for PWM channel 2. We can then use PWM channel 1 to control one pair of the transistors of the H-bridge, and the PWM channel 2 to control the other pair of the transistors of the H-bridge.

Once we know the actual position, we can digitally calculate the actual (estimated) speed of the shaft. Then, if we have a programmed (desired) position or velocity, we can determine the PWM output based on a PID type control algorithm, that is PD algorithms

```
// Assume
//    xd - the desired position variable, programmed.
//    vd - the desired speed variable, programmed.
//     x - the measured position
//     v - calculated speed based on measured position
//    Kp - proportional gain of the PD control algorithm: a
//         constant.
//    Kd - derivative gain of the PD control algorithm: a
//         constant
//    Ki - integral gain control
//    ui - integral control


// The PD closed loop control position control algorithm

      PWM_Out = Kp * (xd - x ) + Kd * (v - vd)


// If the objective is only to control speed, but not the
//   position, then the PWM output calculation should not be
//    function of position information, i.e. a proportional
//    closed loop speed control algorithm.

      PWM_Out = Kd * (v -vd)


 // PID algorithm for position control loop
 //     Initialize integrator term on startup;
 //     Set  "first_call = true ;" on startup.

      if (first_call == true)
          ui= 0.0 ;
          first_call = false;
      endif

      ui = ui + Ki * (xd -x )


      PWM_Out = Kp * (xd - x ) + Kd * (v - vd) + ui ;
```

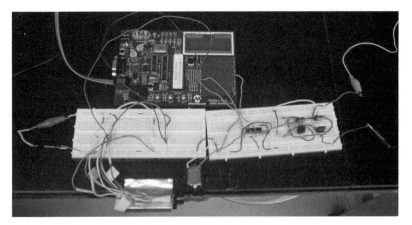

FIGURE 11.35: Picture of DC motor closed loop position control experiment.

Procedure

1. Decide on the hardware interface method of the opto-interrupters, that is the electrical signals coming from the "home-made" incremental encoder, to the PIC microcontroller. It is highly recommended that a microcontroller with incremental encoder (QDEC) interface peripheral is used, such as the PIC 18F4331 microcontroller.

2. Keep the motor current (or voltage) control and PWM circuit as in the previous lab (Figure 11.35), that is the H-bridge amplifier (power stage) and the PIC microcontroller interface to it via PWM output channels.

3. Assemble the circuit.

4. Implement software to measure actual position and speed.

5. Program a closed loop PID control algorithm that can be used either for closed loop position control or closed loop speed control. The program or user should select which mode (closed loop position or speed control mode) to operate and update frequency.

6. Program desired position and speed control trajectories as a function of time in the control algorithm and test the closed loop position control. For instance, command 1/4 rev rotations in forward and reverse directions, command slow and fast speeds in forward and reverse direction.

7. In closed loop position control mode, while holding current position (no change in the desired position), try to disturb the rotor position, does the motor react to keep its current position?

8. Demonstrate that you can control the motor position or velocity to any desired value and direction.

9. What is the smallest positioning accuracy you can achieve with this system?

MATLAB®, SIMULINK®, STATEFLOW, AND AUTO-CODE GENERATION

MATLAB® (and its tools such as Simulink®, Stateflow) is the standard software tool used for control system design by engineers. In addition to MATLAB®, the following tools are used as part of control system development, proto-type deployment, and testing: Simulink®, Stateflow, and an auto-code generator tool called Similink Coder (formerly Real Time Workshop (RTW) and Stateflow Coder (SFC)). Programs for control and simulation are written in MATLAB® as text files using MATLAB® script language, which is similar to C language. Simulink® and Stateflow are graphical tools to define the control logic. Simulink® Coder is the tool to automatically generate C-code (ANSI/ISO C and C++ code) for a target embedded hardware ECM (electronic control module).

Simulink® is the main graphical modeling tool for control system design. Stateflow is a complementary tool that is most suitable for modeling event-driven supervisory logic aspects of a control software, whereas Simulink® is more suitable for more mathematical aspects of it, such as PID control algorithms. The Simulink® Coder auto-code generation tool is used to automatically convert Simulink®, Stateflow, and MATLAB® files to C-language files, compiled with a C-compiler, and deployed on a target microcontroller hardware. There are many, and growing, numbers of target microcontroller/DSP hardware supported by MATLAB® and its tools. One of the most general purposes of those is xPC hardware. xPC is an industrial PC/104 bus based controller hardware architecture. It is a flexible and general purpose real-time controller hardware platform for development purposes, with a lesser degree of ruggedness compared to commercial ECMs. MATLAB® and its tools support the xPC hardware through the "xPC Target" software tool that allows the Simulink® Coder to generate C-code for the xPC target hardware. The xPC hardware runs the DOS operating system and the xPC Target Kernel as its real-time operating system.

A.1 MATLAB® OVERVIEW

MATLAB® is a general purpose modeling and simulation software, widely used in the world by control engineers. Students can use this introduction as a starting point, and get more details on specific points using the online help for MATLAB®. The MATLAB® software package is an integrated software tool with many components and new components are being added continuously. These components are:

1. **MATLAB®** being the main front end tool, text based programming environment. In addition to these basic MATLAB® components, there are **Toolboxes** that can be added to increase its capabilities. For instance, the Control Systems Toolbox can

Mechatronics with Experiments, Second Edition. Sabri Cetinkunt.
© 2015 John Wiley & Sons, Ltd. Published 2015 by John Wiley & Sons, Ltd.
Companion Website: www.wiley.com/go/cetinkunt/mechatronics

be added to MATLAB® to extend its function libraries for that purpose. Similar toolboxes exist for Simulink®, Stateflow, and other software components.

2. **Simulink**® for graphical block diagram modeling of dynamic systems and control algorithms.

3. **Stateflow** for event-driven system and supervisory control logic modeling (also with graphical interface).

4. **Simulink**® **Coder (SC)** for auto-code generation in C-language from Simulink®, Stateflow, and MATLAB® files (formerly called Real Time Workshop (RTW) and Stateflow Coder (SFC)). The **Embedded Coder** tool generates auto-code in C for specific microcontrollers and DSPs. The **MATLAB**® **Coder** tool generates auto-code in C from the MATLAB® text m-function files.

5. **xPC-Target** software for the PC-based embedded hardware tool called xPC-hardware. There are other embedded target microcontroller software and hardware combinations supported, such as Real-time Windows environment to run real-time code on a PC with relevant I/O cards. For other specific commercial ECMs (i.e., for engine control, transmission control etc.), MATLAB® provides "target ECM hardware" specific tools which work with Simulink® Coder (SC), Embedded Coder, and MATLAB® Coder in order to generate C-code automatically for that target ECM.

6. **C-Compiler** to compile the generated code for the target embedded controller hardware. MATLAB® installation includes the **LCC C-Complier**.

Figure A.1 shows the relationship between MATLAB® and its main "add-on" tools. In the discussion that follows, we used the term MATLAB® or MATLAB®/Simulink® to refer to this whole suite of tools in the MATLAB® package. If testing on actual embedded controller hardware is not needed and simulations are sufficient for a given purpose, then only the MATLAB®, Simulink®, and Stateflow package components are needed. It should be noted that the supervisory and event driven logic can also be coded in MATLAB® and Simulink® alone without the use of Stateflow. However, the Stateflow environment makes

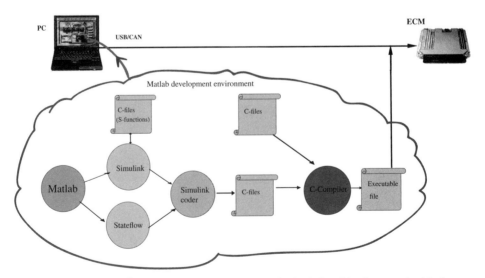

FIGURE A.1: MATLAB®, its add-on tools and their typical relationships in an embedded control software development.

it more convenient and easy to code the event driven nature of logic. Many examples are provided under the

MATLAB® > Help > Demos

menu with example categories in automotive, aerospace, and heating systems.

The main benefit of the MATLAB®/Simulink® environment over the standard high level programming languages such as C/C++ are:

1. faster program development due to reduced detail in defining data structures (i.e., in MATLAB®, we do not have to declare data types before usage) and easy to use library functions (including I/O functions),

2. good graphical communication of the model structure in Simulink® and Stateflow as oppose to text based programming language description,

3. auto-code generation tools for real-time implementation.

In addition, the MATLAB®/Simulink® environment can interface with external user written C functions. However, use of external functions should be balanced with the difficulty they impose and maintaining the advantages of the MATLAB® environment.

MATLAB® is started by double clicking on the MATLAB® short-cut icon on the desktop under Windows development environment. The standard prompt of MATLAB® when it starts and opens its user interface window, called the **"Command Window"**, is

```
>>
```

The user interacts with MATLAB® from the command window. In addition, the window has menu items which will allow the user to perform typical file I/O, print, and setup operations, and get help on any MATLAB® topic.

The MATLAB® development environment has seven basic windows. Each window is dockable and can be disconnected from the MATLAB® window and treated as a separate window. Each window can be opened/closed using the View menu check mark next to each window. MATLAB®'s user interface is modified in each version a little bit. The current version may have a slightly different user interface than described here.

The main windows are (Figure A.2):

1. Command Window: the window through which the user enters MATLAB® commands interactively.

```
>> clc ;   % Clears the screen of the Command Window.
```

2. Command History Window: used to store and reuse previously used interactive MATLAB® commands. It is useful during program debugging.

3. Workspace Window: this window shows all of the variables defined in the MATLAB® workspace and their values.

```
>> clear ;  % Clears (deletes) all data variables defined from the
               MATLAB® Workspace.
>> who  ;   % Displays the currently defined data variables
            %  (without their size info) in the MATLAB® Workspace.
>> whos  ;  % Displays the currently defined data variables
            %(with their size info) in the MATLAB® Workspace.
```

4. Current Directory Window: directory navigation window. Using the icons on this window, the user can quickly change the current directory. By default, MATLAB®

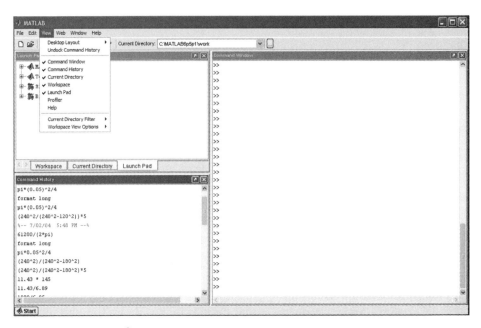

FIGURE A.2: MATLAB® environment and main user interface windows.

saves/accesses the files from the Current Working Directory. If it cannot find a file in the Current Working Directory, then it searches the deirectories defined in the "Path."

5. Launch Pad Window: to start different components of MATLAB®, such as Simulink®, StateFlow, and so on.

6. Profiler: The profile function is used to debug and optimize M-files by tracking their execution time. For each function in the M-file, the Profiler records information about execution time, number of calls, parent functions, child functions, code line hit count, and code line execution time.

7. Help Window: a separate window that allows access to a complete online help on all of the MATLAB® tools.

MATLAB® is used in one of two modes:

1. Interactive mode: the user types the commands in the "Command Window" and gets results after executing that command.

2. Programming mode: the user writes programs in MATLAB® language (also called MATLAB® script) and saves them in M-files. The name of the M-file is typed at the command window to run the programs.

A.1.1 Data in MATLAB® Environment

In MATLAB®, a data variable is created by giving it a name and assigning a value to that name by placing it on the left hand side of an assignment statement. For instance, the following example shows how to create a variable named "x" and assign a value to it. By default, it has data type of **double**.

```
>> x = 1.25 ;
```

Unlike in high level languages, such as C and Fortran, we do not need to declare the data type of a variable before it is used.

The main data structure in MATLAB® is a "matrix." A scalar is considered a 1x1 matrix. A vector is considered a row matrix. MATLAB® does not require the data type of a variable or matrix be declared. By default, all numerical data is treated as `double`.

The percent sign, %, is used to indicate the beginning of comments. The semi column " ; " at the end of a line suppresses the echo of the data to the screen. Three dots are used to indicate that the command line continues in the next line.

Variable names must start with a character, and are case sensitive, so x and X are different variable names.

Help on any topic can be obtained by

```
>>  help
>>  help topic
```

The display accuracy of the numerical data can be controlled by

```
>> format   short  ;
>> format   long   ;
>> format   short e ;    % to display in scientific notation
>> format   long  e  ;
>> format   bank   ;    % Display two decimals only, i.e. 5.40
>> format   rat    ;    % Display in fractional form,  i.e. 10/3
```

Matrix data is defined as follows,

```
>>   A = [  1      2      3
            4      5      6
            7      8      9  ]         ;
```

The same matrix, A, can also be defined with the following syntax,

```
>>  A = [ 1    2    3 ;  4    5    6 ;   7    8    9 ]  ;
```

Instead of spaces between matrix elements, comma "," can be used,

```
>>  A = [ 1,2,3 ; 4,5,6  ; 7,8,9]   ;
```

A specific element of a matrix, (k,l) - row k, column l, is referenced as,

```
>>  x = A(k,l) ;
>>  A(k,l) = 5.0  ;
>>
```

A row, or a column or a sub-matrix of a matrix can be referenced (for reading and writing)

```
>>  row_k  = A(k, : )    ;
>>  col_l  = A(:, l )    ;
>>  B   = A(1:k, 10:20)  ;     %  B is a sub-matrix of A.
```

Note that MATLAB® uses variable names "i" and "j" for representing $\sqrt{-1}$ imaginary number for complex variable definitions. In order to make sure we do not change the definition of built in variables "i" and "j", we should not use "i" and "j" in our programs as data variables, such as index number to vectors and matrices. When a "clear" statement is executed, the original definitions of the variables are restored.

```
>> clear
>> i
```

```
ans =

       0 + 1.0000i
>> j

ans =

       0 + 1.0000i
>> i=2.0

i =

    2
>> v= 5 + 10*i

v =

    25

>> i

i =

    2

>> clear
>> i

ans =

       0 + 1.0000i

>>
```

A large vector can be assigned as follows using a starting value, increment value, and final value,

```
>>  x =  0 :  1 : 100   ;   %  x=[0   1  2  . . . . . 100];
```

By default, vectors are row vectors of (1xn). To change a vector into a column vector, simply use the transpose operator,

```
>> x = x' ;   % Original x is a row vector.
              % After this statement it is a column vector.
```

Defining matrices with rows and columns as follows is often needed, as shown below. Notice how we transpose the row vectors into column vectors in the definition of the matrix

```
>> Angle_Degrees = [ 0   90   180 270   360]

Angle_Degrees =

    0    90    180    270    360

>> Angle_Radians = (pi/180) * Angle_Degrees

Angle_Radians =

        0    1.5708    3.1416    4.7124    6.2832

>> Angle_Matrix = [Angle_Degrees' , Angle_Radians' ]
```

```
Angle_Matrix =

        0         0
  90.0000    1.5708
 180.0000    3.1416
 270.0000    4.7124
 360.0000    6.2832

>>
```

The default numerical data type in MATLAB® is double precision floating point data, **double** in C-language. Data can be defined and converted to integer data type using data conversion functions. MATLAB® supports 8-bit, 16-bit, 32-bit, and 64-bit signed and unsigned integer data types. The data conversion function examples are shown below,

```
k1=int8(-10.3)
l1=int16(-1000.4)
m2=uint32(1000000.5)
n2=uint64(-100000000.6)

>>
k1 =   -10
l1 =   -1000
m2 =    1000001
n2 =    0

>>
```

Characters and strings (array of characters) are stored as vectors of ASCII code for each character using a single quote to assign to a string variable. For instance

```
myTitle='Plot of Torque versus Speed' ;
size(myTitle)

>>
ans =

    1     27
```

The character array variable (equivalently called the string variable) "myTitle" is a vector, where each element of the vector stores the ASCII code of the characters. The "size()" function counts the number of elements in this variable. The conversion between character representation in ASCII code and numerical representations can be performed as follows,

```
char(97)      % Prints the character corresponding to
                ASCII code 97:  a
double('a')   % ..........ASCII code corresponding to
                character 'a'

S = ['1 2'
     '3 4'] ;

X = str2num(S)

S = num2str(X) ;
```

```
>>
    ans =

        a

    ans =

        97
    X =

        1       2
        3       4

    S =

    1 2
    3 4
```

The basic arithmetic operators, +,-,∗,/ are defined for scalars as well as vectors and matrices. In C++ terminology, the +, -, ∗, / operators have been overloaded to handle scalar, vector, and matrix data objects. In MATLAB®, we can add two matrices like scalars,

```
>>  A = [   1       2 ;     3       4  ]  ;
>>  B = [   1       2 ;     3       4  ]  ;
>>  C = A + B   ;
```

Element-by-element algebraic operations are also defined using the following operators:

```
    .+   .−   .∗ ./
```

The following is a standard matrix multiplication operator,

```
>>  C = A ∗ B   ;
```

Whereas, the following multiplies the A and B matrices element by element and assigns it to the C matrix

```
>>  C = A .∗ B   ;    %   C(i,j) = A(i,j) ∗ B(i,j)
```

MATLAB® automatically defines the following variables,

```
>> pi
>> eps
>> inf
>> NaN
>> i
>> j
```

where "pi" is π, "eps" is a very small number which is $2^{-52} \approx 2.2204\,E^{-16}$, and "inf" represents the infinity. "NaN" is used to represent a condition that data is not a number. The maximum and minimum numbers that MATLAB® can represent as integers and floating points can be determined on a given computer as follows. Any mathematical operation that results in a number beyond this range will result in overflow (larger than maximum representable number) or underflow (smaller than the smallest representable number) errors.

```
>> intmax

ans =

  2147483647

>> intmin
```

```
ans =
 -2147483648
>> realmax
ans =
  1.7977e+308
>> realmin
ans =
  2.2251e-308
>>
```

For complex numbers, the constant "i" and "j" are both defined as square root of -1, $\sqrt{-1}$, when used in the context of complex variables.

```
>>   z = 2+3*i
>>   z = x+i*y

>>   z = 2+3j
>>   z = x+j*y
>>
>>   z=complex(2,3)   ; % equivalent to z = 2 + 3*i
>>
>>   x = real(z)      ; % Assign real parf of z to x
>>   y = imag(z)      ; % Assign imaginary parf of z to y
>>
>>   w = conj(z)      ; %  if z = 2 + 3* i, then  w = 2 - 3 * i
```

Note that "i" and "j" are commonly used as loop counters in programming languages such as C and Fortran. In MATLAB®, if in a program we use "i" or "j" for loop counters or for any other purpose on the left hand side of an assignment statement, then we have effectively redefined the meaning of it, and hence can no longer be used as a complex number $\sqrt{-1}$ within the scope of its definition. In other words, if "i" and "j" are redefined in an M-script file with global visibility, it will effect the redefinition of the "i" and 'j' as complex number globally. However, if 'i' and 'j' are redefined in a function as local variable, it will only effect the definition locally within the scope of that function. M-script files, M-function files, and variable scopes are discussed later in this chapter.

A.1.2 Program Flow Control Statements in MATLAB®

The following logic and flow control statements, coupled with operators (arithmetic, logical, and relational operators), allows us to code logic into our MATLAB® programs in the form of loops (iterative and conditional loops) and decision blocks,

```
% Loops:  for and while constructs.

for  k=1:10       % Iterative loop: execute the loop 10 times.
    ...
    statements    %    for i= start_value: increment_value :
                          end_value
    ...           %    increment_value is 1 by default if omitted
end

%     Conditional loop:
%     Execute the loop while the (condition) is true or non-zero
```

```
while (condition)
   ...
   statements
   . . .

end
```

The "continue" and "break" statements are defined for "for" and "while" loops. The "continue ; " statement skips the rest of the block and continues the next iteration of the loop in the "for" or "while" loop. The "break " statement breaks out of the current loop.

```
Decision Blocks:    if and   switch   constructs.

if (expression1)
    statements1
%      execute the first block for which (expressionX) is true
%      or non-zero
      .   .   .
elseif (expression2)
    statements2
      .   .   .
elseif (expression3)
    statements3
   .   .   .

%  (repeat elseif (exp) as many times as necessary)
else
    statementsN
      .   .   .
end


switch  (switch_expr)
  case  (case_expr )
%  Execute the block for which case_expr = switch_expr
          statements
            ...
  case {case_expr1,case_expr2,case_expr3,...}
          statements
            ...
  otherwise
          statements
            ...
end
```

Example for "switch" construct,

```
method = 'Bilinear';

 switch lower(method)
    case {'linear','bilinear'}
       disp('Method is linear')
    case 'cubic'
       disp('Method is cubic')
    case 'nearest'
      disp('Method is nearest')
    otherwise
      disp('Unknown method.')
 end
```

The "otherwise" part is executed only if none of the preceding case expressions match the switch expression. Only the statements between the matching "case" and the next "case," "otherwise," or "end" are executed. Unlike C, the "switch" statement does not fall through, so "break" statements are not necessary.

The various operators used in MATLAB® are summarized below.

Arithmetic operators,

```
A+B
A-B
A*B      A.*B
A/B      A./B
A\B      A.\B
A^B      A.^B
A'       A.'
```

Relational operators,

```
A < B
A > B
A <= B
A >= B
A == B              (equal ?)
A ~= B              (not equal ?)
```

Logical operators,

```
A & B     ;  AND element-by-element (bit-wise) evaluation
A | B     ;  OR  element-by-element (bit-wise) evaluation

A && B    ;  AND  logical
A || B    ;  OR   logical

~A        ;  NOT
xor(A,B)  ;  Exclusive OR,
             logical 1 (TRUE) where either A or B, but not
             both, is non-zero.
```

A.1.3 Functions in MATLAB®: M-script files and M-function files

There are two different ways to call functions in MATLAB®:

1. run another M-file: script file,

2. call another MATLAB® function in an M-file: function file.

To run another M-file, called a script file, just include the name of that file in the program M-file. It is equivalent to including the content of that M-file at that location. The data in the M-file is global and accessible in the MATLAB® workspace as well as the current workspace. One M-script file can include and execute another M-script file, and in turn that file can include yet another M-script file. The nesting can continue as part of the program logic.

Script M-file example: Matrix1.m

```
% Filename:  Matrix1.m
%  Example of an M-script file.
%  Typing the file name in MATLAB® command window or another
   MATLAB® M-file is
```

```
% equivalent to running the code in this file as if they were
  typed explicity at that location.

          A = [ 1 2 ; 3  4 ] ;
          B = [ 5 6 ; 7  8 ] ;
          C =  A + B  ;
          Matrix2 ;
          F = [ 50  60 ; 70  80] ;
          G = E + F ;
```

Script M-file example: Matrix2.m

```
% Filename:  Matrix2.m
%  Example of an M-script file.
%  This file is executed by the "Matrix2" line in the Matrix1.m
%  script file.

          D = [ 10 20 ; 30  40 ] ;
          E=   C +  D ;


   >> Matrix1   ;  % to run the Matrix1.m script file.
                   %  Then Matrix1.m script file runs Matrix2.m
                      script file.
```

Calling a **"user-written MATLAB® function"** follows the function call rules similar to the rules used in high level programming languages such as C and Fortran: it accepts input arguments in the function call and returns data. The data defined locally in the function are local to the function file. Unlike the script M-files, they are not globally accessible. Likewise, the function can access only the data variables passed through the function call argument list. It cannot access the variables in the workspace. The only way to override that is to define **"global"** variables. Good programming practice suggests that we use M-function files with controlled local data scope, and minimize the use of M-script files with global data scope or "global" statement that can lead to unintended data over-writes. The syntax of a user-written MATLAB® function M-file is as follows (filename is the function-name with extension ".m"),

Filename: function_name.m

```
function   [Y,Z] = function_name(input_1, input_2)
 % Input variable to this function from the caller: input_1,
   input_2, ...
 % Output variable from this function to the caller: Y, Z
 %
 %   Local data
 %
 %   Logic
 %
 %   I/O

          Y= 5.0 * input_1 ;
          Z= 10.0 * input_2 ;

 %  Return (output) variables Y and Z must appear on the left
    hand side of
 %  assignment operator.
 %
 %  This is the end of the function.
```

MATLAB® Example for M-Function File

```
% Filename: Example1.m
%
%
  x=[0.0:0.1:2*pi];

  y = My_Function1(x) ;

  plot(x,y) ;  %  This will plot the sine function on Figure 1.
  grid on;

% End of file


  function  z1 = My_Function1(u1)
   % Input variable to this function from the caller: input_1,
     input_2, ...
   % Output variable from this function to the caller: Y, Z
   %
   %   Local data
   %
   %   Logic
   %
   %   I/O

      z1 = My_Function2(u1) ;  % Let's illustrate that function
                                calls can be nested.

   %  Return (output) z1 must appear on the left hand side of
   %  assignment operator.
   %
   %  This is the end of the function.

  function  y1 = My_Function2(x1)
   % Input variable to this function from the caller: input_1,
     input_2, ...
   % Output variable from this function to the caller: Y, Z
   %
   %   Local data
   %
   %   Logic
   %
   %   I/O

      y1 = sin(x1) ;

   %  Return (output) y1 must appear on the left hand side of
   %  assignment operator.
   %
   %  This is the end of the function.
```

The result is shown in Figure A.3.

Global variables can be defined in MATLAB® for data variables to be shared across functions without having to pass it in the input–output arguments. However, a global data definition violates the "encapsulation" principle in programming. Therefore, it is not recommended for good programming practice. However, if it must be used, the global variables are declared **both in M-script file (i.e., the M-script file which holds your**

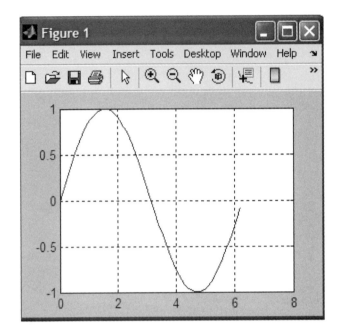

FIGURE A.3: Output plot of the MATLAB® M-function file example.

main program or in the Command Window of MATLAB®) and in the function that uses it.

```
%  Main script file name: filename.m
%

global gravity

gravity = 9.81 ;

 ....


_____
%  File name: myFunction1.m

function z = myFunction1(time)
  global gravity

 ...

return
```

Built-in Functions in MATLAB® Environment MATLAB® provides a rich library of built-in functions to provide convenience to the programmer for data manipulation. It is the simplicity of data handling with this rich set of built-in functions and the simplicity of I/O handling that makes MATLAB® an attractive computational engineering tool.

The following built-in functions are useful in data generation

```
>>  A  =  zeros(5,5) ;
            % all elements of  A matrix, 5x5, are zero
```

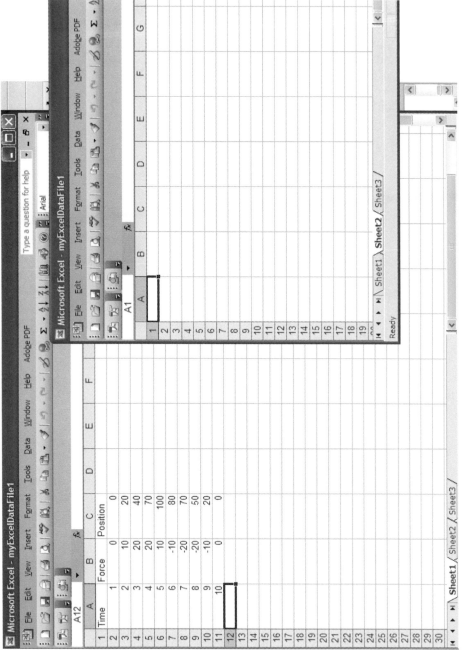

FIGURE A.4: Excel data file I/O example: the content of Sheet1 and Sheet2 of the Excel data file at the beginning.

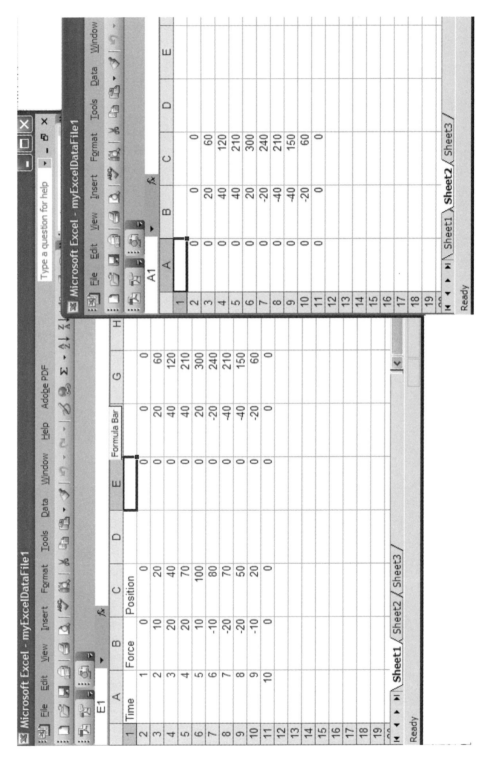

FIGURE A.5: Excel data file I/O example: the content of Sheet1 and Sheet2 of the Excel data file after writing to it.

```
>>   B  =  ones(6,4) ;
             %  all elements of B  matrix, 6x4, are  ones.

>>   Y  =  eye(n)  ;
             %  Returns (creates) an identity matrix: nxn

>>   Y  =  eye(m,n) ;
             %  .............. mxn
```

Matrices can be concatenated to build larger size matrices or elements can be deleted to build smaller matrices,

```
>> B = [ A,    A+2 ;   A+2,   A ] ;
>>
>> A(:,2)=[]   ;   % Delete second  column of matrix A
>>
>> C = B(2:4, 3:6) ; % Assign sub-set of B to C: rows 2,3,4
                        and colums 3, 4, 5, 6 of
>>                   % B are assigned to C
>>
>> v=linspace(0,10,6)  %  Create an array v: start at 0,
                        end at 10,  6 numbers evenly spaced

 v =

    0     2     4     6     8    10

>> v=logspace(0,3,4)  ; % First two arguments are powers of 10,
                        % third  argument is the number of
                          elements
                        % in the range, in Logarithmic scale.

 v =

          1          10         100        1000
```

Various matrix properties are calculated using MATLAB® library functions, such as determinant, eigenvalues of a matrix, rank of a matrix, inverse of a matrix

```
>>  Value  =  det(A)    ;      %  Determinant of matrix A,
>>  Lambda =  eig(A)    ;      %  Eigenvalues of matrix  A.
>>  n      =  rank(A)   ;      %  Rank of matrix A
>>  Ainv   =  inv(A)    ;      %  inverse of A:  A^-1
```

Some commonly used built-in mathematical funtions in MATLAB® are as follows:

```
>> y = sqrt(x) ;  % Square root of x assigned to y
>> y = exp(x) ;   % Exponential function
>> y = abs(x) ;   % Absolute value of x assigned to y
>> y = log(x) ;   %  y = ln(x).  Log(x) is logarithmic function
                     with base 'e'.
>> y = log10(x) ;   %  y = log(x).  Log10(x) is logarithmic
                     function with base '10'.
>>
>> x = 5.4  ;
>> y = round(x) ; % Round to nearest integer:  y = 5
>> y = fix(x) ;   % Round to integer towards zero:  y = 5
>> y = ceil(x) ;  % Round to integer towards infinity :
                     y = 6
>> y = floor(x) ; % Round to integer  towards negative infinity:
                     y = 5
```

```
>> y = rem(x,z) ; % Reminder function:  y = x - fix(x/z)
>> y = sign(x)  ; % returns 1 if x>0, 0 if x=0, -1 if x< 0.

>>               % Trigonometric functions
>> y = sin(x) ;   %  Sin function
>> y = cos(x) ;   %  Cosine function
>> y = tan(x) ;   %  Tangent function
>> y = cot(x) ;   %  Cotangent function

>> y = asin(x) ;  %  Arcsin (inverse Sin) function
>> y = sinh(x) ;  %  Hyperbolic sin function
>> y = asinh(x) ; %  Inverse hyperbolic Sin function
                  %  Similar functions for cos, tan, cot.
```

Polynomials, their roots, and multiplication of polynomials are handled by the coefficients data vector. Let us consider a polynomial as follows:

$$a_0 s^n + a_1 s^{n-1} + a_2 s^{n-2} + \cdots + a_n = 0 \qquad (A.1)$$

The roots of the polynomial can be found as follows

```
>>   c = [ a_0  a_1  a_2 ... a_n] ; % Defines the coefficients of
                                    the polynomial

>>   r = roots(c);   %  r is the vector that has the roots of the
                        polynomial.

>>   c = poly(r) ;   %  poly() does the opposite of roots() --
                        given the roots,
                     %  it calculates the coefficients of the
                        corresponding polynomial.

>>   c3 = conv(c1, c2) ; % multiplies two polynomials, and
                         % returns the coefficients of the
                           resulting polynomial.
                         % (Length of c3) = (length of c1 + length
                           of c2 - 1)

>>   y1 = interp1(x,y,x1); % given (x,y) vector pair,
                          % find interpolated value(s) of y1 at x1
                          %  using linear (default) interpolation.
                          %  If x1 is scalar, y1 is scalar. If x1
                            is vector, so is y1.
```

A.1.4 Input and Output in MATLAB®

MATLAB® has most of the features of a high level programming language, such as C, plus the additional convenience of easily handling graphical plotting of data and automatic declaration of data type. Unlike in C, the data type of a variable does not need to be declared in advance of its usage. MATLAB® automatically handles that. In any programming environment, we need to be able to input–output the data to-and-from the program and apply logic to it.

Data can be input to a MATLAB® environment in one of the following ways:

1. From the keyboard, using "input" function.
2. From a previously saved "∗.mat" file, using "load."

3. From an M-file where data is defined following the MATLAB® syntax (this is the most common form of inputing data in MATLAB® environment), using the "filename" in the program or in command window.

4. From any file in a format supported by MATLAB® such as "filename.txt" or "filename.dat," including Excel spreadsheet data file, "filename.xls."

```
>> n = input ('Enter value for n : ') ;   % Inputs numeric data
                                             to variable 'n'.
```

To enter a vector or matrix, use the same notation that we use in m-script file or Command Window, i.e.

```
>>  v = input ('Enter value for v: ') ;   % Inputs numeric data
                                            for vector v

   [ 1   2   3] ;
>>  matrix1 = input ('Enter value for Matrix1: ') ;   % Inputs
                                 % numeric data for vector v

   [  1  2   3 ;
      4  5   6 ;
      7  8   9   ] ;
```

The entered data are assigned to the variable on the left side of the assignment statement.

Formatted input and output can be accomplished using **fscanf()** and **fprintf()** functions, respectively, which are similar to the same named functions in C language. Before a file can be accessed for read/write operations, it must be opened with the **fopen()** function. When the read/write operations are completed for the current application, the file should be closed using the **fclose()** function. As a result, there are typically four functions used for formatted file I/O operations,

```
>> name = input ('Enter string data for name ', 's') ;
              % enters the string data (specified by 's') into
                variable 'name'

>> filename    ;
    %  executes the filename.m
    %  all the data defined in the m-file becomes part
    %  of the MATLAB® workspace and
    %  therefore accessible to the program.

 >> load filename
    %  loads the variables saved in the filename.mat file
    %  which was saved before.

>>   load('filename.dat', v)
      % Loads ASCII data from the file to variable v

>> save
    %  saves all workspace variables to  matlab.mat file in
       binary format.

>> save filename
    %  save all of the workspace variables in the filename.mat
       in binary format.
```

```
>> save filename  var1  var2   -ascii  -double -tab
   %  saves the selected variables in ascii format in the
      filename specified.
   %  if no variable name is provided, all of the variables
   %  in workspace is saved.
```

Data can be displayed in text format on the screen, plotted in graphical form, or saved to a file,

```
x
   %  Not terminating a line in m-file (or in command
      window) ";" displays
   %  the numerical result of the current line on the
      screen including the name of the variables

 disp(x)
   % Displays the value of x without printing the name of
     the variable

 disp('Message');
   %   display the text string 'Message'
```

Formatted printing to screen or file can be done using the **fprintf()** function, which is almost identical to the **fprintf()** function used in C language.

```
A = [ 1  2  3 ;
      4  5  6 ]  ;

 [row,col]=size(A) ;


 fprintf('The matrix has %4.0f rows and \t %4.0f columns \r\n',
 row, col) ;
   %    the '%5d'    means the data for 5 digit integer format,
   %    the '%4.0f'  means the data is in decimal  notation,
   %                 4 characters long, 0 digits after the decimal
                     point.

 myTitle = 'Result: Force versus Time' ;

 fprintf('%s \n', myTitle) ;
```

The result is

```
>>
The matrix has    2 rows and       3 columns

Result: Force versus Time
>>
```

The

```
\n
```

format specifier adds a carriage-return after printing "myTitle" on the screen. In the above example, "fprintf()" statements print on the screen.

The fprintf() output can also be directed to a file, as follows,

```
x = 0:.1:1;
y = [x ; exp(x)];
fHandle = fopen('exponential.txt','wt');
fprintf(fHandle,'%6.2f  %12.8f\n',y);
fclose(fHandle);
```

Data can be input from different file formats that MATLAB®'s built-in functions support. Below is a list of typical file formats supported by MATLAB® built-in functions.

File Format	Filename extensions	MATLAB® Built-in Functions
Text File	*.dat, *.txt, *.mat	y= load(' .. ') save('....', y) fopen() fclose() fscanf() fprintf()
Excel File	*.xls	y=xlsread('..') xlswrite('..',y)
Image File	*.bmp, *.jpg, *,jpeg	WL=imread('wheel_loader.jpg') ; % to read image(WL); % to display imwrite(WL,'MWL.jpg') % to write
Audio File	*.wav, *.au	[Adata,Af]=wavread('myAudio1.wav') ; % to input sound(Adata,Af) ; % to play it wavwrite(Adata,Af, Ares,'myAudio2.wav') % to % write it
Video File	*.avi	getframe() movie()

where *Af* is sampling frequency in Hz and *Ares* is the number of bits per sample (resolution of sound data). In sound file of type "*.wav," the magnitude of the data should be between -1.0 and 1.0 which represents the amplitude or the volume of the sound.

A useful MATLAB® built-in function for user interface using monitor and keyboard is the **menu**() function. The **menu**() function is used to display a window on the screen and provides a number of choices, whichever one is selected by the user, the result is returned to the variable on the assignment statement as an integer, indicating the choice. The general format of the "menu()" function is as follows,

```
choice = menu('menu_header_message', 'item1 text', 'item2
text', ... )
```

and an example of usage is given below.

```
choice = menu('Simulation choices:', 'Case 1', 'Case 2',
        'Case 3')

switch choice
  case 1
     disp('Case 1 is selected');
  case 2
     disp('Case 2 is selected');
  case 3
    disp('Case 3 is selected');
end
```

Example: Working with Excel Files Data from an Excel file can be entered to a MATLAB® variable by specifying the (Figures A.4 and A.5)

1. Excel filename,
2. Worksheet name in the file,
3. optionally a specified range of rows and columns in the spreadsheet and assign it to a MATLAB® variable.

Let us assume that we have an Excel file with three colums of data, first column represents the "time," the second column represents the "force," and the third column represent the "position". At the very top of the file, in row 1, we have the text information on top of colums 1, 2, 3 as "time," "force," "position," respectively. Our objective is to read the numeric data interactively, plot the results, and modify the data, and then write the modified data by appending it to colums next to the existing data on the same worksheet. Plot the data in "position versus time" and "force versus time," then multiply the "force" data by 2 and position data by 3, save the new data into the same file into columns 5, 6, 7 as "time," "force," "position." Before running this program, make sure Excel program does not have the data file "myExcelDataFile1.xls" open. The host PC should have the Excel program installed in order to run the program properly.

```
% Filename; myExcelFile1.m

 disp('Reading Excel Data File...') ;
 myData = xlsread('c:\temp\myExcelDataFile1.xls',-1);
 disp('Completed reading...');

 subplot(221)
 plot(myData(:,1), myData(:,2))
 ylabel('Position');
 xlabel('Time');
 subplot(222)
 plot(myData(:,1), myData(:,3))
 ylabel('Force');
 xlabel('Time');

 myData2(:,2) = 2.0 .* myData(:,2) ;
 myData2(:,3) = 3.0 .* myData(:,3) ;
 subplot(223)
 plot(myData(:,1), myData2(:,2))
 ylabel('Position');
 xlabel('Time');
 subplot(224)
 plot(myData(:,1), myData2(:,3))
 ylabel('Force');
 xlabel('Time');

 disp('Writing Excel Data File...');
 xlswrite('c:\temp\myExcelDataFile1.xls', myData2, 'Sheet1', 'E2') ;
 xlswrite('c:\temp\myExcelDataFile1.xls', myData2, 'Sheet2', 'A2') ;
 disp('Competed writing...');
```

The ease of plotting in MATLAB® is one of the most attractive features of it. In plotting data, we need to open a figure window, divide the window into subplot areas if desired, define the title, labels of x and y axes, scales of x and y axes, and for each x-y pair of data we can specify the line type and color. These are illustrated in the following example.

```
% open a new figure window
%   2x3 grid of sub-plot areas in the figure,
```

```
%     current target is number 1 spot

        Figure(1);
        Subplot(2,3,1);

% add plot title, x and y axis labels, axis scales, grid

        Title('My  X-Y Plot');
        Xlabel('time(sec)');
        Ylabel('Position (meters)');
        Axis([xmin  xmax  ymin ymax]) ;
        Grid (on);

% plot type: linear in both axes, data vector pairs
%     (x1, y1),   (x2, y2), and plot line specs

        Plot(x1,y1,'-Or', x2,y2,'-+b');

% line specs defined with the ' ...' following data pair definition:
%              - line type:    solid line '-', dashed '-', dotted ':',
%                              dash-dot '-.'    ...
%              - line symbol:  circle 'o' , plus '+' ,  x-mark x,
%                              square 's', diamond 'd'   ...
%              - line color:   red 'r', black 'k', blue 'b', green 'g',
%                              yellow 'y', cyan 'c', magneta 'm'...
```

Plotting data in logarithmic scale, such as only *x*-axis in log scale, only *y*-axis in log scale or both in log scale is accomplished by the following statements instead of the "plot"

```
semilogx(x1, y1);   % x-axis is in Log scale
semilogy(x1, y1);   % y-axis is in Log scale
loglog(x1,y1);      % Both x and y axes are in Log scale
```

Sometimes it is necessary to generate a frequency data vector that is spaced logarithmically instead of linearly as shown below

```
wlinear = 0.0: 1.0 : 100.0 ;
wlog    = logspace(-3,3,100);
                    % generate frequencies
                    % in the range of 10^-3 to 10^+3, 100
                    data points
                    % logaritmically spaced.
```

Likewise, three-dimensional plots are obtained either as 3D line plots or 3D surface plots. 3D line plots are done by

```
plot3(x,y,z) ;
```

where *x*, *y*, and *z* all have the same number of elements and *z* value plotted for each *x,y* pair (Figure A.6).

```
>>
>> x = [1 2 3   4 ];
>> y = [1 2 3   4] ;
>> z = [0 2 10 50] ;
>> plot3(x,y, z) ;
>> grid on
>>
```

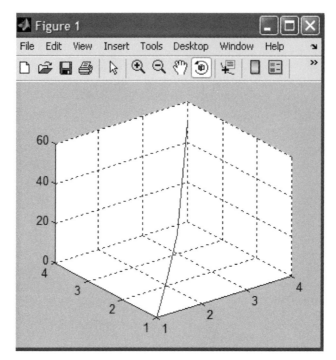

FIGURE A.6: 3D line plot example.

3D surface plots are done by

```
mesh(x1,y1,z1)   ; % 3-d plots with lines
surf(x1,y1,z1)   ; % 3-d plots with surface
contour(x1,y1,z1) ; % Generates a contour plot
surfc(x1,y1,z1)  ; % 3-d plots with surface and contour
                     combined
```

where x1 and y1 form a 2D matrix defining the *x*-*y* grid's coordinates. These data are typically created from row vectors x and y using the

```
[x1, y1] = meshgrid(x,y)
```

function. The x1 vector is rows of the x vector, and the y1 vector is the columns of the y vector. For example, to evaluate the function $x \cdot e^{(-x^2 - y^2)}$ over the range $-1 < x < 1$, $-1 < y < 1$,

```
        clc ;
        clf ;
        [X,Y] = meshgrid(-1:.5:1, -1:.5:1)
        Z = X .* exp(-X.^2 - Y.^2)
        surf(X,Y,Z)
>>

 X =

    -1.0000   -0.5000        0   0.5000   1.0000
    -1.0000   -0.5000        0   0.5000   1.0000
    -1.0000   -0.5000        0   0.5000   1.0000
    -1.0000   -0.5000        0   0.5000   1.0000
    -1.0000   -0.5000        0   0.5000   1.0000
```

```
Y =

    -1.0000    -1.0000    -1.0000    -1.0000    -1.0000
    -0.5000    -0.5000    -0.5000    -0.5000    -0.5000
          0          0          0          0          0
     0.5000     0.5000     0.5000     0.5000     0.5000
     1.0000     1.0000     1.0000     1.0000     1.0000


Z =

    -0.1353    -0.1433          0     0.1433     0.1353
    -0.2865    -0.3033          0     0.3033     0.2865
    -0.3679    -0.3894          0     0.3894     0.3679
    -0.2865    -0.3033          0     0.3033     0.2865
    -0.1353    -0.1433          0     0.1433     0.1353
```

The following is an example of four different three-dimensional surface plots (Figure A.7).

```
x = [-2.0: 0.1 : 2.0];
y = [-2.0: 0.1 : 2.0];
[x1,y1] = meshgrid(x,y) ;
```

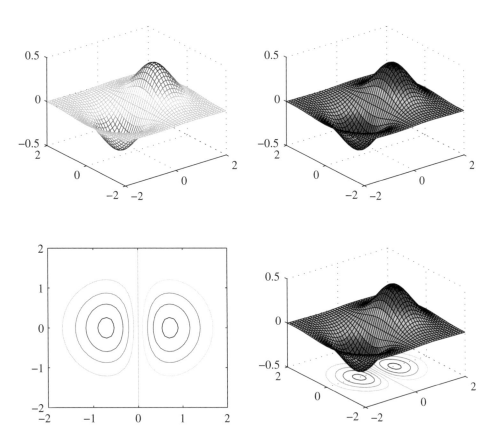

FIGURE A.7: 3D surface plot examples: using "mesh," "surf," "contour," and "surfc" functions.

```
z1= x1 .*  exp(-x1.^2 -  y1.^2) ;

figure(1)
subplot(2,2,1)
mesh(x1,y1,z1)
subplot(2,2,2)
surf(x1,y1,z1)
subplot(2,2,3)
contour(x1,y1,z1)
subplot(2,2,4)
surfc(x1,y1,z1)
```

The viewing angle of each 3D plot can be changed using the interactive tool on the plot window using the mouse or can be programmed using the "view()" function,

```
az_angle = 45 ;
el_angle = 45 ;
view(azimuth_angle, elevation_angle) ;
```

Plots can be printed from the **File** − > **Print** menu of the plot window. In addition, in the Windows environment, the screen can also be captured using the "PrtScr" key on the keyboard, which copies the content of the screen to memory. Then the content can be copied to a file using "Microsoft Paint" or a similar program. The file can be saved in bit mapped format.

Animation can be done by storing images into an array of frames, and displaying these frames in a sequence using "getframe" and "movie" functions.

```
for j=1:60
      plot(.....);
      F(j) = getframe;
end

movie(F,10)  ; % play the frames stored in F for 10 times.
               % Once the movie frames are save to a matrix,
                  i.e. F,
               % the animation can be replayed anytime.
```

Here is an example of animation by first calculating and saving each frame, then playing them as an animation (Figure A.8).

```
% Animation example

  clear
  clc
  clf

  x= 0 : pi/50 : 2*pi ;
  y = x ;
  [X,Y] = meshgrid(x,y);

  z = 20*sin(X)+cos(Y);
  h = surf(z) ;

  axis tight ; %  sets the axis limits to the range of the data.
  set(gca, 'nextplot','replacechildren');  %  'gca' is the current
                                               axis handle
                                           %  next plot replaces the
                                              previous plot, without
                                              rescaling the axis.
```

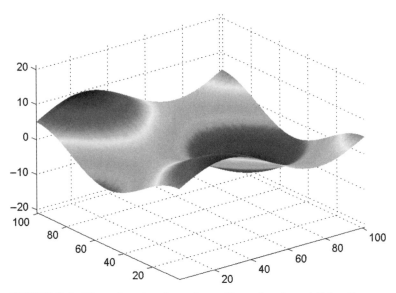

FIGURE A.8: 3D movie animation using **getframe()** and **movie()** functions.

```
shading interp ;   % color shading uses 'interpolation' method based
                     on the value of the plotted data
colormap(jet) ;    % Color map (number range versus color spectrum):
                     jet =  blue - green - yellow- red range.

m=1 ;
for k = 0 : pi/50: 2*pi
   z = (sin(X)+cos(Y)).*8*sin(k) ;
   set(h,'Zdata',z);
   M(m)=getframe; % create and save each frame
   m=m+1 ;
end

movie(M,3) ; % play the movie 3 times.
```

Other related MATLAB® functions are:

```
axis, bar, grid, hold on,  legend, line, LineSpec,
loglog, plotyy, plot3, semilogx, semilogy,  xlabel,
xlim, ylabel, ylim, zlabel, zlim, stem, set.
```

A.1.5 MATLAB® Toolboxes

The power of MATLAB® comes from its extensibility through libraries called "Toolboxes," grouped into different functional categories. For instance all of the control system design functions are grouped into a "Control Systems Toolbox." Similarly, all of the signal processing related functions are grouped into a "Signal Processing Toolbox." There are many such toolboxes and new toolboxes are being added continuously for different application fields. A typical control engineer needs the "Signal Processing Toolbox" and "Control Systems Toolbox." There are also toolboxes that provide functions specifically for system identification, neural networks, fuzzy logic, power electronics, statistics, and so on.

A.1.6 Controller Design Functions: Transform Domain and State-Space Methods

A transfer function is defined with numerator and denominator coefficients. A transfer function description can be converted between three different ways of expressing it: (i) numerator and denominator polynomials form, (ii) poles-zeros and gain form, (iii) partial fraction expansion form (poles, residues, and DC gain).

```
          num(s)
H(s)  =  -------
          den(s)
```

```
             (s-z1)(s-z2)...(s-zn)
H(s)  =   K  --------------------
             (s-p1)(s-p2)...(s-pn)
```

```
num(s)         R(1)        R(2)              R(n)
----   =    --------- + --------- + ... + --------- + K(s)
den(s)      s - P(1)    s - P(2)          s - P(n)
```

These functions are applied in *s*-domain as well as *z*-domain transfer functions.

```
numG = [ 1 ] ;                  % Numerator coeficients
                                %   of the transfer function.
denG = [ 1   2    4] ;          % Denominator ......
                                % ..........................
Gs  = tf(numG,denG) ;           %  Gs defines the transfer funtion.

[z,p,k]      = tf2zp(numG, denG);  % convert to zero-pole-gain form
                                   % from the polynomial num(s)/den(s)
                                     form

[numG, denG] = zp2tf(z,p,k);       % do the reverse

[r,p,k] = residue(numG, denG) ;    % convert  from num(s)/den(s) form
                                   % to Partial Fraction Expansion form

[num, den] = residue(r,p,k) ;      % convert from partial fraction
                                     expansion form
                                   % to numerator and denominator
                                     polynomial form

p = pole(Gs)  ;                    % Given G(s), get poles,
[z, k] = zero(Gs)   ;              % .......... get zeros and gain
[num,den]=zp2tf(z,p,k) ;           % obtain the num(s)/den(s) form
```

If we define complex variables "s" and "z" as follows for Laplace and Z-transforms, then we can express the transfer functions in symbolic form,

```
s = TF('s') ; %  specifies the transfer function H(s) = s
                     (Laplace variable).

z = TF('z',TS) ; %  specifies H(z) = z with sample time TS.
```

You can then specify transfer functions directly as rational expressions in *s* or *z*, for example,

```
s = tf('s');
G = (s+1)/(s^2+3*s+1) ;
z = tf('z');
H = (z+1)/(z^2+4*z +2) ;
```

Similar functions are available to convert between different forms of transfer function and state-space representations as follows.

```
[numG, denG] = ss2tf(A, B, C, D);
[A,B,C,D]    = tf2ss(numG, denG);

[z, p, k]    = ss2zp(A, B, C, D);
[A, B, C, D] = zp2ss (z,p,k) ;

sys = ss(A,B,C,D) ; % creates a SS object SYS representing
                    % the continuous-time state-space model

p = eig(A) ;        % Eigenvalues of A.

z = tzeros(sys) ;   % Transmission zeros of the corresponding
                    % transfer function between input and output.

 dx/dt = Ax(t) + Bu(t)
 y(t)  = Cx(t) + Du(t)
```

You can set D=0 to mean the zero matrix of appropriate dimensions.

```
sys = ss(A,B,C,D,Ts) ; % creates a discrete-time state-space
                         model with
                       % sample time Ts (set Ts=-1
                       % if the sample time is undetermined).
```

Finding the discrete equivalents of continuous system transfer functions using various approximation methods can be done as follows in transform and state-space form

```
sysD= c2d(sys, 0.01, 'tustin') ; % given 'sys' for continuous
                                   system,
                                 % sampling period = 0.01, use
                                   'Tustin' method
                                 % to obtain 'sysD' for
                                   discrete system equivalent.

Gs = tf(num,den);          % Define a G(s)
Gz = c2d(Gs, T,'zoh') ;    % Obtain ZOH equivalent of it in
                             z-domain G(z) for
                           % sampling period T.

zgrid ;    % draws the constant damping ratio and constant
             natural frequency
           % contours on z-plane

zgrid([],[]) ; % draws the |z|=1 unit circle.
```

Root locus related functions are illustrated with an example below.

```
num = [1] ;
den = [ 1   5    2    6] ;
sys = tf(num, den);
K = 0.0 : 0.1 : 100 ;
rlocus(sys,K) ;          % plot root locus for standard closed
                           loop case,
                         % for the given range of K values.
rltool(sys) ;        % interactively  plots root locus and
                       calculates gain and pole
                           % locations seleted at the points where
                             mouse is clicked.
```

Frequency response related functions are illustrated in the example below.

```
num = [1] ;
den = [ 1   5    2    6] ;
sys = tf(num, den);

[mag, phase, w] = bode(sys) ; % Get Bode plot data
loglog(w, mag);                % plot magnitude plot
semilogx(w, phase);            % plot phase plot

nyquist(sys) ;                 %  Plot Nyquist (polar) plot

[GM, PM, wgm, wpm] = margin(G) ; % Obtain Gain Margin, Phase
                                   margin
                                 % and the freqencies at those
                                   locations.
```

Given the A, B, C matrices of state-space representation, controllability and observability matrices can be obtained as follows,

```
Wc = ctrb(A,B);  % Calcualate controllability matrix
Wo = obsv(A,C);  % ........... observability........

nc = rank(Wc) ;    % Get the rank of controllability matrix
value1 = det(Wc) ; % ....... determinant of .............

no = rank(Wo) ;    %  Same for observability matrices.
value2 = det(Wo) ;
```

Given the A, B, C, D matrices of state-space representation, and desired pole locations for closed loop system p_c, the state feedback gain can be calculated by

```
K = acker(A, B, p_c) ;
K = place(A, B, p_c) ;
```

For the observer design, given the observer poles, p_e,

```
L = (acker(A', C', p_e))' ;
L = (place(A', C', p_e))' ;
```

MATLAB® Functions to Simulate Linear Dynamic Systems   MATLAB® has three convenient functions to simulate **linear dynamic** systems. They are

1. impulse(sys,t) function for impulse response of the system for a given time duration,
2. step(sys,t) function for unit magnitude step function response,
3. lsim(sys,u,t) function to simulate the response to an arbitrary input function defined in vector "u."

The example MATLAB® script file below illustrates their use. Note that these functions are usable only for dynamic systems defined by linear, constant coefficient differential equations. Figure A.9 shows the results of the simulation.

```
% Simulating Linear Systems response using MATLAB® Functions
%   impulse()  and step() and lsim()

% Linear system definition:
  wn=2*pi*1.0;
  psi=0.5;
```

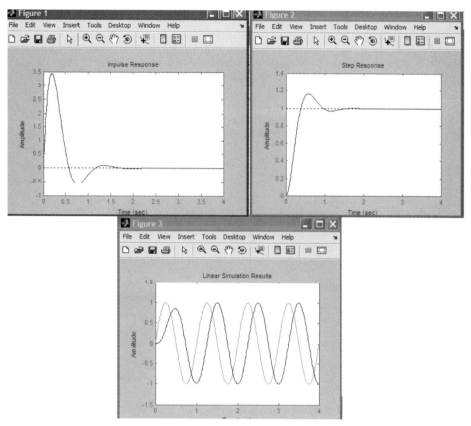

FIGURE A.9: Simulation results of a second-order system for impulse, step, and a user-defined input functions using the MATLAB® functions **impulse, step**, and **lsim**.

```
num=[wn^2];
den=[1 2*psi*wn  wn^2];
sys1=tf(num,den);

% Impulse response

t=[0:0.01:4.0];
figure(1);
impulse(sys1,t);

% Step response

t=[0:0.01:4.0];
figure(2);
step(sys1,t)

% Response to a general input

t=[0:0.01:4.0];
u= 1.0 * sin(2*pi*t);
figure(3);
lsim(sys1,u,t) ;
```

MATLAB® provides the following functions for the numerical solution of nonlinear differential equations.

MATLAB® Function ode45() for ODE Solution MATLAB® provides numerical integration functions to solve differential equations, some of which are as follows:

```
[t, x]=  ode45('ODE_FunctionName', t_vector, x_initial)
[t, x]=  ode23('ODE_FunctionName', t_vector, x_initial)
```

where "ODE_FunctionName" is the name of the M-function file that has the differential equations that describes the dynamics of the system, t_{vector} can be either $[t_0, t_f]$ beginning and ending time interval of the simulation and the algorithm decides on the number of intermediate solution points or it can be a vector of time points at which the solution is to be obtained and returned at the $[t, x]$ output vector, $x_{initial}$ is the initial conditions for states at the beginning time t_0. The difference between different numerical integration methods (above functions, ode45(), ode23()) are about their efficiency and suitability in dealing with different types of dynamic systems, such as "stiff" numerical problems. A "stiff" numerical problem is one that has low and very high frequency (sudden changes in the dynamics, i.e., impact problems) content in the solution.

A.2 SIMULINK®

Simulink® is a graphical modeling and simulation software tool which runs as part of the MATLAB® package. In MATLAB®, a dynamic system is modeled using M-files which holds the model description in text file format written in MATLAB® language. In Simulink®, a system is modeled using graphical block diagrams and the connections between the blocks. Therefore, the graphical representation of the model is easier to understand and facilitates better communication between different users working on the same model.
A model in Simulink® is:

1. a collection of blocks,
2. their inter-connections, and
3. the parameters of each block.

A typical model development using Simulink® involves the following steps:

1. Given a mathematical and logical description of a dynamic system and control algorithm, build the model of the system using Simulink® blocks from the block library. This involves selecting a number of blocks from the Simulink® block library, and making the connections between them.

2. Setup the parameters of each block. Double-clicking on a block brings up the user interface menu for the block which allows us to setup the parameters of the block, i.e. the value of the gain for a "Gain" block. Numerical values can be constants directly typed in the user interface dialog box or can be variables where the values of the variables are assigned in the MATLAB® Workspace by either directly making assignment statements in the MATLAB® Command Window or placing the assignment statements into a script-M file (the recommended method), i.e. filename "Filename_Parameters.m."

3. Setup the simulation conditions on the "Model Window's Simulation" menu item,
   (a) select the integration algorithm related parameters, simulation time interval,
   (b) select the variables to be saved and plotted and connect them to the appropriate Simulink® blocks (Scope, To Workspace, etc.).

4. Run the simulation in Simulink® and analyze the results. Simulink® also has a built in "Debugger" which is accessed from the Tools → Debugger menu item of the model window.

Simulink® is a numerical modeling and time domain simulation software that allows us to mix linear, nonlinear, analog, and digital systems in the model using interconnected blocks. In Simulink®, the dynamic model and control algorithm are modeled using interconnected blocks. The user interface is a graphical one. In order to simulate the system response for given input conditions, inputs are connected to various source blocks (i.e., function generator block, step input block). The response is recorded by connecting various output signals to sink blocks (i.e., scope block). The simulation start time, stop time, sampling time and integration method are selected in the **Simulink® > Simulation** setup window. The Simulink® model simulates an *analog PD controller* for the mass-force system. The controller is sampled at the same rate that the integration algorithm uses for the solution of the mass-force model. In Simulink® models, the input–output relationships are described by interconnected blocks. The individual blocks may represent linear or nonlinear functions between its input and output, transfer functions in Laplace transform form for analog systems and z-transforms form for digital systems. Input functions for the simulated case are represented in time domain form. The responses of the system are also time domain functions.

Simulink® is started by typing "simulink" in the MATLAB® prompt, or placing that command in the "Startup.m" file, or it can be launched from the "Launch Pad Window" of MATLAB®.

```
>> simulink
```

Once Simulink® starts, it brings up the **"Simulink® Library Browser"** window (Figure A.10). A new model is developed by selecting **File → New** menu item and saved by selecting **File → Save As** menu. The default name for a newly opened model file is "Untitled.mdl." A Simulink® model is saved in a file with ".mdl" extension. All the hierarchical levels of the model are saved in one file.

To build a model, different blocks are selected from the Simulink® "Block Library" and copied over to the "Model Window," in one of two ways:

1. copy/paste menu items of the Edit menu item, or

2. a block can be selected from the "Block Library" and dragged over to the "Model Window," which is the preferred method.

Then the connections between the blocks are made to define the relationship between blocks to form the complete model.

Preferences in Simulink® provides a window of many menu items to configure the Simulink® environment. For most applications, the default settings work well. It is accessed from the **File > Preferences** menu item.

The parameters of a block can be modified by double clicking on the block once it is in the model window. A menu comes up associated with the block. The parameters can be set to either to

1. a constant value, or

2. a variable to be defined in the MATLAB® workspace later on.

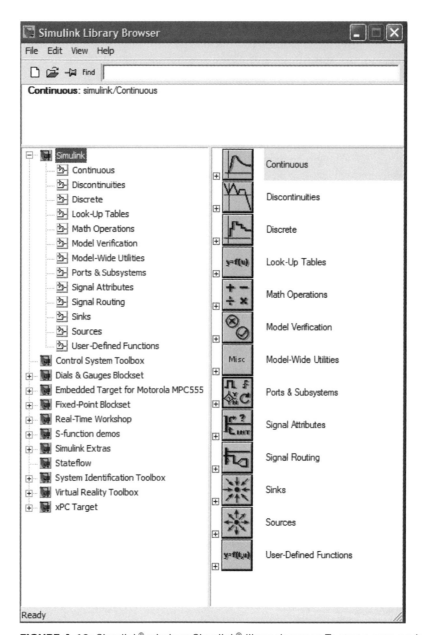

FIGURE A.10: Simulink® window: Simulink® library browser. To start a new model file or open an existing model file, use the "File" menu. Then pick, drag, and drop component blocks from this library window.

The Simulink® block library is grouped into functional categories such as Sources, Sinks, Connections, Math, Signals, and so on. In addition to the standard blocks provided with Simulink®, application specific block libraries can be added as "Block Library Toolboxes" similar to the toolboxes of MATLAB®.

Blocks: From the Simulink® Library Browser window, a new model is created by

    File → New → Model

menu selection. This creates a model window with name "Untitled." Now we can pick different blocks from the Simulink® Library Browser window, drag them into the "Untitled" model window and build a model. Help can be obtained on any library block by right-clicking on it, and selecting "Help" from the "context-sensitive menu." Basic properties of a block, such as orientation of the block and so on, can be accessed through the block's context-sensitive properties menu. Each block has a name which is by default placed under the block. The names are strictly for the benefit of the user and are not used as a variable reference in the Simulink® operations. The names of the blocks must be unique, in other words, the same name cannot be used for more than one block. The parameters of a block are accessed by double-clicking on it. The model is saved in the "Untitled" model window menu:

File → Save As: enter filename (extension is ".mdl").

Once a name is specified for the model and saved, the window title for the model window changes from "Untitled" to the specified file name.

Connections: After selecting the blocks for the model, the next step is to draw the proper connections between them (Figure A.11).

- In order to draw *straight line* connections from the output of one block to the input of another block, simply click the mouse at the output point and drag the mouse to the input point of the other block.

- In order to draw a *perpendicular break* point, simply click the mouse in the interme-diate point, then continue the drawing.

- In order to draw *diagonal lines*, hold the "Shift" key while drawing.

- To define a *branch point* on a line, place the mouse at the point on the line to create the branch point and hold the "Ctrl" key plus click the left mouse button.

- In order to insert a block over an existing line, just move and drop the block over the line. If the block is not inserted into the line, delete the line, insert the block, and make the line connections again.

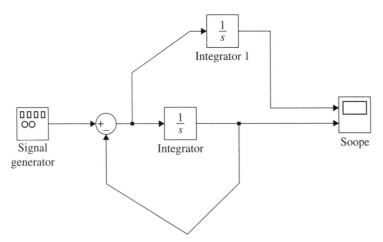

FIGURE A.11: Drawing connections in Simulink®: straight and perpendicular lines, diagonal lines, creating break points on a line.

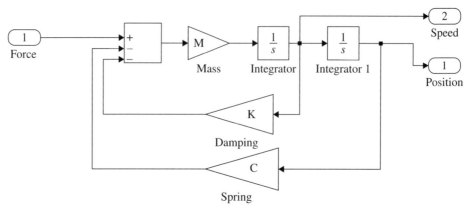

FIGURE A.12: Simulink® model of a mass-spring-damper system.

- In order to delete a line, select the line with the mouse, and hit "Delete" key or use the Edit → Delete menu item.
- To remove a connection without deleting it, select it, pick the edge of the connection, and move it.
- To make a straight line into a non-straight line, select the line, pick a point on the line, and drag it while holding the "Shift" key.

Grouping of Blocks: Simulink® supports block *hierarchy*. A block can be made of smaller sub-blocks. The hierarchy can be many layers deep. An existing group of blocks can be grouped into a single "sub-system block." In order to group a set of blocks into a "sub-system block," select them with the mouse in a rectangular select box (left click to define the corner of the bounding rectangle, then drag the move to include the blocks desired in the sub-system, and release the move), then right click to bring up the context-sensitive menu, then select from menu

Edit → Create SubSystem

The set of blocks will now be replaced by a sub-system block. The new block will show the input ports and output ports of the grouped set and hide the details. In order to look into the details of the sub-system block (lower level hierarchy details), double click on the sub-system block. It will open a new window with the details of the block (Figures A.12 and A.13).

Grouping of blocks is a very useful tool in designing Simulink® models with clear graphical and logical organization. Good programming practice dictates that the systems should be divided into sub-systems, and then those sub-systems should be divided into

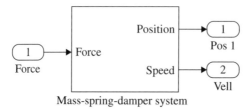

Mass-spring-damper system

FIGURE A.13: The mass-spring-damper system is grouped into a sub-system block showing the input and output ports, hiding the internal details of it.

further sub-systems in a logical tree structure. The sub-system definitions should be a representation of physical reality, as opposed to programming short cuts, in order to have good re-usability and maintainability for the model.

Masking: A related topic to grouping is the "masking" of a block. Masking allows the user to define a custom user interface to access the block parameters. It is like a custom "wrapper" around the block (Figure A.14). Using "masking," we can provide

1. the user more descriptive parameter information about the block parameters, and
2. hide the block details.

If a sub-system block is *masked*, double-clicking on the block brings up the user interface that shows the parameters of the block that the user can change. It does not show the logical details of the block. If a sub-system block is not masked but grouped, then double-clicking on the block brings up a new window showing the logical details of that block. In order to mask a block or sub-system block, select it first. Then, select the sub-system,

Edit → **Mask Subsystem** to "mask" a subsystem
Edit → **Unmask Subsystem** to "unmask" an already "masked" subsystem

Under this selection, we need to define the following tabs for masking

```
Icon
Parameters
Initialization
Documentation
```

Block masks can be activated and deactivated. Once a mask is defined, deactivating it does not lose the mask information. The mask information is still saved and can be activated by the user at any time. The masked block parameters can be assigned values as constants in this interface or as variables where the variables would later be assigned values in the MATLAB® workspace. "Icon" allows us to define a bit mapped image file to be displayed on the masked block (file formats supported: bmp, jpg, tif),

```
image ( imread ('filename.jpg'))
```

The "Icon" feature can be very useful to visually describe the model by attaching an image to the block to display, and it is highly recommended.
    Another useful tool in Simulink® models it to have a button that you can double click to execute a MATLAB® command, such as running an M-file. The Simulink® block used for that function is in

"Model-Wide Utilities: Model Info Block"

Copy this block to the model window. Right-click the mouse to bring up the block properties menu. You can enter any desired text to describe (document) information about the model.
    Let us connect the PD controller (Figure A.13) and the mass-spring-damper model (Figure A.14) and simulate the system response. The equations described by the Simulink® graphics are

$$m\ddot{x}(t) = u(t) \ ; \quad mass - force\ dynamics \tag{A.2}$$

$$u(t) = k_\mathrm{p}(x_\mathrm{d}(t) - x(t)) - k_\mathrm{v}\dot{x}(t) \ ; \quad PD\ control \tag{A.3}$$

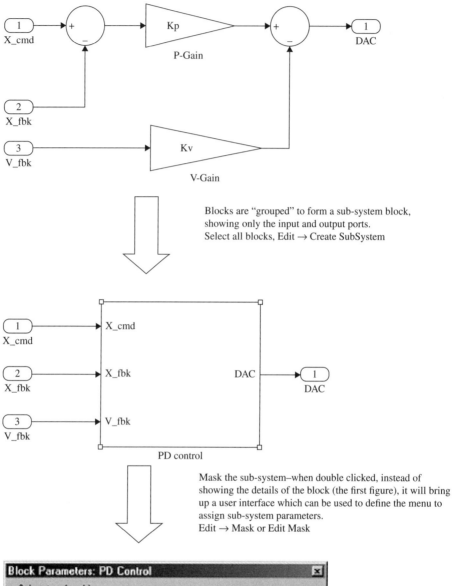

Blocks are "grouped" to form a sub-system block, showing only the input and output ports. Select all blocks, Edit → Create SubSystem

Mask the sub-system–when double clicked, instead of showing the details of the block (the first figure), it will bring up a user interface which can be used to define the menu to assign sub-system parameters. Edit → Mask or Edit Mask

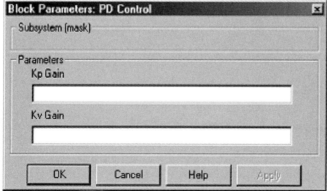

FIGURE A.14: Grouping and masking of Simulink® blocks. This example shows the grouping and masking for a PD control block diagram.

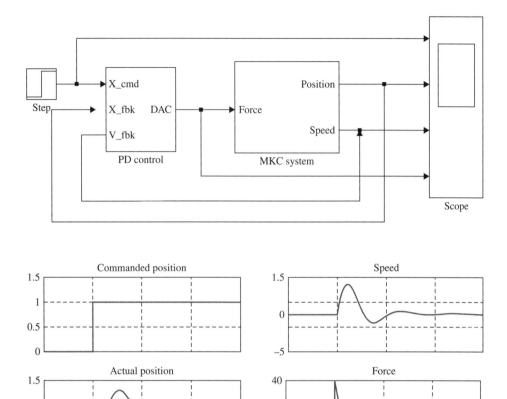

FIGURE A.15: PD control and mass-spring-damper system model and simulation results using Simulink®.

where $x_d(t)$ is the desired (commanded) position signal, $x(t)$ is the actual position signal. For the simulations, we use the following parameters:

$$m = 1.0 \tag{A.4}$$

$$K_p = (2 \cdot \pi)^2 \tag{A.5}$$

$$K_v = 0.7 \cdot 2 \cdot \pi \tag{A.6}$$

A simulation case result is shown in Figure A.15.

A.2.1 Simulink® Block Examples

Here we will examine some of the most commonly used Simulink® blocks.

The *integrator* block and *unit time delay* are shown below (Figure A.16). Integrator, $\frac{1}{s}$ is used to integrate a signal, that is the output is the integral of the input. The integrator block is used in analog control systems. In addition to its input, an initial condition must be specified.

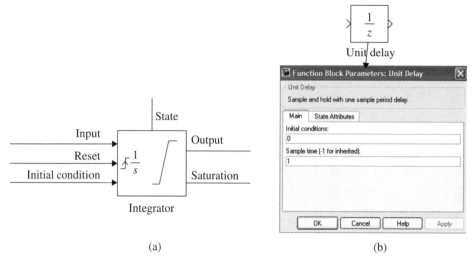

(a)   (b)

FIGURE A.16: (a) Integrator block, (b) unit time delay block.

Unit time delay, $\frac{1}{z}$, is used to delay the input signal by one sampling period. Unit time delay is used in digital control systems. In addition to its input, initial condition and sampling time period must be specified.

Inport and *Outport* are the "input" and "output" port connections for Simulink® sub-systems so that they can connect to other sub-systems and external I/O. The figure below (Figure A.17) shows a sub-system with two inports (In1, In2) and one outport (Out1). The labels of the ports can be changed to anything desired.

Unused input/output ports should be terminated with the "Ground block" (outputs zero to an input terminal otherwise would be left unconncted) and "Terminal block" (terminates an output otherwise would be left unconnected) as a good programming practice (Figure A.18).

Logical and relational operators provide TRUE (1) or FALSE (0) output based on the inputs. The logical operator block property can be set to (Figure A.19),

AND, OR, XOR, NOT, NAND, NOR

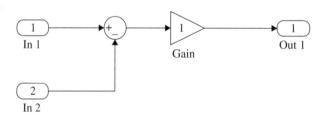

FIGURE A.17: Inport and outport blocks for Simulink® sub-systems.

Ground block   Terminal block

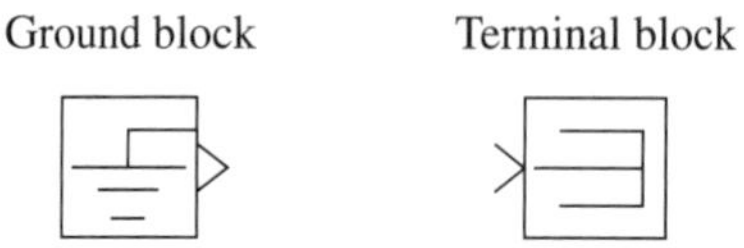

FIGURE A.18: Ground block used to terminate the open input point. Terminal block is used to terminate the open output point.

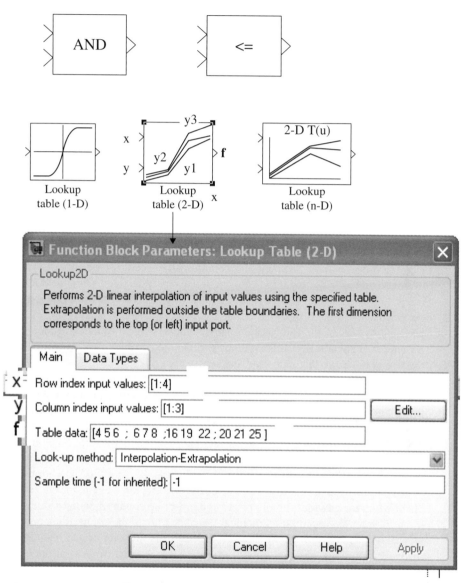

FIGURE A.19: Simulink® blocks for logical operators (i.e., AND), relational operators (comparator i.e., <=), and look-up tables.

The relational operator property can be set to

$$== , = , > , < , >= , <=$$

Look-up tables and data interpolation based on a table are commonly used in control applications. Simulink® provides one-dimensional (i.e., $f(x)$), two-dimensional (i.e. $f(x, y)$), and multi-dimensional look-up tables that can be used for interpolation. A single dimensional table is simply two data vectors: row-vector (independent variable values, x), and the corresponding function ($f(x)$) values with the same dimension. Two-dimensional tables have two variables: row-vector (i.e., 1xm) and column-vector (i.e., 1xn), then the table must be a matrix of dimension mxn. The row-vector variable is connected to the x-port and the column vector variable is connected to the y-port of the 2D look-up table interpolator. The

Enabled block   Triggered block

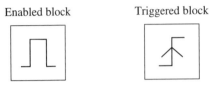

FIGURE A.20: Enable and trigger blocks to make a sub-system enabled and triggered.

block displays the table data as follows: the row vector is treated as the x-axis, and the column vector is a parameter, and there is a curve for each column-vector in the 2D plot.

Conditionally executed sub-systems can be implemented in one of three different ways:

1. enabled,
2. triggered, and
3. enabled and triggered sub-systems.

In order to make a sub-system conditionally executed, just include one of, or both of, the "enabled" and "triggered" blocks in the sub-system and setup the parameters (Figure A.20). A sub-system can be made *enabled* by placing an "enable" block in it, *triggered* by placing a "trigger" block in it, or it can be both *enabled* and *triggered* by placing both blocks. When the user adds "enable" and/or "trigger" blocks inside a sub-system, Simulink® automatically adds an input port for the "enable" and "trigger" signals on the top line of the block that represents the sub-system.

The enabled block makes the sub-system execute during the time the enable signal is larger than zero. The triggered subsystem makes the sub-system execute once on every true trigger signal condition. These two can be combined so that an "enabled and triggered" sub-system would execute once on the trigger signal if it is also already enabled.

Simulink® initializes the states in the "enabled" and "triggered" states at the beginning of the simulation based on the user choices (initial conditions are set to zero if they are not explicitly defined by the user). We can call these the "start-up" condition of the states. When the sub-system is disabled and re-enabled/re-triggered again, the states can either be "held" from the previous run or "reset" to the start-up condition initial values.

In order to simplify the graphical visual display of the connections between blocks, we can use "Goto" and "From" blocks (Figure A.21). In complicated models, the number of connection lines and their routing can get very crowded for visual understanding. In order to simplify that, a signal can be connected to a "Go to" block in one part of the model, and that signal can be picked-up using a "From" block in another part of the model. That way, long routing of the connection lines is eliminated.

In order to make the graphical display easier when a large number of I/O lines is involved, Mux/Demux and Bus-Creater/Bus-Selector blocks can be used (Figure A.22).

Data can be input/output from/to signal sources (i.e., function generator), signal sinks (i.e., scope), MATLAB® workspace and files (Figure A.23). The "Signal Builder" block is a particularly useful signal source tool. It can be used to define custom signal groups rather quickly and interactively for simulation input signals. Signal Builder provides a graphical

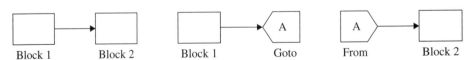

Block 1   Block 2   Block 1   Goto   From   Block 2

FIGURE A.21: Goto and From blocks to simplify line connections in Simulink® graphical model description.

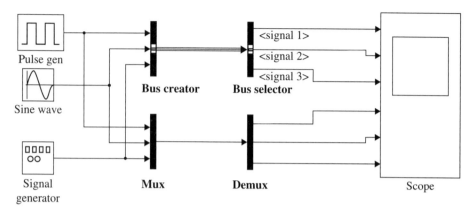

FIGURE A.22: Connection signal management blocks to simplify graphical display: Mux, Demux, Bus Create, and Bus Selector blocks used in an example.

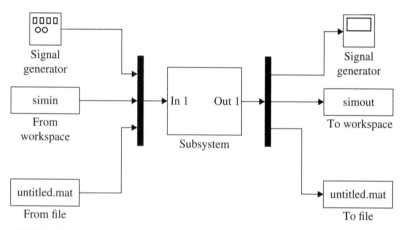

FIGURE A.23: Input sources and output destinations for the data of a Simulink® model.

interface to define groups of signals (i.e., three different groups of signals), then within each group define individual signals (i.e., input functions $r_1(t), r_2(t), r_3(t)$). Each signal (such as $r_1(t)$) function can be defined using the main interface tools (such as constant, step, pulse, rectangular, triangular signal type).

There are three main "function" blocks in the Simulink® block library (Figure A.24):

1. 'MATLAB function block" (like $sin(), exp()$, but not an expression) which accepts vector input and provides vector output. It allows selection of a single MATLAB® function. It does not allow typing an expression.

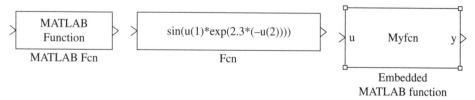

FIGURE A.24: Function blocks in Simulink®: MATLAB® Function (a single MATLAB® function such as "sin()" or 'cos()'). The Fnc() allows an expression (such as "2 ∗ sin(x) - 5 ∗ cos(y)"). The Embedded MATLAB® Function is linked to an M-function script file and is the most general purpose block among them where any desired logic can be implemented.

2. "Fcn function and expression block" which operates on scalar or vector inputs but provides a scalar output. This block allows an expression made of MATLAB® functions and operators and input data.

3. "Embedded MATLAB Function Block" which can accept vector input and provide vector output and implement any function desired. Inputs can be signal inputs in the Simulink® environment or variables from the MATLAB® workspace. Outputs show up at the output port of the Embedded MATLAB® Function Block. Parameters can be defined so that they are visible to the block from the MATLAB® Workspace using **Tools > Edit Data/Ports**. Inputs and Outputs show up both on the Embedded MATLAB® Function Block and also in the call arguments of the function in MATLAB® script. This block is basically a link to a MATLAB® M-function file such as the one shown below, and can be very effectively used for simulation of complicated dynamic systems. When the dynamics of a system get complicated, it is easier to describe it in MATLAB® script language (or in C-language using the S-functions) than express it graphically in Simulink®. In such cases, the overall structure of the model can be designed in Simulink®, and individual complicated sub-systems defined in the "Embedded MATLAB® Function Block" in MATLAB® script language.

```
function y = Myfcn(u)
 % This block supports an embeddable subset of the MATLAB
    language.
 % See the help menu for details.

  y = 2.0 * ( u + 1.0 );

return
```

Simulink® provides sub-system blocks that implement the equivalent five (5) blocks for

1. loops: for, while, do-while,

2. conditional blocks: if, switch-case.

In each case, the block is executed within a given simulation sampling period as long as the condition is satisfied or the iteration count is not larger than the maximum value set.

1.
```
  for (i=begin : increment : end)  % See "for loop" blocks
    ...
    ...
  end
```

The "for" block parameters are set in the menu of the block. The number of input and output port signals and the logic inside the block can be decided by the user, that is one can define three input signals (in1, in2, in3) and two output signals (out1, out2) and some logic in between them. The block is executed for the number of times the "for block" specifies within the current sampling period. This is equivalent to the for-loop in C language.

2.
```
  while(condition)           % See "while" block
    ...
    ...
  end
```

The while loop has two signals to determine its execution: "Condition" and "IC" ports. As long as "Condition" is TRUE and number of iterations is less than the maximum number

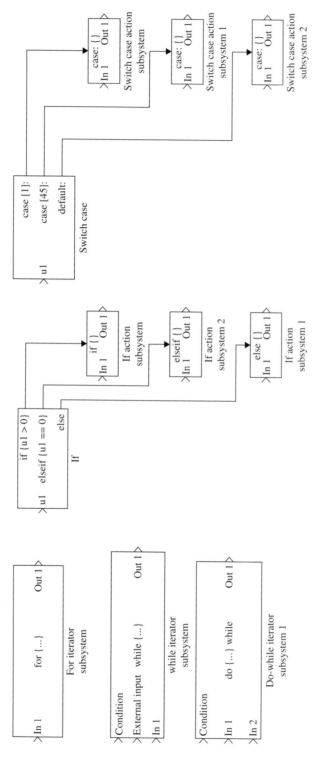

FIGURE A.25: Decision blocks: loops (for, while, do-while) and conditional (if, switch) blocks.

of the iteration set, the block executes. After each execution, "Condition" is checked. The "IC" port condition is checked only at the beginning. If the "IC" condition is FALSE, the block is not executed.

Again, as many inports and outports can be added in the sub-system and the desired logic can be defined. There is a parameter for this block named "maximum number of iterations." It is advisable to set this parameter to a finite value. Simulink® executes the block endlessly as long as the "while (condition)" is TRUE. Unless it is certain that the "while (condition)" will become FALSE at some point in the simulation, a maximum number of iterations should be specified to avoid endless loops in this block.

3.
```
do                        % See "Do-while" block
   ...
   ...
while(condition)
```

The sub-system block is executed at least once. After that, the status of the "Condition" input port is checked. If true, the sub-system block is executed again, until the "Condition" is false. As many input ports and output ports can be defined, as well as any desired logic defined inside the sub-system block. The number of iterations (execution) within a sampling period follow the same rules for the "while block."

4.
```
if (condition)    % See "if" and "if-action" blocks
   ...
elseif
   ...
elseif
   ...
else
   ...
end
```

5.
```
switch case (variable)    % See "Switch case" and "Action
                                 blocks"
   case 1:
     ....
   case 2:
     ....
   case 3:
     ....
   default:
     ....
end
```

"If" and "Switch" statement blocks allow us to select a block from a set of blocks to execute. Each "If" and "Switch" block is accompanied by "Action" blocks. There must be as many action blocks as there number of sections to "if - elseif - elseif - else" and "Switch case" statements.

Example Let us consider the pendulum problem and its nonlinear differential equation which describes the relationship between the torque at the base and the angular position of the pendulum (Figure A.26). We will simulate the response of the pendulum as a result of

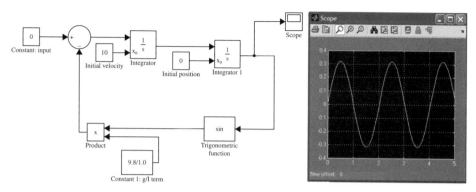

FIGURE A.26: Nonlinear dynamic model of a pendulum and simulation result in Simulink®.

zero input, non-zero initial velocity, and zero initial position. The equation to build a model of in Simulink® is

$$\ddot{\theta}(t) + \frac{g}{l} \cdot \sin(\theta(t)) = u(t) \tag{A.7}$$

$$\theta(t_0) = 0.0 \tag{A.8}$$

$$\dot{\theta}(t_0) = 1.0 \tag{A.9}$$

$$u(t) = 0.0; \; t \geq t_0 \tag{A.10}$$

Let us simulate this for $t_0 = 0.0\,$s, and for a time period of $t = 0.0$ to $5.0\,$s. Figure A.26 shows the Simulink® model that represents this equation and the simulation result for this condition. Since this is a second-order system (highest derivative of the dependent variable in the differential equation is two), we use two integrators in the Simulink® representation. Notice the use of initial conditions for position and velocity as constant inputs to the integration blocks.

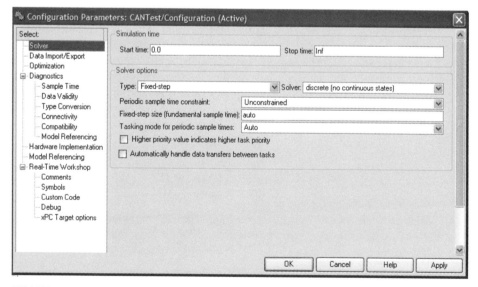

FIGURE A.27: Configuration of Parameters in Simulink® for a particular model simulation. Before running a simulation, **Simulation > Configuration Parameters > Solver (and other options as needed)** should be configured appropriately for the model.

A.2.2 Simulink® S-Functions in C Language

S-functions are the most powerful method of extending the capabilities of Simulink®. S-functions can be written in MATLAB® script, C, C++, Fortran and Ada languages. Here, we briefly discuss the S-functions written in C language only. S-functions created in C language are called "MEX-File S-Functions." Third party companies add value to Simulink® through the use of MEX-File S-functions, that is adding hardware device drivers and custom algorithms. S-functions are also supported by the auto-code generation tools such as the Real Time Workshop (RTW) or Simulink® Coder (SC). The user interface to S-functions can be "masked" to customize its dialog box with the user.

S-functions are used to:

1. add new algorithmic functionalities to Simulink® that are not provided by standard Simulink® blocks,

2. develop hardware device drivers for Simulink®,

3. include existing C-functions into Simulink® blocks.

There are two main steps to using S-functions in Simulink® (see Online Help for further details):

1. Create an S-function MEX-file.

2. Add the **"Simulink® Library Browser > User-Defined Functions > S-Function"** block to your Simulink® model, associate it with the desired S-function MEX-file you created, and setup its parameters, if any.

These two steps are integrated into the **"Simulink® Library Browser > User-Defined Functions > S-Function Builder"** block where the user interface dialog of this block guides the user to create the S-function MEX-file as well as associate it with that block in the Simulink® model.

Example   Consider the liquid level in a tank and its control system shown in Figure 1.4. Let us further consider a computer controlled version of the system: the mechanical levers are replaced by a level sensor, a digital controller, and a valve that is actuated by a solenoid. In-flow rate to the tank is controlled by the valve. The valve is controlled by a solenoid. The input to the solenoid is the current signal from the controller and the output of the solenoid is proportional force. The force generated by the solenoid is balanced by a centering spring. Hence, the valve position or opening of the orifice is proportional to the current signal. The flow rate through the valve is proportional to the valve opening. Assuming simplified linear relationships, the input–output relationship for the valve can be expressed as

$$F_{\text{valve}}(t) = K_1 \cdot i(t) \tag{A.11}$$

$$= K_{\text{spring}} \cdot x_{\text{valve}}(t) \tag{A.12}$$

$$Q_{\text{in}}(t) = K_{\text{flow}} \cdot x_{\text{valve}}(t) \tag{A.13}$$

$$= K_{\text{flow}} \cdot \frac{1}{K_{\text{spring}}} \cdot K_1 \cdot i(t) \tag{A.14}$$

$$= K_{\text{valve}} \cdot i(t) \tag{A.15}$$

where $K_{\text{valve}} = K_{\text{flow}} \cdot K_1 / K_{\text{spring}}$, which is the valve gain between current input and flow rate through the valve.

The liquid level in the tank is a function of the rate of in-flow, rate of out-flow, and the cross-sectional area of the tank. The time rate of change in the volume of the liquid in the tank is equal to the difference between in-flow rate and out-flow rate,

$$\frac{d(Volume\ in\ Tank)}{dt} = (In\text{-}flow\ rate) - (Out\text{-}flow\ rate) \tag{A.16}$$

$$\frac{d(A \cdot y(t))}{dt} = Q_{in}(t) - Q_{out}(t) \tag{A.17}$$

$$A\frac{dy(t)}{dt} = Q_{in}(t) - Q_{out}(t) \tag{A.18}$$

The Q_{in} is controlled by the valve to be between zero and maximum flow that can go through the valve, $[0, Q_{max}]$. The Q_{out} is a function of the liquid level and the orifice geometry at the outlet. Let us approximate the relationship as a linear one, that is the higher the liquid height is, the larger the out-flow rate,

$$Q_{out}(t) = \frac{1}{R} \cdot y(t) \tag{A.19}$$

where R represent the orifice restriction as the resistance to flow. Then, the tank dynamic model can be expressed as

$$A\frac{dy(t)}{dt} + \frac{1}{R} \cdot y(t) = Q_{in}(t) \tag{A.20}$$

Let us consider a practical ON/OFF type controller with hysteresis. The controller either fully turns ON or turns OFF the valve depending on the error between the actual and measured liquid level. In order to make sure the controller does not switch the valve ON/OFF at high frequency due to small changes in the liquid level, a small amount of hysteresis is added in the control function. This type of controller is called a relay with hysteresis and is commonly used in many automatic control systems such as home temperature control, liquid level control ($[e_{max}, e_{max}]$ range). In Simulink®, the controller function is implemented with a hysteresis block. In mathematical terms, the controller function is

$$e(t) = y_d(t) - y(t) \tag{A.21}$$

$$i(t) = Relay_{Hysteresis}(e) \tag{A.22}$$

The relay control function with hysteresis where the hysteresis band is $[e_{max}, e_{max}]$ range. Flow rate can vary linearly between zero and maximum flow rate as a function of current signal. Since current signal is either zero or maximum value, the flow rate will be either zero or maximum flow.

$$Q_{in}(t) = K_{valve} \cdot i(t) \tag{A.23}$$

$$= Q_{max} \ ; \ \text{when } i(t) = i_{max} \tag{A.24}$$

$$= 0 \ ; \ \text{when } i(t) = 0 \tag{A.25}$$

Let us simulate the liquid level control system for the following conditions. The system parameters are $e_{max} = 0.05$, $i_{max} = 1.0\,A$, $Q_{max} = 1200\,l/min = 20\,l/s = 0.02\,m^3/s$, $A = 0.01\,m^2$, $R = 500\,m/m^3/s$. Consider the case that the desired liquid height is $y_d(t) = 1.0\,m$ which is commanded as a step function, and the initial height of the liquid is zero (empty tank). Figure A.28 shows the Simulink® model and simulation results.

Example   Consider a room or furnace temperature control system (Figure 1.8). We need to consider room temperature, outside temperature (cold) and a heater. The heater is controlled by a relay type controller with hysteresis. The room temperature is initially at the same temperature as the outside temperature. The controller is set to increase the room temperature to a higher level. The heater is controlled to regulate the room temperature. As

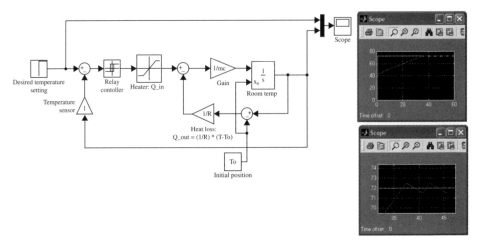

FIGURE A.28: Model and simulation of a liquid level control system.

the room temperature increases and becomes larger than the outside temperature, there is a heat loss from the room to the outside. The net added heat rate to the room is the difference between the heat generated by the heater and the heat loss to the outside since outside temperature is colder. The temperature rise in the room is function of this difference and the size of the room. The heat loss is a linear function of the inside and outside room temperatures. Net heat added (or lost) to the room will result in the temperature change,

$$(Net\ Heat\ Added\ to\ Room) = (Heat\ in\ Rate) - (Heat\ out\ Rate) \qquad (A.26)$$

$$Q_{net} = Q_{in} - Q_{out} \qquad (A.27)$$

$$mc\frac{dT}{dt} = Q_{in} - \frac{1}{R}(T - T_o) \qquad (A.28)$$

where mc is the heat capacitance of the room which is function of the room size, R is the resistance of the heat transfer from walls due to the temperature difference. The effective resistance to heat transfer between room and outside is function of the type of dominant mode of heat transfer (conduction, convection, radiation) as well as the size and insulation type of walls. T and T_o are inside and outside temperatures, respectively.

Let us simulate the room temperature control system for the following conditions; desired temperature $T_d = 72\,°F$, initial temperature $T_o = 42\,°F$, allowed error in room temperature in the hysteresis function of the relay control system, $e_{max} = 0$, maximum heat-in rate $Q_{max} = 100$, $R = 100$, $mc = 1.0$. Initially the room is assumed to be at the same temperature as the outside temperature. After entering the room, that is one second later, the temperature is commanded to be $T_d = 72\,°F$. Figure A.29 shows the Simulink® implementation of the model and simulation results.

Example Consider the web tension control system shown in Figure 1.7. The wind-off roll is driven by another part of the machinery where the speed $v_1(t)$ is dictated by other considerations. The wind-up roll is driven by an electric motor. This motor is required to run in such a way that the tension in the web (F) is maintained constant and equal to a desired value (F_d). So, if the wind-off roll speeds up, the wind-up roll is suppose to speed up. Similarly, if the wind-off roll slows down, the wind-up roll is suppose to slow down quickly. The wind-off roll speed is given as an external input and is not under our control. The wind-up speed is our controlled variable. Our objective is to minimize the tension error, $e(t) = F_d(t) - F(t)$.

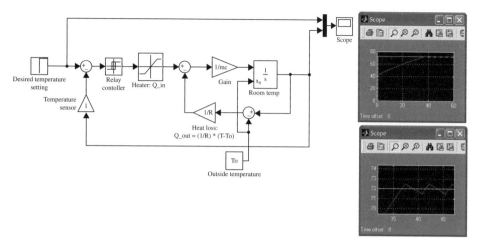

FIGURE A.29: Model and simulation of furnace or room temperature control system.

The tension in the web will be determined by the difference between the integral of $v_1(t)$ and $v_2(t)$,

$$y(t) = y(t_0) + \int_{t_0}^{t} (v_2(t) - v_1(t))dt \tag{A.29}$$

$$F(t) = F_0 + k \cdot y(t) \tag{A.30}$$

If initially the web tension is adjusted so that when $y = y_0$, the tension $F = F_0 = 0$ by proper calibration, then we can express the tension as function of change in $y(t)$,

$$\Delta y(t) = y(t) - y(t_0) \tag{A.31}$$

$$= \int_{t_0}^{t} (v_2(t) - v_1(t))dt \tag{A.32}$$

$$\Delta Y(s) = \frac{1}{s} \cdot (V_2(s) - V_1(s)) \tag{A.33}$$

$$F(t) = k \cdot \Delta y(t) \tag{A.34}$$

$$F(s) = \frac{k}{s} \cdot (V_2(s) - V_1(s)) \tag{A.35}$$

The control system that controls the $v_2(t)$ is a closed loop control system and is implemented using an analog controller. Let us consider the dynamics of the amplifier and motor as a first-order filter, that is the transfer function between the commanded speed $w_{2,cmd}$ to actual speed w_2,

$$\frac{w_2(s)}{w_{2,cmd}(s)} = \frac{1}{\tau_m s + 1} \tag{A.36}$$

where τ_m is the first-order filter time constant for the amplifier and motor. The corresponding linear speeds are

$$v_{2,cmd}(t) = r_2 \cdot w_{2,cmd}(t) \tag{A.37}$$

$$v_2(t) = r_2 \cdot w_2(t) \tag{A.38}$$

Let us consider a proportional controller,

$$w_{2,cmd}(t) = K_p \cdot (F_d(t) - F(t)) \tag{A.39}$$

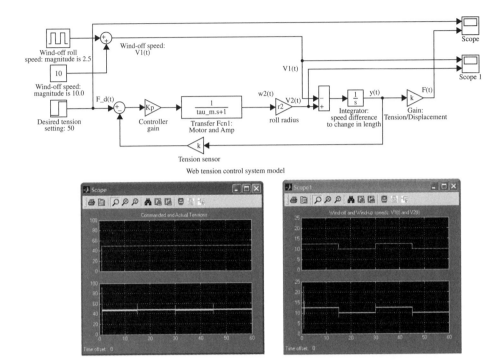

FIGURE A.30: Model and simulation of web tension control system. The top figure is the Simulink® model of the tension control system. The left plot shows the commanded tension on the top and the actual tension at the bottom. The right plot shows the wind-off and wind-up speeds.

Figure A.30 shows the model and simulation conditions in Simulink®. The parameters of the system used in the simulation are,

$$k = 10\,000\,\text{N/m} \tag{A.40}$$

$$K_p = 10\,\text{m/s/m} \tag{A.41}$$

$$\tau_m = 0.01\,\text{s} \tag{A.42}$$

$$r_2 = 0.5\,\text{m} \tag{A.43}$$

We will simulate a condition where $v_1(t)$ has a step change from its nominal value for a period of time.

$$v_1(t) = 10.0 + 2.5 f_1(t) \tag{A.44}$$

$$F_d(t) = 50 \cdot step(t - 1.0) \;\; ; \;\; step\,function\,starts\,at\,1.0\,\text{s} \tag{A.45}$$

where $f_1(t)$ represents a square pulse function with a period of $T = 30\,\text{s}$. The interested reader can easily experiment with different control algorithms as well as different process parameters, that is different roll diameter values r_2.

A.3 STATEFLOW

Stateflow (SF) is the graphical modeling software as part of MATLAB® that is used to model *event-driven systems*. Control logic or process dynamic models, which have various modes of operation and are event-driven and supervisory in nature, are best expressed in

Stateflow rather than in Simulink®. In typical usage, a SF model is a block in a Simulink® model. The SF model is called the *chart*. The chart is used to represent various modes of operation.

The mathematical modeling paradigm in Stateflow is based on the concept of operating modes. The operating modes are called the **states**, in which various logics are implemented and are called **actions**. The **transitions** between these states are based on some **events** and/or **conditions**.

The programmer needs to express the problem at hand in these terms;

1. states,
2. actions,
3. transitions and junctions,
4. conditions,
5. events,
6. data.

Events and **data** are non-graphical objects of Stateflow, whereas the others are graphical objects.

Stateflow is started from MATLAB® command window by typing (Figure A.31)

```
>> stateflow
```

This brings up the "sflib" (Stateflow library) window. The development tools for Stateflow include the

1. **Stateflow Editor**
2. **Model Explorer**, and
3. **Debugger**.

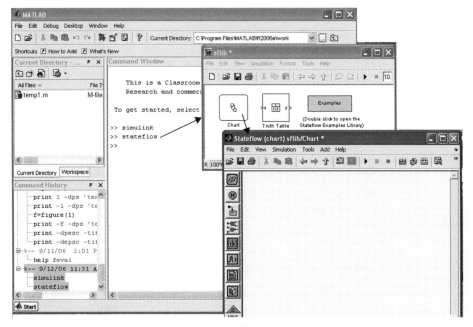

FIGURE A.31: Stateflow windows in MATLAB®.

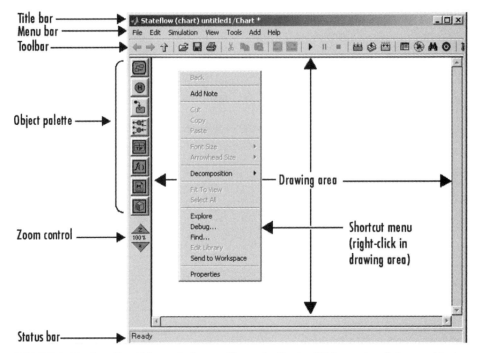

Title bar
Menu bar
Toolbar
Object palette
Zoom control
Status bar
Drawing area
Shortcut menu
(right-click in
drawing area)

FIGURE A.32: Stateflow Editor window and its tools. On the "Object palette" the graphical components of the Stateflow chart are displayed: state, history junction, default transition, connective junction, truth table, function call (may be used to implement "Actions"), embedded MATLAB® functions (may be used to implement "Actions"), and box for graphical grouping.

The **Stateflow Editor** is used to graphically build the SF model, using objects of states, transitions, junctions, conditions, and actions. **Model Explorer** is used to define the non-graphical components of a SF chart; events and data. The **Debugger** is used to debug the SF model during simulation, and also allows *animation* of states in order to visually assist the developer evaluate the operation of the SF chart.

By double clicking on the "Chart" icon, we bring up the "StateFlow Editor" window, where a SF chart is designed (Figure A.32). A pop-up menu can be displayed and content changed for any SF chart component by pointing on the component and right-clicking the mouse. Stateflow "Chart" and "Truth Table" blocks can also be accessed under Simulink® toolboxes, named "Stateflow." The Stateflow Editor window and its various menu options are shown in Figures A.33 and A.34. In the **File > Chart Properties**, select **Classic, Inherited** and enable C-bit operations. Checking this box instructs the Stateflow to interpret C bitwise operators ($\sim$, &, |, ^, >>, and so on) in action language statements and encode them as C bitwise operations. If this box is not selected & and | are interpreted as logical operators.

A Stateflow chart may be triggered at each sampling period or by *events*. The source and nature of each event can be defined in the SF chart. In control applications, the most common source of trigger method for Stateflow charts are:

1. periodic timer, such as every 10 ms,
2. external events (interrupts); such as status change of an ON/OFF input signal, or the value of an analog signal going beyond a certain range.

The whole logic, or the chart, is a collection of states plus the *actions* and *conditions* that may be executed in each state. Each mode can be considered as a *state*. The actions and conditions are also used to define the *transitions* between states.

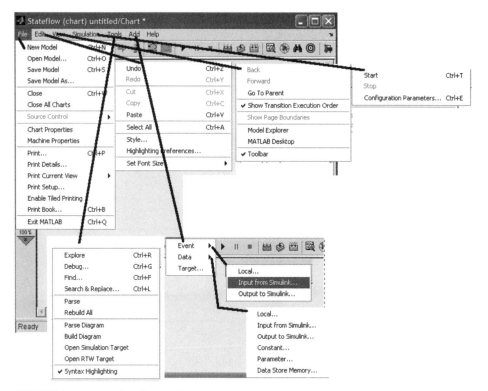

FIGURE A.33: Stateflow Editor Menu.

The SF state concept supports *hierarchy:* states can be grouped in a hierarchy. The group of states can be exclusive (also called OR states, meaning only one of the states in the group can be active at any given time, i.e., a switch state can be either ON or OFF, not both at the same time) or parallel (also called AND states, meaning the states in the parallel group can be active concurrently during the same sampling period). For instance, assume we have ON and OFF states for a FAN motor and a HEATER. The FAN can be ON and the HEATER can be ON at the same time. The states of FAN motor and HEATER motor are parallel states. The states of FAN motor (ON state and OFF state) are exclusive (OR states), clearly meaning that a FAN motor cannot be ON and OFF at the same time. The same thing applies for the HEATER motor.

Grouping states is a graphical visual aid, it does not make any difference to the logic of the chart. It simply hides the sub-states in the hierarchy, making the visual display less crowded. To group a state and hide its sub-states graphically, right-click on the state, and then select

Make Contents > Grouped

You must ungroup a superstate to select objects within the superstate (grouped state). To ungroup a state, double-click it or its border and select

Make Contents > Ungrouped

Once the code is auto-code generated, both Simulink® and StateFlow graphical programs are converted to C language and then compiled/linked into an executable code.

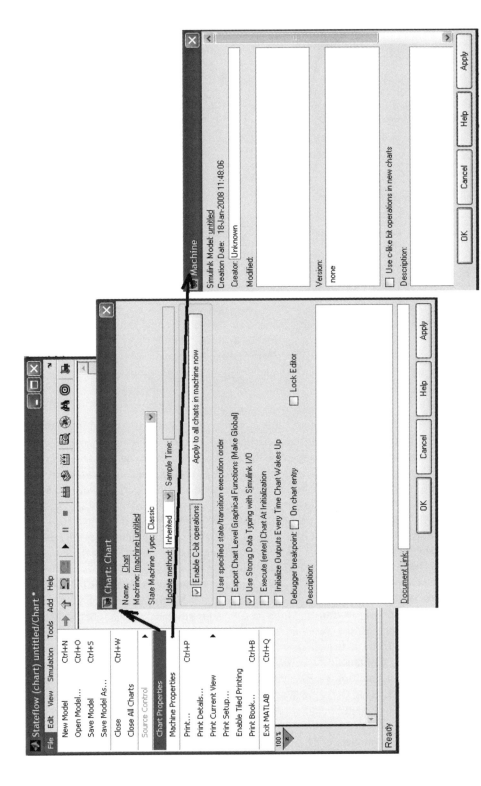

FIGURE A.34: Stateflow Editor Menu: File > Chart Properties and File > Machine Properties.

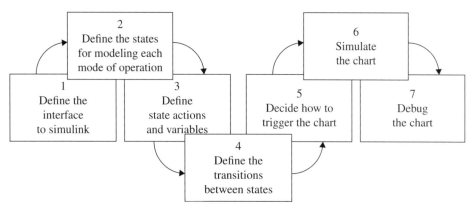

FIGURE A.35: Typical steps involved in designing a Stateflow chart as part of a Simulink® model.

The typical steps involved in developing a SF chart is shown in Figure A.35.

Step 1: Input and outputs between Stateflow (SF) and Simulink®. The first step in designing a typical SF chart is to plan it as part of a Simulink® model, and then decide on the input–output data and events between the SF block and the Simulink® model. The SF chart interacts with Simulink® through its input and output signals. The input and output signals can be data or events needed by SF and Simulink®.

In Simulink®, select Stateflow from the Simulink® Library Browser, then double-click on it. This will bring up the Stateflow blocks: Chart, Truth Table. Select "Chart" and drag-and-drop it in the Simulink® model. Now double-click on the "Chart" block in the Simulink® model to edit its content using Stateflow Editor.

In order to define I/O between SF and Simulink®, use

Stateflow Editor > Add > Data/Event > Input from Simulink®

Stateflow Editor > Add > Data/Event > Output to Simulink®

The user then will need to fill in the data in the user interface provided, that is, provide the name of the data variable, data type, scope, and so on. Then, Stateflow adds input and output ports for each "Input from Simulink®" and "Output to Simulink®" data definitions. The port numbers are assigned in the order they are created, but can be changed by explicitly specifying the port number in the definition menu.

Step 2: States, their hierarchy and decomposition. How many states are needed, their hierarchy in terms of super-states and sub-states (parent-child relationship), and whether they are exclusive (OR-states) or parallel (AND-states, concurrent) should be decided. Consider the operation of your control logic or process model, and decide on the modes of operation. In general, each mode of operation is assigned a state (Figure A.36). In a state hierarchy, among the collection of OR states, only one of them can be active at a given time, whereas multiple AND states can be active concurrently (although they are executed still sequentially). The order of the sequential execution of parallel states (AND states) can be left to default or specified by the designer. In the SF editor, exclusive (OR) states are displayed with solid line rectangles, whereas parallel (AND) states are displayed with dotted line rectangles. In the order to define the states as parallel in a hierarchy, right-click the mouse inside the super-state where the parallel states will be defined, then select

Decomposition > Parallel (AND)

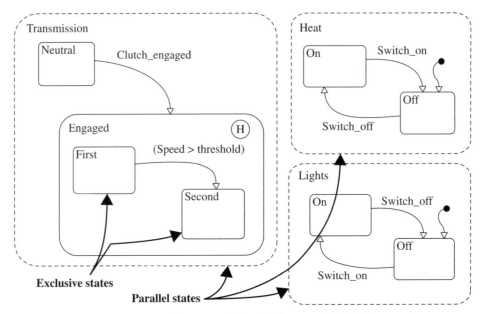

FIGURE A.36: Decomposition of states: exclusive (OR) or parallel (AND) states. At a super-state, only one of the exclusive states can be active, whereas multiple parallel states (AND states) can be concurrently active.

Step 3: Define State Actions and Data Variables. The hierarchy of states describes the type structure of the logic. The actual logic, that is the actions, and the data acted on in each state defines the details of the algorithm. Actions can be present either in the state logic or in the transitions.

There are four different categories of actions in each state (Figure A.37):

```
entry:
        action1();   % executed once when the state is activated

during:              % executed  every sampling period while the
                        state is active
        action21();
        action22();

exit:                % executed once when transition from that
                        state to another state
        action3();   % occurs and this state is 'exited'

on_"EventName1":     % executed on every occurance of the named
                        event
        action41();
        action42();

on_"EventName2":     % executed on every occurance of the named
                        event
        action43();
        action44();
        action45();
```

When a state is "active", some of the actions are executed depending on the condition of the state. For instance, on the first entry to the state, the actions in the "entry:" block are

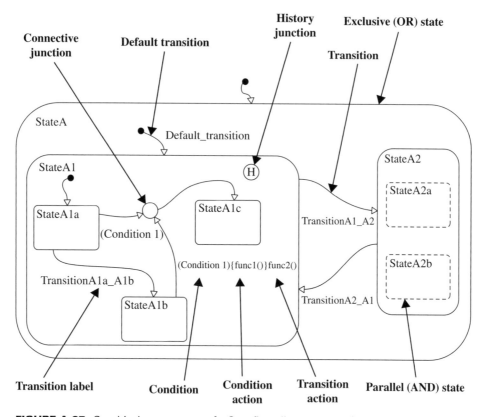

FIGURE A.37: Graphical components of a Stateflow diagram: state (parallel and exclusive), transitions (normal transitions, default transitions), actions, junctions (connective and history types), conditions.

executed. After that, the actions in the "during:" block are executed during the following sampling periods. When a transition occurs from the currently active state, then the actions in the "exit:" block are executed before the program transitions to another state. Multiple "on_EventName" blocks can be defined. This allows us to define a unique set of actions for each event.

In addition to actions, we may need to define data variables (local or persistent) in each state to keep track of data. It is also possible that we only need the input and output data, and no local (or persistent) data may be needed.

Step 4: Define the Transitions Between States. So far we have defined a state hierarchy and actions within those states. We also need to define the logic to make the transitions between those states.

Transitions are the logical flow paths between different states. To simplify the logic, it is advisable to exercise modular programming practice and plan transitions between states at the same hierarchy level. Otherwise, the transition logic may look like a "spaghetti code" which means it is difficult to understand and maintain. Each transition may have a *condition* which must be TRUE for the transition to occur, and *an action* that is executed during the transition before the control is passed to the next state. These transition conditions and actions are expressed with a notation on the transition path in text format. At every hierarchy of states, there should be a **default transition** defined. This transition would be used in case of ambiguity over which state to activate. This often happens on start-up conditions. For every level of exclusive OR-states, there should be a **default transition** defined as a good programming practice. Although this is strictly not required, it is a good programming practice to clearly define which state will be activated under unclear conditions, such as on power-up. A transition can support two different **actions**

- a condition action, and
- a transition action.

Transition labels are defined in the following syntax (Figure A.37),

```
event [condition]{condition_action}/transition_action
```

where *event* defines the event that triggers the transition, [condition] must be true in order for the transition to take place. Then the two sets of actions are performed, if defined. You can specify some, all, or none of the parts of the transition label. Use "..." to continue on the next line in defining the transition labels.

Connection junctions are used to provide multiple branch paths for a transition between more than two states. They are equivalent to **if-elseif-else** decision blocks (Figure A.37).

History junctions provide the information on the most recent active sub-state in a super-state or a chart. When a transition occurs to a super-state which has its sub-states and history junction, the transition occurs to the sub-state that was active most recently, as a result of the history junction component in the model. The history junction remembers the last active sub-state. In other words, the history junction overrides the default transition if they both exist in the same state hierarchy (Figure A.37). The history junction applies only at the hierarchy in which it is placed.

Conditions are boolean logical expressions that must result in a true or false result. For a transition to occur, the transition condition must be TRUE.

Events and *data* are non-graphical components of the SF chart. Events are defined using the menu item in the SF editor,

Stateflow > Add > Event

Events can be defined to trigger a SF chart, trigger a transition to occur, trigger an action to be executed. Data is a signal or local information needed and logic operations are applied to. Data need to be defined in SF charts using menu item in SF editor,

Stateflow > Add > Data

These non-graphical elements of a state can be viewed and displayed by using the **Model Explorer**. To do so, point to the state, right-click, and select **Explorer** from the pop-up menu.

Step 5: Define Triggers to the Stateflow Chart. A Stateflow chart can be executed at a specified rate or can be triggered by an external signal. Triggers are defined as a part of events. In Stateflow Editor, use

Stateflow > Add > Event > Input from Simulink®

to define a trigger event from Simulink® to the SF chart. The signal source and the conditions at which it causes a trigger (i.e., rising-edge, falling-edge, or both) can be defined.

Step 6: Simulate. A SF chart is normally simulated as part of the Simulink® model. However it can also be simulated with its own simplified input and output for debugging purposes.

> **Stateflow > Simulation > Configuration Parameters**

Step 7: Debug. Graphical animation tools can be enabled to display the active or inactive states as debugging tools. Breakpoints can be set at various points in the SF, such as at the wake-up of the chart, the entry of a state, event occurrences. To set breakpoints, enable animation, and set the speed of animation, select

> **Stateflow Editor > Tools > Debug**

This brings up a new window with options to set breakpoints, enable/disable animation and the speed of the animation. Most common breakpoints are set at the "Chart Entry" and "State Entry" by checking those option boxes. Then the program execution can be controlled with "Start," "Stop" "Step" "Break" buttons.

A.3.1 Accessing Data and Functions from a Stateflow Chart

MATLAB®, Simulink®, and Stateflow work together to model and simulate controlled dynamic systems. Therefore, these three components must be able to access data in each others' workspace in order to apply logic to it. MATLAB® maintains all its data variables in its "Workspace." Data sharing between Simulink® and Stateflow is generally accomplished by the input and the output ports in the block diagram. This is the recommended method of exchanging data between MATLAB®, Simulink®, and Stateflow. It shows clearly in graphical form what data is passed between Simulink® and Stateflow. It is that graphical display of information and logic that makes Simulink® and Stateflow attractive programming tools. Below are also other methods of accessing data in MATLAB® workspace from a Stateflow chart.

Accessing MATLAB® Data and Built-in Functions from the SF Chart
One way to access MATLAB® data and functions from Stateflow is to use the "ml" namespace operator and "ml" function. Using the "ml" namespace operator we can access data in the MATLAB® workspace. The "ml" acts like an object name, followed by a dot operator and variable name in the MATLAB® workspace. For instance, let us assume there is a variable

x

in the MATLAB® workspace and we want to assign it to variable in Stateflow,

```
 a
```

```
a = ml.x  ;  % Stateflow variable 'a' is assigned
            %  to the value of MATLAB® workspace variable 'x'.
```

Likewise, we can assign a Stateflow variable to a MATLAB® workspace variable. If the MATLAB® workspace variable was not created before, it will be automatically created. The following accomplishes that,

```
ml.x = a  ;  % MATLAB® workspace variable 'x'
             % is assigned to the value of Stateflow variable 'a'.
```

```
% If MATLAB® workspace variable does not exit,
% it is created automatically.
```

We can use the same operator to access MATLAB® built-in functions:

```
a = ml.sin(ml.x)  ;  % Stateflow variable 'a' is assigned to
                     % the value of Sin of MATLAB® workspace
                     % variable 'x'.

b = ml.sin(a)    ;  % .................'b' ..............
                     % .......... Sin of Stateflow variable 'a'.
```

Notice the difference between the above two examples: in the first one the data variable 'x' in the MATLAB® workspace is used for the sine function. In the second example, the data variable 'a' in the Stateflow is used in the sine function.

Function Calls from a Stateflow Chart: Embedded MATLAB® Functions

An "Embedded MATLAB® Function" can be added by selecting the "Embedded MATLAB® Function" icon on the left pane of the Stateflow Editor, and dragging and dropping it in the chart, then typing the name of the function. Then double-click on the box just created to bring up the text editor and edit the content of the function. Input and output arguments from the function can be any MATLAB® data type of the form scalar, vector, or matrix. This is a very effective and convenient way to code "actions" that are more than a single line.

Figure A.38 shows the windows on how to add an Embedded MATLAB® Function in a Stateflow chart. The function is used to define an action.

1. We draw the transitions and actions which calls an embedded MATLAB® function in the usual way.
2. Then use the "Embedded MATLAB® Function" icon on the left pane of the Stateflow Chart, copy it into the chart.
3. Then type the name of the function in the created rectangle.
4. Double-clicking on the rectangle block for the function brings up the **Embedded MATLAB® Editor**.
5. From the Embedded MATLAB® Editor, use **Tool** > **Model Explorer**, then edit the input–output data type and size, as well as local variables as needed.
6. Then simply type the logic of the function into the editor.

Figure A.39 shows the window for the 'Model Explorer" that allows us to edit the data for the function.

If the embedded MATLAB® function calls yet other function(s), that function(s) can be included in the same file below the embedded MATLAB® function code.

Function Calls from a Stateflow Chart: User Written C Functions

User-written C functions can be called from Stateflow chart's transition actions and state actions (Figure A.40). The C-function call is made just like in a C language program, that is

```
function_name(argument1, ... ) ;
```

The string parameters are passed using quotes, such as "string1" or "Hello" in the argument list. Function calls can be nested, that is one function call can include another function call in its argument list (Figure A.40). Arguments can be passed by an address

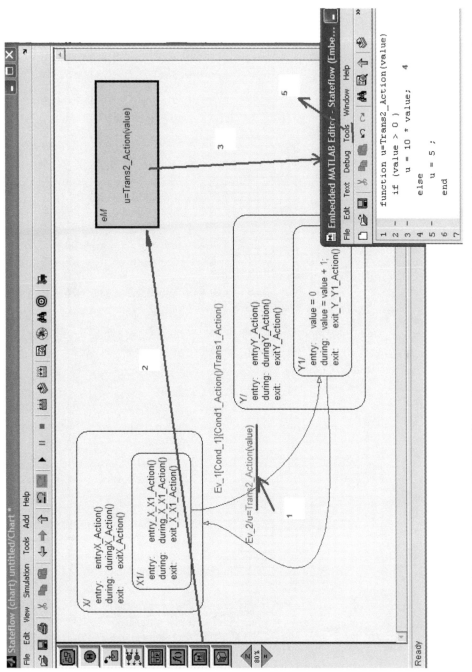

FIGURE A.38: Adding an Embedded MATLAB® Function in a Stateflow chart: 1. Type the name of the function in an action (i.e., a transition action). 2. Using the **"Embedded MATLAB® Function"** icon from the Stateflow Chart Editor, copy one to the Chart, and type the name on it. 3. Double-click on the **"Embedded MATLAB® Function"** rectangle to bring up the text editor for the function. 4. From the **Embedded MATLAB® Editor**'s tool menu, select **Model Explorer,** and define the input–output data (as well as local data) to the function. 5. Type in the logical code in the text file.

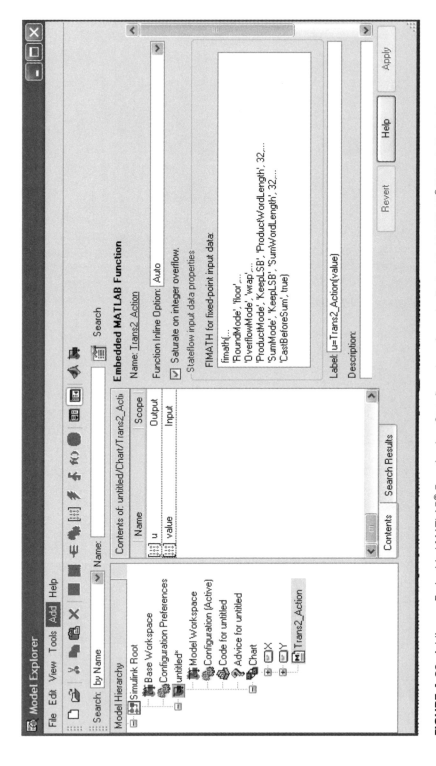

FIGURE A.39: Adding an Embedded MATLAB® Function in a Stateflow chart: from the Embedded MATLAB® Editor's Tools Menu, select Model Explorer, and define the input–output data as well as local variables to the function.

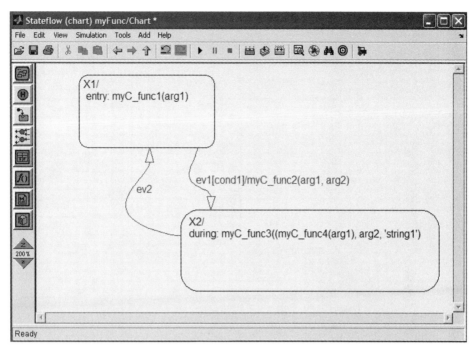

FIGURE A.40: Calling User-Written C-functions from Stateflow Charts: calls from transition actions and state actions.

using pointer notation. However, data from "Input from Simulink®" should not be passed by an address.

```
f(&x) ;       % Call to function f(), pass data by address
                 (reference)
g(&(z[0])) ; %  Passing array z[0], z[1], ...
g(&(z[1])) ; %  Passing array z[1], z[2], ...
```

Example: Embedded MATLAB® Functions in Stateflow Chart   The purpose of this example is to illustrate how to include "Embedded MATLAB® Functions" in a SF model. Figures A.41 and A.42 show the user interface windows for the development.

Figure A.41 shows the windows '1,' '2,' and '3' for MATLAB®, Simulink®, and Stateflow respectively. Then from Simulink® we start a model, shown in window '4.' In this model, we include a Stateflow chart. Double-clicking on the SF chart in the model, brings up the Stateflow Editor window '5.' In this case we define two states, X and Y for the SF model, and two transitions between them. In each state, and for each transition, we define an action, which calls an "Embedded MATLAB® Function." Hence, we have four "Embedded MATLAB® Functions" for this example. In order to make the function calls, simply click on the state X, and type the name of the function to be called. In this case, the actions are defined for the "during:" stage of the state's status. Repeat the same for state Y and transitions.

Then pick-and-drag four copies of the "Embedded MATLAB® Function" icon from the left pane into the Stateflow Editor window. Then select each one-by-one, and type the name of the function on the box. Then double-click on each one-by-one (Figure A.42). This will bring up the "Embedded MATLAB® Function Editor." Now, enter your code here. The editor already has the function declaration line. The editor shows one file at a time.

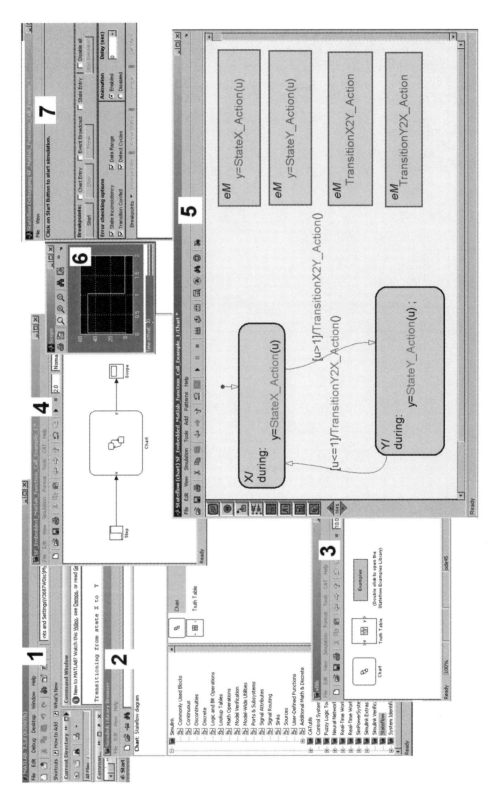

FIGURE A.41: Adding an Embedded MATLAB® Function in a Stateflow chart: one of two (1/2) sets of windows.

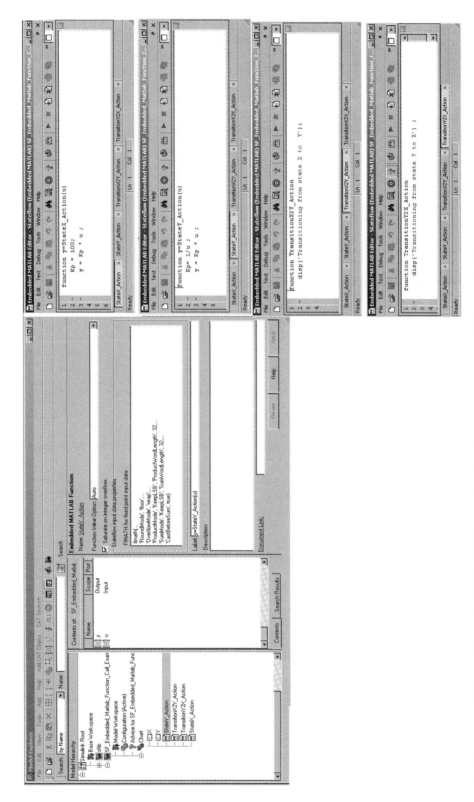

FIGURE A.42: Adding an Embedded MATLAB® Function in a Stateflow chart: two of two (2/2) sets of windows.

For convenience, we copied all the files for illustration purposes here in the figure. Note that this is a "MATLAB® Function" of an embedded kind. Only the data visible to it are passed in its argument list. The "ml" operator we used in the Stateflow graphical model file is not available to access MATLAB® workspace data. If we need to access MATLAB® workspace data which are not passed by the function call arguments, then we must use the "global" declaration. For instance, in the function "StateX_Action(u)," we define the Kp as a local variable. The Kp of the MATLAB® workspace can not be accessed via "ml.Kp." If it is desired to access the Kp of the MATLAB® workspace instead of using a local Kp, then the "global" declaration must be used.

Example: Use of C-Functions in Stateflow Chart   The purpose of this example is to illustrate how to incorporate C-functions into Stateflow charts. Simply follow the order of activities in the dialog windows as shown in Figures A.43 and A.44. There are four C-functions which are called from the Stateflow chart;

```
StateX_Action(u) ;
TransitionY2X_Action() ;
StateY_Action(u);
TransitionX2Y_Action() ;
```

The source code for these four C-functions is in the file named

```
StateFlow_C_Function_Example.c
```

The declarations (include statements) code for each function is typed directly into the dialog box.

Example: Liquid Level in a Tank Control Problem with Multiple Modes of Controller Operation   For the purpose of illustrating how a typical Stateflow chart can be used as part of a Simulink® model, let us assume that the controller logic is now different than the one implemented in the Simulink® example. The new controller we want to implement here has three modes of operation. In each mode, a different control algorithm is implemented with different parameters. A supervisory control logic is to monitor the sensor signal for the measured liquid height and the desired height data, then decide on an output value that selects one of three modes of operation. Using that information, the Simulink® part of the code is to execute the selected control algorithm.

The supervisory control logic will be as follows, expressed in MATLAB® text language, and will be implemented in Stateflow. For the purpose of illustration, we select a simple supervisory controller that can equally be implemented in MATLAB® or Simulink® with ease. However, as supervisory logic gets more complicated, the Stateflow programming model provides advantages in ease of programming.

```
% Supervisory control logic to select controller mode of operation,

    err_th1 = 10.0 ;    % Percentage of liquid height error,
                          threshold value 1
    err_th2 = 25.0 ;    % Percentage of liquid height error,
                          threshold value 2

    err = ((hd - h)/hd )* 100   ; % Current error, as percentage of
                                    desired height.
```

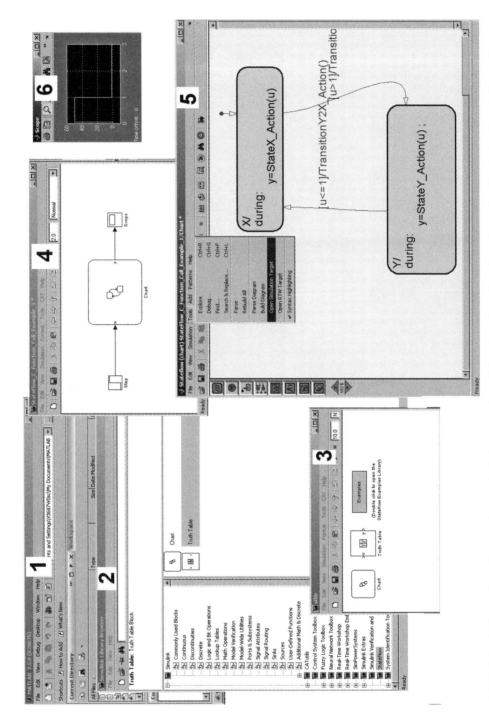

FIGURE A.43: Example to illustrate how to include user-written C-functions from Stateflow Charts for Simulation Target: Illustrative figure 1 of 2. Windows 1 through 6 show the development process involved in order.

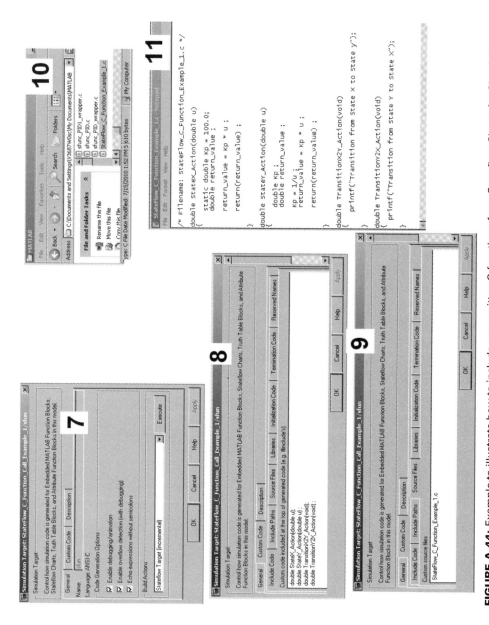

FIGURE A.44: Example to illustrate how to include user-written C-functions from Stateflow Charts for Simulation Target: Illustrative figure 2 of 2. Windows 7 through 11, including the location of the C-source file in the directory, how it is specified in the dialog window of the Stateflow Editor, and the C-source file content.

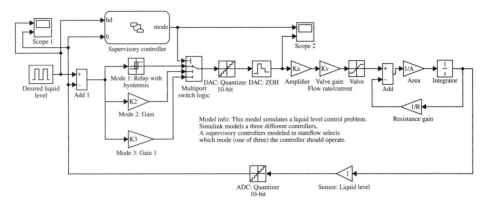

FIGURE A.45: Liquid level control example using a Stateflow chart for supervisory control logic.

```
if (err < err_th1)    %  Supervisory control logic: select
                          controller mode.
    control_mode = 1 ;
elseif (err >= err_th1 and err < err_th2 )
    control_mode = 2 ;
elseif (err >= err_th2)
    control_mode = 3 ;
end
```

Figures A.45 and A.46 show the new Simulink® block diagram of the liquid level control system as well as the Stateflow chart for this very simple supervisory controller. A simulation result is shown in Figure A.47. The simulation was run for the following

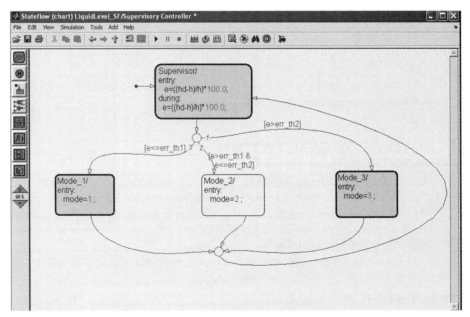

FIGURE A.46: Liquid level control example using a Stateflow chart for supervisory control logic: Stateflow chart details. This Stateflow chart implements the supervisory logic shown in the MATLAB® text programming language above.

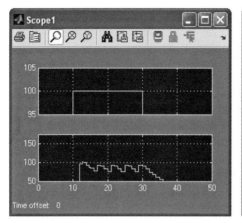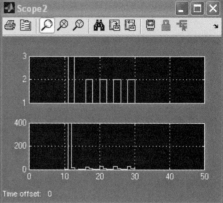

FIGURE A.47: Simulation results for the liquid level control example using a Stateflow chart for supervisory control: Scope 1 top: Desired Liquid Level, Scope 1 bottom: Measured Liquid Level, Scope 2 top: Supervisory Controller Output: Control Mode (1, 2, 3). Scope 2 bottom: Control Signal to Valve Amplifier.

parameters (i.e., this data can be placed in a MATLAB® script M-file, and executed before a simulation is run),

```
err_band = 1.0 ; % for the Relay function.
K2 = 1.0 ;        % Controller mode 2 gain
K3 = 4.0 ;        % Controller mode 3 gain
Ka = 1.0 ;        %  Amplifier current/voltage gain
Kv = 1.0 ;        %  Valve flowrate/currrent gain
Qin_max = 100.0 ; % Valve saturation values: maximum flow rate
                    out of valve
Qin_min = 0.0 ;   % Minimum flow rate: valve closed, so zero.
A = 1.0 ;         % Tank cross sectional area
R = 10.0 ;        % Outflow rate resistance as function of
                    liquid
                  %  height: Q_out = (1/R) * h
```

Notice the change in the controller output discontinuity as the supervisory controller switches from one controller mode to another.

A.4 AUTO CODE GENERATION

One of the most powerful, versatile, and widely used MATLAB® features is automatic code generation from model for a target embedded controller for real-time implementation. The auto-code generation approach to generate a real-time code to run on an embedded electronic control module (ECM) has already largely replaced the manual coding in C and assembly languages in industry.

For a real-time implementation on a target ECM, a typical Simulink®/StateFlow algorithm should be modified to remove all non-real time components as follows.

1. Remove all input and output components (Function generator, Display, Scope, etc.) which are included for the purpose of analysis, and debug (Figure A.48).

2. Connect Simulink® I/O driver blocks (such as ADC, DAC, DIO, CAN blocks) to the input/output ports. The external I/O will be handled by these blocks.

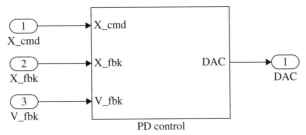

Simulation version

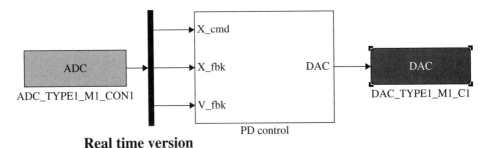

Real time version

FIGURE A.48: An example model in Simulink®: Simulation version (non-real time) and real-time version with appropriate I/O software drivers for the target ECM (electronic control module) to be used by the auto-code generation tools.

3. Provide initial values in the MATLAB® workspace for the parameters and variables so that these initialization values can be included in the generated auto-code.

There are three auto-code generation tools in MATLAB®:

1. **Simulink® Coder**,
2. **Embedded Coder**, and
3. **MATLAB® Coder**.

The main auto-code generation tool is the **Simulink® Coder**. The **MATLAB® Coder** generates C/C++ code from MATLAB® language files. The **Embedded Coder** generates optimized code specific to various microcontrollers and digital signal processors, making use of the Simulink® Coder and MATLAB® Coder tools in the process. Once the auto-code generation tools are configured for a target embedded micro controller, they can be saved and re-used without having to deal with any configuration issues.

Let us consider the following target microcontroller for auto-code generation: Microchip dsPIC33F. For development purposes, the following hardware and software tools are needed:

1. Microchip dsPIC33F Starter Kit which has a development board with the target microcontroller. The board also has a built-in debugger.
2. PC with USB interface to communicate with the development board.

The additional software tools needed for this auto-code generation for the target microcontroller are (Figure A.1):

1. Microchip dsPIC33F Blockset for Simulink® (which would be integrated into the Simulink® environment in Figure A.1).

2. Microchip C Compiler for dsPIC33F (see Figure A.1).

3. Microchip X IDE (Integrated Development Environment), that is for downloading programs to the microcontroller development board.

The minimum required MATLAB® tools needed to support the auto-code generation for the Microchip dsPIC33F target microprocessor are

1. MATLAB®,

2. Simulink®,

3. Simulink® Coder,

4. Embedded Coder.

REFERENCES

[1] Mori, T., "Mechatronics", Yasakawa Internal Trademark Application Memo, 21.121.01, July 12, 1969.

[2] Harashima, F., Tomizuka, M., Fukuda, T., "Mechatronics – What is it, why and how?", *IEEE/ASME Transactions on Mechatronics*, Vol. 1, No. 1, pp. 1–4, 1996.

[3] *National Geographic Magazine*, January 2010 issue, pp. 38–53.

[4] Bryson, A.E., Jr., *Control of Spacecraft and Aircraft*, Princeton, N.J., Princeton University Press, 1994.

[5] Churchill, R.V., Operational Mathematics, Third Edition, McGraw-Hill, 1972.

[6] Ogata, K., *Modern Control Engineering*, Prentice Hall, 1990.

[7] Hartenberg, R.S., Denavit, J., *Kinematic Synthesis of Linkages*, McGraw-Hill, New York, 1964.

[8] Craig, J.J., *Introduction to Robotics: Mechanics and Control*, Third Edition, Addison Wesley, 2004.

[9] Paul, R., *Robot Manipulators*, The MIT Press, Cambridge, MA, 1981.

[10] Orin, D.E., Schrader, W.W., "Efficient Computation of the Jacobian for Robot Manipulators," *International Journal of Robotics Research*, Vol., No. 4, pp. 66–75, 1984.

[11] U.S. Coast Gaurd Navigation Center, `http://www.navcen.uscg.gov`.

[12] Zogg, J-M., GPS Basics, `www.u-blox.com`.

[13] Predko, M., *Programming Robot Controllers*, McGraw Hill, 2003.

[14] *Fluid Power Designers' Lightning Reference Handbook*, Eighth Edition, Berendsen Fluid Power, 1990.

[15] Vickers Inc., Subsidiary of Eaton Corp., *Industrial Hydraulics Manual*, 1999. www.eatonhydraulics.com.

[16] Kenjo, T., Nagamori, S., *Permanent-Magnet and Brushless DC Motors*, Oxford Science Publications, 1985.

[17] Wilson, C.S., "Universal Commutation Algorithm Adapts Motion Controller for Multiple Motors," Proceedings of PCIM 1989, Intertec Communications, pp. 348–360.

[18] Articulated Trucks Users Manual, Volvo Construction Equipment, 2008, `http://www.volvoce.com`.

FURTHER READINGS

Aardema, J., Koehler, D.W., System and Method for Controlling Independent Metering Valve, US Patent 5,947,140, Sept. 7, 1999.

Adams, H.W., *Aircraft Hydraulics*, McGraw Hill, 1943.

American Society of Mechanical Engineers, www.asme.org.

Bertoline, G.R., Wiebe, E.N., *Technical Graphics Communication*, Third Edition, McGraw Hill, 2003.

Bishop, R., *Basic Microprocessor and the 6800*, Hayden Book Co., 1979.

Bosch Automotive Handbook, 6th Edition, Professional Engineering Publishing, 2004.

Brady, R.N., *Modern Diesel Technology*, Prentice Hall Inc., 1996.

Brey, B.B., *The Intel Microprocessors*, Sixth Edition, Prentice Hall, 2003.

Mechatronics with Experiments, Second Edition. Sabri Cetinkunt.
© 2015 John Wiley & Sons, Ltd. Published 2015 by John Wiley & Sons, Ltd.
Companion Website: www.wiley.com/go/cetinkunt/mechatronics

Cetinkunt, S., Chen, C., Egelja, A., Muller, T., Ingram, R., Pinsopon, U., Method and System for Selecting Desired Response of an Electronic Controlled Sub System, US Patent 6,330,502, December 11, 2001.

Clark, D.C., "Selection and Performance Criteria for Electrohydraulic Servodrives," Technical Bulletin 122, Moog Inc.

Cleveland Clinic Web Page, ClevelandClinic.com.

Cobo, M., Ingram, R., Reiners, E.A., Wiele, M.F.,V, Positive Flow Control system, US Patent 5,873,244, February 23, 1999.

Cogdell, J.R., *Foundations of Electrical Engineering*, Prentice Hall, 1990.

Cox, R.A., *Technician's Guide to Programmable Controllers*, Third Edition, Delmar Publishing, Albany, NY, 1989.

DC Motors, Speed Controls, Servo Systems, Electro-Craft Corp, 1980.

Design Engineers Handbook, Volume 1-Hydraulics, Motion Control Technology Series, Parker Hannifin Corp.

Deutsche Institute fur Normung, www2.din.org.

Dorf, R.C., Bishop, R.H., *Modern Control Systems*, Pearson – Prentice Hall, 2001.

Doughman, G., *Programming the Motorola M68HC12 Family*, Anna Books, 2000.

El-Rabbany, A., *Introduction to GPS: The Global Positioning System*, Second Edition, Artech House, 2006.

Figliola, R.S., Beasley, D.E., *Theory and Design for Mechanical Measurements*, John Wiley, 1995.

Ford, W., Topp, W., *MC 68000: Assembly Language and Systems Programming*, D.C. Heath and Company, 1987.

Fortney, L.R., *Principles of Electronics*, Harcourt, Brace, Jovanovich Publishers, 1987.

Franklin, G.F., Powell, J.D., Emami-Naeini, A., *Feedback Control of Dynamic Systems*, Addison Wesley, 1994.

Franklin, G.F., Powell, J.D., Workman, M., *Digital Control of Dynamic Systems*, Addison Wesley, 1998.

Graphical Symbols for Fluid Power Diagrams, American National Standard, ANSI Standard y32.10–1967, 1967.

Green, W.L., *Aircraft Hydraulic Systems*, John Wiley, 1985.

Grimheden, R. N., Hansen, M, "Mechatronics – the evolution of an academic discipline in engineering education", *International Journal of Mechatronics*, Vol. 15, pp. 179–1996, 2005.

Hanselman, D., *Brushless Permanent Magnet Motor Design*, The Writers' Collective, 2003.

Heisler, H., *Advanced Engine Technology*, Society of Automotive Engineers, 1995.

Heywood, J.B., *Internal Combustion Engine Fundamentals*, McGraw Hill, 1988.

Horowitz, P., Hill, W., *The Art of Electronics*, Cambridge Press, Second Edition, 1989.

International Standards Organization, www.iso.org.

Iovine, J., *PIC Microcontroller Project Book*, McGraw Hill, 2000.

Jung, W.G., *IC Op-Amp Cookbook*, Third Edition, Prentice Hall.

Keller, G., *Aircraft Hydraulic Design*, Applied Hydraulics Edition, 1957.

Kenjo, T., *Electric Motors and Their Controls*, Oxford Science Publications, 1991.

Kenjo, T., Sugawara, A., *Stepping Motor and Their Microprocessor Controls*, Oxford Science Publications, 1994.

Kennedy, M., *The Global Positioning System and GIS: An Introduction*, Ann Arbor Press, Inc., 1996.

Klafter, R.D., Chmielewski, T.A., Negin, M., *Robotic Engineering: An Integrated Approach*, Prentice Hall, 1989.

Labrosse, J.J., *Embedded Systems Building Blocks*, R&D Publications, 2000.

Making the Choice: Selecting and Applying Piston, Bladder and Diaphragm Accumulators, Brochure 1660-USA, www.parker.com/accumulator.

Manring, N., *Hydraulic Control Systems*, John Wiley and Sons, 2005.

Mazidi, M.A., Mazidi, J.G., *The 80x86 IBM PC and Compatible Computers*, Prentice Hall, 2003.

Merritt, H.E., *Hydraulic Control Systems*, John Wiley and Sons, 1967.

Miller, T.J.E., *Switched Reluctance Motors and Their Control*, Magna Physics Publishing and Clarendon Press, Oxford Science Publications, 1993

National Fluid Power Association, www.nfpa.com.

Neal, T.P., "Performance Estimation for Electrohydraulic Control Systems," Technical Bulletin 126, Moog Inc.

Neese, W., *Aircraft Hydraulic Systems*, Krieger Ed., Third Edition, 1991.

Newton, K., Steeds, W., Garrett, T.K., *The Motor Vehicle*, Twelfth Edition, SAE International, 1996.

Newton, K., Steeds, W., Garrett, T.K., *The Motor Vehicle, SAE International*, 12th Edition, 1996.

Norton, R.L., *Design of Machinery*, McGraw Hill, Second Edition, 1999.

Novotny, D.W., Lipo, T.A., *Vector Control and Dynamics of AC Drives*, Clarendon Press, 2000.

Peatman, J.B., *Design with PIC Microcontrollers*, Prentice Hall, 1998

Pippenger, J.J., Hicks, T.G., *Industrial Hydraulics*, McGraw Hill, 1970.

Ramshaw, R., van Heeswijk, R.G., *Energy Conversion: Electric Motors and Generators*, Saunders College Publishing, 1990.

Raymond, E.T., Chenoweth, C.C., *Aircraft Flight Control Actuation System Design*, SAE Press, 1993.

Sciavicco, L., Siciliano, B., *Modelling and Control of Robot Manipulators*, Springer Verlag; Second Edition, 2000.

Shigley, J.E., Mischke, C.R., Budynas, R.G., *Mechanical Engineering Design*, Seventh Edition, McGraw Hill, 2004.

Society of Automotive Engineers, www.sae.org.

Spong, M.W., Vidyasagar, M., *Robot Dynamics and Control*, John Wiley & Sons, January 1989.

Technical Bulletin 98, Design Guidelines for Electric Multi Disc Clutches and Brakes, The Carlyle Johnson Machine Co, www.cjmco.com.

Thompson, J.E., Campbell, R.B., *Manual for Aircraft Hydraulics*, Aviation Press, 1942.

Tsui, J.B., *Fundamentals of Global Positioning System Receivers: A Software Approach*, John Wiley & Sons, Inc., 2000.

Uhlir, P., Kubiczek, Z., "3-Phase AC Motor Control with V/Hz Speed Open Loop Using DSP56F80X," Motorola, Semiconductor Application Note, AN1911/D, 2001.

Valvano, J.W., *Embedded Microcomputer Systems: Real Time Interfacing*, Brooks/Cole, 2000.

Vaughan, N.D., Gamble, J.B., "The Modelling and Simulation of a Proportional Solenoid Valve," ASME Winter Annual Meeting, Nov. 25–30, 1990, 90-WA/FPST-11.

SUPPLIERS OF MECHATRONIC SYSTEMS AND COMPONENTS

ABB Control Inc. 1206 Hatton Road 76302 Wichita Falls, TX, Phone: (940) 397 7000 FAX: (940) 397 7085 http://www.abb.com.

Analog Devices, Two Technology Way, PO Box 280, Norwood MA 02062, www.analogdevices.com.

Bosch-Rexroth Corp., PO Box 25407, Lehigh Valley, PA 18002-5407, Phone: (610) 694-8246 FAX: (610) 694-8266, www.boschrexroth.com.

Digi-Key, 701 Brooks Ave., PO Box 677, Tief River Falls, MN 56701-0677, Phone: 800-344-4539 FAX: 218-681-3380 http://www.digikey.com.

Encoders and various sensors, Dynapar Corp., http://www.dynapar-encoders.com.

Fairchild Semiconductor Corp., 313 Fairchild Drive, Mountain View, CA 94003 www.fairchild.com.

Garmin International, http://www.garmin.com/.

GPS Basics Tutorial, Trimble Navigation, http://www.trimble.com.

Husco Corporation, www.husco.com.

HYDAC Technology Corporation, HYDRAULIC Division, 445 Windy Point Drive, USA-Glendale Heights, IL 60139, Phone: +1-630 - 5 45-08 00, FAX: +1-630 - 5 45-00 33, www.hydac.com.

Hydraforce Inc., 500 Barclay Blvd., Lincolnshire, IL 60069 USA, Phone: 847-793-2300, FAX: 847-793-0086, http://www.hydraforce.com.

Magellan Systems Corp., http://www.magellangps.com/.

Magnet Schultz Solenoids, www.magnetschultz.com.

Mitsubishi Electric Automation, Inc., 500 Corporate Woods Parkway, Vernon Hills, Illinois 60061, Phone: 847-478-2100, Email: marcomm@meau.mea.com, http://www.mitsubishielectric.com.

Moog Controls Inc., Industrial Division, East Aurora, NY 14052, Phone: (716) 655-3000, FAX: (716) 655-1803, http://www.moog.com.

Motorola Semiconductor Prodcuts PO Box 20912, Phoenix, AZ 85036 www.motorola.com.

MTS, Temposonic Sensor, 3001 Sheldon Dr, Cary, NC 27513, Phone: 919-677-0100, http://www.mtssensors.com

National Semiconductor Products, Inc., 2900 Semiconductor Drive PO Box 58090 Santa Clara CA 95052.

Newark Electronics 4801 N. Ravenswood Chicago, Illinois 60640, Phone: 773-784-5100 FAX: 773-907-5339 http://www.newark.com.

Novtel, http://www.novatel.ca.

Omega, Sensors and Measurement Systems, http://www.omega.com.

Omron Electronics LLC, One East Commerce Drive, Schaumburg, IL 60173 USA, Phone: 847-843-7900, FAX: 847-843-8081, http://oeiweb.omron.com/oei/.

Parker Hannifin Corp., Hydraulic Valve Division, 520 Ternes Ave, Elyria, OH, 44035, USA, Phone: 440-366-5200, FAX: 440-366-5253, www.parker.com/hydraulicvalve.

PC Based Data Acquisition Prodducts, National Instruments, www.ni.com.

Peerless-Winsmith Inc., 172 Eaton Street PO Box 530 Springville, NY 14114, Phone: 716-592-9311 http://www.winsmith.com.

Piezoelectric accelerometers, PCB Piezotronics Inc., http://www.pcb.com.

Precision Industrial Components Corp. 86 Benson Road, PO Box 1004 Middlebury, CT 06762-1004 Phone: 800-243-6125 FAX: 203-758-8271 http://www.pic-design.com

Rockwell Automation Corporate Headquarters, Allen-Bradley US Bank Center, 777 East Wisconsin Avenue, Suite 1400, Milwaukee, Wisconsin 53202 USA, Phone: (414) 212-5200, http://www.rockwellautomation.com/, http://www.ab.com.

Sauer-Danfoss Co, 2800 E. 13th Street, Ames, IA 50010, USA, Phone: 515-239-6000, FAX: 515-239-6618, www.sauer-danfoss.com.

Siemens Energy and Automation 1901 N. Roselle Rd. Suite 220 Schaumburg, IL 60195, Phone: 847-310-5900 FAX: 847-310-6570 http://www.sea.siemens.com.

Sokkia Corporation, http://www.sokkia.com/.

Sun Hydraulics, 1500 West University Parkway, Sarasota, Florida 34243, Phone: 941-362-1200, FAX: 941-355-4497, www.sunhydraulics.com.

Texas Instruments, PO Box 3640, Dallas TX 75285, www.ti.com.

The Oilgear Company, P.O. Box 343924, Milwaukee, WI 53234-3924, Phone: 414-327-1700, www.oilgear.com.

Thomson Industries, Inc. 2 Channel Drive Port Washington, NY 11050, Phone: 800-544-8466 www.thomsonindustries.com.

SUPPLIERS OF INDUSTRIAL ROBOTS

ABB Inc (Asea Robots), US Head Office, 501 Merritt 7, Norwalk, CT 06851, Phone: 203 750 2200, FAX: 203 750 2263 abb.com/robotics.

Adept Technologies Inc, 3011 Triad Drive, Livermore, CA 94551, Phone: 925-245-3400, FAX: 925-960-0452, www.adept.com.

EPSON Factory Automation/Robotics (formerly Seiko), 18300 Central Avenue, Carson, CA 90746, Phone: (562) 290-5910 FAX: (562) 290-5999, robots.epson.com.

Fanuc Robotics (formerly GMF Robotics), fanucrobotics.com, 3900 W. Hamlin Road Rochester Hills, MI 48309-3253, Phone: 800-47-ROBOT (or 1-800-477-6268).

Kawasaki Robotics (USA), Inc., 28059 Center Oaks Court, Wixom, Michigan 48393, Phone: 248-305-7610, FAX: 248-305-7618, kawasakirobots.com.

Mitsubishi Electric Automation Inc, a Mitsubishi Company, 500 Corporate Woods Parkway, Vernon Hills, IL 60061, USA, Phone: 847-478-2100, FAX: 847-478-2396, www.meau.com.

Motoman Inc., a Division of Yaskawa Electic Company, motoman.com, 805 Liberty Lane, West Carrollton, Ohio 45449, Phone: 937-847-6200, FAX: 937-847-6277.

INDEX

Mechatronics with Experiments, Second Edition. Sabri Cetinkunt.
© 2015 John Wiley & Sons, Ltd. Published 2015 by John Wiley & Sons, Ltd.
Companion Website: www.wiley.com/go/cetinkunt/mechatronics